Jenny Stanford Series on Biocatalysis
Volume 4

Pharmaceutical Biocatalysis

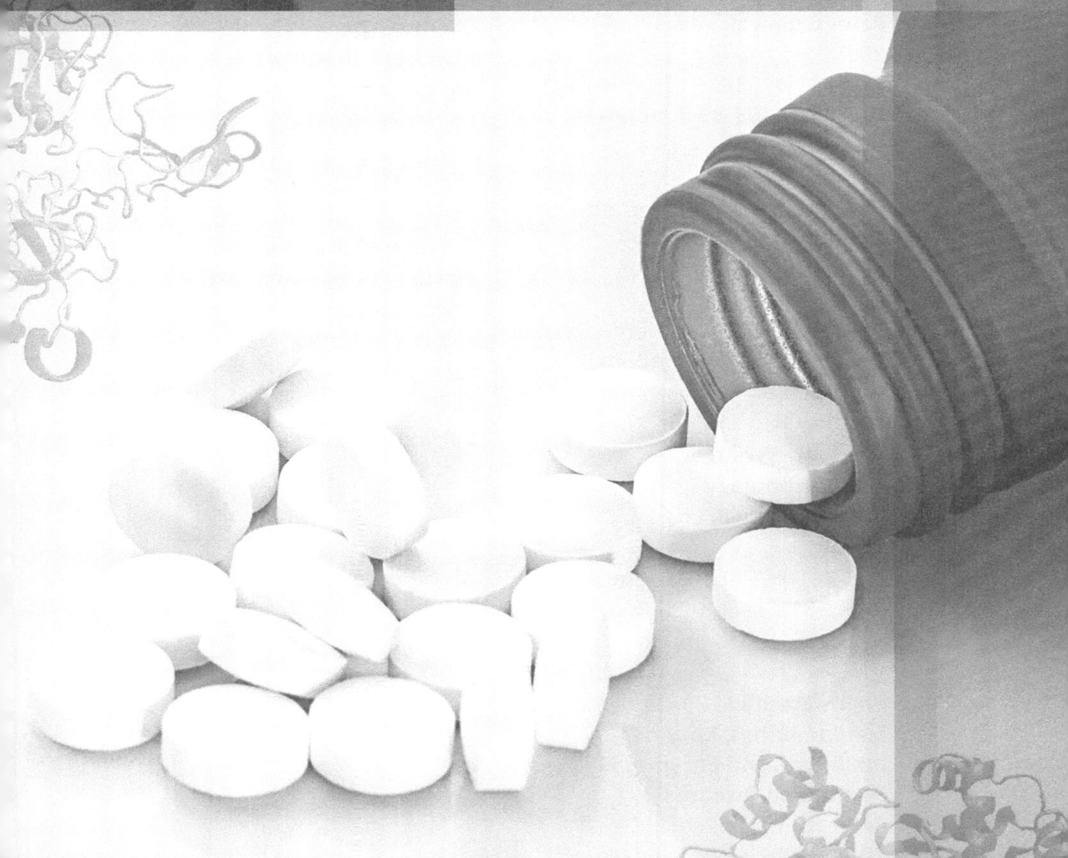

Jenny Stanford Series on Biocatalysis

Series Editor
Peter Grunwald

Titles in the Series

Published

Vol. 1
Industrial Biocatalysis
Peter Grunwald, ed.
2015
978-981-4463-88-1 (Hardcover)
978-981-4463-89-8 (eBook)

Vol. 2
Handbook of Carbohydrate-Modifying Biocatalysts
Peter Grunwald, ed.
2016
978-981-4669-78-8 (Hardcover)
978-981-4669-79-5 (eBook)

Vol. 3
Biocatalysis and Nanotechnology
Peter Grunwald, ed.
2017
978-981-4613-69-9 (Hardcover)
978-1-315-19660-2 (eBook)

Vol. 4
Pharmaceutical Biocatalysis: Fundamentals, Enzyme Inhibitors, and Enzymes in Health and Diseases
Peter Grunwald, ed.
2019
978-981-4800-61-7 (Hardcover)
978-0-429-29503-4 (eBook)

Forthcoming

Vol. 5
Pharmaceutical Biocatalysis: Chemoenzymatic Synthesis of Active Pharmaceutical Ingredients
Peter Grunwald, ed.
2019

Vol. 6
Pharmaceutical Biocatalysis: Biotransformations, Novel Therapeutics, Natural Sources, and Degradation
Peter Grunwald, ed.
2020

Jenny Stanford Series
on Biocatalysis
Volume 4

Pharmaceutical Biocatalysis

Fundamentals, Enzyme Inhibitors, and Enzymes in Health and Diseases

edited by

Peter Grunwald

JENNY STANFORD
PUBLISHING

Published by

Jenny Stanford Publishing Pte. Ltd.
Level 34, Centennial Tower
3 Temasek Avenue
Singapore 039190

Email: editorial@jennystanford.com
Web: www.jennystanford.com

British Library Cataloguing-in-Publication Data
A catalogue record for this book is available from the British Library.

Pharmaceutical Biocatalysis: Fundamentals, Enzyme Inhibitors, and Enzymes in Health and Diseases

ISBN 978-981-4800-61-7 (Hardcover)
ISBN 978-0-429-29503-4 (eBook)

Contents

Preface

The relations between biocatalysis and pharmaceutical sciences are manifold. Outstanding aspects are the excellent properties of biocatalysts, which are among others their high chemo-, regio-, and enantioselectivities, making them particularly suitable for the synthesis of safe chiral drug intermediates and drugs. Furthermore, several (engineered) enzymes find increasing application for the treatment of important and in part serious diseases, such as fibrinolytic disorders, cancer, or the rare lysosomal storages diseases. Drug discovery also comprises the isolation of therapeutic enzymes and drugs from natural sources such as microorganisms and plants; the blockbuster anti-cancer drug paclitaxel is an early example. Finally, enzymes and microorganisms are indispensable in connection with the degradation of pharmaceuticals including antibiotics that entered the ecosystem with potentially hazardous consequences. These issues are discussed in volumes 4 to 6 of this Series on Biocatalysis. The present volume imparts fundamentals, and introduces into the fields of therapeutic enzymes and enzyme inhibitors by means of relevant examples.

The first chapter is about some fundamental aspects such as the world market of drugs, regulatory authorities with a focus on the US FDA and the EMA, including quality standards, rare diseases and orphan drugs, therapeutic enzymes and enzyme inhibitors, randomized clinical trials, cost of drug development and therapies, and future developments in different pharmaceutical disciplines. In Chapter 2, Duarte Miguel Prazeres then provides insights into the different roles enzymes play as ancillary materials in the downstream processing of biopharmaceuticals to dissociate and lyse cells, remove impurities, or to perform structural modifications in target molecules until the end-product specifications are met. The third chapter, authored by Andrea Angeli and Claudiu Supuran, deals with the therapeutic use of carbonic anhydrase inhibitors as diuretic drugs, anti-glaucoma and anti-tumor agents, and anti-epileptics, together with the multiple drug interactions of these inhibitors. It is followed by a chapter on fibrinolytic enzymes for thrombolytic therapy, in which Swaroop Kumar and Sabu

Abdulhameed describe the advantage of many plasminogen activators and plasmin-like enzymes over anti-coagulants and anti-platelet drugs for the treatment of thrombolytic diseases such as acute myocardial infarction and stroke. In Chapter 5, Khushboo Gulati and Krishna Poluri give a comprehensive overview of enzyme engineering approaches in connection with various protein therapeutics, including monoclonal antibodies, glycoproteins, enzymes, cytokines, hormones, coagulation factors, interleukins, and protein scaffolds.

Bioactive peptides are widely used in areas such as antimicrobials, treatment of cancer, hypertension, and skin diseases, and in connection with the modulation of immunity as well as cellular mechanisms. In Chapter 6, Jaison Jeevanandam, Ashish Kumar Solanki, Shailza Sharma, Prabir Kumar Kulabhusan, Sapna Pahil, and Michael K. Danquah describe apart from their many pharmaceutical applications the different synthesis methods, including the rDNA technology, as well as the relevant purification and characterization procedures. Based on respective metabolic pathways and reactive oxygen species (ROS)- and reactive nitrogen species (RNS)-dependent enzymatic reactions, Marco Fernandes, Alisha Patel, and Holger Husi (Chapter 7) introduce the reader to the versatile relationships between oxidative stress and cardiovascular disease (CVD), the types of ROS in CVD, or the maintenance of redox homeostasis, and discuss options to target ROS via pharmacological intervention. The amino acid deprivation cancer therapy relies on the fact that cancers often require high concentrations of one or more amino acid residues that are essential for their development, survival, and spread. Several successful examples of the enzymatic amino acid deprivation therapies using enzymes such as arginine deiminase, methionine-γ-lyase, or L-asparaginase in the treatment of several types of tumors are the topic of Chapter 8, authored by Carla S. S. Teixeira, Henrique S. Fernandes, Sérgio F. Sousa, and Nuno M. F. S. A. Cerqueira. Mammalian carboxylesterases catalyze the cleavage of a large number of structurally diverse ester- or amide-containing substrates, which is why they play key roles in both endogenous and xenobiotic metabolism; Li-Wei Zou and Guang-Bo Ge discuss in Chapter 9 the structural and catalytic properties of these enzymes and their tissue distribution and biological function together with the importance of carboxylesterase inhibitors to alter their activity for therapeutic purposes with the aim to improve the outcomes of patients administrated ester drugs.

Chapters 10 to 12 are devoted to treatment options for neurodegenerative diseases. It is well known that abnormal enzyme activities may be associated with pathologies. An example is monoamine oxidase MAO; elevated MAO B is observed in neurodegenerative diseases such as Alzheimer's disease. Rona R. Ramsay discusses in Chapter 10 the molecular aspects of the activity and inhibition of the FAD-containing MAO and gives examples of (tight-binding) reversible and irreversible MAO inhibitors used as drugs to treat, for instance, Parkinson's disease. Ireen Denya, Sarel F. Malan, and Jacques Joubert then report in Chapter 11 about multi-functional monoamine oxidase and cholinesterase inhibitors for the treatment of Alzheimer's Disease. For the multi-target drug ligand (MTDL) approach, drugs are often designed via linking two properly selected pharmacophores through a spacer, and subsequent optimization of the hybrid molecules for the two targets by generating analogs based on structure–activity relationship studies. Similarly, in Chapter 12, Hayrettin Ozan Gülcan and Ilkay Erdogan Orhan summarize the neurodegenerative characteristics of Alzheimer's disease, the validated targets, and the current drugs used for the treatment of the disease, together with actual approaches on multi-target ligand design studies.

Aldehyde oxidases are a group of enzymes with important functions in the context of drug and xenobiotic metabolism. In Chapter 13, Cristiano Mota, Teresa Santos-Silva, Mineko Terao, Enrico Garattini, Maria João Romão, and Silke Leimkühler focus on the significance of these enzymes in xenobiotic and drug metabolism, and for the design and development of new drugs. As several tumor types express aldehyde oxidases, they may also serve to activate anti-tumor pro-drugs and contribute to increased tumor selectivity. Aldehyde oxidases exert overlapping activities with cytochrome P-450 mono-oxygenases (CYP450s). Sanjana Haque, Yuqing Gong, Sunitha Kodidela, Mohammad A. Rahman, Sabina Ranjit, and Santosh Kumar provide a general background on CYP enzymes and discuss the importance of wild-type CYP enzymes in the synthesis of novel drugs and drug metabolites and associated limitations, the engineering of CYP enzymes for enhanced activity, stability, and specificity to overcome these limitations, and the advantages of using CYP enzymes in the synthesis of drugs and drug metabolites. Finally, Chin Eng Ong, Amelia Dong, Boon Hooi Tan, and Yan Pan elucidate the role of CYP enzymes in extra-hepatic tissues (heart, brain, kidney, etc.) and give an overview of the role of CYP enzymes in organ homeostasis with special focus on vasculature and cardiovascular systems, and an outlook for therapeutic strategies in vascular and cardiovascular diseases.

Many disease-related proteins have been identified during the past decades, including oncogenic proteins that drive tumor growth. These are often not druggable by conventional means, e.g., by small-molecule inhibitors, and require alternative treatment strategies. One such approach described by Norihito Shibata, Nobumichi Ohoka, Takayuki Hattori, and Mikihiko Naito in Chapter 16 is the use of so-called SNIPERS or PROTACS, small hybrid molecules with two ligands connected by a linker. One ligand targets the oncogenic protein and the other one, E3 ubiquitin, ligases in the cells, resulting in ubiquitylation and subsequent degradation of the targeted oncogenic protein via the ubiquitin-proteasome system. Chapter 17, authored by Rolf Marschalek, provides an overview of the many resistance mechanisms of tumor cells and furthermore informs in different subsections about the development of solid and hematological tumors, general treatment strategies and patient outcome, the concept of cancer stem cells, and cancer immune therapies.

The concept of biocatalytic nanoreactors for medical purposes is discussed by Oscar González-Davis, Kanchan Chauhan, and Rafael Vazquez-Duhalt in Chapter 18. This chapter deals with the various nanobioreactor design strategies and the use of nanomaterials, such as biocompatible nanoparticles and nanodevices, as carriers for enzymes with medical relevance to treat enzyme deficiency disorders or to transport enzymatic activity to cancer cells. The employment of these nanoreactors is still in its infancy but shows enormous potential, as demonstrated by the authors with different examples. In Chapter 19, Pravin Ambure and Kunal Roy report about computer-aided drug design for the identification of MTDLs (see above) in complex diseases, such as cancer, multiple sclerosis, osteoporosis, and neurodegenerative disorders. They show that the most promising way to improve the quality of results in this polypharmacological approach is to combine structure-based and ligand-based studies. In article chapter about the development of improved therapeutics through a glycan-“designer” approach (Chapter 20), Natalia G. Bednarska reviews efforts that have been made in the field of protein–glycan coupling technology to modify protein/peptide immunogenicity through glycosylation with the intention of enhancing the anti-microbial and immunogenic properties. The different types of glycosylation are discussed together with its effects on peptides, proteins, and other therapeutics, as well as the impact of glycoconjugates on the cells of the immune system, and the rational design of glycotherapeutics.

The final chapter of this volume, authored by Andreas Sebastian Klein, Thomas Classen, and Jörg Pietruszka, is on biocatalysis as a resourceful methodology for complex syntheses (selective catalysis, cascades, and biosynthesis). The authors illustrate various modes of utilizing biocatalysis, including biosynthesis, which can be applied for the production of pharmaceuticals or other high-value products. This is demonstrated in addition to numerous other examples with the semisynthesis approaches towards the industrial production of the anti-cancer drug Taxol and the anti-malarial drug artemisinin. This chapter is a suitable introduction to volume 5 of this Series on Biocatalysis, where the (chemo)enzymatic synthesis of active pharmaceutical ingredients is dealt with.

Peter Grunwald

Chapter 1

Pharmaceuticals: Some General Aspects

Peter Grunwald

Department of Physical Chemistry, University of Hamburg,
D-20146 Hamburg, Germany

grunwald@chemie.uni-hamburg.de

1.1 Introduction

In this chapter, apart from brief general remarks, some aspects of enzymes and enzyme inhibitors employed for therapeutic applications will be discussed. The topics are, among others, the following:

- The world market of these drugs
- Regulatory authorities (mainly the US FDA and the EMA)
- Quality standards
- Rare diseases and orphan drugs
- Therapeutic enzymes and enzyme inhibitors
- Cost of drug development and therapies
- Future developments

The term "drug" is used here in connection with recombinant therapeutic proteins (for the world market of therapeutic proteins,

Pharmaceutical Biocatalysis: Fundamentals, Enzyme Inhibitors, and Enzymes in Health and Diseases
Edited by Peter Grunwald

ISBN 978-981-4800-61-7 (Hardcover), 978-0-429-29503-4 (eBook)
www.jennystanford.com

see Research and Markets, 2016) as well as chemically synthesized small molecules that have been approved for the treatment and prevention of diseases. Not included are diagnostic imaging agents as, e.g., Netspot (detection of rare neuroendocrine tumors) or Axumin (detection of recurrent prostate cancer) both approved by the FDA in 2016, and other compounds used for diagnostic purposes. The targets of small molecule drugs are proteins to which they bind to exert therapeutic efficacy; enzymes involved in drug metabolism are not defined here as drug target. The WHO Collaborating Centre for Drug Statistics Methodology classifies drugs by The Anatomical Therapeutic Chemical (ATC) Classification System (ATC and ATCvet) with respect to the organ or system, on which they act, their therapeutic, pharmacological, and chemical properties, etc., including, e.g., defined daily drug doses (WHOATC 2018).

1.2 General Remarks

Enzyme inhibitor drugs are used to treat chronic diseases and disorders such as cardiovascular diseases (angina and myocardial infarction, heart failure, hypertensive heart and rheumatic heart disease, etc.), gastrointestinal diseases, Diabetes Type 2, pain, fever, inflammation, neurodegenerative diseases such as Parkinson's and Alzheimer's, asthma and COPD, psoriasis and psoriatic arthritis, different infectious diseases (trypanosomiasis, leishmaniasis, tuberculosis, malaria, acquired immunodeficiency syndrome, etc.), a variety of different types of cancers (see below), some of the rare lysosomal storage diseases, life style related conditions (erectile dysfunction, benign prostatic hyperplasia, alopecia) and many others.

Pharmaceutical enzymes may be categorized as enzymes in replacement therapy, enzymes in cancer treatment, enzymes for fibrinolysis, enzymes to cure digestive disorders and enzymes that are used topically for various treatments (Yari et al., 2017). Modern DNA technology together with other approaches such as PEGlation or glycoengineering generates enzymes with extended half-life and reduced or no immunogenicity.

The Food and Drug Administration (FDA or USFDA) a federal agency of the United States Department of Health and Human

Services is—together with several FDA organizations among them the Center for Drug Evaluation and Research (CDER)—responsible for controlling apart from cosmetics, animal foods and feed, tobacco products, dietary supplements and veterinary products, new drug application (new molecular entities, NMEs) and prescription, over-the-counter pharmaceutical drugs, generic drugs, vaccines, biopharmaceuticals, etc. An NME is a drug with an active moiety that is approved by the FDA as a single ingredient drug or as part of a combination product or marketed in the US for the first time; hence, treatment with an NME is often tantamount to a new therapy for patients.

Its European counterpart is the European Medicines Agency (EMA), which recently moved from London to Amsterdam. The function of the EMA (2017, 2018) is to protect "public and animal health in 28 EU Member States, as well as the countries of the European Economic Area, by ensuring that all medicines available on the EU market are safe, effective and of high quality; the EMA serves a market of over 500 million people living in the EU." A European medicines regulatory network aims among others on reducing the administrative burden through the centralized authorization procedure, translating into medicines reaching patients faster. The EMA has a number of committees for Medicinal Products for Human Use, Pharmacovigilance Risk Assessment, for Medicinal Products for Veterinary Use, for Orphan Medicinal Products, Herbal Medicinal Products, for Advanced Therapies, and a Paediatric Committee.

According to more recent developments, the FDA offers an Expedited Program for speeding up the availability of drugs that treat serious diseases: *Fast track* (the aim is to facilitate the development, and expedite the review of drugs to treat serious conditions, filling an unmet medical need), *breakthrough therapy* (substantial improvement over available therapy), *priority review* (FDA's goal is to take action on an application within 6 months), and *accelerated approval* (allows drugs to be approved based on a surrogate endpoint) are the 4 ways applied singly or in conjunction in which the FDA cooperates with the industry (see Table 1.1) in developing drugs (FDA, 2018; 2018a). This resulted in an increase of the speed at which new drugs are and have been developed and reviewed, and is expressed by the fact that 78% and 87% of drug applications were approved by the FDA in the first review cycle in 2014 and 2015, respectively, compared with just 58% between 2008 and 2012 (Tworkoski, 2016).

Table 1.1 Drugs approved by the FDA in 2016

API	Drug Name	FDA-approved use	Expedited program	Company
Elbasvir & grazoprevir	Zepatier	Chronic hepatitis C virus	BT	MSD SHARP & DOHME GMBH
Brivaracetam	Briviact	Partial onset seizures		UCB (Union Chimique Belge)
Obiltoxaximab	Anthim	Inhalational anthrax		Elusys Therapeutics, Inc.
Ixekizumab	Taltz	Plaque psoriasis		Eli Lilly & Company
Reslizumab	Cinqair	Severe asthma		Teva Pharmaceuticals
Defibrotide sodium	Defitelio	Hepatic veno-occlusive disease	PR, OD	Jazz Pharmaceuticals
Venetoclax	Venclexta	Chronic lymphocytic leukemia	AA, BT, PR, OD	AbbVie Inc.
Pimavanserin	Nuplazid	Hallucinations & delusions	BT, PR	Acadia Pharmaceuticals Inc.
Atezolizumab	Tecentriq	Urothelial carcinoma	AA, BT, PR	Genentech
Daclizumab	Zinbryta	Multiple sclerosis		Biogen, Inc.
Obeticholic acid	Ocaliva	Chronic liver disease	AA, FT, OD	Intercept Pharmaceuticals, Inc.
Fluciclovine F 18; imaging agent	Axumin	To detect recur-rent prostate cancer		Blue Earth Diagnostics, Ltd.
Ga 68 dotatate; imaging agent	NETSPOT	To detect neuroendocrine tumors	PR, OD	Advanced Accelerator Applications USA, Inc.

API	Drug Name	FDA-approved use	Expedited program	Company
Sofosbuvir & velpatasvir	Epclusa	All six major forms of hepatitis C virus	PR	Gilead Sciences, Inc.
Lifitegrast ophthalmic solution	Xiidra	Dry eye disease		Shire US Inc.
Lixisenatide	Adlyxin	Improve glycemic control		Sanofi-Aventis U.S. LLC
Eteplirsen	Exondys 51	Duchenne muscular dystrophy	AA, FT, PR, OD	Sarepta Therapeutics
Olaratumab	Lartruvo	Soft tissue sarcoma	AA, BT, FT, PR, OD	Eli Lilly & Company
Crisaborole	Eucrisa	Atopic dermatitis		Anacor Pharmaceuticals, Inc.
Bezlotoxumab	Zinplava	Decrease the risk of *Clostridium difficile* infection		Merck Sharp & Dohme Corp.
Rucaparib	Rubraca	Ovarian cancer	AA, BT, PR, OD	Foundation Medicine, Inc.
Nusinersen	Spinraza	Spinal muscular atrophy	FT, PR, OD	Ionis Pharmaceuticals

Note: AA, Accelerated approval; PR, Priority review status, BT, Breakthrough therapy designation; FT, Fast track designation; OD, Orphan drug designation; API, Active Pharmaceutical ingredient.

The understanding of cancer pathogenesis has significantly improved in the recent past which has led to new therapeutic concepts such as targeted agents designed to inhibit molecular pathways that are crucial for tumor growth, or immunotherapeutic approaches (Vanneman and Dranoff, 2012; Gotwals et al., 2017; Banchereau and Palucka, 2018) including immune checkpoint inhibitors (Dine et al., 2017). Another recent example is Yescarta, a chimeric antigen receptor (CAR) T cell therapy, approved as the second gene therapy by the FDA for treating diffuse large B-cell lymphoma (DLBCL, the most common type of non-Hodgkin lymphoma; FDA, 2018c); Yescarta has received the orphan drug status (see next section). This has led to an improved survival of patients suffering from certain types of cancer and finds expression, too, in a trend towards an increased share of oncologic drugs as part of the number of drugs approved per year as shown in the following figure (see also Centerwatch, 2018).

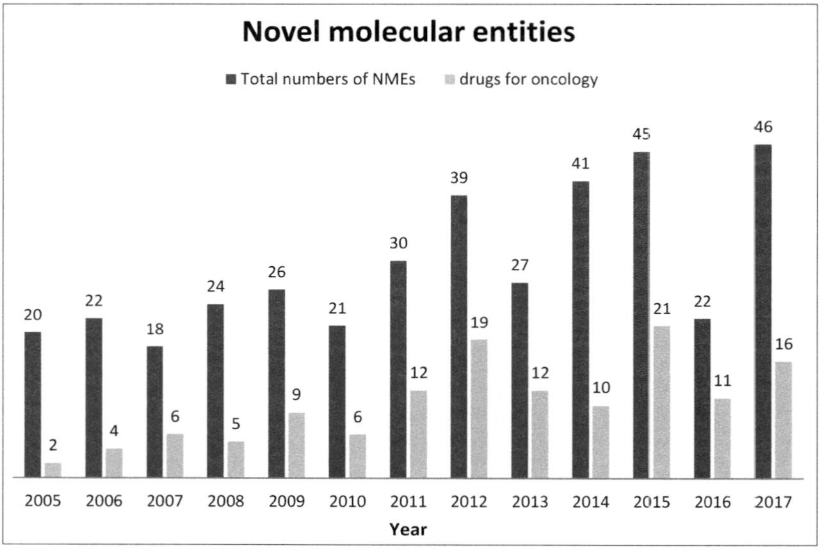

1.3 Orphan Drugs and Rare Diseases

According to the European Medicines Agency the number of rare/orphan diseases is between 5,000 and 8,000, and 27 million to 36 million patients in the EU are suffering from such diseases of which most (~80%) are of genetic origin. The NORD's (National Organization for Rare Disorders, Inc.) database contained details for more than 1,200 rare diseases by 2018. The orphan drug status has been introduced in various countries, among them the USA (for an *Orphan Drug Modernization Plan* see FDA, 2017a, 2017b, 2017c), Japan, Singapore, Australia and the EU (since 2000), to foster the development of drugs for treating rare diseases by a set of commercial incentives. The prerequisites for receiving this status are different. In the USA (orphan medicinal product designation) a rare disease is one that affects <200,000 individuals whereas in the EU (orphan drug designation) the respective limit is 5 of 10,000 inhabitants, and in Australia just about 1 per 10,000. In the EU, the orphan status of a drug is linked to a rare disease that is life-threatening and for which no alternative method of treatment (or prevention) exists.

Rare diseases include spinal muscular atrophy, lysosomal storage disorders, patent ductus arteriosus (PDA), familial adenomatous polyposis (FAP) and cystic fibrosis that occur at birth or in childhood whereas other rare diseases among them certain types of cancer (renal-cell carcinoma, glioma, or acute myeloid leukemia) appear during adulthood. The EMA works together in this field with international organizations such as the FDA or the Japanese Ministry for Health, Labour and Welfare (EMA, 2017a; Mariz et al., 2016).

Pharmaceutical companies developing orphan drugs benefit from a variety of advantages among them reduced regulatory fees, financial support for clinical trials, and protection from competition once the medicine enters the market; the duration of marketing exclusivity for an approved drug is 7 years (USA) and 10 years, respectively. On the other hand, clinical studies on orphan drugs require a larger network of clinical operations in more than one country in order to recruit enough patients (Ku, 2015). The current status of orphan drug development in the EU and the US has been reviewed by Aronson (2006), Hall and Carlson (2014), and more recently by Giannuzzi et al. (2017), and Mezher (2017).

Currently there are 98 orphan drugs approved by the EMA; 45 additional drugs lost their orphan status after 10-year market exclusivity incentive (vfa, 2018) but most of them are still at the market. Table 1.2 (modified from vfa, 2018b) shows a selection of orphan drugs currently approved in the EU.

The three latest orphan medical products designated by the EMA early 2018 (EMA, 2018b) via accelerated assessment after the European Medicines Agency's Committee for Medicinal Products for Human Use (CHMP) recommended marketing approval are Crysvita (burosumab), Prevymis (letermovir), and Jorveza (budesonide). The active substance of Crysvita (Kyowa Hakko Kirin Co. Ltd., Kyowa Kirin International PLC and Ultragenyx Pharmaceutical) is burosumab, a recombinant fully humanized

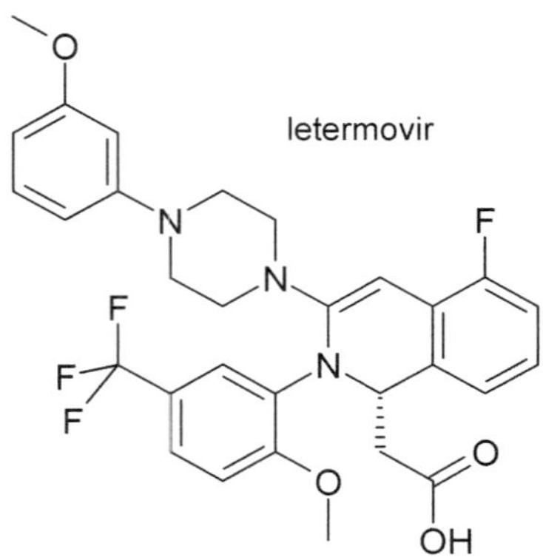
letermovir

monoclonal IgG1 antibody that binds to and inhibits the activity of fibroblast growth factor 23. The drug is indicated for the treatment of X-linked hypophosphataemia (XLH, a chronic progressive musculoskeletal disorder) with radiographic evidence of bone disease in children 1 year of age and older. The antiviral agent Prevymis (letermovir), a non-nucleosidic, inhibitor is used for prophylaxis of cytomegalovirus (CMV) infection and disease in adults and acts by inhibiting the human CMV viral terminase by a mechanism that remains to be elucidated (Ligat et al., 2018). CMV may cause life-threatening infections in immuno-compromised patients and serious congenital malformations. Prevymis is marketed by Merck & Co., Inc. (or Merck Sharp & Dohme (kurz MSD)) and was developed by AiCuris up to the clinical phase 2b. AiCuris received from Merck €30 million in addition to €105 million that became due after approval of Prevymis by the FDA in November 2017. Jorveza (with the glucocorticosteroid budesonide), marketed by Dr. Falk Pharma GmbH, Germany, is a medicine used to treat adults with the rare inflammatory disease eosinophilic esophagitis, a designation stemming from the fact that in the tissue of the esophagus in patients with this disease numbers of white blood cells called eosinophils are very high. Budesonide acts by preventing antigen-stimulated secretion of proinflammatory signal molecules such as thymic stromal lymphopoeitin (TSLP, a protein belonging to the cytokine family and produced mainly by non-hematopoietic cells (its expression is linked not only to eosinophilic esophagitis but also to other diseases, e.g., asthma, inflammatory arthritis, atopic dermatitis, etc.), interleukin-13 (the IL-13 protein is associated primarily with the induction of airway diseases), and eotaxin-3 (or chemokine (C-C motif) ligand 26 (CCL26) in the esophageal epithelium. Eotaxin-3 is a small cytokine belonging to the CC chemokine family and chemotactic for eosinophils (and basophils); for the regulatory role of this human chemokine in connection with attracting eosinophils and basophils, see Petkovic et al. (2004).

Table 1.2 Some of the currently approved orphan drugs by the EMA

Disease	Patients (EU)	API	Name	Company	Approval
Morbus Fabry	1,500	Agalsidase alfa Agalsidase beta	Replagal® Fabrazyme®	Shire Genzyme	08-2001[a]
Hemophilia D	10,000	Albutrepenonacog	Idelvion®	CSL Behring	05-2016
Morbus Pompe	6,850	Alglucosidase alfa	Myozyme®	Genzyme	03-2006[a]
Pulmonary hypertension	103,000	Ambrisentan	Volibris®	GlaxoSmithKline	04-2008
Lambert-Eaton myasthenic syndrome	5,300	Amifampridin	Firdapse®	BioMarin Europe	12-2009
Acute promyelocytic leukemia	41,000	Arsenic oxide	Trisenox®	Teva	03-2002[a]
Hypophosphatasia	500	Asfotase alfa	Strensiq®	Alexion	08-2015
Duchenne muscular dystrophy	20,000	Ataluren	Translarna®	PTC Therapeutics	07-2014
Merkel cell carcinoma	21,000	Avelumab	Bavencio®	Merck	09-2017
Acute myeloid leukemia	101,000	Azacitidin	Vidaza®	Celgene	12-2008
Acute lymphocytic leukemia	92,000	Blinatumomab	Blincyto®	Amgen	11-2015

(Continued)

Table 1.2 (*Continued*)

Disease	Patients (EU)	API	Name	Company	Approval
Chronic myeloid leukemia	81,000	Bosutinib	Bosulif®	Pfizer	03-2013
Hodgkin's lymphoma	51,300	Brentuximab vedotin	Adcetris®	Takeda	12-2012
Multiple myeloma	205,000	Daratumumab	Darzalex®	Janssen	05-2008
Myasthenia gravis	191,000	Eculizumab	Soliris®	Alexion	08-2017
Morbus Gaucher Typ 1	15,000	Eliglustat	Cerdelga®	Genzyme	01-2015
Mukopolysaccharidosis Typ VI A	1,500	Elosulfase alfa	Vimizim®	BioMarin Europe	04-2014
Mukopolysaccharidosis Typ II	1,000	Idursulfase	Elaprase®	Shire	01-2007[a]
Pulmonary arterial hypertension	112,000	Iloprost	Ventavis®	Bayer	09-2003[a]
Acute lymphatic leukemia	20,000	Inotuzumab ozogamicin	Besponsa®	Pfizer	06-2017
Multiple myeloma	205,000	Ixazomib	Ninlaro®	Takeda	11-2016
Cushing Syndrome	46,000	Ketoconazol	Ketoconazole HRA®	Laboratoire HRA Pharma	11-2014

[a]Drugs that lost their orphan status after 10 years but are still on the market.

Finally, Strimvelis (GlaxoSmithKline) that was granted orphan status on August 2005 should be mentioned as the second gene therapy approved by the EMA after Glybera for treating the ultrarare (about 15 patients per year in Europe) ADA-SCID (Severe Combined Immunodeficiency due to Adenosine Deaminase Deficiency). This inherited condition is caused by a mutation in the gene responsible for expressing the enzyme adenosine deaminase (ADA) essential for maintaining healthy lymphocytes to fight off infections. Strimvelis is produced from CD34+ cells extracted from the patient's bone marrow cells; an ADA gene is inserted into these CD34+ cells by means of a genetically modified retrovirus. After the autologous CD34+ enriched cell fraction is given as an infusion into a vein in a dose depending on the bodyweight of the patient, CD34+ cells start to grow in the bone marrow and produce normal lymphocytes with the ability to synthesize functional ADA (EMA, 2016; Aiuti et al., 2017).

1.4 The World Market of Pharmaceuticals, Therapeutic Enzymes, and Enzyme Inhibitors

The sales of the global pharmaceutical industry were estimated at about US\$ 967 × 10^9 in 2016 (~US\$ 1.2 × 10^{12} in 2017); thereof 46% account for sales in the US. The top 10 drugs in terms of sales are compiled in Table 1.3.

This ranking will change in the future among others due to patent expirations (e.g., top-selling drugs like AbbVie's Humira and Pfizer's Lyrica). Although the pharmaceutical market is dominated by small molecule drugs, 9 of the 10 top selling pharmaceutical products are biologics and only Lenalidomide is a derivative of thalidomide. Since 2002, biologics increased from 11% of the global sales to about 20% (\$212 × 10^9) in 2017. This trend is confirmed by a study of Yu et al. (2017) during a 9-year period in the USA with 415,405 patients among others suffering from inflammatory bowel disease (Crohn's disease; 188,842 patients) and ulcerative colitis (195,183 patients). For instance, in the case of CD patients the proportion of patients using biologics increased

Table 1.3 Top 10 global prescription drugs in 2016

Trade name	Generic name	Disease	Mechanism of action	Marketing	Sales[a] (US$·$10^9$)	Patent expired (*USA)
Humira	Adalimumab	Rheumatoid arthritis, psoriatic arthritis, ankylosing spondylitis, Crohn's disease, ulcerative colitis, etc.	TNFα inhibitor	AbbVie	16.078	2016 *2018
Harvoni	Ledipasvir/ sofosbuvir	Hepatitis C	NS5 A/B poly-merase inhibitor	Gilead Science	9.081	2030
Enbrel	Etanercept	rheumatoid arthritis, psoriatic arthritis, plaque psoriasis, ankylosing spondylitis	TNFα inhibitor	Amgen, Pfizer, Takeda Pharm.	8.874	expired
Rituxan	Rituximab	Autoimmune diseases, different cancers, non-Hodgkin's lymphoma	Among others: induces apoptosis of CD20+ cells	Roche	8.583	2013 *2016
Remicade	Infliximab	Crohn's disease, ulcerative colitis, plaque psoriasis, ankylosing spondylitis, etc.	TNFα inhibitor	Janssen Biotech, etc.	7.829	2015 *2018

Trade name	Generic name	Disease	Mechanism of action	Marketing	Sales[a] (US$·$10^9$)	Patent expired (*USA)
Revlimid	Lenalidomide	Multiple myeloma, Myelodysplastic syndromes	Induces tumor cell apoptosis	Celgene	6.974	2024 *2027
Avastin	Bevacizumab	Various types of cancer	Angiogenesis inhibitor	Roche-Genentech	6.752	2019 *2022
Herceptin	Trastuzumab	breast cancer (HER2 receptor positive)	Downregulation of HER2	Roche	6.751	2014 *2019
Lantus	Insulin glargin	Diabetes mellitus Type 1/2	Binds to Insulin-like GF	Sanofi Aventis	6.054	expired
Prevnar 13	Pneumococcal 7-valent conjugate vaccine	Develops immunity to pneumococcal disease such as Pneumonia or Meningitis		Pfizer	5.718	*2026

[a]Numbers from Craig W. Lindsley (2017).

from 21.8% (2007) to 43.8% (2015). Biosimilars and non-original biologics are steadily increasing their share in the pharmaceutical market. For rituximab alone, there are 25 companies from all over the world involved in approval or in development of corresponding products, among them *Amgen*, USA, *Apotex (Apobiologix)*, Canada, *AryoGen Pharmed*, Iran, *Biocad*, Russia, *Boehringer Ingelheim*, Germany, *Celltrion/Hospira* (Pfizer), South Korea/USA, *Celltrion/Teva Pharmaceutical Industries*, South Korea/Israel, *Dr Reddy's Laboratories*, the *Hetero Group*, *Intas Biopharmaceuticals*, and *Reliance Life Sciences* from India, *JHL Biotech*, Taiwan, *Shanghai Henlius Biotech*, China, *Samsung BioLogics*, South Korea, and others (GaBI, 2017).

The current ranking of the top 10 pharmaceutical companies in terms of sales is as follows: Pfizer ($\$48 \times 10^9$), Novartis ($\$44.5 \times 10^9$), Roche ($\$39.8 \times 10^9$), Sanofi ($\$39.7 \times 10^9$), Merck ($\$37.9 \times 10^9$), GlaxoSmithKline ($\$33.5 \times 10^9$), Johnson & Johnson ($\$33.2 \times 10^9$), Gilead ($\$30.1 \times 10^9$), Abbvie ($\$26 \times 10^9$), and AstraZeneca ($\$22.7 \times 10^9$).

As outlined in Vol. 5, Chapter 20, the conventional ways in which enzymes are inhibited by small-molecule inhibitors, are competitive, noncompetitive, uncompetitive or by mixed-type inhibition. More recently, (reversible) covalent inhibitors have also become important in connection with development of new drugs. Of the total of all drugs approved more than 40% are estimated to act as enzyme inhibitors; by far most of them work by a competitive mechanism, followed by noncompetitive inhibitors whereas uncompetitive inhibitors are rather rare.

According to a study by Research and Markets Ltd., Dublin, Ireland (2017), the global market for enzyme inhibitors should reach $179.9 billion by 2022 from $168 billion in 2017 at a compound annual growth rate (CAGR) of 1.4%. This study covers the market of North America, Europe, Japan, Australia etc., and the emerging markets (BRICS and MIST countries, also termed pharmerging markets) including countries such as India, China, South Korea, Taiwan, Africa, New Zealand, Canada, Latin America and others. The opportunities and challenges for the pharma industry planning to expand or already active in these emerging markets have been reviewed by Tannoury and Attieh (2017). Currently these fast growing markets account for almost one-quarter-to-one-half of the revenues for big pharma with an average sales growth between 2010 and 2014 in China of about 88% followed by Brazil (33%) and Russia (19%) as reported by Gautam and Pan (2016).

Among the drug categories considered are acetylcholine esterase inhibitors (AChE Is: Alzheimer's (Parkinson's)), 5α-Reductase inhibitors (5-AR Is: benign prostatic hyperplasia, lower urinary tract symptoms), xanthine oxidase inhibitors (XO Is: hyperuricemia), or monoamine oxidase inhibitors (MAO Is: depression, bipolar disorders, Parkinson's). Drugs that entered the market more recently include integrase inhibitors (HIV-1 infection), proprotein convertase subtilisin kexin 9 inhibitors (PCSK9 Is: lowering LDL cholesterol levels), poly ADP ribose polymerase inhibitors (PARP Is: several types of cancer such as ovarian, colorectal, prostate cancer, etc.), phosphodiesterase type 4 inhibitors (PDE4 Is: various inflammatory conditions such as COPD, psoriasis and psoriatic arthritis, etc.), sodium–glucose cotransporter 2 inhibitors (SGLT-2 Is: diabetic nephropathy, and dipeptidyl peptidase 4 inhibitors (DPP-4 Is/gliptins: diabetes mellitus type 2).

One of the main reasons for the optimistic estimates concerning the development of the *global market for proteins* are the advances in recombinant DNA technology in the recent past, enabling the production of desired commercial enzymes in high quantities.

According to GBI Research's analysis (a business unit of GlobalData, UK) the global market for *therapeutic proteins* is driven by the increasing demand of monoclonal antibodies, but also insulins, interferons, growth hormones, colony stimulating factors, erythropoietins, and blood factors. The market was estimated at $93 billion in 2010 to reach $141.5 billion in 2017, which means a CAGR of 6.2% within this period (GBI Research, 2016).

A new report by Allied Market Research (2018) expects the *Specialty Enzymes Market* to reach $947.5 million, globally, by 2020 with a CAGR of 9.4% during 2015–2020. Among the different application segments, the pharmaceutical sector will grow fastest (CAGR of 9.7%).

Industrial enzymes accounted for 56.9% of the overall market whereas pharmaceuticals valued at $1.63 billion in 2015 were dominant in the group of speciality enzymes, followed by research and biotechnology, diagnostics, and biocatalysts (Grand View Research, 2018).

The Market Reports forecasted the global Enzyme Replacement Therapy market to be valued at $7,400 million in 2017; they expect that this value will more than double by the end of 2023 ($15,800 million) which is tantamount to a CAGR of 13.4% between 2017 and 2023.

1.5 Drug Development Process, Risks, and Costs of Drug Development, and Prescription Drug Prices

At the beginning of a drug project, many questions have to be cleared such as whether there are diseases urgently needing the approval of a new drug, whether there are new findings from basic research in this field or whether an existing prescription drug should be replaced by the one with less side effects and whether it is realistic to expect that compulsory/private health insurances or patients pay for such a drug when it reaches the market, to name but a few.

In the United States, the drug development process proceeds via the following five steps (FDA, 2018; 2018a):

1. Discovery and development
2. Preclinical Research
3. Clinical Research, Phases 1 to 3
4. FDA Review
5. FDA Post-Market Safety Monitoring

Step 1 starts from many potential compounds for treating a certain disease of which only some few remain left after first testing. These are further investigated with respect to their absorption and distribution and how they are metabolized and excreted (ADME), and apart of mechanism of action, criteria are dosage and kind of administration, side effects and possible toxicity, interaction with other drugs and advantage over similar drugs, and whether certain groups of people are affected differently. The costs for this first step that aims to answer basic questions about a drug's safety, add up to $US 196×10^6 within a period of 1.5 years.

Research in step 2 is devoted to investigations (pharmacodynamics, pharmacokinetics, ADME, toxicology including animal experiments) into aspects related to whether the drug in question has no harmful effects so that it can be tested on humans, highly regulated by FDA's *Good Laboratory Practice for Non-clinical Laboratory Studies*. The costs amount to \$US 122 × 10^6 in 1.5 years.

Step 3, which consists of three parts, is initiated by the design of clinical trials, the submission of an Investigative New Drug (IND) application containing results from steps 1 and 2 to the FDA, and finally the FDA approval—possibly with suggestions how to improve the clinical trial. Costs as well as duration of this intermediate phase are comparatively low.

Clinical research phase studies consist of the 3 parts Phase 1, Phase 2, and Phase 3. *Phase 1* is performed with 20 to 100 healthy volunteers or people with the disease for which the drug is designated. If the drug is an oncologic one Phase 1 is conducted with patients suffering from the respective type of cancer. The results provide first answers to questions such as how the drug interacts with the body, what are the consequences of dose increase, e.g., concerning side effects, and what kind of administration should be preferred. The costs for phase 1 that last 1.6 years on average are about \$US 169 × 10^6 and approximately 70% of the drugs investigated move to the next phase.

In *Phase 2* the number of study participants with the disease for which the drug is being developed amounts to several hundred. Phase 2 studies serve to gather additional safety data, to formulate research questions more precisely and to develop additional problem-related research methods. At the end of this phase that costs \$US 402 × 10^6 during 2.5 years, ca. 33% of the drugs enter phase 3.

Phase 3 with total costs of ~\$US 536 × 10^6 (2.5 years) is conducted with 300 to 3000 participants with the respective disease designed such that questions can be answered as to whether the application of a drug offers a treatment benefit (assessment of effectiveness), and whether long-term or rare side effects have possibly been overlooked in the previous phases.

In some cases, a *Phase 4* trial also known as *Postmarketing Surveillance Trial* is conducted (particularly after approval under FDA Accelerated Approval Program) with several thousands of volunteers to additionally test drug safety and efficacy. In Table 1.4, the data associated with the development of a new drug are compiled (see also Palgon, 2017; Tufts CSDD, 2017).

Table 1.4 Estimated costs for developing a drug

Stage	Direct costs (US$ × 10^6)	Duration (years)
Discovery	196	1.5
Preclinical	122	1.6
Phase 1	169	1.6
Phase 2	402	2.5
Phase 3	536	2.5
Phase 4[a]	312	
Registration	11	1.3
Total	1,748	11

[a]Lasts all the time the drug is at the market.

In addition investors consider so-called *opportunity costs* that is the money they could have received from investments in other projects; based on an 11% cost of capital opportunity costs sum up to roughly $US 1.2 × 10^9 so that the costs for drug development amounts to a total of about $US 2.9 × 10^9 (see, e.g., Palgon, 2017). This $2.9 billion figure has been critically discussed by Avorn (2015), and DiMasi et al. (2016).

1.5.1 Drug Development in Europe and Germany

In **Europe**, the process of drug development is similar. This holds for the long way from screening large compound libraries, chemical modification of possible hits and further improvement of their properties concerning the interaction with the target in question and the following preclinical phase (toxicological tests using cell cultures and/or via animal experiments) as well as for the three (four) successive phases. Until a new drug receives approval takes about 13 years and the costs for a pharmaceutical company amount to about US$ 2 billion; this includes all projects that failed and the above mentioned opportunity costs.

Within the EU new pharmaceuticals to treat diseases must be tested with under-aged persons, too, in the case that the incidence for this disease is not restricted to adults. According to the *Paediatric Committee* of EMA studies with underaged persons normally start after phase II with adults is completed. Exceptions are life-threatening diseases which allow contemporaneous studies with adults and underaged persons or diseases that only occur with the latter; phase I studies with healthy minors are ineligible. If a drug is envisaged for men and women, studies have to be performed with members of both sexes—at least in phase II. For reasons of security women of child-bearing age have to apply safe methods of contraception. In the EU member states, the rights, safety, and wellbeing of human subjects involved in a trial is protected by the Ethics Committee; in addition, discovery and preclinical development raise questions of animal ethics. Ethical aspects in drug development have been recently discussed by Arányi (2017).

In **Germany** nearly one half of the member companies of the vfa (German Association of Research-Based Pharmaceutical Companies) are involved in research on active ingredients or galenics; these companies that in part are also running headquarters in the USA, France, UK, Japan and Switzerland coordinate, from their German offices, their clinical studies in Germany as well as occasionally in other countries and spend around €5.5 billion per year on research and development in Germany alone. This underlines that Germany belongs to the leading research locations for internationally active pharmaceutical companies.

Research in German pharmaceutical companies focuses on drug development for treating cancer, inflammatory diseases, cardio-vascular illnesses, metabolic disorders (e.g., diabetes type 2), Alzheimer's disease, drug formulations and devices for drug administration; further important topics are infectious, respiratory and gynecological disorders and the development of new antibiotics. Germany is one of the leading countries for producing biopharmaceuticals; more than 360 companies, among them many small ones, are involved in medical biotechnology (vfa, 2018a). An example of one such company is Evotec AG, founded in 1993 by the German biophysical chemist Manfred Eigen, who was awarded the Nobel Prize in Chemistry (1967, together with Ronald George Wreyford Norrish and George Porter) for his studies on the kinetics of extremely fast chemical reactions using relaxation methods. Evotec employed 1,238 people by end of 2016 with sales amounting to €160 million. The company, a drug discovery and development partnership company, is active in developing therapeutics for neuroscience, diabetes and complications of diabetes, pain and inflammation, oncology and infectious diseases. Apart from its headquarters in Hamburg, Germany, the company runs operation sites in Abingdon and Manchester (UK), Toulouse (France), Göttingen and Munich (Germany) and South San Francisco, Branford and Princeton (USA).

Concerning the number of industry-sponsored clinical trials, Germany ranges second after the USA and is followed by the UK, France, and Canada, whereas in countries such as India and Brazil the number of trials is comparatively small. Most of these studies are performed on a multinational basis and the same standards must be met in all countries taking part (vfa, 2018b).

1.5.2 Risks in Drug Development and Mitigation Strategies

Challenges concerning investments in developing drugs are among others the tight regulation of such processes by approval authorities and the long development period of at least 12 years until drug approval. Further characteristics of drug development processes are R&D investments of several 100 million US$/€ together with the fact that 99% of all drug discovery projects will not result in an approved product with the highest clinical attrition rates being observed in phase II (see e.g., Hay et al., 2014). Kannt and Wieland (2016; see also literature cited therein) made proposals for managing risks in drug discovery. They recommend a more translational approach in drug discovery based on solid clinical evidence of target relevance, the availability of suited biomarkers for target engagement, and an in-depth understanding of the link between target and disease; these criteria should be applied very early in the drug discovery process. A positive example of drug development in a comparatively short time are the neutralizing mAbs (e.g., alirocumab and evolocumab, approved by the US FDA and the EMA in 2015) against protease proprotein convertase subtilisin/kexin type 9 (PCSK9) (see literature cited by Kannt and Wieland, 2016) leading to a distinct reduction in atherogenic cholesterol levels with a good safety profile for patients that do not respond optimally to statin therapy (Ito and Santos, 2017). PCSK9 mutations were found in humans that cause familial hypercholesterolemia (overproduction of apoB100) via a gain-of-function mechanism (evidence of target relevance). In contrast, in humans with a loss-of-function mutation in their *PCSK9* gene the LDL-C levels are very low; LDL-C serving as a biomarker enabled a rapid proof of concept. In addition, exploring a new therapeutic target is accompanied by an intensive literature search, however, the majority of these published studies could not be reproduced under standardized conditions which means that replication of such data is indispensable before starting a drug development project (Kannt and Wieland, 2016).

1.5.3 Randomized Clinical Trials and Beyond

Randomized clinical trials (RCTs)—for an evidence-based, minimum set of recommendations for reporting randomized trials see the CONSORT statement (http://www.consort-statement.org/consort-statement/)—represent strong evidence and are normally considered the "gold standard" of clinical trials assessing pharmaceutical drug and medical device safety and efficacy. Carefully planned experimental framework serving to compare a treatment with a control frequently using a placebo control and patients carefully screened based on precise clinical criteria are characteristics of RCTs.

It is, however, meanwhile widely acknowledged that despite its excellent internal validity, there is often a gap between the scientific evidence generated on drugs/medical devices during clinical trials to support their approval and real-world environments (Sherman et al., 2017). This is why real-world data (RWD) and real-world evidence (RWE) are playing an increasing role in health care decisions beyond randomized clinical trials. Traditional phase 3 RCTs and post-launch observational studies provide evidence of a drug effect under ideal circumstances but not about effectiveness, which means true benefit to patients in routine practice. Reasons that limit the clinical applicability of RCT results may be among others that patients and practitioners in RCTs are different from those in routine practice, an under-representation of older patients and those with comorbidity, detection of a clinically modest effect that does not apply to less selected patients, or a limited ability to detect rare and chronic toxicities, e.g., those emerging after trial completion, relevant for patients with comorbidity. Booth and Tannock (2014) discussed the relative strengths and limitations of randomized controlled trials and population-based observational research and came to the result that both—if well-designed—can serve in a complementary way to ensure that the results of clinical trials translate into concrete benefits in the general population.

FDA uses RWD and RWE to monitor post-market safety and adverse events and to make regulatory decisions; the health care community is using these data to support coverage decisions and to develop guidelines and decision support tools for use in clinical practice, and medical product developers are using RWD and RWE to support clinical trial designs (e.g., large simple trials, pragmatic clinical trials) and observational studies to generate innovative, new treatment approaches (FDA, 2018b; 2017d). Both RWD and RWE should contribute to an acceleration concerning the development of scientific evidence for medical products; the 21st Century Cures Act places additional focus on the use of these types of data to support regulatory decision making (NIH, 2016). The FDA defines real-world evidence as "the clinical evidence regarding the usage and potential benefits or risks of a medical product derived from analysis of RWD." It is of importance to evaluate treatment effectiveness in daily settings to guide clinical decision-making and answer scientific questions. Real-world data come (apart from those derived from RCTs and phase IV RCTs) from sources such as electronic health records (EHRs), insurance claims and billing activities, product and disease registries, patient-related activities in out-patient or in-home use settings, or health-monitoring devices and include data derived from RCTs and phase IV RCTs. An assessment and analysis of these large amounts of data has been eased due to advances in data management. According to this rather new trend the FDA recently released a document entitled "Use of Real-World Evidence to Support Regulatory Decision-Making for Medical Devices"; it clarifies how to "evaluate real-world data to determine whether it may be sufficiently relevant and reliable to generate the types of real-world evidence that can be used in FDA regulatory decision-making for medical devices" (FDA). For the many benefits RWE provides, including improved health outcomes, reduction of medical costs, better identification of medication side-effects and contraindications, increased personalization of medical care, or discover new indications for existing treatments, see Jadhav (2017). Zuidgeesta et al. (2017) provided an introduction to the topic "Pragmatic Trials and Real World Evidence. Concerning the EMA's contributions to address these developments, see, e.g., Cave and Cerreta (2017), and EMA (2017, 2018a).

American Health & Drug Benefits (AHDB) and The Peer Reviewed Forum for Real-World Evidence in Benefit Design are trademarks of Engage Healthcare Communications, LLC. The independent, peer-reviewed journal founded in 2008 publishes studies dealing with the incorporation of Real-World Evidence in patient care, and offers a forum for stakeholder collaboration toward the promotion of value-based healthcare. Furthermore it "provides benefit design makers the integrated information they need to devise benefit designs and make coverage decisions that stand up to today's special healthcare delivery and business needs" (AHDB, 2018).

1.5.4 Prescription Drug Prices

In the following Section the topic *costs for prescription drugs* is discussed in more detail with examples related to the respective situations in the USA, Europe, and Germany. In most countries all over the world, these costs are high and mainly justified by the enormous expenditure of R&D in connection with the development of new drugs, which, however, does not remain uncontradicted. According to Waxman et al. (2017), Prasad et al. (2017), Schumock and Vermeulen (2017), Saltz (2016), and others, the reasons responsible for this situation in the United States (which can, however, be transferred to a different extent to many other countries, too) are, among others, the following:

- Natural monopolies for drugs that enable manufacturers to significantly increase prices.
- Spending on marketing routinely exceeds R&D spending.
- Lack of robust competition among manufacturers of generic drugs.

- Lack of price competition among biologics and biosimilars.
- Monopolistic pricing practice in the generic market.
- Anticompetitive behavior by some manufacturers.
- New and expensive speciality medications and cancer drugs.
- Patients, providers, and payers lack information about the comparative effectiveness of drugs.
- Large payers may be prevented by law from negotiating the price of drugs.
- The pharmaceutical distribution system does not make essential pricing information available to patients, providers, and payers.

Examples are shown in Table 1.5 with the monthly treating costs of cancer which are particularly high, as are those for many orphan drugs. Prices of these drugs in the USA are consistently higher (in some cases several 100%) than those in, e.g., Norway, despite similar median monthly incomes of the population. Carrera et al. (2018) termed this problem *Financial Toxicity of Cancer Treatment.*

Table 1.5 US FDA-approved oral cancer drugs in 2016 to 2017 and costs of treatment in the US (after Carrera et al., 2018)

Generic name	Brand name	Cancer type	Treatment costs per month (US$)
Brigatinib	Alunbrig	NSCLC, ALK	12,868.76
Cabozantinib	Cabometyx	Renal cell carcinoma, FLT3, KIT, VEGFR	15,156.59
Enasidenib	Idhifa	Acute myeloid leukemia, IDH2	25,141.37
Neratinib	Nerlynx	Breast cancer, HER2 positive	10,613.75
Ribociclib	Kisqali	Breast cancer, HER2 negative	8,476.31
Rucaparib	Rubraca	Ovarian cancer, BRCA	20,162.74

Abbreviations: ALK, anaplastic lymphoma kinase; FLT3, fms–related tyrosine kinase 3; HER2, human epidermal growth factor receptor 2; IDH, isocitrate dehydrogenase; KIT, stem cell factor receptor; NSCLC, non-small cell lung cancer; VEGFR2, vascular endothelial growth factor receptor 2; BRCA, BReast CAncer.

On the other hand, these tremendous costs raise questions concerning the average benefit derived from such treatments which often remains marginal. The American Society of Clinical Oncology (ASCO) defined clinically meaningful outcomes for progression-free survival (PFS) and overall survival (OS) as a relative improvement of 25% and an absolute increase of 2.5 months in PFS and/or OS compared with that achieved with standard treatments. Kumar et al. (2016), in an analysis of 47 FDA anticancer drugs approved between April 2014 and February 2016, found that only 9 (19%) met these modest definitions set by ASCO (see also Prasad et al., 2017, for other studies concerning cost-effectiveness analyses of anti-cancer drugs).

The total revenue of the global pharmaceutical market is $1.05 trillion; nearly half of it—roughly $515 billion—comes from the United States and Canada. The 2015 profit margin that Forbes estimated for the healthcare technology industry was 3.21%, making it by far the most profitable industry of all (Speights, 2016).

According to the Centers for Medicare & Medicaid Services, CMS (2018), (part of the Department of Health and Human Services), national health expenditure (NHE) in the USA grew 4.3% to $3.3 trillion in 2016, or $10,348 per person, and accounted for 17.9% of Gross Domestic Product (GDP). These numbers include *medicare* (a Federal health insurance program that pays for hospital and medical care for elderly and certain disabled Americans), *medicaid* (governmental health and medical services program for certain individuals and families with low incomes and few resources) and out-of-pocket spendings, hospital as well as physician and clinical services expenditures, and private health insurance spending. Prescription drug spending in 2016 amounted to $328.6 billion which is about 17% of overall personal health care services. The largest shares of total health spending came from the federal government (28.3%), the households (28.1%), the private business (19.9%), state and local governments (16.9%), and other private revenues (6.7%). Under the current legal conditions, national health spending is expected to grow at an average rate of 5.5% per year for 2017–2026 and to reach $5.7 trillion by 2026. Per capita personal health care spending varied significantly by state of residence and ranged in 2014 from $5,982 in Utah to $11,064 in Alaska.

To address these increases in costs related to prescription drugs, the services of pharmacy benefit managers (PBMs) have been utilized in the USA. The three major PBMs (Express Scripts, CVS Health, and OptumRx) manage the drug benefits for approximately 95% of the US population. A PBM is a third-party administrator of prescription drug programs for commercial health plans, self-insured employer plans, Medicare Part D (medicare prescription drug benefit—a United States federal government program to subsidize the costs of prescription drugs) plans, the Federal Employees Health Benefits Program, and state government employee plans. Historically, PBMs were established to serve as mediators for processing prescription medication claims for insurance companies and plan sponsors (a company or employer that sets up a healthcare or retirement plan) for a small fee as equivalent. Meanwhile, PBMs are involved in activities like drug utilization review and drug plan formulary development. They further have an active influence on decisions concerning the pharmacies included in the prescription drug plan's network, and the amount of reimbursement network pharmacies receive for their services. In other words, they determine in which pharmacy a given drug is available for an insured person as well as the drug's price. PBMs nowadays are often multi-enterprises that in addition operate speciality pharmacies, large mail order facilities and act as wholesale traders. PBMs have become extremely profitable and make money by rebates that are not or only a part shared back to the patient but kept as part of an "administrative fees" by the so-called pharmacy spread in a way that obviously lacks transparency. Pharmacy spread means charging a health plan potentially much more than the PBM pays the network pharmacy for filling a prescription and the PBM pockets the difference. Hence, the activities of PBMs contribute to rising costs in the public health sector (Hoffman-Eubanks, 2017; Rumore and Vogenberg, 2017).

Kesselheim et al. (2016) discussed *The High Cost of Prescription Drugs in the United States* together with its origins and prospects for reform: Per capita prescription drug spending in the US exceeds that in all other countries in particular driven by brand-name drug prices which increased in recent years at rates significantly above the consumer price index; in 2013, per capita spending on these drugs with $858 was more than twice as high compared with an average of $400 for 19 other industrialized nations. High drug prices can be set by companies mainly due to monopoly rights awarded upon Food and Drug Administration approval and by patents. The access to generic drugs after patent expiration dates is the main factor of decreasing drug prices but their availability is delayed by numerous business and legal strategies. An opinion that seems to be generally accepted in the USA as well as in other countries is that the R&D cost for developing drugs do not justify the high drug prices (see also above).

In an international comparison between the USA, Canada, Japan, and 21 European countries, pharmaceutical expenditures vary significantly. Expenditures per capita in USD (adjusted for purchasing power) are 1,034 (USA), 761 (Canada), 756 (Japan), 678 (Germany); the lowest expenditures are in Denmark with USD 288 per capita. As a percentage of gross domestic product, Hungary with 2.26% is followed by Japan (2.12%), Slovakia (2.01%), the USA (1.95%), and Canada (1.78%); Germany with 1.55% ranges in the middle of the comparison scale. Concerning values as a percentage of total healthcare expenditures the first four countries are Hungary (30.6%), Slovakia (26.5%), Poland (21.6%), and Japan (21.%); Canada (17.5%) ranks in the middle of the table, Germany (14.1%) at position 14 and the USA (11.9%) at position 18. Denmark again is bottom of the table with 6.3%.

In most EU member states and other countries, tax burdens for pharmaceuticals are reduced or altogether waived. According to a WHO/HAI Review Series (joint World Health Organization (WHO)/Health Action International (HAI) Project on Medicine Prices and Availability) on Pharmaceutical Pricing Policies and Interventions; Working paper 5: Sales taxes on medicines (Creese, 2014), in 34 out of 50 states of the USA prescription medicines (or publicly reimbursed medicines) are exempted from value-added taxes (VAT); in some states this also holds for non-prescription medicines. Other states exempt drugs from surtaxes on the sales tax rate.

In Europe, VAT rates for prescription medicines are between zero (United Kingdom, Ireland, Malta, Sweden) and values of 15% (Czech Republic) and above (Germany, 19%; Bulgaria, 20%; Denmark and Norway, 25%; and Iceland with the highest rate of 25.5%).

Low taxes are raised in France (2.1%), Switzerland, 2.5%), Luxemburg (3%), and Spain (4%). In the middle of the VAT scale range Poland, Serbia, Turkey (8%), Estonia and Romania (9%), Slovenia (9.5%) and Finland, Italy, Austria, and Slovakia with 10%.

The pricing structure in the different EU countries varies significantly, too, in the EU member states. The retail price for prescription or reimbursed pharmaceuticals is composed of manufacturer, wholesaler, and pharmacy shares and taxes and discounts. The later value (with combined statutory rebates for manufacturers and pharmacies) amounts to 32% of the end price in Germany which is the highest such rate in Europe. The shares of manufacturers, wholesalers and pharmacies are 50%, 4%, and 14%, respectively. The other extreme is Sweden, where the corresponding values are 80%, 3%, and 17%, followed by Ireland (70%, 5%, 25%) with taxes and discounts being zero. For more detailed information, see the vfa statistics (vfa, 2015).

1.5.4.1 Cost reduction: public–private drug partnerships

The combination of increasing costs of drug development and largely capped budgets for medication requires additional models for a more effective and less costly drug development (Yildrim et al., 2016). To respective strategies belongs—apart from *adaptive trial designs* and the *use of Big Data* in drug development—*drug repurposing* which means finding a new use for an "old" drug is one of several possibilities. An earlier example is sildenafil, a selective type 5 phosphodiesterase inhibitor, initially developed to treat angina and then repurposed as a life-style drug to treat erectile dysfunction; meanwhile it is also an agent against pulmonary arterial hypertension and furthermore strong evidence exists that PDE5 inhibitors exert a reduction in cardiovascular events and mortality (Hutchings, 2018). Another example for drug repurposing is metformin (Glucophage), a common oral anti-diabetic medication. An increasing number of investigations and clinical trials have demonstrated positive results for the use of metformin in anti-cancer therapy (e.g., prostate cancer therapy; Chae et al., 2016; Whitburn et al., 2017). The advantage of drug repurposing is that aspects like safety, efficacy, and toxicity have already been extensively studied which saves money and time with respect to gaining approval for a specific indication by the FDA and/or the EMA (Hernandez et al., 2017). Lee and Kim (2016) reviewed perspectives of drug repurposing in drug discovery and development in neuropsychiatric disorders.

Finally, the public–private drug partnerships (PPPs) (Vaudano, 2013; Gottwald et al., 2016; de Vrueh and Crommelin, 2017; Denne and Davis, 2018), or multiple stakeholder partnerships including apart from universities and industries regulatory and health technology assessment agencies, patient organizations, health foundations or charities should be mentioned here. In a Milken Institute review (the Milken Institute is focused on accelerating medical solutions), Melissa Stevens (2018) stated that "more capital, and more of it directed to early-stage research, would do much to increase medical R&D productivity." A rather new instrument to overcome private risk aversion of market investors and pharmaceutical companies in drug development is the entrance of venture philanthropy into life-science finance. Philanthropic Investors are disease-specific public charities, private foundations and individual philanthropists or families. Examples of such non-profit organizations from the United States are the Juvenile Diabetes Research Foundation, Leukemia & Lymphoma Society with its Therapy Acceleration Program, the National MS Society (Association for Advancement of Research on Multiple Sclerosis) and the Cystic Fibrosis Foundation. This organization supported the development and use of five FDA-approved therapies among them Ivacaftor (Kalydeco); in 2014 it received $3.3 billion for the sale of royalty rights to drugs developed with its financial support. A further example is BioMotiv a mission-driven development company associated with The Harrington Project for Discovery & Development, a national initiative centered at University Hospitals Cleveland Medical Center. The company's focus is accelerating breakthrough discoveries from research institutions into therapeutics for patients (http://www.biomotiv.com/). In this connection product development partnerships such as Medicines for Malaria Venture (MMV) and the Drugs for Neglected Diseases Initiative should also be mentioned These collaborations with public and private organizations help to mitigate drug development risks (see also IVESTOPEDIA, 2018) and use public and philanthropic funds to invest in early-stage research in cooperation with both the pharmaceutical industry and academic research organizations. An example from Europe is *The Innovative Medicines Initiative*

(IMI), a large-scale public–private partnership between the European Commission and the European Federation of Pharmaceutical Industries and Associations (EFPIA).

1.6 Concluding Remarks

Biocatalysis plays an important role in pharmaceutical and medical science concerning its various applications. Enzymes as well as whole cells are employed for the synthesis of chiral active pharmaceutical ingredients or key intermediates thereof as a sustainable alternative to traditional chemical methods. Several chapters of this volume address this topic. A number of enzymes, most of them nowadays produced by recombinant techniques and optimized with respect to their properties by directed evolution methods, are used for therapeutic purposes (Baldo, 2015). According to Yari et al. (2017), they may be subdivided

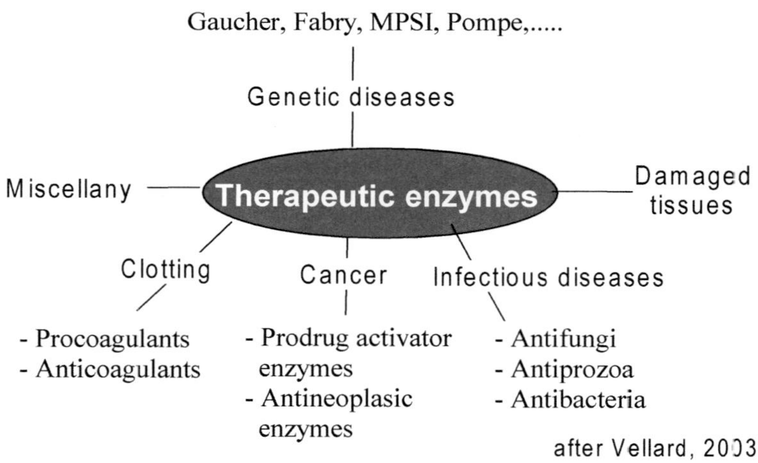

into enzymes used in enzyme replacement therapy, in cancer treatment (e.g., for amino acid depletion therapy in cancer (Cantor et al., 2012)), enzymes for fibrinolysis, and those used topically for various treatments. A categorization proposed by Vellard (2003) is shown opposite. Several of these applications are discussed in Vol. 6 of this Series on Biocatalysis. The importance of therapeutic enzymes is underlined by the fact that the International Pharmaceutical Federation (FIP) founded an *International Commission on Pharmaceutical Enzymes* already in 1960. The safety of treatments based on therapeutic enzymes used for enzyme replacement therapy in lysosomal storage disorders has been discussed by Broomfield et al. (2016) with respect to immunological mechanisms involved in the development of anti-drug antibodies (see also the FDA *Investigational Enzyme Replacement Therapy Products: Nonclinical Assessment Guidance for Industry* http://www.fda.gov/Drugs/GuidanceComplianceRegulatoryInformation/Guidances/default.htm, 2015).

On the other hand, the development of enzyme inhibitors correcting, e.g., metabolic imbalances is an area of research in biochemistry and pharmacology (Balbaa and El Ashry, 2012) gaining increasing actuality and recent examples have been described in this and other chapters of *Pharmaceutical Biocatalysis*. Research has also been performed during the last decade on developing drugs containing more than one functionality into the same molecule to give multi-target-directed ligands (MTDLs) as, e.g., those interacting with acetylcholinesterases and monoamine oxidases for treating neurodegenerative diseases (Ramsay and Tipton, 2017; and literature cited therein). MTDLs-based treatments are one of the most followed approaches in the active central nervous system (CNS) therapeutic area. Wang et al. (2016) published their results about the employment of AChE inhibitors-based MTDLs as a novel pharmacological approach for Alzheimer's disease therapy; they observed among others inhibition of Aβ aggregation, regulation of Aβ procession, reduced activation of platelet-activating factor (PAF) receptor (caused, e.g., by damaged cerebral endothelial cells), a chelation of metal ions, and an inhibition of monoamine oxidase B (MAO-B). The opposite scheme shows an example of Tacrine-Resveratrol fused hybrids as multi-target-directed ligands against Alzheimer's disease developed by Jeřábek et al. (2017). Tacrine is a cholinesterase inhibitor drug and the stilbenoid Resveratrol

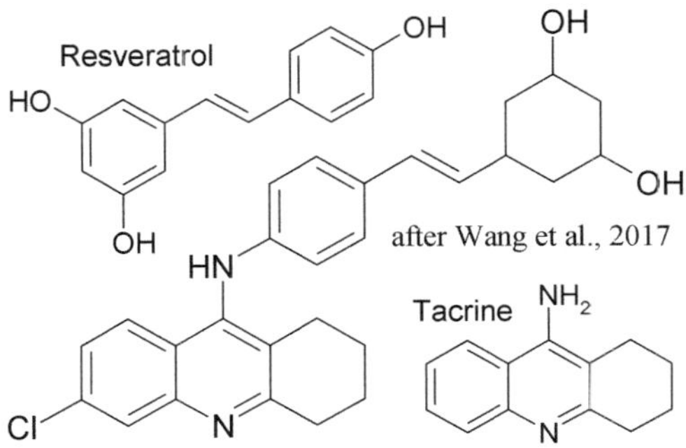

is assumed (albeit not uncontradicted) to have antioxidant and anti-neuroinflammatory activities. The MTDL drug comprising structural features of both compounds exerted an inhibitory effect on human acetylcholinesterase at μmolar concentrations and effectively modulated Aβ self-aggregation in vitro. In addition; moreover, it showed significant anti-inflammatory and immunomodulatory properties in neuronal and glial AD cell models and it could be predicted with high probability that this MTDL is able to cross the blood-brain barrier and exerts low cytotoxicity on primary neurons.

Copeland et al. (2007) provided examples of how a detailed understanding of the catalytic mechanism of the enzyme in question is a prerequisite for identifying and optimizing small molecules as potent and selective inhibitors able to interact with conformationally distinct forms of the enzyme. The functional elements of the epigenomic machinery (writers, erasers, and readers; discussed also in Vols. 5 and 6 of this Series on Biocatalysis), aberrations in the control of cellular gene expression, diseases based thereon and small molecules that modulate epigenetic processes have been described by Pachaiyappan et al. (2014). Key examples of epigenetic dysregulation in tumors together with an analysis of prospects for discovery and development of novel cancer agents that target epigenetic proteins are summarized in a review published by Campbell and Tummino (2014). A useful source of drug and disease information is KEGG (http://www.kegg.jp/ or http://www.genome.jp/kegg/) the Kyoto Encyclopedia of Genes and Genomes; it is an encyclopedia of genes and genomes developed in 1995. Apart from the databases PATHWAY, GENES, COMPOUND and ENZYME with which KEGG started it now contains DRUG (a comprehensive collection of approved drugs, drug target and drug metabolism information) and DISEASE databases—containing disease entries, each consisting of a list of disease genes, carcinogens, pathogens and other environmental factors (Kanehisa et al., 2017).

Deviations of a patient's response to an administered drug from the normal behavior in presence of a second one may be a result of pharmocodynamic or/and pharmacokinetic drug-drug interactions (DDI; Srinivas et al., 2017). Pharmacodynamic interactions are coined by the pharmacological response changed through agonism or antagonism. Agonists are pure ones when they bind to the same (main) locus of the receptor with an effect similar to the main drug but partial ones when binding to a secondary locus—such an agonist provokes the same effect but with a lower intensity. However, an antagonist produces an opposite effect by binding to the receptor's main locus in a competitive or an uncompetitive way. Pharmacodynamic DDIs include the occurrence of unwanted signal transduction mechanisms as side effect. In contrast, pharmacokinetic interactions include those where the availability of a drug is altered by induction or inhibition of metabolic enzymes or transporters involved in drug absorption, distribution, metabolism or excretion leading to a behavior of the drugs being different to that when administered individually (Kohler et al., 2000). Metabolism-based DDIs resulting in induction or inhibition of CYP 450 enzymes are among the most dangerous ones as estimated, e.g., by Pirmohamed and Park (2003) in their paper about Cytochrome P450 enzyme polymorphisms and adverse drug reactions.

The efficacy of a drug (as well as its toxicity) depends among others on the cytochrome P450 mediated drug metabolism (oxidative drug metabolizing enzymes; see also other chapters in *Pharmaceutical Biocatalysis* volumes). Of the more than 50 CYP450 enzymes, CYP1A2, CYP2C9, CYP2C19, CYP2D6, CYP3A4, and CYP3A5 enzymes metabolize 90% of drugs, and of these P450 2D6 is responsible for the metabolism of almost 25% of drugs among them different types of analgesics, anti-arrhythmic agents, antidepressant and β-blockers. The response to standard doses of a drug may be different from patient to patient due to the fact that CYP 450 enzymes exhibit genetic variability (polymorphism). Those with two copies of wild-type genetic alleles are normal so-called "extensive" metabolizers whereas persons with two copies of variant alleles are "poor" metabolizers, and those with one variant and one wild-type allele show reduced enzyme activity (Lynch and Price, 2007). Flockhart (2007) provided a table with lists of drugs in columns under the designation of specific cytochrome P450 isoforms with published evidence (see cited literature) that the respective drug is metabolized, at least in part, via that isoform. In 2007 the FDA approved the first genotype test (Roche AmpliChip cytochrome P450 genotyping test) a DNA microarray that enables the detection of 29 polymorphisms of CYP2D6 and two of CYP2C19 from a blood sample. Investigations into cytochrome P450 2D6 genotype assessment and phenotype prediction using the Roche test have been reported by Rebsamen et al. (2009).

Other drug-metabolizing biocatalysts are flavin-containing monooxygenases (FMOs). They act mechanistically different from CYP enzymes in that they catalyze the oxidation of organic compounds using NADPH and molecular oxygen. UDP-glucuronosyl transferase (UGT) enzymes, as well as aldehyde oxidase, γ-glutamyl transpeptidase, cathepsin B, and ADP-ribosyltransferase, as examples of non-P450- and non-UGT-mediated metabolism have been reviewed by Fan et al. (2016). The UGT1A and 2B subfamilies are involved in terminating the biological actions and enhancing the renal elimination of lipophilic drugs from all therapeutic classes (Rowland et al., 2013). UDP-glucuronosyl transferase (UGT) enzymes, glutathione transferases (GSTs), sulfotransferases, and *N*-acetyltransferases are Conjugative DMEs. For example UGT uses UDP-glucuronic acid as a donor to catalyze the glucuronidation of the drug (UDP-glucuronic acid + drug $\rightarrow$ UDP + drug-glucuronide) resulting in products with enhanced solubility, whereas GSTs catalyze the formation of thioether conjugates between glutathione and reactive xenobiotics.

Concerning future prospects of enzyme inhibitors as drugs, there is a never-ending need to develop safe, efficient, and affordable new treatment strategies as alternatives to existing ones. These efforts will be supported by using in vitro systems and in silico models to understand pharmacogenetic causes of variability in drug disposition (Brian et al., 2016). Another important aspect is the further improvement of existing structure-based or ligand-based methods and tools of computer-aided drug design and discovery which increases the hit rate of novel drug compounds due to using a much more targeted search than traditional HTS and combinatorial chemistry (Sliwoski et al., 2014). Furthermore, one of the core areas in drug development should be the neglected diseases The Drugs for Neglected Diseases initiative (DNDi) is a drug research and development (R&D) organization that is developing new treatments for neglected patients.

References

AHDB, http://www.ahdbonline.com/about-us (2018).

Aiuti A., Roncarolo M. G., Naldini L., *EMBO Mol. Med.*, **9** (2017), 737–740.

Allied Market Research, https://www.alliedmarketresearch.com/press-release/specialty-enzymes.html (2018).

Arányi P., *J. Pharmacol. Med. Chem.*, **1** (2017), 29–31.

Aronson J. K., *Br. J. Clin. Pharmacol.*, **61** (2006), 243–245.

Avorn J., *N. Engl. J. Med.*, **372** (2015), 1877–1879.

Balbaa M., El Ashry E. S. H., *Biochem. Physiol.*, **1** (2012), 103.

Baldo B. A., *BioDrugs*, **29** (2015), 31–55.

Banchereau J., Palucka K., *Nat. Rev. Clin. Oncol.*, **15** (2018), 9–10.

Booth C. M., Tannock I. F., *Br. J. Cancer*, **110** (2014), 551–555.

Brian W., Tremaine L. M., Arefayene M., de Kanter R., Evers R., Guo Y., Kalabus J., Lin W., Loi C.-M., Xiao G., *Pharmacogenomics*, **17** (2016), 615–631.

Broomfield A., Jones S. A., Hughes S. M., Bigger B. W., *J. Inherit. Metab. Dis.*, **39** (2016), 499–512.

Campbell R. M., Tummino P. J., *J. Clin. Invest.*, **124** (2014), 64–69.

Cantor J. R., Panayiotou V., Agnello G., Georgiou G., Stone E. M., *Methods Enzymol.*, **502** (2012), 291–319.

Carrera P. M., Kantarjian H. M., Blinder V. S., *CA: Cancer J. Clinicians*, https://doi.org/10.3322/caac.21443 (2018).

Cave E., Cerreta F., https://www.google.com/search?q=EMA%2Breal+world+evidence&ie =utf-8&oe=utf-8&client=firefox-b, (2017).

Centerwatch, https://www.centerwatch.com/drug-information/fda-approved-drugs/therapeutic-area/12/oncology (2018).

Chae Y. K., Arya A., Malecek M.-K., Shin D. S., Carneiro B., Chandra S., Kaplan J., Kalyan A., Altman J. K., Platanias L., Giles F., *Oncotarget*, **7** (2016), 40767–40780.

CMS, 2018: SOURCE: Centers for Medicare & Medicaid Services, Office of the Actuary, National Health Statistics Group.

Copeland R. A., Harpel M. R., Tummino P. J., *Exp. Opin. Ther. Targets*, **11** (2007), 967–978.

Creese A., www.haiweb.org/medicineprices/policy/index.html, haiweb.org/wp.../5_WHOHAI_Policy-Brief_Taxes_2015.pdf, (2014).

de Vrueh R. L., Crommelin D. J., *Pharm. Res.*, **34** (2017), 1985–1999.

Denne S. C., Davis J. M., *Pediatr. Res.*, **83** (2018), 401–402.

DiMasi J. A., Grabowski H. G., Hansen R. W., *J. Health Econ.*, **47** (2016), 20–33.

Dine J., Gordon R., Shames Y., Kasler M. K., Barton-Burke M., *Asia Pac. J. Oncol. Nurs.*, **4** (2017), 127–135.

EMA, EMA/CHMP/272303/2016 *Rev*, **1**, (2016).

EMA, http://www.ema.europa.eu/ema/index.jsp?curl=pages/news_and_events/news/2017/05/news_detail_002743.jsp&mid=WC0b01ac058004d5c1 (2017).

EMA, http://www.ema.europa.eu/ema/index.jsp?curl=pages/medicines/landing/orphan_search.jsp (2017a).

EMA, www.ema.europa.eu/docs/en_GB/.../07/WC500251842.pdf (2018).

EMA, https://www.ema.europa.eu/en/news/harnessing-potential-real-world-data-through-learning-healthcare-system (2018a).

EMA, https://www.ema.europa.eu/en/medicines/human/EPAR/crysvita (2018b).

Fan P. W., Zhang D., Halladay J. S., Driscoll J. P, Khojasteh S. C., *Drug Metab. Dispos.*, **44** (2016), 1253–1261.

FDA, http://www.accessdata.fda.gov/scripts/opdlisting/oopd (2017)

FDA, https://www.fda.gov/NewsEvents/Newsroom/PressAnnouncements/ucm581216.htm (2017a).

FDA, FDA's Orphan Drug Modernization Plan (2017b).

FDA, https://www.fda.gov/ForIndustry/DevelopingProductsforRareDiseasesConditions/HowtoapplyforOrphanProductDesignation/ucm588744.htm (2017c).

FDA, Use of Real-World Evidence to Support Regulatory Decision-Making for Medical Devices—Guidance for Industry and Food and Drug Administration Staff (2017d).

FDA, https://www.fda.gov/forpatients/approvals/fast/default.htm (2018).

FDA, Fast Track, Breakthrough Therapy, Accelerated Approval, Priority Review; https://www.google.de/search?q=fast+track%2Bfda&ie=utf-8&oe=utf-8&client=firefox-b&gfe_rd=cr&dcr=0&ei=Ki1WWrOGN_Hw8AfykZyoDw (2018a).

FDA, https://www.fda.gov/ScienceResearch/SpecialTopics/RealWorldEvidence/default.htm (2018b).

FDA, https://www.lymphoma.org/aboutlymphoma/nhl/dlbcl/dlbclfdaupdates/ (2018c).

Flockhart D. A., Drug Interactions: Cytochrome P450 Drug Interaction Table. Indiana University School of Medicine (2007).

GaBI, http://www.gabionline.net/Biosimilars/General/Biosimilars-of-rituximab (2017).

Gautam A., Pan X., *Drug Discov. Today*, **21** (2016), 379–384.

GBI Research, http://www.gbiresearch.com/report-store/market-reports/archive/therapeutic-proteins-market-to-2016-high-demand-for-monoclonal-antibodies-and-erythropoietins-will-drive-the-market (2016).

Giannuzzi V., Conte E., Landi A., Ottomano S. A., Bonifazi D., Baiardi P., Bonifazi F., Ceci A., *Orphanet J. Rare Dis.*, **12** (2017), 64.

Gottwald M., Becker A. Bahr I., Mueller-Fahrnow A., *Arch. Pharm., (Weinheim)*, **349** (2016), 692–697.

Gotwals P., Cameron S., Cipolletta D., Cremasco V., Crystal A., Hewes B., et al., *Nat. Rev. Cancer*, **17** (2017), 286–301.

Grand View Research, https://www.grandviewresearch.com/press-release/global-enzymes-market (2018).

Hall A. K., Carlson M. R., *Intractable Rare Dis. Res.*, **3** (2014), 1–7.

Hay M., Thomas D. W., Craighead J. L., Economides C., Rosenthal J., *Nat. Biotechnol.*, **32** (2014), 40–51.

Hernandez J. J., Pryszlak M., Smith L., Yanchus C., Kurji N., Shahani V. M., Molinski S. V., *Front. Oncol.*, **7** (2017), 273.

Hoffman-Eubanks B., http://www.pharmacytimes.com/news/the-role-of-pharmacy-benefit-mangers-in-american-health-care-pharmacy-concerns-and-perspectives-part-1 (2017).

Hutchings D. C., Anderson S. G., Caldwell J. L., Trafford A. W., *Heart*, doi: 10.1136/heartjnl-2017-312865 (2018).

Ito M. K., Santos R. D., *J. Clin. Pharmacol.*, **57** (2017), 7–32.

IVESTOPEDIA, https://www.investopedia.com/ask/answers/032315/what-are-primary-risks-associated-investing-drugs-sector.asp (2018).

Jadhav S., http://www.clinicalinformaticsnews.com/2017/1/13/using-real-world-data-to-enhance-clinical-trials.aspx (2017).

Jeřábek J., Uliassi E., Guidotti L., Korábečný J., Soukup O., Sepsova V., Hrabinova M., Kuča K., Bartolini M., Peña-Altamira L. E., Petralla S., Monti B., Roberti M., Bolognesi M. L., *Eur. J. Med. Chem.*, **127** (2017), 250–262.

Kanehisa M., Furumichi M., Tanabe M., Sato Y., Morishima K., *Nucleic Acids Res.*, **45** (2017), D353–D361.

Kannt A., Wieland T., *Naunyn-Schmiedeberg's Arch. Pharmacol.*, **389** (2016), 353–360.

Kesselheim A. S., Avorn J., Sarpatwari A., *JAMA*, **316** (2016), 858–871.

Kohler G. I., Bode-Böger S. M., Busse R., Hoopmann M., Welte T., Böger R. H., *Int. J. Clin. Pharmacol. Ther.*, **38** (2000) 504–513.

Ku M. S., *J. Food Drug Anal.*, **23** (2015), 595–608.

Kumar H., Fojo T., Mailankody S., *JAMA Oncol.*, **2** (2016), 1238–1240.

Lee H.-M., Kim Y., *Schizophr. Res. Treatment*, **2016** (2016), 6378137.

Ligat G., Cazal R., Hantz S., Alain S., *FEMS Microbiol. Rev.*, **42** (2018), 137–145.

Lindsley C. W., *ACS Chem. Neurosci.*, **8** (2017), 1635–1636.

Lynch T., Price A., *Am. Fam. Physician*, **76** (2007), 391–396.

Mariz S., Reese J. H., Westermark K., Greene L., Goto T., Hoshino T., Llinares-Garcia J. Sepodes B., *Nat. Rev. Drug Discov.*, **15** (2016), 440–441.

Mezher M., http://www.raps.org/Regulatory-Focus/News/2017/01/19/26647/Updated-FDA-and-EMA-in-2016-A-Look-at-the-Numbers/ (2017).

NIH, https://www.nih.gov/research-training/medical-research-initiatives/cures (2016).

NORD, https://rarediseases.org/for-patients-and-families/information-resources/rare-disease-information/ (2018).

Pachaiyappan B. M., Woster P. M, *Bioorg. Med. Chem. Lett.*, **24** (2014), 21–32.

Palgon G., Liaison Technologies; https://www.liaison.com/blog/2017/05/02/cost-drug-development/ (2017).

Petkovic V., Moghini C., Paoletti S., Uguccioni M., Gerber B., *J. Biol. Chem.*, **279** (2004), 23357–23363.

Pirmohamed M., Park B. K., *Toxicology*, **192** (2003), 23–32.

Prasad V., De Jesús K., Mailankody., *Nat. Rev. Clin. Oncol.*, **14** (2017), 381–390.

Ramsay R. R., Tipton K. F., *Molecules*, **22**, 1192 (2017); doi:10.3390/molecules22071192.

Rebsamen M. C., Desmeules J., Daali Y., Chiappe A., Diemand A., Rey C., Chabert J., Dayer P., D Hochstrasser D., Rossier M. F., *Pharmacogenom. J.*, **9** (2009), 34–41.

Research and Markets, https://www.researchandmarkets.com/research/2bgjgx/protein (2016).

Research and Markets, https://www.researchandmarkets.com/research/9hhc8d/179_9_billion?w=5 (2017).

Rowland A., Miners J. O. Mackenzie P. I., *Int. J. Biochem. Cell Biol.*, **45** (2013), 1121–1132.

Rumore M. M., Vogenberg F. R., *Pharm. Ther.*, **42** (2017), 330–335.

Saltz L. B., *JAMA Oncol.*, **2** (2016), 19–21.

Schumock G. T., Vermeulen L. C., *Pharmacotherapy*, **37** (2017), 9–11.

Sherman R. E., Davies K. M., Robb M. A., Hunter N. L., Califf R. M., *Nat. Rev. Drug Discov.*, **16** (2017), 297–298.

Sliwoski G., Kothiwale S., Meiler J., Lowe Jr E. W., *Pharmacol. Rev.*, **66** (2014), 334–395.

Speights K., https://www.fool.com/investing/2016/07/31/12-big-pharma-stats-that-will-blow-you-away.aspx (2016).

Srinivas M., Thirumaleswara G., Pratima S., *Peertechz J. Med. Chem. Res.*, **3** (2017), 001–011.

Stevens M., http://www.milkenreview.org/articles/financing-high-risk-medical-research (2018).

Tannoury M., Attieh Z., *Curr. Ther. Res.*, **86** (2017), 19–22.

The Market Reports, https://www.themarketreports.com/report/global-enzyme-replacement-therapy-market-2018-by-manufacturers-countries-type-and-application-forecast-to-2023 (2018).

Tufts CSDD, http://csdd.tufts.edu/news/complete_story/tufts_csdd_rd_cost_study_now_published (2017). Tufts Center for the Study of Drug Development (Tufts CSDD, 2017).

Tworkoski, Record Numbers of FDA-Approved Drugs: Recent Trends; http://www.impactpharma.com/blog/record-numbers-of-fda-approved-drugs/ (2016).

Vanneman M., Dranoff G., *Nat. Rev. Cancer*, **12** (2012), 237–251.

Vaudano E., *Comput. Struct. Biotechnol. J.*, **6** (2013), e201303017.

Vellard M., *Curr. Opin. Biotechnol.*, **14** (2003), 444–450.

Vfa, https://www.vfa.de/embed/statistics-2015.pdf (2015).

Vfa, https://www.vfa.de/de/arzneimittel-forschung/datenbanken-zu-arzneimitteln/orphan-drugs-list/_p9#liste-medikamente (2018).

Vfa, https://www.vfa.de/embed/bcg-vfa-bio-biotech-report-2018.pdf (2018a).

Vfa, https://www.vfa.de/en/en-amf-standortfaktoren.htm (2018b).

Wang X. L., Li T., Li J. H., Miao S Y., Xiao X. Z., *Molecules,* **22** (2017), pii: E1529.

Wang Y., Wang H., Chen H.-Z., *Curr. Neuropharmacol.,* **14** (2016), 364–375.

Waxman H., Corr B., Martin K., et al., *Getting to the Root of High Prescription Drug Prices: Drivers and Potential Solutions,* The Commonwealth Fund (2017).

Whitburn J., Edwards C. M., Sooriakumaran P., *Curr. Urol. Rep.,* **18** (2017), 46.

WHO ATC, https://ec.europa.eu/cefdigital/wiki/pages/viewpage.action?pageId=52609352 (2018).

Yari M., Ghoshoon M. B., Vakili B., Ghasemi Y., *Curr. Pharm. Biotechnol.,* **18** (2017), 531–540.

Yildirim O., Gottwald M., Schüler P., Michel M. C., *Front. Pharmacol.,* **7** (2016), 461.

Yu H., MacIsaac D., Wong J. J., Sellers Z. M., Wren A. A., Bensen R., Kin C., Park K. T., *Aliment. Pharmacol. Ther.,* **47** (2017), 364–370.

Zuidgeesta M. G. P., Goetz I., Groenwold R. H. H., Irving E., van Thiel G. J.M.W., *J. Clin. Epidemiol.,* **88** (2017), 7e13.

Chapter 2

Use of Enzymes in the Downstream Processing of Biopharmaceuticals

Duarte Miguel F. T. Prazeres
iBB—Institute for Bioengineering and Biosciences, Department of Bioengineering, Instituto Superior Técnico, Universidade de Lisboa, Portugal
miguelprazeres@tecnico.ulisboa.pt

2.1 Introduction

The manufacturing of biopharmaceuticals encompasses a string of activities that are designed with the goal of producing a certain mass or biological activity of a target product with specific quality attributes. These activities can be grouped into two stages: the upstream processing, which comprises the generation of cell lines/strains, the establishment of cell banks, inoculum preparation and cell cultivation, and the downstream processing, which includes all the unit operations required to purify the product up to a point where pre-defined, end-product specifications are met (Heinzle et al., 2006; Jungbauer, 2013).

Pharmaceutical Biocatalysis: Fundamentals, Enzyme Inhibitors, and Enzymes in Health and Diseases
Edited by Peter Grunwald

ISBN 978-981-4800-61-7 (Hardcover), 978-0-429-29503-4 (eBook)
www.jennystanford.com

The downstream processing is indisputably the most challenging stage in biopharmaceutical manufacturing (Gronemeyer et al., 2014). The selection of a train of recovery and purification operations requires an adequate knowledge of the properties of the target molecule and associated impurities. Then, one must examine the different unit operations available (e.g., microfiltration, centrifugation, extraction, precipitation, chromatography, etc.), and sequence them in such a way that the final product meets the expected quality (Heinzle et al., 2006). Ideally, the overall process should have a small number of high-yield steps and lengthy operations should be avoided, so that processing costs and complexity are reduced. In general, manufacturing processes with a larger number of operations are required to obtain highly pure biopharmaceuticals, but this often comes with a high cost in terms of yield (Gronemeyer et al., 2014). Additionally, any reagents used should be considered safe *a priori*, to expedite both validation and approval by regulatory agencies.

A range of process additives are typically used during the downstream processing to facilitate product isolation and purification. For example, sodium hydroxide and detergents are used to lyse cells (Chamsart and Karnjanasorn, 2007; Berlowska et al., 2015), urea and guanidine hydrochloride facilitate the solubilization of inclusion bodies (Singh and Panda, 2005), mercaptoethanol, L-arginine and a number of additives aid in protein refolding (Singh and Panda, 2005; Yamaguchi et al., 2013), salts, alcohols and polymers promote precipitation (Sim et al., 2012; Yoshikawa et al., 2012) and surfactants like Triton-X, Tween or Polysorbate are common as purification aids (Roberts, 2008; Shukla and Hinckley, 2008; Moleirinho et al., 2018). These process additives are relatively simple from a molecular point of view ($M_w < 10^3$ Da) and relatively inexpensive when cost is considered on a mass basis. Occasionally, however, complex macromolecules like enzymes may be included in specific steps of the downstream processing (Prazeres, 2016).

The rationale for the use of enzymes as ancillary materials during the downstream processing is to explore their high specificity and ability to function under mild pH/temperature conditions that are compatible with the biomolecules being purified. The operation of an enzyme-assisted step encompasses the addition of a purified enzyme preparation, usually in a soluble form, to a process stream. Here, the right conditions (e.g., temperature, pH, co-factors, viscosity) must be present for the enzyme to act appropriately on its substrate. Nevertheless, if the appropriate environment exists, enzyme activity can be extremely valuable in different contexts. A review of the literature on the topic shows that the process steps that take advantage of enzyme activity in vitro can be clustered into three main groups (Fig. 2.1) according to the specific goal pursued: (i) cell dissociation and lysis, (ii) impurity clearance, and (iii) structural modifications (Prazeres, 2016).

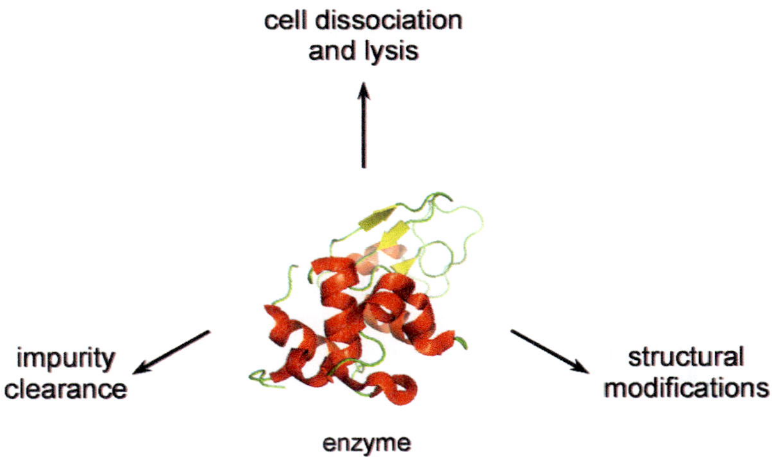

Figure 2.1 Overview of enzyme-assisted steps in the downstream processing of biopharmaceuticals.

The majority of the enzymes used in these categories are hydrolases, which fall within the Enzyme Commission (EC) enzyme class 3[1] (Schomburg et al., 2014). This class of enzymes catalyzes the cleavage of hydrolytic bonds such as C–O, C–C, and C–N in a biomolecule to yield two smaller molecules. In the following sections, an outline of enzyme-assisted process steps within each of these categories is presented (Table 2.1). The focus is on operations that have gained industry acceptance and/or are routinely used in bioprocess engineering research.

[1]The Enzyme Commission classification is used throughout this chapter.

2.2 Cell Dissociation and Lysis

Enzymes can be used as process additives very early in a biopharmaceutical manufacturing process to tackle specific challenges linked to cell dissociation and cell lysis (Table 2.1). The mass/total activity of enzyme involved may be high since process volumes at such an earlier stage of a process are typically very large.

2.2.1 Cell Dissociation

The majority of mammalian cells that are cultured in vitro in petri dishes, flasks, and bioreactors adhere to the substrate surfaces by a sequence of steps that includes an initial attachment, flattening and spreading (Sagvolden et al., 1999; Khalili and Ahmad, 2015). Furthermore, cell to cell adhesion also occurs as proliferation proceeds. Integrin receptors and heterodimeric transmembrane proteins play a central role in this process of cell-substrate and cell-cell adhesion (Sagvolden et al., 1999). Following proliferation and once growth is halted, for example at a certain degree of confluence, cells must be dissociated from each other and from the surfaces of the containers or microcarriers used as supports, for further processing or passaging to a second stage of proliferation. The use of enzymes in these harvesting and sub-culturing operations is widespread and well known to researchers working with mammalian cells. These operations rely essentially on trypsin (EC 3.4.21.4), a serine endopeptidase found in the digestive system of many vertebrates, which preferentially hydrolyzes proteins by cleaving bonds next to arginine and lysine residues (Stroud et al., 1974). Trypsin is used specifically to hydrolyze the proteins involved in cell adherence, and thus promote disaggregation and detachment from the substratum (Gori, 1964; Codinach et al., 2016). The procedure is called trypsinization and can be found in processes used to manufacture recombinant proteins (Spearman et al., 2005), viral vectors (Fernandes et al., 2013), advanced therapy medicinal products (Koller et al., 1993; Stute et al., 2005; Kami et al., 2013; Codinach et al., 2016), vaccines (Wu et al., 2005), or other cell culture–derived medicinal products (Sun and Zhang, 2007; Sinacore et al., 2000).

Table 2.1 Examples of enzyme-assisted process steps in the downstream processing of biopharmaceuticals

Category	Application	Enzymes	References
Cell dissociation and lysis	Lysis and disruption of microbial cells for intracellular product release	Lysozyme, glucanases, mannanases, chitinases	Andrews and Asenjo, 1987; Salazar and Asenjo, 2007; Middelberg, 1995; Vogels and Kula, 1992; Zhu et al., 2005; Zhu et al., 2007
	Detachment of adherent cells from culture vessels for passage	Trypsin	Spearman et al., 2005; Fernandes et al., 2013; Koller et al., 1993; Stute et al., 2005; Kami et al., 2013; Codinach et al., 2016; Wu et al., 2005; Sun and Zhang, 2007; Sinacore et al., 2000
Impurity clearance	Clearance of nucleic acids from process streams arising from mammalian cell culture	DNase/RNase	Kamen and Henry, 2004; Konz et al., 2005; Segura et al., 2012; Guo et al., 2012; Bandeira et al., 2012; Milián and Kamen, 2015; Weigel et al., 2014; Steppert et al., 2016
	Digestion of plasmid DNA vectors following transient transfection of cells in viral particle production	DNase	Sastry et al., 2004; Grassman et al., 2008; Shaw et al., 2012
	Clearance of host cell (*E. coli*) RNA in plasmid DNA purification	RNase	Varley et al., 1998; Prazeres et al., 1998; Voss et al., 2006
	Digestion of linear DNA fragments in plasmid DNA purification	Exonuclease	Balagurumoorthy et al., 2008
	Digestion of nucleic acids and proteins in capsular polysaccharide purification	DNase, trypsin, pronase E, proteinase K, nargase	Gonçalves et al., 2003; Gonçalves et al., 2007; Takagi et al., 2008; Albani et al., 2015

(*Continued*)

Table 2.1 (*Continued*)

Category	Application	Enzymes	References
Structural modifications	Removal of affinity or solubility tags from protein fusions via hydrolysis of peptidic bonds	Trypsin, TEV, Factor Xa, enteropeptidase, R3C, thrombin, DAPase, carboxypeptidase A and B	Arnau et al., 2006; Unal et al., 2008; Waugh, 2011
	Conversion of porcine insulin into human insulin by transpeptidation	Trypsin	Markussen, 1982; Markussen et al., 1983
	Generation of insulin via cleavage of recombinant proinsulin	Trypsin, carboxypeptidase B	Nilsson et al., 1996; Kjeldsen, 2000
	Generation of antibody fragments by proteolytic digestion of whole antibodies	Papain, pepsin	Adamczyk et al., 2000; Cresswell et al., 2005; Thillaivinayagalingam et al., 2007; Jones and Landon, 2002; Kumpalume et al., 2002; Boushaba et al., 2003
	Linearization of miniplasmid impurities in minicircle vector purification	Restriction endonuclease	Darquet et al., 1999
	Modification of topology of miniplasmid impurities in minicircle vector purification	Nicking endonuclease	Alves et al., 2016
	Glycosylation of recombinant proteins produced in prokaryotic hosts	Endoglycosidases	Takegawa et al., 1995; Meynial-Salles and Combes, 1996; Yamamoto and Takegawa, 1997; Yamamoto, 2001

From a technical point of view, trypsinization involves addition of the enzyme (e.g., at 0.05% to 0.25%) and an incubation period at 37°C that can last from 2 to 20 min, depending on the enzyme concentration used (Koller et al., 1993; Stute et al., 2005). Once cells have been detached, trypsin activity is arrested usually by low temperature or by the addition of serum proteins (Koller et al., 1993). These precautions are taken to ensure that the proteolytic activity of trypsin does not damage the cells by cleaving cell surface receptors, membrane and chromatin proteins, which could impair cell function (Maizel et al., 1975; Huang et al., 2010). For example, a proteome study found that the expression of 36 proteins changed significantly following trypsin treatment (Huang et al., 2010). Furthermore, as a result of trypsinization, apoptosis-related proteins were up-regulated and growth- and metabolism-related proteins were down-regulated (Huang et al., 2010).

Native trypsin from bovine or porcine pancreas is widely available for trypsinization. The enzyme is extracted from pools of frozen pancreatic glands and purified further by precipitation or chromatography. The final product is an ill-defined mixture of α- and β-trypsin with traces of chymotrypsin (Merck Millipore, 2015). A long incubation at low pH is usually included together with a final pathogen inactivation step such as gamma irradiation, e-beam irradiation, or UV irradiation (EMA, 2013; Melnick and Wallis, 1976). In spite of these precautions, a recurrent concern is the possibility that animal-derived trypsin may introduce adventitious agents in the target product during manufacturing, which could pose a risk to humans (Berger et al., 2005; Victoria et al., 2010; EMA, 2013). For instance, at least 55 porcine virus species are widespread among pigs that are very difficult to eliminate due to their high resistance to physicochemical treatment (Marcus-Sekura et al., 2011). Furthermore, a significant number of these viruses can replicate in human and primate (e.g., Vero) cells. The potential for problems to occur has been demonstrated, for example in the case of the paediatric vaccine Rotarix, which was found to be contaminated with porcine circovirus originating from trypsin lots derived from pigs (Baylis et al., 2011). Thus, regulatory agencies strongly recommend that trypsin of animal origin is appropriately selected, tested and treated to inactivate and/or remove adverse biological agents before it is used to manufacture biopharmaceuticals (EMA, 2013). As a response to these concerns, recombinant versions of the enzyme were commercially developed specifically to be used in cell dissociation and vaccine production and processing (Novozymes, 2013; Merck Millipore, 2015). These enzymes are manufactured under cGMP standards in facilities that comply with non-animal origin requirements and can thus be used in the production of biopharmaceuticals (Merck Millipore, 2015). An alternative strategy, encouraged by the European Medicines Agency (EMA), would be to use plant or invertebrate trypsins (EMA, 2013; Muhlia-Almazán et al., 2008).

2.2.2 Cell Lysis

The release of biopharmaceuticals produced intracellularly by microbial cells (e.g., proteins, plasmids) requires the inclusion of a lysis and disruption step right after cells are recovered from fermentation broths. While mechanical processes (e.g., bead milling and homogenization) and chemical based methods (e.g., alkaline lysis) are often chosen, the use of enzymes in this early step of product isolation is also well established, dating back to the early years of modern Biotechnology (Andrews and Asenjo, 1987). A range of enzymes have been studied at lab scale for this purpose, which include chitinases, glucanases, mannanases, glycosidases and endopeptidases. Furthermore, formulations of lytic enzymes are commercialized by many companies to aid researchers in the isolation of proteins. Among the lytic enzymes available, the glycosidase lysozyme (peptidoglycan *N*-acetylmuramoylhydrolase, EC 3.2.1.17) is by far the most important when it comes to the process scale lysis of cells (Andrews and Asenjo, 1987; Salazar and Asenjo, 2007). Chicken lysozyme from hen egg white is a relatively small protein (~14 kDa). The ability of the enzyme to hydrolyze the β-1,4-glycosidic bond between units of *N*-acetylglucosamine and *N*-acetylmuramic acid in the peptidoglycan of bacterial cell wall is conferred by a catalytic site with a glutamic acid and an aspartic acid (Held and van Smaalen, 2014). This specificity makes lysozyme an excellent choice when it comes to disrupting Gram-positive cells like yeast or *Bacillus* that have a thick peptidoglycan layer in their cell wall (Knorr et al., 1979; Ibrahim et al., 1994). Gram-negative bacteria, on the other hand, are less susceptible because lipopolysaccharides (LPS) in the outer layer bind and inhibit lysozyme (Ohno and Morrison, 1988), effectively shielding the thin peptidoglycan cell wall from the enzymatic action (Middelberg, 1995). Nevertheless, lysozyme can be made more effective if assisted by chelating agents like EDTA (Middelberg, 1995) or used in combination with high pressure homogenization, wet milling (Vogels and Kula, 1992) and high temperatures (van Hee et al., 2004; Zhu et al., 2005; Zhu et al., 2007). The majority of lysozyme preparations commercially available are derived from hen egg white. While this is not an issue for research applications, in a biopharmaceutical manufacturing scenario the animal origin of lysozyme poses safety and consistency challenges (Merck Millipore, 2013). Alternatively, a recombinant human lysozyme (~15 kDa) produced in transgenic rice, purified by extraction and cation exchange (Wilken and Nikolov, 2006) and specifically manufactured under cGMP standards to allow its use in the production of biopharmaceuticals is available from Merck Millipore (CellPrime® rLysozyme, Merck Millipore, 2013). The use of rice as a "cell factory" to produce protein-based additives like lysozyme for biopharmaceutical manufacturing (as well biopharmaceuticals and nutraceuticals for human consumption) is very attractive given the Generally Regarded As Safe (GRAS) status of rice (Huang et al., 2002). Furthermore, the recombinant human lysozyme displays more than three times greater specific activity compared to chicken lysozyme and hence a lower concentration is needed to lyse an equivalent amount of bacteria. Additionally, the higher pI of human lysozyme (10.9) when compared with the pI of chicken lysozyme (9.2) may facilitate the subsequent removal of the enzyme from the biopharmaceutical containing stream (Merck Millipore, 2013).

2.3 Impurity Clearance

The use of enzymes to clear cell-derived impurities from process streams is relatively common in the downstream processing of biopharmaceuticals (Table 2.1). The goal is to degrade either nucleic acids or proteins into smaller fragments, which are easier to remove from the target biopharmaceutical in subsequent unit operations.

2.3.1 Nucleic Acid Clearance

The presence of cellular DNA from continuous mammalian cell lines in human biopharmaceuticals is one of the key points addressed by regulatory agencies. The concern is that residual fragments of foreign DNA can encode or harbor oncogenes and infectious agents, or inadvertently insert into the genome of the recipient, thus hypothetically inactivating tumor suppressor genes or activating oncogenes (WHO, 1998; Yang et al., 2010; Yang, 2013). The acceptable limits of residual DNA in biopharmaceuticals have thus been set by the World Health Organization (WHO) and by the Food and Drug Administration (FDA) to be of the order of 10 ng per parenteral dose and 100 ng per oral dose (WHO, 1998; FDA, 2010). While the risks posed by bacterial DNA are smaller when compared with those posed by mammalian cell DNA, specifications for genomic DNA are also commonly used in the case of recombinant proteins (WHO, 2013), plasmid vectors (Prazeres, 2011) or capsular polysaccharides (WHO, 2000; Takagi et al., 2008) produced by microbial hosts. In view of these regulatory requirements, it is clear that the downstream processing of biopharmaceuticals must address this issue, for example by including unit operations that decrease the amount of residual DNA and/or reduce its size to below the size of a functional gene (approximately 200 base pairs) (Butler et al., 2009; FDA, 2010). Although the safety concerns associated with residual DNA do not extend to RNA with the same pertinence, the level of RNA impurities should also be reduced below a specific limit. Apart from improving purity and safety of biopharmaceuticals, clearance of nucleic acids may also contribute to reducing the high viscosity associated with high-molecular-weight DNA (Boynton et al., 1999), the fouling of separation media (Siu et al., 2006; 2007), the complexation of recombinant protein biopharmaceuticals with DNA (Zabriskie and DiPaolo, 1988) and the clumping of cells and agglomeration of viral particles, which is usually induced by adhesion of nucleic acids (Zhang et al., 2001).

An excellent way to reduce the content and size of DNA in a process stream is to use a deoxyribonuclease (DNase)-based degradation step. Endonucleases are an excellent choice due to their ability to cleave internal phosphodiester bonds present between nucleotides (Benedik and Strych, 1998). Upon complete hydrolysis, DNA macromolecules are reduced to 5′-monophosphate-terminated oligonucleotides with 3–5 bases (Rangarajan and Shankar, 2001), which are then relatively easy to remove from the target biopharmaceutical (Gousseinov et al., 2014). One of the most popular nuclease preparation available in the market for applications in the pharmaceutical and biotechnological industry is Benzonase.[2] The key component in the preparation is a recombinant endonuclease normally expressed in *Serratia marcescens* (EC 3.1.30.2). A homodimer of 30 kD subunits, the enzyme requires divalent cations to function (with preference for Mg^{2+}) and acts as a sugar nonspecific hydrolase that cleaves both RNA and DNA (Benedik and Strych, 1998). Nuclease activity is displayed across a broad range of pH (6–10) and temperatures (0 to 42°C). The enzyme is highly active when compared to other nucleases (e.g., staphylococcal nuclease, DNase I) and shows no preference for a particular base or nucleic acid structure, cleaving single-stranded, double-stranded, linear, circular, or supercoiled structures alike (Benedik and Strych, 1998). Another endonuclease preparation that has been suggested is Pulmozyme[3] (Grassman et al., 2008; Shaw et al., 2012), a pharmaceutical grade recombinant human DNase I (EC 3.1.21.1) produced in Chinese hamster ovary (CHO) cells that has received approval by the FDA (Shire, 1996). This 37 kDa enzyme uses a single-strand nicking mechanism to hydrolyze double-stranded DNA nonspecifically and is also dependent on Mg^{2+} (Parsiegla et al., 2012). While the Pulmozyme preparation was originally developed for the treatment of cystic fibrosis, its use in the downstream processing of biopharmaceuticals has been considered advantageous as compared to Benzonase due to its pharmaceutical grade and to the fact that it is not of bacterial origin (Shaw et al., 2012).

---

[2]EMD Millipore, Benzonase® endonuclease, Darmstadt, Germany, 2013.

[3]Pulmozyme® is a proprietary name from Genentech.

As in the case of DNA, the content of single-stranded RNAs in a process stream can be reduced very effectively using endoribonucleases (RNAse). RNase A from bovine pancreas, for example, will cleave the 3′-end of unpaired C and U residues, yielding 3′-phosphorylated products (Cuchillo et al., 2011). The four disulfide linkages of this small enzyme (13.7 kDa, EC 3.1.27.5) are responsible in part for its remarkable stability at elevated temperatures and under harsh chemical conditions (Raines, 1998). This is duly explored in many cases by complementing the enzyme action with strongly alkaline environments that contribute to RNA degradation (Malek et al., 2000).

Excellent examples of enzyme-based DNA/RNA removal can be found in the mammalian/insect cell-based manufacturing of adenovirus (Kamen and Henry, 2004; Konz et al., 2005; Segura et al., 2012), adeno-associated virus (Zolotukhin et al., 1999; Chahal et al., 2007; Guo et al., 2012), lentivirus (Bandeira et al., 2012), influenza virus (Milián and Kamen, 2012; Weigel et al., 2014) and virus-like particles (Steppert et al., 2016). In the majority of these examples, DNase and/or RNase treatment is usually performed prior to or after cell clarification to reduce cell nucleic acid content. In some cases, nuclease treatment can be carried out at the end of the process, after chromatographic purification (Bandeira et al., 2012; Green et al., 2002). Nucleases can also be used to digest plasmid DNA vectors if a transient transfection strategy is used for viral particle production (Sastry et al., 2004; Grassman et al., 2008; Shaw et al., 2012).

Nuclease treatment is also found in the downstream processing of biopharmaceuticals produced in microbial cells. A very interesting illustration is found in the purification of the capsular polysaccharides that are used to manufacture vaccines against *Haemophilus infuenzae*, *Streptococcus pneumoniae* and *Neisseria meningitidis* (Gonçalves et al., 2003; Gonçalves et al., 2007; Takagi et al., 2008; Albani et al., 2015). Here, enzymatic treatment with endonucleases has been advocated as the most efficient strategy to achieve the WHO requirement for nucleic acid content inferior to 1% dry weight (WHO, 2000). For example, the use of a Benzonase digestion step (2–4 h incubation at 37°C and 100 rpm) combined with tangential ultrafiltration has been reported to achieve 50–70-fold purification of capsular polysaccharides relatively to nucleic acids (Gonçalves et al., 2003;

Gonçalves et al., 2007; Takagi et al., 2008). It is worth noting that this endonuclease digestion is the first step of an enzymatic treatment that includes also digestion of proteins by proteases (see Section 2.3.2, Protein Clearance).

Another good example of the use of nucleases in the downstream processing of microbial cell-based biopharmaceuticals is seen in the lab-scale isolation of plasmid DNA from the producer *E. coli* cells. RNase A, in particular, is very popular due to its excellent ability to remove RNA from *E. coli* cells (Varley et al., 1998; Prazeres et al., 1998). The beneficial impact of RNase clearance in the performance of subsequent chromatographic operations is well known to those who have worked in plasmid DNA purification. However, regulatory concerns with the animal origin of the benchmark RNAse A from bovine pancreas have contributed to the banning of the enzyme from the process scene. An obvious way to circumvent this problem is to use recombinant RNAses, as proposed for example by Voss et al. (2006), who demonstrated the feasibility of using RNase Ba from *Bacillus amyloliquefaciens* as an alternative to bovine RNase A in the downstream processing of plasmid DNA. Alternatively, many native microbial RNases have been described that could be used for RNA clearance (Hames and Demir, 2015) Surprisingly, neither one of these solutions has been pursued by biopharmaceutical manufacturers.

While not as popular as endonucleases, exonucleases (i.e., enzymes that cleave nucleotides one at a time from the end of the nucleic acid chain), have also proved valuable in the downstream processing context. For example, Balagurumoorthy et al. (2008) describe an application in the lab scale purification of plasmid DNA where a combination of two exonucleases is used specifically to digest linear DNA fragments that originate from the shearing/degradation of genomic DNA or from plasmid linearization. The enzymes do not degrade intact supercoiled plasmid molecules since these are covalently closed, and thus do not present a free terminal available for hydrolysis.

2.3.2 Protein Clearance

A few reports describe the use of enzymes to degrade protein impurities from process streams that contain non-protein biopharmaceuticals. Here, endopeptidases that act by breaking peptide bonds of nonterminal amino acids are the preferred choice for digesting proteins. An interesting example is again found in the downstream processing of capsular polysaccharides used to manufacture vaccines (Gonçalves et al., 2003; Gonçalves et al., 2007; Takagi et al., 2008; Albani et al., 2015). In sequence with the endonuclease digestion of nucleic acids described above (see Section 2.3.1, Nucleic Acid Clearance), protein impurities are degraded below the recommended level of 1% dry weight (WHO, 2000) using a series of proteases, which may include proteinase K (EC 3.4.21.64), trypsin (EC 3.4.21.4), nargase (EC 3.4.21.62) and pronase E (EC 3.4.24.31) (Tanizaki et al., 1996; Gonçalves et al., 2007). The proteases are added sequentially after Benzonase treatment (see above) with 2 h interval and incubated at 37°C and 100 rpm (Gonçalves et al., 2003; Takagi et al., 2008). The use of a combination of proteases broadens the specificity of action of the individual enzymes thus decreasing the size of the resulting peptide fragments. Combined with ultrafiltration, this step can yield 2.6- to 20-fold purification of capsular polysaccharides relatively to proteins (Gonçalves et al., 2003; Gonçalves et al., 2007; Takagi et al., 2008). Furthermore, it is clearly advantageous when compared with the protein extraction with chloroform/butanol or phenol used traditionally in the manufacturing of capsular polysaccharides (Gotschlich et al., 1969).

2.4 Targeted Structural Modifications

In the applications reviewed above, enzyme activity is used essentially to hydrolyze biomacromolecules from producer cells like peptidoglycan, proteins and nucleic acids. However, more sophisticated applications can be found where a very precise alteration of the structure of a precursor or impurity is required (Table 2.1).

2.4.1 Affinity Tag Removal

A specific application where enzymes are used to modify the structure of a precursor of the target molecule is the removal of affinity tags (Arnau et al., 2006; Waugh, 2011). These tags, which are usually composed of a number of exogenous amino acid sequences, are added by inserting the corresponding coding sequences next to those that code for the protein of interest. Tags that have a high affinity for a specific ligand like the common hexa-histidine (HHHHHH) tag (Hoffmann and Roeder, 1991) or that contain an epitope recognized by immobilized antibodies like the FLAG (DYKDDDDK) (Hopp et al., 1988; Knappik and Pluckthun, 1994) and Streptag II (WSHPQFEK) (Skerra and Schmidt, 2000; Schmidt and Skerra, 2007) tags, are commonly used to facilitate purification. The high affinity of the corresponding tag-protein fusions to their ligands makes it possible to purify them by affinity chromatography with immobilized nickel (His tag), antibody (FLAG tag) and streptavidin (Streptag II) matrices (Arnau et al., 2006; Unal et al., 2008). In other cases, tags like glutathione S-transferase (GST) (Ren et al., 2003), maltose binding protein (MBP) (Guan et al., 1988) and elastin-like polypeptides (ELP) (Kowalczyk et al., 2014) are introduced to improve the solubility of the target protein. Proteins fused with these tags are also easier to purify due to their ability to bind to specific matrices (e.g., glutathione–Sepharose for GST, amylose for MBP) or to undergo temperature-induced aggregation (ELPs) (Arnau et al., 2006).

Whichever is the reason for introducing a tag, removing it is an imperative in the case of therapeutic recombinant proteins designed for human administration to guarantee that the final product is structurally identical to the authentic native product (Lauritzen et al., 2013). Here, proteases are used to precisely clip away the tag from the target protein by hydrolyzing the peptidic bond at a specific given location within the fusion, ideally without leaving any residues behind (Unal et al., 2008). Exoproteases like carboxypeptidase A (EC 3.4.17.1), carboxypeptidase B (EC 3.4.17.2) and dipeptidyl-aminopeptidase I (DAPase, EC 3.4.14.1) or endoproteases like tobacco etch virus protease (EC 3.4.22.44), Factor Xa (EC 3.4.21.6), thrombin (EC 3.4.21.5), enteropeptidase (EC 3.4.21.9) and rhinovirus 3C proteases (EC 3.4.22.28) have all be used to remove tags. The choice of the appropriate enzyme will depend on the exact genetic design of the fusion and location of the tag (Waugh, 2011). A critical aspect is to guarantee that the enzyme will not cleave within the protein of interest and that non-specific digestion of the target protein is not an issue (Unal et al., 2008). The tag-removal process is typically performed right after the affinity chromatography step by incubating the isolated fusion protein with the selected protease under appropriate conditions and at an enzyme to target ratio that may vary between 0.1% to 10% (wt/wt) (Unal et al., 2008). Next, both the cleaved tag and the tag-free protease must be separated from the target molecule, a task that is usually carried out by adding an extra chromatography step (Unal et al., 2008). In most cases, the lack of residual enzyme activity in the final product will have to be demonstrated.

2.4.2 Precursor Maturation

A number of processes used to manufacture therapeutic proteins like insulin and antibody fragments call for the production of a precursor molecule, which is then enzymatically converted into the target molecule by a protease. The case of insulin is particularly interesting in this regard. Already before the advent of recombinant DNA technology, a commercial process was devised to enzymatically convert porcine insulin purified from pancreatic extracts into human insulin (Markussen, 1982; Markussen et al., 1983). At the molecular level, this requires replacing an alanine residue at the C terminal of the B chain of porcine insulin for the threonine that is characteristic of human insulin. This was accomplished by promoting a trypsin catalyzed transpeptidation in aqueous-organic medium in the presence of a large excess of threonine ester. The process followed on with the chromatographic removal of trypsin (gel filtration) and unconverted porcine insulin (anion-exchange), and with the cleavage of the ester group in threonine (Markussen, 1982; Markussen et al., 1983; Walsh, 2005). Eventually, human insulin started being produced by recombinant DNA technology (Kehoe, 1990; Walsh, 2005). One of the manufacturing strategies devised involves the expression of a single proinsulin polypeptide chain, intracellularly by *E. coli* (Nilsson et al., 1996), or of proinsulin-like precursors by *S. cerevisiae* (Thim et al., 1986). Following recovery and purification of the proinsulin or proinsulin-like precursors, the C-peptide is excised using mixtures of trypsin and carboxypeptidase B (Nilsson et al., 1996) or via trypsin-mediated transpeptidation (Kjeldsen, 2000).

Another interesting example is found in the manufacturing of recombinant trypsin, an enzyme that is required as an ancillary material in the therapeutic recombinant insulin production mentioned above, as well as in biopharmaceutical manufacturing processes that use adherent cells (Section 2.3.1) (Muller et al., 2010; Merck Millipore, 2015). The enzyme is manufactured in the yeast *Pichia pastoris* as an inactive trypsinogen precursor, which is then converted into a mixture of α- and β-trypsin by a controlled auto-catalytic cleavage (Merck Millipore, 2015). In this process, small amounts of trypsin molecules (which may be added to accelerate the process) act as an activator of trypsinogen, releasing additional trypsin molecules that in turn act on more trypsinogen until substrate depletion. While the rate of trypsin production is very slow at first, it increases rapidly due to the auto-catalytic nature of the process (Vernon, 1913; Neurath and Dreyer, 1955; Muller et al., 2010).

Antibody fragments for therapeutic and diagnostic applications can be obtained by controlled proteolytic digestion of precursor, whole antibodies (mono- or polyclonal). For example, digestion with papain (~23 kDa, EC 3.4.22.2), a cysteine protease from papaya (*Carica papaya*), yields two 50 kDa Fab fragments and one 50 kDa Fc fragment, as long as the reaction is carried out in the presence of cysteine (Adamczyk et al., 2000; Cresswell et al., 2005; Thillaivinayagalingam et al., 2007). However, a stable and consistent product may be difficult to obtain by papain digestion and operating parameters should be optimized for the different types and classes of antibodies (Bennet et al., 1997). Digestion with papain can be used to yield a bivalent F(ab′)2 fragment and an Fc′ fragment if cysteine is absent from the reaction mixture (Bennet et al., 1997). Likewise, F(ab′)2 fragments can also be obtained by the targeted attack of pepsin (EC 3.4.23.1), an aspartic protease from the gastric mucosa of pigs, below the hinge region of antibodies (Jones and Landon, 2002; Kumpalume et al., 2002; Boushaba et al., 2003). However, unlike papain, pepsin cleaves the Fc fragment into small molecular weight molecules, a situation that facilitates the subsequent purification of the antigen-binding F(ab′)2 fragment (Boushaba et al., 2003).

The porcine origin of the pepsin used is a concern here, as was the case of porcine trypsin described in Section 2.2.1.

2.4.3 Structural Modification of Impurities

Precise enzyme-catalyzed structural modifications have also targeted key impurities with the goal of facilitating their separation from the target molecule. Two notable examples can be found in the case of the very challenging purification of minicircle (MC) DNA vectors. MCs are produced in vivo in *Escherichia coli* by first replicating a parental plasmid (PP) backbone and then promoting an intramolecular site-specific recombination between two strategically placed multimer resolution sites (Simcikova et al., 2016). This generates a replication-deficient MC, which contains the expression cassette with therapeutically useful transgene, and a miniplasmid (MP), which contains the prokaryotic replication segment used for PP amplification. The separation of MCs from the MP impurities is very difficult to perform since the two species are very similar in terms of structure and size (Alves et al., 2016).

In one of the strategies suggested to tackle the MC/MP separation, an in vitro restriction with two type II restriction endonucleases (EC 3.1.21.4) is used after in vivo recombination, cell lysis and DNA isolation, to cut only the MP and un-recombined PP, but not the corresponding MC (Darquet et al., 1999). This greatly facilitates the subsequent separation and purification of supercoiled MC by CsCl ethidium bromide (CsCl-EtBr) density gradient centrifugation (Darquet et al., 1999).

In the second example, the purification of MC is tackled by exploring the in vitro enzymatic action of Nb.BbvCI (Alves et al., 2016; Silva-Santos et al., 2019), an artificial nicking endonuclease derived from the natural, R.BbvCI type IIA restriction endonuclease (EC 3.1.21.4) of *Bacillus brevis* (Wei et al., 2016). Nb.BbvCI was engineered by mutating one of the catalytic subunits of the heterodimeric Rb.BbvCI (31 kDa + 32 kDa) in such a way that it recognizes a 7-base pair asymmetric sequence on DNA substrates and introduces a single cut in one of the DNA chains (50′-CCTCAGC-3′/GCTGAGG → CCTCAGC/GC^TGAGG) (Heiter et al., 2005). At the early stage of vector design, a target site for Nb.BbvCI is strategically placed in the MP part of the PP backbone.

The manufacturing process then starts with *E. coli* culture and recombination of PPs into target MC, followed by cell harvesting and alkaline lysis, precipitation with isopropanol and ammonium sulfate and diafiltration/concentration by microfiltration. In the subsequent in vitro digestion step, Nb.BbvCI cleaves one of the strands of the MPs and of non-recombined PPs. The introduction of a nick in these species modifies their topology from supercoiled (sc) to open circular (oc). Sc MCs, on the other hand, remain unaffected since they are devoid of the recognition site. Hydrophobic interaction or multimodal chromatography is finally used to purify sc MCs from MPs, oc MCs and oc PPs (Alves et al., 2016; Silva-Santos et al., 2019).

2.4.4 Glycosylation

The physicochemical and pharmacological properties of many protein biopharmaceuticals in vivo are dependent in many cases on a proper glycosylation of the molecules (Solá and Griebenow, 2010). Indeed, the glycan pattern of these so-called glycoproteins may directly influence the stability, bioactivity, and immunogenicity of a protein (Solá and Griebenow, 2010; Zhang et al., 2016). Monoclonal antibodies provide a striking example of this situation, but many other recombinant proteins can be found that require appropriate glycosylation to function appropriately (Higel et al., 2016; Zhang et al., 2016). Accordingly, regulatory agencies require an adequate analysis and control of the glycan variants of glycoprotein biopharmaceuticals to ensure product quality (Zhang et al., 2016). Both eukaryotic and prokaryotic cells have the ability to glycosylate proteins post-expression, albeit with slightly different mechanisms and resorting to sets of enzymes that may vary from specie to specie (Dell et al., 2010). This means that when dealing with the production of glycoprotein drugs, host-expression systems and process conditions must be judiciously chosen to guarantee that the glycosylation pattern of the proteins produced in vivo is the one that maximizes its efficacy (Brooks, 2006). While in vivo glycosylation of recombinant proteins during expression is the norm, in principle nonglycosylated protein forms could be glycosylated in vitro using enzymes and glycan moieties according to the required glycan pattern (Meynial-

Salles and Combes, 1996; Yamamoto, 2001; Khmelnitsky, 2004). For example, the formation of the glycosidic bond between the polypeptide and chemically synthesized glycan parts can be carried out by specialized enzymes like endoglycosidases (Kobata, 2013). While the main function of endoglycosidases is to cleave glycan groups from glycoproteins or to hydrolyze the internal glycosidic bonds in sugar chains, they are also able to transfer oligosaccharides via transglycosylation reactions (Khmelnitsky, 2004; Kakizaki and Takagaki, 2008). A number of examples can be found in the literature where this strategy has been explored using endo-β-*N*-acetylglucosaminidases (EC 3.2.1.96) from different origins, e.g. Endo-A (Takegawa et al., 1995), Endo-M (Yamamoto, 2001) and Endo-F (Yamamoto and Takegawa, 1997). While this strategy is not pursued actively by the industry, it could open up the way for the synthesis of homogeneously glycosylated proteins with glycan side chains at natural or unnatural glycosylation sites (MacMillan et al., 2001).

2.5 Discussion

2.5.1 Points to Consider

A number of technical and non-technical points should be addressed when considering the use of enzymes in the downstream processing of biopharmaceuticals (Table 2.2). From a technical perspective, the first challenge is to find an enzymatic activity that is appropriate to meet the process goal at hand. The stability and activity of the selected enzyme should be compatible with the process stream characteristics (e.g., pH, temperature) and the need for specific co-factors or additives must be addressed. In principle, one should aim for the highest specific activity possible, to minimize the amount of enzyme required and/or guarantee that the duration of the enzymatic step is compatible with the process schedule. Additionally, it is important to check if the activity of the enzyme is partially or completely inhibited by the presence of product-related impurities in the process stream. This is more likely to occur at the earlier stages of the downstream processing, when the impurity burden is highest.

From a process point of view, decisions will have to be made regarding the positioning of the enzyme step within the downstream processing train. In some cases, the position is determined by the task at hand, e.g., an enzyme-assisted cell lysis must be performed at the start of the process and tag removal only makes sense after affinity chromatography, or by the specificity of the enzyme/process, e.g., enzyme inhibition may require the inclusion of the step after inhibitory impurities have been cleared. In other cases, however, there may be room for placing the enzyme step at different locations. The choice should then be made by balancing economic and regulatory aspects. For example, while in principle it is preferable to move the step toward the process end where process volumes are lowest to minimize

Table 2.2 Points to address when considering the use of enzymes in the downstream processing of biopharmaceuticals

Points to address	Description
Enzyme characteristics	Are stability and activity compatible with the process stream characteristics (e.g., pH, temperature)? Is inhibition by product-related impurities a possibility? Does the enzyme require specific co-factors? Is the duration of the enzymatic step compatible with the process schedule?
Process	Where should we position the enzyme step in the DSP train? What operating variables do we need to control? Is the enzyme step reproducible? Can we remove the enzyme from the final product? How easy is it to validate enzyme removal?
Availability	Is there a purified enzyme preparation commercially available from a trusted source? Are the amounts available compatible with the enzyme-assisted step being considered?
Regulatory aspects	Are the enzymes derived from an animal source?
Economics	What is the cost of the enzyme preparation? Is the enzyme step cost effective?

the amounts involved, the regulatory need to remove the enzyme from the final product indicates that the step must followed by one or two unit operations (e.g., tangential flow filtration and chromatography). The need to validate the lack of residual enzyme activity in the final product must also be considered. The non-technical aspects underlying the decision are more obvious, but, nevertheless, as critical as the technical points discussed above (Table 2.2), since a lack of enzyme availability, regulatory compliance and cost effectiveness will certainly deter decision makers from including an enzymatic step within a manufacturing process.

2.5.2 Pros and Cons

While a number of enzyme-assisted process steps have gained industry acceptance (e.g., the trypsin/carboxypeptidase C excision of the C-peptide from proinsulin (Walsh, 2005)), the fact that more of them have not left the realm of academic labs and entered into the biopharmaceutical downstream processing practice is intriguing. This is even more puzzling given the vast number of enzyme activities at our disposal that could be explored for purification purposes. Arguments against the use of enzymes in biopharmaceutical downstream processing usually point to (i) a lack of cost effectiveness and commercial availability and (ii) hurdles associated with regulatory approval.

The lack of cost effectiveness/commercial availability argument is compelling, especially in those cases where the enzyme in question has not been developed for biopharmaceutical manufacturing. However, it is also evident that if a clear demand exists from the biopharmaceutical industry for those products, technologies are available nowadays that could be mobilized to produce process-scale amounts of enzymes at a reduced cost. After all, enzymes are produced at a cost low enough so that they can be incorporated into products like food additives (e.g., chymosin), detergents (e.g., subtilisin), and textile aids (e.g., pectinase, cellulase). The highest purity required for enzymes used in the manufacturing of pharmaceuticals is of course a major difference. However, the fact that the use of enzymes could improve downstream processing by reducing the impact of impurities in chromatography operations, decreasing the number of process steps and associated labor, or improving final product quality, should compensate for the higher cost of such enzymes. If one takes the example of protein A, for example, it seems obvious that the use of expensive mass separating agents can be accommodated by companies producing high-value products like biopharmaceuticals.

Additionally, a final decision on whether to use or not an enzyme such should only be taken on the basis of a professional cost analysis such as the one provided by modeling tools like SuperProDesigner (Intelligen Inc., Scotch Plains, NJ, USA), as exemplified for the case of insulin (Petrides et al., 1995). In this case study, the cost associated with ~2 kg of enzymes required to produce 1500 kg of insulin represented less than 1.5% of the total costs of raw materials. This interesting example warns us that enzyme costs may not be as significant as one might anticipate.

The issues related to regulatory approval are also critical, in particular those that are rooted on the concerns raised by the EMA and FDA with the animal origin of some of the most interesting enzymes (e.g., lysozyme, trypsin, pepsin, RNase) used in downstream processing (Berger et al., 2005; EMA, 2013). One obvious way to circumvent the stringent regulations in these cases is to resort to animal-component-free recombinant enzymes (Novozymes, 2013; Merck Millipore, 2013; 2015; Voss et al., 2006). For example, recombinant trypsin and lysozyme manufactured under cGMP standards to allow their use in the production of biopharmaceuticals are already available in the market (Novozymes, 2013; Merck Millipore, 2013; 2015). Thus, there is no reason why other animal-derived enzymes, if judged valuable (e.g., pepsin, RNAse A) cannot follow the same path. Clearly, the biopharmaceutical industry should be able to reap the benefits of enzyme activity and diversity, particularly in the downstream processing area, in a more systematic and rational way.

2.6 Conclusions

Enzymes can be used in vitro as ancillary materials during the downstream processing of biopharmaceuticals. A number of relevant examples have been described where enzymatic action is explored to dissociate and lyse cells, remove impurities and perform structural modifications in target molecules. However, the fact that enzymes are not used more often in the downstream processing of biopharmaceuticals is surprising if one considers the number of enzymatic activities that could be explored for this purpose. A critical analysis of issues related to cost effectiveness, commercial availability and regulatory hurdles, indicates that the potential of enzyme activity and diversity ought to be explored more systematically by the biopharmaceutical industry in general, and by the downstream processing area in particular.

Acknowledgment

Funding received by iBB-Institute for Bioengineering and Biosciences from FCT-Portuguese Foundation for Science and Technology (grant UID/BIO/04565/2013) and from Programa Operacional Regional de Lisboa 2020 (Project N. 007317) is acknowledged.

References

Adamczyk M., Gebler J. C., Wu J., *J. Immunol. Meth.*, **237** (2000), 95–104.

Albani S. M. F., Silva M. R., Fratelli F., Junior C. P. C., Iourtov D., Cintra F. O., Takagi M., Cabrera-Crespo J., *Carbohyd. Polym.*, **116** (2015), 67–73.

Alves C. P. A., Šimčíková M., Brito L., Monteiro G. A., Prazeres D. M. F., *J. Chrom. A.*, **1443** (2016), 136–144.

Andrews B. A., Asenjo J. A., *Trends Biotechnol.*, **5** (1987), 273–277.

Arnau J., Lauritzen C., Petersen G. E., Pedersen J., *Protein Expr. Purif.*, **48** (2006), 1–13.

Balagurumoorthy P., Adelstein S. J., Kassis A. I., *Anal. Biochem.*, **381** (2008), 172–174.

Bandeira V., Peixoto C., Rodrigues A. F., Cruz P. E., Alves P. M., Coroadinha A. S., Carrondo M. J. T., *Hum. Gene Ther. Meth.*, **23** (2012), 255–263.

Baylis S. A., Finsterbusch T., Bannert N., Blümel J., Mankertz A., *Vaccine*, **29** (2011), 690–697.

Benedik M. J., Strych U., *FEMS Microbiol. Lett.*, **165** (1998), 1–13.

Bennett K. L., Smith S. V., Truscott R. J. W., Sheil M. M., *Anal. Biochem.*, **245** (1997), 17–27.

Berger C. N., Le Donne P., Windemann H., *Biologicals*, **33** (2005), 1–7.

Berlowska J., Dudkiewicz M., Kregiel D., Czyzowska A., Witonska I., *Enzyme Microb. Technol.*, **75–76** (2015), 44–48.

Boushaba R., Kumpalume P., Slater N. K. H., *Biotechnol. Prog.*, **19** (2003), 1176–1182.

Boynton Z. L., Koon J. J., Brennan E. M., Clouart J. D., Horowitz D. M., Gerngross T. U., Huisman G. W., *Appl. Environ. Microbiol.*, **65** (1999), 1524–1529.

Brooks S. A., *Exp. Rev. Proteom.*, **3** (2006), 345–359.

Butler M. D., Kluck B., Bentley T., *J. Chromatogr. A*, **1261** (2009) 6938–6945.

Chahal P. S., Aucoin M. G., Kamen A., *J. Virol. Methods*, **139** (2007), 61–70.

Chamsart S., Karnjanasorn T., *Biotechnol. Bioeng.*, **96** (2007), 471–482.

Codinach M., Blanco M., Ortega I., Lloret M., Reales L., Coca M. I., Torrents S., Doral M., Oliver-Vila I., Requena-Montero M., Vives J., Garcia-López J., *Cytotherapy*, **18** (2016), 1197–1208.

Cresswell C., Newcombe A. R., Davies S., Macpherson I., Nelson P., O'Donovan K., Francis R., *Biotechnol. Appl. Biochem.*, **42** (2005), 163–168.

Cuchillo C. M., Nogués M. V., Raines R. T., *Biochemistry*, **50** (2011), 7835–7841.

Darquet A., Rangara R., Kreiss P., Schwartz B., Naimi S., Delaère P., Crouzet J., Scherman D., *Gene Ther.*, **6** (1999), 209–218.

Dell A., Galadari A., Sastre F., Hitchen P., *Int. J. Microbiol.*, **2010** (2010), 148178.

EMA, EMA/CHMP/EWP/814397/2011 (2013).

FDA, Food and Drug Administration, Guidance for Industry: Characterization and Qualification of Cell Substrates and Other Biological Materials Used in the Production of Viral Vaccines for Infectious Disease Indications, Rockville, (February, 2010).

Fernandes P., Peixoto C., Santiago V. M., Kremer E. J., Coroadinha A. S., Alves, P. M., *Gene Ther.*, **20** (2013), 353–360.

Gonçalves V. M. M., Massaldi H., Giordano R. C., Tanizaki M. M., *Biotechnol. Appl. Biochem.*, **37** (2003), 283–287.

Gonçalves V. M, Takagi M., Carmo T. S., Lima R. B., Albani S. M. F., Pinto J. V., Zangirolami T. C., Giordano R. C., Tanizaki M. M., Cabrera-Crespo J., Simple and efficient method of bacterial polysaccharides purification for vaccines production using hydrolytic enzymes and tangential flow ultrafiltration, in *Communicating Current Research and Educational Topics and Trends in Applied Microbiology*, Mendez-Vilas A., ed., Formatex (2007).

Gori G. B., *Appl. Microbiol.,* **12** (1964), 115–121.

Gotschlich E. C., Liu T. Y., Artenstein M. S., *J. Exp. Med.,* **129** (1969), 1349–1365.

Gousseinov E., Kools W., Pattnaik P., *BioProcess Int.*, **12** (2014), 59–64.

Grassman E., Higashimoto T., Rohrbach J., Hall D., Williams D. A., Reeves L., Malik P., *Blood*, **112** (2008), 4626.

Green A. P., Huang J. J., Scott M. O., Kierstead T. D., Beaupre I., Gao G. P., Wilson J. M., *Hum. Gene Ther.*, **13** (2002), 1921–1934.

Gronemeyer P., Ditz R., Strube J., *Bioengineering,* **1** (2014), 188–212.

Guan, C., Li P., Riggs P. D., Inouye H., *Gene*, **67** (1988), 21–30.

Guo P., El-Gohary Y., Prasadan K., Shiota C., Xiao X., Wiersch J., Paredes J., Tulachan S., Gittes G. K., *J. Virol. Methods.*, **183** (2012), 139–146.

Hames E. E., Demir T., *World J. Microbiol. Biotechnol.,* **31** (2015), 1853–1862.

Heinzle E., Biwer A., Cooney C., Development of bioprocesses, in *Development of Sustainable Bioprocesses,* Heinzle E., Biwer A., Cooney C., eds., John Wiley & Sons, Chichester (2006).

Heiter D. F., Lunnen K. D., Wilson G. G., *J. Mol. Biol.*, **348** (2005), 631–640.

Held J., van Smaalen S., *Acta Cryst.*, **D70** (2014), 1136–1146.

Higel F., Seidl A., Sörgel F., Friess W., *Eur. J. Pharm. Biopharm.*, **100** (2016), 94–100.

Hoffmann A., Roeder R. G., *Nucleic Acids Res.*, **19** (1991), 6337–6339.

Hopp T. P., Prickett K. S. Price V. L., Libby R. T., March C. J., Cerretti P. D., Urdal D. L., Conlon P. J., *Bio/Technology*, **6** (1988), 1204–1210.

Huang H.-L., Hsing H.-W., Lai T.-C., Chen Y.-W., Lee T.-R., Chan H.-T., Lyu P.-C., Wu C.-L., Lu Y.-C., Lin S.-T., Lin C.-W., Lai C.-H., Chang H.-T., Chou H.-C., Chan H.-L. *J. Biomed. Sci.*, **17** (2010), 36.

Huang J., Nandi S., Wu L., Yalda D., Bartley G., Rodriguez R., Lonnerdal B., Huang N., *Mol. Breeding*, **10** (2002), 83–94.

Ibrahim H. R., Yamada M., Matsushita K., Kobayashi K., Kato A., *J. Biol. Chem.*, **269** (1994), 5059–5050.

Jones R. G. A., Landon J., *J. Immunol. Meth.*, **263** (2002), 57–74.

Jungbauer A., *Trends Biotechnol.,* **13** (2013), 479–492.

Kakizaki I., Takagaki K., Endoglycosidases (glycosaminoglycans), in *Experimental Glycoscience; Glycochemistry*, Taniguchi N., Suzuki A., Ito Y., Narimatsu H., Kawasaki T., Hase S., eds., Springer Japan, Tokyo (2008).

Kamen A., Henry O., *J. Gene Med.*, **6** (2004), S184–S192.

Kami D., Watakabe K., Yamazaki-Inoue M., Minami K., Kitani T., Itakura Y., Toyoda M., Sakurai T., Umezawa A., Gojo S., *BMC Biotechnol.*, **13** (2013), 102.

Kehoe J. A. The story of biosynthetic human insulin, in *Frontiers in Bioprocessing*, Sikdar S. K., Bier M., Todd O., eds., CRC Press, Boca Raton (1990), pp. 45–50.

Khalili, A. A., Ahmad, M. R., *Int. J. Mol. Sci.*, **16** (2015), 18149–18184.

Khmelnitsky Y. L., *J. Mol. Catal. B: Enzymatic*, **31** (2004), 73–81.

Kjeldsen T., *Appl. Microbiol. Biotechnol.*, **54** (2000), 277–286.

Knappik A., Pluckthun A., *Biotechniques*, **17** (1994), 754–761.

Knorr D., Shetty K. J., Kinsella J. E., *Biotechnol. Bioeng.*, 21 (1979), 2011–2021.

Kobata A., *Proc. Jpn. Acad., Ser. B*, **89** (2013), 97–117.

Koller M. R., Emerson S. G., Palsson, B. O., *Blood*, **82** (1993), 378–384.

Konz J. O., Lee A. L., Lewis J. A., Sagar S. L., *Biotechnol. Prog.*, **21** (2005), 466–472.

Kowalczyk T., Hnatuszko-Konka K., Gerszberg A., Kononowicz A. K. *World J. Microbiol. Biotechnol.*, **30** (2014), 2141.

Kumpalume P., Boushaba R., Jones R. G. A., Slater N. K. H., *Trans. IChemE*, **80** (2002), 80–97.

Lauritzen C., Nygaard G. E., Pedersen J., Arnau J., Römer U., Schäfer F., Reliable, Large-scale cleavage of tags from affinity-purified biopharmaceuticals, in *Modern Biopharmaceuticals: Recent Success Stories*, Knäbleir J., ed., Wiley-VCH Verlag GmbH & Co. KGaA, Weinheim, Germany (2013).

Macmillan D., Bill R. M., Sage K. A., Fern D., Flitsch S. L., *Chem. Biol.*, **8** (2001), 133–145.

Maizel A., Nicolini C., Baserga R., *J. Cell Physiol.*, **86** (1975), 71–81.

Malek J. A., Shatsman S. Y., Akinretoye B. A., Gill J. E., *BioTechniques*, **29** (2000), 250–252.

Marcus-Sekura C., Richardson J. C., Harston R. K., Dane N., Sheets R. L., *Biologicals* **39** (2011), 359–369.

Markussen J., US patent 4,343,898 (1982).

Markussen J., Damgaard U., Jørgensen K. H., Sørensen E., Thim, L., *Acta Med. Scand.*, **213** (1983), 99–105.

Melnick J. L., Wallis C., *Dev. Biol. Stand.*, **37** (1976), 77–82.

Merck Millipore, Enhanced Protein Recovery From *E. coli* Using CellPrime™ Recombinant Human Lysozyme, Technical Brief, Billerica, MA (2013).

Merck Millipore, CellPrime® rTrypsin Recombinant trypsin, Technical Brief, Billerica, MA (2015).

Meynial-Salles I., Combes D., *J. Biotechnol.*, **46** (1996), 1–14.

Middelberg A. P. J., *Biotechnol. Adv.* **13** (1995), 491–551.

Milián E., Kamen A. A., *BioMed Research Int.* (2015), Article ID 504831.

Moleirinho M. G., Rosa S., Carrondo M. J. T., Silva R. J. S., Hagner-McWhirter A., Ahlén G., Lundgren M., Alves P. M., Peixoto C., *Curr. Gene Ther.*, **18** (2018), 366–374.

Muhlia-Almazán A., Sánchez-Paz A., García-Carreño F. L., *J. Comp. Physiol. B*, **178** (2008), 655–672.

Muller R., Glaser S., Geipel F., Thalhofer J.-P., Rexer B., Schneider C., Ratka M., Ronning S., Eckstein H., Giessel C., US patent 7,666,629 B2 (2010).

Neurath H., Dreyer W. J., *Discuss. Faraday Soc.*, **20** (1955), 32–43.

Nilsson J., Jonasson P., Samuelsson E., Ståhl S., Uhlén M., *J. Biotechnol.*, **48** (1996), 241–250.

Novozymes. rTrypsin. Product Data Sheet. (2013).

Ohno N., Morrison, D. C., *J. Biol. Chem.*, **264** (1988), 4434–4441.

Parsiegla G., Noguere C., Santell L., Lazarus R. A., Bourne Y., *Biochemistry*, **51** (2012), 10250–10258.

Petrides D., Sapidou E., Calandranis J., *Biotechnol. Bioeng.*, **48** (1995), 529–541.

Prazeres D. M. F., Product Specifications and Quality Control, in *Plasmid Biopharmaceuticals: Basics, Applications and Manufacturing*, Prazeres D. M. F., ed., John Wiley & Sons, Hoboken (2011).

Prazeres D. M. F., *Pharm. Bioprocess.*, **4** (2016), 91–95.

Prazeres D. M. F., Schluep T., Cooney C., *J. Chromatogr. A.*, **806** (1998), 31–45.

Raines R. T., *Chem. Rev.*, **98** (1998), 1045–1065.

Rangarajan E. S., Shankar V., *FEMS Microbiol. Rev.*, **25** (2001), 583–613.

Ren L., Chang E., Makky K., Haas A. L., Kaboord B., Qoronfleh M. W., *Anal. Biochem.*, **322** (2003), 164–169.

Roberts P. L., *Biologicals*, **36** (2008), 330–335.

Sagvolden G., Giaever I., Pettersen E. O., Feder, J., *Proc. Natl. Acad. Sci. U. S. A.*, **96** (1999), 471–476.

Salazar O, Asenjo J. A., *Biotechnol. Lett.*, **29** (2007), 985–994.

Sastry L., Xu Y., Cooper R., Pollok K., Cornetta K., *Hum. Gene Ther.*, **15** (2004), 221–226.

Schmidt T. G. M., Skerra A., *Nature Protocols*, **2** (2007), 1528–1535.

Schomburg I., Chang A., Schomburg D., *Persp. Sci.*, **1** (2014), 15–23.

Segura M. M., Puig M., Monfar M., Chillón M., *Hum. Gene Ther. Meth.*, **23** (2012), 182–197.

Shaw A., Bischof A., Jasti A., Ernstberger A., Hawkins T., Cornetta K., *Hum. Gene Ther. Meth.*, **23** (2012), 65–71.

Shire S. J., *Pharm. Biotechnol.*, **9** (1996), 393–426.

Shukla A. A., Hinckley P., *Biotechnol. Prog.*, **24** (2008), 1115–1121.

Silva-Santos A. R., Alves C. P., Monteiro G. A., Azevedo A. M., Prazeres D. M. F., *Sep. Pur. Technol.*, **212** (2019), 161–170.

Sim S.-L., He T., Tscheliessniga A., Mueller M., Tan R. B. H., Jungbauer A., *J. Biotechnol.*, **157** (2012), 315–319.

Simcikova M., Alves C. P., Brito L., Prather K. L. J., Prazeres D. M. F., Monteiro G. A., *Appl. Biotechnol. Microbiol.*, **100** (2016), 6725–6737.

Sinacore M. S., Drapeau D., Adamson S. R., *Mol. Biotechnol.*, **15** (2000), 249–257.

Singh S. M., Panda A. K., J. *Biosci. Bioeng.*, **99** (2005), 303–310.

Siu S. C., Boushaba R., Liau J., Hjorth R., Titchener-Hooker N. J., *J. Chem. Technol. Biotechnol.*, **82** (2007), 871–881.

Siu S. C., Boushaba R., Topoyassakul V., Graham A., Choudhury S., Moss G., Titchener-Hooker N. J., *Biotechnol. Bioeng.*, **95** (2006), 714–723.

Skerra A., Schmidt T. G. M., *Methods Enzymol.*, **326** (2000), 271–304.

Solá R. J., Griebenow K., *BioDrugs*, **24** (2010), 9–21.

Spearman M., Rodriguez J., Huzel N., Butler M., *Biotechnol. Prog.*, **21** (2005), 31–39. 10.1021/bp0498084.

Steppert P., Burgstaller D., Klausberger M., Berger E., Aguilar P. P., Schneider T. A., Kramberger P., Tover A., Nöbauer K., Razzazi-Fazeli E., Jungbauer A., *J. Chromatogr. A*, **1455** (2016) 93–101.

Stroud, R. M., Kay, L. M., Dickerson, R. E., *J. Mol. Biol.*, **83** (1974), 185–208.

Stute N., Holtz K., Bubenheim M., Lange C., Blake F., Zander A. R., *Exp. Hematol.*, **32** (2005), 1212–1225.

Sun X., Zhang Y., *Biotechnol. Lett.*, **29** (2007), 697–701.

Takagi M., Lima R. B., Albani S. M. F., Zangirolami T. C., Tanizaki M. M., Cabrera-Crespo J., *J. Ind. Microbiol. Biotechnol.*, **35** (2008), 1217–1222.

Takegawa K., Tabuchi M., Yamaguchi S., Kondo A., Kato I., Iwahara S., *J. Biol. Chem.*, **270** (1995), 3094–3099.

Tanizaki M. M., Garcia L. R., Ramos J. B., Leite L. C. C., Hiss H., Furuta J. A., Cabrera-Crespo J., Raw, I., *J. Microbiol. Meth.*, **27** (1996), 19–23.

Thillaivinayagalingam P., O'Donovan K., Newcombe A. R., Keshavarz-Moore E., *J. Chromatogr. B*, **848** (2007), 88–96.

Thim L., Hansen M. T., Norris K., Hoegh I., Boel E., Forstrom J., Ammerer G., Fiil N. P., *Proc. Natl. Acad. Sci. U. S. A.*, **83** (1986), 6766–6770.

Unal O., Scholtyssek M., Chelur D., Strickler J., *BioPharm Int.*, **Jun 02** (2008), issue 5.

van Hee P., Middelberg A. P. J., van der Lans R. G. J. M., van der Wielen L. A. M., *Biotechnol. Bioeng.*, **88** (2004), 100–110.

Varley D. L., Hitchkock A. G., Weiss A. M. E., Horler W. A., Cowell R., Peddie L., Sharpe G. S., Thatcher D. R., Hanak J. A. J., *Bioseparation*, **8** (1998), 209–217.

Vernon H. M., *J. Physiol.*, **47** (1913), 325–338.

Victoria J. G., Wang C., Jones M. S., Jaing C., McLoughlin K., Gardner S., Delwart E. L., *J. Virol.*, **84** (2010), 6033–6040.

Vogels G., Kula M.-R., *Chem. Eng. Sci.*, **47** (1992), 123–131.

Voss C., Lindau D., Flaschel E., *Biotechnol. Prog.*, **22** (2006), 737–744.

Walsh G., *Appl. Microbiol. Biotechnol.*, **67** (2005), 151–159.

Waugh S., *Protein Exp. Purif.*, **80** (2011), 283–293.

Wei H., Zhao G., Hu T., Tang S., Jiang J., Hu B., Guan Y., *Sci. Rep.*, **6** (2016), 32560.

Weigel T., Solomaier T., Peuker A., Pathapati T., Wolff M. W., Reichl U., *J. Virol. Methods*, **207** (2014), 45–53.

WHO, *Biologicals*, **26** (1998), 175–193.

WHO, Recommendations for the production and control of Haemophilus infuenzae type b conjugate vaccines. *WHO Tech. Rep. Ser.*, **897** (2000), 27–56.

WHO, Guidelines on the quality, safety, and efficacy of biotherapeutic protein products prepared by recombinant DNA technology, *Replacement of Annex 3 of WHO Technical Report Series*, No. 814 (2013).

Wilken N. R., Nikolov, Z. L., *Biotechnol. Prog.*, **22** (2006), 745–752.

Wu F., Reddy K., Nadeau I., Gilly J., Terpening D., Clanton D. J., *Cytotechnology*, **49** (2005), 95–107.

Yamaguchi S., Yamamoto E., Mannen T., Nagamune T., *Biotechnol. J.*, **8** (2013), 17–31.

Yamamoto K., *J. Biosci. Bioeng.*, **92** (2001), 493–501.

Yamamoto K, Takegawa K., *Trends Glycosci. Glyc.*, **9** (1997), 339–354.

Yang, H., *PDA J. Pharm. Sci. Technol.*, **67** (2013), 155–163.

Yang H., Zhang L., Galinski M., *Vaccine*, **28** (2010), 3308–3311.

Yoshikawa H., Hirano A., Arakawa T., Shiraki K., *Int. J. Biol. Macromol.*, **50** (2012), 865–871

Zabriskie D. W., DiPaolo M., *Biotechnol. Bioeng.*, **32** (1988), 100–104.

Zhang L., Luo S., Zhang B., *MABS*, **8** (2016), 205–215.

Zhang S., Thwin C., Wu Z., Cho T., US patent application 6194191 B1, Feb 27 (2001).

Zhu K., Jin H., He Z., Zhu Q., Wang B., *Nat. Protoc.*, **1** (2007), 3088–3093.

Zhu K., Jin H., Ma Y., Ren Z., XiaoC., He Z., Zhang F., Zhua Q., Wang B., *J. Biotechnol.*, **118** (2005), 257–264.

Zolotukhin S., Byrne B. J., Mason E., Zolotukhin I., Potter M., Chesnut K., Summerford C., Samulski R. J., Muzyczka N., *Gene Ther.*, **6** (1999), 973–985.

Chapter 3

Therapeutic Use of Carbonic Anhydrase Inhibitors and Their Multiple Drug Interactions

Andrea Angeli and Claudiu T. Supuran

University of Florence, NEUROFARBA Department,
Sezione di Scienze Farmaceutiche,
Via Ugo Schiff 6, 50019 Sesto Fiorentino (Florence), Italy

claudiu.supuran@unifi.it

3.1 Introduction

Carbonic anhydrases, also known as carbonate dehydratases (CAs, EC 4.2.1.1), are ubiquitous metalloenzymes present throughout most living organisms prokaryotes and eukaryotes (Supuran, 2008). These enzymes are evolved in a very interesting and intricate example of convergent/divergent evolution phenomenon in seven evolutionarily unrelated gene families: α-, β-, γ-, δ-, ζ-, η-and θ-CAs (Neri and Supuran, 2011). The α-CAs were present in vertebrates, bacteria, algae, and cytoplasm of green plants (Supuran, 2012), the β-CAs were predominantly in bacteria, algae, and chloroplasts of both mono and dicotyledons

Pharmaceutical Biocatalysis: Fundamentals, Enzyme Inhibitors, and Enzymes in Health and Diseases
Edited by Peter Grunwald

ISBN 978-981-4800-61-7 (Hardcover), 978-0-429-29503-4 (eBook)
www.jennystanford.com

(Capasso and Supuran, 2015), the γ-CAs were isolated in Bacteria, cyanobacteria, and Archaea (Supuran, 2016), the δ-, ζ-, and θ-CAs are present in marine diatoms (Capasso and Supuran, 2015. Kikutani et al., 2016), whereas the η-CAs in protozoa belong to the Plasmodium spp. (Del Prete et al., 2014). All these enzymes catalyze the reversible hydration of carbon dioxide to bicarbonate ion and proton, although this reaction may occur without a catalyst at physiological pH values it is too slow to meet metabolic needs of most organisms on earth. This catalytic mechanism follows two steps (Supuran, 2008): The first one, in the hydration direction, is the nucleophilic attack of a Zn^{2+}-bound hydroxide ion on CO_2 with consequent formation of HCO_3^-, which is then displaced from the active site by a water molecule. The last step, which is rate limiting, regenerates the catalytically active Zn^{2+}-bound hydroxide ion through a proton transfer reaction from the Zn^{2+}-bound water molecule to an exogenous proton acceptor or to an active site residue as represented in Fig. 3.1.

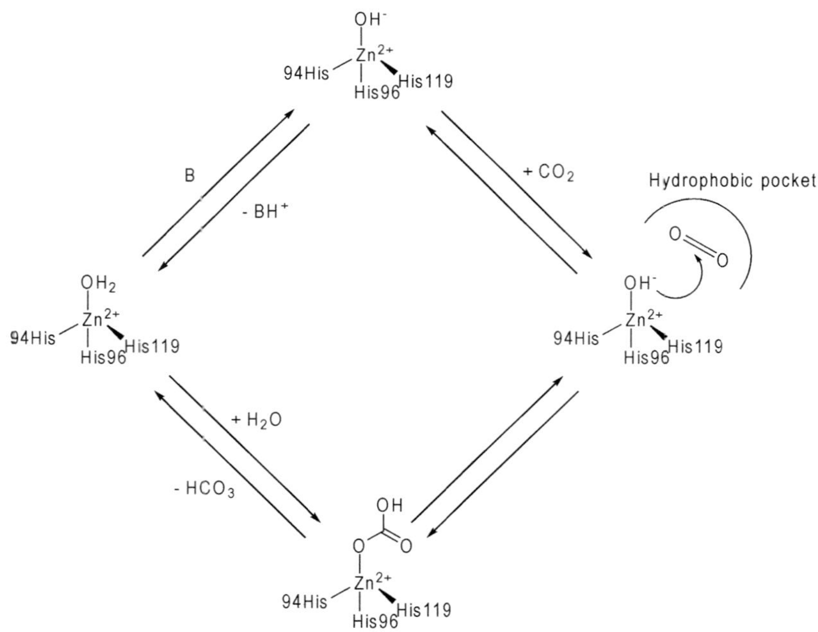

Figure 3.1 Catalytic mechanism of reversible hydration of CO_2 to HCO_3^- and proton.

Only α-CAs have been reported in vertebrates and, to date, in humans 16 α-CA isozymes or CA-related proteins have been described with different catalytic activity, subcellular localization, and tissue distribution (Hilvo et al., 2005; Thiry et al., 2006; Supuran et al., 2003). Only twelve isoforms showed catalytic activity, among these, they provided five cytosolic forms (CA I, CA II, CA III, CA VII and CA XIII), five membrane-bound isozymes (CA IV, CA IX, CA XII, CA XIV and CA XV), two mitochondrial forms (CA VA and CA VB), and one secreted CA isozyme (CA VI). Several studies demonstrated the important roles of CAs in a variety of physiological processes, and showed that abnormal levels or activities of these enzymes have been often associated with different human diseases (Suparan, 2008; Suparan et al., 2009). Specifically hCA I is found in many tissues and it is involved in retinal and cerebral edema, in this manner, its inhibition may be a valuable tool for fighting these conditions (Gao et al., 2007). hCA II is involved in several diseases, such as glaucoma, edema, epilepsy, and probably altitude sickness (Mincione et al., 2008; Suparan, 2008; De Simone et al., 2009). hCA III is involved in the oxidative stress, characterizing a lot of inflammatory diseases (Barreiro and Hussain, 2010; Brancaccio et al., 2010). hCA IV is surely a drug target for several pathologies, including glaucoma (together with hCA II and XII), retinitis pigmentosa and stroke (Matsui et al., 1996; Tang et al., 2006; Datta et al., 2009). The mitochondrial isoforms hCA VA and VB are targets for obtaining new antiobesity agents (De Simone and Supuran, 2007; Supuran et al., 2008), whereas CA VI is implicated in cariogenesis (Kiveläet al., 1999). CA VII has been noted for its contributions to epileptiform activity together with CA II and XIV (Ruusuvuori, et al., 2004). hCA IX is now a well-established marker of disease progression in many types of hypoxic tumors, and recently its inhibition has been shown to be associated with a significant inhibition of the growth of both primary tumors and metastases (Guler et al., 2010; De Simone and Supuran, 2010; Pastorekova et al., 2006). Finally, hCA XIV is involved in epileptogenesis and, similarly to hCA IV, in some retinopathies, thus, it may be a drug target for innovative agents useful in the management of such disorders (Shah et al., 2005; Ogilvie et al., 2007). Inhibitors of carbonic anhydrase (CAIs) are in clinically use since 1950s, but to date, no CAI showed selectivity for a specific isoform. Given this situation, patients often receive several or more drugs at the same time and, not surprising that an interaction may occur between them. For this reason, this chapter lays down the aspects of drug-drug interactions involving in the clinically used CAIs which all belong to the sulfonamide or sulfamate class (Supuran, 2008; Neri and Supuran, 2011).

3.2 First- and Second-Generation Clinically Used Sulfonamides and Their Drug Interactions

The systemic carbonic anhydrase inhibitor, acetazolamide (**1**), was initially discovered as a diuretic agent (Supuran, 2008; Neri and Supuran, 2011) and introduced in 1954 into ophthalmology as a treatment for glaucoma and, since then, has been widely used for this purpose. At first, its clinical use as diuretic drug is limited to development of metabolic acidosis (Kassamali and Sica, 2011). However, this feature can afford some potential clinical benefit for disease states such as metabolic alkalosis (Faisy et al., 2010; Mazur et al., 1999) nephrolithiasis (Sterrett et al., 2008), rhabdomyolysis (Subbaramaiah et al., 2010; Davidov et al., 2006), contrast-induced nephropathy (CIN) (Pakfetrat et al., 2009), sleep apnea (Nakayama et al., 2002; Angeli and Supuran, 2018), and high-altitude erythropoiesis (Leaf and Goldfarb, 2007; Richalet et al., 2008). In subsequent years the systemic side effects of acetazolamide led to the development of new CAIs such as brinzolamide **3** and dorzolamide **4** (Fig. 3.2) with their high water solubility are administered via the topical route directly into the eye, making them less prone to drug-drug interactions and systemic side effects (Carta et al., 2012; Masini et al., 2013).

The main drawback of these drugs was non-selective CAI, which indiscriminately inhibits the most of human Carbonic anhydrase isoforms (Supuran, 2008; Neri and Supuran, 2011; Carta et al., 2012; Masini et al., 2013) and present a large number of interactions with drugs belonging to a variety of pharmacologic classes (Anderson et al., 1978; Sweeney et al., 1986; Hazouard et al., 1999; Cappon et al., 2005; Yablonski et al., 1988; Rodicio and Hernando, 1977; Beermann et al., 1975; Scozzafava et al., 2006; Sato et al., 1981; Sato et al., 1983; Keogh et al., 1988; Tabbara et al., 1998; Ruokoniemi et al., 2009; Scott et al., 1990; Bell et al., 1997; Stankiewicz et al., 2010; Rulo et al.,1997; Kennedy et al., 1985; Ali et al., 2015; Asiedu et al., 2010).

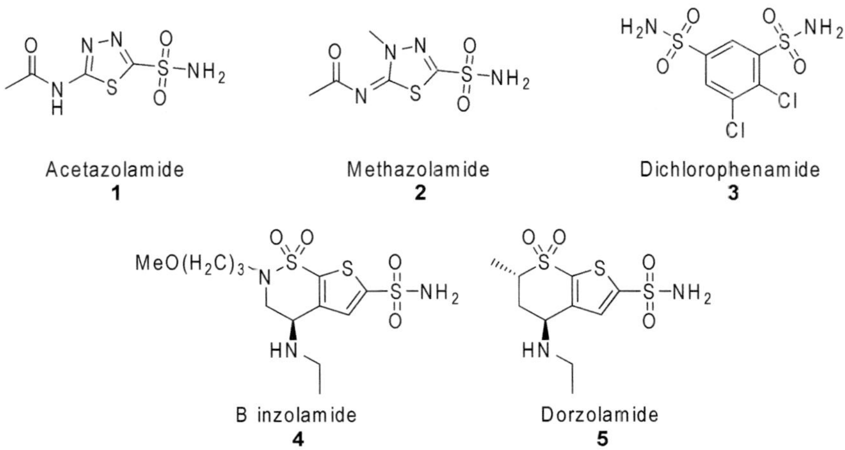

Figure 3.2 Clinically used classical carbonic anhydrase inhibitors 1–5.

Clinically adverse drug interactions have been found when the CAIs and nonsteroidal anti-inflammatory drugs (NSAIDs) are co-prescribed (Anderson et al., 1978). The ability of salicylate (aspirin) to inhibit significantly the renal tubular secretion of the acetazolamide is well known and an important cause of severe toxicity in glaucoma patients receiving these drugs concomitantly (Sweeney et al., 1986). The metabolic acidosis enhanced by acetazolamide caused in turn an enhanced of non-ionized form of salicylate which crosses the biological membrane more rapidly and better than the ionized form of salicylate. In this way, Non-ionized salicylic acid passes from the blood stream into the central nervous system increased cerebral toxicity. Furthermore, it was also enhances the free form of acetazolamide leading to an increased toxicity of both drugs (Hazouard et al., 1999). This particular toxicity can be extended to all NSAIDs such as ketoprofen, ibuprofen, diflunisal and fenbufen that significantly depressed the renal clearance of acetazolamide (Hazouard et al., 1999; Cappon et al., 2005; Yablonski et al., 1988). On the other hand, co-administration of acetazolamide during the action of the other diuretics incorporating a primary sulfonamide moiety such as furosemide, resulted in potentiation of inhibition, suggesting a different mechanism of action (Rodicio and Hernando, 1977). The changes observed in potassium excretion are of great interest. Separately, furosemide or acetazolamide produced kaliuresis, but when, furosemide was administered during acetazolamide diuresis potassium excretion was reduced despite the sharp rise in sodium excretion (Rodicio and Hernando, 1977; Beermann et al., 1975). However, the additive CA inhibitory effects of the two drugs may lead to electrolyte abnormalities, dehydration, kidney stones and other side effects typical of systemically acting CAIs (Scozzafava et al., 2006).

Another potentiating effect of first and second generation of sulfonamides are observed against barbiturates and more in general with antiepileptic drugs (Sato et al., 1981). The concomitantly use of acetazolamide with barbiturates (barbital, mephobarbital, metharbital, and phenobarbital) increased the antiepileptic activity of them. The mechanism of action is complicated and not completely understood, on one side, acetazolamide indicate a common antiepileptic effect due inhibition of carbonic anhydrase isoforms present in the brain, on the other side, acetazolamide may influenced the pharmacodynamics of antiepileptic drugs contributing overall to increase the effects on brain (Sato et al., 1983).

An additional important effect of acetazolamide we found also on cyclosporine clearance. Cyclosporine has been used for the treatment of organ transplation, autoimmune diseases and uveitis that may have an increase in intraocular pressure. When administered in conjunction with cyclosporine, acetazolamide has been shown to cause 5-fold increase in cyclosporine trough blood levels as well as pronounced nephrotoxicity and neurotoxicity although the mechanism of this effect does not seem to be related to urinary alkalinization (Keogh et al., 1988; Tabbara et al., 1998; Ruokoniemi et al., 2009). Acetazolamide is a teratogenic agent, inducing limb malformations possibly by reducing the embryonic intracellular pH (Scott et al., 1990). This effect was exacerbated when acetazolamide was combined with amiloride (Bell et al., 1997) and is probably present with all other CAIs of the sulfonamide type (Scozzafava et al., 2006).

Sulfonamides drug interactions have been found also with positive pharmacologic interactions. Topical carbonic anhydrase inhibitor dorzolamide with prostamides and prostaglandin analogues (bimatoprost) had additive effect on the reduction of intraocular pressure (IOP) with different mechanisms of hypotensive action (Stankiewicz et al., 2010; Rulo et al., 1997). The combination of mitomycin C and dorzolamide showed an increased activation of pro-apoptotic and anti-proliferative signals of Thioredoxin-Interacting Protein (TXNIP). Dorzolamide was able to increase the capability of mitomycin C to reduce the solid tumor weight after their combination. This synergistic effect augmented mitomycin C in the active form and increasing its DNA adduct formation (Kennedy et al., 1985) leading to decrease of solid tumor weight (Ali et al., 2015). In the field of neuropathic pain, spinal co-administration of acetazolamide and midazolam act synergistically to reduce neuropathic allodynia after peripheral nerve injury. These findings indicate that the combined use of CA inhibitors and benzodiazepines may be effective in the clinical management of neuropathic pain in humans (Asiedu et al., 2010). However, newer-generation sulfonamide CAIs seem to have a much better effect than acetazolamide, without the need of drug combination regimens (Carta et al., 2015).

3.3 Sulfonamides with Diuretic Action: Bumetanide, Hydrochlorothiazide, Furosemide, Indapamide, and Chlortalidone

Diuretic drugs, which show an effect also independently of their CA inhibitory action (Supuran, 2012), have been in clinical use for decades and widely employed for controlling hypertension (Padilla et al., 2007; Smith and Ashiya, 2007; Elliott, 2007) alone or in combination with other drugs such as β-adrenergic receptor antagonists (Kharwade and Bhushan, 2008), cholesterol-lowering agents (Sasmal et al., 2010), inhibitors of neutral peptidase (Straub et al., 2010). The primarily sulfonamide moiety acting as excellent zinc-binding group for the CAs active site and, their inhibition are responsible for some of the therapeutic effects of the drugs or for some side effects observed with it (Fig. 3.3). However, to date, no clinical trials are being conducted to understand in detail the poly-pharmacological aspects related to CAs inhibition of these drugs. However, the presence of a zinc binding group as sulfonamide moiety induce strong CA inhibitory activities and, only recently re-evaluated against all the mammalian active CA isoforms (Temperini et al., 2008; 2009; 2009a).

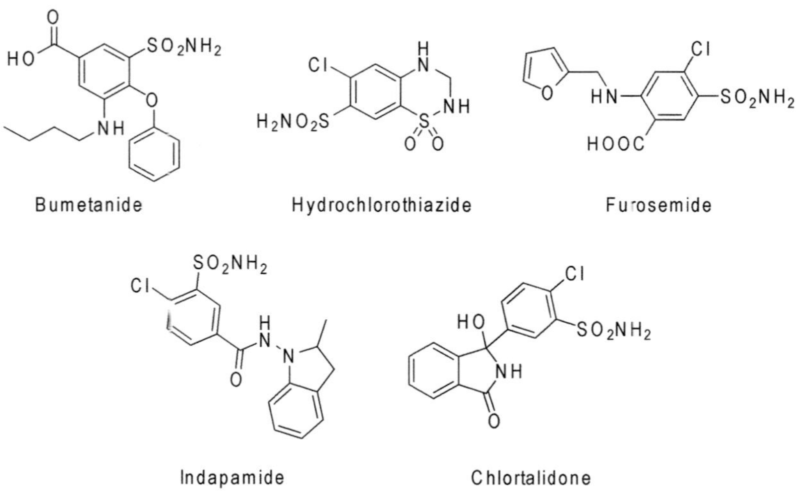

Figure 3.3 Clinically used sulfonamides with diuretic action.

The common drug-interaction are typical of sulfonamides reported above such as with NSAIDs, where, the nonsteroidal anti-inflammatory drugs (such as aspirin and indomethacin) competition for a common secretory mechanism, could mediate the blunting of furosemide and the other drugs reported in Fig. 3.3 (Brater, 1986). The same situation we found when these compounds are co-administered with barbiturates, cyclosporine and lithium drugs producing a remarkable increase the toxicity of them (Sato et al., 1981; Höcherl et al., 2004; Finley et al., 1995; Dzhala et al., 2008). Furosemide showed, recently, ototoxicity in patients taking Phosphodiesterase type 5 (PDE-5) inhibitors such as sildenafil (Skeith et al., 2013). Although loop diuretics such as furosemide are not widely appreciated for their potential synergistic effects in terms of ototoxicity causing cochlear injury and hearing loss, particularly in combination of known ototoxic drugs such as aminoglycosides (Hirose and Sato, 2011) and co-administration also with cisplatin (McAlpine and Johnstone, 1990). Furosemide, finally, may decrease renal elimination of Methotrexate that following the same renal tubular secretion and could potentiate its toxicity as well as the toxicity of furosemide (Nierenberg and Mamelok, 1983).

Bumetanide and indapamide showed antagonistic effects when using with probenecid producing natriuresis and hyperreninemia. These effects are not due to a direct action on sodium excretion but are probably secondary to its inhibitory effect on renal tubular secretion of diuretic drugs (Opie, 2012; Holland and Williamson, 1979). This interaction, however, can be used to increase brain levels of bumetanide and decrease its elimination, which could have therapeutic potential in the treatment of brain disorders (Töllner et al., 2015). On the other hand, absorption of Hydrochlorothiazide is impaired in the presence of anionic exchange resins such as Cholestyramine. Single doses of it bind the Hydrochlorothiazide and reduce its absorption from the gastrointestinal tract (Hunninghake and Hibbard, 1986; Hunninghake et al., 1982).

Finally, the concomitant use of corticosteroids and most of these sulfonamide drugs increased the risk of hypokalemia since both classes of drugs excrete potassium: from one side, via mineralocorticoid effects, and the other via diuretic effects of sulfonamide molecules (Seale and Compton, 1986).

3.4 Sulfonamides/Sulfamates as Antiepileptic Drugs (AEDs): Sulthiame, Topiramate, and Zonisamide

Two sulfonamide derivatives (sulthiame **6** and zonisamide **8**) and one sulfamate compound (topiramate **7**) are current used as antiepileptic drugs and act as highly efficient CAIs (Fig. 3.4)

(Supuran, 2008; De Simone et al., 2008). The mechanisms by which these drugs exert their antiepileptic effects are complex and involved several different pathway (Milburn-McNulty et al., 2015; Bialer et al., 2004; Leppik, 2004), but probably inhibition of CA isoforms present in the brain (CA II, III, IV, VA, VII, VIII, XII and XIV) caused a selective increase of the cerebral blood flow, with the concomitant raising of the carbon dioxide partial pressure (Supuran, 2008; 2012; 2016). As a consequence, CA inhibitors are useful in the treatment of conditions associated with increased intracranial pressure, as well as different neurological/neuromuscular pathologies such as epilepsy. The three drugs are also subject to some drug-drug interaction, especially with other AEDs, since in many patients combination of several such drugs are used (Milburn-McNulty et al., 2015; Bialer et al., 2004; Leppik, 2004).

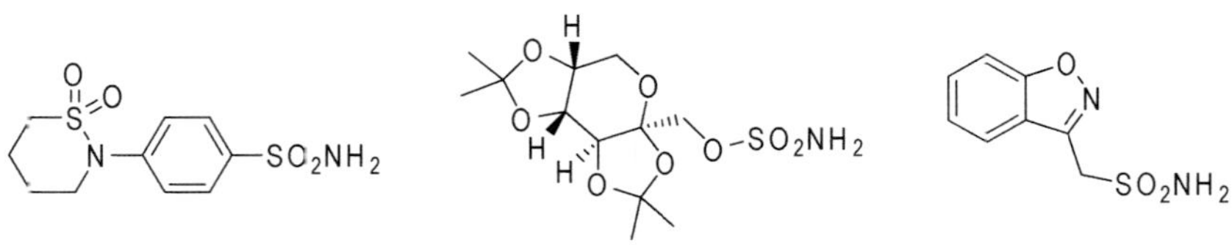

Sultiame **6** Topiramate **7** Zonisamide **8**

Figure 3.4 Sulfonamide and sulfamate derivatives used as AEDs 6–8.

Sulthiame has been used since the early 1960s as an effective anticonvulsant agent in combination with other AED drugs for major epilepsy. For this reason, the incidence of adverse effects caused by drug-drug interaction has been high in several studies especially with other AED drugs. A well-known drug interaction among sulthiame and phenytoin was reported (Houghton and Richens, 1974). Phenytoin is mainly metabolized by CYP2C9 to several inactive metabolites, and CYP2C19 is also involved in their metabolism. The mechanism by which sulthiame inhibits this drug metabolism is not clearly understood but was probably due to displacement of phenytoin from its binding to proteins, for example, from red cells (Houghton and Richens, 1974a, Hansen et al., 1968). This interaction may account entirely or in part for the anticonvulsant action of sulthiame and, in second place, it may lead to intoxication in patients who have been receiving phenytoin without adverse effect for a number of years. These findings may explain several clinical investigations showing that sulthiame was not very active per se but it was effective in improving the therapeutic effect of other anticonvulsant drugs. Whether the increased concentration of phenytoin in the brain after sulthiame association is due to a metabolic inhibition or to a redistribution modification due the tissue acidosis induced by sulthiame remains to be established. In the same manner, when sulthiame and clobazam are combined the concentration may increase, potentially leading to both enhanced antiseizure activity and aggravation of adverse effects (Yamamoto et al., 2014).

Topiramate was licensed for clinical use in 1995 as adjunctive treatment of partial seizures and primary generalized tonic-clonic seizures. Topiramate has multiple mechanisms of action including over Carbonic anhydrase inhibition, the inhibition of voltage-sensitive sodium and calcium channels, enhancement of GABA-mediated activity and inhibition of kainite/AMPA-type glutamate receptors (Kawasaki et al., 1998). The drug-interaction, also for this drug, has been high especially with other AED drugs. Carbamazepine and phenytoin enhances the metabolism of topiramate in similar manner of sulthiame (Contin et al., 2002; Adin et al., 2004). Furthermore, the metabolic clearance of topiramate was extent to other barbital drugs and primidone. An important drug interaction with topiramate is that with oral contraceptives such as ethinylestradiol. Thus, sulfonamide compound increases the oral clearance of ethinylestradiol and may compromise contraception control (Doose et al., 2003). Other important interaction was observed between topiramate and lithium drugs where co-administration enhances the toxicity of this one (Abraham and Owen, 2004). However, the low dosage (under 200 mg/day) of topiramate exhibited no or little (with lithium, propranolol and other drugs) pharmacokinetic interactions with these drugs (Bialer et al., 2004).

Zonisamide was a potent CAI and licensed for clinical use since 2005 as antiepileptic drug (AED). Zonisamide exerts its pharmacological effects by multiple complementary mechanisms, including over carbonic anhydrase inhibition, block voltage-gated sodium channels (providing activity against partial-onset seizures) (Schauf, 1987), and inhibit T-type calcium channels (providing activity against absence seizures) (Rogawski and Löscher, 2004). One of the principal pathways of zonisamide inactivation is reductive metabolism by CYP3A4. As a result, it is important to assess the potential for zonisamide to alter the pharmacokinetics of concomitant AEDs, and vice versa (Sills and Brodie, 2007). The interaction between carbamazepine zonisamide is well recognized. Both drugs are metabolized by CYP3A4 and carbamazepine is a potent inducer of this isoenzyme. Thus, carbamazepine significantly increases zonisamide elimination via induction of CYP3A4. However, caution must be exercised if the enzyme-inducing agents are subsequently withdrawn, as this will lead to a slow increase in zonisamide plasma levels and the potential development of toxicity (Patsalos, 2013). Zonisamide was widely used not only for the treatment of epilepsy, but also for the management of a variety of non-epileptic neurological conditions and psychiatric disorders. Zonisamide was approved as adjunctive therapy with levodopa in previously treated adult patients with Parkinson's disease (Yang and Perry, 2009). Its mechanism of action in Parkinson's disease is still poor known. However, recent reports showed zonisamide as a weakly inhibitor of MAO-B (Binda et al., 2011) and it activates dopamine synthesis exhibiting an increased levels of dihydroxyphenylalanine (DOPA; a dopamine precursor) (Okada et al., 1992; Okada and Kaneko, 1995).

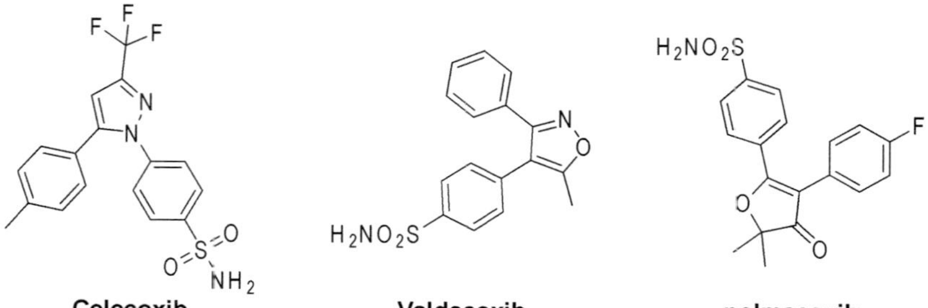

Figure 3.5 Dual CA/COX inhibitors: celecoxib, valdecoxib, and polmacoxib.

3.5 Celecoxib, Valdecoxib, and Polmacoxib as CA/COX-2 Inhibitors

Several selective inhibitors of cycloxigenase-2 (COX-2) had inside their scaffold a sulfonamide moiety (Fig. 3.5). Thus, in the last years was reported in literature that they exhibited also potent CAIs against most of the human CA isoforms involved in crucial physiologic functions (Supuran, 2008; Skarke et al., 2012; Weber et al., 2004; Di Fiore et al., 2006).

Celecoxib is the first selective COX-2 inhibitor to be approved for clinical practice in the treatment of patients with rheumatoid arthritis, osteoarthritis and for management of the pain of these conditions (Skarke et al., 2012). However, this class is putative gastrointestinal safety advantages compared with other NSAIDs. Celecoxib is metabolized primarily by CYP2C9 and may be increased or decreased by CYP2C9 modifiers such as fluconazole (200 mg once/day), fluvastatin, and zafirlukast. When, it was administered concomitantly with celecoxib, plasma level of coxib drug increased two folds (Garnett, 2001). It also inhibits CYP2D6 and may increase serum concentrations for a substantial number of agents that are substrates for CYP2D6, including many antidepressants (selective serotonin reuptake inhibitors, tricyclic antidepressants), antifungals, antipsychotics (flecainide, propafenone, haloperidol, etc.), analgesics such as codeine, and β-blockers (Werner et al., 2003). Significant interactions were reported for the administration of celecoxib with lithium. In fact, it decreases the clearance of lithium by inhibiting synthesis of renal prostaglandins, thereby elevating its plasma levels and lead to toxic effects (Gunja et al., 2002; Slørdal et al., 2003). Celecoxib showed an antagonistic drug-interaction with many anti-cancer drugs as doxorubicin (El-Awady et al., 2011) and platin derivative drugs such as cisplatin or oxaliplatin (Bijman et al., 2008; Chen et al., 2013; Kong et al., 2016).

The antagonism between celecoxib and platin drugs were accompanied with reduced caspase-3 activity and Pt-DNA adduction formation due to reduced drug influx, which may antagonize apoptosis pathways possibly reducing cisplatin efficacy (Dent and Grant, 2001).

On the other hand, co-administration of celecoxib with docetaxel or sorafenib produced an enhancement of the antitumor effects (Shaik et al., 2006; Rosendahl et al., 2012). It has been also reported that celecoxib can attenuate to a limited extent the antiplatelet effects of aspirin (Ruzov et al., 2016).

There are few reports on interactions between the other two coxib drugs: valdecoxib and polmacoxib. Valdecoxib is approved by the FDA for the treatment of rheumatoid arthritis (RA), osteoarthritis (OA), and primary dysmenorrhea since 2001 (Gotta, 2002). Polmacoxib is still in clinical trials (Phase II) for the same pathologies (Skarke et al., 2012).

However, valdecoxib and polmacoxib are a substrate of CYP3A4 and CYP2C9 and subject to drug-drug interactions typical for the entire class, such as for example with diuretics (acetazolamide, furosemide, hydrochlorothiazide), lithium or AEDs drugs discussed above (Chavez et al., 2003).

3.6 Inhibitors of Carbonic Anhydrases as Antitumor Agents

Hypoxia is a key component of tumor microenvironment with strong effects on tumor phenotype and cancer progression. This particular condition leads to molecular changes in the transcriptome and proteome of hypoxic cells, which induce genes and proteins that facilitate survival of cells. The hypoxia-induced transcriptional program is principally activated by the hypoxia-inducible factor (HIF), a master transcription factor consisting of two subunits, each existing in at least two isoforms: α and β one (Ratcliffe, 2013). An important gene activated by HIF1α is carbonic anhydrase (Carnero and Lleonart 2016) and in particular the overexpression of hCA IX. This enzyme, with its catalytic mechanism, can help the tumor to adapt in hypoxic stress condition neutralizing the acidic microenvironment (Neri and Supuran, 2011; Carnero and Lleonart 2016). In the last twenty years, hCA IX ((McIntyre et al., 2012; Hussain et al., 2007) but also hCA XII (Haapasalo et al., 2008) isoforms have been shown to be the target of many carbonic anhydrase inhibitors (CAIs) because their overexpression in various tumors is correlated with poor survival and progression (Svastová et al., 2004; Said et al., 2013; Dubois et al., 2009). Recently, new approaches, in order to provide a more targeted and efficient form of cancer therapy, were developed by inhibiting these two enzymes with small molecule compounds (Fig. 3.6) or antibodies.

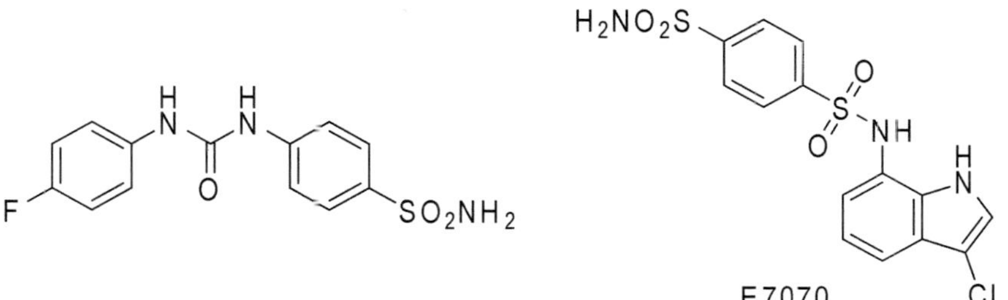

Figure 3.6 Small molecule CA inhibitors in Phase I/II clinically trials as anticancer agents.

Monotherapy approach is still a very common treatment modality for many different forms of cancer, but the combination of two or more therapeutic treatments to specifically target cancer-inducing or cell-sustaining pathways is a cornerstone of cancer therapy (Yap et al., 2013).

There are few reports regarding the interactions of these agents with other drugs. An interesting study showed SLC-0111 co-administrated with temozolomide increased the therapy against glioblastoma. This combination works in a synergistic or additive manner to decreased the glioblastoma survival and reduced its chemoresistance (Boyd et al., 2017). Moreover, this compound with Gemcitabine is under investigation in a clinical trial for its synergistic effects against Metastatic Pancreatic Ductal Cancer (NIH, 2018). In this field, also acetazolamide, which exhibited a good inhibition for hCA IX and XII showed antitumor effects. It enhances doxorubicin toxicity and reduces melphalan toxicity in cell lines that highly expressed hCA IX under anoxic conditions, such as HT29 and MDA435 CA9/18 cell lines (Gieling et al., 2013). Also the co-administration among acetazolamide and sulforaphane for the treatment of bronchial carcinoid enhanced the efficacy compared to monotherapy of each agent. The enhanced efficacy was attributed to increased inhibitory effects on clonogenic and invasive ability in vitro, and increased tumor growth inhibitory effects in vivo (Mokhtari et al., 2013; Islam et al., 2016).

Indisulam (E7070) is another small molecule as anticancer agent that is currently being evaluated in phase II clinical studies (Haddad et al., 2004). The precise mechanism of action has not been determined, since it showed multiple mechanisms of action such as inhibitor of hCA IX and XII (Supuran, 2003) and multiple cell-cycle checkpoints (Ozawa et al., 2012) Also for this compound there are few reports regarding the interactions of these agents with other drugs. A phase I clinical trial reported the excessive hematological toxicity with relatively high exposure to carboplatin (Dittrich et al., 2007). In addition, the plasma protein binding of acenocoumarol was reduced at higher concentrations of E7070. This is caused by potential inhibition of cytochrome CYP2C9 that is the major enzyme involved in the metabolism of acenocoumarol. This could be lead to hypoprothrombinemia (van den Bongard et al., 2004).

Monoclonal antibodies (mAbs) are examples of new strategies to prevent or treat cancer (White et al., 2001). The G250 mAb and its humanized, chimeric, and bispecific variants have become the most important imaging and therapeutic tools targeting hCA IX-expressing tumors. Chimeric G250 (Girentuximab) has very recently completed Phase III trial as an adjuvant immunotherapy of patients with non-metastatic renal cell cancer (ccRCC) (Winum and Capasso, 2016). This chimeric antibody has also recently completed clinical Phase II trials in combination therapy with cytokines (IL-2 and IFN) demonstrating clinical benefit in patients with progressive metastatic RCC (Siebels et al., 2011; Ullrich, 2004). Radionuclide conjugated mAbs have been studied extensively for the treatment of solid tumors. Killer isotopes (α or β emitters such as ^{131}I, ^{90}Y, ^{177}Lu, and ^{211}At) can be effective used for a therapeutic effect (Divgi et al., 2013; Stillebroer et al., 2013; Brouwers et al., 2005). Finally, a phase I/II Clinical trials with radiolabeled murine mAb CA IX (mG250) in patients with RCC have demonstrated selective and specific delivery of mAb to renal cancer sites (Stillebroer et al., 2007).

3.7 Conclusions and Repurposing Old Drugs for New Disease Approaches

Recently, the revisitation of these old drugs, mentioned above, may have relevance for the drug design of CAIs with diverse pharmacological applications. In sight of this, obesity has become one of the most global health problems and can be simply defined as a physiological status derived from a disruption of the equilibrium between energy intake and energy expenditure. The reasons of the energetic unbalance are multiple and complex. One possible pathway involves the two mitochondrial hCA isoforms, hCA VA and hCA VB. These isoforms are involved in lipogenesis and their inhibition leads to diminished fatty acid biosynthesis (Dodgson and Cherian, 1989; Lynch et al., 1995). They are expressed only in the mitochondria and localized in the liver for hCA VA (Fujikawa-Adachi et al., 1999; Dodgson et al., 1980), whereas CA VB is more widely distributed in the organism (Shah et al., 2000). In fact, Clinical observation on patients treated with topiramate and zonisamide for neurological pathologies showed a marked body weight reduction (Gordon and Price, 1999). This "side effect" could be supported by inhibition of the two mitochondrial CAs and, therefore, lipogenesis. Thus different patents, in the last 10 years, claiming the association of topiramate and zonisamide with different sympathomimetic agent as new anti-obesity approach or obesity-related pathologies (Najarian, 2000; KozachuK, 2002; Hebebrand et al., 2000). In this context, one important obesity-related pathology is diabetes and subsequently the alteration of CAs activity of isoforms involved in the biosynthesis of glucose and lipids, i.e., CA II and CA VA/VB (Supuran, 2008). Ghosh et al. (2016) reported the potential CA activity in human erythrocytes as biomarker for selectively distinguishing type 1 Diabetes from type 2. In fact, some early evidences suggest that the changes in carbonic anhydrases activities in erythrocytes may be an initial step of altered metabolism in diabetes mellitus (Gambhir et al., 1997). However, the development of selective mitochondrial CA inhibitors indeed represents a key point in the treatment of metabolic related pathologies (Supuran, 2008).

Recently, new approaches against neuropathic pain involved the application of carbonic anhydrase inhibitors. In this field was demonstrated that CA inhibition may represent a new and effective strategy (Supuran, 2008). One possible pathway engaged CAIs such as acetazolamide that reduce the bicarbonate-dependent depolarization of GABAA receptors when the function of KCC2 is compromised showing analgesic effects in animal models of neuropathic pain. In fact, intrathecal administration of acetazolamide rapidly restored an analgesic effect for the KCC2 blockers, suggesting an important role of CA activity in regulating GABAA-mediated analgesia, but also, Spinal administration also led to a profound reduction in the mouse formalin pain test, indicating that spinal CA inhibition produced analgesia when primary afferent activity is driven by chemical mediators (Asiedu et al., 2010; 2014). Moreover, the same study showed that spinal co-administration of acetazolamide and midazolam acted synergistically to reduce neuropathic allodynia after peripheral nerve injury (Asiedu et al., 2010). Consequently, several clinically drugs reported in this chapter, such as methazolamide, topiramate, zonisamide, and celecoxib were investigated in various neuropathic pain models in animals, and reported to be effective in relieving pain by several groups (Snedecor et al., 2014; Sun et al., 2014; Rezende et al., 2009). Supported by these studies, a new class of sulfonamide compounds was recently reported and tested in vivo, in animal models of neuropathic pain induced by oxaliplatin, showing promising results (Carta et al., 2015; Angeli et al., 2018). However, a main unresolved issue remains: Which CA isoforms are being targeted for such applications, considering the fact that many of them are present in the central nervous system and the ones present in the peripheral nervous system are still unknown.

References

Abraham G., Owen J., *J. Clin. Psychopharmacol.*, **24** (2004), 565–567.

Adin J., Gomez M. C., Blanco Y., Herranz J. L., Armijo J. A., *Ther. Drug Monit.*, **26** (2004), 251–257.

Ali B. M., Zaitone S. A., Shouman S. A., Moustafa Y. M., *Naunyn Schmiedebergs Arch. Pharmacol.*, **388** (2015), 1271–1282.

Anderson C. J., Kaufman P. L., Sturm R. J., *Am. J. Ophthalmol.*, **86** (1978), 516–519.

Angeli A., Di Cesare Mannelli L., Lucarini E., Peat T. S., Ghelardini C., Supuran C. T., *Eur J. Med. Chem.*, **154** (2018), 210–219.

Angeli A., Supuran C. T. *Expert Opin. Ther. Pat.*, **16** (2018), 1–5.

Asiedu M. N., Mejia G. L., Hübner C. A., Kaila K., Price T. J., *J. Pain*, **15** (2014), 395–406.

Asiedu M., Ossipov M. H., Kaila K., Price T. J., *Pain*, **148** (2010), 302–308.

Barreiro E., Hussain S.N. A., *Antioxid. Redox Signal.*, **12** (2010), 417–429.

Beermann B., Hellström K., Lindström B., Rosén A., *Clin. Pharmacol., Ther.*, **17** (1975), 424–432.

Bell S. M., Schreiner C. M., Resnick E., et al., *Reprod. Toxicol.*, **11** (1997), 823–831.

Bialer M., Doose D. R., Murthy B., Curtin C., Wang S. S., Twyman R. E., Schwabe S., *Clin. Pharmacokinet.*, **43** (2004), 763–780.

Bijman M. N., Hermelink C. A., van Berkel M. P., Laan A. C., Janmaat M. L., Peters G. J., Boven E., *Biochem. Pharmacol.*, **75** (2008), 427–437.

Binda C., Aldeco M., Mattevi A., Edmondson D. E., *J. Med. Chem.*, **54** (2011), 909–912.

Boyd N. H., Walker K., Fried J., Hackney J. R., McDonald P. C., Benavides G. A., Spina R., Audia A., Scott S. E., et al., *JCI Insight*, **2** (2017), pii: 92928.

Brancaccio, P., Lippi, G., Maffulli, N., *Clin. Chem. Lab. Med.*, **48** (2010) 757–767.

Brater D. C., *Am. J. Med.*, **80** (1986), 62–77.

Brouwers A. H., Buijs W. C., Mulders P. F., et al., *Clin. Cancer Res.*, **11** (2005), 7178s–786s.

Capasso C., Supuran C. T., *J. Enzyme Inhib. Med. Chem.*, **30** (2015), 325–332.

Cappon G. D., Fleeman T. L., Cook J. C., et al., *Drug Chem. Toxicol.*, **28** (2005), 409–421.

Carnero A., Lleonart M., *Bioessays*, **38** (2016), S65-S74.

Carta F., Supuran C. T., Scozzafava A., *Expert Opin. Ther. Pat.*, **22** (2012) 79–88.

Carta F., Di Cesare Mannelli L., Pinard M., Ghelardini C., Scozzafava A., McKenna R., Supuran C. T., *Bioorg. Med. Chem.*, **23** (2015), 1828–1840.

Chavez M. L., DeKorte C. J., *Clin. Ther.*, **25** (2003), 817–851.

Chen M., Yu L., Gu C., Zhong D., Wu S., Liu S., *Cancer Lett.*, **329** (2013), 189–196.

Contin M., Riva R., Albani F., Avoni P., Baruzzi A., *Ther. Drug Monit.*, **24** (2002), 332–337.

Datta R., Waheed A., Bonapace G., Shah G. N., Sly W. S., *Proc. Natl. Acad. Sci. U. S. A.*, **106** (2009), 3437–3442.

Davidov T., Hong J. J., Malcynski J. T., *J Trauma*, **61** (2006), 213–215.

De Simone G., Di Fiore A., Supuran C. T., *Curr. Pharm. Des.*, **14** (2008), 655–660.

De Simone G., Scozzafava A., Supuran C. T., *Chem. Biol. Drug Des.*, **74** (2009), 317–321.

De Simone G., Supuran C. T., *Biochim. Biophys. Acta*, **1804** (2010), 404–409.

De Simone G., Supuran C. T., *Curr. Top. Med. Chem.*, **7** (2007), 879.

Del Prete S., Vullo D., Fisher G. M., Andrews K. T., Poulsen S. A., Capasso C., Supuran C. T., *Bioorg. Med. Chem. Lett.*, **15** (2014), 4389–4396.

Dent P., Grant S., *Clin. Cancer. Res.*, **7** (2001), 775–783.

Di Fiore A., Pedone C., D'Ambrosio K., et al., *Bioorg. Med. Chem. Lett.*, **16** (2006), 437–442.

Dittrich C., Zandvliet A. S., Gneist M., Huitema A. D., King A. A., Wanders J., *Br. J. Cancer.*, **96** (2007), 559–566.

Divgi C. R., Uzzo R. G., Gatsonis C., et al., *J. Clin. Oncol.*, **31** (2013), 187–94.

Dodgson S. J., Cherian K., *Am. J. Physiol.*, **257** (1989), E791–E796.

Dodgson S. J., Forster R. E. II, Storey B. T., Mela L., *Proc. Natl. Acad. Sci. U. S. A.*, **77** (1980), 5562–5566.

Doose D. R., Wang S.-S., Padmanabhan P., et al., *Epilepsia*, **44** (2003), 540–549.

Dubois L., Lieuwes N. G., Maresca A., Thiry A., Supuran C. T., Scozzafava A., et al., *Radiother. Oncol.*, **92** (2009), 423–428.

Dzhala V. I., Brumback A. C., Staley K. J., *Ann. Neurol.*, **63** (2008), 222–235.

El-Awady R. A., Saleh E. M, Ezz M., Elsayed A. M., *Toxicol. Appl. Pharmacol.*, **255** (2011), 271–286.

Elliott W. J., *Curr. Probl. Cardiol.*, **32** (2007), 201–259.

Faisy C., Mokline A., Sanchez O., et al., *Intensive Care Med.*, **36** (2010), 859–863.

Finley P. R., Warner M. D., Peabody C. A., *Clin. Pharmacokinet.*, **29** (1995), 172–191.

Fujikawa-Adachi K., Nishimori I., Taguchi T., Onishi S., *J. Biol. Chem.*, **274** (1999), 21228–21233.

Gambhir K. K., Oates P., Verma M., Temam S., Cheatham W., *Ann. N. Y. Acad. Sci.*, **827** (1997) 163–169.

Gao B. B., Clermont A., Rook S., Fonda S. J., Srinivasan V. J., Wojtkowski M., Fujimoto J. G., Avery R. L., Arrigg P. G., Bursell S. E., Aiello L. P., Feener E. P., *Nat. Med.*, **13** (2007), 181–188.

Garnett W. R., *Pharmacotherapy*, **21** (2001), 1223–1232.

Ghosh C., Mandal S. Banik G. D., Maity A., Mukhopadhyay P., Ghosh S., Pradhan M., *Sci. Rep.*, **6** (2016), 35836.

Gieling R. G., Parker C. A., De Costa L. A., et al., *J. Enzyme Inhib. Med. Chem.*, **28** (2013), 360–369.

Gordon A., Price L. H., *Am. J. Psychiatry*, **156** (1999), 968–969.

Gotta A., *Curr. Opin. Invest. Drugs.*, **3** (2002), 240–245.

Guler O. O., De Simone G., Supuran C. T., *Curr. Med. Chem.*, **17** (2010), 1516–1519.

Gunja N., Graudins A., Dowsett R., *Intern. Med. J.*, **34** (2002), 494.

Haapasalo J., Hilvo M., Nordfors K., Haapasalo H., Parkkila S., Hyrskyluoto A., Rantala I., Waheed A., Sly W. S., et al., *Neuro Oncol.*, **10** (2008), 131–138.

Haddad R. I., Weinstein L. J., Wieczorek T. J., Bhattacharya N., Raftopoulos H., Oster M. W., Zhang X., et al., *Clin. Cancer. Res.*, **10** (2004), 4680–4687.

Hansen J. M., Kristensen M., Skovsted L., *Epilepsia*, **9** (1968), 17–22.

Hazouard E., Grimbert M., Jonville-Berra A. P., et al., *J. Fr. Ophtalmol.*, **22** (1999), 73–75.

Hebebrand J., Antel J., Preuschoff U., et al., WO07821 (2000).

Hilvo M., Tolvanen M., Clark A., Shen B., Shah G. N., Waheed A., Halmi P., Hanninen M., Hamalainen J. M., Vihinen M., Sly W. S., Parkkila S., *Biochem. J.*, **392** (2005), 83–92.

Hirose K., Sato E., *Hear. Res.*, **272** (2011), 108–116.

Höcherl K., Kees F., Krämer B. K., Kurtz A., *Kidney Int.*, **65** (2004), 2071–2080.

Holland S. D., Williamson H. E., *Proc. Soc. Exp. Biol. Med.*, **161** (1979), 299–302.

Houghton G. W., Richens A., *J. Neurol. Neurosurg. Psychiatry*, **37** (1974), 275–281.

Houghton G. W., Richens A., *Br. J. Clin. Pharmacol.*, **1** (1974a), 59–66.

Hunninghake D. B., Hibbard D. M., *Clin. Pharmacol. Ther.*, **39** (1986), 329–334.

Hunninghake D. B., King S., LaCroix K., *Int. J. Clin. Pharmacol. Ther. Toxicol.*, **20** (1982), 151–154.

Hussain S. A., Ganesan R., Reynolds G., Gross L., Stevens A., Pastorek J., Murray P. G., Perunovic B., Anwar M. S., et al., *Br. J. Cancer.*, **96** (2007), 104–109.

Islam S. S., Mokhtari R. B., Akbari P., Hatina J., Yeger H., Farhat W. A., *Target Oncol.*, **11** (2016), 209–227.

Kassamali R., Sica D. A., *Cardiol. Rev.*, **19** (2011), 276–278.

Kawasaki H., Tancredi V., D'Arcangelo G., Avoli M., *Brain Res.*, **807** (1998), 125–134.

Kennedy K. A., McGurl J. D., Leondaridis L., Alabaster O., *Cancer. Res.*, **45** (1985), 3541–3547.

Keogh A., Esmore D., Spartt P., Sardie E., McClskey P., *Transplantation*, **46** (1988), 478–479.

Kharwade P., Bhushan I., US0107726 (2008).

Kikutani S., Nakajima K., Nagasato C., Tsuji Y., Miyatake A., Matsuda Y., *Proc. Natl. Acad. Sci. U. S. A.*, **113** (2016), 9828–9833.

Kivelä J., Parkkila S., Parkkila A. K., Rajaniemi H., *Caries Res.*, **33** (1999), 178–184.

Kong Y., Gu C., Zhong D., Zhao X., Lin Q., Wang K., Xun T., Yu L., Liu S., *Eur. J. Pharm. Sci.*, **81** (2016), 137–148.

KozachuK W. E., WO03915 (2002).

Leaf D. E., Goldfarb D. S., *J. Appl. Physiol.*, **102** (2007), 1313–1322.

Leppik I. E., *Seizure*, **13** (2004), Suppl 1, S5–S9.

Lynch C. J., Fox H., Hazen S. A., et al., *Biochem. J.*, **310** (1995), 197–202.

Masini E., Carta F., Scozzafava A., et al., *Expert Opin. Ther. Pat.*, **23** (2013), 705–716.

Matsui H., Murakami M., Wynns G. C., Conroy C. W., Mead A., Maren T. H., Sears M. L., *Exp. Eye Res.*, **62** (1996), 409–417.

Mazur J. E., Devlin J. W., Peters M. J., et al., *Crit. Care Med.*, **27** (1999), 1257–1261.

McAlpine D., Johnstone B. M., *Hear Res.*, **47** (1990), 191–203.

McIntyre A., Patiar S., Wigfield S., Li J. L., Ledaki I., Turley H., Leek R., Snell C., Gatter K., Sly W. S., et al., *Clin. Cancer Res.*, **18** (2012), 3100–3111.

Milburn-McNulty P., Powell G., Sills G. J., et al., *Cochrane Database Syst. Rev.* (2015) CD009472.

Mincione F., Scozzafava A., Supuran C. T., *Curr. Pharm. Des.*, **14** (2008), 649–654.

Mokhtari R. B., Kumar S., Islam S. S., Yazdanpanah M., Adeli K., Cutz E., Yeger H., *BMC Cancer*, **13** (2013), 378.

Najarian T., WO76493 (2000).

Nakayama H., Smith C., Rodman J., et al., *Am. J. Respir. Crit. Care Med.*, **165** (2002), 1251–1259.

Neri D., Supuran C. T., *Nature Rev. Drug Discov.*, **10** (2011), 767–777.

Nierenberg D. W., Mamelok R. D., *Arch. Dermatol.*, **119** (1983), 449–450.

NIH: https://clinicaltrials.gov/ct2/show/NCT03450018 (2018).

Ogilvie J. M., Ohlemiller K. K., Shah G. N., Ulmasov B., Becker T. A., Waheed A., Hennig A. K., Lukasiewicz P. D., Sly W. S., *Proc. Natl. Acad. Sci. U. S. A.*, **104** (2007), 8514–8519.

Okada M., Kaneko S., *Epilepsy Res.*, **22** (1995), 193–205.

Okada M., Kaneko S., Hirano T., et al., *Epilepsy Res.*, **13** (1992), 113–119.

Opie L. H., *S. Afr. Fam. Pract*, **54** (2012), S23-S25.

Ozawa Y., Kusano K., Owa T., Yokoi A., Asada M., Yoshimatsu K., *Cancer Chemother. Pharmacol.*, **69** (2012), 1353–1362.

Padilla M. C., Armas-Hernandez M. J., Hernandez R. H., et al., *Am. J. Ther.*, **14** (2007), 154–160.

Pakfetrat M., Nikoo M. H., Malekmakan L., et al., *Int. Urol. Nephrol.*, **41** (2009), 629–634.

Pastorekova S., Parkkila S., Zavada J., *Adv. Clin. Chem.*, **42** (2006), 167–216.

Patsalos P. N., *Clin. Pharmacokinet.*, **52** (2013), 927–966.

Ratcliffe P. J., *J. Physiol.*, **591** (2013), 2027–2042.

Rezende R. M., Dos Reis W. G., Duarte I. D., Lima P. P., Bakhle Y. S., de Francischi J. N., *Pain*, **142** (2009), 94–100.

Richalet J. P., Rivera-Ch M., Maignan M., et al., *Am. J. Respir. Crit. Care Med.*, **177** (2008), 1370–1376.

Rodicio J. L., Hernando L., *Rev. Esp. Fisiol.*, **33** (1977), 113–118.

Rogawski M. A., Löscher W., *Nat. Rev. Neurosci.*, **5** (2004), 553–564.

Rosendahl A. H., Gundewar C., Said K., Karnevi E., *Pancreatology*, **12** (2012), 219–226.

Rulo A. H., Greve E. L., Hoyng P. F., *Ophthalmology*, **104** (1997), 1503–1507.

Ruokoniemi P., Tertti R., Paalosmaa-Puusa P., et al., *NDT Plus*, **2** (2009), 298–299.

Ruusuvuori E., Li H., Huttu K., Palva J. M., Smirnov S., Rivera C., Kaila K., Voipio J., *J. Neurosci.*, **24** (2004), 2699–2707.

Ruzov M., Rimon G., Pikovsky O., Stepensky D., *Br. J. Clin. Pharmacol.*, **81** (2016), 316–326.

Said H. M., Hagemann C., Carta F., Katzer A., Polat B., Staab A., Scozzafava A., Anacker J., Vince G. H., Flentje M., Supuran C. T., *Bioorg. Med. Chem.*, **21** (2013), 3949–3957.

Sasmal B. K., Reddy B. P., Nasare V. D., et al., US0068269 (2010).

Sato J., Nioka M., Owada E., et al., *J. Pharmacobiodyn.*, **4** (1981), 952–960.

Sato J., Nioka M., Owada E., Ito K., Murata T., *J. Pharmacobiodyn.*, **6** (1983), 295–300.

Schauf C. L., *Brain Res.*, **413** (1987), 185–188.

Scott W. J., Duggan C. A., Schreiner C. M., et al., *Toxicol. Appl. Pharmacol.*, **103** (1990), 238–254.

Scozzafava A., Mastrolorenzo A., Supuran C. T., *Expert Opin. Ther. Pat.*, **16** (2006), 1627–1664.

Seale J. P., Compton M. R., *Med. J. Aust.* **144** (1986), 139–142.

Shah G. N., Hewett-Emmett D., Grubb J., et al., *Proc. Natl. Acad. Sci. U. S. A.*, **97** (2000), 1677–1682.

Shah G. N., Ulmasov B., Waheed A., Becker T., Makani S., Svichar N., Chesler M., Sly W. S., *Proc. Natl. Acad. Sci. U. S. A.*, **102** (2005), 16771–16776.

Shaik M. S., Chatterjee A., Jackson T., Singh M., *Int. J. Cancer*, **118** (2006), 396–404.

Siebels M., Rohrmann K., Oberneder R., et al., *World J. Urol.*, **29** (2011), 121–126.

Sills G., Brodie M., *Epilepsia*, **48** (2007), 435–441.

Skarke C., Alamuddin N., Lawson J. A., et al., *Clin. Pharmacol. Ther.*, **91** (2012), 986–993.

Skeith L., Yamashita C., Mehta S., Farquhar D., Kim R. B., *J. Popul. Ther. Clin. Pharmacol.*, **20** (2013), e128–e131.

Slørdal L., Samstad S., Bathen J., Spigset O., *Br. J. Clin. Pharmacol.*, **55** (2003), 413–414.

Smith R. E., Ashiya M., *Nat. Rev. Drug Discov.*, **6** (2007), 597–598.

Snedecor S. J., Sudharshan L., Cappelleri J. C., Sadosky A., Mehta S., Botteman M., *Pain Pract.*, **14** (2014), 167–184.

Stankiewicz A., Wierzbowska J., Siemiatkowska A., Fuksinska B., Robaszkiewicz J., Zegadlo A., Ehrlich R., Siesky B., Harris A., *Br. J. Ophthalmol.*, **94** (2010), 1307–1311.

Sterrett S. P., Penniston K. L., Wolf J. S. Jr, et al., *Urology,* **72** (2008), 278–282.

Stillebroer A. B., Boerman O. C., Desar I. M., et al., *Eur. Urol.*, **64** (2013), 478–485.

Stillebroer A. B., Oosterwijk E., Oyen W. J., Mulders P. F., Boerman O. C., *Cancer Imaging*, **7** (2007), 179–188.

Straub M., Witte K., Ziegler D., et al., US0323012 (2010).

Subbaramaiah M. T., Sapsford D., Banham-Hall E., *Anaesth. Intensive Care,* **38** (2010), 398.

Sun Y. J., Chen Y., Pang C., Wu N., Li J., *Acta Pharmacol. Sin.*, **35** (2014), 41–47.

Supuran C. T., *Expert Opin. Investig. Drugs,* **12** (2003), 283–287.

Supuran C. T., *Nature Rev. Drug Discov.,* **7** (2008), 168–181.

Supuran C. T., *J. Enzyme Inhib. Med. Chem.*, **27** (2012), 759–772.

Supuran C. T., *Biochem. J.*, **473** (2016), 2023–2032.

Supuran C. T., Casini A., Mastrolorenzo A., et al., *Mini Rev. Med. Chem.*, **4** (2004), 625–632.

Supuran C. T., Di Fiore A., De Simone G., *Expert Opin. Emerg. Drugs,* **13** (2008), 383–392.

Supuran C. T., Scozzafava A., Casini A., *Med. Res. Rev.*, **23** (2003), 146–189.

Supuran C. T., Winum J.-Y., *Drug Design of Zinc-Enzyme Inhibitors: Functional, Structural, and Disease Applications,* eds., Wiley: Hoboken, NJ (2009); Part II.

Supuran C. T., *Curr. Pharm. Des.,* **14** (2008), 641–648.

Svastová E., Hulíková A., Rafajová M., Zat'ovicová M., Gibadulinová A., Casini A., Cecchi A., et al., *FEBS Lett.*, **577** (2004), 439–445.

Sweeney K. J., Chapron D. J., Brandt J. L., Gomolin I. H., Feig P. U., Kramer P. A., *Clin. Pharmacol. Ther.*, **40** (1986), 518–524.

Tabbara K. F., Al-Faisal Z., Al-Rashed W., *Arch. Ophthalmol.*, **116** (1998), 832–833.

Tang Y., Xu H., Du X., Lit L., Walker W., Lu A., Ran R., Gregg J. P., Reilly M., Pancioli A., Khoury J. C., Sauerbeck L. R., Carrozzella J. A., Spilker J., Clark J., Wagner K. R., Jauch E. C., Chang D. J., Verro P., Broderick J. P., Sharp F. R. J., *Cereb. Blood Flow Metab.*, **26** (2006), 1089.

Temperini C., Cecchi A., Scozzafava A., et al., *Bioorg. Med. Chem. Lett.*, **18** (2008), 2567–2573.

Temperini C., Cecchi A., Scozzafava A., et al., *J. Med. Chem.*, **52** (2009), 322–328.

Temperini C., Cecchi A., Scozzafava A., et al., *Bioorg. Med. Chem.*, **17** (2009a), 1214–1221.

Thiry A., Dognè, J. M., Masereel B., Supuran C. T. *Trends Pharmacol. Sci.*, **27** (2006), 566–573.

Töllner K., Brandt C., Römermann K., Löscher W., *Eur. J. Pharmacol.*, **746** (2015), 167–173.

Ullrich S., Warnaar O. S., WO2004002526 (2004).

van den Bongard H. J., Sparidans R. W., Critchley D. J., Beijnen J. H., Schellens J. H., *Invest. New Drugs*, **22** (2004), 151–158.

Weber A., Casini A., Heine A., et al., *J. Med. Chem.*, **47** (2004), 550–557.

Werner U., Werner D., Rau T., et al., *Clin. Pharmacol. Ther.*, **74** (2003), 130–137.

White C. A., Weaver R. L., Grillo-Lopez A. J., *Annu. Rev. Med.*, **52** (2001), 125–145.

Winum J. Y., Capasso C., *Expert Opin. Ther. Pat.*, **23** (2016), 757–760.

Yablonski M. E., Maren T. H., Hayashi M., et al., *Am. J. Ophthalmol.*, **106** (1988), 332–336.

Yamamoto Y., Takahashi Y., Imai K., Mogami Y., Matsuda K., Nakai M., Kagawa Y., Inoue Y., *Epilepsy Behav.*, **34** (2014), 124–126.

Yang L. P., Perry C. M., *CNS Drugs.*, **23** (2009), 703–711.

Yap T. A., Omlin A., de Bono J. S., *J. Clin. Oncol.*, **31** (2013), 1592–1605.

Chapter 4

Fibrinolytic Enzymes for Thrombolytic Therapy

Swaroop S. Kumar and Sabu Abdulhameed

Department of Biotechnology and Microbiology, School of Life Sciences, Kannur University, Thalassery Campus, Kannur, Kerala 670661, India

drsabu@gmail.com

4.1 Introduction

Coronary thrombosis had been identified as fatal in the early 19th century itself but little was known about its cause and risk factors. It was rather regarded as a medical curiosity, and for many years it was believed to cause sudden death every time. Until the dawn of the 20th century, pathophysiology of acute myocardial infarction (AMI) was unclear. In 1901, Krehl proposed that thrombosis does not always cause immediate death (Krehl, 1901). Later it was found that recovery from AMI is possible and bed rest for post infarction recovery was advocated (Herrick, 1912). During that period, morphine was recommended to relieve pain after AMI (Parkinson and Bedford, 1928). Thus, the

Pharmaceutical Biocatalysis: Fundamentals, Enzyme Inhibitors, and Enzymes in Health and Diseases
Edited by Peter Grunwald

ISBN 978-981-4800-61-7 (Hardcover), 978-0-429-29503-4 (eBook)
www.jennystanford.com

management of AMI was palliative than curative during the earlier times. Aspirin was reported to prevent myocardial infarction (MI) during the 1950s and used widely for preventing primary and secondary MIs though it has not been proved useful for acute phases of infarction (Craven, 1950, 1953). Coronary care units (CCU) were introduced in the early 1960s for having continuous cardiac monitoring and treating fatal arrhythmias, thereby reducing the mortality rate due to MI (Julian, 1961).

Whenever an injury or trauma occurs, the blood flow through the vessels should be immediately arrested to prevent the loss. This is achieved by hemostasis, characterized by equilibrium between blood coagulation and fibrinolysis. The hemostatic system includes primary hemostasis (platelet plug formation), secondary hemostasis (coagulation), and tertiary hemostasis (fibrinolysis). Primary hemostasis initiates through vascular spasm or local vaso-constriction occurring immediately after injury and thereby reducing the blood flow through the vessel followed by platelet aggregation to form platelet plug. Coagulation factors normally exist as pro-enzymes and circulate through blood stream along with pro-coagulant and anticoagulant factors such as von Willebrand factor (vWF), tissue factor, platelet activating factors, tissue factor pathway inhibitor, endothelium-derived relaxing factor or nitric oxide (NO), tissue plasminogen activator (t-PA) and prostacyclin (Lane et al., 2005; Pearson, 1994). Once trauma occurs, the above factors get activated. Coagulation or secondary hemostasis usually referred as coagulation cascade initiates a series of enzyme reactions which consists of intrinsic and extrinsic pathways (Macfarlane, 1964). Chief coagulating enzyme thrombin, formed from Factor Xa initiates fibrin polymerization and thereby blood clotting (Kovalenko et al., 2017). However, removal of clot after wound healing is also important in order to ensure the proper blood circulation. Tertiary hemostasis is characterized by the dissolution of clot or fibrinolysis controlled by fibrinolytic system. Plasmin is the enzyme that degrades clot, and it is formed form plasminogen with the aid of plasminogen activators. However, this is regulated by plasminogen activator inhibitor 1 (PAI-1), thrombin-activatable fibrinolysis inhibitor (TAFI), α2-antiplasmin and α2-macroglobulin, etc. (Stassen et al., 2004).

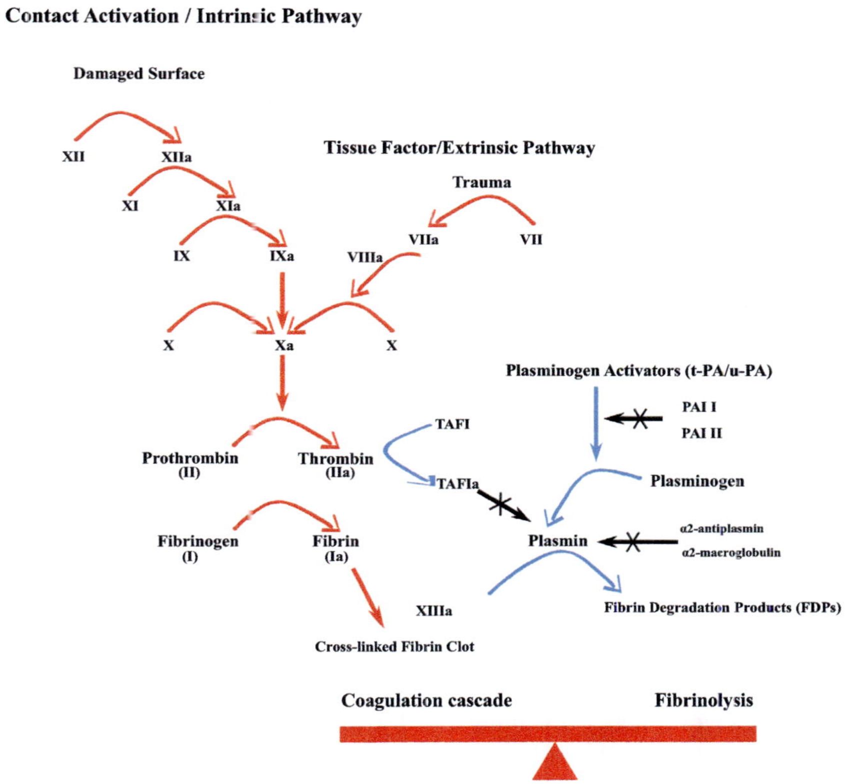

Figure 4.1 Normal hemostasis- equilibrium between coagulation and fibrinolysis.

Normal functioning of the circulatory system is achieved through hemostatic balance between fibrin formation and degradation. Improper accumulation of blood clot inside the vessel is called thrombus which is usually removed by plasmin. Nevertheless, when the balance shifts toward the inappropriate accumulation of fibrin, normal oxygen supply towards cells will be disturbed causing tissue damage, leading to thrombolytic diseases such as acute myocardial infarction and stroke (Arnout et al., 2006). According to WHO report 17.7 million deaths (below 70 years old) are due to heart diseases representing 31% of total mortality occurred globally during the year 2015 (WHO, 2017). Among these 7.4 million were due to coronary heart disease while

6.7 million were due to stroke. Cardiovascular conditions such as acute myocardial infarction, ischemic heart disease, and stroke often occur due to blockage of blood flow and many factors play a role in these obstructions. Removal of fibrin clot is essential for proper management of thrombolytic diseases. Cardiovascular disease management often involves the use of anticoagulants such as warfarin and heparin, antiplatelets such as dipyridamol and aspirin, and removal of blood clot either by surgery or by using thrombolytic enzymes.

4.2 Status of Anticoagulants and Antiplatelets

4.2.1 Heparin and Derivatives

Anticoagulants are known to inhibit blood clotting and were used in preventing MI from the early 20th century. A fat-soluble anticoagulant was discovered while searching for procoagulants from dog liver in 1915 by McLean under the guidance of Howell. Later, Howell and Holt isolated fat-soluble anticoagulant heparin which was claimed to be distinct from the anticoagulant molecule obtained by McLean. The term heparin was coined from the Greek word "hepar," meaning liver, by Howell indicating the source of isolation (Howell and Holt, 1918). However, the credit of discovery was disputed since McLean claimed himself that he discovered heparin (Marcum, 2000). Heparin is a glycosaminoglycan inhibiting blood clotting. Although it was discovered a century ago, it took long time for its clinical application. However, earlier studies conducted by Mayo Clinic, Minnesota, USA, showed that heparin causes side effects such as headache, fever, and nausea (Mason, 1924). Later, researchers from various parts of the world worked on the purification and improvement of heparin preparation and thereby eliminating the side effects to a great extent.

Murray et al. (1937) showed that heparin could prevent thrombus formation in traumatized veins of dogs. Later it was used as anticoagulant in humans in the year 1937. It enhanced the clotting time during 2 h infusion (Murray et al., 1937). Several reports expressed successful use of heparin (McLean and Johnson, 1946; Lange et al., 1945), but clinical practice of the molecule reflected certain shortcomings such as heparin induced thrombocytopenia (HIT), bleeding complications, etc. which have to be constantly monitored as well as multiple and regular administration of heparin on a daily basis is essential (Cohen, 1999). Despite these limitations, heparin still remains as one of the widely used cardiovascular drugs, though it is the oldest therapeutic used as anticoagulant. Apart from heparin, low-molecular-weight heparins (LMWHs) derived from heparin by its depolymerization also find their application as anticoagulants. They unveil better therapeutic index with lower frequency of HIT in comparison with heparin. But, LMWHs are not appropriate for management of already established HIT, since the antibodies produced against heparin react with them. Each LMWH is distinct in its molecular weight and therapeutic properties (Bick and Fareed, 1999; Cohen, 1999). On the other hand, LMWHs also need to be injected and the major route of clearance is renal. Hence, the half-life of the molecule is higher in patients with renal failure leading to its accumulation (Hirsh et al., 2001). A few of these LMWHs include enoxaparin, dalteparin, nadroparin, tinzaparin, certoparin, reviparin, ardeparin, parnaparin, and bemiparin.

4.2.2 Coumarin Derivatives

Anticoagulants like coumarin derivatives were discovered in 1939 by the identification of dicumarol from spoiled sweet clover hay which is a vitamin K antagonist by Link and Campbell (Stahmann et al., 1941). Further studies on coumarin derivatives led to the discovery of warfarin which was initially used as rodenticide and later approved for clinical use as anticoagulant. It was the first oral thrombin inhibitor and found useful in preventing embolic strokes (Aguilar and Hart, 2005). Limitations of these vitamin K antagonists were bleeding complications, interaction with food, necrosis and hair loss (Dantas et al., 2004; Ansell et al., 2008; Pirmohamed, 2006). Even though other coumarin derivatives such as phenprocoumon and acenocoumarol also were used as anticoagulants, warfarine is the most common vitamin K antagonist in practice and remains as affordable in cardiovascular disease management.

4.2.3 Thrombin Inhibitors

Several thrombin inhibitors have been developed during the past couple of decades and proved their efficacy as anticoagulants. Hirudin is one of the most important, naturally occurring thrombin inhibitor molecule isolated from leach. Administration of hirudin has been associated with increased risk of bleeding as well as formation of non-neutralizing antibodies in patients (Hoppensteadt et al., 2008). Later, many thrombin inhibitors with better efficacy and therapeutic potential have been developed such as argatroban, bivalirudin, lepirudin, and dabigatran etexilate. The first used thrombin inhibitor is argatroban and it is now widely used in Japan. It was recommended as an alternate anticoagulant for patients suffering from HIT and its clinical use significantly reduced bleeding complications in comparison to heparin (Lewis et al., 2001, 2003). Bivalirudin is a bivalent reversible inhibitor and when compared to heparin and LMWHs, they declined the bleeding complications by almost 50%, whereas the efficacy remained same for all of them (Carswell and Plosker, 2002; Ahrens et al., 2007). Another thrombin inhibitor lepirudin was found to be marginally superior to heparin and more suitable for patients with previously reported HIT though continuous monitoring is required here also (Lubenow et al., 2004). Dabigatran etexilate is an oral prodrug that would get converted into active dabigatran, a direct thrombin inhibitor, upon intestinal absorption (Lee and Ansell, 2011). All those thrombin inhibitors described here are FDA approved for preventing various cardiovascular diseases.

4.2.4 Factor Xa Inhibitors

Factor Xa is a serine protease and since it is part of prothrombinase complex, it carries crucial role in blood coagulation. Factor Xa inhibitors are another class of anticoagulant drugs reducing thrombin formation. These small molecules bind with active site of factor X (Xa), which in turn inhibits the formation of thrombin. They demonstrate varying structural and functional diversity and can be either naturally occurring or synthetic molecule and chemically it can be peptide, protein, or heparin saccharidic sequences (Fareed et al., 1999; Leone et al., 2004; Bauer, 2006;

Samama et al., 1999). Their mode of action could be either direct by binding to factor Xa or indirect through binding with antithrombin III (AT), leading to Factor Xa inactivation. The interaction of Xa inhibitors could be reversible or irreversible (Harenberg and Fenyvesi, 2004; Walenga et al., 2002). Fondaparinux is a heparin pentasaccharide, the FDA approved this for its use after surgery and also for initial management of thrombotic disorders such as pulmonary embolism and deep vein thrombosis trials (Hoppensteadt et al., 2008). Another indirect-acting FXa inhibitor is Idraparinux (SANORG-34006), which is a hyper methylated variant of fondaparinux and showed more affinity towards antithrombin III (AT) (PERSIST investigators, 2004). Rivaroxaban was an orally active, direct thrombin inhibitor and extensively studied for its use in preventing venous thromboembolism during knee and hip replacement (Turpie et al., 2005; Eriksson et al., 2006). A few other direct Factor Xa inhibitors include otamixaban, apixaban, edoxaban, etc. Although most of the Factor Xa inhibitors are either under clinical trial or in different developmental stages, still they are considered safer than thrombin inhibitors (Hoppensteadt et al., 2008).

4.2.5 Antiplatelet Drugs

Platelet activation is initiated by vWF interacting with collagen and platelet surface glycoprotein receptors resulting in its adherence to vascular sub-endothelium. Activated platelets further release thromboxane A2 (TxA2), thrombin, and adenosine diphosphate (ADP). These series of events eventually lead to the expression of platelet surface glycoproteins (GP IIb-IIIa) subsequently leading to platelet aggregation and thrombus formation (Oprea and Popescu, 2013). Antiplatelet drugs are used for the prevention and management of thrombotic disorders. The first antiplatelet drug used was aspirin, though its mechanism of action was unresolved at the time of use. Later it was found that platelet-connective tissue reaction can be considerably damaged by aspirin uptake. This inhibits the platelet plug formation and thereby prolongs the clotting time (Weiss and Aledort, 1967). Further studies revealed that antiplatelet effect of aspirin is largely due to its irreversible inactivation of cyclooxygenase enzyme, primarily cyclooxygenase 1 (COX1). Inhibition of prostaglandin synthase/

cyclooxygenase enzyme impedes the prostaglandin produced by platelets. Prostaglandin (PH_2) is the substrate for the synthesis of the TxA2 by thromboxane-A synthase (Vane, 1971; Smith and Willis, 1971; Oprea and Popescu, 2013) and thus, COX1 inhibition by aspirin is accountable for its use as antiplatelet drug. Another class of antiplatelet therapeutics is ADP receptor inhibitor, which blocks binding of ADP to its specific receptor on blood platelet surface. Ticlopidine was the earliest ADP receptor antagonist, later discontinued due to its adverse effects such as thrombocytopenia. Successively a series of ADP receptor inhibitors are developed like clopidogrel, prasugrel, and ticagrelor. Clopidogrel is a $P2Y_{12}$ receptor inhibitor which irreversibly blocks binding of ADP. Prasugrel is another irreversible inhibitor of $P2Y_{12}$ receptor. Ticagrelor on the other hand mainly blocks $P2Y_1$ receptor and it partially inhibits $P2Y_{12}$ receptor too. The inhibitory mechanism is instead reversible here. A third class of antiplatelet drugs belongs to glycoproteins (GP IIb-IIIa) antagonists. GP IIb-IIIa are chief integrins responsible for platelet aggregation and subsequent thrombus formation. Abxicimab, tirofiban, and eptifibatide are a few GP IIb-IIIa inhibitors that are FDA approved and antagonism exhibited by these molecules are reversible (Oprea and Popescu, 2013).

4.3 Thrombolytic Drugs

Despite all the advancements in the area, anticoagulants and antiplatelet drugs are with significant shortcomings such as bleeding complications, but are in practice for a longer period of time. However, the major drawback of this therapy is that these drugs cannot act on an existing clot. Demand for clot buster drugs was there for a longer time. Enzyme therapies are becoming more significant during the current scenario. Fibrinolytic enzyme therapy is one among the many potential applications of therapeutic enzymes. Accidental discovery of streptokinase paved the way for the development of thrombolytic therapy. Streptokinase was regarded as a wonder drug since its discovery and still a drug of choice in most of the developing countries.

The major advantage of treatment using fibrinolytic enzymes over anticoagulants and antiplatelets is that they could act upon an existing clot. The primary focus of these enzymatic actions is on the protein (fibrin) either directly or indirectly. They are often called clot busting enzymes and categorized into two types based on the mechanism of action. The first category is plasminogen activators; converting plasminogen to active plasmin that cleaves the fibrin clot formed, e.g. tissue-type plasminogen activator (t-PA) and urokinase. They are widely used for the treatment of cardiovascular diseases. t-PA was approved in 1996 by the United States Food and Drug Administration (USFDA) for intravenous injection against thrombosis. Recombinant tissue plasminogen activator alteplase (Activase®) against cardiovascular diseases is the first genetically engineered enzyme to get approved by the USFDA. Many plasminogen activators such as alteplase, reteplase, tenecteplase, urokinase, streptokinase, and anistreplase are available for clinical use and have been commonly used in treating cardiovascular diseases for the past five to six decades (Kotb, 2014). The second category of thrombolytic enzyme is plasmin-like enzymes that are direct-acting fibrinolytic enzymes which do not require the activation of plasminogen. Instead, they can perform clot dissolution by acting upon it, e.g., plasmin, nattokinase, and lumbrokinase.

4.3.1 Plasminogen Activators as Thrombolytic Drugs

Plasminogen activators, also called indirect thrombolytic drugs are serine proteases that stimulate the formation of plasmin from plasminogen. They are classified into three categories, namely, first-, second-, and third-generation plasminogen activators. First-generation plasminogen activators (streptokinase and urokinase) were less fibrin specific and caused bleeding complications. Second-generation plasminogen activators were developed to increase fibrin specificity and thereby reducing bleeding complications. Although it reduced the complications associated with first-generation plasminogen activators, they still possess risk of bleeding. Third-generation plasminogen activators are mostly genetically engineered t-PA variants (except staphylokinase and desmoteplase) to achieve superior properties such as fibrin selectivity, improved half-life, and prolonged renal clearance time. The development of plasminogen activators as thrombolytic drugs is discussed below.

4.3.1.1 First-generation plasminogen activators as thrombolytic drugs

With the discovery of streptokinase, a new line of treatment for cardiovascular diseases was started, i.e., dissolution of clot using thrombolytic enzymes. First-generation plasminogen activators convert circulating plasminogen to active plasmin which in turn cleaves fibrin as well as other proteins such as clotting factors, fibrinogen, etc. The first-generation thrombolytics thus are not fibrin specific and lead to systemic bleeding complications but offered a significant hope in the new therapeutic approach towards cardiovascular diseases (Flemmig and Melzig, 2012).

4.3.1.1.1 Streptokinase

First-generation fibrinolytic enzyme therapy began with the discovery of streptokinase during the 1930s. Streptococci a beta hemolytic bacterium was known to produce a fibrinolytic substance which was isolated from wound exudates (Tillett and Garner, 1933). This substance was earlier named streptococcal fibrinolysin and later named as streptokinase by Christensen. Plasminogen in human plasma gets activated by these kinases to active plasmin which is responsible for fibrinolysis (Christensen, 1945). Further studies proved that plasmin could digest not only fibrin but fibrinogen too. Plasmin inhibitors prevented fibrinolysis in normal course. Partially purified streptokinase was used to study the therapeutic potential in earlier studies (Christensen, 1947). However, the prospective of using streptokinase (SK) as thrombolytic drug was acknowledged only with the experiment of Johnson and Tillett (1952) when they successfully dissolved intravascular clots in the marginal ear veins of rabbit. Later thrombolysis in femoral artery by streptokinase was reported (Sherry, 1954).

Large-scale purification of human plasminogen was achieved by Kline (1953). This led to the direct use of plasmin as clot buster. Intravenous administration of streptokinase-activated human plasmin dissolved clots formed in canine vessels (Cliffton et al., 1953). The major concern was whether direct use of plasmin or use of plasminogen activators was beneficial. The mechanism of action by streptokinase turned out to be a two-step process. Initially it formed a complex with human plasma factor and converted the plasma factor to plasminogen activator and finally this plasminogen activator converted plasminogen to plasmin (Davies et al., 1964; Sherry, 1954). Direct use of plasmin dissolved not only the clot but also other proteins such as clotting factors and fibrinogen. Thus, it was assumed that plasminogen activators are better to be used as therapeutic agent rather than circulating plasmin (Alkjaersig et al., 1959; Fletcher et al., 1959b). It was established that streptokinase infusion via the intravenous route after AMI significantly reduced mortality (Fletcher et al., 1958; Fletcher et al., 1959a). Streptokinase as cardiovascular therapeutic was found promising since it was the only drug that could improve the condition after AMI at that time. Dissolution of intracoronary clots by streptokinase was proved for first time by Ruegsegger and colleagues (1959).

Correspondingly, the adverse effects of SK were reported earlier by Tillet and Sherry (1949). Usage of SK leads to pyrogenic reaction associated with symptoms like headache, malaise, arthralgia, and occasionally nausea (Tillet and Sherry, 1949). However, it was impossible to produce streptokinase with reduced pyrogenic reaction (Sherry, 1981). During those days, the purity of streptokinase attained was only 10% (Sikri and Bardia, 2007). In order to reduce the side effects, further purification of SK was carried out. Purification of SK was achieved and the molecular weight was determined to be 47 kDa (De Renzo et al., 1967). It is a single polypeptide composed of 415 amino acids (Jackson and Tang, 1982). Continuous evolution of SK has been achieved through various attempts. With advances in recombinant DNA technology, cloning and expression of streptokinase was made possible to yield better preparations (Malke et al., 1984; Muller et al., 1989; Wong et al., 1994). Recombinant SK (rSK) thus produced was superior with lesser antigenicity compared to wild-type SK (Collen et al., 1996).

A large number of clinical trials were performed involving both animals and humans by various groups to determine the safety and therapeutic potential of SK. Some of these studies were representing only a smaller population (Koren et al., 1985). Extensive studies involving large number of populations were conducted during the early 1980s. GISSI (Gruppo Italiano per la Sperimentazione della Streptochinasi nell'Infarto Miocardico) was the first among them. In this study of 11,806 patients in 176 coronary care units were monitored for a period of 17 months. Survival rate was more for patients who received SK during early onset of AMI (GISSI, 1986).

The scenario changed after the introduction of t-PA for treating AMI, and finding the best thrombolytic agent among SK and t-PA was the new challenge. t-PA become popular in developed countries for its superior action and t-PA as thrombolytic agent will be discussed later. But SK is still the drug of choice in the case of developing nations and poor countries due to its reduced cost. Streptase®, Kabikinase®, Heberkinasa®, etc., are a few of the commercially available streptokinase formulations of which Heberkinasa® is a recombinant streptokinase (rSK) (Kumar and Abdulhameed, 2017).

4.3.1.1.2 Urokinase plasminogen activator (u-PA)

Macfarlane and Pilling (1947) reported the fibrinolytic activity of normal urine for the first time. Urine dilutions of 1/2000 to 1/4000 dissolved blood clots. Urokinase was later purified from urine (Ploug and Kjeldgaard, 1957). Studies of adult human kidney have revealed that primary culture of adult kidney cells produced plasminogen activator. Apart from hepatic cells, primary cultures of other tissues (both prenatal and post natal) such as ureter, renal blood vessels, bladder, and lung too produced plasminogen activators, which were antigenically similar to Urokinase (Bernik and Kwaan, 1967, 1969). Urokinase is released as an inactive single-chain precursor molecule of 55 kDa and proteolytic enzymes like plasmin cleaves the precursor giving rise to active form of urokinase (Kasai et al., 1985; Petersen et al., 1988; Bernik, 1973). The proteolytic cleavage at Lys157 gives the active urokinase, which is a disulfide linked heterodimer of A chain with a molecular weight of 20 kDa and B chain with molecular weight 34 of kDa. The A chain consists of an epidermal growth factor-like domain and a kringle domain, whereas B chain is the catalytic unit (Kasai et al., 1985; Appella et al., 1987; Patthy et al., 1984). Plasma is devoid of urokinase inhibitors and hence it is freely available at the site of action.

4.3.1.2 Second-generation plasminogen activators

First-generation thrombolytics were nonspecific towards fibrin (Flemmig and Melzig, 2012). Side effects such as bleeding complication demanded more specific thrombolytic drugs. Second-generation thrombolytics were developed to overcome the shortcomings of first-generation plasminogen activators.

4.3.1.2.1 Development of t-PA as a thrombolytic agent

Earlier it was reported that fibrinokinase, a plasminogen activator is present in animal tissues and isolation of the enzyme with chaotropic agents were described by Stage and Astrup (1952). Later it was named as tissue type plasminogen activator (t-PA). Purification of these plasminogen activators was attempted from many sources. Extraction of human t-PA was achieved with high purity, though the yield was as low as 1 mg of t-PA from 5 kg uterine tissue. Later it was found that t-PA is formed in vascular endothelial cells and then released in to circulatory system (Rijken et al., 1979, 1980). However, cloning and expression of human t-PA was achieved which resolved the difficulties with respect to low yield and cost of production. t-PA is a serine protease of 527 amino acids with a molecular weight of 70 kDa (Ranby et al., 1982; Pennica et al., 1983). It is secreted as single polypeptide which is cleaved by plasmin into a two-chain form, linked together by disulfide bonds. The C-terminal light chain carries the catalytic serine protease domain while the N-terminal heavy chain comprises an epidermal growth factor domain, a finger domain and two kringle domains (Rijken and Groeneveld, 1986). Plasminogen interacts with kringle 2 domain alone, while fibrin binds to kringle 2 domain and fibronectin type III finger domain of t-PA, pointing towards the stimulatory effect of fibrin on t-PA (van Zonneveld et al., 1986). From the kinetic studies of t-PA also, it was found that the activation of plasminogen by the enzyme is remarkably enhanced in the presence of fibrin. Thus its specificity towards fibrin is substantiated (Hoylaerts et al., 1982). Alteplase is a recombinant t-PA approved in 1996 by the FDA, and it is the first recombinant therapeutic enzyme to get FDA approval. Since then it is in clinical practice. Alteplase is available in market under the trade name Activase® (Kumar and Abdulhameed, 2017). It is used for treating acute ischemic stroke, pulmonary embolism, and acute myocardial infarction and is regulated to be used within 3 h after ischemic stroke. Also, it has a short half-life of 4–8 min and is unsuitable for bolus injections (Ali et al., 2014). t-PA is clinically used as therapeutic for acute ischemic stroke even though it is often associated with the risk of cerebral hemorrhage and edema (Kaur et al., 2004).

Studies such as GUSTO (Global Utilization of Streptokinase and Tissue Plasminogen Activator for Occluded Coronary Arteries) (GUSTO Angiographic Investigators, 1993), Third International Study of Infarct Survival Collaborative Group, ISIS-3 (ISIS-3, 1992), and GISSI-2 (Gruppo Italiano per so Studio della Sopravvienza nell'Infarto Miocardico) (GISSI-2, 1990) were used to evaluate the therapeutic potential of SK to t-PA. All these studies did not show any significant difference in death rate for those received either streptokinase or t-PA. However, follow-up of one year showed significant reduction in mortality for those treated with t-PA according to GUSTO. t-PA was having a slight edge over streptokinase in the case of patients either having chance to receive drug at the earliest, or at younger age, or earlier treated with streptokinase for AMI (Gershlick and More, 1998).

4.3.1.2.2 Prourokinase (scu-PA)

Prourokinase or single-chain urokinase-type plasminogen activator (scu-PA) is an inactive precursor form of urokinase and in a surface-controlled process, plasmin associated with fibrin network cleaves the single-chain prourokinase into two-chain urokinase which in turn converts plasminogen to plasmin. Major advantage of scu-PA is that it shows specificity towards fibrin unlike native urokinase (Bernik, 1973; Fleury and Angles-Cano, 1991). Pro-UK showed increased half-life of more than 24 h in circulation compared to urokinase and complete lysis of fibrin was attained within 1.5 h whereas urokinase required 3 h for the fibrinolysis (Zamarron et al., 1984). Saruplase or prourokinase (scu-PA) is a naturally occurring glycoprotein carrying 411 amino acids. Recombinant production of prourokinase (r-proUK) can be achieved in either bacterial system (*E. coli*) or mammalian cell line. Prolyse® (Abbott Laboratories) is a recombinant prourokinase (r-proUK) yet to be approved by the FDA and production is achieved in murine hybridoma cell line. It was found effective in stroke and myocardial infarctions (Sun et al., 1998; Ouriel, 2002). These recombinant prourokinases have undergone a large number of clinical studies. SESAM (Study in Europe with Saruplase and Alteplase in Myocardial Infarction) revealed that saruplase was safe and effective as altplase in acute myocardial infarction (Bar et al., 1997). Another study, PROCAT (Prolyse in Acute Cerebral Thromboembolism) showed the recanalization efficacy in acute thromboembolic strokes (del Zoppo et al., 1998). COMPASS (Comparative Trial of Saruplase Versus Streptokinase) study showed comparable 30-day mortality rate for both saruplase and streptokinase (Tebbe et al., 1998). According to PROCAT II intra-arterial administration of pro-urokinase within 6 h after the acute ischemic stroke considerably enhanced clinical outcome at 90 days (Furlan et al., 1999).

4.3.1.2.3 Anistreplase (APSAC)

Anistreplase (anisoylated plasminogen streptokinase activator complex or APSAC) is an equimolar mixture of streptokinase and plasminogen. The active site of plasminogen is protected from plasmin inhibitors by reversible acylation with p-amidinophenyl p-anisate hydrochloride (APAN). Upon injection, deacylation starts by hydrolysis of p-anisate group leaving active plasminogen-streptokinase complex. APSAC is more stable with improved half-life and fibrin binding capacity (Sherry, 1990). APSAC is a readily available plasminogen activator and shows more specificity towards plasminogen bound to clot (Smith et al., 1981). ISIS-3 reported that APSAC and streptokinase showed similar results in mortality. Also, it is comparable to streptokinase in terms of antigenicity and safety (Sherry, 1990; Munger and Forrence, 1990; ISIS-3, 1992). Eminase® is a commercially available anistreplase formulation. However, its usage was discontinued in the USA recently (Ali et al., 2014).

4.3.1.3 Third-generation plasminogen activators (by protein engineering)

Third-generation plasminogen activators with improved properties were developed for thrombolytic therapy. Most of these classes of drugs are genetically engineered t-PA variants with improved resistance to inhibitors, more safe upon administration, prolonged half-life, etc. Their increased clearance rate makes them suitable for bolus administration. A few non-t-PA variants such as staphylokinase and desmoteplase also belong to this category.

4.3.1.3.1 Reteplase (recombinant plasminogen activator, r-PA, BM 06.022)

Reteplase is an unglycosylated shortened single-chain deletion variant of t-PA, recombinantly produced in *E. coli* and having a molecular weight of 39 kDa. It consists of 355 of 527 amino acids from native t-PA; precisely 1–3 and 176–527 amino acids of t-PA. Thus reteplase includes only catalytic protease domains along with kringle 2 domain while it lacks epidermal growth factor domain, finger domain and kringle 1 domains of t-PA (Kohnert et al., 1992; Sturzebecher et al., 1992). The molecules attained enhanced half-life from 4 to 15 min as it lacks kringle 1 domain which is responsible for quick renal clearance (Martin et al., 1991b; Flemmig and Melzig, 2012). Absence of fibrin binding finger domain decreases the binding affinity of the molecule, although kringle 2 domain interacts with fibrin (Kohnert et al., 1992; Martin et al., 1991a). The mortality rate and hemorrhagic stroke for 30 days are similar for reteplase and alteplase (GUSTO III Investigators, 1997). Reteplase is commercially available as Retavase® (Centocor, Inc.) and Rapilysin® (Roche).

4.3.1.3.2 Tenecteplase (TNK-tPA)

Tenecteplase is a recombinant mutant variant of t-PA. It is a 65 kDa glycoprotein with 527 amino acids. The mutations include Asp103Thr, Glu117Asn and 4 alanine substitutions (Lys296Ala, His297Ala, Arg298Ala, and Arg299Ala). These point mutations increase the fibrin specificity by 14-fold and resistance to plasminogen activator inhibitor 1 (PAI-1) by 80-fold (Smalling, 1996). Mutations in kringle 1 domain prolongs the renal clearance time and thus increases the half-life to about 10–24 min (Cannon et al., 1997). This makes it suitable for bolus administration. Clinical trials such as TIMI 10B (Thrombolysis in Myocardial Infarction) trial, ASSENT-1 and ASSENT-2 (Assessment of Safety and Efficacy of a New Thrombolytic agent) trials reported similar efficacy and safety while used for treatment of AMI. However, bleeding rate was lesser compared to altepase (Zeymer and Neuhaus, 1999; Cannon et al., 1998; Van de Werf et al., 1999a, 1999b). Trade name of tenectplase is TNKase™ (Genentech, Inc.) and was approved by USFDA in 2000 for treatment of myocardial infarction.

4.3.1.3.3 Lanoteplase/novel plasminogen activator (n-PA)

Lanoteplase is a novel truncated mutant variant of t-PA created by removing finger domain and epidermal growth factor domain. Apart from that, it has point mutation at kringle domain Asn117Gln. This in turn results in deletion of glycosylation site. These modifications increase the renal clearance time and the plasma half-life of the molecule turns out to be 37–45 min. It is also suitable for bolus administration. The enzyme is a single polypeptide with a molecular weight of 53.6 kDa and recombinant production is achieved in Chinese Hasmster Ovary (CHO) cell lines (Larsen et al., 1989, 1991; Nordt et al., 1991; Collen et al., 1988). In comparison with alteplase higher patency rate was shown by lanetoplase for 90 min while similar mortality rate was shown at 30 days (den Heijer et al., 1998; InTIME-II investigators, 2000). However, mild bleeding complications with hemorrhagic stroke were observed more than that found for alteplase (Verstraete, 2000).

4.3.1.3.4 Pamiteplase/Solinase

Pamiteplase is another shortened mutant of t-PA, devoid of kringle 1 domain (residues 92–173) along with point mutation Arg274Glu. This variant is resistant to cleavage by plasmin and having a half-life of about 30–47 min (Flemmig and Melzig, 2012). In comparison with t-PA, much higher fibrinolysis was shown by pamiteplase under an in vivo study using canine coronary thrombosis model (Kawasaki et al., 1993). Another trial reported comparable results for t-PA and pamiteplase, where TIMI-3 flow rate was 25% at 30 min and 50% after 60 min for pamiteplase while it was, respectively 16% and 48% for t-PA (Flemmig and Melzig, 2012). Recombinant production is achieved in CHO cell lines like lanoteplase (Yokota and Tanji, 2008).

4.3.1.3.5 Duteplase (met-t-PA)

Duteplase in another variant of t-PA with a single point mutation Val245Met and recombinant production is achieved in CHO cell lines. Duteplase is a nonglycosylated, two-chain form as opposed to the single-chain form of alteplase. According to ESPRIT study, incidence of patency in infarct-related coronary artery was achieved in 70% patients at 90 min with weight-adjusted duteplase infusion along with oral aspirin and intravenous heparin. Safety of the molecule was similar to that of alteplase (Kalbfleisch et al., 1992; Malcolm et al., 1996).

4.3.1.3.6 Monteplase (E6010)

Another mutated variant of t-PA is monteplase where a point mutation has been achieved in epidermal growth factor domain (Cys84Ser). However, extended half-life of about 23 min makes the mutant suitable for bolus administration (Suzuki et al., 1991; Kawai et al., 1997). TIMI-3 flow rate was 69% at 60 min post monteplase treatment for AMI (Flemmig and Melzig, 2012). Higher rate of early recanalization was reported for the molecule in comparison to tisokinase, a native t-PA variant in a randomized double-blind study conducted in Japan. Also, fatal bleeding was not observed during the trial (Kawai et al., 1997).

4.3.1.3.7 Amediplase (MEN 9036)

Amediplase is a chimeric enzyme of 365 amino acids with kringle 2 domain of t-PA (1–3 and 176–275 residues) and the protease domain of single-chain urokinase, scu-PA (159–411). Molecular weight of the protein is 39.9 kDa and recombinant production is achieved in CHO cell lines (Agnelli et al., 1993; Muller et al., 1994). Amediplase showed 10 times lesser binding affinity towards fibrin in comparison with alteplase under internal and external clot lysis methods. However, clot penetration ability of the molecule was superior to t-PA (Rijken et al., 2004). However, amediplase has shown slightly better activity than tenecteplase and scu-PA in another external clot lysis method (Guimaraes et al., 2006).

4.3.1.3.8 Desmoteplase (Bat-PA, DSPAα1, v-PA)

Desmoteplase is third-generation plasminogen activator, a non-t-PA variant. Desmodus salivary plasminogen activators (DSPAS) is a group of four enzymes isolated from common vampire bat, *Desmodus rotundus* and desmoteplase is one among them, called DSPAα1 or Bat-PA (Kratzschmar et al., 1991). It is a single-chain 50 kDa protein which could be produced in both CHO cell lines and insect cells by recombinant techniques. It has an extended terminal half-life when compared with other thrombolytic enzymes such as t-PA (Hildebrand et al., 1996). DSPAα1 contains a single kringle domain quite similar to the kringle 1 domain of t-PA but it is devoid of any lysine binding site. Apart from kringle domain, it also contains a signal peptide, an epidermal growth factor (EGF) domain, a serine protease domain, and a finger domain. Also the protease domain lacks plasmin cleavage site. The bimolecular rate constant of plasminogen activation was 13000 for DSPAα1 and 72 for t-PA in the presence of fibrin versus fibrinogen (fibrin selectivity), proving it to be highly fibrin specific (Bringmann et al., 1995). Tranexamic acid is a synthetic lysine analog that can inhibit desmoteplase, therefore possible for managing any bleeding complications, occurring during the usage of the enzyme (Niego et al., 2008). The canine model of arterial thrombosis showed desmoteplase to be better in thrombolysis compared to t-PA (Mellott et al., 1992). Intravenous administration directly into brain causes excitotoxic injury leading to brain damage in case of t-PA, whereas desmoteplase does not encourage excitotoxic injury and thus proves to be a better therapeutic for ischemic stroke (Reddrop et al., 2005). DSPAα1 definitely has therapeutic advantages over t-PA and preclinical studies have already showed better specificity and selectivity towards fibrin. Most thrombolytic drugs have a time window of less than 3–4.5 h. However, desmoteplase is supposed to demonstrate better clinical advantages by extending the therapeutic window of patients with ischemic stroke. Desmoteplase in acute ischemic stroke (DIAS) trial; a phase II trial showed that the administration of this drug within 3–9 h after the onset of the symptoms of stroke has got a clinical benefit with use of DSPAα1 (Hacke et al., 2005). However, the DIAS-2 failed to prove any significant advantage of the drug within 3–9 h time window after onset of stroke (Hacke et al., 2009). According to DIAS 3 and DIAS 4 trials, acceptable safety profile was observed for DSPAα1, but no evidence to extend therapeutic window up to 9 h for ischemic stroke was found (Albers et al., 2015; Kummer et al., 2016). Satisfactory reperfusion efficacy was shown by desmoteplase after 3 h since onset of AIS; however, more studies with well-designed clinical trials are needed to be evaluated for the effectiveness of the drug to improvise the therapeutic window in treating AIS patients

(Li et al., 2017). Although intracranial hemorrhages are low for desmoteplase, it did not showed any neurological improvement or survival rate improvement (Hacke et al., 2009; Albers et al., 2015; Kummer et al., 2016).

4.3.1.3.9 Staphylokinase (SAK)

Staphylokinase is another non-t-PA variant belonging to the third-generation plasminogen activators. It is an extracellular protein produced by *Staphylococcus aureus*, which forms 1:1 stoichiometric complex with plasmin or plasminogen and activates more plasminogen molecules. SAK is a monomer of 136 amino acids without any disulfide bond and molecular weight is about 15.5 KDa. SAK complexes with plasminogen molecules that are bound to partially degraded fibrin [Sakharov et al., 1996]. α2-antiplasmin inhibits plasminogen-SAK complex in the absence of fibrin, however, in presence of fibrin the lysine binding domain of the complex is occupied, preventing the inhibition by α2-antiplasmin and there is a 4-fold increment in activity with presence of fibrin, thus more specificity towards fibrin (Lijnen et al., 1991).

Recombinant production of SAK is achieved in *Escherichia coli*, *Bacillus subtilis* and yeast *Hansenula polymorpha* (Sako, 1985; Gerlach, 1988; Moussa et al., 2012). Recombinant wild-type SAK (wt-SAK) showed more fibrinolysis and less fibrinogenolysis compared to that of streptokinase (Collen and Van de Werf, 1993). Yet, therapeutic application of wt-SAK is associated with antigenicity and hence its repetitive use as a drug is ineffective (Vanderschueren et al., 1997). Antigenic epitopes can be removed without affecting the activity, by site directed mutagenesis (Laroche et al., 2000). Since then many site directed mutants were produced with reduced immunogenicity. Many of the in vivo studies and clinical trials showed the therapeutic efficacy and safety of recombinant SAK over t-PA. It did not show any systemic fibrinogenolysis or allergic reactions during the earlier studies (Vanderschueren et al., 1995a, 1995b, 1996, 1997; Armstrong et al., 2000, 2003).

Two hybrid proteins, SAK-RGD-K2-Hir and SAK-RGD-K2-Hirul, carrying fibrinolytic C-terminal part of recombinant staphylokinase (r-SAK), fibrin binding kringle 2 domain of tissue-type plasminogen activator (t-PA), antiplatelet aggregating tripeptide sequence Arg-Gly-Asp and anticoagulant hirudin (65 amino acids) or hirulog (hirudin analogue of 20 amino acids), were produced by cloning and expression in pESP-3 expression vector. In vitro studies proposed these fusion proteins to be better thrombolytic agents than recombinant wt-SAK. However, SAK-RGD-K2-Hirul did not show any clinical benefit over SAK-RGD-K2-Hir and it also caused bleeding complications (Szemraj, 2005, 2011).

The major disadvantage of plasminogen activators (PAs) is high cost and side effects such as bleeding, nausea, and hemorrhage. Upon systemic administration, plasminogen activators (PAs) exceed the plasminogen activator inhibitors (PAIs) and start dissolving the clot. Apart from clot dissolution, they also disrupt the vascular plug formed at the site of injury. Thus bleeding complications are inevitable in treatment with plasminogen activators that could lead to serious situations such as intracranial hemorrhage (Gore et al., 1991; Marder, 1979; Marder and Sherry, 1988). Their disadvantages necessitated the discovery of better thrombolytic agents.

4.3.2 Direct-Acting Plasmin-Like Thrombolytic Drugs

Research on plasminogen activators resulted in development of recombinant and chimeric protein production. Significant improvement in fibrin specificity, safety, extended plasma life, etc., has been achieved so far. But still these thrombolytic agents possess risk of bleeding complication, allergic reaction, and repercussion. Hence, search for new and better drugs continues. Catheter based local administration of plasminogen activators fastens the process of thrombolysis, yet systemic bleeding cannot be avoided completely. Direct-acting fibrinolytic enzymes were reported from various sources which could bypass plasminogen activation and disintegrate fibrin clot. They were isolated from non-microbial as well as microbial sources.

4.3.2.1 Non-microbial thrombolytic enzymes

Non-microbial thrombolytic enzymes include plasmin, variants of plasmin, fibrinolytic proteases from earthworm, snake venom, plant, and algae. Few among them offered hope as cardiovascular therapeutic; however, further clinical studies are required to determine their safety and efficacy. Even though reported with thrombolytic potential, further therapeutic usage of plant and algal proteases were not reported so far.

4.3.2.1.1 Plasmin

Plasmin is the fibrin degrading enzyme which can be administered without any bleeding complication and allergic reactions. Systematically administered plasmin often gets inhibited by α2-antiplasmin, thereby losing the therapeutic benefit (Marder, 2011). But when given locally, plasmin does not get blocked by its inhibitors and will lead to fibrinolysis. Later it gets inhibited by α2-antiplasmin when it is released to circulation after thrombolysis (Marder and Novokhatny, 2010, Marder, 2011). In vivo animal models and clinical trials suggested the clinical benefit and safety of plasmin as a cardiovascular drug that could accomplish fibrinolysis (Hoefer et al., 2009; Sadeghi et al., 2003; Jahan et al., 2008; Shlansky-Goldberg et al., 2008). Plasmin is a heterodimer with an N-terminal heavy chain of molecular weight 65 kDa and a C-terminal light chain of molecular weight 25 kDa linked together by two disulfide bonds. Heavy chain contains 5 kringle domains and a 77 residue N-terminal peptide, whereas light chain carries a serine protease domain. It binds to fibrin through lysine binding sites at kringle domain, chiefly at kringle 1 domain (Wiman et al., 1979; Robbins et al., 1967; Sottrup-Jensen et al., 1978; Violand and Castellino, 1976; Wiman, 1977, Wiman and Wallen, 1977; Lucas et al., 1983; Wu et al., 1994).

4.3.2.1.2 Miniplasmin

Numerous variants of plasmin have been constructed for therapeutic applications. Miniplasmin is one among them and it is a truncated mutant obtained from plasminogen by proteolysis using porcine pancreatic elastase. It contains kringle 5 domain and serine protease domain of plasmin (Sottrup-Jensen et al., 1978). Kringle 5 domain mediates binding of fibrin to miniplasmin and inhibition of this protein is lower compared to that of plasmin (Wiman et al., 1978). Recombinant production of miniplasminogen is achieved in *E. coli* using T7 expression system and activated to form miniplasmin by urokinase (Moroz, 1981). However, the kinetic data for thrombolysis by miniplasmin is contradictory, which is showing higher and lower efficiency in two separate studies (Moroz, 1981; Kolev et al., 1996). Using the canine model of femoral artery thrombosis, it was observed that the occurrence of reperfusion was much higher upon miniplasmin administration, when compared to rt-PA. Also, the miniplasmin-treated group left without any reocclusion, whereas 20% reocculsion was shown by the rt-PA treated group. This points towards the lower incidence of hemorrhage while administering miniplasmin over rt-PA (Fu et al., 2008).

4.3.2.1.3 Microplasmin

Microplasmin is another plasmin variant generated by autolysis of plasmin at high alkaline pH (pH 11) and is a 29 kDa protein. It is a heterodimer made up of "A" chain with 31 amino acids and "B" chain with 230 amino acids (Wu et al., 1987). Microplasminogen can be recombinantly produced in *Pichia pastoris* and *E. coli* (Nagai et al., 2003; Ma et al., 2007). Bioengineering significantly reduces the overall production cost. On the other hand, it does not bind with fibrin since it is devoid of all the kringle domains of plasmin. Also, the rate of inhibition by α2-antiplasmin is very low with respect to native plasmin (Nagai et al., 2003). In vivo animal models and clinical trials proved its safety to be used as a therapeutic since it was not causing any hemorrhage (Suzuki et al., 2004; Chen et al., 2007; Thijs et al., 2009). Microplasmin (ocriplasmin) also finds application as therapeutic for vitreomacular adhesion. It causes vitreomacular detachment of fibronectin and laminin (Chen et al., 2008; de Smet et al., 2009). Ocriplasmin received FDA approval for treating vitreomacular adhesion and is commercially available under trade name JETREA®. However, its use as a cardiovascular drug is yet to be validated by further clinical studies.

4.3.2.1.4 Delta plasmin

The third and most important plasmin variant is Delta plasmin, which is a truncated mutant formed by removing krinlgle 2–5 domains leaving only kringle 1 domain attached to the serine protease domain of plasmin. Recombinant production of the

Delta plasminogen [Delta(K2-K5)Pg] was carried out in *E. coli* T7 expression system and found functionally similar to human plasminogen. It was converted to active Delta plasmin [Delta (K2-K5) Pm] by plasminogen activators such as t-PA, urokinase and streptokinase. Even though it showed fibrinolytic efficacy equivalent to plasmin (Hunt et al., 2008) further validation is required.

Plasmin and its variants are serine proteases that directly dissolve clot, but get inhibited by serine protease inhibitors (SERPINS) in plasma. Most common serpin in circulation is α2-antiplasmin, which immediately inhibits plasmin upon systemic administration. Plasmin, however, does not show bleeding complications since it gets inhibited rapidly in circulation. However, catheter-based local administration of plasmin is more effective in dissolving thrombus (Marder et al., 2001). Delivery of thrombolytics through sophisticated catheter-based system limited their use (Marder and Novokhatny, 2010).

Common serine proteases such as trypsin, chymotrypsin and elastase are capable of dissolving thrombi (Sarkar, 1960; Sherry, 1954; Komorowicz et al., 1998). Both plasmin and plasminogen activators (PAs) are also chymotrypsin like serine proteases. However, the broad substrate specificity of these common proteases makes them less suitable for therapeutic applications (Madison et al., 1995). The SERPINS found in circulation is another important factor affecting the efficacy of serine proteases while performing administration. One among them is α1-antitrypsin (α1-PI), which is found in abundance and characterized by its ability to inhibit serine proteases such as plasmin and thrombin. They also could directly impede trypsin, chymotrypsin and elastases. Even though reactivation of these proteases by α2-macroglobulin is possible, the slow rate of reactivation is also a limiting factor (Aubry and Bieth, 1977). Thus, systemic administration of these proteases is not recommendable for treating cardiovascular disorders. Limitations of enzymes involved in plasma fibrinolytic system (plasmin, t-PA) and common proteases (trypsin, chymotrypsin and elastase) demand administration of fibrinolytic proteases from external sources. Few of the external non-microbial sources include earthworms *Eisenia fetida*, *Lumbricus rubellus* (Mihara et al., 1991; Roch, 1979), snake venoms (Markland, 1998), *Catharsius molossus* (Ahn et al., 2003), *Spirodela polyrhiza* (Choi and Sa, 2001), etc.

4.3.2.1.5 Snake venom thrombolytic enzymes

Another important source for fibrinolytic enzymes is venom of snakes. Fibrolase, is a fibrinolytic enzyme isolated from snake *Agkistrodon contortrix*. It is a zinc metalloprotease of 203 amino acids having molecular weight of approximately 29 kDa. A truncated mutant of 201 amino acids is produced from fibrolase called alfimeprase with improved fibrinolytic properties (6-fold higher activity than plasminogen activators such as t-PA and urokinase) without bleeding complications upon administration. Two deletions and one substitution changes EQRFPQR of fibrolase to SFPQR in alfimeprase. Animal models as well as Phase I and II clinical studies proved efficacy and safety of the drug upon administration (Randolph et al., 1992; Deitcher et al., 2006). Lebetase is another direct-acting fibrinolytic enzyme isolated from *Vipera lebetina*. It is a single-chain polypeptide of 214 amino acids constituting 23.7 kDa and is a metalloprotease (inhibited by EDTA). Lebetase achieves fibrinogenolysis by rapid dissolution of Aα-chain together with slow lysis of Bβ-chain. It appears to be safer since it showed only lower hemorrhagic activity (Siigur and Siigur, 1991).

4.3.2.1.6 Earthworm thrombolytic enzymes

Earthworms are a rich source of proteases with thrombolytic potential. Initial characterization of earthworm fibrinolytic enzyme was performed from *Lumbricus rubellus* and named lumbrokinases (LK) representing a group of protease enzymes (Mihara et al., 1991). *Eisenia fetida, E. andrei L. bimastus,* etc., are a few other earthworm species identified as fibrinolytic enzyme producers. They produce multiple thrombolytic proteases. *L. rubellus* produced around six lumbrokinase variants and *E. fetida* yields 7 variants of fibrinolytic enzymes (EFE a-g) (Cho et al., 2004; Wang et al., 2003). These enzymes show more specificity towards fibrin in comparison with plasminogen activators such as t-PA and urokinases. Lumbrokinases are direct-acting fibrinolytic enzymes, also showing functions like activation of plasminogen, reduction of platelet aggregation, blood thinning, etc. Molecular mass of these enzymes ranges 20–35 kDa and they retain their activity over a wide range of pH (pH 1–11) (Pan et al., 2010). In vivo animal models revealed their potential to be used for oral administration (Lee et al., 2007; Yan et al., 2010). Recombinant production of this enzyme can be achieved in *pichia pastoris* (Ge et al., 2005). LK has its main advantage as a safer therapeutic enzyme with lower risk of bleeding disorders (Tjandrawinata et al., 2016); but yet to get approval from the FDA. However, Boluoke® is a commercially available LK approved by the China Food and Drug Administration. *E. fetida* enzymes belongs to serine proteases with molecular mass ranges from 24.5 to 29.5 kDa and few of them were quite similar to lumbrokinases (Wang et al., 2003).

4.3.2.1.7 Plant-based thrombolytic enzymes

Herbal formulations are known for therapeutic usage since it is widely used in traditional medicines. Plants are a rich source of proteolyitc enzymes, especially plant latex. Crude preparations of plants are reported to show in vitro clot dissolution and they show considerable stability over a wide range of temperature and pH. *Fagonia Arabica, Bacopa monnieri, Leea indica, Leucas aspera, Clausena suffruticosa, Urena sinuate, and Trema orientalis* are few among the plants producing thrombolytic enzymes (Prasad et al., 2007; Rahman et al., 2013; Emran et al., 2015). Latex enzymes from various plants such as *Calotropis gigantea, Wrightia tinctoria and Synadenium grantii, Cynanchum puciflorum, Asclepias curassavica,* and *Pergularia extensa* showed procoagulant activity (clot inducing) as well as clot dissolving activity. These enzymes are either serine proteases or cysteine proteases. Although the protease from *C. gigantea* was found to be most potent thrombolytic agent, it showed high risk of hemorrhage also, whereas the enzymes from *S. grantii* and *W. tinctoria* showed no hemorrhage (Rajesh et al., 2007; Shivaprasad et al., 2009). Cysteine protease enzyme from fruits of *Bromelia hieronymi, B. balansae,* and *Pseudananas macrodontes* was evaluated for its thrombolytic potential; it was found that at low concentration, it showed procoagulant activity, and as concentration increased, it showed fibrinolytic activity (Errasti et al., 2016). Extracts from *Litchi chinensis* were shown to have antiplatelet, thrombolytic, and anticoagulant potential. The extract showed increase in fibrinolytic activity in a dose dependent manner (Sung et al., 2012). Fibrinolytic proteases are also reported from latex of *Cryptostegia grandiflora* and *Plumeria rubra* (Viana et al., 2013). A fibrinogenolytic enzyme Eumiliin was purified and characterized from the latex of *Euphorbia milii* var. *hislopii*. Eumiliin is a 30 kDa cysteine protease and showed stability over a wide range of temperature and pH. It lost complete activity at 80°C and predominantly degraded Aα chain of fibrinogen (Fonseca et al., 2010). A serine protease, EuP-82, with fibrinogenolytic potential was isolated form *Euphorbia* cf. *lactea*. It is a homodimer and retains its stability over a wide range of pH and temperature (Siritapetawee et al., 2015). Hirtin is a 34 kDa serine protease isolated from the latex of *Euphorbia hirta* with both fibrinolytic as well as fibrinogenolytic potential. It was stable between pH 4–8 and temperature from 30 to 70°C. Hirtin also showed more affinity towards Aα chain of fibrinogen during fibrinogenolysis (Patel et al., 2012). Another fibrinolytic protease, AMP-48, was isolated from the latex of *Artocarpus heterophyllus*. It is a serine protease of molecular weight 48 kDa hydrolyzing fibrinogen and readily dissolving α subunit followed by partial hydrolysis of β and γ subunits just like other fibrinolytics of plant origin (Siritapetawee et al., 2012). Some of the plant fibrinolytic enzymes along with their properties are summarized in Table 4.1.

4.3.2.1.8 Algal thrombolytic enzymes

Thrombolytic enzymes are isolated from algal sources too. Seaweeds from the genus *codium* such as *Codium intricatum, C. latum and C. diaricatum* are reported to produce multiple fibrinolytic enzymes. CIP I (*C. intricatum* protease I) and CIP II (*C. intricatum* protease II) are two direct-acting fibrinolytic enzymes isolated from *C. intricatum.* Molecular mass of CIP I and CIP II were approximately 20 kDa under denaturing conditions. Protease isolated from *C. latum* designated as CLP (*C. latum* protease) also showed fibrinolytic potential. Molecular weight of the enzyme was 23 KDa and N terminal sequencing confirms it to be distinct from CIP. A 31 KDa thrombolytic enzyme CDP (*C. diaricatum* protease) was isolated from C. *diaricatum.*

Table 4.1 Fibrinolytic enzymes produced by plants and their characteristics

Organism	Enzyme	Type	Molecular weight, optimum temperature and pH	Mode of action	Reference
Euphorbia milii var. hislopii	Eumiliin	Cysteine protease	30 kDa pH 8.0	Preferentially cleaved Aα chain of fibrinogen	Fonseca et al., 2010
Euphorbia cf. lactea	EuP-82	Serine protease	35 kDa (monomeric form) 35°C, pH 11	cleaved all subunits of human fibrinogen	Siritapetawee et al., 2015
Euphorbia hirta	Hirtin	Serine Protease	34 kDa 50°C, pH 7.2	Preferentially cleaved Aα chain of fibrinogen followed by Bβ and γ chains	Patel et al., 2012
Artocarpus heterophyllus	AMP-48	Serine protease	48 kDa 55–60°C, pH 8	Preferentially cleaved α followed by partial hydrolysis of β and γ subunits	Siritapetawee et al., 2012

Although the fibrinolytic enzymes from these seaweeds are diverse they showed almost similar properties. They are serine proteases and predominantly hydrolyzed Aα chain of fibrinogen while prolonged incubation cleaved all subunits (Matsubara et al., 1998, 1999, 2000) Fibrinolytic enzymes from algae with their properties are listed in Table 4.2.

Table 4.2 Fibrinolytic enzymes produced by algae and their characteristics

Organism	Enzyme	Type	Molecular weight, and pH	Mode of action	Reference
Codium intricatum	CIP	Serine protease	20 kDa pH 8–9	Preferentially cleaved Aα chain followed Bβ and γ chains	Matsubara et al., 1998
Codium latum	CLP	Serine protease	23 kDa pH 10	Preferentially cleaved Aα chain followed Bβ chain and γ chain to end	Matsubara et al., 1999
Codium diaricatum	CDP	Serine protease	31 kDa pH 9	Cleaved Aα chain, very little affinity towards Bβ chain and γ chain	Matsubara et al., 2000

4.3.2.2 Microbial thrombolytic enzymes

Although fibrinolytic enzymes were reported from various sources, most of the fibrinolytic enzymes reported were from microbial sources. A few of the microbial producers include fungi such as *Aspergillus ochraceus* (Batomunkueva and Egorov, 2001), *Fusarium pallidoroseum* (El-Aassar, 1995), *Pleurotus ostreatus* (Choi and Shin, 1998), algae such as *Codium* sp. (Matsubara et al., 1998) and bacteria such as *Streptomyces* sp. (Wang et al., 1999), *Pseudomonas* sp. (Wang et al., 2009), *Staphylococcus* (Gerheim, 1948) and *Bacillus* sp. (Sumi et al., 1987; Peng et al., 2003; Ko et al., 2004). Major advantages of microbial enzymes are its ease of production, lesser bleeding complications, oral administration, etc. (Goldhaber and Bounameaux, 2001; Tough, 2005). Past couple of decades witnessed the discovery of thrombolytic agents from numerous microorganisms and this elicited the medical interest in development of better therapeutic agents from them with lesser or no side effects. Direct-acting plasmin-like fibrinolytic enzymes from microbial sources are discussed below.

4.3.2.2.1 Fungal thrombolytic enzymes

Fungi are a rich source of proteolytic enzymes and some of them are having fibrinolytic potential too. Here we categorize fungus under microbial producers though micro and macro fungi are discussed. The ability of fungi to grow on solid substrates such as agro-industrial waste residues made them more suitable for economic production of fibrinolytic enzymes. *Aspergillus* species are most widely occurring fungal group and few of them turn out to be thrombolytic enzyme producers. Protease enzyme produced by *Aspergillus ochraceus* 513 was reported to have fibrinolytic potential and subsequent purification steps yielded a 36.5 kDa protease with both fibrinolytic as well as anticoagulant potential (Batomunkueva and Egorov, 2001). *Aspergillus oryzae* KSK-3 produced a 30 kDa serine protease with thrombolytic potential. It remained stable up to temperatures of up to 50°C and pH ranging from 4 to 9 (Shirasaka et al., 2012). There are numerous reports of fungal fibrinolytic enzymes of *fusarium* species (Tao et al., 1997, 1998). A thrombolytic enzyme producing fungus *Fusarium oxysporum* was successfully grown on a solid substrate rice chaff (Tao et al., 1997, 1998). *Fusarium pallidoroseum* produced a metalloprotease with plasmin like activities (El-Aassar, 1995). *Fusarium BLB* isolated from hibiscus leaf produced an alkaline serine protease with fibrinolytic potential. This 27 kDa protein exhibited pH tolerance and retained its activity between pH 2.5 to 11.5 (Ueda et al., 2007). Fu-P is a fibrinolytic enzyme isolated from *Fusarium* sp. CPCC 480097. It is a chymotrypsin like enzyme, with a molecular weight of 28 kDa and has high affinity towards α-chain but showing only low affinity towards β and γ chain during fibrinogenolysis (Wu et al., 2009). Fungal strains such as *Cochliobolus lunatus* and *Penicillium chrysogenum* H9 were also reported to produce fibrinolytic enzymes (Abdel-Fattah and Ismail, 1984; El-Aassar et al., 1990). *Rhizopus chinensis*-12 is another fungal producer, of a hydrosulfuryl-metalloprotease with fibrinolytic activity. The enzyme is a single polypeptide of 18 kDa according to SDS-PAGE results and showed similar affinity towards α, β and γ chains of fibrinogen (Xiao-Lan et al., 2005). Purification of a fibrinolytic metalloprotease was achieved from the mycelium of *Perenniporia fraxinea*. The 42 kDa protein thus obtained showed affinity towards α-chain of fibrin (Kim et al., 2008).

Thermophilic fungus *Oidiodendron flavum* produced a 22 kDa fibrinolytic enzyme which showed stability over a pH range of 6–11 and completely lost its activity at 70°C (Tharwat, 2006). PTEFP is another fungal fibrinolytic enzyme extracted from *Paecilomyces tenuipes*. It is a serine protease with a molecular mass of 14 kDa. PTEFP showed affinity towards Aα-chain only during fibrinogenolysis (Kim et al., 2011). A 52 kDa metalloprotease with fibrionolytic potential as well as anticoagulant activity was isolated from the mycelium of *Ganoderma lucidum* (Choi and Sa, 2000). Properties of thrombolytics from fungi excluding those from mushrooms are given in Table 4.3.

Mushrooms are also known producers of thrombolytic enzymes including some edible mushrooms. Mushrooms belonging to the genus *Cordyceps* such as C. *sinensis* and C. *militaris* yield direct-acting plasmin-like fibrinolytic enzymes. C. *sinensis* produce a 31 kDa serine protease called CSP capable of hydrolyzing Aα chain of fibrinogen. C. *militaris* produced another fibrinolytic enzyme, CMase is with a molecular mass of 27.3 kDa and it is a metalloprotease (Li et al., 2007; Cui et al., 2008). *Flammulina velutipes* proteases were reported having thrombolytic potential. The organism produced FVP-I, a 37 kDa metalloprotease showing higher degree of affinity for α-chain during fibrinolysis (Park et al., 2007). Zn metalloprotease isolated from *Pleurotus ostreatus* dissolves fibrin as well as Bβ and γ chains of fibrinogen (Choi and Shin, 1998) whereas *Pleurotus eryngii* produce a 14 kDa serine protease with specificity towards Aα-chain of fibrinogen during hydrolysis (Cha et al., 2010). Medicinal mushroom *Schizophyllum commune* is also reported to have fibrinolytic activity. The enzyme isolated was a metalloprotease with a molecular mass of 18.2 kDa (Lu et al., 2010). Another metalloprotease enzyme, TSMEP1 with fibrinolytic activity was isolated from fruiting bodies of *Tricholoma saponaceum* (Kim and Kim, 2001). AMMP isolated from mycelium of *Armillaria mellea* is also a fibrinolytic

Table 4.3 Fibrinolytic enzymes produced by fungi (excluding mushrooms) and their characteristics

Microorganism	Enzyme	Type	Molecular weight, optimum temperature and pH	Mode of action	Reference
Aspergillus ochraceus 513	—	—	36.5 kDa	—	Batomunkueva and Egorov, 2001
Aspergillus oryzae KSK-3	—	Serine protease	30 kDa 50°C, pH 6.0	—	Shirasaka et al., 2012
Fusarium pallidoroseum	—	Metalloprotease	40°C, pH 7.0	—	El-Aassar, 1995
Fusarium sp. BLB	—	Serine protease	27 kDa 50°C, pH 9.5	—	Ueda et al., 2007
Fusarium sp. CPCC 480097	Fu-P	Serine metalloprotease	28 kDa 45°C, pH 8.5	Cleaves α-chain fibrinogen, low affinity towards β and γ chain	Wu et al., 2009
Cochliobolus lunatus	—	—	pH 6.98	—	Abdel-Fattah and Ismail, 1984
Rhizopus chinensis-12	—	Hydrosulfuryl-metalloprotease	18 kDa 45°C, pH 10.5	Similar affinity towards α, β and γ chains of fibrinogen	Xiao-lan et al., 2005
Perenniporia fraxinea	—	Metalloprotease	42 kDa 35–40°C, pH 6.0	Cleaves α-chain of fibrinogen followed by β and γ chains	Kim et al., 2008
Oidiodendron flavum	—	—	22 kDa 45–55°C, pH 8.0		Tharwat, 2006
Paecilomyces tenuipes	PTEFP	Serine protease	14 kDa 35°C, pH 5.0	Quickly dissolved Aα chain of fibrinogen, No affinity towards β and γ chains	Kim et al., 2011
Ganoderma lucidum	—	Metalloprotease	52 kDa pH 7.0–7.5	cleaves Aα and Bβ chain of fibrinogen	Choi and Sa, 2000

Table 4.4 Fibrinolytic enzymes produced by fungi (mushrooms) and their characteristics

Microorganism	Enzyme	Type	Molecular weight, optimum temperature and pH	Mode of action	Reference
Cordyceps sinensis	CSP	Serine protease	31 kDa 40°C pH 7.0	Hydrolyses Aα chain of fibrinogen	Li et al., 2007
Cordyceps militaris	CMase	metalloprotease	27.3 kDa 25°C, pH 6.0	—	Cui et al., 2008
Flammulina velutipes	FVP-I	metalloprotease	37 kDa 20–30°C, pH 6.0	Higher degree of affinity for α-chain followed by β and γ chain	Park et al., 2007
Pleurotus ostreatus	—	Zn metalloprotease	12 kDa pH 7.5–8.0	Cleaves aα, Bβ and γ chains of fibrinogen	Choi and Shin, 1998
Pleurotus eryngii	—	Serine protease	14 kDa 40°C, pH 5.0	specificity towards Aα-chain of fibrinogen followed by Bβ chain	Cha et al., 2010
Schizophyllum commune	—	Metalloprotease	21.32 kDa 45°C, pH 7.4		Lu et al., 2010
Tricholoma saponaceum	TSMEP1	Metalloprotease	18 kDa 55°C, pH 7.5	cleaves Aα and Bβ chain of fibrinogen, no affinity for γ chain	Kim and Kim, 2001
Armillaria mellea	AMMP	Metallprotease	21 kDa 33°C, pH 6.0	Preferentially cleaved Aα chain	Lee et al., 2005
Fomitella fraxinea	FFP1	Serine protease	32 kDa 40°C, pH 10.0	Aα and Bβ chain of fibrinogen, no affinity for γ chain	Lee et al., 2006
Fomitella fraxinea	FFP2	Metalloprotease	42 kDa 40°C, pH 5.0	Aα chain and Bβ chain of fibrinogen	Lee et al., 2006
Lyophyllum shimeji	—	Serine protease	21 kDa 37°C pH 8.0	Preferentially cleaves Aα chain of fibrinogen, followed by Bβ and γ chain	Moon et al., 2014

metallprotease of 21 kDa (Lee et al., 2005). *Fomitella fraxinea* is another fibrinolytic enzyme producer contributing two potential enzymes, a 32 kDa serine protease called FFP1 (*F. fraxenia* proteases 1) and 42 kDa metalloprotease, FFP2 (*F. fraxenia* proteases 2) (Lee et al., 2006). *Lyophyllum shimeji*, an edible mushroom, also has thrombolytic potential and the enzyme derived from it is 21 kDa serine protease which performs rapid Aα fibrinogenolysis, followed by Bβ and γ chain hydrolysis (Moon et al., 2014). The enzymes with thrombolytic potential from mushrooms are shown in Table 4.4.

4.3.2.2.2 Bacterial thrombolytic enzymes

Bacteria are the major fibrinolytic enzyme producers and they are the most preferred sources also. It is anticipated that bacterial proteins are suitable for oral administration (Sumi et al., 1990; Omura et al., 2004). Another advantage of bacterial enzymes is ease for large-scale production. Plasminogen activators such as streptokinase and staphylokinase of bacterial origin were previously discussed, while a large number of direct-acting plasmin like enzymes also were reported from bacterial sources.

4.3.2.2.2.1 *From Bacillus sp.*

Most of the bacterial producers belong to the genus *Bacillus* and nattokinase isolated from *Bacillus* sp. turned to be the most prospective fibrinolytic enzyme among them. Nattokinase was initially isolated from *Bacillus subtilis natto* (Sumi et al., 1987). Various other bacterial strains were also reported to produce nattokinase enzyme and it will be discussed separately later. Apart from Nattokinase several other fibrinolytic enzymes were isolated from *Bacillus* sp. Subtilisin QK1 and subtilisin QK2 are two serine fibrinolytic proteases having a molecular weight of 42 and 28 kDa, respectively, isolated from B. *subtilis* QK02 (Ko et al., 2004). Bafibrinase, a thrombolytic enzyme of 32.3 kDa is produced by *Bacillus* sp. strain AS-S20-I (Mukherjee et al., 2012) and a 31 kDa thrombolytic enzyme is produced by *B. subtilis* BK-17 (Jeong et al., 2001). Subtilisin BSF1 from *Bacillus subtilis* A26 and BAF1 produced by *Bacillus amyloliquefaciens* An6 are two serine proteases with molecular weight of 28 and 30 kDa, respectively, both having fibrinolytic potential (Agrebi et al., 2009, 2010).

Thrombolytic activity along with plasminogen activation was exhibited by two proteolytic enzymes having molecular weight 29 and 29.5 kDa produced by *Bacillus halodurans* IND18 and *Bacillus cereus* IND1, respectively (Vijayaraghavan et al., 2016b; Vijayaraghavan and Vincent, 2014). A 20.5 kDa metalloprotease was purified from *Bacillus subtilis* K42 which also showed fibrinolytic as well as anticoagulant potential (Hassanein et al., 2011). URAK enzyme produced by *Bacillus cereus* NK1 [Deepak et al., 2010] and bacillopeptidase DJ-2 (bpDJ-2) from *Bacillus* sp. DJ-2 are two other thrombolytic enzymes from the *Bacillus* species (Choi et al., 2005). A few fibrinolytic enzymes with their properties are summarized in Table 4.5.

4.3.2.2.2.2 *From Streptomyces sp.*

Second largest genus of bacterial fibrinolytic enzyme producers is *Streptomyces.* Thermophilic *Streptomyces megasporus* strain SD5 produced a chymotrypsin-like serine peptidase of molecular weight 35 KDa with fibrinolytic potential (Chitte and Dey, 2000). Another member of the genus, *Streptomyces omiyaensis* yielded a thrombolytic serine protease, SOT (Uesugi et al., 2011). Low-molecular-weight serine metalloprotease of 18 kDa has been isolated from *Streptomyces* sp. CS624 and named as FES624. It quickly dissolved Aα, Bβ and γ chain of fibrinogen (Mander et al., 2011), whereas a 20 kDa fibrinolytic serine protease SFE1 of *Streptomyces* sp. XZNUM 00004 showed preference towards Aα chain (Ju et al., 2012). FP84 is another *Streptomyces* derived fibrinolytic enzyme with a molecular weight of 35 kDa contributed by *Streptomyces* sp. CS684. It is a serine metalloprotease with affinity towards Bβ chain during fibrinogenolysis (Simkhada et al., 2010). SW-1 is a fibrinolytic enzyme isolated from *Streptomyces* sp. Y405 and it was found to be a 30 kDa serine metallo protease with a lysine binding site (Wang et al., 1999). FSP3, a 44 kDa serine protease was extracted from *Streptomyces* sp. P3. It is direct-acting thrombolytic serine protease which also acts as plasminogen activator (Cheng et al., 2015). *Streptomyces* sp. DPUA 1576 produced a fibrinolytic serine protease of 39 kDa that could digest Aα-chain, Bβ-chain, and γ-chain during fibrinogenolysis (Silva et al., 2016). A few fibrinolytic enzymes from actinomycetes with their properties are summarized in Table 4.6.

Table 4.5 Fibrinolytic enzymes produced by Bacteria (*Bacillus* sp.) and their characteristics

Microorganism	Enzyme	Type	Molecular weight, Optimum temperature and pH	Mode of action	Reference
Bacillus subtilis HK176	AprE 176	Serine metalloprotease	27 kDa 40°C pH 8.0	Rapid dissolution of Aα chain followed by Bβ of fibrinogen	Jeong et al., 2015
B. subtilis CH3-5	Apr E2	Serine protease	29 kDa 40°C, pH 7.0	Rapid dissolution of Aα and Bβ chains of fibrinogen only	Jeong et al., 2014
Bacillus subtilis natto	Nattokinase (NK)	Serine protease	27.7 kDa		Sumi et al., 1987
Bacillus subtilis QK02	Subtilisin QK1	Serine protease	42 kDa	Readily hydrolyzed Aα and Bβ and γ chains of fibrin	Ko et al., 2004
Bacillus subtilis QK02	Subtilisin QK2	Serine protease	28 kDa 55°C pH 8.5	Readily hydrolyzed Aα and Bβ and γ chains of fibrin	Ko et al., 2004
Bacillus sp. strain AS-S20-I	Bafibrinase	Serine protease	32.3 kDa 37°C pH 7.4	Preferentially cleaved Aα and Bβ chains of fibrinogen	Mukherjee et al., 2012
Bacillus subtilis BK-17	Bacillokinase (BK)	Serine protease	31 kDa 50°C pH 7.0–8.0		Jeong et al., 2001 Bek et al., 2005

Microorganism	Enzyme	Type	Molecular weight, Optimum temperature and pH	Mode of action	Reference
Bacillus subtilis A26	Subtilisin BSF1	Serine protease	28 kDa 60°C pH 9.0		Agrebi et al., 2009
Bacillus amyloliquefaciens An6	BAF1	Serine protease	30 kDa 60 °C pH 9.0		Agrebi et al., 2010
Bacillus halodurans IND18	—	—	29.5 kDa 60°C pH 9.0		Vijayaraghavan et al., 2016b
Bacillus cereus IND1	—	—	29.5 kDa 60°C pH 8.0		Vijayaraghavan and Vincent, 2014
Bacillus subtilis K42	K42	Metalloprotease	20.5 kDa 40°C pH 9.4		Hassanein et al., 2011
Bacillus cereus NK1	URAK	—	46 kDa		Deepak et al., 2010
Bacillus sp. DJ-2	Bacillopeptidase DJ-2 (bpDJ-2)	Serine protease	42 kDa 60°C pH 9		Choi et al., 2005

Table 4.6 Fibrinolytic enzymes produced by *Streptomyces* sp. and their characteristics

Microorganism	Enzyme	Type	Molecular weight, Optimum temperature and pH	Mode of action	Reference
Streptomyces sp. Y405	SW-1	Serine metalloprotease	30 kDa pH 8.0	—	Wang et al., 1999
Streptomyces megasporus strain SD5	—	Serine protease	35 kDa		Chitte and Dey, 2000
Streptomyces sp. CS684	FP84	Serine metalloprotease	35 kDa 45°C, pH 7.0–8.0	Preference towards Bβ chain of fibrinogen	Simkhada et al., 2010
Streptomyces omiyaensis	SOT	Serine protease	23 kDa	—	Uesugi et al., 2011
Streptomyces sp. CS624	FES624	Serine metalloprotease	18 kDa 60°C, pH 7.0	Rapid dissolution of Aα, Bβ and γ chain of fibrinogen	Mander et al., 2011
Streptomyces sp. XZNUM 00004	SFE1	Serine protease	20 kDa 35°C, pH 7.8	Preferentially cleaved Aα chain of fibrinogen followed Bβ-chain and γ-chain at the end	Ju et al., 2012
Streptomyces sp. P3	FSP3	Serine protease	44 kDa 50°C, pH 6.5	—	Cheng et al., 2015
Streptomyces sp. DPUA 1576	—	Serine protease	39 kDa 45°C, pH 7.5	Cleaved Aα, Bβ, and γ-chains of fibrinogen	Silva et al., 2016

4.3.2.2.2.3 *Other Bacterial fibrinolytic enzyme producers*

Various bacterial species excluding *Bacillus* sp. and *Streptomyces* sp. are also reported to produce fibrinolytic enzymes although those two genera contribute most of the fibrinolytic enzymes discovered so far. A 63.3 kDa serine protease was isolated from *Paenibacillus polymyxa* EJS-3 named PPFE-I. It was found dissolving Aα-chain of fibrinogen quickly and subsequently Bβ-chain and followed by γ-chain during fibrinogenolysis (Lu et al., 2010). *Paenibacillus* sp. IND8 is also known to produce fibrinolytic enzyme (Vijayaraghavan and Vincent, 2016a). A 50 kDa metalloprotease with thrombolytic effect was reported from *Serratia* sp. RSPB11 and *Serratia* sp. KG-2-1 (Lakshmi and Prakasham, 2013; Taneja et al., 2017). Another fibrinolytic enzyme producer is *Shewanella* sp. IND20 which contributed a 55.5 kDa thrombolytic enzyme (Vijayaraghavan and Vincent, 2015). *Psuedoalteromonas* sp. IND11 yielded a 64 kDa protease which acted as plasminogen activator and disintegrated blood clot directly (Vijayaraghavan et al., 2015). *Treponema denticola* also reported with thrombolytic enzyme production (Rosen et al., 1994). Fibrinolytic enzyme from *Proteus penneri* showed higher affinity towards α-chain followed by β-chain, showing lower affinity towards γ-chain. It leaves partially hydrolyzed γ-chain after fibrinolysis (Jhample et al., 2015). Nattokinase enzyme production was reported from *Pseudomonas* sp. TKU015 with molecular weight of 24 kDa as determined by gel filtration chromatography (Wang et al., 2009). Bacterial fibrinolytic enzymes other than those from *Bacillus* sp. and actinomycetes are shown in Table 4.7.

4.3.2.2.2.4 *Nattokinase (NK) subtilisin NAT*

Fibrinolytic activity of Japanese fermented food natto was first described by Sumi and team. The enzyme extracted was termed nattokinase (Sumi et al., 1987). Subsequent years witnessed studies leading to the characterization and functional analysis of the enzyme (Sumi et al., 1990). Later, nattokinase production was achieved from various bacterial strains. A few among them are *Bacillus subtilis* isolated from Thua nao, a fermented food of Thailand (Inatsu et al., 2006), *Bacillus subtilis* ICTF-1 isolated from marine sources (Mahajan et al., 2012), *Bacillus subtilis* 1A752

Table 4.7 Fibrinolytic enzymes produced by other bacteria (excluding *Bacillus* sp. and *Streptomyces* sp.) and their characteristics

Microorganism	Enzyme	Type	Molecular weight, Optimum temperature and pH	Mode of action	Reference
Paenibacillus polymyxa EJS-3	PPFE-I	Serine protease	63.3 kDa 37°C, pH 7.5	Preferentially cleaved Aα-chain of fibrinogen followed Bβ-chain and γ-chain at the end	Lu et al., 2010
Paenibacillus sp. IND8	—	—	44 kDa 40°C, pH 7.0	—	Vijayaraghavan and Vincent, 2016a
Serratia sp RSPB11	—	Metalloprotease	50 kDa 37°C, pH 9.0	Preferentially cleaved Aα chain of fibrinogen followed Bβ-chain and γ-chain at the end	Lakshmi and Prakasham, 2013
Serratia sp. KG-2-1	—	Metalloprotease	50 kDa 40°C, pH 8.0	—	Taneja et al., 2017
Shewanella sp. IND20	—	—	55.5 kDa 50°C, pH 8.0	—	Vijayaraghavan and Vincent, 2015
Pseudoalteromonas sp. IND11	—	—	64 kDa pH 8.0, 40°C	—	Vijayaraghavan et al., 2015

(Deepak et al., 2003) and *Pseudomonas sp.* TKU015 (Wang et al., 2009). Nattokinase (EC 3.4.21.62) belongs to subtilisin family and also known as subtilisin NAT. It was derived from *Bacillus subtilis natto* encoded by *apr N* gene (Nakamura et al., 1992). It is a 27 kDa serine protease devoid of cysteine and hence lacking disulfide bonds (Fujita et al., 1993; Wang et al., 2009). Fibrinolysis by nattokinase is achieved by different ways. In vivo studies involving oral administration of nattokinase or natto showed slight enhancement in plasma fibrinolysis, thus it turned to be direct-acting fibrinolyic enzyme. Apart from this, another important factor aiding fibrinolysis is that the enzyme degrades plasminogen activator inhibitor-1(PAI-1). Also it increases the production of plasminogen activators and thereby aiding plasmin formation (Sumi et al., 1990; Urano et al., 2001). Hence, it leads to combined fibrinolysis, by NK and plasmin. NK shows more specificity towards cross-linked fibrin and less affinity towards fibrinogen (Fujita et al., 1995c). NK shares immense homology with other subtilisins such as subtilisin E (99.5%) and subtilisin J (99.3%) (Kurosawa et al., 2015), but it exhibits higher substrate specificity towards fibrin in comparison with other subtilisins, making it more suitable for therapeutic applications (Peng et al., 2005). In vitro studies substantiated that NK is a superior thrombolytic enzyme and six times more active than plasmin (Fujita et al., 1995c). NK is the only microbial fibrinolytic therapeutic enzyme whose in vivo studies have been conducted in the best possible manner. Early in vivo studies involving dogs showed that after oral administration, complete clot dissolution was achieved within 5 h and blood circulation was restored (Sumi et al., 1990). Another study proved that after intraduodenal administration, NK was transported across intestinal tract of rat and plasma fibrinogen was hydrolyzed thereafter (Fujita et al., 1995b). On a molar basis, NK was found to be more effective in clot lysis and restoration of arterial blood flow in comparison with plasmin and elastase (Fujita et al., 1995a). NK prevented platelet aggregation, along with reduction in euglobulin clot lysis time (ECLT) and extension of partial thromboplastin time (PATT) upon its oral administration in rats (Park et al., 2012). Clinical trials were performed on three groups viz. healthy group, cardiovascular group (patients with cardiovascular risk factors), and dialysis group (patients undergoing dialysis) and showed that NK reduced fibrinogen, factor VII, and factor VIII in plasma without affecting the blood lipids (Hsia et al., 2009). Trials involving healthy volunteers in Japan proved that fibrinolysis and anti-coagulation was considerably enhanced by single-dose of NK administration (Kurosawa et al., 2015). The effect of nattokinase on North American hypertensive population showed that its consumption reduced blood pressure and vWF (Jensen et al., 2016). NK is under Phase II clinical trial study in the USA for its competence in inhibiting atherothrombosis (Weng et al., 2017). Improvement of NK as a fibrinolytic as well as antithrombotic drug that could significantly recover blood circulation is highly anticipated.

4.3.2.2.2.5 *Serrapeptase/Serralysin*

Serrapeptase (SP) also known as serralysin is a proteolytic enzyme derived from enterobacterium *Serratia* E-15, isolated from silkworm *Bombyx mori* (Miyata et al., 1970). It comprises of 470 amino acids with a molecular weight of approximately 50 kDa (Nakahama et al., 1986) and is a metalloprotease with zinc ion, essential for its activity (Miyata et al., 1971). Later, it was identified from various species of *Serratia, Xenorhabdus, Pseudomonas,* etc. (Massaoud et al., 2011; Wu et al., 2016; Lakshmi and Prakasham, 2013; Louis et al., 1998). It has been used as an anti-inflammatory agent from 1960s. Because of the anti-inflammatory properties along with anti-edemic and analgesic properties, SP frequently finds its application in surgery, gynecology, orthopedics, etc. Also, the enzyme shows fibrinolytic potential making it suitable for treatment of atherosclerosis (Bhagat et al., 2013). It can dissolve only dead and damaged tissue without damaging the living tissue. Thus, removal of deposits from the arteries such as cholesterol, fatty substances, cellular waste products, etc. can be achieved by SP. Elimination of atherosclerotic plaques and blood thinning along with clot dissolution also can be achieved by the enzyme (Kotb, 2013). An in vitro study indicated serralysin from *Serratia* sp. RSPB11 as a direct-acting thrombolytic agent and shown to be more active than plasmin (Lakshmi and Prakasham, 2013). In vivo studies proved intestinal absorption of serrapeptase upon oral administration and it remained in blood stream in active form (Moriya et al., 1994). Rare cases of pneumonitis have been reported as side effects associated with use of SP (Hirahara et al., 1989; Sasaki et al., 2000). Apart from that, not much information is available on its side effects ever since the literatures are available as monographs from pharmaceutical companies. SP has been studied extensively for its anti-inflammatory properties. However, regarding its thrombolytic potential in vivo studies or trials have not been conducted.

4.4 Conclusion

During their development, thrombolytic enzymes have gone through many phases since the discovery of streptokinase. Both plasminogen activators and plasmin-like enzymes were reported with their advantages and disadvantages. More and more thrombolytic enzymes have been discovered from microbial sources in recent times. In spite of all these developments, a thrombolytic enzyme without any serious adverse effects such as bleeding complications, suitable for bolus administration and with lower production costs is not in use currently. In the past few decades, recombinant DNA technology has emerged as a powerful tool in genetic manipulation and can be used for the development of thrombolytic enzymes without systemic bleeding and improved circulation time. Chimeric enzymes (fusion proteins) with improved properties can be achieved through cloning and expression. With cutting-edge technologies, therapeutic enzymes for cardiovascular disease management are expecting new avenues.

References

Abdel-Fattah A. F., Ismail A. M., *Biotechnol. Bioeng.*, **26** (1984), 407–411.

Agnelli G., Pascucci C., Nenci G. G., Mele A., Burgi R., Heim J., *Thromb. Haemost.*, **70** (1993), 294–300.

Agrebi R., Haddar A., Hmidet N., Jellouli K., Manni L., Nasri M., *Process. Biochem.*, **44** (2009), 1252–1259.

Agrebi R., Hmidet N., Hajji M., Ktari N., Haddar A., Fakhfakh-Zouari N., Nasri M., *Appl. Biochem. Biotechnol.*, **162** (2010), 75–88.

Aguilar M. I., Hart R., *Cochrane Database. Syst. Rev.*, **3**, CD001927 (2005).

Ahn M. Y., Hahn B. S., Ryu K. S., Kim J. W., Kim I., Kim Y. S., *Thromb. Res.*, **112** (2003), 339–347.

Ahrens I., Smith B. K., Bode C., Peter K., *Expert. Opin. Drug. Metab. Toxicol.*, **3** (2007), 609–620.

Albers G. W., von Kummer R., Truelsen T., Jensen J. K. S., Ravn G. M., Gronning B. A., Chabriat H., Chang K.-C., Davalos A. E., Ford G. A., Grotta J., Kaste M., Schwamm L. H., Shuaib A., DIAS-3 Investigators., *Lancet. Neurol.*, **14** (2015), 575–584.

Ali M. R., Salim Hossain M., Islam M. A., Saiful Islam Arman M., Sarwar Raju G., Dasgupta P., Noshin T. F., *Sci. World. J.*, 586510 (2014).

Alkjaersig N., Fletcher A. P., Sherry S., *J. Clin. Invest.*, **38** (1959), 1086–1095.

Ansell J., Hirsh J., Hylek E., Jacobson A., Crowther M., Palareti G., Pharmacology and management of the vitamin K antagonists: American College of Chest Physicians Evidence-Based Clinical Practice Guidelines (8th Edition). *Chest,* **133** (2008), 160S–198S.

Appella E., Robinson E. A., Ullrich S.J., Stoppelli M. P., Corti A., Cassani G., Blasi F., *J. Biol. Chem.*, **262** (1987), 4437–4440.

Armstrong P. W., Burton J., Pakola S., Molhoek P. G., Betriu A., Tendera M., Bode C., Adgey A. A., Bar F., Vahanian A., Van de Werf F., *Am. Heart J.*, **146** (2003), 484–488.

Armstrong P. W., Burton J. R., Palisaitis D., Thompson C. R., Van de Werf F., Rose B., Collen D., Teo K. K., *Am. Heart J.*, **139** (2000), 820–823.

Arnout J., Hoylaerts M. F., Lijnen H. R., Haemostasis. In: Moncada S., Higgs A. (eds.) *The Vascular Endothelium II. Handbook of Experimental Pharmacology*, volume 176/II. Springer, Berlin, Heidelberg. pp. 1–41, (2006).

Aubry M., Bieth J., *Clin. Chim. Acta*, **78** (1977), 371–380.

Bar F. W., Meyer J., Vermeer F., Michels R., Charbonnier B., Haerten K., Spiecker M., Macaya C., Hanssen M., Heras M., Boland J. P., Morice M. C., Dunn F. G., Uebis R., Hamm C., Ayzenberg O., Strupp G., Withagen A. J., Klein W., Windeler J., Hopkins G., Barth H., von Fisenne M. J., SESAM Study Group., *Am. J. Cardiol.*, **79** (1997), 727–732.

Batomunkueva B. P., Egorov N. S., *Microbiology*, **70** (2001), 519–523.

Bauer K. A., *Hematology,* **1** (2006), 450–456.

Bek H., Hak-Seob L., Kyung Kae C., Choi Y., Byung Tae C., Min-Jeong S., Ji-Eun K., Eun-Ju R., Man Kyu H., Joo W. H., Young Kee J., *J. Life Sci.*, **15** (2005), 987–993.

Bernik M. B., *J. Clin. Invest.*, **52** (1973), 823–834.

Bernik M. B., Kwaan H C., *J. Lab. Clin. Med.*, **70** (1967), 650–661.

Bernik, M. B., Kwaan, H. C., *J. Clin. Invest.*, **48** (1969), 1740–1753.

Bhagat S., Agarwal M., Roy V., *Int. J. Surg.*, **11** (2013), 209–217.

Bick R.L., Fareed J., *Clin. Appl. Thromb. Hemost.*, **5** (1999), S63–S66.

Bringmann P., Gruber D., Liese A., Toschi L., Kratzschmar J., Schleuning W. D., Donner P., *J. Biol. Chem.*, **270** (1995), 25596–25603.

Cannon C. P., Gibson C. M., McCabe C. H., Adgey A. A., Schweiger M. J., Sequeira R. F., Grollier G., Giugliano R. P., Frey M., Mueller H. S., Steingart R. M., Weaver W. D., Van de Werf F., Braunwald E., *Circulation*, **98** (1998), 2805–2814.

Cannon C. P., McCabe C. H., Gibson C. M., Ghali M., Sequeira R. F., McKendall G. R., Breed J., Modi N. B., Fox N. L., Tracy R. P., Love T. W., Braunwald E., *Circulation*, **95** (1997), 351–356.

Carswell C. I., Plosker C. L., *Drugs*, **62** (2002), 841–870.

Cha W. S., Park S. S., Kim S. J., Choi D., *Bioresour. Technol.*, **101** (2010), 6475–6481.

Chen W., Huang X., Ma X. W., Mo W., Wang W.J., Song H.Y., *Eye*, **22** (2008), 300–307.

Chen F., Suzuki Y., Nagai N., Sun X., Wang H., Yu J., Marchal G., Ni Y., *Radiology*, **244** (2007), 429–438.

Cheng G., He L., Sun Z., Cui Z., Du, Y., Kong Y., *J. Microbiol. Biotechnol.*, **25** (2015), 1449–1459.

Chitte R. R., Dey S., *Lett. Appl. Microbiol.*, **31** (2000), 405–410.

Cho I. H., Choi E. S., Lim H. G., Lee H. H., *J. Biochem. Mol. Biol.*, **37** (2004), 199–205.

Choi H. S., Sa Y. S., *Mycologia.*, **92** (2000), 545–552.

Choi H. S., Sa Y. S., *Biosci. Biotechnol. Biochem.*, **65** (2001), 781–786.

Choi H. S., Shin H. H., *Mycologia.*, **90** (1998), 674–679.

Choi N. S., Yoo K. H., Hahm J. H., Yoon K. S., Hyun B. H. Maeng P. J., Kim S. H., *J. Microbiol. Biotechnol.*, **15** (2005), 72–79.

Christensen L. R., *J. Gen. Physiol.*, **28** (1945), 363–383.

Christensen L. R., *J. Gen. Physiol.*, **30** (1947), 465–473.

Cliffton E. E., Cannamela D. A., Grossi C., *J. Appl. Physiol.*, **6** (1953), 143–150.

Cohen M., *Semin. Hematol.*, **36** (1999), 33–36.

Collen D., Moreau H., Stockx L., Vanderschueren S., *Circulation*, **94** (1996), 207–216.

Collen D., Stassen J. M., Larsen G., *Blood*, **71** (1988), 216–219.

Collen D., Van de Werf F., *Circulation*, **87** (1993) 1850–1853.

Craven L. L., *Ann. West. Med. Surg.*, **4** (1950), 95.

Craven L. L., *Miss. Valley. Med. J.*, **75** (1953), 38–44.

Cui L., Dong M. S., Chen X. H., Jiang M., Lv X., Yan G., *World. J. Microbiol. Biotechnol.*, **24** (2008), 483–489.

Dantas G. C., Thompson B. V., Manson J. A., Tracy C. S., Upshur R. E. G., *BMC Fam. Pract.*, **5** (2004), 15.

Davies M. C., Englert M. E., Derenzo E. C., *J. Biol. Chem.*, **239** (1964), 2651–2656.

De Renzo E. C., Siiteri P. K., Hutchings B. L., Bell P. H., *J. Biol. Chem.*, **242** (1967), 533–542.

de Smet M. D., Valmaggia C., Zarranz-Ventura J., Willekens B., *Ophthalmol. Vis. Sci.*, **50** (2009), 814–819.

Deepak V., Ilangovan S., Sampathkumar M. V., Victoria M. J., Pasha S. P. B. S., Pandian S. B. R. K., Gurunathan S., *Enzyme. Microb. Technol.*, **47** (2010), 297–304.

Deepak V., Kalishwaralal K., Ramkumarpandian S., Babu S. V., Senthilkumar S. R., Sangiliyandi G., *Bioresour. Technol.*, **99** (2008), 8170–8174.

Deitcher S. R., Funk W. D., Buchanan J., Liu S., Levy M. D., Toombs C.F., *Expert. Opin. Biol. Ther.*, **6** (2006), 1361–1369.

del Zoppo G. J., Higashida R. T., Furlan A. J., Pessin M. S., Rowley H. A., Gent M., *Stroke*, **29** (1998), 4–11.

den Heijer P., Vermeer F., Ambrosioni E., Sadowski Z., López Sendón J. L., von Essen R., Beaufils P., Thadani U., Adgey J., Pierard L., Brinker J., Davies R. F., Smalling R. W., Wallentin L., Caspi A., Pangerl A. Trickett L., Hauck C., Henry D., Chew P., *Circulation*, **98** (1998), 2117–2125.

El-Aassar S. A., *Biotechnol. Lett.*, **17** (1995), 943–948.

El-Aassar S. A., El-Badry H. M., Abdel-Fattah A. F., *Appl. Microbiol. Biotechnol.*, **33** (1990), 26–30.

Emran T. B., Rahman M. A., Uddin M. M. N., Rahman M. M., Uddin M. Z., Dash R., Layzu C., *BMC Complement Altern. Med.*, **15** (2015), 128.

Eriksson B. I., Borris L., Dahl O. E., Haas S., Huisman M. V., Kakkar A. K., Misselwitz F., Kälebo P., ODIXa-HIP Study Investigators., *J. Thromb. Haemost.*, **4** (2006), 121–128.

Errasti M. E., Prospitti A., Viana C. A., Gonzalez M. M., Ramos M. V., Rotelli A. E., Caffini N. O., *Blood Coagul. Fibrinolysis.*, **27** (2016), 441–449.

Fareed J., Walenga J. M., Hoppensteadt D., Kaiser B., Jeske W., *Hamostaseologie*, **19** (1999), 55–62.

Flemmig M., Melzig M. F., *J. Pharm. Pharmacol.*, **64** (2012), 1025–1039.

Fletcher A.P., Alkjaersig N., Sherry S., *J. Clin. Invest.*, **38** (1959a), 1096–1110.

Fletcher A. P., Alkjaersig N., Smyrniotis F. E., Sherry S., *Trans. Assoc. Am. Phys.*, **71** (1958), 287–296.

Fletcher A. P., Sherry S., Alkjaersig N., Smyrniotis F. E., Jick S., *J. Clin. Invest.*, **38** (1959b), 1111–1119.

Fleury V., Angles-Cano E., *Biochemistry*, **30** (1991), 7630–7638.

Fonseca K. C., Morais N. C. G., Queiroz M. R., Silva M. C., Gomes M. S., Costa J. O., Mamede C. C. N., Torres F. S., Penha-Silva N., Beletti M. E., Canabrava H. A. N., Oliveira F., *Phytochemistry*, **71** (2010), 708–715.

Fu J., Ren J., Zou L., Bian G., Li R., Lu Q., *Thromb. Res.*, **122** (2008), 683–690.

Fujita M., Hong K., Ito Y., Fujii R., Kariya K., Nishimuro S., *Biol. Pharm. Bull.*, **18** (1995a), 1387–1391.

Fujita M., Hong K., Ito Y., Misawa S., Takeuchi N., Kariya K., Nishimuro S., *Biol. Pharm. Bull.*, **18** (1995b), 1194–1196.

Fujita M., Ito Y., Hong K., Nishimuro S., *Fibrinolysis*, **9** (1995c) 157–164.

Fujita M., Nomura K., Hong K., Ito Y., Asada A., Nishimuro S., *Biochem. Biophys. Res. Commun.*, **197** (1993), 1340–1347.

Furlan A., Higashida R., Wechsler L., Gent M., Rowley H., Kase C., Pessin M., Ahuja A., Callahan F., Clark W. M., Silver F., Rivera F., *JAMA*, **282** (1999), 2003–2011.

Ge T., Sun Z.-J., Fu S.-H., Liang G.-D., *Protein Exp. Purif.*, **42** (2005), 20–28.

Gerheim E. B., *Nature*, **162** (1948), 732.

Gerlach D., Kraft R., Behnke D., *Zentralbl. Bakteriol. Mikrobiol. Hyg. A*, **269** (1988), 314–322.

Gershlick A. H., More R. S., *BMJ*, **316** (1998), 280–284.

GISSI, Gruppo Italiano per la Sperimentazione della Streptochinasi nell'Infarto Miocardico (GISSI), *The Lancet*, **327** (1986), 397–402.

GISSI-2: Gruppo Italiano per lo Studio della Sopravvivenza nell'Infarto Miocardico. *Lancet*, **336** (1990) 65–71.

Goldhaber S. Z., Bounameaux H., *Semin. Vasc. Med.*, **01** (2001), 213–220.

Gore J. M., Sloan M., Price T. R., Randall A. M., Bovill E., Collen D., Forman S., Knatterud G. L., Sopko G., Terrin M. L., *Circulation*, **83** (1991), 448–459.

Guimaraes A. H. C., Barrett-Bergshoeff M. M., Criscuoli M., Evangelista S., Rijken D. C. *Thromb. Haemost.*, **96** (2006), 325–330.

GUSTO Angiographic Investigators, *N. Engl. J. Med.*, **329** (1993), 1615–1622.

GUSTO III Investigators, *N. Engl. J. Med.*, **337** (1997), 1118–1123.

Hacke W., Albers G., Al-Rawi Y., Bogousslavsky J., Davalos A., Eliasziw M., Fischer M., Furlan A., Kaste, M., Lees K. R., Soehngen M., Warach S., *Stroke*, **36** (2005), 66–73.

Hacke W., Furlan A. J., Al-Rawi Y., Davalos A., Fiebach J. B., Gruber F., Kaste M., Lipka L. J., Pedraza S., Ringleb P. A., Rowley H. A., Schneider D., Schwamm L. H., Leal J. S., Söhngen M., Teal P. A., Wilhelm-Ogunbiyi K., Wintermark M., Warach S., *Lancet. Neurol.*, **8** (2009), 141–150.

Harenberg J., Fenyvesi T., *Hamostaseologie*, **24** (2004), 261–278.

Hassanein W. A., Kotb E., Awny N. M., El-Zawahry Y. A., *J. Biosci.*, **36** (2011), 773–779.

Herrick J. B., *JAMA*, **59** (1912), 2015–2020.

Hildebrand M., Bhargava A. S., Bringmann P., Schutt A., Verhallen P., *Fibrinolysis*, **10** (1996), 269–276.

Hirahara K., Saitoh T., Terada I., Uno K., Nagai A., Kioi S., Arakawa M., *Nihon Kyobu Shikkan Gakkai Zasshi*, **27** (1989), 1231–1236.

Hirsh J., Anand S. S., Halperin J. L., Fuster V., *Circulation*, **103** (2001), 2994–3018.

Hoefer I. E., Stroes E. S. G., Pasterkamp G., Levi M. M., Reekers J. A., Verhagen H. J. M., Meijers J. C., Humphries J. E., Rotmans J. I., *J. Vasc. Interv. Radiol.*, **20** (2009), 951–958.

Hoppensteadt D. A., Jeske W., Walenga J., Fareed J., *Semin. Respir. Crit. Care. Med.*, **29** (2008), 90–99.

Howell W., Holt E., *Am. J. Physiol.*, **47** (1918), 328–341.

Hoylaerts M., Rijken D. C., Lijnen H. R., Collen D., *J. Biol. Chem.*, **257** (1982), 2912–2919.

Hsia C. H., Shen M. C., Lin J. S., Wen Y. K., Hwang K. L., Cham T. M., Yang N. C., *Nutr. Res.*, **29** (2009), 190–196.

Hunt J. A., Petteway S. R., Scuderi P., Novokhatny V., *Thromb. Haemost.*, **100** (2008), 413–419.

Inatsu Y., Nakamura N., Yuriko Y., Fushimi T., Watanasiritum L., Kawamoto S., *Lett. Appl. Microbiol.*, **43** (2006), 237–242.

InTIME-II Investigators, *Eur. Heart J.*, **21** (2000), 2005–2013.

ISIS-3, ISIS-3 (Third International Study of Infarct Survival) Collaborative Group, *Lancet*, **339** (1992), 753–770.

Jackson K. W., Tang J., *Biochemistry*, **21** (1982), 6620–6625.

Jahan R., Stewart D., Vinters H. V., Yong W., Vinuela F., Vandeberg P., Marder V. J., *Stroke*, **39** (2008), 1613–1615.

Jensen G. S., Lenninger M., Ero M. P., Benson K. F., *Integr. Blood. Press. Control.*, **9** (2016), 95–10[illegible].

Jeong S. J., Cho K. M., Lee C. K., Kim G. M., Shin J. H., Kim J. S., Kim J. H., *J. Microbiol. Biotechnol.*, **24** (2014), 969–978.

Jeong S. J., Heo K., Park J. Y., Lee K. W., Park J. Y., Joo S. H., Kim J. H., *J. Microbiol. Biotechnol.*, **25** (2015), 89–97.

Jeong Y. K., Park J. U., Baek H., Park S. H., Kong I. S., Kim D. W., Joo W. H., *World. J. Microbiol. Biotechnol.*, **17** (2001), 89–92.

Jhample S. B., Bhagwat P. K., Dandge P. B., *Biocatal. Agric. Biotechnol.*, **4** (2015), 370–379

Johnson A. J., Tillett W. S., *J. Exp. Med.*, **95** (1952), 449–464.

Ju X., Cao X., Sun Y., Wang Z., Cao C., Liu J., Jiang J., *World J. Microbiol. Biotechnol.*, **28** (2012), 2479–2486.

Julian D. G., *The Lancet*, **278** (1961), 840–844.

Kalbfleisch J., Thadani J., LittleJohn J. K., Brown G., Magorien R., Kutcher M., Taylor G., Maddox W. T., Campbell W. B., Perry J., Spann J. F., Vetrovec G., Kent R., Armstrong P. W., *Am. J. Cardiol.*, **69** (1992), 1120–1127.

Kasai S., Arimura H., Nishida M., Suyama T., *J. Biol. Chem.*, **260** (1985), 12382–12389.

Kaur J., Zhao Z., Klein G. M., Lo E. H., Buchan A. M., *J. Cereb. Blood. Flow. Metab.*, **24** (2004), 945–963.

Kawai C., Yui Y., Hosoda S., Nobuyoshi M., Suzuki S., Sato H., Takatsu F., Motomiya T., Kanmatsuse K., Kodama K., Yabe Y., Minamino T., Kimata S., Nakashima M., *J. Am. Coll. Cardiol.*, **29** (1997), 1447–1453.

Kawasaki T., Katoh M., Kaku S., Gushima H., Takenaka T., Yui Y., Kawai C., *Jpn. J. Pharmacol.*, **63** (1993), 9–16.

Kim H. C., Choi B. S., Sapkota K., Kim S., Lee H. J., Yoo J. C., Kim S. J., *Process Biochem.*, **46** (2011) 1545–1553.

Kim, J. H., Kim, Y. S., *Biosci. Biotechnol. Biochem.*, **65** (2001), 356–362.

Kim J. S., Kim J. E., Choi B. S., Park S. E., Sapkota K., Kim S., Lee H. H., Kim C. S., Park Y., Kim M. K., Kim Y. S., Kim S. J., *Mycol. Res.*, **112** (2008), 990–998.

Kline D. L., *J. Biol. Chem.*, **204** (1953), 949–955.

Ko J. H., Yan J. P., Zhu L., Qi Y. P., *Comp. Biochem. Physiol. C. Toxicol. Pharmacol.*, **137** (2004), 65–74.

Kohnert U., Rudolph R., Verheijen J. H., Weening-Verhoeff E. J., Stern A., Opitz U., Martin U., Lill H., Prinz H., Lechner M., *Protein. Eng.*, **5** (1992), 93–100.

Kolev K., Komorowicz E., Owen W. G., Machovich R., *Thromb. Haemost.*, **75** (1996), 140–146.

Komorowicz E., Kolev K., Lerant I., Machovich R., *Circ. Res.*, **82** (1998), 1102–1108.

Koren G., Weiss A. T., Hasin Y., Appelbaum D., Welber S., Rozenman Y., Lotan C., Mosseri M., Sapoznikov D., Luria M. H., *N. Engl. J. Med.*, **313** (1985), 1384–1389.

Kotb E., *Appl. Microbiol. Biotechnol.*, **97** (2013), 6647–6665.

Kotb E., *Biotechnol. Prog.*, **30** (2014), 656–672.

Kovalenko T. A., Panteleev M. A., Sveshnikova A. N., *Biophysics*, **62** (2017), 291–300.

Kratzschmar J., Haendler B., Langer G., Boidol W., Bringmann P., Alagon A., Donner P., Schleuning W. D., *Gene*, **105** (1991), 229–237.

Krehl L., Diseases of the heart muscle and nervous heart disease. Vienna: Alfred Holder (1901).

Kumar S. S., Abdulhameed, S., Therapeutic enzymes, in: Sugathan, S., Pradeep, N. S., Abdulhameed, S. (eds.), *Bioresources and Bioprocess in Biotechnology: Volume 2: Exploring Potential Biomolecules.* Springer Singapore, Singapore, pp. 45–73 (2017).

Kummer R., Mori E., Truelsen T., Jensen J. K. S., Gronning B. A., Fiebach J. B., Lovblad K. O., Pedraza S., Romero J. M., Chabriat H., Chang K. C., Davalos A., Ford G. A., Grotta J., Kaste M., Schwamm L. H., Shuaib A., Albers G. W., *Stroke*, **47** (2016), 2880–2887.

Kurosawa Y., Nirengi S., Homma T., Esaki K., Ohta M., Clark J. F., Hamaoka T., *Sci. Rep.*, **5** (2015), 11601.

Lakshmi B. P., Prakasham R. S., *Int. J. Biol. Macromol.*, **61** (2013), 479–486.

Lane D. A., Philippou H., Huntington J. A., *Blood.*, **106** (2005), 2605–2612.

Lange K., Boyd L. J., Loewe L., *Science*, **102** (1945), 151–152.

Laroche Y., Heymans S., Capaert S., De Cock F., Demarsin E., Collen D., *Blood*, **96** (2000), 1425–1432.

Larsen G. R., Metzger M., Henson K., Blue Y., Horgan P., *Blood*, **73** (1989), 1842–1850.

Larsen G. R., Timony G. A., Horgan P. G., Barone K. M., Henson K. S., Angus L. B., Stoudemire J. E., *J. Biol. Chem.*, **266** (1991), 8156–8161.

Lee C. J., Ansell J. E., *Br. J. Clin. Pharmacol.*, **72** (2011), 581–592.

Lee C. K., Shin J. S., Kim B. S., Cho I. H., Kim Y. S., Lee E. B., *Arch. Pharm. Res.*, **30** (2007), 475–480.

Lee J. S., Hyung-Suk B., Sang-Shin P., *J. Microbiol. Biotechnol.*, **16** (2006), 264–271.

Lee S. Y., Kim J. S., Kim J. E., Sapkota K., Shen M. H., Kim S., Chun H. S., Yoo J. C., Choi H. S., Kim M. K., Kim S. J., *Expr. Purif.*, **43** (2005), 10–17.

Leone G., Rossi E., Leone A. M., De Stefano V., *Curr. Med. Chem. Cardiovasc. Hematol. Agents*, **2**, (2004), 311–326.

Lewis B. E., Wallis D. E., Berkowitz S. D., Matthai W. H., Fareed J., Walenga J. M., Bartholomew J., Sham R., Lerner R. G., Zeigler Z. R., Rustagi, P. K., Jang I. K., Rifkin S. D., Moran J., Hursting M. J., Kelton J. G., *Circulation*, **103** (2001), 1838–1843.

Lewis B. E., Wallis D. E., Leya F., Hursting M. J., Kelton J. G., Argatroban-915 Investigators, *Arch. Int. Med.*, **163** (2003), 1849–1856.

Li H., Hu Z., Yuan J., Fan H., Chen W., Wang S., Zheng S., Zheng Z., Zou G., *Phytother. Res.*, **21** (2007), 1234–1241.

Li X., Ling L., Li C., Ma Q., *Medicine*, **96** (2017), e6667.

Lijnen H. R., Van Hoef B., De Cock F., Okada K., Ueshima S., Matsuo O., Collen D., *J. Biol. Chem.*, **266** (1991), 11826–11832.

Louis D., Bernillon J., Wallach J. M., *Biochim. Biophys. Acta Protein Struct. Mol. Enzymol.*, **1387** (1998), 378–386.

Lu C. L., Chen S., Chen S. N., *J. Food. Drug Anal.*, **18** (2010a), 69–76.

Lu F., Lu Z., Bie X., Yao Z., Wang Y., Lu Y., Guo Y., *Thromb. Res.*, **126** (2010b), 349–355.

Lubenow N., Eichler P., Lietz T., Farner B., Greinacher A., *Blood*, **104** (2004), 3072–3077.

Lucas M. A., Fretto L. J., McKee P. A., *J. Biol. Chem.*, **258** (1983), 4249–4256.

Ma Z., Lu W., Wu S., Chen J., Sun Z., Liu J.-N., *Biotechnol. Lett.*, **29** (2007), 517–523.

Macfarlane R. G., *Nature*, **202** (1964), 498–499.

Macfarlane R. G., Pilling J., *Nature*, **159** (1947), 779.

Madison E. L., Coombs G. S., Corey D. R., *J. Biol. Chem.*, **270** (1995), 7558–7562.

Mahajan P. M., Nayak S., Lele S. S., *J. Biosci. Bioeng.*, **113** (2012), 307–314.

Malcolm A., Keltai M., Walsh M., *Eur. Heart. J.*, **17** (1996), 1522–1531.

Malke H., Ferretti J. J., *Proc. Natl. Acad. Sci. U. S. A.*, **81** (1984), 3557–3561.

Mander P., Cho S. S., Simkhada J. R., Choi Y. H., Yoo J. C., *Process. Biochem.*, **46** (2011), 1449–1455.

Marcum J. A., *J. Hist. Med. Allied. Sci.*, **55** (2000), 37–66.

Marder V. J., *Ann. Int. Med.*, **90** (1979), 802–808.

Marder V. J., *J. Thromb. Haemost.*, **1** (2011), 364–373.

Marder V. J., Landskroner K., Novokhatny V., Zimmerman T. P., Kong M., Kanouse J. J., Jesmok G., *Thromb. Haemost.*, **86** (2001), 739–745.

Marder V. J., Novokhatny V., *J. Thromb. Haemost.*, **8** (2010), 433–444.

Marder V. J., Sherry S., *N. Engl. J. Med.*, **318** (1988), 1512–1520.

Markland F. S., *Toxicon*, **36** (1998), 1749–1800.

Martin U., von Mollendorff E., Akpan W., Kientsch-Engel R., Kaufmann B., Neugebauer G., *Clin. Pharmacol. Ther.*, **50** (1991a), 429–436.

Martin U., von Möllendorff E., Akpan W., Kientsch-Engel R., Kaufmann B., Neugebauer G., *Thromb. Haemost.*, **66** (1991b), 569–574.

Mason E. C., *J. Lab. Clin. Med.*, **10** (1924), 203–206.

Massaoud M. K., Marokházi J., Venekei I., *Biochim. Biophys. Acta Proteins Proteomics*, **1814** (2011), 1333–1339.

Matsubara K., Hori K., Matsuura Y., Miyazawa K., *Phytochemistry*, **52** (1999), 993–999.

Matsubara K., Hori K., Matsuura Y., Miyazawa K., *Comp. Biochem. Physiol. B Biochem. Mol. Biol.*, **125** (2000), 137–143.

Matsubara K., Sumi H., Hori K., Miyazawa K., *Comp. Biochem. Physiol. B Biochem. Mol. Biol.*, **119** (1998), 177–181.

McLean J., Johnson A. B., *Surgery*, **20** (1946), 324–336.

Mellott M. J., Stabilito I. I., Holahan M. A., Cuca G. C., Wang S., Li P., Barrett J. S., Lynch J. J., Gardell S. J., *Arterioscler. Thromb.*, **12** (1992), 212–221.

Mihara H., Sumi H., Yoneta T., Mizumoto H., Ikeda R., Seiki M., Maruyama M., *Jpn. J. Physiol.*, **41** (1991), 461–472.

Miyata K., Maejima K., Tomoda K., Isono M., *Agric. Biol. Chem.*, **34** (1970), 310–318.

Miyata K., Tomoda K., Isono M., *Agric. Biol. Chem.*, **35** (1971), 460–467.

Moon S. M., Kim J. S., Kim H. J., Choi M. S., Park B. R., Kim S. G., Ahn H., Chun H. S., Shin Y. K., Kim J. J., Kim D. K., Lee S. Y., Seo Y. W., Kim Y. H., Kim C. S., *J. Biosci. Bioeng.*, **117** (2014), 544–550.

Moriya N., Nakata M., Nakamura M., Takaoka M., Iwasa S., Kato K., Kakinuma A., *Biotechnol. Appl. Biochem.*, **20** (1994), 101–108.

Moroz L. A., *Blood*, **58** (1981), 97–104.

Moussa M., Ibrahim M., El Ghazaly M., Rohde J., Gnoth S., Anton A., Kensy F., Mueller F., *BMC. Biotechnol.*, **12** (2012), 96.

Mukherjee A. K., Rai S. K., Thakur R., Chattopadhyay P., Kar S. K., *Biochimie*, **94** (2012), 1300-1308.

Muller D., Domon B., Karas M., van Oostrum J., Richter W. J., *Biol. Mass. Spectrom.*, **23** (1994), 330–338.

Muller J., Reinert H., Malke H., *J. Bacteriol.*, **171** (1989), 2202–2208.

Munger M. A., Forrence E. A., *Clin. Pharm.*, **9** (1990), 530–540.

Murray D. W. G., Jaques L. B., Perrett T. S., Best C. H., *Surgery*, **2** (1937), 163–187.

Nagai N., Demarsin E., Van Hoef B., Wouters S., Cingolani D., Laroche Y., Collen D., *J. Thromb. Haemost.*, **1** (2003), 307–313.

Nakahama K., Yoshimura K., Marumoto R., Kikuchi M., Lee I. S., Hase T., Matsubara H., *Nucleic Acids Res.*, **14** (1986), 5843–5855.

Nakamura T., Yamagata Y., Ichishima E., *Biosci. Biotechnol. Biochem.*, **56** (1992), 1869–1871.

Niego B., Horvath A., Coughlin P. B., Pugsley M. K., Medcalf R. L., *Blood Coagul. Fibrinolysis.*, **19** (2008), 322–324.

Nordt T. K., Moser M. Kohler B., Kubler W., Bode C., *Thromb. Haemost.*, **82** (1991), 121–123.

Omura K., Hitosugi M., Kaketani K., Zhu X., Maeda H., Tokudome S., *Biofactors*, **22** (2004), 185–187.

Oprea A. D., Popescu W. M., *Br. J. Anaesth.*, **111** (2013), i3–i17.

Ouriel K., *Ann. Vasc. Surg*, **16** (2002), 797–804.

Pan R., Zhang Z. J., He R. Q., *Appl. Environ. Soil Sci.*, **2010** (2010), 294258.

Park K. J., Kang J. I., Kim T. S., Yeo I. H., *Prev. Nutr. Food. Sci.*, **17** (2012), 78–82.

Park S. E., Li M. H., Kim J. S., Sapkota K., Kim J. E., Choi B. S., Yoon Y. H., Lee J. C., Lee H. H., Kim C. S., Kim S. J., *Biosci. Biotechnol. Biochem.*, **71** (2007), 2214–2222.

Parkinson J., Bedford D. E., *The Lancet*, **211** (1928), 4–11.

Patel G. K., Kawale A. A., Sharma A. K., *Physiol. Biochem.*, **52** (2012), 104–111.

Patthy L., Trexler M., Vali Z., Banyai L., Varadi A., *FEBS Lett.*, **171** (1984), 131–136.

Pearson J. D., *Clin. Haematol.*, **7** (1994), 441–452.

Peng Y., Huang Q., Zhang R., Zhang Y., *Comp. Biochem. Physiol. B Biochem. Mol. Biol.*, **134** (2003), 45–52.

Peng Y., Yang X., Zhang Y., *Appl. Microbiol. Biotechnol.*, **69** (2005), 126–132.

Pennica D., Holmes W. E., Kohr W. J., Harkins R. N., Vehar G. A., Ward C. A., Bennett W. F., Yelverton E., Seeburg P. H., Heyneker H. L., Goeddel D. V., Collen D., *Nature*, **301** (1983), 214–221.

PERSIST investigators., *J. Thromb. Haemost.*, **2** (2004), 47–53.

Petersen L. C., Lund L. R., Nielsen L. S., Dano K., Skriver L., *J. Biol. Chem.*, **263** (1988), 11189–11195.

Pirmohamed M., *Br. J. Clin. Pharmacol.*, **62** (2006), 509–511.

Ploug J., Kjeldgaard N. O., *Biochim. Biophys. Acta*, **24** (1957), 278–282.

Prasad S., Kashyap R. S., Deopujari J. Y., Purohit H. J., TaoriG. M., Daginawala H. F., *BMC Complement. Altern. Med.*, **7** (2007), 36.

Rahman M. A., Sultana R., Bin Emran T., Islam M. S., Rahman M. A., Chakma J. S., Rashid H., Hasan C. M. M., *BMC. Complement. Altern. Med.*, **13** (2013), 25.

Rajesh R., Shivaprasad H. V., Gowda C. D., Nataraju A., Dhananjaya B. L., Vishwanath B. S., *Planta Med.*, **73** (2007), 1061–1067.

Ranby M., Bergsdorf N., Pohl G., Wallen P., *FEBS Lett.*, **146** (1982), 289–292.

Randolph A., Chamberlain S. H., Chu H. L., Retzios A. D., Markland F. S., Masiarz F. R., *Protein Sci.*, **1** (1992), 590–600.

Reddrop C., Moldrich R. X., Beart P. M., Farso M., Liberatore G. T., Howells D. W., Petersen K. U., Schleuning W. D., Medcalf R. L., *Stroke*, **36** (2005), 1241–1246.

Rijken D. C., Barrett-Bergshoeff M. M., Jie A. F. H., Criscuoli M., Sakharov D. V., *Thromb. Haemost.*, **91** (2004), 52–60.

Rijken D. C., Groeneveld E., *J. Biol. Chem.*, **261** (1986), 3098–3102.

Rijken D. C., Wijngaards G., Welbergen J., *Thromb. Res.*, **18** (1980), 815–830.

Rijken D. C., Wijngaards G., Zaal-de Jong M., Welbergen J., *Biochim. Biophys. Acta*, **580** (1979), 140–153.

Robbins K. C., Summaria L., Hsieh B., Shah R. J., *J. Biol. Chem.*, **242** (1967), 2333–2342.

Roch P., *Dev. Comp. Immunol*, **3** (1979), 599–608.

Rosen G., Naor R., Kutner S., Sela M. N., *Infect. Immun.*, **62** (1994), 1749–1754.

Ruegsegger P., Nydick I., Hutter R. C., Freiman A. H., Bang N. U., Cliffton E. E., Ladue J. S., *Circulation*, **19** (1959), 7–13.

Sadeghi S., Marder V. J., Stewart D., Kong M., Humphries J., Baumbach G. A., Jesmok G., *J. Thromb. Haemost.*, **1** (2003), 2621–2625.

Sakharov D. V., Lijnen H. R., Rijken D. C., *J. Biol. Chem.*, **271** (1996), 27912–27918.

Sako T., *Eur. J. Biochem.*, **149** (1985), 557–563.

Samama M., Walenga J., Kaiser B., Fareed J., Specific factor Xa inhibitors, in Verstraete, M., Fuster, V., Topol, E. (eds.), *Cardiovascular Thrombosis: Thrombocardiology*, Lippincott-Raven, Brussels, Belgium (1999), pp. 173–188.

Sarkar N. K., *Nature*, **185** (1960), 624–625.

Sasaki S., Kawanami R., Motizuki Y., Nakahara Y., Kawamura T., Tanaka A., Watanabe S., *Nihon Kokyuki Gakkai Zasshi*, **38** (2000), 540–544.

Sherry S., *J. Clin. Invest.*, **33** (1954), 1054–1063.

Sherry S., *Am. Heart J.*, **102** (1981), 1134–1138.

Sherry S., *Clin. Cardiol.*, **13** (1990), 3–10.

Shirasaka N., Naitou M., Okamura K., Fukuta Y., Terashita T., Kusuda M., *Mycoscience*, **53** (2012), 354–364.

Shivaprasad H. V., Riyaz M., Venkatesh Kumar R., Dharmappa K. K., Tarannum S., Siddesha J. M., Rajesh R., Vishwanath B. S., *J. Thromb. Thrombolysis*, **28** (2009), 304–308.

Shlansky-Goldberg R. D., Matsumoto A. H., Baumbach G. A., Siegel J. B., Raabe R. D., Murphy T. P., Deng C., Ray Dawkins J., Marder V. J., *J. Thromb. Haemost.*, **6** (2008), 944–950.

Siigur E., Siigur J., *Biochim. Biophys. Acta*, **1074** (1991), 223–229.

Sikri N., Bardia A., *Tex. Heart. Inst. J.*, **34** (2007), 318–327.

Silva G. M. M., Bezerra R. P., Teixeira J. A., Silva F. O., Correia J. M., Porto T. S., Lima-Filho J. L. Porto A. L. F., *Acta Amazon.*, **46** (2016), 323–332.

Simkhada J. R., Mander P., Cho S. S., Yoo J. C., *Process Biochem.*, **45** (2010), 88–93.

Siritapetawee J., Sojikul P., Klaynongsruang S., *Plant Physiol. Biochem.*, **92** (2015), 30–38.

Siritapetawee J., Thumanu K., Sojikul P., Thammasirirak S., *Biochim. Biophys. Acta*, **1824** (2012), 907–912.

Smalling R. W., *Am. J. Cardiol.*, **78** (1996), 2–7.

Smith J. B., Willis A. L., *Nat. New Biol.*, **231** (1971), 235.

Smith R. A., Dupe R. J., English P. D., Green J., *Nature*, **290** (1981), 505–508.

Sottrup-Jensen L., Claeys H., Zajdel M., Petersen T., Magnusson S., The primary structure of human plasminogen: isolation of two lysine binding fragments and one mini plasminogen (MW 38000) by elastase catalyzed specific limited proteolysis, in Davidson J. F., Rowan R. M., Samama M. M., Desnoyers P. C. (eds.), *Progress in Chemical Fibrinolysis and Thrombolysis*. Raven Press, New York, (1978), pp. 191–209.

Stage A., Astrup T., *Nature*, **170** (1952), 929.

Stahmann M. A., Huebner C. F., Link K. P., *J. Biol. Chem.*, **138** (1941), 513–527.

Stassen J. M., Arnout J., Deckmyn H., *Curr. Med. Chem.*, **11** (2004), 2245–2260.

Sturzebecher J., Neumann U., Kohnert U., Kresse G. B., Fischer S., *Protein Sci.*, **1** (1992), 1007–1013.

Sumi H., Hamada H., Nakanishi K., Hiratani H., *Acta Haematol.*, **84** (1990), 139–143.

Sumi H., Hamada H., Tsushima H., Mihara H., Muraki H., *Experientia*, **43** (1987), 1110–1111.

Sun A. L., Hua Z. C., Yao J., Yang Y. H., Yin D. Q., *Biochem. Mol. Biol. Int.*, **46** (1998), 479–486.

Sung Y. Y., Yang W. K., Kim H. K., *Mol. Med. Rep.*, **5** (2012), 721–724.

Suzuki Y., Nagai N., Collen D., *J. Thromb. Haemost.*, **2** (2004), 1617–1621.

Suzuki S., Saito M., Suzuki N., Kato H., Nagaoka N., Yoshitake S., Mizuo H., Yuzuriha T., Yui Y., Kawai C., *J. Cardiovasc. Pharmacol*, **17** (1991), 738–746.

Szemraj J., Walkowiak B., Kawecka I., Janiszewska G., Buczko W., Bartkowiak J., Chabielska E., *J. Thromb. Haemost.*, **3** (2005), 2156–2165.

Szemraj J., Zakrzeska A., Brown G., Stankiewicz A., Gromotowicz A., Grędzinski T., Chabielska E., *Pharmacol. Rep.*, **63** (2011), 1169–1179.

Taneja K., Bajaj B. K., Kumar S., Dilbaghi N., *3 Biotech.*, **7** (2017), 184.

Tao S., Peng L., Beihui L., Deming L., Zuohu L., *Biotechnol. Lett.*, **19** (1997), 465–467.

Tao S., Peng L., Beihui L., Deming L., Zuohu L., *Bioprocess Eng.*, **18** (1998), 379–381.

Tebbe U., Michels R., Adgey J., Boland J., Caspi A., Charbonnier B., Windeler J., Barth H., Groves R., Hopkins G. R., Fennell W., Betriu A., Ruda M., Mlczoch J., *J. Am. Coll. Cardiol.*, **31** (1998), 487–493.

Tharwat N., *Biotechnology*, **5** (2006), 160–165.

Thijs V. N. S., Peeters A., Vosko M., Aichner F., Schellinger P. D., Schneider D., Neumann-Haefelin T., Rother J., Davalos A., Wahlgren N., Verhamme P., *Stroke*, **40** (2009), 3789–3795.

Tillett W. S., Garner R. L., *J. Exp. Med.*, **58** (1933), 485–502.

Tillett W. S., Sherry S., *J. Clin. Invest.*, **28** (1949), 173–190.

Tjandrawinata R. R., Yunaidi D. A., Susanto L. W., *Drug Res. (Stuttg)*, **66** (2016), 293–299.

Tough J., *Nurs. Stand.*, **19** (2005), 55–64.

Turpie A. G. G., Fisher W. D., Bauer K. A., Kwong L. M., Irwin M. W., Kalebo P., Misselwitz F., Gent M., OdiXa-Knee Study Group., *J. Thromb. Haemost.*, **3** (2005), 2479–2486.

Ueda M., Kubo T., Miyatake K., Nakamura T., *Appl. Microbiol. Biotechnol.*, **74** (2007), 331.

Uesugi Y., Usuki H., Iwabuchi M., Hatanaka T., *Enzyme Microb. Technol.*, **48** (2011), 7–12.

Urano T., Ihara H., Umemura K., Suzuki Y., Oike M., Akita S., Tsukamoto Y., Suzuki I., Takada A., *J. Biol. Chem.*, **276** (2001), 24690–24696.

Van De Werf F., Adgey J., Ardissino D., Armstrong P. W., Aylward P., Barbash G., Betriu A., Binbrek A. S., Califf R., Diaz R., Fanebust R., Fox K., Granger C., Heikkila J., Husted S., Jansky P., Langer A., Lupi E., Maseri A., Meyer J., Mlczoch J., Mocceti D., Myburgh D., Oto A., Paolasso E., Pehrsson K., Seabra-Gomes R., Soares-Piegas L., Sùgrue D., Tendera M., Topol E., Toutouzas P., Vahanian A., Verheugt F., Wallentin L., White H., *Lancet*, **354** (1999a), 716–722.

Van de Werf F., Cannon C. P., Luyten A., Houbracken K., McCabe C. H., Berioli S., Bluhmki E., Sarelin H., Wang-Clow F., Fox N. L., Braunwald E., *Am. Heart J.*, **137** (1999b), 786–791.

van Zonneveld A. J., Veerman H., MacDonald M. E., van Mourik J. A., Pannekoek H., *J. Cell. Biochem.*, **32** (1986), 169–178.

Vanderschueren S., Barrios L., Kerdsinchai P., Van den Heuvel P., Hermans L., Vrolix M., De Man F., Benit E., Muyldermans L., Collen D., *Circulation*, **92** (1995a), 2044–2049.

Vanderschueren S., Collen D., van de Werf F., *Thromb. Haemost.*, **76** (1996), 541–544.

Vanderschueren S., Dens J., Kerdsinchai P., Desmet W., Vrolix M., De Man F., Van den Heuvel P., Hermans L., Collen D., Van de Werf F., *Am. Heart J.*, **134** (1997), 213–219.

Vanderschueren S., Stockx L., Wilms G., Lacroix H., Verhaeghe R., Vermylen J., Collen D., *Circulation*, **92** (1995b), 2050–2057.

Vane J. R., *Nat. New Biol.*, **231** (1971), 232–235.

Verstraete M., *Am. J. Med.*, **109** (2000), 52–58.

Viana C. A., Oliveira J. S., Freitas C. D. T., Alencar N. M. N., Carvalho C. P. S., Nishi B. C., Ramos M. V., *Blood Coagul. Fibrinolysis*, **24** (2013), 386–392.

Vijayaraghavan P., Raj F., Vincent S. G. P, *Int. J. Biol. Chem.*, **9** (2015), 11–20.

Vijayaraghavan P., Vincent S. G. P., *Biomed. Res. Int.*, **2014** (2014), 725064.

Vijayaraghavan P., Vincent S. G. P., *Biotechnol. Rep.*, **7** (2015), 135–142.

Vijayaraghavan P., Vincent S. G. P., Arasu M. V., *SIJBS.*, **2** (2016a), 434–444.

Vijayaraghavan P., Vincent S. G. P., Arasu M. V., Al-Dhabi N. A., *Electron. J. Biotechnol.*, **20** (2016b), 1–8.

Violand B. N., Castellino F. J., *J. Biol. Chem.*, **251** (1976), 3906–3912.

Walenga J. M., Jeske W. P., Samama M. M., Frapaise F. X., Bick R. L., Fareed J., *Expert. Opin. Investig. Drugs.*, **11** (2002), 397–407.

Wang C., Du M., Zheng D., Kong F., Zu G., Feng Y., *J. Agric. Food Chem.*, **57** (2009a), 9722–9729.

Wang F., Wang C., Li M., Gui L., Zhang J., Chang W., *Biotechnol. Lett.*, **25** (2003), 1105–1109.

Wang J., Wang M., Wang Y., *Chin. J. Biotechnol.*, **15** (1999), 83–89.

Wang S. L., Chen H. J., Liang T. W., Lin Y. D., *Process Biochem.*, **44** (2009b), 70–76.

Weiss H. J., Aledort L. M., *Lancet*, **2** (1967), 495–497.

Weng Y., Yao J., Sparks S., Wang K. Y., *Int. J. Mol. Sci.*, **18** (2017), 523.

WHO, 2017. World Health Organization fact sheet fs317 Cardiovascular diseases (CVDs); updated May (2017).

Wiman B., *Eur. J. Biochem.*, **76** (1977), 129–137.

Wiman B., Boman L., Collen D., *Eur. J. Biochem.*, **87** (1978), 143–146.

Wiman B., Lijnen H. R., Collen D., *Biochim. Biophys. Acta*, **579** (1979), 142–154.

Wiman B., Wallen P., *Thromb. Res.*, **10** (1977), 213–222.

Wong S. L., Ye R., Nathoo S., *Appl. Environ. Microbiol.*, **60** (1994), 517–523.

Wu B., Wu L., Chen D., Yang Z., Luo M., *J. Ind. Microbiol. Biotechnol.*, **36** (2009), 451–459.

Wu D., Li P., Zhou J., Gao M., Lou X., Ran T., Wu S., Wang W., Xu D., *Int. J. Biol. Macromol.*, **93** (2016), 98–106.

Wu T. P., Padmanabhan K. P., Tulinsky A., *Blood Coagul. Fibrinolysis*, **5** (1994), 157–166.

Wu H. L., Shi G. Y., Wohl R. C., Bender M. L., *Proc. Natl. Acad. Sci. U.S.A.*, **84** (1987), 8793–8795.

Xiao-lan L., Lian-xiang D., Fu-ping L., Xi-qun Z., Jing X., *Appl. Microbiol. Biotechnol.*, **67** (2005), 209–214.

Yan X. M., Kim C.-H., Lee C. K., Shin J. S., Cho I. H., Sohn U. D., *Korean J. Physiol. Pharmacol.*, **14** (2010), 71–75.

Yokota M., Tanji Y., *Biochem. Eng. J.*, **39** (2008), 297–304.

Zamarron C., Lijnen H. R., Van Hoef B., Collen D., *Thromb. Haemost.*, **52** (1984), 19–23.

Zeymer U., Neuhaus K. L., *Curr. Opin. Cardiol.*, **14** (1999), 392.

Chapter 5

Role of Engineered Proteins as Therapeutic Formulations

Khushboo Gulati[a] **and Krishna Mohan Poluri**[a,b]

[a]*Department of Biotechnology, Indian Institute of Technology Roorkee, Roorkee 247667, Uttarakhand, India*
[b]*Centre for Nanotechnology, Indian Institute of Technology Roorkee, Roorkee 247667, Uttarakhand, India*

gulati.khushboo9@gmail.com, krishfbt@iitr.ac.in/mohanpmk@gmail.com

5.1 Introduction

Proteins, string of amino acids, play life-sustaining functions. They act as molecular transporters, sensors, signaling molecules, maintain cell shape, aid cell movement, and actively participate in the synthesis of other biomolecules (Goodsell, 1991). The secret to perform such diverse variety of functions by the protein is hidden in their amino acid sequence. Such a sequence of amino acids in turn encodes for the unique structural architecture of the protein. The three-dimensional or quaternary structure of the protein provides unique binding pockets, charged surfaces or catalytic sites that confer the unique functionality to the protein. Any alteration in the structure of the protein can either modulate

Pharmaceutical Biocatalysis: Fundamentals, Enzyme Inhibitors, and Enzymes in Health and Diseases
Edited by Peter Grunwald

ISBN 978-981-4800-61-7 (Hardcover), 978-0-429-29503-4 (eBook)
www.jennystanford.com

or completely disrupt its functions (Voet and Voet, 2011). This implies that nature has bestowed on us the proficiency to play with the amino acid sequences and assemble them in a way that they perform a desired function. With this idea of rewiring proteins, a novel research area known as "protein engineering" has flourished.

Protein engineering is a rapidly evolving field that has made major strides in the era of designing novel proteins with the desired fold and function. The field has gained momentum as a result of development of recombinant DNA technology, high-throughput screening approaches, and supercomputing techniques (Brannigan and Wilkinson, 2002). Protein engineering approaches have been employed to imbibe several key features such as specificity, stability, solubility, activity, and environmental stability in the proteins. Hence, several diverse varieties of proteins have been designed for their various industrial and biomedical applications. Indeed, this field has also made a remarkable progress in the era of "Protein Therapeutics" (Leader et al., 2008; Caravella and Lugovskoy, 2010; Kimchi-Sarfaty et al., 2013).

In the current era of modern medicines, protein therapeutics has gained an upper hand as a result of advancements in protein engineering techniques. The global protein therapeutics market was $140,109 million in 2016 and is expected to reach to $217,591 million in 2023 at an annual growth rate (AGR) of 10.9% from 2016 to 2023 (Sumant and Shaikh, 2017). Novel protein drugs are being discovered, and efforts are in progress to improve their efficacy. Protein therapeutics are being attracted owing to their tendency to treat the previously untreatable diseases such as infertility, chronic renal failure, and dwarfism (Johnson-Leger et al., 2006). Further, high binding selectivity and specificity allow protein therapeutics to act on a particular step in disease pathology in contrast to small molecule drugs that broadly target both harmful and protective immune responses instinctively thereby, causing adverse side effects (Rutgeerts et al., 2006). Protein therapeutics can act in several different modes depending on the origin of the disease. First, if the disease is caused by some undesirable extracellular molecule, then enzyme therapeutics can degrade those molecules specifically. Second, if the disease is due to the deficiency of a particular protein, then protein therapeutics can be given in order to recover the health. Thirdly, if any discrepancy in the signaling pathways or inappropriate immune responses are responsible for a disease, then protein therapeutics can act as inhibitors or activators of the particular cell surface receptor (Wen et al., 2010).

Proteins in their natural form cannot be directly used as therapeutics for the treatment of disease as they do not possess optimal properties such as physical and chemical stability, specificity, activity, and affinity. There are several limitations of natural proteins to be used as therapeutics such as (1) their short half-life in the body, hence requiring frequent dosage, (2) higher production cost owing to their shorter shelf life, (3) their rapid digestion in intestine influencing therapeutic outcomes, (4) protein drugs causing immunogenicity, implying that patients develop immune responses against protein drugs, that sometimes result in miserable situations (Kefalides, 1998; Barbosa and Celis, 2007; De Groot and Moise, 2007; De Groot and Scott, 2007). Hence, there is a high need to precisely design or engineer the protein therapeutics that meets all the requirements of an optimal therapeutic.

Protein engineering has provided us various valuable tools that aid in designing novel or improving the existing protein therapeutics in several aspects including their activity, specificity, affinity, pharmacokinetics, reduced immunogenicity, improved productivity, and stability. This chapter focuses briefly on the overview of protein engineering approaches that have revolutionized the protein therapeutics era. Further, it discusses various protein therapeutics either in clinical use or under development that are tailored using protein engineering principles.

5.2 Protein Engineering Approaches

Smart protein therapeutics with controlled mechanism of actions are being designed using protein engineering tool box (Fig. 5.1). Protein engineering approaches hold great potential to fine-tune the desired properties of protein therapeutics, to attenuate their undesirable features and to imbibe novel activities in them. These protein designing approaches are majorly distributed among two main strategies. They include: (1) directed evolution/experimental-based strategies (Cobb et al., 2012; Cobb et al., 2013a; Cobb et al., 2013b) and (2) computational/rational designing of proteins. These approaches are being exploited either alone or in conjunction with each other in order to achieve the desirable pharmacokinetics and pharmacodynamics of protein therapeutics (Gulati and Poluri, 2016; Poluri and Gulati, 2017).

5.2.1 Directed Evolution

Directed evolution (DE) is a strategy to engineer proteins based on the nature's theme of evolution. However, DE evolves proteins with desirable traits in few days or months in comparison to millions of years for them to evolve under natural selection pressures. DE aims at exploring the sequence space which has not been scrutinized by the nature. DE has emerged as a versatile tool to tailor or alter the desired properties in numerous proteins for the wide range of biological applications with desired physicochemical and pharmacokinetic properties. The DE methodologies have not only flourished in the era of protein designing but are also being employed to engineer novel pathways, viruses, operons, and even whole microbes (Cobb et al., 2012; Cobb et al., 2013a; Cobb et al., 2013b).

DE is an iterative two-step process that involves creation of diverse libraries of proteins followed by high-throughput screening to obtain the protein of interest. Such loop process will continue until the desired protein will be obtained. Hence, the process of DE depends majorly on two things namely; one is the creation of good protein libraries, and second is the adoption of an efficient screening strategy. Creating good protein libraries implies that the library will contain large number of protein variants with refurbished properties that can be easily and rapidly screened for the aspired protein variants (Arnold, 1998; Bloom and Arnold, 2009; Cobb et al., 2012).

Methods for the generation of libraries for the DE of proteins have been segregated into two categories namely, (1) **Asexual methods** and (2) **Sexual methods**. The choice of method (asexual/sexual) depends on several factors including how much information is available about the parent protein and the extent of novelty that has to be imbibed in the protein (Chen and Arnold, 1993; Stemmer, 1994a; Stemmer, 1994b; Moore and Arnold, 1996; You and Arnold, 1996; Dalby, 2011). Further, the availability and accessibility of various resources and techniques also play a significant role in selecting the appropriate strategy.

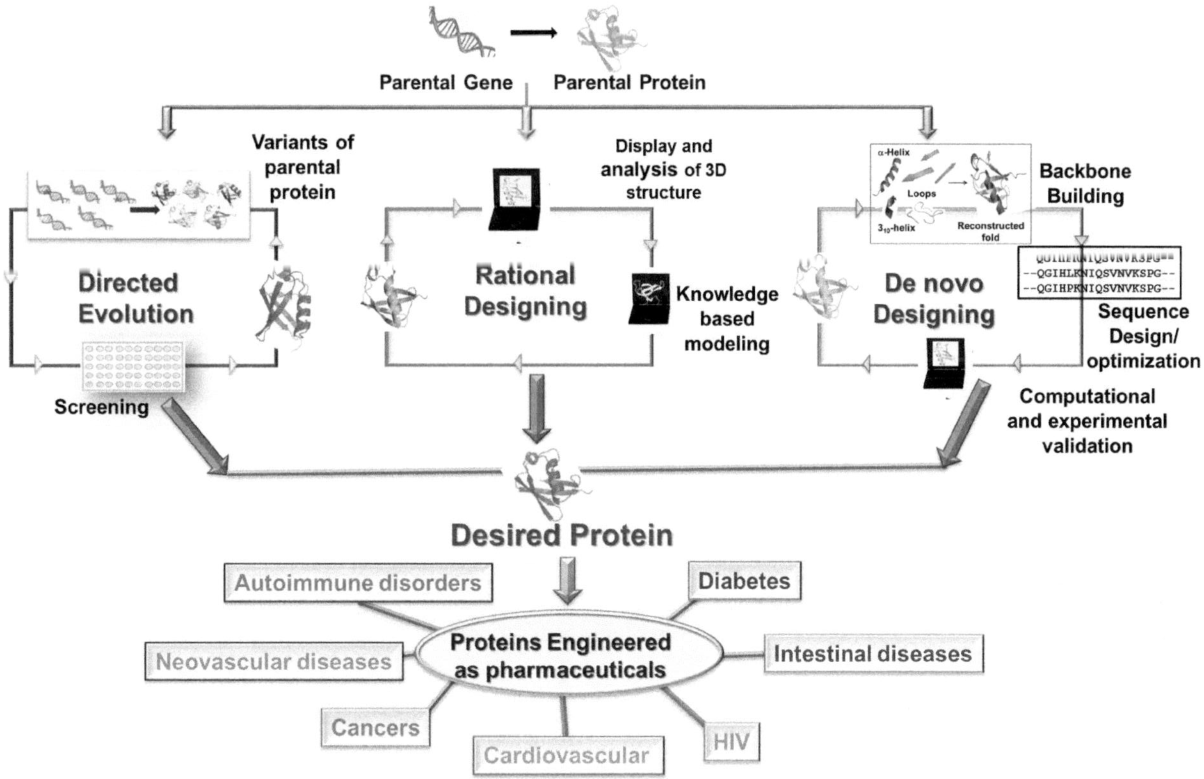

Figure 5.1 Schematic highlighting the protein engineering strategies namely directed evolution, rational designing, and de novo designing to engineer desired protein with pharmaceutical potential to treat numerous diseases.

5.2.1.1 Asexual methods

Asexual methods rely on a single parental gene for the creation of variant libraries with evolved structural, functional, or stability features. Asexual methods are broadly classified into two groups: (a) random mutagenesis and (b) focused mutagenesis.

(A) Random mutagenesis

It is the most rigorous method used for the creation of libraries of mutant proteins. Different mutations like insertion, deletion, transversions, and transitions, etc., can be introduced into the genes randomly using several methods. This include chemical mutagenesis (Deshler, 1992; Labrou, 2010), error prone PCR (Cadwell and Joyce, 1992; Patrick et al., 2003) or mutator strains (Greener et al., 1996). Chemicals including sodium bisulfite, ethyl methane sulfonate, hydroxylamine, mitomycin C, hydroxylamine, etc., are reported as chemical mutagens that can randomly mutate the DNA. Biased nature of chemicals towards particular amino acid substitution and no control over the mutation rate limits the usage of this method (Deshler, 1992; Lai et al., 2004; Liu et al., 2009; Labrou, 2010). Error Prone PCR is one of the most common techniques that uses Taq DNA polymerase lacking 3′-5′ exonuclease activity, as a low fidelity agent (Tindall and Kunkel, 1988). Mutator strain method is based on transforming the parent plasmid in a strain that lacks the DNA repair mechanism. *Epicurian coli XL1*-red is the most commonly employed strain which lacks DNA repair pathways to introduce errors in the plasmid DNA during replication cycle (Greener et al., 1996). Several other methods have also been developed like dual approach to random chemical mutagenesis (DuARCheM) (Mohan and Banerjee, 2008), rolling circle error-prone PCR (Fujii et al., 2014), targeting glycosylases to embedded arrays for mutagenesis (TaGTEAM) (Finney-Manchester and Maheshri, 2013), mutagenesis by random insertion and deletion (Murakami et al., 2002), transposon-based random mutagenesis (Tee and Wong, 2013), random mutagenesis methods altering the target DNA length (Kipnis et al., 2012; Fujii et al., 2013), that are described elsewhere (Tee and Wong, 2013).

(B) Focused mutagenesis

In contrast to random mutagenesis, focused mutagenesis involves generation of small libraries that can be readily screened. Smaller libraries are the result of mutations that are focused only to protein hot spot residues. Hot spots correspond to the residues which have phenomenal functional/structural role. Site saturation mutagenesis (SSM) is the most commonly employed focused mutagenesis method which tries all the 20 amino acids against the targeted amino acid in a single shot. The two most commonly employed SSM methods are: (1) site directed mutagenesis (SDM) that uses complementary primer pairs containing the mutation site (Matsumura and Rowe, 2005; Alcalde et al., 2006). (2) Overlap extension PCR that employs two pairs of complementary primers with two rounds of PCR to introduce the mutation at the targeted site (Gratz and Jose, 2008). Other focused mutagenesis methods such as cassette mutagenesis (Reidhaar-Olson and Sauer, 1988), sequence saturation mutagenesis (SeSAM) (Wong et al., 2008), single-primer reactions in parallel (SPRINP) (Edelheit et al., 2009), megaprimed and ligase-Free (Tseng et al., 2008), Ω-PCR (Chen et al., 2013), PFunkel (Firnberg and Ostermeier, 2012), ominchange (Dennig et al., 2011), OSCARR (Hidalgo et al., 2008), trimer-dimer mutagenesis (Gaytan et al., 2009), synthetic saturation mutagenesis (Patwardhan et al., 2009), Amber codon saturation mutagenesis have also been developed and are being employed for the creation of variant libraries (Shozen et al., 2012).

Regardless of the various advantages of random mutagenesis methods, several limitations of these methods like restriction to the usage of single parental sequence, lower mutational frequency, and exhausting screening strategies derived the alternative strategy to overcome these pitfalls. The other strategy which is based on recombination was originated by Stemmer and his coworkers (Stemmer, 1994a; Stemmer, 1994b).

5.2.1.2 Sexual methods

These methods engage in vitro recombination of homologous/non-homologous parental sequences. The result of such recombination events will inherit the characteristic features from all the recombining parental sequences (Stemmer, 1994a; Stemmer, 1994b). Sexual methods are also broadly divided into categories depending on the extent of homology of the parental DNA sequences. They include: (a) homologous recombination and, (b) non-homologous recombination methods.

(A) Homologous recombination

Homologous recombination techniques are based on high sequence homology of the parental gene sequences. The resultant chimeric gene shows an amalgamation of characteristic features of the combining parental sequences. The homologous recombination methods have been classified into in vitro and in vivo homologous recombination methods. DNA shuffling is the most prominent in vitro homologous recombination technique developed by Stemmer and coworkers (Stemmer, 1994b). Later advancements in the DNA shuffling method resulted in improvised schemes such as family shuffling (Crameri et al., 1998), and DOGs (degenerate oligonucleotide gene shuffling) (Gibbs et al., 2001). Other in vitro homologous recombination methods include: random priming in vitro recombination (RPR) (Shao et al., 1998), truncated metagenomic gene-specific PCR (TMGS-PCR) (Wang et al., 2010), staggered extension process (StEP) (Zhao et al., 1998), random chimeragenesis on transient templates (RACHITT) (Coco, 2003), synthetic shuffling (Ness et al., 2002). The in vivo homologous recombination methods include: cloning performed in yeast (CLERY) (Abecassis et al., 2003), Mutagenic organized recombination process by homologous in vivo grouping (MORPHING) (Gonzalez-Perez et al., 2014), and phage assisted continuous evolution (PACE) (Esvelt et al., 2011).

(B) Non-homologous recombination

Non-homologous recombination methods have also been developed to produce chimeric proteins via recombination of parental sequences which are non-homologous. Non-homologous parental sequences imply that they share very little sequence identity. Such non-homologous sequences cannot be recombined using the above-mentioned homologous recombination methods. Exon shuffling is the one of the non-homologous recombination methods developed by Kolkman and Stemmer (Kolkman and Stemmer, 2001). This method essentially requires mixing of exons from different protein genes via recombination at the intervening regions known as introns. Chimeric protein will exhibit the features of all the recombining exons (Kolkman and Stemmer, 2001). Further, diversity can be inculcated in these chimeras by insertion, deletion or substitution during the recombination event. Few other non-homologous recombination techniques such as incremental truncation for the creation of hybrid enzymes (ITCHY) (Ostermeier et al., 1999), sequence homology independent protein recombination (SHIPREC) (Sieber et al., 2001), Degenerate Homo-duplex Recombination (DHR) (Coco et al., 2002), Random Multi-recombinant PCR (RM-PCR) (Tsuji et al., 2001), golden gate shuffling (GGS) Recombination (Engler et al., 2009), Integron (Bikard et al., 2010), Y-ligation-based shuffling (YLBS) (Kitamura et al., 2002) have also been developed to generate chimeric proteins with novel functionalities.

5.2.2 Computational Designing

Directed evolution methods have gained great success in designing the proteins. However, keeping in mind the tedious efforts involved in creation of huge libraries and screening of those libraries, scientific community devised a more efficient way to capture the desired proteins. In this modern computing age with huge advancements in bioinformatics tools, databases, servers, computational designing of proteins has taken the lead to create novel proteins with fancy folds meeting the desirable structural and functional expectations. Computational designing amalgamates all the available knowledge on similar proteins to engineer the protein of interest, which is subjected to several experimental assessments for the desirable features like specificity, activity, stability, etc. (Saven, 2011; Tiwari et al., 2012; Khan et al., 2016). The two major strategies for computational designing of proteins are (a) rational design and (b) de novo design. Rational designing involves the recreation of existing protein which implies inculcating the desired characteristic in the available protein. In contrast, De novo designing involves the creation of proteins from scratch. Hence, de novo designing exploits the knowledge gained about protein's structural and functional features during the rational designing of proteins (Lutz, 2010).

Both the approaches have been successfully employed to design novel proteins with fancy folds decorated with desired traits.

5.2.2.1 Rational designing

Shaping the desired protein computationally requires the detailed information about the structure and functions of the desired protein. All such information is concealed in the sequence of protein. Thus, it is imperative to perform an exhaustive scan of all the related protein sequences to decipher all the information in terms of evolutionary perspectives, coevolving and hot spot residues that are vital for maintaining the structural/functional activities of protein. To disentangle all this information, numerous computational methods have been devised.

Multiple Sequence Alignment (MSA)

It is the simplest approach to computationally design a protein if the structural information is not available for the target protein. Several algorithms have been developed to align large number of sequences within a few seconds (Pei, 2008; Dagelaite et al., 2013). The alignment of target protein with its homolog will discern the hot spots in the protein which are important for the function of the protein. Hence, the hot spots can be targeted for the mutations to modulate the functional activities/stability of the proteins. Several groups across the globe are being exploiting MSA to determine the residues specific for the stability, activity, functionality of the proteins, and to target them for creation of proteins with newer functions (Steipe et al., 1994; Watanabe et al., 2006; Sullivan et al., 2011).

Coevolutionary analysis

The word "Coevolution" itself indicates the simultaneous evolution. This coevolution implies that there are several residues in the protein which are quiet far away from each other in the sequence; however, they come closer during the structure formation (Lovell and Robertson, 2010). Such residues are involved in making essential interactions to hold the structural fold of the protein. Decoding coevolving residues in the protein aids in figuring out the ingrained elements in the protein that exert huge influence on the structural and functional aspects of the protein (Horner et al., 2008). Several research groups are focusing on identification of coevolving residues and targeting them for mutations to inculcate novelties in the existing protein (Socolich et al., 2005; Yip et al., 2008; Chen et al., 2011).

Structure-based designing of proteins

Elucidating the three-dimensional structure of protein is imperative to decipher the functional characteristics of protein. In many cases, it has been reported that proteins share similar structure despite huge difference in their sequence. Hence, structure of a protein can be used as a template to redesign the protein (Choi et al., 2009; Qi et al., 2012; Choi et al., 2015). Although structures of several proteins have been obtained, there is still a large chunk of proteins whose structures are not known due to several experimental restraints such as expression of protein in insoluble fraction, difficulty in formation of crystal, poor solubility/precipitation. Hence, researchers have taken a lead to overcome this limitation by acquiring the structure of proteins in silico by computational methods like homology modeling (Xiang, 2006), protein threading (David et al., 2000), fragment-based methods (Zhang, 2007), and ab initio methods (Bonneau and Baker, 2001). These computational methods are based on the fact that structures of proteins are evolutionarily more conserved in contrast to their sequences.

5.2.2.2 De novo designing

De novo designing is a bit sluggish as no prior information is available about the target protein. De novo designing strategy is based on fitting the appropriate sequence onto a designed structural fold and, hence, described as inverse protein folding problem (Samish et al., 2011; Woolfson et al., 2015). Constituting a protein using de novo design essentially comprises three main parts: **(1) Backbone building**: This is the first step in designing the protein from scratch. Several approaches are available to build a backbone. One simple way is to congregate the paragon secondary structural elements (helices and sheets, etc.,) with allowed torsion angles and interconnecting loops. Spacing between the secondary structure elements and loops should be optimized by aligning the target backbone structure onto the structure of protein closely resembling the target protein structure (Murzin et al., 1994; Kuhlman et al., 2002; Offredi et al., 2003). **(2) Sequence design and optimization**: This is an important step in de novo designing of the protein. It involves designing the sequence which best fits the structure. Several methods are available to explore the large sequence space for the target backbone structures. Some of these methods involve rotamer libraries which contain the conformations that are most likely accessible by an amino acid. All these methods focus on identification of low energy sequences with structural specificity (Mendes et al., 2002; Dunbrack, Jr. 2002; Lassila et al., 2006; Fung et al., 2008). **(3) Validation of selected sequence and structure pair**: once sequences are optimized for the target backbone structure, these sequences are assessed in silico for their ability to reach the targeted structural fold. After in silico evaluation of protein structures, the sequence will be recombinantly cloned and the three-dimensional structure of the target protein will be validated using one of the following structural biology techniques (NMR spectroscopy/X-ray crystallography/cryo electron microscopy, etc.,) (Summa et al., 2002; Liang et al., 2009; Joh et al., 2014).

5.2.3 Combinatorial Approach

Bioinformaticians have eased the process of designing the protein by writing advanced algorithms, creation of numerous databases, and servers. Such computational efforts not only advanced the in silico designing of proteins, but also provided relaxation to those experimentalists who screen large libraries of targets under directed evolution methods. Combinatorial approach is the hybrid approach that amalgamates the computational designing and directed evolution to develop concise libraries and to efficiently screen those libraries for desired functional variants (Saraf et al., 2005; Pantazes et al., 2007; Frith et al., 2008; Lane and Seelig, 2014).

Table 5.1 List of some of the proteins engineered as therapeutics that are either approved by FDA or under different phases of clinical trials

Protein engineered	Generic name/ brand name	Disease/condition	Reference
		Antibodies	
Anti-epidermal growth factor receptor	Cetuximab	Metastatic colorectal cancer	(Wong, 2005)
Bevacizumab	Avastin	Cancer treatment	(Kazazi-Hyseni et al., 2010)
Brentuximab vedotin	Adcetris	Hodgkin lymphoma	(Scott, 2017a)
Ado-trastuzumab emtansine	Kadcyla	Breast Cancer	(Lambert and Chari, 2014)
Catumaxomab	Removab®	EpCAM-positive carcinomas	(Linke et al., 2010)
Trastuzumab Natalizumab	Herceptin Tysabri	Breast Cancer, Relapsing multiple sclerosis	(Vogel et al., 2002; Miller et al., 2003)
Ranibizumab285	Lucentis	Neovascular age-related, macular degeneration	(Brown et al., 2006)
		Enzymes	
Agalsidase beta	Fabrazyme®	Fabry Disease	(Borgwardt et al., 2013)
Taliglucerase	Elelyso™	Type I Gaucher disease	(Haddley, 2012)
Velaglucerasealfa	VPRIV™	Type I Gaucher disease	(Hughes et al., 2015)
Laronidase	Aldurazyme®	Hurler Syndrome	(Jameson et al., 2013)
Galsulfase	Naglazyme™	MPS VI (Maroteaux–Lamy syndrome)	(Brunelli et al., 2016)

(Continued)

Table 5.1 (*Continued*)

Protein engineered	Generic name/ brand name	Disease/condition	Reference
Idursulfase	Elaprase®	MPS II (Hunter disease)	(Burrow and Leslie, 2008)
PEGylated asparaginase	Oncaspar®	Acute lymphoblastic leukemia (ALL)	(Dinndorf et al., 2007)
PEG-arginine deiminase	ADI-PEG 20	Melanoma	(Ott et al., 2013)
rh DNase I	Pulmozyme®	Cystic fibrosis	(Thomson, 1995)
		Cytokines	
PEGylated colony stimulating factor	Pegfilgrastim (PEG-rmetHuG-CSF, Neulasta)	Neutropenia	(Crawford, 2003; Molineux, 2004)
IFN-α-2a	Pegasys®	Chronic hepatitis C	(Mangia et al., 2017)
Albinterferon alpha-2b (alb-IFN)	Zalbin	Chronic hepatitis C	(Subramanian et al., 2007)
Denileukin diftitox	ONTAK	Cutaneous-T-cell Lymphoma	(Foss, 2006)
rh IL-2	Proleukin®	Renal cell carcinoma	(Marshall et al., 2003)
		Hormones	
Insulin	Humilin®	Diabetes	(Landgraf and Sandow, 2016)
Insulin	Humalog® (Insulin lispro)	Diabetes	(Thrasher et al., 2015a)

Protein engineered	Generic name/ brand name	Disease/condition	Reference
Insulin	NovoLog® (Insulin aspart)	Diabetes	(Muchmore, 2018)
Insulin	Apidra® (Insulin glulisine)	Diabetes	(Garnock-Jones and Plosker, 2009)
Insulin	Lantus® (Insulin glargine)	Diabetes	(Goykhman et al., 2009)
Insulin	Levemir® (Insulin detemir)	Diabetes	(Vigneri et al., 2010)
rh GH		Chronic renal failure, Turner syndrome, idiopathic short stature	(Bajpai and Menon, 2005; Reddy et al., 2017)
		GPCR targeting antibodies	
Mogamulizumab	Poteligeo®	Relapsed or refractory CCR4 + adult T-cell leukemia-lymphoma (ATL)	(Subramaniam et al., 2012)
PA14 antibody	PRO-140	HIV infection	(Thompson, 2018)
Erenumab	AMG 334	Migraine	(Ashina et al., 2017)
Vantictumab	OMP-18R5	Solid tumors, metastatic breast cancer	(Jo and Jung, 2016)
		Cardiovascular therapeutics	
Etanercept	Enbrel®	Rheumatoid arthritis	(Haraoui and Bykerk, 2007)

(Continued)

Table 5.1 (*Continued*)

Protein engineered	**Generic name/ brand name**	**Disease/condition**	**Reference**
Infleximab	Remicade®	Rheumatoid Arthritis, Crohn's disease	(Jay and Lee, 2013)
Tocilizumab	RoActemra®	Rheumatoid arthritis	(Scott, 2017h)
		Coagulation factors	
rh FVIII (rFVIII)	Adynovate®	Hemophilia A	(Dunn et al., 2018)
Turoctocog alfa	NovoEight®	Hemophilia A	(Tiede et al., 2015)
human-cl *rh*FVIII/ simoctocog alfa	Nuwiq®	Hemophilia A	(Klukowska et al., 2018)
Efraloctocog alfa	Eloctate®	Hemophilia A	(Chowdary et al., 2016)
		Protein scaffolds	
Ecallantide	Kalbitor®	Hereditary angioedema	(Farkas and Varga, 2011)
Anticalin (Angiocal)	PRS-050	Advanced solid tumors	(Mross et al., 2013; Gille et al., 2016)
Trivalent, bispecific nanobody	ALX-0761	Psoriasis	(Torres et al., 2016)

rh, recombinant human; cl, cell line; CGRP receptor, (calcitonin gene-related peptide receptor); GH, growth hormone; IL, interleukin.

5.3 Protein Therapeutics

The protein engineering strategies described in the previous sections have transformed the protein therapeutics era as several different classes of protein therapeutics with heightened specificity, stability, activity, pharmacokinetics, pharmacodynamics, yields, and reduced immunogenicity have been engineered (Table 5.1). They include monoclonal antibodies, glycoproteins, enzymes, cytokines, hormones, coagulation factors, interleukins, MHCs, colony stimulating factors, etc. Various protein scaffolds have also been designed that are endowed with desirable biomedical properties.

5.3.1 Antibodies

Antibodies or immunoglobulins (Ig) are the important members of immune system that neutralize the antigens by recognizing their epitopes. Antigen binding site is located in hypervariable regions or complementarity determining regions (CDRs) that mainly involve first 110 amino acids of heavy and light chains of antibodies. Arrangement of variable regions results in diversity among the antibodies (Owen et al., 2012). Several methods have been developed to engineer antibodies with high specificity and affinity (He and Zhu, 2015). Regions including fragments of antigen binding (Fab) comprising VH, VL, CH1, and CL domains and Fragment of crystallization (Fc region) consisting of hinge, CH2, and CH3 domains are the main targets to engineer novel antibodies. Several strategies are also being employed to increase the in vivo half- life of antibodies. Eculizumab is the first antibody with engineered Fc region that was approved in 2007 (Czajkowsky et al., 2012; Wang et al., 2018). In several cases, monoclonal antibodies are not as effective as therapeutic agents, hence, antibody scaffolds are being used to incorporate the specific activity to develop them as successful therapeutic candidates (Gebauer and Skerra, 2009).

Lippow et al. employed an iterative computational design method based on single mutants and electrostatic binding to increase the affinity of antibodies. They improved the affinity by 10-fold in the anti-epidermal growth factor receptor drug cetuximab (Erbitux), and by 140-fold in anti-lysozyme model antibody D44.1. High affinity mutants were also engineered for bevacizumab (Avastin), and anti-fluorescein antibody 4-4-20 (Lippow et al., 2007; Kazazi-Hyseni et al., 2010).

Antibodies are being considered for drug carriers as ADCs (antibody–drug conjugates). Drugs are conjugated with antibodies through a sensitive linker. This construct when administered will propel/release the drug selectively towards the target location such as cancer cells. The choice of monoclonal antibodies (mABs) depends on two parameters: (1) mABs should be highly selective towards binding to tumor cells; (2) mABs should exhibit low cross reactivity with healthy cells. Highly potent drugs are required as lower quantities of ADCs are administered. Linker should also be highly stable to prevent the prior release of drug. Mostly ADCs are internalized into the cells through receptor mediated endocytosis before the release of the drug. Now days, non-internalizing ADCs and vascular targeting ADCs are being developed. Variety of responsive linkers are being exploited in ADCs that include (1) hydrazones, which are cleavable at low pH conditions of endosome and lysosome, (2) disulfides that release the drug in the reducing environment of cytosol, (3) peptides liable to cancer specific proteases. Non-cleavable linkers like thioethers are also being used that release the drug after the enzymatic degradation of ADC in the lysosome or endosome (Alley et al., 2010; Casi and Neri, 2012; Teicher and Doroshow, 2012).

Ado-trastuzumab emtansine (Kadcyla) and brentuximab vedotin (Adcetris) are two ADCs that are being used for the treatment of hematological malignancies (Hodgkin lymphoma) and solid tumor (breast cancer), respectively (Beck and Reichert, 2014; Scott, 2017a). Brentuximab vedotin is a conjugate of a monoclonal antibody targeting CD30, which is expressed only during Hodgkin lymphoma. It is connected via cathepsin to anti-mitotic drug monomethyl auristatin E (Beck and Reichert, 2014; Scott, 2017a). Ado-trastuzumab emtansine is composed of anti-Her2-mAb Trastuzumab antibody linked to cytotoxic drug mertansine via thioether linkage (Beck and Reichert, 2014; Lambert and Chari, 2014). Variety of drugs can be conjugated at multiple locations in monoclonal antibody to further enhance their actions (Tobin et al., 2014).

Another strategy that is being employed for the construction of antibody-based drugs includes bispecific antibodies, which unites the specificities of two antibodies and hence can simultaneously respond to different antigens. Catumaxomab (Removab®, anti-EpCAM x anti-CD3) was the first bispecific antibody. It was approved in Europe (2009) to cure malignant ascites in patients with EpCAM-positive carcinomas (Linke et al., 2010). Bispecific antibodies provides an advantage of targeting the two mediators of the disease simultaneously; hence, they are more effective therapeutic agents as compared to those that target the single disease mediator (Marvin and Zhu, 2005).

Antibodies are further improved by enhancing the performance of their effector functions such as ADCP (antibody-dependent cellular phagocytosis), ADCC (antibody-dependent cellular cytotoxicity), and CDC (complement-dependent cytotoxicity). This is an effective strategy in case of oncology targets where the destruction of target cells is required (Kubota et al., 2009).

5.3.2 Enzymes as Therapeutics

Enzymes are the catalysts that aid in carrying out various complex chemical reactions. Enzymes are being used as therapeutics since long time. Previously, proteolytic enzymes were employed as supplements in the treatment of gastrointestinal related disorders. Nowadays, enzymes are being exploited in the treatment of various types of cancers and numerous other diseases. Collagenase, a collagen degrading enzyme breaks peptide bonds in collagen. Microbial collagen is being used in debridement of ulcers and burns (Ramundo and Gray, 2008; Ramundo and Gray, 2009). Collagenase is also used for enzymatic fasciotomy owing to its specificity for collagen. Collagenase derived from *Clostridium histolyticum* is used as nonsurgical treatment for Dupuytren's contracture, as approved by US FDA (Coleman et al., 2014). Pancreatic exocrine insufficiency (PEI) is a condition related to pancreatic diseases such as chronic pancreatitis, and cystic fibrosis. This disease leads to the weight loss, malnutrition, morbidity, and mortality. Pancreatic enzyme replacement therapy (PERT) is used to address the pathophysiological conditions in PEI (Nakajima et al., 2012). PERT is also used in patients suffering from cystic fibrosis (Somaraju and Solis-Moya, 2014). Many other diseases including genetic disorders are being treated by the strategy of ERT. ERT has been approved by FDA. Some notable ERT's include Fabrazyme® (agalsidase beta) for Fabry Disease (Borgwardt et al., 2013); Elelyso™ (taliglucerase) (Haddley, 2012), VPRIV™ (velaglucerasealfa) (Hughes et al., 2015) for type I Gaucher disease; Aldurazyme® (laronidase) for MPS I (Hurler, Hurler-Scheie, or Scheie syndrome) (Jameson et al., 2013); Lumizyme® (alglucosidasealfa) for Glycogen Storage Disease type II (Pompe disease); Naglazyme™ (galsulfase) for MPS VI (Maroteaux-Lamy syndrome) (Brunelli et al., 2016), Elaprase® (idursulfase intravenous) for MPS II (Hunter disease) (Burrow and Leslie, 2008). ERT approach is not devoid of limitations, it is also associated with high production cost and lower bioavailability. Hence, continuous efforts are being made in these directions to circumvent these limitations.

Asparaginase is an aminohydrolase which converts aspargine to aspartic acid and ammonia that eventually leads to cell death. Asparginase is used in treatment of acute lymphoblastic leukemia (ALL). PEGylated asparaginase with lower immunogenicity, and longer half-life compared to native asparagine makes it an ideal drug for the treatment of ALL (Rizzari et al., 2013). Similarly, PEGylated asparginase (Oncaspar®, Enzon) is used for treating leukemia (Dinndorf et al., 2007; Pasut et al., 2008a). Other PEGylated enzyme "PEG-arginine deiminase (ADI-PEG 20)" is under clinical trials for curing melanoma, carcinoma, and specific types of tumors (Pasut et al., 2008a; Pasut et al., 2008b; Ott et al., 2013). PEGylation of enzymes is carried out to improve their pharmacokinetic profiles and to lower their immunogenicity (Pasut et al., 2008a). DNA and RNA enzymes are also being used for the treatment of diseases in which gene products are involved. DNase obtained from *Nocardia* and *Bacillus sp.* exhibits mucolytic activity; hence, it is used for the treatment of chronic bronchitis. Recombinant human DNase I (Pulmozyme®, Genentech) or dornase alfa, is used against cystic fibrosis. Recombinant human DNase improves the lung function as it degrades the viscous mucus formed by the DNA (Thomson, 1995). Enzyme therapeutics are being exploited in many other disorders related to cardiovascular, viral, intestinal, and hereditary. Several of them are still in different phases of clinical trials (Vellard, 2003). Both protein engineers and biotechnologists are attempting to produce safer and cheaper enzyme therapeutics with improved efficacies and specificities. Enzymes are also being given along with other drugs to exert synergistic effects for the treatment of diseases and sometimes to reduce the adverse effects of the particular drug.

5.3.3 Cytokines and Their Receptors

Cytokines are the small proteins that mediate communication between the cells and the external environment, by interacting with the cell surface receptors. Cytokine family comprises many different varieties of proteins that include interferons, interleukins, and colony stimulating proteins, tumor necrosis factor, and chemokines. They are essentially secreted by white blood cells (leukocytes), and also by other variety of cells including epithelial, and endothelial cells in response to external stimuli (Turner et al., 2014). Cytokines actively participate in numerous biological activities such as organogenesis, oncogenesis, inflammation, wound healing and many other immune system related functions (Turner et al., 2014).

Owing to their differential roles in several biological processes, cytokines are marked as potent drugs against various cancers, inflammatory, infectious and autoimmune diseases (Vilcek and Feldmann, 2004). Several of the recombinant cytokines are being used for the treatment of a handful of diseases. Some of the approved therapeutic agents include; recombinant IL-2 for metastatic melanoma, recombinant IFN-2α for hairy cell leukemia. IFN-α con-1 for chronic hepatitis C (Melian and Plosker, 2001). Several of the engineered/recombinant cytokine protein therapeutics for different ailments are in clinical trials as summarized by Lombardi et al. (Lombardi et al., 2013).

Cytokines also exert many adverse effects during their exploitation as therapeutics due to their inherent properties such as pleotropic, short serum half-life. Cytokines are pleotropic which implies that they act on more than one cell type causing conflicting effects. Hence, there is a need to engineer cytokines with high specificity, and extended half-life. One of the strategies employed to increase the half-life of therapeutic cytokines is PEGylation. This involves conjugation of PEG chains to cytokine/chemokine either enzymatically or chemically. It prolongs the life of cytokines by preventing them from proteases and by increasing their hydrodynamic size (Harris and Chess, 2003; Jevsevar et al., 2010). PEG-G-CSF (pegfilgrastim) is PEGylated colony stimulating factor which has been approved for the treatment of neutropenia (Crawford, 2003). Two PEGylated cytokines IFN-α-2a (Pegasys®) and IFN-α-2b (Pegintron®) are employed for treating chronic hepatitis C (Perry and Jarvis, 2001; Bukowski et al., 2002).

Further, properties of cytokine therapeutics can also be improved by fusion with other proteins. Cytokines fused with Fc, transferrin, or albumin have been shown to possess an increased half-life. The activity of cytokines has been incremented by fusing cytokines with cytokine agonists. Enhancement in the cytotoxicity of cytokine has been observed in complexes formed by cytokines and bacterial lysins or lytic Fc. Fusion of cytokine with specific antibody also aids in mediating an appropriate delivery of cytokines (Lombardi et al., 2013). Subramanian et al. reported an increase in half-life of IFN-2a when complexed with the polypeptide human albumin. The fusion complex albinterferon alpha-2b (alb-IFN) proposed for the treatment of chronic hepatitis C is under clinical trials (Subramanian et al., 2007).

Bai et al. engineered a complex G-CSF-Tf containing the granulocyte colony-stimulating factor (G-CSF) in fusion with transferrin (Tf). Subcutaneous administration of this complex in mice resulted in an effective increase in absolute neutrophil counts as compared to the condition when the G-CSF was administered alone. The complex also resulted in myelopoietic effect that lastest for three days. This implies that fusion of recombinant proteins with transferrin is a promising technique for the future protein drugs (Bai et al., 2005). The researchers further improved the oral efficacy of the fusion protein by adding the linker peptide (helix linker H4-2) between the G-CSF and Tf domains. Addition of such a linker significantly improved their in vivo myelopoietic effect (Bai and Shen, 2006). A fusion conjugate of IL-2 and EDT (enzymatically active domain of diphtheria toxin) was approved as an engineered protein therapeutic as it is successful in treating cutaneous-T-cell lymphoma (Foss, 2006).

Therapeutic cytokines with improved pharmacological properties are also being engineered by simple mutagenesis. Addition, removal, or substitution of specific amino acids exert large effects on the activities of proteins. Proleukin® is a recombinant human IL-2 that is administered to treat metastatic melanoma and renal cell carcinoma (RCM). The stability of IL-2 was enhanced by the removal of extra surface cysteines by substituting them with serines. Such a focused mutagenesis by analyzing the 3D structure of IL-2 prevented the non-specific aggregation of IL-2 (Marshall et al., 2003). Bishop et al. reengineered granulocyte colony-stimulating factor to enhance its thermo stability. They increased the propensity of residues to form antiparallel α-helical bundles via a rational structural approach to enhance its thermal stability, while retaining the biological activity of the protein (Bishop et al., 2001).

5.3.4 Hormones

Patients suffer from many diseases as a result of deficiency in the endogenous proteins. Protein hormones such as insulin and human growth hormone belong to this category of endogenous proteins. Insulin deficiency causes type I diabetes. Hence, Insulin is given exogenously to the patients suffering from diabetes. Insulin is the first human recombinant protein therapeutic that is also known as Humulin®. It was prepared by Eh Lilly at Genentech and was approved by the US FDA in 1982 (Goeddel et al., 1979). Humulin® was used to replace the natural insulin which is absent in diabetic patients. With advancements in proteins engineering, several of insulin analogs have been designed that are classified into two categories, namely (1) fast-acting analog formulations for bolus injection before meals, and (2) basal analog formulations intended for once-a-day injections (Berenson et al., 2011). Insulin lispro, insulin aspart, and insulin glulisine are the fast-acting insulin analogs that are safe and effective in multi-injection regimens.

Insulin lispro was engineered in which the C-terminal lysine and proline residues of insulin B-chain were swapped. This mutation completely knocked out the insulin hexamer formation without affecting its receptor binding efficiency, thus allowing the presence of higher concentration of insulin monomers in postprandial injections. Insulin lispro was first approved in the United States in 1996 (Thrasher et al., 2015b). Insulin aspart is the fast-acting insulin analog that gets absorbed quickly in the blood. It starts its action in several minutes and hence can be taken by the patients just before the meal (Muchmore, 2018). Insulin glulisine, marketed as Apidra, is a rapidly acting insulin analogue in which the amino acid aspargine at third position in the B-chain was swapped with lysine and the lysine residue at 29th position in B-chain was replaced by glutamic acid. Its receptor binding properties are same as the human insulin but it appears faster in the blood. Insulin glulisine also follows flexible administration, that is, it can be administered immediately before or after the meal (Garnock-Jones and Plosker, 2009). Insulin glargine is marketed as Lantus, is most widely used as long-acting insulin analogue. It contains two basic arginine residues at the end of the B-chain that shifts the isoelectric point to neutrality, hence resulting in precipitation when injected subcutaneously in an unbuffered pH-4 formulation form. The precipitation results in its prolonged actions (Goykhman et al., 2009). Insulin detemir sold by Novo-Nordisk as Levemir, contains a prosthetic fatty acyl group on Lys 29 of B-chain, which causes its binding to serum albumin and results in circulating depot. This analog also exhibits more stable hexameric structure in comparison to natural insulin (Vigneri et al., 2010). Levemir is administered twice a day, as its actions are not as prolonged as that of Lantus.

Another well-known hormone replacement therapy is growth hormone therapy. Recombinant human growth hormone is being used for the treatment of dwarfism in children and adults, and also used as replacement therapy in the children deficient in growth hormone. Growth hormone is administered subcutaneously at higher doses to improve the height outcome. GH is also being used in other disorders including Turner Syndrome, chronic renal failure, idiopathic short stature, intrauterine growth restriction, and idiopathic short stature, etc. (Bajpai and Menon, 2005). Cox et al. engineered long-acting mono-PEGylated growth hormone analogs which exhibit complete bioactivity, enhanced potency, and prolonged half-life in vivo. This analog was developed by substituting the cysteine residue at third position with threonine, followed by the chemical modification of cysteine residue by the PEG chains. PEG-T3C showed 8-fold longer half-life than GH when administered subcutaneously in rats (Cox et al., 2007). The authors further focused on increasing its stability in solution to make it amenable to other delivery systems. They applied the computational approach to increase the thermostability of GH. The engineered GH contained 6–10 mutations, showed improved pharmacokinetics, stability, and long half-life (Filikov et al., 2002). Such a strategy can also be applied to other protein therapeutics to improve their pharmacokinetic profiles.

5.3.5 G-Protein Coupled Receptor (GPCR) Targeting Antibodies

GPCRs are the signaling molecules which transfer the information from external environment to the intercellular environment via effector molecules. They interact with a number of molecules including nucleotides, peptides, lipids, proteins, small organic molecules, photons, and ions and participate in numerous biological processes such as proliferation and migration of cells, homeostasis, etc., Binding of GPCRs with external ligands induces the conformational change in the transmembrane region of GPCR, thereby resulting in the change in the G-proteins, and arrestins. Because of its multifunctional roles in cell, any deterioration in its conformation, expression, or activity results in numerous diseases like cancers, infection, inflammation, and other cardiac diseases (Rosenbaum et al., 2009). Hence, GPCR is an important target for many therapeutic drugs. Drugs specifically designed for targeting GPCRs are the analogs of natural ligands that will act either as agonist or antagonist (Jo and Jung, 2016; Nieto and McDonald, 2018). Antibodies are recognized as the most potent molecules as compared to any other small molecule drugs to regulate the actions of GPCRs. Several obstacles are being faced during the development of anti-GPCR antibodies which includes: difficulty in preparing functionally active GPCR antigens, highly variable GPCR extracellular region, lack of efficient antibody screening strategies, and limited exposure of GPCR epitope (Jo and Jung, 2016). A glycoengineered antibody mogamulizumab (Poteligeo) targeting CCR4 was approved in Japan against relapsed or refractory CCR4 + adult T-cell leukemia–lymphoma (ATL) (Beck and Reichert, 2012). Various other GPCR antibodies are in clinical trials that include AMG-334, AMG-477 targeting CGRPR and GCGR, respectively. A summary of anti-GPCR antibodies that are under different phases of clinical trials is provided elsewhere by Jo and Jung (Jo and Jung, 2016).

5.3.6 Cardiovascular Therapeutics

Numerous strategies are being followed for heart repair, including cell-based therapies, gene therapy, and silencing methods. Recently, protein therapeutics have spread their roots as cardiovascular therapeutics. They hold various advantages over cell therapies for cardiac repair. They are cheaper and more target oriented, and also have high scalability. Protein therapeutics are majorly being designed to induce cardiac regeneration. They exhibit several discrepancies such as poor bioavailability, unwanted bio-distribution, and pharmacokinetics profiles. Hence, the continuous efforts of protein engineering are being exploited to design protein-based cardiovascular therapeutics with desirable characteristics (Jay and Lee, 2013). Several proteins such as tumor necrosis factor-α, natriuretic peptides, fibroblast growth factor (FGF), insulin like growth factor-1 (IGF-1), vascular endothelial growth factor (VEGF), stromal cell–derived factor-1a, granulocyte colony-stimulating factor (G-CSF), Neuregulin-1β (NRG), and erythropoietin can serve as potential candidates for cardiovascular regeneration after myocardial infarction (Segers and Lee, 2010).

5.3.7 Coagulation Factors

Blood clotting factors are being engineered to overcome the bleeding disorders such as hemophilia. Recombinant Factor VIII (rFVIII) was given first in March 1987 to the hemophilic patient and after five years it became available for clinical usage (White et al., 1989; Pipe, 2005). Recombinant Factor IX (rFIX) and recombinant factor VIIa (rVIIa) were later developed for the treatment of hemophilia A and B. Few of the clotting factors have been already approved, and some are under clinical trials (Pipe, 2005; Key and Negrier, 2007). Recombinant activated protein C (APC) has been developed and was reported to reduce the mortality in patients suffering from severe sepsis. This recombinant protein also showed high bleeding risk due to its high anticoagulant activity; hence, efforts were made to reduce its anticoagulant activity to overcome bleeding risk. Two variants of APC were designed by site directed mutagenesis. In one variant, the two arginine residues at 229th and 230th position were mutated to alanine and in other variant three adjacent lysines at positions 191, 192, and 193 were mutated to alanine. Both the variants held back their apoptotic activity and evidenced a decrease in anticoagulant activity. These two APC variants can serve as future engineered therapeutics (Mosnier et al., 2004).

5.4 Protein-Based Scaffolds for Therapeutic Applications

Novel next-generation protein therapeutics are being designed that are based on protein scaffolds. These scaffolds provide the binding site for the target molecules as provided by antibodies for binding antigens. Protein scaffolds exhibit a specific structural fold. Several factors are kept in mind while choosing scaffolds. The scaffold should be small, highly expressible, soluble, stable, and monomeric. A variety of therapeutics were developed using the basic protein scaffolds by introducing diversity in their extracellular loops that aid in recognizing the target (Gill and Damle, 2006).

Several of the protein therapeutics have already been developed using different protein scaffolds and are under clinical trials. Some of them include: Affibodies® from the synthetic Z domain of staphylococcal protein, Adnectins® from type III fibronectin domains, Avimers® from LDLR-A modules, Anticalins® from lipocalins, engineered Kunitz Domains, and DARPins® from ankyrin repeat proteins. One of the synthetic scaffolds, namely ecallantide (Kalbitor®), is a 60-amino-acid-long protein based on kunitz domain and has been approved as a therapeutic that inhibits plasma kallikrein. It was selected by phage display technology and was grown in *Pichia pastoris* strain of yeast (Zuraw et al., 2010). Some of these scaffolds are described in detail in the following sections.

5.4.1 Monobodies

Monobodies are the protein scaffolds based on Fibronectin type III domain. Fibronectin is a large protein that interacts with the extracellular matrix proteins and regulates cell-cell communication. Structurally, it is a large protein containing two 250 kD subunits connected by disulfide bridges. These subunits contain the repeats of three domains (type I, II, III). The first monobody was engineered using the tenth repeat of type III domain of Fibronectin (FNfn10) against ubiquitin using phage display system. FNfn10 is devoid of disulfide bonds, and also contains 7 β-strands and 3 surface loops that can be used to add diversity to the monobodies. Their structural simplicity makes them compatible with any screening method such as phage display, yeast two-hybrid system, and peptide-ribonucleic acid. They can be expressed in large quantities in bacteria and are highly thermo stable (Koide and Koide, 2007). Such properties of monobodies make them ideal candidates for therapeutic protein scaffolds. Monobodies have been engineered for the numerous proteins including EGFR (Hackel et al., 2012), IL-23 (Tang et al., 2012), Ableson (Abl) kinase SH2 domain (Wojcik et al., 2010), etc.

5.4.2 SH3 Domains/Fynomers

Fynomers are the protein scaffolds designed based on the SH3 domain of Fyn enzyme. Fyn enzyme is a tyrosine kinase (belongs to Src family) containing 5 anti-parallel β-strands connected by two loops (Musacchio et al., 1992; Musacchio et al., 1994). Src domain involves in interaction with many proteins associated with cell signaling, cytoskeleton structures, and membrane receptors. Hence, efforts are being made to add diversity to the fynomers by mutagenesis in the two flexible loops. Fyn SH3 domains are also devoid of cysteine residues, hence do not form disulfide bonds and therefore can be expressed efficiently in bacteria. Additional features of higher stability and their occurrence in monomeric form makes fynomers potent therapeutic scaffolds. Fynomers known as 2C1, specific to human IL-17A were designed. They bind to IL-17A with nano-molar affinity. Continuous efforts are being made to fuse the fynomers to antibodies to make them bi-specific or tri-specific or increase their binding affinity to the target protein. 2C1 was fused to the Fc region of antibody using a four amino acid linker. The fynomer-Fc fusion had shown pico-molar binding affinity to IL-17A. Hence, these fusion constructs of fynomers with 100 fold improvement in inhibition of IL-17A, represent the ideal drug candidates for treating IL-17A mediated diseases (psoriasis, psoriatic arthritis and rheumatoid arthritis, etc.) (Silacci et al., 2014). Bispecific FynomAb COVA208 has been designed that reduced the levels of HER2, HER3, and EGFR and thereby showed strong inhibitory action on the downstream signaling of HER2 pathway (Brack et al., 2014).

5.4.3 Lipocalins/Anticalins

Lipocalins are the family of 160–180 amino acid residue proteins that share high structural similarity. Lipocalins are mainly involved in transfer of molecules such as hormones, vitamins, steroids, and other metabolic products. Structurally, lipocalins contain eight antiparallel β-strands that are packed in cylindrical manner. At the N-terminus, the β-sheet region is extended by a coiled polypeptide whereas at the C-terminus it is packed against C-terminal helix and a stretch of amino acids in the extended conformation. On the broader side of the β-barrel, the β-strands are connected by the loops that are variable, which can be mutagenized for binding target proteins (Schlehuber and Skerra, 2005). The binding pocket of Lipocalin protein from *Pieris brassicae* was reengineered to bind to fluorescein. Fluorescein binding lipocalins were designed by mutating the 16 residues at the binding site formed by the four external loops. The fluorescein binding lipocalin variants were selected by bacterial phage display method. Variants showed the nano-molar binding affinity to fluorescein (Beste et al., 1999). Anticalin with picomolar binding affinity for prostate–specific membrane protein (PSMA) was engineered to diagnose the prostate carcinoma or neovasculature of solid tumors (Barinka et al., 2016). Anticalin specific for binding VEGF-A was engineered using the human tear lipocalin.

The anticalin PRS-050 variant showed pico-molar binding affinity to VEGF-A, and blocked its functional activities, including the mitogenic signaling and proliferation of human endothelial cells. The plasma half-life of anticalin was further extended by PEGylation. As blockage of VEGF-A is advantageous in numerous neovascular diseases, anticalin can serve as future potential therapeutic (Gille et al., 2016).

5.4.4 Nanobodies/VHH Domains

Single variable domain from antibody heavy chain (HCAbs) is known as nanobody. Nanobodies are of nanometer size and also contain mutations that prevent the interaction of heavy chain with light chain. These mutations include V42F, G49E, L50R, and W52G that keep the HCAbs more stable and soluble in aqueous environment. Nanobodies are alternatively known as VHH domains. Structurally, they are small, 15 kDa in size, and comprise two β-sheets, three variable loops, and a disulfide bond. Nanobodies are highly stable both chemically and thermally compared to antibodies. Hence, they can be produced to reside in more harsh conditions (Hassanzadeh-Ghassabeh et al., 2013). Nanobodies can be designed for the targets for which the antibodies are not accessible (Muyldermans, 2013). A Belgian company "Ablynx" is designing nanobodies for treating number of diseases that include: psoriasis, rheumatoid arthritis, inflammatory, neurological disorders, and oncology (Steeland et al., 2016).

5.4.5 DARPins

DARPins are designed ankyrin repeat proteins that are made of ankyrin repeats which are 33 amino acids long. An ankyrin repeat is structurally made of two antiparallel α-helices connected by β-turn. More than 3000 proteins contain these ankyrin repeats, which imply their biological functional importance. Ankyrin repeats are known to interact with number of biological macromolecules, hence, gained importance by protein engineers to mend these ankyrin repeats for the targets of their interest. Ankyrin repeat is 3.5 kDa in size and hence the protein scaffolds designed will also be relatively smaller ranging ~14–21 kDa, depending on total number of ankyrin repeats (Pluckthun, 2015). DARPins have been engineered and selected against maltose binding protein (MBP) and mitogen activated protein kinase (MAPK) using phage display screening strategy (Zahnd et al., 2010). They are also engineered for epithelial cell adhesion molecule (Martin-Killias et al., 2011), HER2 (Zahnd et al., 2010), and VEGF-A (Stahl et al., 2013). Loop DARPins have been engineered in which an extra loop is added between the helical regions to generate excess binding surface area (Schilling et al., 2014). DARPins are also being used as drug conjugates (Simon et al., 2013), and as small interfering RNA transporters (Arcuri et al., 1998).

5.4.6 Affibodies

Affibodies are the small protein fragments obtained from immunoglobulin binding region B-domain (~58 amino acids) of staphylococcal protein A. Structurally, affibodies are folded in a three-helix bundle and it has been reported that they follow the fastest folding kinetics. B-domain is thermally very stable. To increase its chemical stability, some of the residues were mutated, which resulted in the formation of "Z-domain." The resulting Z-domain maintained its affinity for Fc part of antibody but lost its affinity for Fab part of the antibody (Nilsson et al., 1987). Affibody molecules with different specificities were designed using the randomization of 13 amino acids in the first and second helices that constitute the Fc binding surface of Z-domain. Affibody molecules for specific protein targets were then screened using biopanning technique (Nord et al., 1997). Affibody molecules specific to different proteins have been engineered that include IL-8, CD28, transferrin, tumor necrosis factor-a, fibrinogen insulin, gp-120, EGFR, HER2, IgA, IgM, IgE, and human serum albumin. Affibody molecules engineered for all these proteins showed binding affinity ranging from μM to pM range. Affibody molecules due to their small size can also be fused to other molecules and can still maintain their size smaller than the conventional antibody (Nygren, 2008; Lofblom et al., 2010).

Z-doamin affibody specific to Taq DNA polymerase was generated through a combinatorial library of three-helix bundle. To further increase the affinity of affibody, a hierarchical library was generated by stochastic evaluation of 6 amino acids in one of helices that bind to Taq DNA polymerase. Affibody with a binding affinity of 30–50 nM was obtained for Taq DNA polymerase (Gunneriusson et al., 1999). Affibodies particular to proteins that are highly specific for different types of cancers are being developed to detect different malignancies. Affibody specific to HER2 has been engineered that showed binding affinity of 22 pmol/L to HER2 and hence can be used to detect the expression of HER2 in the cells using gamma camera (Orlova et al., 2006). Fluorescently labeled affibodies are being produced to function as biosensors. The triple-labeled affibody molecule has already been shown to detect unlabeled human IgG and IgA (Engfeldt et al., 2005). Affibody molecules are also being exploited in numerous biotechnological, pharmaceutical, and diagnostic applications which are discussed in detail in a review by Löfblom et al. and Nygren (Nygren, 2008; Lofblom et al., 2010).

5.4.7 Avimers

Avimer has derived its name from the fusion of two words "avidity" and "multimer" because they have an ability to bind to multiple epitopes of their target protein (Silverman et al., 2005). Avimers are the protein scaffolds generated using A-domain, a poly peptide chain of ~35 residues long (~4 kDa). A-domain comprises β–α–β secondary structures which are stabilized by three disulfide bonds. Presence of three disulfide bonds aids in holding the A-domain in U-type fold. Human proteins such as VLDL (very-low-density lipoprotein) and HDL (high-density lipoprotein) contain more than 200 A-domains. A-domains contain 12 highly conserved residues that are required for holding its structure. The other 25 amino acids can be varied to create the avimers that are specific for different target proteins. Efforts were made to design the avimers that block the pro-inflammatory cytokine IL-6 (Silverman et al., 2005), and bi-specific avimer that imitates the function of FGF21 (Smith et al., 2013).

5.4.8 Knottins and Cyclotides

Cystine knot mini proteins (Knottins) are 30–50-amino-acid-long small proteins comprising a canonical cysteine knot. Structurally, knottins contain three antiparallel β-strands and are stabilized by three disulfide bonds arranged in a unique fashion. These bonds are formed between cys1 to cys 4, cys2 to cys5, and cys3 to cys6. The disulfide bond between cys3 to cys6 crosses a macrocycle formed by other two disulfide bonds and a backbone peptide bond. Together, these bonds form a cystine knot structure. Such a structure is responsible for higher thermal, chemical, and proteolytic stability of knottins. Cyclotides are a class of cystine knot proteins that undergo head to tail cyclization due to the presence of an extra loop in their structural architecture. Both knottins and cyclotides serve as promising candidates for potential therapeutic applications. External loops of both knottins and cyclotides make them amenable to various amino acid substitutions and also to the addition of various amino acids, rendering them structurally stable (Moore et al., 2012).

Kimura et al. engineered *Ecballium elaterium* trypsin inhibitor (EETI-II) as integrin binding agents. They created a yeast displayed library of EETI-II by substituting 6 residues of trypsin binding loop with 11 amino acids comprising integrin binding motif (R-G-D) and randomized the terminal residues. Engineered knottin peptides showed high binding affinity for αvβ3, αvβ5, and α5β1 integrins. These engineered integrin binding knottin peptides are of great medical importance as they can bind to all three receptors that are co-expressed in tumor cells. Jiang et al. engineered peptides that can bind alpha(V)beta(3) integrin with high affinity and specificity by exploiting agouti related protein (a 4kD cysteine knot peptide) as a scaffold (Jiang et al., 2010). Kimura et al. have also developed knottin peptides that are labeled with both near-infrared fluorescence (NIRF) and positron emission tomography (PET) imaging agents. The resulting dual labeled knottin peptides are being proposed to be used for multimodality imaging and for the detection of deeply located tumors in the body (Kimura et al., 2010).

5.4.9 Kringle Domain

Kringle domains (KD) are 70–80 amino acids long and are mostly present in blood plasma proteins (proteases, cofactors, growth factors, apolipoproteins). Around 39 KDs are found in human proteome. Structurally, KD comprises two antiparallel β-sheets, and three disulfide bonds. The core structure of KD is conserved among all the proteins containing KD domains. Around 40 amino acids constituting these surface exposed loops are highly variable and therefore are involved in interaction with various biomolecules such as other proteins, lipids, or some small molecules (Lee et al., 2010a). Lee et al. generated KD variants that are anatagonists for human TNFα. Additionally, they have also designed KD variants that are agonists of death receptor 4 (DR4) and death receptor 5 (DR5) with antitumor activity (Lee et al., 2010b). Further, Lee et al. generated bi-specific kringle domains that can bind to both DR4 and DR5. Kringle domains provide an advantage over other scaffolds for the development of bispecific scaffolds within a single domain (Lee et al., 2011).

5.4.10 Kunitz Domain

Kunitz domain is found in numerous proteases, including bovine pancreatic trypsin inhibitor (BPTI), human pancreatic secretory trypsin inhibitor (PSTI), and ecotin (periplasmic *E. coli* protease inhibitor). Kunitz domains are also involved as ion channel blockers as they mainly inhibit serine proteases. Structurally, kunitz domains are 60 amino acids long that are arranged in form of a mixture of α-helices and β-sheets. The core structure of the Kunitz domain is stabilized by three disulfide bonds that also make the structure compact and protect it from proteases (Ranasinghe and McManus, 2013). Lehmann et al. engineered small protein known as Ecallantide (DX-88) based on Kunitz domain using phage display technology. It acts as an inhibitor of plasma kallikrein that plays a major role in contact cascade to produce bradykinin (Lehmann, 2008). Dennis et al. designed Kunitz domain variants from Alzheimer's amyloid beta-protein precursor inhibitor (APPI), to inhibit the association of human tissue factor-Factor VIIa complex (TF.FVIIa) (Dennis and Lazarus, 1994).

5.5 Concluding Remarks

This chapter presented various protein engineering strategies in detail and their major contributions in engineering or redesigning the proteins for pharmaceutical and therapeutic research.

As the global survey indicated that the pharmaceutical market in the next two decades will be replaced with protein/biomolecule-based therapeutics, these engineering strategies can be envisioned as foremost approaches to formulate next-generation therapeutics against several genetic, infectious, inflammatory, and autoimmune diseases. Although several protein formulations have been designed and approved in market, the design and generation of recombinant protein-based pharmaceuticals is in its infancy. This research area of protein design and engineering will further bloom with contemporary schemes of synthetic biology to produce protein therapeutics with outstanding clinical outcomes.

Acknowledgements

KMP acknowledges the support of DBT-IYBA fellowship–BT/07/IYBA/2013-19, SERB-SB/YS/LS-380/2013 and SERB-EMR/2016/001756, Government of India.

References

Abecassis V., Pompon D., Truan G., *Methods Mol. Biol.*, 231 (2003), 165–173.

Alcalde M., Zumarraga M., Polaina J., Ballesteros A., Plou F. J., *Comb. Chem. High Throughput. Screen.*, 9 (2006), 719–727.

Alley S. C., Okeley N. M., Senter P. D., *Curr. Opin. Chem. Biol*, 14 (2010), 529–537.

Arcuri F., Sestini S., Paulesu L., Bracci L., Carducci A., Manzoni F., Cardone C., Cintorino M., *Mol. Cell Endocrinol.*, 141 (1998), 13–20.

Arnold, F. H., Design by directed evolution. *Acc. Chem. Res.*, 31, 125–131. 1998.

Ashina M., Dodick D., Goadsby P. J., Reuter U., Silberstein S., Zhang F., Gage J. R., Cheng S., Mikol D. D., Lenz R. A., *Neurology*, 89 (2017), 1237–1243.

Bai Y., Ann D. K., Shen W. C., *Proc. Natl. Acad. Sci. U. S. A.*, 102 (2005), 7292–7296.

Bai Y., Shen W. C., *Pharm. Res.*, 23 (2006), 2116–2121.

Bajpai A., Menon P. S., *Indian J. Pediatr.*, 72 (2005), 139–144.

Barinka C., Ptacek J., Richter A., Novakova Z., Morath V., Skerra A., *Protein Eng. Des. Sel.*, 29 (2016), 105–115.

Beck A., Reichert J. M., *MAbs.*, 4 (2012), 419–425.

Beck A., Reichert J. M., *MAbs.*, 6 (2014), 15–17.

Berenson D. F., Weiss A. R., Wan Z. L., Weiss M. A., *Ann. N. Y. Acad. Sci.*, 1243 (2011), E40–E54.

Beste G., Schmidt F. S., Stibora T., Skerra A., *Proc. Natl. Acad. Sci. U. S. A.*, 96 (1999), 1898–1903.

Bikard D., Julie-Galau S., Cambray G., Mazel D., *Nucleic Acids Res.*, 38 (2010), e153.

Bishop B., Koay D. C., Sartorelli A. C., Regan L., *J. Biol. Chem.*, 276 (2001), 33465–33470.

Bloom J. D., Arnold F. H., *Proc. Natl. Acad. Sci. U. S. A.*, 106 Suppl 1 (2009), 9995–10000.

Bonneau R., Baker D., *Annu. Rev. Biophys. Biomol. Struct.*, 30 (2001), 173–189.

Borgwardt L., Feldt-Rasmussen U., Rasmussen A. K., Ballegaard M., Meldgaard L. A., *Clin. Genet.*, 83 (2013), 432–438.

Brack S., et al., *Mol. Cancer Ther.*, 13 (2014), 2030–2039.

Brannigan J. A. Wilkinson A. *J., Nat. Rev. Mol. Cell Biol.*, 3 (2002), 964–970.

Brown D. M., Kaiser P. K., Michels M., Soubrane G., Heier J. S., Kim R. Y., Sy J. P., Schneider S., *N. Engl. J. Med.*, 355 (2006), 1432–1444.

Brunelli M. J., Atallah A. N., da Silva E. M., *Cochrane. Database. Syst. Rev.*, 3 (2016) CD009806.

Bukowski R. M., Tendler C., Cutler D., Rose E., Laughlin M. M., Statkevich P., *Cancer*, 95 (2002), 389–396.

Burrow T. A., Leslie N. D., *Biologics.*, 2 (2008), 311–320.

Cadwell R. C., Joyce G. F., *PCR Methods Appl.*, 2 (1992), 28–33.

Caravella J. Lugovskoy A., *Curr. Opin. Chem. Biol*, 14 (2010), 520–528.

Casi G., Neri D., *J. Control. Release*, 161 (2012), 422–428.

Chen K., Arnold F. H., *Proc. Natl. Acad. Sci. U. S. A.*, 90 (1993), 5618–5622.

Chen Z., Meyer W., Rappert S., Sun J., Zeng A. P., *Appl. Environ. Microbiol.*, 77 (2011), 4352–4360.

Chen L., Wang F., Wang X., Liu Y. G., *Plant Cell Physiol.*, 54 (2013), 634–642.

Choi Y., Hua C., Sentman C. L., Ackerman M. E., Bailey-Kellogg C., *MAbs.*, 7 (2015), 1045–1057.

Choi S. H., Kim H. S., Lee E. Y., *Biotechnol. Lett.*, 31 (2009), 1617–1624.

Chowdary P., Fosbury E., Riddell A., Mathias M., *J. Blood Med.*, 7 (2016), 187–198.

Cobb R. E., Chao R., Zhao H., *AIChE J.*, 59 (2013a), 1432–1440.

Cobb R. E., Si T., Zhao H., *Curr. Opin. Chem. Biol.*, 16 (2012), 285–291.

Cobb R. E., Sun N., Zhao H., *Methods*, 60 (2013b), 81–90.

Coco W. M., *Methods Mol. Biol.*, 231 (2003), 111–127.

Coco W. M., Encell L. P., Levinson W. E., Crist M. J., Loomis A. K., Licato L. L., Arensdorf J. J., Sica N., Pienkos P. T., Monticello D. J., *Nat. Biotechnol.*, 20 (2002), 1246–1250.

Coleman S., Gilpin D., Kaplan F. T., Houston A., Kaufman G. J., Cohen B. M., Jones N., Tursi J. P., *J. Hand Surg. Am.*, 39 (2014), 57–64.

Cox G. N., Rosendahl M. S., Chlipala E. A., Smith D. J., Carlson S. J., Doherty D. H., *Endocrinology*, 148 (2007), 1590–1597.

Crameri A., Raillard S. A., Bermudez E., Stemmer W. P., *Nature*, 391 (1998), 288–291.

Crawford J., *Semin. Oncol.*, 30 (2003), 24–30.

Czajkowsky D. M., Hu J., Shao Z., Pleass R. J., *EMBO Mol. Med.*, 4 (2012), 1015–1028.

Dagelaite J., Driscoll A. O', Sleator Roy D. *ISRN Biomathematics.*, 2013, (2013), article ID 615630.

Dalby P. A., *Curr. Opin. Struct. Biol.*, 21 (2011), 473–480.

David R., Korenberg M. J., Hunter I. W., *Pharmacogenomics.*, 1 (2000), 445–455.

De Groot A. S. Moise L., *Curr. Opin. Drug Discov. Devel.*, **10** (2007), 332–340.

De Groot A. S. Scott D. W., *Trends Immunol.*, **28** (2007), 482–490.

Dennig A., Shivange A. V., Marienhagen J., Schwaneberg U., *PLoS One.*, 6 (2011), e26222.

Dennis M. S., Lazarus R. A., *J. Biol. Chem.*, 269 (1994), 22137–22144.

Deshler J. O., *Genet. Anal. Tech. Appl.*, 9 (1992), 103–106.

Dinndorf P. A., Gootenberg J., Cohen M. H., Keegan P., Pazdur R., *Oncologist.*, 12 (2007), 991–998.

Dunbrack R. L., Jr., *Curr. Opin. Struct. Biol.*, 12 (2002), 431–440.

Dunn A. L., Ahuja S. P., Mullins E. S., *Haemophilia.* (2018).

Edelheit O., Hanukoglu A., Hanukoglu I., *BMC. Biotechnol.*, 9 (2009), 61.

Engfeldt T., Renberg B., Brumer H., Nygren P. A., Karlstrom A. E., *Chembiochem.*, 6 (2005), 1043–1050.

Engler C., Gruetzner R., Kandzia R., Marillonnet S., *PLoS. One*, 4 (2009) e5553.

Esvelt K. M., Carlson J. C., Liu D. R., *Nature*, 472 (2011), 499–503.

Farkas H., Varga L., *Clin. Cosmet. Investig. Dermatol.*, 4 (2011), 61–68.

Filikov A. V., Hayes R. J., Luo P., Stark D. M., Chan C., Kundu A., Dahiyat B. I., *Protein Sci.*, 11 (2002), 1452–1461.

Finney-Manchester S. P., Maheshri N., *Nucleic Acids Res.*, 41 (2013), e99.

Firnberg E., Ostermeier M., *PLoS. One*, 7 (2012), e52031.

Foss F., *Semin. Oncol.*, 33 (2006), S11–S16.

Frith M. C., Saunders N. F., Kobe B., Bailey T. L., *PLoS. Comput. Biol.*, 4 (2008), e1000071.

Fujii W., Kano K., Sugiura K., Naito K., *PLoS. One*, 8 (2013), e59801.

Fujii R., Kitaoka M., Hayashi K., *Methods Mol. Biol.*, 1179 (2014), 23–29.

Fung H. K., William J. W., and Floudas, C. A., *Ind. Eng. Chem. Res.*, 47 (2008), 993–1001. 2008.

Garnock-Jones K. P., Plosker G. L., *Drugs*, 69 (2009), 1035–1057.

Gaytan P., Contreras-Zambrano C., Ortiz-Alvarado M., Morales-Pablos A., Yanez J., *Nucleic Acids Res.*, 37 (2009), e125.

Gebauer M., Skerra A., *Curr. Opin. Chem. Biol.*, 13 (2009), 245–255.

Gibbs M. D., Nevalainen K. M., Bergquist P. L., *Gene*, 271 (2001), 13–20.

Gill D. S., Damle N. K., *Curr. Opin. Biotechnol.*, 17 (2006), 653–658.

Gille H., Hulsmeyer M., Trentmann S., Matschiner G., Christian H. J., Meyer T., Amirkhosravi A., Audoly L. P., Hohlbaum A. M., Skerra A., *Angiogenesis*, 19 (2016), 79–94.

Goeddel D. V., Kleid D. G., Bolivar F., Heyneker H. L., Yansura D. G., Crea R., Hirose T., Kraszewski A., Itakura K., Riggs A. D., *Proc. Natl. Acad. Sci. U. S. A.*, 76 (1979), 106–110.

Gonzalez-Perez D., Molina-Espeja P., Garcia-Ruiz E., Alcalde M., *PLoS. One*, 9 (2014), e90919.

Goykhman S., Drincic A., Desmangles J. C., Rendell M., *Expert. Opin. Pharmacother.*, 10 (2009), 705–718.

Gratz A., Jose J., *Anal. Biochem.*, 378 (2008), 171–176.

Greener A., Callahan M., Jerpseth B., *Methods Mol. Biol.*, 57 (1996), 375–385.

Gulati K., Poluri K. M., *Recent Pat. Biotechnol*, 10 (2016), 235–263.

Gunneriusson E., Nord K., Uhlen M., Nygren P., *Protein Eng.*, 12 (1999), 873–878.

Hackel B. J., Neil J. R., White F. M., Wittrup K. D., *Protein Eng. Des. Sel.*, 25 (2012), 47–57.

Haddley K., *Drugs Today (Barc.)*, 48 (2012), 525–532.

Haraoui B., Bykerk V., *Ther. Clin. Risk Manag.*, 3 (2007), 99–105.

Harris J. M., Chess R. B., *Nat. Rev. Drug Discov.*, 2 (2003), 214–221.

Hassanzadeh-Ghassabeh G., Devoogdt N., De P. P., Vincke C., Muyldermans S., *Nanomedicine (Lond)*, 8 (2013), 1013–1026.

He L., Zhu J., *Curr. Opin. Virol.*, 11 (2015), 103–112.

Hidalgo A., Schliessmann A., Molina R., Hermoso J., Bornscheuer U. T., *Protein Eng. Des. Sel.*, 21 (2008), 567–576.

Horner D. S., Pirovano W., Pesole G., *Brief. Bioinform.*, 9 (2008), 46–56.

Hughes D. A., et al., *Am. J. Hematol.*, 90 (2015), 584–591.

Jameson E., Jones S., Wraith J. E., *Cochrane. Database. Syst. Rev.,* (2013), CD009354.

Jay S. M., Lee R. T., *Circ. Res.*, 113 (2013), 933–943.

Jevsevar S., Kunstelj M., Porekar V. G., *Biotechnol. J.*, 5 (2010), 113–128.

Jiang L., Kimura R. H., Miao Z., Silverman A. P., Ren G., Liu H., Li P., Gambhir S. S., Cochran J. R., Cheng Z., *J. Nucl. Med.*, 51 (2010), 251–258.

Jo M., Jung S. T., *Exp. Mol. Med.*, 48 (2016), e207.

Joh N. H., Wang T., Bhate M. P., Acharya R., Wu Y., Grabe M., Hong M., Grigoryan G., DeGrado W. F., *Science*, 346 (2014), 1520–1524.

Johnson-Leger C., Power C. A., Shomade G., Shaw J. P., Proudfoot A. E., Expert. *Opin. Biol Ther.*, 6 (2006), 1–7.

Kazazi-Hyseni F., Beijnen J. H., Schellens J. H., *Oncologist,* 15 (2010), 819–825.

Kefalides P. T., *Ann. Intern. Med.*, 128 (1998), 1053–1055.

Key N. S., Negrier C., *Lancet*, 370 (2007), 439–448.

Khan F. I., Wei D. Q., Gu K. R., Hassan M. I., Tabrez S., *Int. J. Biol. Macromol.*, 85 (2016), 48–62.

Kimchi-Sarfaty C., Schiller T., Hamasaki-Katagiri N., Khan M. A., Yanover C., Sauna Z. E., *Trends Pharmacol. Sci*, 34 (2013), 534–548.

Kimura R. H., Miao Z., Cheng Z., Gambhir S. S., Cochran J. R., *Bioconjug. Chem.*, 21 (2010), 436–444.

Kipnis Y., lus-Gur E., Tawfik D. S., *Protein Eng. Des. Sel.*, 25 (2012), 437–444.

Kitamura K., Kinoshita Y., Narasaki S., Nemoto N., Husimi Y., Nishigaki K., *Protein Eng.*, 15 (2002), 843–853.

Klukowska A., Szczepanski T., Vdovin V., Knaub S., Bichler J., Jansen M., Dzhunova I., Liesner R. J., *Haemophilia*, 24 (2018), 595–603.

Klukowska A., Szczepanski T., Vdovin V., Knaub S., Bichler J., Jansen M., Dzhunova I., Liesner R. J., *Haemophilia.* (2018).

Koide A., Koide S., *Methods Mol. Biol.*, 352 (2007), 95–109.

Kolkman J. A., Stemmer W. P., *Nat. Biotechnol.*, 19 (2001), 423–428.

Kubota T., Niwa R., Satoh M., Akinaga S., Shitara K., Hanai N., *Cancer Sci.*, 100 (2009), 1566–1572.

Kuhlman B., O'Neill J. W., Kim D. E., Zhang K. Y., Baker D., *J. Mol. Biol.*, 315 (2002), 471–477.

Labrou N. E., *Curr. Protein Pept. Sci.*, 11 (2010), 91–100.

Lai Y. P., Huang J., Wang L. F., Li J., Wu Z. R., *Biotechnol. Bioeng.*, 86 (2004), 622–627.

Lambert J. M., Chari R. V., *J. Med. Chem.*, 57 (2014), 6949–6964.

Landgraf W., Sandow J., *Eur. Endocrinol.*, 12 (2016), 12–17.

Lane M. D., Seelig B., *Curr. Opin. Chem. Biol.*, 22 (2014), 129–136.

Lassila J. K., Privett H. K., Allen B. D., Mayo S. L., *Proc. Natl. Acad. Sci. U. S. A.*, 103 (2006), 16710–16715.

Leader B., Baca Q. J., Golan D. E., *Nat. Rev. Drug Discov.*, 7 (2008), 21–39.

Lee C. H., Park K. J., Kim S. J., Kwon O., Jeong K. J., Kim A., Kim Y. S., *J. Mol. Biol.*, 411 (2011), 201–219.

Lee C. H., Park K. J., Sung E. S., Kim A., Choi J. D., Kim J. S., Kim S. H., Kwon M. H., Kim Y. S., *Proc. Natl. Acad. Sci. U. S. A.*, 107 (2010a), 9567–9571.

Lee C. H., Park K. J., Sung E. S., Kim A., Choi J. D., Kim J. S., Kim S. H., Kwon M. H., Kim Y. S., *Proc. Natl. Acad. Sci. U. S. A.*, 107 (2010b), 9567–9571.

Lehmann A., *Expert. Opin. Biol. Ther.*, 8 (2008), 1187–1199.

Liang H., Chen H., Fan K., Wei P., Guo X., Jin C., Zeng C., Tang C., Lai L., *Angew. Chem. Int. Ed. Engl.*, 48 (2009), 3301–3303.

Linke R., Klein A., Seimetz D., *MAbs*, 2 (2010), 129–136.

Lippow S. M., Wittrup K. D., Tidor B., *Nat. Biotechnol.*, 25 (2007), 1171–1176.

Liu P., Hong Y., Lin Y., Fu X., Lin L., Li C., Meng X., Liu Z., *Anal. Biochem.*, 388 (2009), 356–358.

Lofblom J., Feldwisch J., Tolmachev V., Carlsson J., Stahl S., Frejd F. Y., *FEBS Lett.*, 584 (2010), 2670–2680.

Lombardi R. V., Roome B., Christ D., *Antibodies*, 2 (2013), 426–451.

Lovell S. C., Robertson D. L., *Mol. Biol. Evol.*, 27 (2010), 2567–2575.

Lutz S., *Curr. Opin. Biotechnol.*, 21 (2010), 734–743.

Mangia A., et al., *Ann. Gastroenterol.*, 30 (2017), 327–343.

Marshall S. A., Lazar G. A., Chirino A. J., Desjarlais J. R., *Drug Discov. Today*, 8 (2003), 212–221.

Martin-Killias P., Stefan W., Rothschild S., Pluckthun A., Zangemeister-Wittke U., *Clin. Cancer Res.*, 17 (2011), 100–110.

Marvin J. S., Zhu Z., *Acta Pharmacol. Sin.*, 26 (2005), 649–658.

Matsumura I., Rowe L. A., *Biomol. Eng.*, 22 (2005), 73–79.

Melian E. B., Plosker G. L., *Drugs*, 61 (2001), 1661–1691.

Mendes J., Guerois R., Serrano L., *Curr. Opin. Struct. Biol.*, 12 (2002), 441–446.

Miller D. H., Khan O. A., Sheremata W. A., Blumhardt L. D., Rice G. P., Libonati M. A., Willmer-Hulme A. J., Dalton C. M., Miszkiel K. A., O'Connor P. W., *N. Engl. J. Med.*, 348 (2003), 15–23.

Mohan U., Banerjee U. C., *Chembiochem.*, 9 (2008), 2238–2243.

Molineux G., *Curr. Pharm. Des.*, 10 (2004), 1235–1244.

Moore S. J., Leung C. L., Cochran J. R., *Drug Discov. Today Technol.*, 9 (2012), e1–e70.

Moore J. C., Arnold F. H., *Nat. Biotechnol.*, 14 (1996), 458–467.

Mosnier L. O., Gale A. J., Yegneswaran S., Griffin J. H., *Blood*, 104 (2004), 1740–1744.

Mross K., Richly H., Fischer R., Scharr D., Buchert M., Stern A., Gille H., Audoly L. P., Scheulen M. E., *PLoS. One*, 8 (2013), e83232.

Muchmore D. B., *J. Diabetes Sci. Technol.*, 12 (2018), 152–154.

Murakami H., Hohsaka T., Sisido M., *Nat. Biotechnol.*, 20 (2002), 76–81.

Murzin A. G., Lesk A. M., Chothia C., *J. Mol. Biol.*, 236 (1994), 1369–1381.

Musacchio A., Noble M., Pauptit R., Wierenga R., Saraste M., *Nature*, 359 (1992), 851–855.

Musacchio A., Wilmanns M., Saraste M., *Prog. Biophys. Mol. Biol.*, 61 (1994), 283–297.

Muyldermans S., *Annu. Rev. Biochem.*, 82 (2013), 775–797.

Nakajima K., Oshida H., Muneyuki T., Kakei M., *Core. Evid.*, 7 (2012), 77–91.

Ness J. E., Kim S., Gottman A., Pak R., Krebber A., Borchert T. V., Govindarajan S., Mundorff E. C., Minshull J., *Nat. Biotechnol.*, 20 (2002), 1251–1255.

Nieto G. A., McDonald P. H., *Cell Signal.*, 41 (2018), 65–74.

Nilsson B., Moks T., Jansson B., Abrahmsen L., Elmblad A., Holmgren E., Henrichson C., Jones T. A., Uhlen M., *Protein Eng.*, 1 (1987), 107–113.

Nord K., Gunneriusson E., Ringdahl J., Stahl S., Uhlen M., Nygren P. A., *Nat. Biotechnol.*, 15 (1997), 772–777.

Nygren P. A., *FEBS J.*, 275 (2008), 2668–2676.

Offredi F., et al., *J. Mol. Biol.*, 325 (2003), 163–174.

Orlova A., et al., *Cancer Res.*, 66 (2006), 4339–4348.

Ostermeier M., Nixon A. E., Benkovic S. J., *Bioorg. Med. Chem.*, 7 (1999), 2139–2144.

Ott P. A., et al., *Invest. New Drugs*, 31 (2013), 425–434.

Owen J., Punt J., Stranford S., *Kuby Immunology*. Freeman W. H., 7th ed, 2012.

Pantazes R. J., Saraf M. C., Maranas C. D., *Protein Eng. Des. Sel.*, 20 (2007), 361–373.

Pasut G., Sergi M., Veronese F. M., *Adv. Drug Deliv. Rev.*, 60 (2008a), 69–78.

Pasut G., Sergi M., Veronese F. M., *Adv. Drug Deliv. Rev.*, 60 (2008b), 69–78.

Patrick W. M., Firth A. E., Blackburn J. M., *Protein Eng.*, 16 (2003), 451–457.

Patwardhan R. P., Lee C., Litvin O., Young D. L., Pe'er D., Shendure J., *Nat. Biotechnol.*, 27 (2009), 1173–1175.

Pei J., *Curr. Opin. Struct. Biol.*, 18 (2008), 382–386.

Perry C. M., Jarvis B., *Drugs*, 61 (2001), 2263–2288.

Pipe S. W., *J. Thromb. Haemost.*, 3 (2005), 1692–1701.

Pluckthun A., *Annu. Rev. Pharmacol. Toxicol.*, 55 (2015), 489–511.

Poluri K. M., Gulati K., *Protein Engineering Techniques Gateways to Synthetic Protein Universe*. Springer Brief, ISBN 978-981-10-2732-1. (2017).

Qi X., Guo Q., Wei Y., Xu H., Huang R., *Biotechnol. Lett.*, 34 (2012), 339–346.

Ramundo J., Gray M., *J. Wound. Ostomy. Continence. Nurs.*, 35 (2008), 273–280.

Ramundo J., Gray M., *J. Wound. Ostomy. Continence. Nurs.*, 36 (2009), S4–S11.

Ranasinghe S., McManus D. P., *Dev. Comp. Immunol.*, 39 (2013), 219–227.

Reddy D. V., Sreedevi P., Arun G., Rao P. S., *Indian J. Endocrinol. Metab.*, 21 (2017), 378–381.

Reidhaar-Olson J. F., Sauer R. T., *Science*, 241 (1988), 53–57.

Rizzari C., Conter V., Stary J., Colombini A., Moericke A., Schrappe M., *Curr. Opin. Oncol.*, 25 Suppl 1 (2013), S1–S9.

Rosenbaum D. M., Rasmussen S. G., Kobilka B. K., *Nature,* 459 (2009), 356–363.

Rutgeerts P., Van A. G., Vermeire S., *Aliment. Pharmacol. Ther.,* 23 (2006), 451–463.

Samish I., MacDermaid C. M., Perez-Aguilar J. M., Saven J. G., *Annu. Rev. Phys. Chem.,* 62 (2011), 129–149.

Saraf M. C., Gupta A., Maranas C. D., *Proteins,* 60 (2005), 769–777.

Saven J. G., *Curr. Opin. Chem. Biol.,* 15 (2011), 452–457.

Schilling J., Schoppe J., Pluckthun A., *J. Mol. Biol.,* 426 (2014), 691–721.

Schlehuber S., Skerra A., *Drug Discov. Today,* 10 (2005), 23–33.

Scott L. J., *Drugs,* 77 (2017a), 435–445.

Scott L. J., *Drugs,* 77 (2017b), 1865–1879.

Segers V. F., Lee R. T., *J. Cardiovasc. Transl. Res.,* 3 (2010), 469–477.

Shao Z., Zhao H., Giver L., Arnold F. H., *Nucleic Acids Res.,* 26 (1998), 681–683.

Shozen N., Watanabe T., Hohsaka T., *J. Biosci. Bioeng.,* 113 (2012), 704–709.

Sieber V., Martinez C. A., Arnold F. H., *Nat. Biotechnol.,* 19 (2001), 456–460.

Silacci M., Baenziger-Tobler N., Lembke W., Zha W., Batey S., Bertschinger J., Grabulovski D., *J. Biol. Chem.,* 289 (2014), 14392–14398.

Silverman J., et al., *Nat. Biotechnol.,* 23 (2005), 1556–1561.

Simon M., Frey R., Zangemeister-Wittke U., Pluckthun A., *Bioconjug. Chem.,* 24 (2013), 1955–1966.

Smith R., et al., *PLoS. One,* 8 (2013), e61432.

Socolich M., Lockless S. W., Russ W. P., Lee H., Gardner K. H., Ranganathan R., *Nature,* 437 (2005), 512–518.

Somaraju U. R., Solis-Moya A., *Cochrane. Database. Syst. Rev.* (2014), CD008227.

Stahl A., et al., *Angiogenesis.,* 16 (2013), 101–111.

Steeland S., Vandenbroucke R. E., Libert C., *Drug Discov. Today,* 21 (2016), 1076–1113.

Steipe B., Schiller B., Pluckthun A., Steinbacher S., *J. Mol. Biol.,* 240 (1994), 188–192.

Stemmer W. P., *Proc. Natl. Acad. Sci. U. S. A.,* 91 (1994a), 10747–10751.

Stemmer W. P., *Nature,* 370 (1994b), 389–391.

Subramanian G. M., Fiscella M., Lamouse-Smith A., Zeuzem S., McHutchison J. G., *Nat. Biotechnol.,* 25 (2007), 1411–1419.

Subramaniam J. M., Whiteside G., McKeage K., Croxtall J. C., *Drugs*, 72 (2012), 1293–1298.

Sullivan B. J., Durani V., Magliery T. J., *J. Mol. Biol.*, 413 (2011), 195–208.

Sumant O., Shaikh S., Protein therapeutics market overview, *Allied Market Research* (2017) (https://www.alliedmarketresearch.com/protein-therapeutics-market).

Summa C. M., Rosenblatt M. M., Hong J. K., Lear J. D., DeGrado W. F., *J. Mol. Biol.*, 321 (2002), 923–938.

Tang C., Chen S., Qian H., Huang W., *Immunology*, 135 (2012), 112–124.

Tee K. L., Wong T. S., *Biotechnol. Adv.*, 31 (2013), 1707–1721.

Teicher B. A., Doroshow J. H., *N. Engl. J. Med.*, 367 (2012), 1847–1848.

Thompson M. A., *Curr. Opin. HIV. AIDS* 13 (2018), 346–353.

Thomson A. H., *J. R. Soc. Med.*, 88 Suppl 25 (1995), 24–29.

Thrasher J., Bhargava A., Rees T. M., Wang T., Guzman C. B., Glass L. C., *Endocr. Pract.*, 21 (2015a), 247–257.

Thrasher J., Bhargava A., Rees T. M., Wang T., Guzman C. B., Glass L. C., *Endocr. Pract.*, 21 (2015b), 247–257.

Tiede A., Klamroth R., Oldenburg J., *Hamostaseologie.*, 35 (2015), 364–371.

Tindall K. R., Kunkel T. A., *Biochemistry*, 27 (1988), 6008–6013.

Tiwari M. K., Singh R., Singh R. K., Kim I. W., Lee J. K., *Comput. Struct. Biotechnol. J.*, 2 (2012), e201209002.

Tobin P. H., Richards D. H., Callender R. A., Wilson C. J., *Curr. Drug Metab.*, 15 (2014), 743–756.

Torres T., Romanelli M., Chiricozzi A., *Expert. Opin. Investig. Drugs*, 25 (2016), 751–754.

Tseng W. C., Lin J. W., Wei T. Y., Fang T. Y., *Anal. Biochem.*, 375 (2008), 376–378.

Tsuji T., Onimaru M., Yanagawa H., *Nucleic Acids Res.*, 29 (2001), E97.

Turner M. D., Nedjai B., Hurst T., Pennington D. J., *Biochim. Biophys. Acta*, 1843 (2014), 2563–2582.

Vellard M., *Curr. Opin. Biotechnol.*, 14 (2003), 444–450.

Vigneri R., Squatrito S., Sciacca L., *Acta Diabetol.*, 47 (2010), 271–278.

Vilcek J., Feldmann M., *Trends Pharmacol. Sci.*, 25 (2004), 201–209.

Voet D., and Voet J. G., *Biochemistry*. John Wiley & Sons Inc, Hoboken NJ. 4th ed. (2011).

Vogel C. L., et al., *J Clin. Oncol.*, 20 (2002), 719–726.

Wang X., Mathieu M., Brezski R. J., *Protein Cell*, 9 (2018), 63–73.

Wang Q., Wu H., Wang A., Du P., Pei X., Li H., Yin X., Huang L., Xiong X., *J. Biol. Chem.*, 285 (2010), 41509–41516.

Watanabe K., Ohkuri T., Yokobori S., Yamagishi A., *J. Mol. Biol.*, 355 (2006), 664–674.

White G. C., McMillan C. W., Kingdon H. S., Shoemaker C. B., *N. Engl. J. Med.*, 320 (1989), 166–170.

Wojcik J., Hantschel O., Grebien F., Kaupe I., Bennett K. L., Barkinge J., Jones R. B., Koide A., Superti-Furga G., Koide S., *Nat. Struct. Mol. Biol.*, 17 (2010), 519–527.

Wong S. F., *Clin. Ther.*, 27 (2005), 684–694.

Wong T. S., Roccatano D., Loakes D., Tee K. L., Schenk A., Hauer B., Schwaneberg U., *Biotechnol. J.*, 3 (2008), 74–82.

Woolfson D. N., Bartlett G. J., Burton A. J., Heal J. W., Niitsu A., Thomson A. R., Wood C. W., *Curr. Opin. Struct. Biol.*, 33 (2015), 16–26.

Xiang Z., *Curr. Protein Pept. Sci.*, 7 (2006), 217–227.

Yip K. Y., Patel P., Kim P. M., Engelman D. M., McDermott D., Gerstein M., *Bioinformatics.*, 24 (2008), 290–292.

You L., Arnold F. H., *Protein Eng.*, 9 (1996), 77–83.

Zahnd C., et al., *Cancer Res.*, 70 (2010), 1595–1605.

Zhang Y., *Proteins*, 69 Suppl 8 (2007), 108–117.

Zhao H., Giver L., Shao Z., Affholter J. A., Arnold F. H., *Nat. Biotechnol.*, 16 (1998), 258–261

Zuraw B., Yasothan U., Kirkpatrick P., *Nat. Rev. Drug Discov.*, 9 (2010), 189–190.

Chapter 6

Synthesis of Bioactive Peptides for Pharmaceutical Applications

Jaison Jeevanandam,[a] Ashish Kumar Solanki,[b] Shailza Sharma,[c] Prabir Kumar Kulabhusan,[d] Sapna Pahil,[e] and Michael K. Danquah[f]

[a]*Department of Chemical Engineering, Curtin University CDT 250 Miri, Sarawak 98009, Malaysia*
[b]*Department of Medicine, Medical University of South Carolina, Charleston 29208, USA*
[c]*Department of Rheumatology and Immunology, Medical University of South Carolina, Charleston 29208, USA*
[d]*Department of Chemistry and Biomolecular Science, University of Ottawa, Ontario k1N6N5, Canada*
[e]*Department of Medical Microbiology, Postgraduate Institute of Medical Education and Research, Sector 12, Chandigarh 160012, India*
[f]*Chemical Engineering Department, University of Tennessee, Chattanooga, TN 37403, USA*

michael-danquah@utc.edu

6.1 Introduction

The short chains of amino acid monomers are grouped together with peptide bond linkages to form a beneficial biological entity

Pharmaceutical Biocatalysis: Fundamentals, Enzyme Inhibitors, and Enzymes in Health and Diseases
Edited by Peter Grunwald

ISBN 978-981-4800-61-7 (Hardcover), 978-0-429-29503-4 (eBook)
www.jennystanford.com

called peptide. These peptides are classified into dipeptide, which is the shortest peptide, tripeptide, and polypeptide based on the number of amino acids required for peptide formation. Peptides contain less than 50 amino acids and are different from proteins which contain functional polypeptides bound with ligands or genetic material. Also, peptides are highly beneficial in molecular biology as they allow the formation of antibiotic peptides without the requirement of purified protein (Muheem et al., 2016). Recently, peptides are used in mass spectroscopy for the identification of protein structure and function, by analyzing sequences and masses of peptide (Aebersold and Mann, 2016). Moreover, inhibitory peptides are utilized to inhibit the proteins related to diseases and are broadly utilized as a curative agent of deadly diseases (Risitano et al., 2014). The synthesis procedure of peptides determines their properties and thus, they are classified based on the method of synthesis used for peptide formation. Bioactivity is an important property of peptides and thus bioactive peptides are widely used in biomedical pharmaceutical applications. Ribosomal, non-ribosomal, peptones, milk-based and peptide fragments are the sub-classes of bioactive peptides that possess enormous applicational importance (Korhonen et al., 2006).

Recently, bioactive peptides of animals, plants, and microorganism origins are significantly utilized in the treatment and prevention of cardiovascular diseases, cancer and infections (Cicero et al., 2017). Likewise, antimicrobial peptides with enhanced bioactivity are used to inhibit the growth of harmful microbes and prevent the widespread of their associated diseases (Haney et al., 2017). Hence, researchers are utilizing bioactive peptides as an alternate to conventional pharmaceutical entities to develop peptide-based treatment approaches. Thus, the current chapter is an overview of current synthesis methods that are used to synthesize bioactive peptides as they are the determining factor that elevates their bioactivity. Further, the procedures employed for the refinement, characterization and formulation of bioactive peptides in recent times and their pharmaceutical applications are also presented. Additionally, the existing challenges and the bioactive peptides in the future of pharmaceutical industry are also discussed.

6.2 Synthesis of Bioactive Peptides

The synthesis of bioactive peptides is highly significant in determining the biological properties of these peptides. The synthesis procedures that are conventionally and recently available for peptide production are shown in Fig. 6.1.

6.2.1 Chemical Synthesis

The chemical-based fabrication approach is the conventional approach to produce synthetic peptides that have been developed over years with the help of efficient coupling reagents.

Basic principle: Chemical synthesis of peptides involves the mechanism of protection and activation. The activation of the carboxylic acid (RCOOH) by chemical reagents in the amino acid generates active acyl group (R-CO-) that forms strong peptide bonds, followed by nucleophilic α-amino group attack (H_2N-R). The protection of the sensitive functional clusters that are not included in the formation of peptide bond (Machado et al., 2004) are envisioned to enhance chemical selectivity that are essential for the precise sequence of peptide synthesis (David and Luis, 2001). There are three methods of chemical-based peptide synthesis and they are solution, solid phase, and sequential fragment condensation.

6.2.1.1 Synthesis in solution

This is the classical convergent fragment condensation technique used to produce small peptides in solution. In convergent synthesis, peptides are sequentially synthesized individually and then linked in solution. The main advantage of this technique is the specific discrimination between the reactants and the product. Therefore, the intermediate short peptides formed in solution can be easily purified after every synthesis step and can be recombined to generate large peptides after de-protection. Other advantages of this method are that it takes less time and is relatively less expensive. However, this method has certain disadvantages such as poor solubility, lack of advanced methods to purify protected peptides and racemization of activated amino acids in peptides preferably at the c-terminal of histidine and cysteine amino acids (Kent, 1988; Perez Espitia et al., 2012).

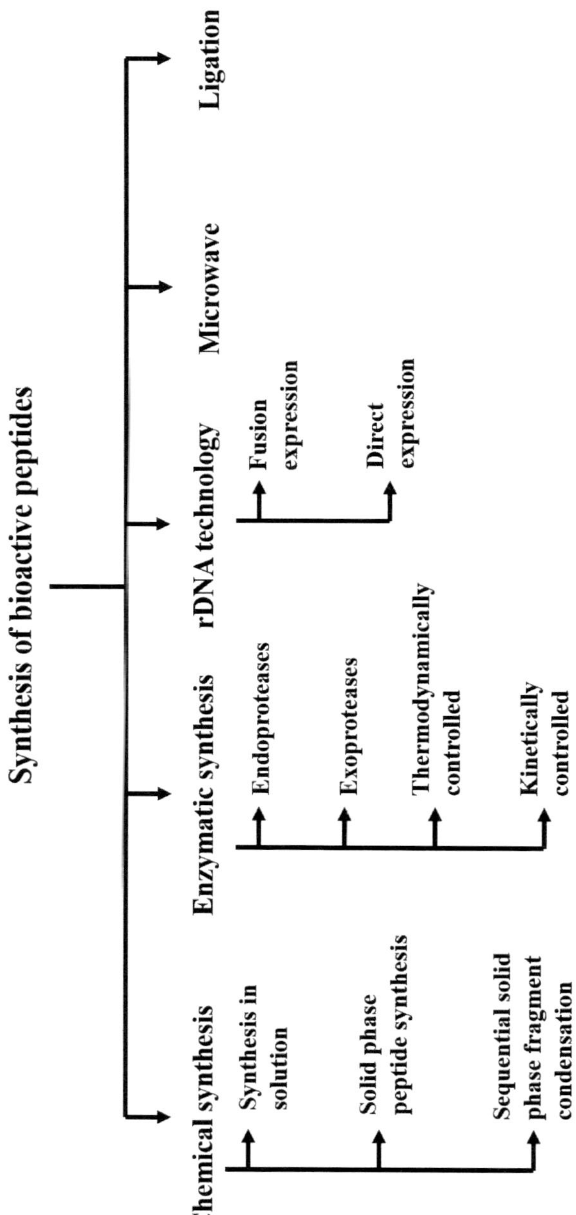

Figure 6.1 Synthesis approaches to produce bioactive peptides.

6.2.1.2 Solid-phase peptide synthesis (SPPS)

This is the simplest and highly preferred method of generating synthetic peptides on a solid support. The peptide is generated on solid matrix by sequential deprotection and coupling, and this sequence is reiterated until the anticipated product is attained. SPPS is advantageous over solution-based peptide synthesis as this method is automated, highly specific, and rapid and its solid matrix-peptide link helps to overcome the limitation of poor solubility of peptide intermediates in solution (Kent, 1988; Perez Espitia et al., 2012). The SPPS method requires three steps to form synthetic peptides such as solid matrix-amino acid coupling, deprotection of Nα-protected amino acid and coupling of amino acids.

(a) *Solid matrix-amino acid coupling*: The c-terminal of Nα-protected amino acid is initially coupled with the firm support by acylation via a linker moiety to form a peptide-linker-resin complex.
(b) *Deprotection of Nα-protected amino acid*: The Nα-amino assembly is then de-protected by chemical methods to allow the linkage of next amino acid in desired peptide.
(c) *Amino acid-amino acid coupling*: The chain of peptides is elongated by sequential addition of amino acids via establishment of amide bond amongst the incoming amino acid with free Nα-protected carboxyl group. Later, the amino acid with free amino group are linked with the solid matrix (Nilsson et al., 2005). In the final step, the peptide–linker–resin complex is cleaved to obtain the desired peptide.

The variation in SPPS lies in the protecting groups, matrix and coupling reagents used; and methodology used to synthesize and cleave the final peptide from the matrix. Generally, chemical synthesis uses two protocols such as Boc and Fmoc which is named on the basis of the protecting group type used for the protection of the tert-butyl and sensitive Nα-amino group of certain amino acid side chains (David and Luis, 2001). The differences in the two protocols are depicted in Table 6.1.

Table 6.1 Evaluation of the Boc and Fmoc protocols for chemical-based peptide synthesis

Properties	Boc	Fmoc
Protecting group	Tert-butyl oxycarbonyl	9-fluorenylmethyloxycarbonyl
Chemo-selectivity	Low	High
Deprotection	Trifluoroacetic acid (TFA)	Piperidine
Synthesis time	Extra neutralization step increases synthesis time	Less
Side chain protecting groups	Ester, ether, and urethane derivatives based on benzyl alcohol; OBzl, Bzl, Tos and Formyl	Ether, ester, and urethane derivatives based on t-butanol; Pbf, Trt, Boc, and tBu
Side chain deprotection	Hydrofluoric acid	Trifluoroacetic acid
Sensitivity	Not recommended for acid-sensitive peptides	Preferred method for the synthesis of routine peptides

Pbf, 2,2,4,6,7-pentamethyl-dihydrobenzofuran-5-sulfonyl; Trt, triphenyl; Boc, tert-butyloxycarbonyl; tBu, t-butyl ether; Bzl, benzyl; OBzl, benzyl ester; Tos, toluenesulfonyl.

6.2.1.3 Sequential condensation of solid phase fragments

The desired peptide is fabricated via condensation of the peptide fragments prepared by stepwise SPPS on a solid support. The fragments are initially protected and synthesized by the stepwise procedure with a solid phase. This was later freed from the solid matrix in the protected form and finally condensed on a resin support. The method has the ability to overcome the difficulty with the insolubility for the resin bound peptide intermediates (Kent, 1988).

6.2.2 Enzymatic Synthesis

This method is highly beneficial in the synthesis of short peptides using enzymes. The development of peptide bond is mediated by the enzymes such as immobilized or free proteases. The process is advantageous over chemical peptide synthesis as it is more stereo and regioselective. Therefore, this method requires minimum protection to the amino acids in the side chain, unlike chemical synthesis. The enzymes utilized for the formation of peptide bond are designated based on their specificity against the amino acids present at the bond cleavage point (Perez Espitia et al., 2012: Kumar and Bhalla, 2005).

6.2.2.1 Classification of proteolytic enzymes used for peptide synthesis

6.2.2.1.1 Endo- and exoproteases

The endoproteases enzymes possess the ability to catalyze and cleave terminal or penultimate peptide bond. They are further classified into amino and carboxypeptidases based on the cleavage of enzymes at amino or carboxy terminus. On the other hand, endoproteases are the enzymes that cleave the internal peptide bond. They are further classified based on the amino acids involved in catalytic mechanism which includes serine proteases, threonine proteases, cysteine proteases, metalloproteases, and aspartic proteases.

6.2.2.1.2 Mechanism of peptide synthesis by proteases

The proteases are employed for peptide bond synthesis by several mechanisms such as transpeptidation and reverse hydrolysis of amides (Boeriu et al., 2010; Machado and Saraiva, 2005). The synthetic reactions catalyzed by proteases are either:

(a) *Thermodynamically controlled*: The peptide is synthesized by reversing the hydrolysis of the peptide bonds that are catalyzed by proteases. The acyl intermediate formed from carboxylic acid represents the rate-limiting step for the synthesis and any proteases can be employed for the thermodynamically controlled fabrication of peptides (Jakubke et al., 1985).

(b) *Kinetically controlled*: This method requires an acyl donor activated as a nitrile, amide, or ester which upon binding the enzyme forms an acyl-enzyme intermediate complex by transpeptidation. In the next step, the complex is deacylated by the nucleophilic attack which is either a peptide or an amino acid that is protected in alpha-carboxyl group leading to the development of novel peptide bonds. The kinetics of the nucleophilic reaction determines the success of the synthesis (David and Luis, 2001). It is further noted that only serine and cysteine proteases are used for the kinetically controlled peptide fabrication, unlike thermodynamically controlled synthesis (Guzmán et al., 2007).

The main disadvantage of enzymatic peptide synthesis is the broad specificity of the proteases (Schellenberger et al., 1991) leading to uncontrolled hydrolytic reactions on substrates and synthesized peptides in solution (Guzmán et al., 2007). However, advancements have been made in the protocols recently to overcome the disadvantages associated with reaction media, solubility, and specificity of proteases by medium engineering, biocatalyst engineering and substrate engineering (Lombard et al., 2005). The most common proteases and their substrates which are used for the synthesis of peptides are itemized in Table 6.2.

Table 6.2 Common proteases and their substrates for peptide synthesis

Class	Enzyme	Substrate sequence
Serine proteases	Chymotrypsin	-Trp (Tyr, Leu, Phe,Met)**Xaa
	Dipeptidyl peptidase IV	-Xaa-Pro(Ala)**Xaa
	Carboxypeptidase Y	-Xaa**Xaa-OH
	Elastase	-Ala (Met, Ser, Phe) **Xaa
	Trypsin	-Arg (Lys)**Xaa
Cysteine proteases	Papain	-Phe (Val, Leu)-Xaa**Xaa
	Clostripain	-Arg**Xaa
Aspartic proteases	Pepsin	-Phe (Leu, Tyr) ** Leu (Phe)
Metalloproteases	Thermolysin	-Phe (Leu, Gly) **Leu (Phe)
	Collagenase	-Pro-Xaa** Gly-Pro
	Carboxypeptidase B	-Xaa** [Arg-Lys]-OH

Xaa: any amino acid; **: site of cleavage.

6.2.3 rDNA Technology

Recombinant protein technologies, with the help of genetic engineering, have immense scope of applications in peptide synthesis. The biggest advantage of the rDNA technology over the traditional peptide chemistry includes the ease of peptide synthesis with sophisticated complications, higher immunogenicity or active biological conformation (Perez Espitia et al., 2012; Li, 2011). This technology poses great implications in the production of high quality and pure antimicrobial peptides and are mostly cost effective and less time consuming, as compared to other peptide synthesis methods (Wang et al., 2011).

Principle: The method employs the cloning and recombinant expression of the desired peptide in hosts such as bacteria, yeast, and plants. More than one peptide can be expressed together. Generally, *E. coli* is utilized as the host of expression for peptide production. It is noteworthy that there are two methods to express peptides in the host expression system; namely fusion and direct expression.

6.2.3.1 Fusion expression

The major drawback of the technology is that the peptides expressed are naturally toxic to the host and nascent peptides are more prone to proteolytic damage inside the host system. The recombinant peptides are expressed as fusion peptides to alleviate their toxicity and degradation (Perez Espitia et al., 2012). The peptide is fused to the carrier protein which mimics the peptide's natural peptide segment and shields it from proteolytic degradation in the host system and minimizes its toxicity. The fused segment is later cleaved by either enzymes or chemicals (Wang et al., 2011). Thioredoxin, glutathione transferase (GST) and small ubiquitin-related modifier (SUMO) (Li, 2011) are the widely utilized carrier proteins used in the expression of fusion.

6.2.3.2 Direct expression

In direct expression, the recombinant peptide is expressed without any carrier tag. The peptide is tagged only to the His-tag and recognition site to remove tag post expression where the His-tag sufficiently shields the host from the peptide toxicity. Different methods, such as low temperature and co-expression with other proteins namely thioredoxin, elevates peptide solubility, are expressed (Li, 2011).

Also, different methods are used to increase the expression and efficacy of the bioactive peptides in rDNA technology such as

(1) Codon optimization: Usually, the native peptide genes are poorly translated in *E. coli* and the bias in codon results in the deprived countenance of the desired product. The native codons are replaced by codon optimization which are better expressed in *E. coli*. This can be performed by either the synthetic approach or using the modified, commercially available bacterial strains that express rare codons.
(2) By introducing unnatural amino acids in the peptide chain by 4-base pair codons increases flexibility of peptides (Young and Schultz, 2010).
(3) Heterologous expression in plant-based expression system is also an efficient way to express bioactive peptides with proper post-translational modifications. Lack of bacterial and yeast proteases in the plant host significantly increases the expression of stable bioactive peptides (Li, 2011).
(4) High-throughput peptide purification strategies could result in the high yield of properly folded and active peptide (Vriens et al., 2014).

6.2.4 Other Novel Peptide Synthesis Methods

The common methods of peptide synthesis discussed in the previous sections are effective for enhanced peptide production. However, there exist challenges in the fabrication of certain sequences of peptides, mainly because of

(i) lesser rate of accomplishment with peptide complexity;
(ii) improved chances of aggregation either inter- and/or intramolecular;
(iii) inclined probability for the development of secondary structure and protecting groups that are hindered by

steric arrangements which generate sequence for early termination. Thus, there are few novel or alternative techniques that have been developed in recent times as a solution to overcome these challenges.

6.2.4.1 Utilizing microwave energy for peptide production

Usually, microwave technology is combined with SPS to resolve the limitations of traditional synthesis procedures and to improve the recent fabrication methods as well as the products of synthetic peptides.

Principle: An oscillating electric or magnetic field is used to agitate the polar molecules or ions and the heat generated is utilized in reactions such as coupling and deprotection during peptide synthesis process, as well as the step that involves cleavage after synthesis for eliminating the final product of peptide and group that protects side chains from the resin.

Advantage: Numerous ionic and polar species such as deprotection bases, group of terminal amines, activators of polar or ionic species and polar solvents are present in the reaction which results in the elevation of temperature that helps to break the accumulated chains and permits for easier contact with the peptide growth chain. This assists in increasing peptide purity, greater reproducibility, higher yield and the capability to fabricate distinct and long peptides which might have difficulty in steric coupling or might form β-sheet structures (Palasek et al., 2006; Vanier, 2013). Moreover, the versatile nature of microwave-assisted technology enables it to perform manipulation of side chains, adjustment of C-terminal peptides and the fabrication of various peptidomimetics for diverse biomedical applications.

Disadvantage: Dehydration and racemization issues may arise due to coupling of cysteine and histidine.

6.2.4.2 Ligation

Smaller peptide chains are ligated to form longer peptides through chemical and enzyme-mediated technologies. This peptide fragment linking process is called "fragment condensation." However, there are limitations associated with this technique, some of which are the racemization and formation of other interfering by-products.

Development in the field of peptide ligation leads to a unique method called native chemical ligation (NCL). This method leads to the establishment of a native bond between Cysteine from N-terminal and thioester from C-terminal with two segments of unprotected peptides. After an initial exchange of chemo-selective thiol-thioester, the reaction progresses to form the desired peptide bond by an irreversible intramolecular S,N-acyl shift (Dawson et al., 1994).

Advantages: Higher yield, high chemo-selectivity, absence of racemization in C-terminal and neutral aqueous media usage which is perfect for unprotected peptide fragment solubility.

Disadvantages: The automated Fmoc SPS with thioester functionality is incompatible and lower quantity of Cysteine in natural peptides are the main drawbacks of the ligation process.

6.2.4.3 Recent developments

Cysteine-free NCL, serine/threonine and α-ketoacid-hydroxylamine ligation strategies are established as a response to the disadvantages of NCL (Noisier and Albericio, 2015). Similarly, ligation via enzymes are recently developed to exploit the survival of ligases from natural peptide which acts as a superior tool for labeling, selective and effectual coupling *N*-terminal and macro-cyclization of peptides. Ligases from peptides including butelase, sortase, subtiligase, trypsiligase, omniligase, and peptiligase offers exclusive opportunities for ligation based on the scope of wide substrate and specificity (Schmidt et al., 2017). Ligation by enzymes are applied unaided or in combination with chemical-based methods of ligation to attain advanced levels of precision and complexity.

6.3 Purification, Characterization, and Nanoformulation of Biopeptides

6.3.1 Purification

Post-synthesis, purification of a biopeptide is one of the most time-consuming stages. The physical and chemical characteristics are exploited for the separation of biopeptides which include specific interactions of covalent or non-covalent bond, polarity, solubility, molecular size and charge. In case of a complex biopeptide mixture, a combination of two or more techniques, which include gel filtration (based on molecular size), ion-exchange chromatography (based mainly on the charge), thin layer electrophoresis (based on both size and charge) and thin-layer chromatography (based on polarity) may be required to resolve them (Agyei et al., 2016). Thus, this section is attributed to discuss the most prominent and recent techniques that are used in the purification of peptide—especially bioactive peptides.

Table 6.3 Summary of the major peptide synthesis methods

Process	Specifications
Chemical Synthesis	
(A) Liquid Phase	Better for large-scale synthesis of smaller peptides with 8 or less amino acids.
	Precise production of single product
	Scale-up production is cheap and easy
	Easy detection of side reaction formation
	Appropriate for ligation technology
	Tedious to purify
	Non-automated or manual
	Long cycle of production
	The protection of the first C-terminal AA requires additional step
(B) Solid Phase	Appropriate for small scale production of long peptides
	Enhanced purification
	Automatic
	Shorter cycle of production
	Produces products in mixture
	Scale-up production is expensive
	Linkage and cleavage steps are required
Enzymatic Synthesis	More stereo and regioselective
	Efficient for more specific products
	Higher control of thermodynamic and kinetic parameters during synthesis

(*Continued*)

Table 6.3 (*Continued*)

Process	Specifications
Microwave assisted	Faster reactions
	Better/higher peptide yield and purity
	Efficient synthesis of difficult sequences of peptides
	Uniform and targeted heating of samples
	Improved reproducibility
	Elevates the racemization or dehydration chances with certain AAs
	Less requirement of solvents and reagents
	Greener solvent
Recombinant	Superior for the large-scale production of immunogenic or bioactive peptides
	Synthetic amino acids are flexibly introduced
	Numerous complications in solubility

6.3.1.1 High-performance liquid chromatography (HPLC) for biopeptide purification

Biopeptide purification employing HPLC utilizes the differences in the size of peptide (size exclusion HPLC), hydrophobicity (RP-HPLC) or net charge (ion exchange HPLC) as a tool for separation. The potential for the separation of an HPLC column can be improved by manipulating the conditions of the mobile phase. Also, a mixed mode approach, named as hydrophilic interaction chromatography (HILIC) exhibited exceptional peptide separation potentials (Arumugam et al., 2018).

6.3.1.1.1 HPLC supports

Silica-based packages are extensively utilized for the key HPLC modes. The silica microparticles are rigid, which facilitates the inclination of mobile phase flow rates. The use of silica-based packing is restricted to 2–8 pH range as silica dissolves in basic eluents. Column packings with elevated performance are introduced to overcome the pH restriction with extensive pH tolerance, based on organic polymers (Rai et al., 2016).

6.3.1.1.2 Size-Exclusion HPLC

Biopeptide separation based on molecular size is a useful early step in the fractionation of different sized complex peptide mixtures. Formerly, the required range of fractionized peptide was based on the ability of commercial column fractionation that was engineered exclusively for the separation of proteins. The size-exclusion column with peptide separation ability was introduced to enhance the quality of HPLC mode for the analysis of peptide in recent years (Arroume et al., 2016).

6.3.1.1.3 Ion exchange HPLC (IEX)

IEX approach is reported to be extensively useful for biopeptide separation, since HPLC that can retain both basic and acidic peptide were introduced. Anion-exchange and cation-exchange HPLC, are employed for peptide separation as negative or positive charged solutes are retained by the respective ion-exchange modes. The charge distribution on the peptide surface regulates the charged group interaction with stationary phase surface. Packaging by exchange of anions commonly consists of primary, secondary, and quaternary amine groups that are linked to a support via absorption or covalent bonds. These packaging with positive charges will interact with peptide residues and α-carboxyl group in C-terminal with negative charge. Likewise, packings with cation-exchange property mostly contain sulfonate or carboxyl groups that are linked with the support matrix. These packings with negative charges will interact with the basic residues and α-amino group in N-terminal with the positive charge (Allen, 1981). Common materials used for ion exchange have been listed in Table 6.4.

Table 6.4 Common ion exchange materials used for separation of biopeptides

Cation exchangers	Anion exchangers
Carboxymethyl-cellulose	Diethylaminoethyl-cellulose
Carboxymethyl-Sephadex	Diethylaminoethyl-Sephadex
Phosphocellulose	Diethyl-(2-hydroxypropyl) aminoethyl-Sephadex
Sulfopropyl-Sephadex	

6.3.1.1.4 Reversed-Phase HPLC (RP-HPLC)

RP-HPLC is another widely utilized procedure for the separation of biopeptides, depending on hydrophobicity and it contains a stationary phase of lower polarity and a highly polarized mobile phase. It is more unique compared to other HPLC modes in both effectiveness and rapidity, which is due to its selective chromatographic ability via alterations in the alignment of mobile phase, comparatively increased recoveries, repeatability and capability to resolve biopeptides from process impurities and degraded products. The volatile mobile phase of RP-HPLC is an ideal characteristic that helps in analytical and preparative separations. Highly acidic HPLC is preferred to possess merits of mobile phase with volatile nature. The volatility of mobile phase is predominantly helpful, while performing purification of a peptide during preparation or during RP-HPLC employment as last mode of purification in a multistep procedure. Lastly, acidity lead to unwanted interactions of ions between residues of amino acid with positive charges and any negatively charged silanol groups, on silica-based packing which is most commonly used for RP-HPLC packing support of peptides. The silica-based support contains covalent bond with functionalities of octadecyl or octyl with hydrophobicity elevated stationary phase that elutes peptides (Spicer et al., 2016; Sarrut et al., 2015).

6.3.1.1.5 Hydrophobic interaction/cation exchange chromatography

Initially, hydrophilic interaction chromatography was introduced to increase solute's hydrophobicity with solute eluted HPLC column which is the opposite behavior of RP-HPLC elution. This technique combines the advantages of two mechanisms such as hydrophilicity or hydrophobicity separation based on the differences in net charge of peptide overlaid (Mant and Hodges, 2008). The hydrophilic solute interaction and the cation exchanging stationary phase is promoted by enhancing the organic modifier. Later, the biopeptides are eluted from the column using salt gradient, whereas group peptide elution is possible by the increment of net positive charge and elution within peptide groups are based on increasing hydrophilicity (Periat et al., 2015).

6.3.1.2 Capillary electrophoresis (CE)

Capillary electrophoresis has also been used for biopeptide separation which separates analytes based on its charge in solution. Numerous separation modes are present in CE which includes capillary isoelectric focusing, capillary zone electrophoresis and micellar electro-kinetic capillary chromatography. Though RP-HPLC is the most favored separation approach for biopeptides, the rapid investigation time, low consumption of sample and reagent, wide detection schemes including fluorescence, UV, ESI-MS and versatile separation mode makes CE an influential alternative to HPLC (Black et al., 2015).

6.3.1.3 Affinity chromatography

Affinity chromatography employs the biological specificity of the biopeptide for its purification. It consists of an immobilized ligand column and the highly specific separation is mediated by biochemical interactions between the ligand and the peptide (Perez Espitia et al., 2012; Sewald and Jakubke, 2015). However, this method is in its infancy stage in biopeptide separation and extensive research will lead to further development in using this method for biopeptide separation.

6.3.2 Characterization Methods

Peptide characterization is an essential part that enables quality control checks on a purified or partially purified peptide. Few modern and sensitive techniques enable physical properties such as mass, size, concentration, amino acid sequence, stability, and structure as a device to address the biophysical peptide characterizations. The methods that are useful to characterize the physical stability of peptides have been summarized in Table 6.5 (D'Addio et al., 2016). Other than these, complimentary methods of characterization are applied to retrieve vital evidence on quality aspects. The use of reverse phase chromatography coupled with mass spectrometry leads to a unique peptide characterization and peptide maps (Valeja et al., 2015). Likewise, mass spectrometric analysis leads to the accurate molecular peptide mass-charge ratio determination and determination of chemical structure with superior resolution and sensitivity (Wilhelm et al., 2014). Different ionization methods are employed during mass spectrometry that involves mass spectrometry ionized by an electrospraying method (ESI-MS), fast atom bombardment mass spectrometry (FAB-MS), liquid chromatography mass spectrometry (LC-MS) or matrix-assisted laser desorption/ionization mass spectrometry (MALDI-MS) that can be used for peptide characterization (Li et al., 2017; Torres-Fuentes et al., 2015) as shown in Table 6.6. Tandem MS has also been employed to analyze any modification that occurred in any amino acid of a peptide or protein (Jedrychowski et al., 2015). Proteolytic peptides are ionized in the gas phase to measure their masses accurately by using electrospray ionization (ESI) or MALDI-MS (Pirrone et al., 2017). Various peptide separations are achieved using electrospray ionization (ESI) LC/MS method due to coupling drawbacks in LC and better tandem mass spectra quality for swift peptide identification (Hernandez-Hernandez et al., 2016).

Table 6.5 Techniques for biophysical characterization of peptide (D'Addio et al., 2016)

	Structure: Secondary
Circular dichroism (CD)	Information of secondary structure
Infrared and Raman spectroscopy	Provides details on secondary, local and specific amino acids
Intrinsic fluorescence	Local aromatic residual structures
Differential scanning calorimetry	Thermal transitions based on temperature and energy
NMR	Atomic resolution details of dynamics and structure
Hydrogen-Deuterium Exchange Mass Spectrometry	Provides insights of flexible conformations from proton exchange kinetics
	Structure: Tertiary
DLS	Average particle size and correlation function
SEC-MALS	Quantification of aggregates, state of oligomers and size of molecules from retention time. Molecular weight is determined by MALS.

Aggregation	
Intrinsic fluorescence	Local aromatic residual structure
DLS	Particle size distribution
SEC-MALS	State of oligomers and size of molecules from time of retention
Nephelometry	Scattering of light due to increased species number or size
Light obscuration	Optical solution density
Cryo TEM	Size, topography and morphology of molecule
Analytical microscopy	Morphological properties of aggregated species
UC	Supernatant analysis provides fractional input of peptides that are soluble in a sample
AUC	Species sizes and their populations
DSC	thermal transitions via temperature and energy
DOSY-NMR	Protein size estimation from rate of diffusion
SAXS	Global shape, flexibility, and molecular weight information

Table 6.6 Ionization methods in mass spectroscopy

Method	Description
ESI-MS	Peptide is dissolved in a solvent and ions are produced using an electrospray in which high voltage is applied to create a fine aerosol.
FAB-MS	Peptide is mixed in matrix which is a chemically protected non-volatile environment and subjected to a high energy atomic beam in a vacuum. Glycerol, thioglycerol, triethanolamine, and several other matrices are used in this method.
LC-MS	A high-energy ionic beam is used to replace atoms and this technique is named as liquid secondary ion mass spectrometry.
MALDI-MS	Peptide is subjected to a nitrogenous laser beam in which the sinapinic acid matrix is utilized for peptide protection. The energy absorbing matrix avoids direct peptide-beam contact to facilitate ionization and vaporization

The parameters of chromatographic conditions and MS have allowed 100% coverage of sequences and highlighted the separation of peptide mapping. Thereby, peptide mapping via mass spectrometer is used to

(a) Approve the precise protein or peptide identity.
(b) Obtain comprehensive characteristics such as *N*- and *C*-terminal peptide confirmation, coverage of highly sequenced peptide maps and substitutions of amino acid.
(c) Identification and screening of post-translational adjustments (Janin-Bussat et al., 2015; Sommella, 2015).

6.3.3 Nanoformulation

Biopeptides are gaining applicational importance in the pharmaceutical sector due to their enhanced bioactivity. The production of active pharmaceutical ingredients has shown tremendous advancements; however, peptide-based drug production is still a significant challenge. There is limited literature available that reports the formulation of peptide-based drug products. This is due to the challenges associated with the peptides such as chemical and physical instability, adopting multiple conformers and tendency to self-associate (Payne and Manning, 2009). The technique of nanoformulation has been developed as a potential entity for improving the stability and delivery of a biopeptide (Matougui et al., 2016). For a drug delivery system, the primary goal is to recover the index of a therapeutic drug by elevating the availability of delivered drug in the target location and reducing its contact with healthy tissues. The initial step for the formulation of biopeptide as a drug includes the comprehensive drug property characterizations and its constancy in diverse formulations. The physicochemical characteristics such as glycosylation, isoelectric point, molecular weight, composition of total amino acids and post-translational alterations of biopeptides are evaluated. These properties with drug behavior in diverse solutions and in vivo conditions help in the initial screening of peptide formulation conditions. Generally, a potential peptide formulation is composed of excipients, buffer components and any essential cofactors, approved by the Food and Drug Administration (FDA). The primary choice for the formulation is to analyze the stability of the biopeptide in a solution with optimum pH and the temperature to estimate their real time stability for shipping or storage (Balcão et al., 2013).

Numerous nanoencapsulations have been established for pharmaceutical applications such as nano-emulsions, liposomes, nanoparticles and hydrogels (Onwulata, 2012). The development of nanoparticles has evolved as one of the most promising aspect to face the barriers during pharmaceuticals delivery (Zhang et al., 2012). A recent study shows that the encapsulated antitumor peptides using CPAP, which is a known anti-proliferation peptide and a post-G1 cell cycle inducer, helps in the targeted peptide delivery for gastric cancer treatment (Wang and Zhang, 2013). The strategies of encapsulation using CPAP peptide include a complicated nanoencapsulation and ionotropic gelation. The first strategy involves the liquid precipitate phase separation, when two hydrophilic colloidal solutions are mixed under appropriate conditions. Strategy for the second one refers to the polyelectrolyte ability of counter ion to form hydrogels (Wang and Zhang, 2013). The efficiency of microencapsulation as compared to nanoencapsulation was about three times more, even though they possess similar polypeptides contents. Moreover, preservation of CPAP against degradation of gastric enzymes after micro or nanoencapsulation as shown in in vitro release tests which leads to slow and controlled release in the intestine.

Antitumor peptides have also been used to form an active tumor cell vaccine with modified genes using the encapsulations (Zhang et al., 2013). Also, nanoformulations have been considered as a potential tool to enhance the stability and delivery for cationic antiviral peptides. A positively charged α-helical analogous peptide p41 was obtained from the HCV protein. These peptides were also incorporated into poly (amino acid)-based anionic block copolymers using electrostatic coupling and used for HIV/HCV co-infection treatment. The nanocomplexes had a size of 35 nm and were highly stable at ionic strength and physiological pH. They retained the antiviral activity in vitro against HCV and HIV, whereas the intrinsic cytotoxicity was decreased (Zhang et al., 2013a). These recent findings validate the potential of the formulations technology to stabilize and deliver a peptide while maintaining their functional activity.

6.4 Pharmaceutical Application of Bioactive Peptides

Bioactive peptides are the next generation of therapeutics considering the rising interest in the peptide-based therapeutics research and pharmaceutical industries. These peptides are known to trigger unique physiological functions such as anti-oxidative, antimicrobial, anti-hypertensive and immunomodulatory actions (Lemes et al., 2016). Moreover, bioactive peptides are superiorly stable, highly bioavailable in body fluids with excellent target affinity and less toxicity, compared to conventional small molecules and drugs, that are produced by almost every living organism as the first line of defense. Figure 6.2 presents the summary of certain important pharmaceutical applications of bioactive peptides.

6.4.1 Antimicrobial Agent

Bioactive peptides with antimicrobial property is a distinct peptide group that is fabricated by various types of cells present in invertebrates, plants, and animal species (Salas et al., 2015). The characteristics of antimicrobial peptides (AMPs) (MW> ~10 KD) are revealed by their microbial cytoplasm membrane interaction. Their size, amino acid composition, hydrophobicity, and cationic charge permit their entry into the anionic membrane bilayers of the pathogenic microbes. In the era of antimicrobial resistance, these bioactive peptides are emerging as an alternative therapeutic agent against pathogens of Gram-positive, Gram-negative and multidrug resistant strains. It has been reported that more than 140 peptide therapeutic agents are currently under clinical trials and approximately 60 drugs are available in the market (Fosgerau and Hoffmann, 2015).

Recently, Pizzo et al. (2018) designed two antimicrobial peptides (IKY31 and IKY23) from the leaves of *Phytolacca dioica* and revealed their bacterial inhibition and anti-biofilm property against planktons and two strains of Gram-negative bacteria. Similarly, lactic acid bacteria that are naturally present in various fermented foods are the main bacteriocin peptide producers (Ahmed et al., 2017) and these peptides are commonly called antibiotics of endogenous origin due to their growth inhibition potential of several pathogenic microorganisms such as *Bacillus cereus, Staphylococcus aureus Listeria monocytogenes*, and *Clostridium botulinum* (Barbosa et al., 2017). There are numerous bacteriocins such as propionicin, nisin and diolococcin that are reported to be formed as a result of bacterial fermentation (Santos et al., 2018). Bioactive peptides are also extant in naturally occurring food products which include milk and dairy products. Milk-based biologically active peptides have elevated interest among researchers because they possess multifunctional properties that aid in modulating and triggering the immuno system simultaneously (Mohanty et al., 2016). Generally, the peptides are obtained from milk via proteolytic, fermentation and enzymatic methods. For instance, Aguilar-Toalá et al. (2017) obtained fractions of peptides and crude extracts (<3 and 3–10 kDa) from milk after fermentation using *Lactobacillus plantarum* bacterial strains and this showed that the crude milk extract possesses higher bioactivity than both peptide fractions with multifunctional properties such as higher anti-inflammatory, antihemolytic, and antioxidant activity. Multidrug resistance is the new route of bacterial pathogenicity towards their host which demands novel anti-bacterial agents. Thus, bioactive peptides derived from milk named as Lactoferrin (Lf) are employed to exhibit antimicrobial activities against bacteria, viruses and fungi by disrupting their cell membranes (Mohanty et al., 2016). Additionally, Vega et al. (2018) synthesized and screened three novel synthetic RRWQWR motif containing AMP based on lactoferricin B (LfcinB). These peptides exhibited a MIC_{50} value between 3.1–198 μM and a minimum bactericidal concentration (MBC) of 25–200 μM against healthcare associated infection causing pathogens. Caseicin is another milk-derived peptide with enhanced antimicrobial activity against *C. sakazakii, S. aureus* and *S. Typhimurium*, and are used as a coating over package material to protect powdered infant food products against microbial activity (Guinane et al., 2015). In the same way, Iglesias-Figueroa et al. (2016) produced recombinant bovine lactoferrin (rbLf) using *Pichia pastoris*, which displayed excellent bacterial growth inhibition properties against *E. coli, S. aureus*, and *P. aeruginosa* after subjected to pepsin digestion (Iglesias-Figueroa et al., 2016).

Bioactive peptides are also called host-defense peptides (HDPs) that are recombinantly produced. Gaglione et al. (2017) demonstrated a broad antimicrobial spectrum against human apolipoprotein B with significant biofilm inhibitory response. Apart from bacteria, these bioactive peptides were also studied against fungi and interference or lysis of fungal cell wall by chitin-targeting AMP was revealed to be the reason for their antifungal activity (El-Azizi et al., 2015). AMPs also possess superior antiviral properties and several researches were focused on revealing their antiviral activity against HSV associated viruses. Marcocci et al. (2018) demonstrated the antiviral activity of an active amphibian-short membrane AMP called Temporin B (TB) against type-1 Herpes simplex virus (HSV-1) infection by interfering specific lifecycle stages of the virus. Another study also revealed the superior antiviral activity of nanoliposome encapsulated LL-37 (human cathelidines) and Indolicidin (cattle) peptides against HSV-1 with reduced cytotoxicity to the cells (Ron-Doitch et al., 2016). Although, AMPs are utilized in diverse applications, there exist certain challenges which include toxicity and protease instability, which obfuscate them for further applications. Hence, modification of the peptide, using small sequences, attachment of alkyl/aromatic group, and incorporation of basic amino acid will enhance their antimicrobial activity against target organisms.

6.4.2 Anticancer Agent

In the medical field, the only dreaded disease without a proper cure is cancer, which is evident from the paucity of clinical trials, in spite of all the research and surveys that have been performed in the last decade. Numerous efforts are being employed in finding an efficient anticancer drug, amongst which bioactive peptides are proving efficient. Accumulated evidence suggested that, cationic bioactive peptides have anticancer properties due to their net negative charge on cancer cell membranes. These peptides have demonstrated promising cytotoxic activity towards various cancer cell lines and inhibit cancer cell growth without any side effects as compared to the conventional chemotherapy (Felício et al., 2017). The major benefits of peptides are, enhanced penetration into tissues, permeability and cell diffusion which leads to the pathway involved in anti-carcinogenesis without genotoxicity

(Blanco-Miguez et al., 2016). The mechanism of action includes cell migration/proliferation hindrance, tumor angiogenesis inhibition, antioxidant activity and apoptosis induction (Tyagi et al., 2015).

Amongst several sources, bioactive peptides of plant origin have attracted much attention. Proteins (lectins) and proteases (cystatin, serpin, and lunasin) from plants have been studied for their potential as anti-cancer agents and the studies revealed that lunasin exhibits side effects of anti-neoplasm towards breast, prostate, skin, leukemia and lymphoma cell lines with diverse mechanisms (Hernandez-Ledesma and Hsieh, 2017). Phyto-peptides, namely lunasin, are found in seeds, legumes, and mainly soya beans; in a recent study, these peptides in the fractions of <5, 5–10, and 10–50 kDa were obtained from soybeans seeds. This study demonstrated the excellent anticancer potential of these phyto-peptides against breast (MCF-7), blood (CCRF-CEM and Kasumi-3), and prostate (PC-3) cancer cell lines. The peptide fractions showed 68.0% inhibition in the cancer cell lines, while the minimum inhibitory concentration (IC_{50}) ranged between 608 and 6786 μg/mL. This peptide was proposed to be useful in alternative cancer therapy, as either nutritional supplements or food products (Rayaprolu et al., 2017). In another study, egg yolk peptides were prepared from the byproducts of egg yolk protein, after lecithin extraction and purification. The filtrated fraction (EYGF-33) significantly inhibited growth of colon cancer cells (Caco-2) without any adverse reactions in normal human epithelial colon cells (HCEC). This significant activity was observed due to the enhanced production of superoxide ions in mitochondria, which is supported by the externalization of phosphatidyl serine (Yousr et al., 2017).

Apart from plants, peptides are also isolated from the animals as well as aquatic organisms. Yu et al. (2014) isolated peptides from goat liver and used it in combination with cisplatin drug against gastric cancer stem cells. They isolated bioactive peptide-3 (ACBP-3) and tested the tumor inhibition property in in vivo and in vitro models which showed that the significant decrement in CD44 (+) cell count depends on the dose of ACBP-3. Also, it is noteworthy that the cisplastin used in combination with ACBP-3 curbed the xenograft-mediated growth of tumor in the in vivo model and the anti-cancer efficacy of cisplastin-peptide combination was more enhanced than the normal cisplatin dosages. Yet another study showed that two different peptides were isolated from the muscles of marine invertebrate called blood clams (*Tegillarca granosa*), which have high nutritive values. These muscle-derived peptides, BCP-A (Trp-pro-pro) and BCP-2 (Gln-Pro), were studied for their anticancer and anti-oxidative properties in which BCP-A peptides are toxic to cancer cells and possess excellent anti-oxidant activity. Also, the study disclosed the good scavenging and lipid peroxidation ability of the peptide with highest anti-cancer activity against PC-3 cells. Thus, these studies suggested that BCP-A (Trp-pro-pro) can be efficiently utilized in cancer treatments (Chi et al., 2015).

The major drawback of biopeptides is lack of bioavailability, which can be improved by nanotechnology-based approaches. Liao et al. (2016) reported a novel fabrication of peptides from walnut (WP-1) that are functionalized with selenium nanoparticles. These peptide-nanoparticle hybrids displayed exceptional selectivity towards normal and cancer cells. The apoptosis induced in human breast adenocarcinoma cells (MCF-7) by these hybrid biopeptides was established by the arrested S-phase accumulation of cells, breakage of DNA, and condensation of nucleus. The cell apoptosis was also established by Fas-associated protein activation and caspases 3, 8, and 9, making them an ideal chemotherapeutic agent for breast cancer. It was ascertained from the recent studies that the bioactive peptides obtained from natural and synthetic approaches have tremendous potential as cancer therapeutic agents. However, their successful clinical trial and optimization of these peptides for specific cancer patients are imperative and warranted to enhance their utilization in the future of cancer treatment.

6.4.3 Antihypertensive Agent

Hypertension is a chronic disease usually characterized as excessive blood pressure that leads to kidney diseases, arteriosclerosis, and strokes (WHO, 2014). Lack of blood pressure control and adverse side effects of anti-hypertensive drugs such as benazepril and perindopril have been used for novel bioactive peptide fabrication. The two key enzymes which regulate the human blood pressure are rennin and angiotensin-converting enzymes (Aluko, 2015). Several bioactive peptides are reported to inhibit angiotensin-I, angiotensin-I converting enzyme (ACE) and Ang II type 1 receptor (AT1) in the cardiovascular system, in addition to augmenting the blood nitric oxide levels. Literature also suggested that bioactive peptides originating from plants, marine organisms, and animals have significant effects on inhibiting the secretion of ACE. Further, natural ACE-inhibitors from invertebrates and marine organism have exhibited less side effects to humans. Several databases, namely, ACEpepDB, BIOPEP, and EROP-Moscow provide valuable information about peptide sequence, inhibitory concentration and molecular weight of the peptides (Lee and Hur, 2007). ACE inhibitor peptides with low molecular weight are obtained from various gastrointestinal enzymes, namely pepsin, proteinases, and trypsin that are used to treat various diseases due to their cell penetration ability. In recent times, Solieri et al. (2015) confirmed that -Pro-Pro and Ile-Pro-Pro peptides derived from fermented milk with *L. casei* PRA205 displayed superior ACE-inhibitory activities as compared to *L. rhamnosus* PRA331. Similarly, Chen et al. (2015) suggested that the novel strain of *L. helveticus* with bioactive ACE-inhibitor from fermented dairy products possesses antihypertensive activity.

Food-derived bioactive peptides also have inhibitory effects on ACE. However, the precise inhibitory mechanism is yet to be identified. Liu et al. (2018) identified a unique peptide-based ACE inhibitor from hazelnut and the cytotoxicity results showed IC_{50} values of 73.06, 15.42, and 249.3 μM, for peptides Ala-Val-Lys-Val-Leu (AVKVL), Tyr-Leu-Val-Arg (YLVR), and Thr-Leu-Val-Gly-Arg (TLVGR), respectively. All these peptides repressed ACE competitively and strong ACE inhibition was observed for YLVR, due to interaction of cation-pi. Furthermore, active peptides from rapeseed protein were obtained by digestion of alcalase and the results revealed that the inhibition of ACE and rennin activities depends on peptide dose. Moreover, in vivo studies of these peptides in hypertensive rats showed maximum systolic blood pressure reduction after 6 h of oral administration (100 mg/kg of body weight) (He et al., 2016). Another study by Ciau–Solís et al. (2018) also exploited alcalase in addition to pepsin for the hydrolysis of proteins obtained from lima beans. The pepsin–pancreatic hydrolysate fraction (>3 kDa) revealed inclined ACE-inhibition activity, while Alcalase®–Flavourzyme® hydrolysate (>3 KDa) showed highest renin-inhibitory activity in vitro. In vivo studies also demonstrated that Alcalase®–Flavourzyme®-derived peptide fractions lowered 64% of systolic blood pressure and 51% of diastolic blood pressure. Thus, these peptides have proved to have potential bioactivity as functional food (Ciau-Solís et al., 2018).

Gelatin, obtained from fish skin, which is a byproduct from the fish processing industry, was separately hydrolyzed to obtain proteases such as bromelain, trypsin, alcalase, and neutrase. Evaluation of ACE-inhibitory activity indicated that trypsin and flavourzyme hydrolysate proteases (>Mw ~ 10 Kd) can act as a potential antihypertensive agent (Ciau-Solís et al., 2018). A similar study was carried out by Salampessy et al. (2015) to produce fractions of fish proteins that are either soluble or insoluble from trevally (*Pseudocaranx* sps.) that are hydrolyzed by bromelain. The IC_{50} of ACE-inhibition was found to be in the range of 1.99 to 3.34 mg/mL for soluble fractions, whereas insoluble protein fractions inhibited ACE at the dosage of 2.45 to 4.74 mg/mL. These peptides indicated that they possess high ACE-inhibition potentials which eventually aid in the treatment of cardiovascular ailments.

6.4.4 Bioactive Peptides for Skin Disease Treatments

The human skin is the outermost covering of the body which forms an initial defensive line against several invading pathogens. The antimicrobial peptides (AMP) are a new and promising strategy to treat skin infections because they show rapid bactericidal activity and no resistance has been reported against them. The medical potential of these bioactive peptides is a well-known phenomenon and their role in modulating biological processes in the skin has been recently identified. It is also noteworthy that the bioactive peptides are used for the treatment of hyper-pigmentation disorders, in anti-aging phenomena, skin infections, wound healing, and skin regeneration processes.

6.4.4.1 Skin infections and biofilms

Various Gram-negative and Gram-positive bacteria are associated with skin infections and drug-resistant biofilms formations. It has been reported that the human antimicrobial peptide named cathelicidin (LL-37), conjugated with silver nanoparticles inhibits skin infections and controls biofilm formation of *P. aeruginosa*. Fungal infections of oral cavity, including oral candidiasis were also reported to be curable with low dosages of AMP (Xia et al., 2017). The short synthetic bioactive peptides RRIKA and RR are found to be a potential microbial growth inhibition agent by the disruption of pore formation and bacterial membranes. Further, these peptides were also effective against clinical and drug-resistant isolates of *S. aureus* that cause skin infections (Pfalzgraff et al., 2018). A peptide (HPA3NT3) isolated from *Helicobacter pylori* exhibited high antimicrobial activity against *Propionibacterium acnes*-mediated skin acne. Numerous modified lipopeptides such as Lauryl-KPV-NH2, Palmitoyl-KPV-NH2, and Palmitoyl-KH-NH2 were also described to be efficient in the treatment of bacterial vaginosis, acne vulgaris, candidiasis, and atopic dermatitis (Kang et al., 2017). Recently, two AMPs, namely, Indolicidin and LL-37, were reported to reduce the spread of Herpes Simplex Virus I, which is responsible for fever blisters and cold sores in the face and mouth (Piotrowska et al., 2017).

6.4.4.2 Skin cancer and hyper-pigmentation

The bioactive peptides are used to treat skin cancers and hyper-pigmentation disorders since the peptides can stimulate and inhibit melanin production. Melanin is a heteropolymer which protects the skin from harmful UV radiation and certain tetrapeptides such as His-D-Phe-Arg-Trp, which have the ability to stimulate melanin production to protect the skin from sunlight. On the other hand, decapeptide-12 (Tyr-Arg-Ser-Arg-Lys-Tyr-Ser-Ser-Trp-Tyr) inhibits biosynthesis of melanin and is extensively used in the treatment of hyper-pigmentation disorders called melasma, as well as an active component in skin-lightening products (Reddy et al., 2012).

6.4.4.3 Skin burns

Burns and their associated wounds are an important public health problem throughout the world. The conventional curative therapies for burn wounds have drawbacks of increasing antibiotic resistance to the common pathogenic bacteria such as *S. aureus*, *A. baumannii*, and *P. aeruginosa*. Thus, antimicrobial peptides with a broad activity spectrum are a capable alternative for surgical wound treatments, burns, and other skin infections. For instance, human AMP cathelicidin (LL-37) controls inflammation and promotes wound healing by stimulating fibroblasts and epidermal cells to form a granulation tissue and thus promoting angiogenesis. Likewise, synthetic peptide LLKKK18 conjugated with dextrins not only was effective in wound healing but also promoted angiogenesis in C57-BL/6 mice. Similarly, Pexiganan, which is a promising derivative of antimicrobial magainin peptide, is efficient against 200 bacteria of Gram-positive and Gram-negative strains. Various synthetic peptides, such as HB107, HB1345, PXL-150, Novispirin G10, IK8L, D2A21, and SHAP1, were also delivered in burn wounds as nano-emulsions, which showed potential antimicrobial property against *P. aeruginosa* in preclinical and clinical trials. Moreover, a synthetic peptide called SHAP-1 promoted wound healing at a very low concentration and is reported to be stable, even when high salt concentrations and proteases are present (Lee et al., 2014). Another synthetic peptide, Novispirin-G10, is reported to possess rapid bacterial inhibition ability in partial-thickness burn wounds that are infected by *P. aeruginosa*. Further, a novel antimicrobial peptide (A3-APO), as either a dimer or a monomer, was found effective against wounds and skin burn infections caused by multi-drug-resistant strains of *Acinetobacter baumannii* (Javia et al., 2018; Heinbockel et al., 2018).

6.4.4.4 Diabetic foot ulcers and other chronic skin wounds

Diabetic foot ulcers and skin ulcers caused by trauma are the major chronic wounds reported worldwide which require medical attention. Various therapies such as antimicrobial dressings, growth factors eluting dressings, and nucleic acid incorporated dressings for chronic wounds are being implemented. Recently, bioactive peptides, insulin, and simvastatin have been explored for such wound healing processes (Anjana et al., 2017). Reyes-Ortega et al. (2015) fabricated a bi-layered structure using pro-adrenomedullin with 20 N-terminal peptides and bemiparin to effectively re-epithelialize ischemic and non-ischemic rabbit wound model. Another study used a novel peptide called Pexiganan in Phase 3 clinical trials and revealed that they are highly utilizable for mild infection treatments caused by diabetic foot ulcer Flamm et al., 2015). Similarly, other synthetic AMPs such as SHAP1 and Pep 19–2.5 were also discovered to be effective for use as antibacterial and chronic wound healing agents by stimulating keratinocyte migration (Gomes et al., 2017). All these studies emphasize that bioactive antimicrobial and natural peptides are highly beneficial in the treatment of skin diseases and as cosmetics.

6.4.5 Bioactive Peptides as Immuno and Cytomodulatory Agents

Bioactive peptides are of great value to the nutraceutical and pharmaceutical industry due to their wide spectrum of bioactivities which include therapeutics, immunomodulatory, and cytomodulatory functions. Antimicrobial peptides (AMP) ease pathogen clearance by regulating immune responses such as inhibiting the expression of pro-inflammatory cytokines by blocking the signal transduction pathways, inducing the production of immune chemokines and modulating the expression of reactive oxygen species (ROS) (Hilchie et al., 2013). They avail as a significant adaptive and innate immune response mediator by acting as chemotactic factors, specific cell-receptor binders, inducing granulocyte recruitment, differentiation of leukocytes and macrophages (Xia et al., 2017). The antimicrobial peptides produced by neutrophils and mast cells not only prevent colonization of pathogens but also stimulate keratinocytes for proliferation and angiogenesis to alter injured tissues (Takahashi and Gallo, 2017). Furthermore, bioactive peptides are used as anti-cancer agents, mediate immunomodulatory effects either by tumor-associated macrophage polarization or by increased infiltration of T-lymphocytes (Zhao et al., 2017).

A study by Chalamaiah et al. (2015) revealed that bioactive peptides obtained from hydrolysis via enzymes of Common Carp (*Cyprinus carpio*) egg (roe) exhibited immunomodulatory functions in BALB/c mice. These peptides lead to the proliferation of spleen lymphocytes, increase splenic cell cytotoxicity and increases the levels of serum IgA. Thus, the carp egg peptides have proved to be capable of improving immunity and have been proposed to be utilized as food and in the pharmaceutical industry, for several purposes. Likewise, two anti-allergic peptides, namely LDAVNR and MMLDF are prepared using *Spirulina maxima* algae and are helpful in reducing the histamine release and intracellular Ca^{2+} elevation for suppressing degranulation of mast-cells. Both peptides are also capable of immunomodulation by reducing the nuclear factor-κB translocation and synthesis of interleukin-4 (Vo et al., 2014). It can be noted that synthetic bioactive peptides exhibit various cytomodulatory effects by activating or inhibiting the receptor-pathways. These peptides aid in inhibiting protein-protein interactions, especially the interaction of kinases with their anchors in the plasma membrane. In T-cells, inhibition of type 1 protein kinase A (PKA) leads to the lymphocyte-specific protein tyrosine kinase (Lck) activation which is negatively regulated by PKA and finally, the T-cell responsiveness is increased by linear peptides that are beneficial in intracellular applications (Zuconelli et al., 2017).

Synthetic peptides that can mimic neural cell adhesion molecules (NCAM) play a vital part in the treatment of neurological disorders by inducing NCAM-downstream signaling pathways. These NCAM-mimicking peptides activate various signaling cascades such as mitogen-activated protein kinases (MEK), extracellular signal-regulated kinases (ERK), p21-activated kinases 1 (Pak1), and focal adhesion kinase (FAK)-dependent pathways in different cell populations (Chu et al., 2018). The peptides which act as natural ligands for cell surface receptors are used to transfer nanoparticles and macromolecules for cellular uptake and exhibit cytomodulatory functions. For example, transferrin peptide serves as a highly significant drug carrier for cancer therapy. It helps drugs penetrate the target cells by transferrin-receptor-mediated endocytosis. Similarly, a cell targeting peptide Pep42 (cyclic 13-mer) possesses high affinity towards glucose-regulated protein 78, which is a marker on malignant cancer cells. Pep42 conjugated with Taxol is used to deliver drugs to cancer cells via receptor-mediated endocytosis and it also induces apoptosis to increase automated melanoma cell death (Mousavizadeh et al., 2017). Also, marine-based bioactive peptides derived from enzymatic hydrolysis of fish proteins from their fillet or muscle such as Hydro-MN and Nutripeptin peptide were useful in decreasing the level of blood glucose and in eliminating type 2 diabetes symptoms (Gevaert et al., 2016). The Angiotensin-I Converting Enzyme (ACE) usually gets converted into a vasoconstrictor Angiotensin II in hypertension patients which eventually results in the elevation of blood pressure. Bioactive peptides were reported to be a superior antihypertensive agent since they act as ACE inhibitors (Gallego et al., 2016). Thus, certain products, namely Calpis, Evolus and BioZate containing bioactive peptides as ACE inhibitors are available in the current pharmaceutical market as antihypertensive drugs (Cheung et al., 2015). Most of the wound healing peptides used for the treatment of chronic wounds were also used as immune and cytomodulatory agents by recruitment of macrophages, neutrophils and monocytes and then activating keratinocyte migration, synthesis of ECM proteins and angiogenesis at the site of infection (Gomes et al., 2017). Extensive research in the direction of bioactive peptides will lead to an increase in the peptide-based commercial products for biomedical and pharmaceutical applications.

6.4.6 Other Novel Pharmaceutical Applications

Biopeptides are also gaining applicational importance just as analgesics, anticoagulants, vaccines and for drug delivery, diagnostics, combating drug resistance, and treatment of neurodegenerative disorders and type 2 diabetes. Recently, several marine-derived peptides and peptide-derived compounds, namely ziconotide and Adcetris were available on the market as analgesics and cancer drugs (Cheung et al., 2015), whereas Conopeptides are used for severe pain management (Schroeder and Craik, 2012). Likewise, peptides are being explored as a new class of anticoagulants. For instance, enzymatically synthesized tripeptide (Z-YQQ) increases the coagulation time by acting on the extrinsic pathway of coagulation cascade (Origone et al., 2018). Peptides are also used as vaccines because they are easy to develop, safe, and effective, compared to conventional vaccines. Peptide vaccines are developed against several diseases causing pathogens such as West Nile virus, Hepatitis C virus, influenza, human papilloma virus, and human immunodeficiency virus (HIV) (Baig et al., 2018; Friedrich et al., 2016). Also, antimicrobial peptides are a promising therapeutic alternative to growing antimicrobial resistance and are highly beneficial in combating drug resistance (Cheung et al., 2015; Pfalzgraff et al., 2016). Further, cell-penetrating peptides (CPP) have been confirmed to possess molecular delivery properties to deliver drugs, nanocarriers, genes, and imaging agents into cells. A peptide-mimetic (LLP2A) is used in tumor detection since it binds to $\alpha4\beta1$-integrin receptor of activated T- and B-lymphomas. CPP also induces tumor cell apoptosis and CPP-cargo complexes are successfully used in treatment of cancer, muscular dystrophy, cardiac disorders, and other diseases (Copolovici et al., 2014). Certain peptides also assist in type 2 diabetes treatment, as they act as potential inhibitors of carbohydrate hydrolyzing enzymes such as alpha-glucosidase and glucose lowering agents by inhibiting dipeptidyl peptidase-IV (e.g. LAPSLPGKPKPD). Certain peptides (KRX-683 and KRX-683) also exhibit anti-diabetic potential by modulating the expression of G-protein-coupled receptors that are responsible for glucose homeostasis (Patil et al., 2015; Khan et al., 2018). Similarly, neurodegenerative disorders occur due to misfolding and aggregation of PolyQ protein stretches and therefore, peptide inhibitors such as QBP1, Neurotrophins, and Aβ (16–20) KLVFF) are under extensive research for the treatment of neurodegenerative disorders, including Parkinson's, Alzheimer's, and Huntington's diseases (Baig et al., 2018).

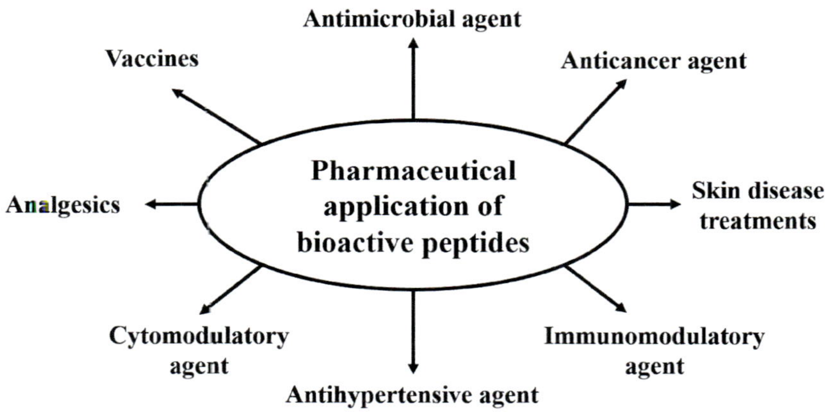

Figure 6.2 Pharmaceutical applications of bioactive peptides.

Bioactive peptides are slowly replacing conventional drug molecules for the treatment of numerous diseases. However, there exist certain challenges in utilizing these bioactive peptides as full-fledged therapeutic entity. In an organic system, it is noteworthy that the impact of bioactive peptide depends on its ability to reach the bloodstream, targeted organ and the injected peptide concentration. There are several labeling approaches to quantify synthesized or extracted bioactive peptides from the natural sources. These approaches have drawbacks such as the requirement of a high sample quantity, exclusive labeling reagents, and monotonous preparation. Further, Cys residues in endogenous Cys thiol peptide groups that are essential to the reaction with isotopic label forming reagents are absent in natural peptides, which is a major setback of labeling techniques (Mora et al., 2017). Also, there is lack of knowledge on natural bioactive peptide generation to predict their bioavailability and bioactivity in living organisms. Even with the development of software, computational simulations and biopeptide databases such as BIOPEP, PEPBANK, or eBASIS, there are some challenges in quantifying biopeptides from complex matrices (Udenigwe, 2014). Thus, extensive research, focusing on these challenges, in future will help to overcome the hurdles in using biopeptides for pharmaceutical applications.

The future of biopeptides in pharmaceutics depends on finding a perfect nanoformulation or a nanocarrier for encapsulating the biopeptides. Nanoencapsulation will counterfeit most of the challenges that are reported to wedge the positive aspects of biopeptides by protecting their stability from biomolecule interaction in body fluids and delivering them at target site (Cheng et al., 2018; Abdelaal et al., 2018). Biopeptides can either be capped with nanoparticles or encapsulated with nanoformulations and both approaches are highly beneficial in enhancing the properties of peptides (Auwal et al., 2018; Hosseini et al., 2018). Recently, certain nanoparticles have been tagged with biopeptides which are useful as labeling agents for precise quantification of peptides from a complex mixture (Carambia et al., 2015). Likewise, metal and metal oxide nanoparticles, especially iron-based magnetic nanoparticles are being used to formulate or bind the biopeptides for magnet-mediated targeted and controlled delivery of biopeptides (Sahoo et al., 2017). Also, virus and virus-like nanoparticles are being used to encapsulate biopeptides to deliver the biopeptides into the target cells which will gain enormous applications in future (Sarkar et al., 2017; Choi et al., 2018). It is also possible in future to formulate medicinal biopeptide as a genetic material in virus, which upon activation generates and multiplies with a specific peptide inside the body until it cures the disease. Thus, there is no doubt that nanoformulated bioactive peptides will capture the market of pharmaceuticals and nutraceuticals as well as dominate conventional medicines and therapeutic methods.

6.5 Conclusion

Bioactive peptides are highly beneficial as a therapeutic entity due to their enhanced biomolecular interactions and bioactivity. The unique biological properties of peptides depend on the synthesis approaches that are used to fabricate them, which includes chemical, enzymatic and r-DNA methods. Along with conventional methods, recent techniques such as microwave and ligation methods were also used for improving the yield of biopeptides. Also, purification and characterization play a substantial role in improving the novelty of each bioactive peptide and utilizes them as drugs for the treatment of specific diseases. These synthesized bioactive peptides that are purified and characterized are widely used in pharmaceutical applications such as antimicrobials, treatment of cancer, hypertension, skin diseases and help in the modulation of immunity as well as cellular mechanisms. Further, bioactive peptides are extensively being researched and processed to be used as analgesics, anticoagulants, vaccines, drug delivery, treatment of neurodegenerative diseases and type 2 diabetes. The profound development of nanotechnology showed that nanoencapsulated peptides can deliver biopeptides to the target site without any stability fluctuations. Thus, these biopeptides have been predicted to occupy the pharmaceutical market as an alternate for conventional medicines in future.

References

Abdelaal I. R., et al., AACR. *Proceedings: AACR Annual Meeting 2018*; April 14–18; Chicago, IL, Abstract 2689 (2018).

Aebersold R., Mann M., *Nature*, **537** (2016), 347.

Aguilar-Toala J. E., et al., *J. Dairy Sci.*, **100** (2017), 65–75.

Agyei D., et al., *Food Biotechnol. Process.*, **98** (2016), 244–256.

Ahmed I., et al., *Food Control.*, **82** (2017), pp. 163–178.

Allen G., Separation and purification of peptides, in *Laboratory Techniques in Biochemistry and Molecular Biology*, (Allen G., ed.) (1981), Elsevier. pp. 73-133.

Aluko R. E., *Ann. Rev. Food Sci. Technol.*, **6** (2015), 235-262.

Anjana J., et al., *Curr. Pharmaceut. Des.*, **23** (2017), 3529–3537.

Arroume N., et al., *Curr. Opin. Food Sci.*, **7** (2016), 101–107.

Arumugam V., et al., *Int. J. Peptide Res. Therap.*, **24** (2018), pp. 13–18.

Auwal S. M., et al., *Sci. Rep.*, **8** (2018), 10411.

Baig M. H., et al., *Biomed. Pharmacother.*, **103** (2018), 574–581.

Balcão V. M., et al., *Food Hydrocoll.*, **32** (2013), 425–431.

Barbosa A. A. T., Mantovani H. C., Jain S., *Crit. Rev. Biotechnol.*, **37** (2017), 852–864.

Black W. A., et al., *Anal. Chem.*, **87** (2015), 6280–6287.

Blanco-Miguez A., et al., *Protein Sci.*, **25** (2016), 1084–1095.

Boeriu C. G., et al., *J. Mol. Catal. B: Enzymatic*, **66** (2010), 33–42.

Carambia A., et al., *J. Hepatol.*, **62** (2015), 1349–1356.

Chalamaiah M., Hemalatha R., Jyothirmayi T., Diwan P. V., et al. *Nutrition*, **31** (2015), 388–398.

Chen Y., et al., *J. Dairy Sci.*, **98** (2015), 5113–5124.

Cheng K., et al., *Nano Lett.*, **18** (2018), 3250–3258.

Cheung R. C. F., Ng T. B., Wong J. H., *Marine Drugs*, **13** (2015), 4006–4043.

Chi C.-F., et al., *J. Funct. Foods*, **15** (2015), 301–313.

Choi J. M., et al., *Mol. Pharm.*, **15** (2018), 2997–3004.

Chu C., et al., *Neurochem. Res.*, **43** (2018), 1–9.

Ciau-Solís N. A., Acevedo-Fernández J. J., Betancur-Ancona D., *J. Sci. Food Agric.*, **98** (2018), 781–786.

Cicero A. F., Fogacci F., Colletti A., *Brit. J. Pharmacol.*, **174** (2017), 1378–1394.

Copolovici D. M., et al., *ACS Nano*, **8** (2014), 1972–1994.

D'Addio S. M., et al., *J. Pharm. Sci.*, **105** (2016), 2989–3006.

David A., Luis R., in *Peptide Antibiotics: Discovery, Modes of Action and Applications* (Dutton M. H. C., McArthur H., Wax R. G., eds.), CRC Press, Florida (2001), pp. 15–45.

Dawson P., et al., *Science*, **266** (1994), 776–779.

El-Azizi M., Farag N., Khardori N., *Ann. Clin. Microbiol. Antimicrob.*, **14** (2015), 21.

Felício M. R., et al., *Front Chem.*, **5** (2017), 5.

Flamm R. K., et al., *Antimicrob. Agents Chemother.*, **59** (2015), 1751–1754.

Fosgerau K., Hoffmann T., *Drug Disc. Today*, **20** (2015), 122–128.

Friedrich B. M., Beasley D. W., Rudra J. S., *Vaccine*, **34** (2016), 5479–5482.

Gaglione, R., et al., *Biochem. Pharmacol.*, **130** (2017), 34–50.

Gallego M., et al., *J. Funct. Foods*, **21** (2016), 388–395.

Gevaert B., et al., *Protein Pept. Lett.*, **23** (2016), 1052–1060.

Gomes A., et al., *Molecules*, **22** (2017), pii: E1743.

Guinane C. M., et al., *Int. Dairy J.*, **49** (2015), 1–7.

Guzmán F., Sonia B., Andrés I., *Electron. J. Biotechnol.*, **10** (2007), doi: 10.2225/vol10-issue2-fullte.

Haney E. F., Mansour S. C., Hancock R. E., in *Antimicrobial Peptides*. Springer (2017), pp. 3–22.

He R., et al., *Food Chem.*, **197** (2016), 1008–1014.

Heinbockel L., et al., *Front. Immunol.*, **9** (2018), 1704.

Hernandez-Hernandez O., et al., *J. Chromatogr. A*, **1428** (2016), 202–211.

Hernandez-Ledesma B., Hsieh C. C., *Crit. Rev. Food Sci. Nutr.*, **57** (2017), 2358–2376.

Hilchie A. L., Wuerth K., Hancock R. E., *Nat. Chem. Biol.*, **9** (2013), 761–768.

Hosseini S. F., Soleimani M. R., Nikkhah M., *Int. J. Biol. Macromol.*, **111** (2018), 730–737.

Iglesias-Figueroa B., et al., *Int. J. Mol. Sci.*, **17** (2016), pii: E902.

Jakubke H.-D., Kuhl P., Könnecke A., *Angew. Chem. Int. Ed.*, **24** (1985), 85–93.

Janin-Bussat M.-C., et al., *J. Chromatogr. B*, **981** (2015), 9–13.

Javia A., et al., *Therap. Deliv.*, **9** (2018), 375–386.

Jedrychowski M. P., et al., *Cell Metab.*, **22** (2015), 734–740.

Kang H.-K., et al., *J. Microbiol.*, **55** (2017), 1–12.

Kent S. B. H., *Annu. Rev. Biochem.*, **57** (1988), 957–989.

Khan H., Nabvi S. M , Habtemariam S., *Life Sci.*, **193** (2018), 153–158.

Korhonen H., Pihlanto A., *Int. Dairy J.*, **16** (2006), 945–960.

Kumar D., Bhalla T. C., *Appl. Microbiol. Biotechnol.*, **68** (2005), 726–736.

Lee S. Y., Hur S. J., *Food Chem.*, **228** (2007), 506–517.

Lee Y. W., et al., *Amino Acids*, **46** (2014), 2333–2343.

Lemes A. C., et al., *Int. J. Mol. Sci.*, **17** (2016), pii: E950.

Li Y., *Protein Expr. Purif.*, **80** (2011), 260–267.

Li H., et al., *Anal. Chem.*, **89** (2017), 2731–2738.

Liao W., et al., *Int. J. Nanomed.*, **11** (2016), 1305–1321.

Liu C., Fang L., Min W., Liu J., Li H., *Food Chem.*, **245** (2018), 471–480.

Lombard C., Saulnier J., Wallach J. M., *Protein Pept. Lett.*, **12** (2005), 621–629.

Machado A., et al., *Química Nova*, **27** (2004), 781–789.

Machado M. F., Saraiva J. M., *Biotechnol. Lett.*, **27** (2005), 1233–1239.

Mant C. T., Hodges R. S., *J. Separ. Sci.*, **31** (2008), 1573–1584.

Marcocci M. E., et al., *Antimicrob. Agents Chemother.*, **62** (2018), pii: e02367-17.

Matougui N., et al., *Int. J. Pharm.*, **502** (2016), 80–97.

Mohanty D., et al., *Int. J. Food Prop.*, **19** (2016), 837–846.

Mora, L., et al., *Trends Food Sci. Technol.*, **69 B** (2017), 306–314.

Mousavizadeh A., et al., *Colloids Surf. B: Biointerfaces*, **158** (2017), 507–517.

Muheem A., et al., *Saudi Pharm. J.*, **24** (2016), 413–428.

Nilsson B. L., Soellner M. B., Raines R. T., *Annu Rev. Biophys. Biomol. Struct.*, **34** (2005), 91–118.

Noisier A. F. M., Albericio F., *Royal Soc. Chem.*, **39** (2015), 1–20.

Onwulata C. I., *Annu. Rev. Food Sci. Technol.*, **3** (2012), 183–202.

Origone A., et al., *Biotechnol Prog.* (2018); doi: 10.1002/btpr.2658.

Palasek S. A., Cox Z. J., Collins J. M., *J. Peptide Sci.*, **13** (2006), 143–148.

Patil P., et al., *Eur. J. Nutrition*, **54** (2015), 863–880.

Payne R. W., Manning M. C., *Innov. Pharm. Technol.*, **28** (2009), 64–68.

Perez Espitia P. J., et al., *Compr. Rev. Food Sci. Food Safety*, **11** (2012), 187–204.

Periat A., Krull I. S., Guillarme D., *J. Separ. Sci.*, **38** (2015), 357–367.

Pfalzgraff A., et al., *Sci. Rep.*, **6** (2016), 31577.

Pfalzgraff A., Brandenburg K., Weindl G., *Front. Pharmacol.*, **9** (2018), 281.

Piotrowska U., Sobczak M., Oledzka E., *Chem. Biol. Drug Des.*, **90** (2017), 1079–1093.

Pirrone G. F., et al., *Anal. Chem.*, **89** (2017), 8351–8357.

Pizzo E., et al., *Biochim. Biophys. Acta-Biomembranes*, **1860** (2018), 1425–1435.

Rai M., et al., *J. Food. Sci. Technol.*, **53** (2016), 3381–3394.

Rayaprolu S. J., et al., *J. Food Sci. Technol.*, **54** (2017), 38–44.

Reddy B., Jow T., Hantash B. M., *Experim. Dermatol.*, **21** (2012), 563–568.

Reyes-Ortega F., et al., *Acta Biomater.*, **23** (2015), 103–115.

Risitano, A. M., Ricklin D., Huang Y., Reis E. S., et al., *Blood*, **123** (2014), 2094–3101.

Ron-Doitch S., et al., *J. Control. Release*, **229** (2016), 163–171.

Sahoo S. K., Misra R., Parveen S., Nanoparticles: A boon to drug delivery, therapeutics, diagnostics and imaging, in *Nanomedicine in Cancer*. Pan Stanford Publishing (2017), pp. 73–124.

Salampessy J., et al., *J. Funct. Foods*, **14** (2015), 716–725.

Salas C. E., et al., *Biomed. Res. Int.*, **2015** (2015), 102129.

Santos J. C. P., et al., *Innov. Food Sci. Emerg. Technol.*, **48** (2018), 179–194.

Sarkar G., et al., Peptide-mediated delivery of active agents across the blood-brain barrier. Google Patents WO2011008823A2 WO Application (2017).

Sarrut M., D'Attoma A., Heinisch S., *J. Chromatogr. A*, **1421** (2015), 48–59.

Schellenberger V., et al., *Biotechnol. Bioeng.*, **38** (1991), 104–108.

Schmidt M., et al., *Curr. Opin. Chem. Biol.*, **38** (2017), 1–7.

Schroeder C. I., Craik D. J., *Future Med. Chem.*, **4** (2012), 1243–1255.

Sewald N., Jakubke H.-D., *Peptides: Chemistry and Biology*, John Wiley & Sons (2015).

Solieri L., Rutella G. S., Tagliazucchi D., *Food Microbiol.*, **51** (2015), 108–116.

Sommella E., et al., *J. Chromatogr. A*, **1375** (2015), 54–61.

Spicer V., et al., *Anal. Chem.*, **88** (2016), 2847–2855.

Takahashi T., Gallo R. L., *Dermatol. Clin.*, **35** (2017), 39–50.

Torres-Fuentes C., et al., *Food Chem.*, **180** (2015), 194–202.

Tyagi A., et al., *Nucleic Acids Res.*, **43** (Database issue) (2015), D837–D843.

Udenigwe C. C., *Trends Food Sci. Technol*, **36** (2014), 137–143.

Valeja S. G., et al., *Anal. Chem.*, **87** (2015), 5363–5371.

Vanier G. S., *Methods Mol. Biol.*, **1047** (2013), 235–249.

Vega S. C., et al., *Front. Microbiol.*, **9** (2018), 329.

Vo T. S., et al., *Mol. Nutr. Food Res.*, **58** (2014), 2226–2234.

Vriens K., Cammue B. P., Thevissen K., *Molecules*, **19** (2014), 12280–12303.

Wang Q., et al., *Biotechnol. Lett.*, **33** (2011), 2121–2126.

Wang X., Zhang X., *Biotechnol. Prog.*, **29** (2013), 681–687.

WHO, *Global Status Report on Noncommunicable Diseases* (2014).

Wilhelm M., et al., *Nature*, **509** (2014), 582–587.

Xia X., et al., *Antonie Van Leeuwenhoek*, **111** (2017), 1–22.

Young T. S., Schultz P. G., *J. Biol. Chem.*, **285** (2010), 11039–11044.

Yousr M. N., et al., *Nutr. Cancer*, **69** (2017), 674–681.

Yu L., et al., *J. Cell Biochem.*, **115** (2014), 697–711.

Zhang J., et al., *Biomaterials*, **34** (2013a), 3846–3857.

Zhang J., et al., *Mol. Immunol.*, **55** (2013), 264–274.

Zhang, X. X., Eden H. S. Chen X., *J. Control. Release*, **159** (2012), 2–13.

Zhao H., et al., *Clin. Epigenet.*, **9** (2017), 4.

Zuconelli C. R., Brock R., Adjobo-Hermans M. J. W, *Curr. Med. Chem.*, **24** (2017), 1862–1873.

Chapter 7

Cardiovascular Disease and Oxidative Stress

Marco Fernandes,[a] Alisha Patel,[a] and Holger Husi[a,b]

[a]*Institute of Cardiovascular and Medical Sciences,*
BHF Glasgow Cardiovascular Research Centre, University of Glasgow, UK
[b]*Department of Diabetes and Cardiovascular Science,*
Centre for Health Science, University of the Highlands and Islands Inverness, UK

holger.husi@glasgow.ac.uk

7.1 Introduction

Cardiovascular disease (CVD) encompasses a multitude of pathological conditions that promote functional impairment of the heart, their structures and the circulatory system. CVD remains the most accountable source of health impairment worldwide. This life-threatening condition is associated with 32.26% of all total global deaths, in which 17.33% are due to ischemic heart disease (IHD), followed by stroke with up-to 10.11%, and by hypertensive heart disease (HHD) that is liable for 1.67% of all death-associated cases (Roth et al., 2017). The Global Burden of Disease estimates that in 2030, CVD-related conditions such as

Pharmaceutical Biocatalysis: Fundamentals, Enzyme Inhibitors, and Enzymes in Health and Diseases
Edited by Peter Grunwald

ISBN 978-981-4800-61-7 (Hardcover), 978-0-429-29503-4 (eBook)
www.jennystanford.com

IHD and stroke will still be ranked as major causes of mortality and disability at a global scale (Mathers and Loncar, 2006).

Some CVD risk factors such as total blood cholesterol, alcohol intake and smoking habits (even though the declining rate has decelerated) followed a declining trend over the last three decades amongst the European population (Wilkins et al., 2017). Nevertheless, prevalence of overweight, obesity, and diabetes is high and the latter have shown a worrying tendency to increase over the last decade, in particular of type 2 diabetes mellitus (T2DM) among children and young adults (Maffi and Secchi, 2017). Diet-related behaviour determinants, such as increased ingestion of fat-based products and a rise of total calorie intake are primary factors that impact in CVD-mortality and morbidity, alongside with high blood pressure prevalence (Palomo et al., 2014). Since low-income countries are more and more emulating and resemble Western-lifestyles, so have the CVD-modifiable cardio-metabolic risk factors in these countries increased, inadvertently boosting the rate of CVD epidemiology (Humayun and Mirza, 2011; Iso, 2011) and set to rise the current CVD prevalence to around 80% as in middle-to-high income countries (Roth et al., 2017).

In the EU economy, CVD is associated with a financial burden of €210 billion per year, in which almost-half of the budget is split into treatment expenses, and costs related with increased work absences and reduced productivity, 26% and 21%, respectively (Wilkins et al., 2017).

Many vascular pathological conditions arise from altered states of the arterial wall and this can lead to disrupted arterial hemodynamics as observed in atherosclerosis, thrombosis and restenosis (Pasterkamp et al., 2000). Many cardiovascular risk factors such as hypertension, hypercholesterolemia, obesity-linked overweight, smoking habits, diabetes and ageing precede the onset of vascular disease, thereby pointing to an underlying causality (Madamanchi et al., 2005). Moreover, oxidative stress appears to be associated with all CVD risk factors and also with the occurrence of the atherosclerotic plaque (Hulsmans and Holvoet, 2010).

Over the last 30 years, it became apparent that many of the observed phenomena in CVD can be attributed to an imbalance of the redox-states within the body. Generally, oxidative stress is caused when a disturbance occurs in the balance between free radical production and the antioxidant defences within a biological system. However, on a molecular level, oxidative stress results from cellular metabolism of external environmental factors, i.e. air pollutants. Several different events occur such as cell signal transduction, activation, proliferation, and altered organ function and all lead to different molecular and physiological outcomes (Hoeschen, 1997). Due to oxidative stress, the molecular outcome results in damage to cell structures and functions, specifically to carbohydrates, nucleic acids, lipids and proteins (Rahal et al., 2014). The physiological outcome causes diseases such as atherosclerosis, hypertension, diabetes, asthma, neurological issues, ischemia, respiratory problems, pulmonary disease, and pulmonary fibrosis (Birben et al., 2012). Oxidative stress is also involved in the aging process and plays important roles in the development of CVD (Cui et al., 2012). Perturbation of the homeostatic redox balance can also occur through environmental factors, such as smoking (Aseervatham et al., 2013), sedentary lifestyle (Al-Gubory, 2014), stress (Limon-Pacheco and Gonsebatt, 2009), and even in effects of space radiation and microgravity (Takahashi et al., 2017).

CVD and oxidative stress are tightly linked to each other. They both have factors that relay on each other to impact further prognosis. All the various CVDs have an underlying similarity: oxidative stress and the role of Reactive Oxygen Species (ROS). ROS are created through different external and internal factors, such as the mitochondrial electron transport chain, NADPH oxidases, xanthine oxidases, arachidonic acid, lipoxygenases, and reactive nitrogen species (RNS), as well as nutritional elements or what type of energy is produced within the body. Due to the multiple problems that ROS can cause to a biological system, redox homeostasis has to be maintained via enzymatic and non-enzymatic antioxidants, and endogenous antioxidants (Cervantes Gracia et al., 2017). Redox maintenance and regulation is critical for any biological system, whether physical, molecular, or cellular that is under pressure for potential ROS overproduction.

Several research methodologies and techniques are being explored to find possible strategies to combat the oxidative stress species associated with CVD. Most strategies focus on targeting ROS via drug-inducement and experimental therapeutic agents (Fearon and Faux, 2009). Oxidative stress can be induced with specific drugs resulting in overproduction of ROS, however, this type of oxidative stress causes more damage to the cell by affecting the cell's DNA, lipids and proteins and eventually results in cell apoptosis. With drug induced oxidative stress the biological system experiences high levels of toxicity towards tissues and organs of the body. In order to combat the toxicity and the excess of ROS production therapeutic agents are being experimented on to observe their reactions to CVD and oxidative stress species within cells and organs (Pashkow, 2011), such as extracts of medicinal plants and many variants of antioxidant molecules that are frequently used in experimental therapeutics.

7.2 The ROS-Scavenging System in Cardiovascular Disease

7.2.1 Chemistry of Free Radical and Non-Radical Species

Reactive oxygen species (ROS) are oxygen (O_2) derivatives that are known to react with intracellular molecular entities such as lipids, proteins, and DNA. Figure 7.1 shows a summarized overview of the various metabolic reactions and molecular species involved in ROS-mediated pathways.

The successive electron acquisition by the superoxide anion ($O_2^{\bullet -}$), the primary form of ROS, and further interactions with other molecules either by metal- or enzyme-based catalyzed reactions leads to the production of secondary ROS (Valko et al., 2007). They can be classified as free radical species such as $O_2^{\bullet -}$, the hydroxyl radical (·OH), and the hydroperoxy radical (HOO·), and in non-radical species such as the hydrogen peroxide (H_2O_2) and hypochlorite (ClO−) (Valko et al., 2007). Superoxide radicals can also interact with nitric oxide (NO) to yield, in a diffusion-controlled reaction, peroxynitrite (ONOO−), a very reactive intermediate with a relatively short half-life, which possesses a remarkable cytotoxic capacity against pathogens and concomitant deleterious functions in disease as

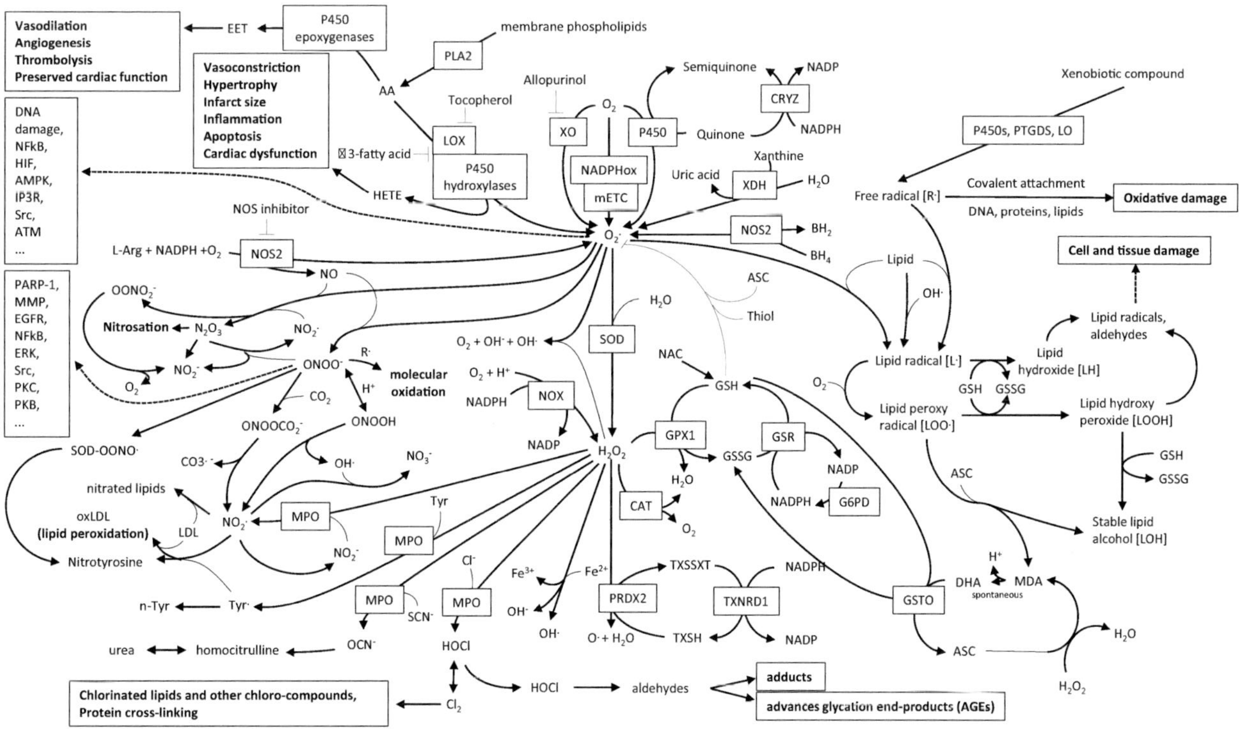

Figure 7.1 Metabolic pathways and ROS- and RNS-dependent enzymatic reactions.

an endogenous toxicant (Radi, 2013). Peroxynitrite belongs to the family of Reactive Nitrogen Species (RNS) are a family of anti-microbial molecules that are derived from NO and $O_2^{\bullet-}$. RNS are produced by enzymatic activity of nitric oxide synthase 2 (NOS2) and NADPH oxidase. Nitric oxide is the considered the most prominent of the RNS family and is found in mitochondria (Bolisetty and Jaimes, 2013). Under physiological conditions, RNS are radicals that are involved in cellular signalling, vasodilation, and the body's immune response (Ozcan and Ogun, 2015). Table 7.1 lists the most common RNS species, however only nitric oxide and nitrogen dioxide are free radicals, whereas peroxynitrite is not since NO and NO_2 molecules are uncharged and contain an unpaired valence electron in the outer shell (Pacher et al., 2007).

Table 7.1 A list depicting reactive nitrogen species (RNS) and their functions and molecular involvements in a biological system

Name	Formula	Function
Nitric oxide	NO	Signalling molecule that modulates the functions of cardiomyocyte, generation of ATP, and contraction of sarcomeres
Peroxynitrite	$ONOO^{\bullet}$	Damages DNA, proteins and other cellular structures
Nitrogen dioxide	NO_2	Produces inflammation to the lungs and absorbs UV light

7.2.2 CVD-Related Conditions and Risk Factors and Their Association with Oxidative Stress

ROS are widely recognised to be involved in a wide-spectrum of human diseases, affecting the cardiovascular system, several conditions related with metabolic disorders, neurodegenerative diseases, cancer, and other ailments, yet only a few antioxidant drugs are currently in use in the clinical setting. Examples of these are Ebselen for the treatment of cerebral infarctions, edaravone for ischemic stroke, alpha-lipoic acid in diabetic neuropathy, flavonoids such as diosmin, hesperidin and oxerutins for chronic venous insufficiency, and other flavonoids as the baicalein and cathechins for therapeutic intervention in osteoarthritis and an L-cysteine precursor, the *N*-acetylcysteine, mostly applied in the treatment of acetaminophen (paracetamol) overdose (Suzuki, 2009), but has pharmacological roles in chronic obstructive pulmonary disease (COPD) (Fowdar et al., 2017), cystic fibrosis (Conrad et al., 2015), and some psychiatric disorders (Dean et al., 2011).

The human redox-machinery in a "non-disease" state is able to counteract the harmful effects of ROS and therefore prevent or even resolve most of the oxidative stress occurrences. However, if the system is subjected to an acute or long-term exposure to oxidant agents, and if an underlying clinical condition exists, this protective process could be diminished or event fully impaired.

Since most disease conditions do not occur alone but exist as a concatenate of dynamically interacting processes and pathways, it is conductive to assess cardiovascular diseases in a more holistic manner, incorporating and considering clinical conditions such as hyperglycaemia, hyperinsulinemia, hypertension and diabetic cardiomyopathy, as well as adding known triggers of oxidative stress and disease-specific key molecules (Fig. 7.2). This integrated view will undoubtedly allow to obtain a global view of the molecular events and trade-offs associated with clinical outcomes that can lead to an increased risk of atherothrombosis.

There is a strong association of obese patients developing type 2 diabetes mellitus (T2DM) (Eckel et al., 2011), thereby, secretion within the adipose tissue of free fatty acids (FFAs) and stimulatory cytokines directly impacts on insulin sensitivity (Kang et al., 2016) (Fig. 7.2). FFA-induced ROS production reduces activation of insulin receptor substrate 1 (IRS1) and the intracellular signal transduction PI3K-Akt pathway, and thereby leads to a reduction of the stimulus of the facilitated glucose transporter member 4 (SLC2A4/GLUT4) (Saltiel and Kahn, 2001). FFA-induces impairment of the PI3K-pathway, reduces Akt activity, and phosphorylation of endothelial NO synthase (eNOS), which in turn results in diminished production of NO, leading to endothelial dysfunction, and vascular remodelling (Fig. 7.2). Consequently, accumulation of ROS induces activation of the transcription factor NF-kappaB, which leads to CD14+ monocytes/macrophages increased release of pro-inflammatory adhesion molecules and cytokines (Wang Y. et al., 2011).

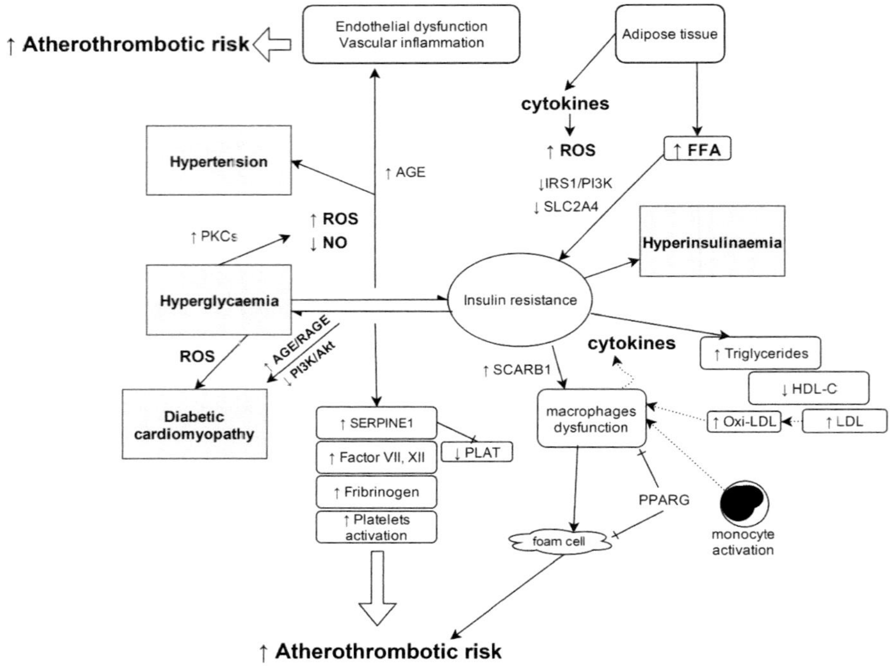

Figure 7.2 Conceptual diagram of the molecular events and trade-offs associated with clinical outcomes that can lead to an increased risk of atherothrombosis. Advanced glycation end products (AGE), advanced glycosylation end product-specific receptor (AGER/RAGE), free fatty acids (FFA), Solute carrier family 2, facilitated glucose transporter member 4 (SLC2A4/GLUT4), high-density lipoprotein cholesterol (HDL-c), low-density lipoprotein (LDL), nitric oxide (NO), protein kinases C (PKCs), insulin receptor substrate 1 (IRS1), peroxisome proliferator-activated receptor gamma (PPARG), plasminogen activator inhibitor 1 (SERPINE1), tissue-type plasminogen activator (PLAT), phosphatidylinositide 3-kinase (PI3K), reactive oxygen species (ROS), scavenger receptor class B member 1 (SCARB1).

Elevation of glucose levels targets, amongst others, the mitochondrial electron transport chain (mETC), which leads to an increased production of $O_2^{\bullet-}$. Concomitantly, consecutive gains in $O_2^{\bullet-}$ levels are achieved by several loop-cycles of ROS inducing protein kinase C isoforms activation (PKCs). This event promotes increased activity of the NADPH oxidase, increased levels of the signalling adapter SHC-transforming protein 1 (SHC1), as well of prostaglandin G/H synthase 1 (PTGS1) levels and thromboxane production, resulting on an compromised NO release (Cosentino et al., 2003; Cosentino et al., 2008) (Fig. 7.1). Mitochondrial ROS (mtROS), in turn, trigger known signalling cascades involved in the pathogenesis of many cardiovascular conditions, which includes advanced glycation end-products (AGEs) and their receptors (RAGEs), PKC's signalling and the hexosamine biosynthesis pathway (HSP) (Muntean et al., 2016) (Fig. 7.2).

Consecutive monocyte differentiation into macrophages and their local accumulation in the adipose tissue is recognized as a key event in insulin resistance (IR) (Patsouris et al., 2014). The local deposition of insulin resistant macrophages up-regulates the scavenger receptor class B member 1 (SCARB1), which promotes foam cell formation and then the establishment of an atherosclerotic lesion (Chistiakov et al., 2017). In case of activation of the peroxisome proliferator-activated receptor gamma (PPARG), the latter described processes can be overturned by transcriptionally regulation of the macrophage insulin signalling activation (Chawla, 2010) (Fig. 7.1). ROS accumulation enduring IR and long-term hyperglycaemia adds-up to both cardiac and structural defects such as endothelial dysfunction and tissue remodelling.

7.3 Types of ROS and Main Sources in CVD

7.3.1 Mitochondrial Electron Transport Chain

The mitochondrial electron transport chain is involved in the transport of electrons and is embedded within the inner mitochondrial membrane. It generates a proton gradient and regenerates electron carriers. Two specific carriers, NADH and FADH2, pass through the electron transport chain, releasing electrons through an oxidation reaction (Turrens, 2003). These electrons are used in catabolic processes to release energy in the form of ATP to the rest of the body (Stein and Imai, 2012). Both NADH and FADH2 are principally produced in the TCA cycle during cellular respiration, and NADH donates electrons in redox reactions when the electron energy is high (Murphy, 2009). Instead of donating, FADH2 feeds the electrons into the transport chain directly through the mitochondrial complex II (Quinlan et al., 2012). The mitochondrial electron chain is a transport for potential ROS to enter any biological system through exchanging, passing, and adding of electrons (Liemburg-Apers et al., 2015).

7.3.2 Nicotinamide Adenine Dinucleotide Phosphate (NADPH) Oxidases

Nicotinamide adenine dinucleotide phosphate-oxidase (NADPHox) is a multi-subunit enzyme complex consisting of up to five subunits, with the NOX enzyme responsible for the conversion of oxygen to superoxide. The five NOX isoforms NOX1-5 (Table 7.2) are involved with immune defence, cell signalling, transporting electrons across membranes (Panday et al., 2015) and can be found within the plasma membrane and the membrane of phagosomes (Meitzler et al., 2014). They are considered to be key producers of ROS in various cells.

In vascular disease, NADPHox isoforms NOX1, NOX2, NOX4 and NOX5 are expressed in cellular components of arteries and veins, with their assembling site being primary in membranes of endothelial cells (ECs) and smooth muscle cells (SMCs), and can also be found in fibroblasts within connective tissue (Giardino et al., 2017). NOX1, NOX2 and NOX4 occur as multimeric protein complexes, thereby the process of electron transfer involves a NOX subunit and different-NOX-type regulatory elements (Drummond and Sobey, 2014). NOX2 regulatory subunits includes cytochrome b-245 light chain (p22phox, CYBA), neutrophil cytosol factor 2 (p67phox, NCF2), p47phox (NCF1), neutrophil cytosol factor 4 (p40phox, NCF4), and ras-related C3 botulinum toxin substrate 1 (RAC1) (Drummond and Sobey, 2014; Konior et al., 2014). NOX1 contains the same regulatory subunits as NOX2, but does not require p40phox to carry on its enzymatic activity (Konior et al., 2014). In vascular smooth muscle cells (VSMCs) an additional enhancer, the polymerase delta-interacting protein 2 (POLDIP2), binds to the NOX4-p22phox complex and increases its catalytic activity by 3-fold (Lyle et al., 2009) whereas NOX5 ability to produce ROS only requires calcium ions for modulation of its catalytic activity (Guzik et al., 2008). Most of the NOX enzymes are characterized by O_2^- production, from the conversion of diatomic oxygen (O_2) using NADPH as electron donors, NOX4 isoform seems to have as main end-product hydrogen peroxide (H_2O_2) (Konior et al., 2014).

Table 7.2 The basic functions and description of NADPH oxidase isoform enzymes that are involved with causes of oxidative stress

	Description	Function
NOX1	Participates in cellular pH regulation, generates superoxide and might conduct H^+ ions	Voltage-gated ion/proton channel activity, protein binding, superoxide-generating NADPH oxidase activity, and oxidoreductase activity
NOX2 (CYBB)	Critical component for membrane-bound oxidase of phagocytes and participates in cellular pH regulation	Flavin adenine dinucleotide binding, heme binding, metal ion binding, protein heterodimerization activity, superoxide-generating NADPH oxidase activity, and voltage gated ion channel activity
NOX3	Role in biogenesis of otolith and detection of gravity	Superoxide-generating NADPH oxidase activity and oxidoreductase activity
NOX4	Involved in redox signalling within vascular cells, capable of damaging nuclear DNA, modulates nuclear activation of ERK1, ERK2, and transcription factor ELK1 and increased activity relatable to NOX1	Electron transfer activity, flavin adenine dinucleotide binding, heme binding, modified amino acid binding, NADPH oxidase activity, nucleotides binding, oxygen sensor activity, and superoxide-generating NADPH oxidase activity
NOX5	Calcium dependent, involved with endothelial generation of ROS, and may regulate redox-dependent processes in lymphocytes and spermatozoa	Calcium ion binding, flavin adenine dinucleotide binding, heme binding, NADP binding, proton channel activity, and superoxide-generating NADPH oxidase activity

NOX homologues are linked to the pathogenesis of hypertension (Meijles and Pagano, 2016; Sedeek et al., 2009), and a knockout of p47phox, a NOX2 regulatory subunit, in apolipoprotein E-deficient (ApoE$^{-/-}$) mice reduced blood pressure responses to angiotensin II and attenuated atherosclerotic lesions (Dikalova et al., 2005; Matsuno et al., 2005). Additionally, mice with disrupted NOX1 show reduced blood pressure in response to angiotensin II (Matsuno et al., 2005). On the other hand, up-regulation of the same enzyme isoform in SMCs lead to overproduction of ROS and an amplified blood pressure response to angiotensin II (Dikalova et al., 2005). Moreover, there is a correlation between increased coronary expression of NADPH oxidases isoforms (p22phox, NOX2, p47phox, and p67phox) with atherosclerosis (Guzik et al., 2006).

Several NOX isoform have also been linked to the regulation of vascular inflammation. NOX1, NOX2 and NOX5 are involved in the regulation of the expression of endothelial adhesion molecules, such as VCAM1, ICAM1 and the P- and E-members of the selectins family (Liu et al., 2011; Sorescu et al., 2004). Augmented expression of those proteins conducts to leukocyte invasion into the sub-endothelial layer, a process that is linked to early onset of atherosclerotic injury or hypertensive vascular dysfunction (Chistiakov et al., 2015; Krieglstein and Granger, 2001). In vascular cells, NOX1, NOX2, NOX4 and NOX5 have been linked to several pathophysiological conditions such as hypertension, atherosclerosis, post-injury restenosis and diabetic-associated vascular lesion (Guzik et al., 2008; Sorescu et al., 2002), and sepsis-induced cardiomyopathy can be prevented by inhibition of NADPH oxidase 2 (NOX2) (Joseph et al., 2017).

Over the years, numerous pharmacological compounds were developed targeting NADPHox (Table 7.3) since their enzymatic activity requires the interaction of multiple subunits, and thus give a wider spectrum of intervention points. NF02, a peptide originally isolated from a flavin reductase (EC 1.5.1.30) of bacteria, inhibits NOX1 in a specific manner, without impacting on NOX2 or xanthine oxidase activity (Mousslim et al., 2017). Another peptide, NoxA1ds, mimics a putative activation domain (Phe-199) of p67phox (NOXA1), and by binding to NOX1, inhibits its activation (Ranayhossaini et al., 2013). This was demonstrated by screening of NoxA1ds effects in human pulmonary artery endothelial cells (HPAEC), where under hypoxia it fully inhibited production of $O_2^{\bullet-}$ and attenuated migration of ECs (Ranayhossaini et al., 2013).

Table 7.3 NADPH antagonists, isoforms-specificity and vascular-cardio effects

Inhibitor	Mode of action	Vascular-cardio effects	IC50s (μM)							References
			NADPH-ox	NOX1	NOX2	NOX3	NOX4	NOX5	XO	
Apocynin	p47phox-binding membrane complex (Stolk et al., 1994); NOX5 inhi. effect can be due to scavenging	Dose-dep. reduction of P-selectin, VCAM1, platelet adhesion, monocyte accumulation in LDL-$R^{-/-}$ (Liu et al., 2013)	10[a]					↓		Baumer et al., 2007; Heumuller et al., 2008; Liu et al., 2013
		Attenuation of ROS and hypertension in rats (Baumer et al., 2007)								
		Prevention of NADPHox expression and activation in fructose-fed animals (Shinozaki et al., 2004) and STZ-diabetes (Olukman et al., 2010)								

(*Continued*)

Table 7.3 (*Continued*)

Inhibitor	Mode of action	Vascular-cardio effects	IC50s (μM)							References
			NADPH-ox	NOX1	NOX2	NOX3	NOX4	NOX5	XO	
DPI	non-selective inhibition of flavin containing oxidases; inhibition of NF-κB, inhibition of ChE activity and Ca^{2+} pump; iodide-specific transporter	Broad spectrum of metabolic action and toxicity, avoid use as vascular NOX inhibitor (O'Donnell et al., 1993)	5.6*							Massart et al., 2013
VAS2870	may not inhibit NOX-dependent ROS production (Gatto et al., 2013); Pan-NADPH-ox inhibitor	Inhibition of PDGF-stimulated NADPH-ox activity in primary VSMCs (ten Freyhaus et al., 2006), and ox-LDL-mediated ROS formation in HUVECs (Stielow et al., 2006) Doubtful effects on NADPH-ox (Gatto et al., 2013)		↓	neutrophils: 10.6; HUVECs: 10; HL-60 cells: 2		↓			Stielow et al., 2006; ten Freyhaus et al., 2006; Wind et al., 2010
VAS3947				0.25[c]	2.00[d]		13.00[a]		ns[e]	Wind et al., 2010

Inhibitor	Mode of action	Vascular-cardio effects	IC50s (μM)							References
			NADPH-ox	NOX1	NOX2	NOX3	NOX4	NOX5	XO	
Fulvene-5	UNK	Blunted cardiac arrhythmia in zebrafish model (Zhang et al., 2014)			↓ (~5.00[c])		↓ (~5.00[c])			Bhandarkar et al., 2009
Naloxone	UNK				2					Wang et al., 2012
Perhexiline	p47phox-binding	Approved prescription drug for angina			1.5-3.6; 3.0[d]				ns[f]	Gatto et al., 2013; Kennedy et al., 2006
Grindelic acid	UNK				>20.0[d]		2.06[c]	>20.0[c]	ns[g]	Kofler et al., 2013
Phenantridinones	UNK						0.17[c]		ns[h]	Borbely et al., 2012
Shionogi	UNK				0.56[d]					Gatto et al., 2013
S17834	activation of AMPK	Reduced expression of VCAM, ICAM1 and E-selectin TNF-α-stimulated; AoAther. in ApoE$^{-/-}$ mice is reduced by 60% (Cayatte et al., 2001)	↓							Cayatte et al., 2001; Zang et al., 2006

(Continued)

Table 7.3 (*Continued*)

			IC50s (μM)							
Inhibitor	**Mode of action**	**Vascular-cardio effects**	**NADPH-ox**	**NOX1**	**NOX2**	**NOX3**	**NOX4**	**NOX5**	**XO**	**References**
AEBSF	p47phox/ p67phox-cytochrome b-559 interaction (9148950)	prevents activation of NADPH-ox in stimulated macrophages and in cell-free systems; Lack of specificity (Wind et al., 2010)	870[a]							Diatchuk et al., 1997
Celastrol	p47phox-p22phox-interface, non-specific	Decreases ROS production by NOX1 inhibition in mice microvascular EDCs (Wu et al., 2009)		0.41[c]	0.59[c]		2.7[c]	3.13[c]	ns[e]	Jaquet et al., 2011
Ebselen	p47phox and p67phox translocation			0.15[c]	0.5[c]		>50.0[c]	0.7[a]	ns[f]	Smith et al., 2012
GKT136901	inhibition of NOX1/4-derived ROS production	Reduction of PPAR-gamma expression; H_2O_2 production and cell proliferation in PAEDCs and SMCs in pulmonary hypertension (Green et al., 2012)		0.16[a,h]	1530[a,b]		0.17[a,b]	0.45[a,b]	>30 000[b]	Garrido-Urbani et al., 2011; Laleu et al., 2010

Inhibitor	Mode of action	Vascular-cardio effects	IC50s (μM)							References
			NADPH-ox	NOX1	NOX2	NOX3	NOX4	NOX5	XO	
GKT137831	UNK	Prevention of atherosclerosis in diabetic ApoE-/- mice (Gray et al., 2013); NOX4/1-mediated effects should be carefully observed		0.14[a,b]	1750[a,b]		0.11[a,b]	0.41[a,b]	>100*	Gaggini et al., 2011; Musset et al., 2012
GSK2795039	competitive inhibition, NADPH competitive			6.18±0.11[a]						Hirano et al., 2015
ML171	UNK	NOX1 inhibition in colon cancer and vascular dysfunction (Gianni, Nicolas, et al., 2010)		0.25[c]	5[c]	3.00[c]	5[c]		5.5	Gianni, Taulet, et al., 2010
17-DMAG	HSP90 inhibitor	reduction of NADPH-ox activity in VSMCs and monocytes; decreased expression of NOX1 and NOXO1	↓	↓						Madrigal-Matute et al., 2012
ACD084	UNK				>5.00[c]		3.08[c]	>5.00[c]	ns[i]	Kofler et al., 2013

(*Continued*)

Table 7.3 (*Continued*)

Inhibitor	Mode of action	Vascular-cardio effects	IC50s (μM)							References
			NADPH-ox	NOX1	NOX2	NOX3	NOX4	NOX5	XO	
DGLA	cyclooxygenase substrate metabolized in macrophages to prostaglandin	Decreases aortic expression of p22phox and gp91phox in ApoE$^{-/-}$ mice (Takai et al., 2009)		↓						Takai et al., 2009
Tyrphostin AG490	JAK2 inhibitor	Reduces activity of aortic NOX1, NOX2 and NOX4 in ApoE-/- mice		↓	↓		↓			Fenyo et al., 2011
11g	p47phox-p22phox-interface			ns	20		ns	ns	ns	Cifuentes-Pagano et al., 2013
11h	p47phox-p22phox-interface			ns	32		ns	ns	ns	Cifuentes-Pagano et al., 2013
XJP-1	modulation of the PI3K/Akt/eNOS pathway	Attenuation of p22phox and p47phox expression in HUVECs			↓					Fu et al., 2013
EXP3179	inhibition of Losartan	Potential inhibition of NADPH-ox in								

Inhibitor	Mode of action	Vascular-cardio effects	IC50s (μM)							References
			NADPH-ox	NOX1	NOX2	NOX3	NOX4	NOX5	XO	
		hypertension (Fortuno et al., 2009); Inhibition of protein kinase C-dependent p47phox subunit phosphorylation			↓					Fortuno et al., 2009
NOXA1ds	NOX1-NOXA1 interface			0.02[a]	ns[a,h]		ns[a,h]	ns[a,h]	ns[h]	Ranayhossaini et al., 2013
NOX2ds-tat	p47phox-p22phox interface; Lack of activity by oral intake	Blocks Ang II-induced O2- and reduces Ang II-induced hypertension (Rey et al., 2001)		ns	0.74		ns	ns	ns	Csanyi et al., 2011; Rey et al., 2001; Shi et al., 1996
		Ameliorates endothelial dysfunction in hypertension Dahl salt-sensitive rats by reducing vascular O2- and ONOO- (Zhou et al., 2006)								
PR-39	binding to SH3-domain of p47phox	Lessens PMN-induced cardiac dysfunction in a IR rat model (Ikeda et al., 2001)			↓					Diatchuk et al., 1997; Ikeda et al., 2001

(*Continued*)

Table 7.3 (*Continued*)

			IC50s (μM)							
Inhibitor	**Mode of action**	**Vascular-cardio effects**	**NADPH-ox**	**NOX1**	**NOX2**	**NOX3**	**NOX4**	**NOX5**	**XO**	**References**
Cluster E	NADPH-binding site				8.4					Dahan et al., 2012
Cluster I	UNK				5.38					Dahan et al., 2012
Pep1	EF-hand-AID interface							30		Tirone et al., 2010
Pep3	EF-hand-AID interface							30		Tirone et al., 2010
Melittin	EF-hand	Inhibits VSMC proliferation by inducing apoptosis via suppression of NF-kB and Akt activation (Son et al., 2006)						0.1		Comte et al., 1983
Tertiary sulphonylureas 23	inhibit NOX4-dependent signalling						27			Xu et al., 2018
Tertiary sulphonylureas 24	inhibit NOX4-dependent signalling						3.7			Xu et al., 2018

Inhibitor	Mode of action	Vascular-cardio effects	IC50s (μM)							References
			NADPH-ox	NOX1	NOX2	NOX3	NOX4	NOX5	XO	
Tertiary sulphonylureas 25	inhibit NOX4-dependent signalling						0.5			Xu et al., 2018
Imipramine blue	inhibits NADPH-mediated-ROS production, PI3K pathway and alters expression of actin regulatory elements		↓							Munson et al., 2012
NF02	no $O_2^{\bullet-}$ scavenging activity			16.7±5.4[a]						Mattioni et al., 1989

Note: Inhibits, but no IC50 determined (↓), no significant (ns), not determined (white/blank space), unknown (UNK), inhibitory constant (*K_i), IC50 values were determined.

Abbreviations: STZ, streptozotocin ; ChE, cholinesterase; EDCs, endothelial cells; CVI, chronic venous insufficiency; SMCs, smooth muscle cells; AoAther., aortic atherosclerosis; NADPH-ox, NADPH oxygenase; PAEDCs, pulmonary artery endothelial cells; IR, ischemia-reperfusion; VSMCs, vascular smooth muscle cells; HUVECs, Human umbilical endothelial cells; CYBA, p22phox; NCF1, p47phox; NOX2, gp91phox; XO, xanthine oxidase; S17834, benzo(b)pyran-4-one derivative; AEBSF, 4-(2-aminoethyl)-benzenesulphonyl fluoride; ML171, 2-acetylphenothiazine; 17-DMAG, 17-dimethylaminoethylamino-17-demethoxygeldanamycin; DGLA, dihomo-gamma-linolenic acid; 11g, N-pentyl substituted analog of (1SR,4RS)-3,3-Dimethyl-1,2,3,4-tetrahydro-1,4-(epiminomethano)naphthalene; 11h, N-methylenethiophene substituted analog of (1SR,4RS)-3,3-Dimethyl-1,2,3,4-tetrahydro-1,4-(epiminomethano)naphthalene; XJP-1, (±)7,8-Dihydroxy-3-methyl-isochromanone-4; gp91ds-tat, Nox2ds-tat; DPI, diphenyleneiodonium; PR-39, proline/arginine-rich antimicrobial peptide; PDGF, platelet-derived growth factor; VAS3947, Novel triazolo pyrimidine; VAS2870, 3-benzyl-7-(2-benzoxazolyl)thio-1,2,3- triazolo[4,5-d]pyrimidine; AMPK, AMP-activated protein kinase.

[a]cell-free assay, [b]Ki exists, [c]over-expressing cells, [d]native cells, [e]no significant inhibition for concentrations of: 30 μM or [f]100 μM, [g]20 μM, [h]10 μM, [i]5 μM.

ML171 (2-acetylphenothiazine) another effective NOX1 inhibitor was discovered by high-throughput screening (HTS) and is effective in the reduction of invadopodia initiation and lessens extracellular matrix (ECM) degradation in human colorectal adenocarcinoma cell lines (Gianni, Nicolas, et al., 2010; Gianni, Taulet, et al., 2010). Additionally, ML171 prevented aldosterone-induced VSMC's remodelling and reduced vascular hyper-contractility in experimental rat-models of severe hypertension and cerebral stroke (Harvey et al., 2017).

GKT137831 and GKT136901 are lead-based derivatives of the pyrazolopyridine class of compounds and show dual inhibitory ability against NOX4 and NOX1 (Aoyama et al., 2012; Laleu et al., 2010). Inhibition of NOX4 by those compounds leads to reduced peroxisome proliferator-activated receptor gamma (PPAR-gamma) expression, induces H_2O_2 production and cell proliferation in human pulmonary artery endothelial cells (PAEDCs) and smooth muscle cells (SMCs) of hypoxia induced human pulmonary hypertension cell lines (Green et al., 2012). Impairment of NOX1-derived ROS with GKT137831 prevented the development of atherosclerotic lesions in double-knockout apolipoprotein E ($ApoE^{-/-}$) diabetic mice (Gray et al., 2013). NOX2 in vivo inhibition was originally explored using the novel 7-azaindole derivative GSK2795039 (Hirano et al., 2015). This novel NOX2 agonist unlike the inhibitory mode of action of diphenyleneiodonium (DPI) has specificity for NOX2 and its pharmacological profile is due to competitive enzymatic inhibition (Hirano et al., 2015).

Apocynin a NADPHox agonist exerts its mode of action by binding to the p47phox subunit (Stolk et al., 1994). Experimental studies with animal models using this inhibitor have shown reduction of ROS and an overall enhancement of the hypertensive state in rats (Baumer et al., 2007), and it could also be demonstrated to lead to an inhibition of both NADPHox expression and activation in streptozotocin (STZ)-diabetes (Olukman et al., 2010) and high fructose-fed mice (Shinozaki et al., 2004). Additionally, apocynin seems to reduce, in a dose-dependent mode, the number of cell adhesion molecules, such as of P-selectin and VCAM1, and also reduce platelet adhesion and monocyte accumulation (Liu et al., 2013). Other promising NOX4 inhibitors such as the recently designed novel tertiary sulphonylureas (23–25) are being discovered using computational approaches, but those need further scientific scrutiny of functional classification and characterization of their pharmacological profile (Xu et al., 2018).

7.3.3 Xanthine Oxidase

Xanthine oxidase (XO), a form of xanthine oxidoreductase, is important in the catabolism process of purines such as uric acid (UA) (Schuchardt et al., 2017). It readily donates electrons to diatomic oxygen yielding O_2^- and H_2O_2. UA is the final end-product of the purine catabolism and is formed from xanthines and hypoxanthines, and XO may be responsible for most of the ROS production (Berry and Hare, 2004). The catalytic reaction on how it generates ROS is shown in three separate, but relatable chemical reactions:

$$\text{Xanthine} + NAD^+ + H_2O \leftrightarrow \text{urate} + NADH$$
$$\text{Hypoxanthine} + NAD^+ + H_2O \leftrightarrow \text{xanthine} + NADH$$
$$\text{Xanthine} + H_2O + O_2 \leftrightarrow \text{urate} + H_2O_2$$

Xanthine oxidase has the ability to catalyse each of these chemical reactions by using oxygen as the electron acceptor on CH and CH_2 groups (Harrison, 2002), and most importantly can produce superoxides (Lee et al., 2014).

Hyperuricemia can be secondary to either an exaggerated production of UA that follows high cellular turnover conditions or, most frequently, to a low renal excretion in patients with impaired renal function. Recent data suggest that serum UA (SUA) at high-normal level is associated with cardiovascular disease risk factors and cardiovascular disease, often being a predictor of incident events. Preliminary data suggest that the reduction of SUA level in subjects with normal-high SUA could prevent at least a part of target-organ damage related to high SUA, especially when xanthine oxidase is selectively inhibited (Borghi et al., 2014; Maia et al., 2015; Schuchardt et al., 2017). Inhibition of uric acid reduces cardiovascular events in the clinical setting (Schuchardt et al., 2017), but it is not well-established if uric acid is a marker of poor prognosis in acute and chronic heart failure (HF) by itself or is an active participant in the disease pathogenesis (Kaufman and Guglin, 2013). Hypoxia-induced of liver cancer cells

(HepG2) and primary mammary epithelial cell (HMECs) lines, under the presence of XO/XD (xanthine dehydrogenase) and aldehyde oxidase (AO) lead to derivative of superoxide species such as NO production (Maia et al., 2015).

Pharmacological intervention targeting this class of proteins has shown very little promise. XO inhibitory drugs, such as Oxypurinol, administered to CVD and hyperuricaemia patients did not shown benefit. More multi-centre clinical trials with controlled randomisation and further patient stratification are required, not only to uncover the connection of uric acid with disease pathogenesis, but also with other associated functions such as assessment of renal function (Zhang J. et al., 2017).

7.3.4 Lipoxygenases

Lipoxygenases (LOXs) constitute a family of non-heme iron-containing enzymes that produce fatty acid hydroperoxides by catalysing the oxidation of poly-unsaturated fatty acids and lipids that contain a *cis,cis*-1,4-pentadiene structure (Neau et al., 2009). It is these hydroperoxides that form a secondary oxidation product. Lipoxygenases are involved in the synthesis of signalling molecules that contribute to structural or metabolic changes within the cell (Brash, 1999). The most relevant members in the CVD context of the lipoxygenase (LOX) family are 5-lipoxygenase (5-LOX) and 12/15-LOX isoforms which are able to convert fatty acids into several lipid mediators using arachidonic acid (AA) as substrate (Kuhn, 2004).

LOXs are expressed in many cellular microenvironments, including epithelium, immune and tumour, which are involved in a diversity of physiological functions and pathological states, such as inflammation, asthma, skin disorders (e.g. ichthyosis), and as well oncogenesis (Mashima and Okuyama, 2015). When 5-LOX oxidizes AA, in addition to the consecutive reduction steps to yield hydroperoxides (-ROOH) and hydroxides (OH^-), they produce key inflammatory mediators, leukotrienes, which transduce pro-inflammatory signals to elicit leukocytes adhesion, chemotaxis and aggregation (Radmark et al., 2007). In animal models of atherosclerosis was observed that targeting leukotriene receptors lessens disease progression (Poeckel and Funk, 2010), in humans this was confirmed by following sections of the population treated with leukotriene receptor antagonists (LTRAs), such as in asthma conditions, and they infer an association with diminished cardiovascular risk (Back et al., 2013; Ingelsson et al., 2012). On the other hand, consecutive lipoxygenation of AA produces lipoxins, A4 (LXA4) and B4 (LXB4), which are anti-inflammatory and transduce resolution of inflammation and restores tissue homeostasis (Chandrasekharan and Sharma-Walia, 2015; Karra et al., 2015). Supplementation with omega-3 was seen to induce inflammation resolution and decrease atherosclerosis (Back, 2017). In atherosclerotic mice, the G protein-coupled lipoxin A4 receptor (ALX)/formyl peptide receptor (FPR2) is activated by lipoxins and peptide agonists, in which FPR2/ALX disruption lessens the size of the atherosclerotic lesion, but promotes plaque instability (Petri et al., 2015). In cardiac tissue of streptozotocin (STZ)-induced diabetic mice, 12/15-LOX isoforms triggered inflammation and oxidative stress, in opposition, Alox15-deficient mice yielded decreased expression of TNF-alpha and nuclear factor (NF)-κB and eradicated STZ-induced reactive oxygen species production (Suzuki et al., 2015). Targeting 12/15-LOX isoforms could lead to a potential treatment of diabetic cardiomyopathy (DCM), a well-known risk factor of the atherogenesis process and heart failure (Kayama et al., 2009; Suzuki et al., 2015).

A number of pharmacological compounds have been used to target LOX. The most promising candidate so far is GSK2190915, a 5-LOX activating protein (FLAP) agonist that inhibits the production of LTB4 and other cysteinyl-derived leukotrienes (Bain et al., 2013; Stock et al., 2011). Results of a Phase I study indicated that there was no clear difference in adverse events between placebo and drug-treated subjects in both cohorts in Western Europe (Trial-EURDACT2007-00484872) and Japan (Trial-NCT00955383) (Bain et al., 2012). 5-LOX induces neutrophilic inflammation by its recruitment in response to LTB4. Importantly, 5-LOX activity is strictly regulated by FLAP though the distribution of 5-LOX in the nucleus. Currently, pharmacological drugs targeting FLAP are under active scientific scrutiny (Mashima and Okuyama, 2015).

7.3.5 Myeloperoxidase

Myeloperoxidase (MPO), a heme protein released by leukocytes, plays a key role in inflammation and oxidative stress. MPO's anti-microbicidal activity is due to its ability to oxidize chloride (Cl^-), bromide (Br^-), iodine (I^-) and thiocyanate (SCN^-) by H_2O_2, thus producing their corresponding hypohalous acids (HOX) (Lazarevic-Pasti et al., 2015). It has become increasingly recognized that MPO performs a very important role as part of the innate immune system through the formation of microbicidal reactive oxidants, whilst it affects the arterial endothelium with a number of mechanisms that include modification of net cellular cholesterol flux and impairment of NO-induced vascular relaxation. Thus, MPO is implicated into both the formation and propagation of atheromatosis and there is substantial evidence that it also promotes ischemia through destabilization of the vulnerable plaque. Numerous studies have added information on the notion that MPO and its oxidant products are part of the inflammatory cascade initiated by endothelial injury and they are significantly overproduced at the site of arterial inflammation. Subsequent studies achieved quantification of this observation showing significant elevations of the systemic levels of MPO in a wide spectrum of cardiovascular disease scenarios with acute coronary syndromes and heart failure being the most studied (Anatoliotakis et al., 2013). There is a positive correlation of MPO plasma levels with severity of coronary artery disease (CAD) within CAD-patients co-exhibiting type 2 diabetes mellitus (T2DM) (Song et al., 2015). In opposition, other study found that five MPO polymorphisms were positively associated with MPO plasma level, but not with cardiovascular mortality (Scharnagl et al., 2014). Other groups have shown the potential role of MPO has as a biomarker for sepsis in the intensive care unit (ICU)(Schrijver et al., 2017) which could aid in the diagnosing of sepsis and/or as predictor of mortality.

MPO is a key molecular mediator in a vast array of clinical conditions, such as in cardiovascular, inflammatory, neurodegenerative, renal and immune-mediated diseases (Lazarevic-Pasti et al., 2015). MPO and its downstream pathways might be attractive targets for therapeutic intervention or even prophylaxis action of these ailments. Development of new efficient and specific inhibitors is undergoing using state-of-the-art virtual screening procedures (VSP), based on ligand-pharmacophore modelling (Soubhye et al., 2017). Oral administration of PF-1355, an MPO inhibitor, to a mice model of myocardial infarction (MI), decreased enzymatic MPO activity, likewise decreased the number of inflammatory cells and decreased left ventricular dilation (LVD). Extending the period of treatment lead to an improved outcome of both cardiac function and remodelling (Ali et al., 2016).

7.3.6 Arachidonic Acid and P450 Hydroxylases

Arachidonic acid (AA) is a poly-unsaturated fatty acid that is important for metabolism, specifically for the synthesis of prostaglandins, thromboxanes and leukotrienes (Elinder and Liin, 2017). It is specific to the body's muscles, liver and brain. AA has the ability control the permeability of cell membranes which then influences different proteins involved with cellular signaling (Piomelli, 1993). AA helps in maintaining cell integrity and vascular permeability, making it important for development and growth upon cell damage occurrence. As a major constituent of membrane phospholipids, it becomes detached from membranes through enzymatic cleavage by phospholipase A2 (PLA2) (O'Donnell et al., 2009), and then can be used as substrate by LOX and P450 hydroxylases to generate superoxide and hydroxyeicosatetraenoic acid (HETE) (Chawengsub et al., 2009).

Experimental evidence from animal and human cell-lines suggests that increased 20-hydroxyeicosatetraenoic acid (20-HETE) levels are liable for ROS production (Guo et al., 2007; Medhora et al., 2008; Zeng et al., 2010) by regulating either in a dependent (Medhora et al., 2008), independent (Guo et al., 2007) manner NADPH oxidase activity or via endothelial nitric oxide synthase (eNOS) uncoupling (Cheng et al., 2008). Additionally, in the clinical setting was observed that excretion of 20-HETE in urine is associated with oxidative-stress in hypertensive patients (Ward et al., 2005). Thus, the role of 20-HETE as an up-stream effector compound, regulating down-stream events in many conditions that lead to, or related to, cardiac dysfunction make this AA derivative a potential drug target (Waldman et al., 2016).

7.4 Maintenance of Redox Homeostasis

Redox homeostasis is the essential balance of a cell's oxidation and reduction systems. Within a cellular response (i.e. signaling, apoptosis, detoxification, repair, and cell cycle arrest), maintaining redox homeostasis is dependent (Ursini et al., 2016). Redox homeostasis is responsible for kick-starting oxidant production and bypassing antioxidant mechanisms. When there is an imbalance of redox homeostasis, ROS levels increase and causes problems for cellular processes (i.e. signaling, adaption, cell cycle arrest, apoptosis, and detoxification repair/cell defense) (Ayer et al., 2014). Maintaining redox homeostasis is essential for cellular processes. In order for redox homeostasis to be effective the redox couple (i.e. NAD^+/NADH and $NADP^+$/NADPH) has to have high concentration levels and reducing at a high capacity (Valko et al., 2007).

7.4.1 Antioxidants

Antioxidants can have endogenous or exogenous sources such as dietary origin, or administered as a therapeutic approach. Endogenous antioxidants include superoxide dismutase (SOD), catalase (CAT), glutathione (GSH), GSH peroxidase (GPx), thioreductase, and uric acid (urate). Dietary sources include vitamins A, B, C, and E, as well as polyphenolic compounds (polyphenols, flavonoids), trace metals and minerals (Fig. 7.3).

Enzymatic antioxidants allow the body to protect itself from ROS. They are critical in preventing lipid peroxidation and maintaining a cell's membrane function (Nimse and Pal, 2015). Primarily, enzymatic antioxidants remove reactive species and their derivatives (Mironczuk-Chodakowska et al., 2018). Alongside enzymatic antioxidants, non-enzymatic antioxidants modulate free radical reactions as well (Bowler and Crapo, 2002). Additionally, there are two types of endogenous antioxidants: non-protein and protein. Non-protein endogenous antioxidants focus on ending free radical chain reactions, the molecular repair of damage done by free radicals, scavenging of ROS and RNS, and sequestration of transitional metal ions (He et al., 2017). The 6 main non-protein endogenous antioxidants are Glutathione, Co-enzyme Q

(CoQ), α-lipoic acid (LA), Ferritin, Uric acid, and Bilirubin (Brown and Griendling, 2015). Protein-based endogenous antioxidants involve enzymatic activity and are considered as the first line of defense against oxidative stress. The three main endogenous proteins are superoxide dismutase (SOD), catalase (CAT), and glutathione peroxidase (Brash, 1999). Table 7.4 lists the various enzymatic, non-enzymatic and endogenous antioxidants, their functions, and subcellular locations.

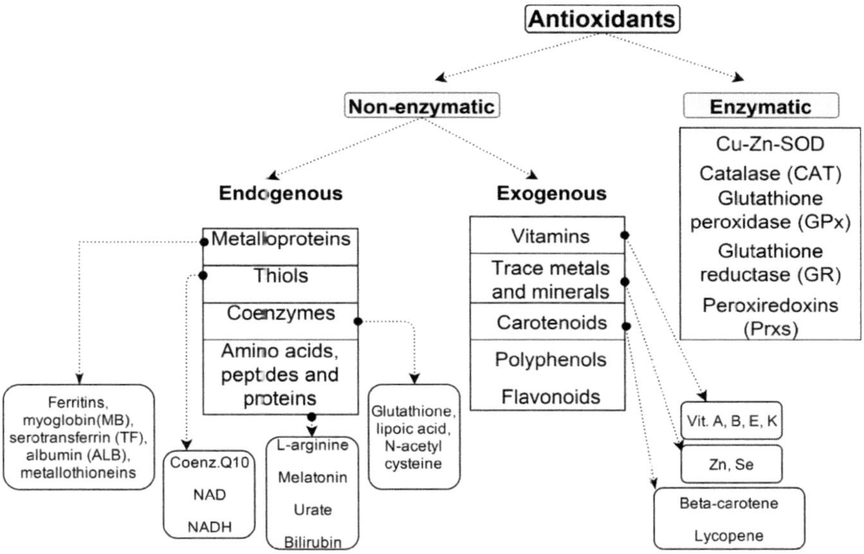

Figure 7.3 Enzymatic and non-enzymatic sources of antioxidants. Nicotinamide adenine dinucleotide (NAD+, NADH), zinc (Zn), selenium (Se).

7.4.2 ROS, Antioxidants, and Signal Transduction

ROS, antioxidants, and signal transduction are connected by both ROS and antioxidants and are affecting a cell's signal transduction directly. The overproduction of ROS leads to oxidative stress which causes damage to cellular DNA, metabolic processes, and cell death (Ott et al., 2007). With each damage that ROS causes on a biological system, antioxidant systems are critical to regulating and maintaining ROS signaling. The levels of ROS

Table 7.4 Enzymatic, non-enzymatic and non-protein endogenous antioxidants, and their molecular functions that are performed when free radicals are present in a biological system

Name	Type of antioxidant	Function	Subcellular location
Superoxide dismutase (SOD)	Enzymatic	Destroys radicals produced within cells that are toxic to biological systems	Peroxisomes, mitochondria, cytosol
Catalase (CAT)	Enzymatic	Protects cells from toxic effects of hydrogen peroxide and promotes cell growth (i.e. T-cells, B-cells, myeloid leukemia, melanoma cells, mastocytoma cells, and fibroblast cells)	Peroxisomes and mitochondria
Ascorbate peroxidase (APX)	Enzymatic	Catalyzes peroxide into water	Peroxisomes, mitochondria, cytosol
Monodehydroascorbate reductase (MDHAR)	Enzymatic	Catalyzes monodehydroascorbate to ascorbate and oxidizing NADH	Mitochondria, cytoplasm
Dehydroascorbate reductase (DHAR)	Enzymatic	Affects cell responsiveness and tolerance to environmental ROS (in plants only)	Mitochondria, cytoplasm
Glutathione reductase (GR)	Enzymatic	Maintains high levels of reduced glutathione in the cytosol	Mitochondria, cytoplasm
Guaiacol peroxidase (GPX)	Enzymatic	Inactivates and hunts of hydrogen and lipid peroxides	Mitochondria, cytoplasm and ER
Ascorbic Acid, Vitamin C (ASC)	Non-enzymatic	Detoxifies H_2O_2 via Ascorbate peroxidase	Cytosol, mitochondria
Reduced Glutathione (GSH)	Non-enzymatic	Acts as a detoxifying co-substrate for enzymes like peroxides, GR and GST	Cytosol, mitochondria

Name	Type of antioxidant	Function	Subcellular location
α-Tocopherol	Non-enzymatic	Guards against and detoxifies products of membrane LPO	Mostly in membranes
Flavonoids	Non-enzymatic	Direct scavenger of H_2O_2 and $^1O_2$ and OH·	Cytosol, mitochondria
Proline	Non-enzymatic	Efficient scavenger of OH· and $^1O_2$ and prevent damages due to LPO	Cytosol, mitochondria
Glutathione (GSH)	Endogenous	An antioxidant defense against ROS and RNS; mitochondrial protection by preventing production of oxidants, scavenging free radicals, or inhibiting oxidant reactivity	
Co-enzyme Q (CoQ)	Endogenous	Energy transduction and ATP production; transports electrons in the respiratory chain; in the mitochondrial layer, it is a redox carrier, antioxidant	
α-lipoic acid (LA)	Endogenous	Delivers antioxidant activity in non-polar and polar mediums and presents an antioxidant effect in its oxidized (LA) and reduced form (DHLA)	
Ferritin	Endogenous	Limits Fe (II) availability to participate in producing ROS	
Uric acid	Endogenous	Scavenge oxygen radicals and protect the erythrocyte membrane from lipid oxidation	
Bilirubin	Endogenous	Prevents ischemic injury to isolated hearts and modulates smooth muscle contractility	

exposed on a cell determines the signal transduction a cell has on a molecular basis (i.e. cell binding) and interaction with other cells (Zhang et al., 2016). Between ROS and antioxidants, cellular signaling is affected directly and can lead to altered gene expression (Birben et al., 2012).

7.4.3 Redox Regulation of Physiological Functions

Redox regulation is important in physiological functions. Redox regulation is involved with transcriptional and translational regulation of encoding antioxidant enzymes (Manganas et al., 2017). Regulation occurs within the cellular inner membrane of the mitochondria. Redox regulation of physiologic functions is important because increased levels of ROS are signalling molecules that can affect physiological functions (i.e. cells, tissues and organs). Redox regulation maintains the balance of ROS levels in order to keep physiological functions at a base level and to keep the control of various cellular functions to prevent diseases from appearing (Grant, 2008).

7.5 Pharmacological Interventions

7.5.1 Drug-Induced Oxidative Stress

Drug induced oxidative stress occurs when certain drugs are administered to individuals to place oxidative stress on biological systems at a cellular level. This can happen either deliberately or, more common, as a result of accidental over-dosing potentially culminating in organ failure. One such example is an excessive consumption of paracetamol, thereby inducing liver failure by a cascade of events after the depletion of cellular glutathione levels (Du et al., 2016). Linked to this, there is a high possibility that an excess of ROS production will occur, resulting in local cellular defense responses to that stressor target (Deavall et al., 2012). The consequence of allowing drug induced oxidative stress to occur is not just the heightened levels of ROS, but the overwhelming pressure on the body's defense system which can result into further damage to the cell. However, induced ROS can help cells by triggering cellular apoptosis which induces the oncogenic phenotype established in cancer cells (Valko et al., 2006). Luckily on a molecular level, damage created by ROS is limited on the factor of metal ions being available as well as the reactivity level of the targeted protein or lipid (Nita and Grzybowski, 2016).

Oxidative stress naturally attacks mitochondrial DNA, which can cause more cellular problems in the long run. Oxidative stress, especially drug induced, modulates the cell directly, and can lead to ROS damaging mitochondrial DNA, consequently resulting in mutations, more oxidative stress, transcriptional errors, and eventually cell death. The amount of oxidative stress the cell is exposed to determine any modifications that could be generated in gene expression (Uttara et al., 2009). The more exposed the cell is to ROS, the more molecular modifications will happen (i.e. pro-survival or pro-apoptotic response pathways) (Trachootham et al., 2008).

Exposing cells to excess oxidative stress, which is normally the case in drug induced therapy, toxicity becomes a problem for numerous tissues and cells. The main targets are DNA, lipids, and protein macromolecules (Rahman, 2007). When the drug is introduced into a cell, oxidative stress is introduced via overproduction of ROS from the mitochondrial electron transport chain. Once the ROS leave the mitochondria there is a chain reaction effect within the cell causing damage to DNA, lipids and proteins.

The DNA is damaged by oxidative stress attacking the guanine nucleotide base by attaching a hydroxyl radical. DNA damage leads to mutations, p53 activation and impairment in structure and function in the cytochrome c release within mitochondria (Chakraborty et al., 2010). Lipids are damaged by ROS and oxidative stress through lipid peroxidation, and the oxidized lipids react with DNA, proteins and macromolecules in the cell. Proteins are damaged by being oxidized, resulting in loss of cellular function and leads to cell apoptosis. Oxidized proteins can lead to protein aggregation and removal of proteasomes (Hohn and Grune, 2014). Exposure of ROS activates MAP kinase activity and detoxification via antioxidants, ascorbic acid, vitamin E, glutathione, and enzymes SOD and catalase (Espinosa-Diez et al., 2015). These molecular reactions occur with ROS exposure, however when oxidative stress is drug induced, the ROS exposure is higher and potent. Drug induced oxidative stress to cells ultimately and continuously results in cell apoptosis.

Alternative to activating a chain reaction to cell apoptosis, drug induced oxidative stress is considered to a mechanism of toxicity for cells in biological systems for various tissues and organs (i.e. liver, cardiovascular and nervous system, kidney, and ear) (Deavall et al., 2012). There are multiple drugs that can cause various levels of oxidative stress damage to a biological system. These drugs and their toxicity specificity are described in Table 7.5.

Table 7.5 Drugs and their toxicity specificities associated with drug induced oxidative stress

Drug	Toxicity specificity	Effect for oxidative stress
Doxorubicin	Cardiac	If reduced to free radical increases ROS in cardiomyocytes. Lipid peroxidation, mitochondrial dysfunction, cell apoptosis
AZT	Skeletal myopathy and cardiac	Increases ROS and NOS (peroxide and peroxynitrate). Overexpression of superoxidase dismutase/catalase does protect against toxicity, cell apoptosis
Diclofenac	Nephrotoxicity and hepatotoxicity	Oxidative stress caused because of redox cycling of intermediates that are derived from hydroxylation, mitochondrial dysfunction
Paracetamol	Hepatotoxicity	Glutathione, mitochondrial dysfunction, inflammation, and activating of pro-apoptotic proteins
Cisplatin	Nephrotoxicity and ototoxicity	Increase in superoxide anion, hydrogen peroxide, and hydroxyl radicals, mitochondrial dysfunction, cell apoptosis
Chlorpromazine	Dermal toxicity	Production of single oxygen radicals and superoxide in response to UVA/B irradiation

7.5.2 Experimental Therapeutic Agents against CVD

Experimental therapeutic agents in pre-clinical drug development focus on a broad spectrum with transition-state inhibitors, stabilizing agents and targeted drugs against various diseases. There are experimental therapeutics focused towards CVD and associated causes of oxidative stress and one experimental therapeutic group of compounds constantly being studied and tested are antioxidants. Yet, there are limitations on experimental therapeutic agents because the pathophysiology for specific CVDs are not addressed and analyzed (Sovari and Dudley, 2012), such as ion channel blockage of sodium channels and limitations on the catheter ablation that is used for tissue destruction for blocking focal activity (Xiao, 2011). Several experimental drugs are being tested as therapeutic agents to lessen the effect of oxidative stress species in various experimental cell-based and animal models of cardiovascular disease (Table 7.6) and their potential preclinical use assessed in many clinical trials (Table 7.7). Currently there are three sources of experimental therapeutic drugs that have been approved in combating oxidative stress species in CVD. Quercetin is a polyphenolic flavonoid found originally in plants. It is a dietary antioxidant that comes from fruits, vegetables, leaves and grains. Rhein is an anthraquinone that is obtained from the rhubarb plant and considered to be one of many medicinal plants. Lipoic acid is an organosulphur compound that is naturally produced within the body and essential for metabolism, though it can be administered from an external source as well.

Quercetin is used as a nutritional supplement found in plants to benefit against several ailments (Zhang et al., 2011). One study focused on whether Quercetin could improve the cardiovascular risk in women with Type 2 diabetes (T2DM), and it was shown that it decreases the systolic blood pressure (Waxman and Ritchie, 1985). In addition, the supplementation of Quercetin led to a reduced serum concentration of tumour necrosis factor-α and interleukin-6. In another study, it was utilized to reduce oxidized plasma low-density lipoprotein (LDL) in overweight subjects with high risk for cardiovascular disease (Schweigerer and Fotsis, 1991). Quercetin could be beneficial in reducing blood pressure and plasma oxidized LDL, however, has no effect on other cardiovascular risk factors, neither on inflammatory markers (Zahedi et al., 2013).

Table 7.6 Experimental therapeutic agents against oxidative stress species in CVD

Agent	Type/class	Source	Effects	References
Quercetin	Polyphenolic flavonoid	Dietary antioxidant	Diminished activation of ERK1/2, PKB, p38-MAPK and JNK induced by H_2O_2; cardioprotection after 24 h exposure and cardiotoxicity after 48 h	(Daubney et al., 2015; Zahedi et al., 2013; Zhang et al., 2011)
Rhein	Anthraquinone	Many medicinal plants	Hepato-, nephron-protective, anti-tumour, anti-inflammatory, and potentially anti-altherosclerotic	(Zhou et al., 2015)
Lipoic acid	Dithiol compound derived from octanoic acid	Naturally occurring	Reduced TBARS levels, DNA damage and blunted expression of adhesion molecules in the endothelium of a hypertension rat model	(Tayebati et al., 2016)
			Inhibits NF-κB activation via antioxidant-independent mechanisms; blunted PTGS2 expression in HUVECs	(Ying et al., 2011)
Curcumin	Phenolic compound	Rhizoma of *Curcuma longa*	Improves endothelial dysfunction and vascular remodeling by elevating NO and decreasing oxidative-stress in a hypertension rat model	(Boonla et al., 2014)
			Inhibits cardiac hypertrophy and recovers cardiovascular function; up-regulation of NCX and eNOS in TransAC model rat	(Bai et al., 2018)
Antioxidin-RL and antioxidin-2	Biomimetic peptides	Skin of plateau frog *Odorrana livida*	Inhibition of intracellular build-up of free radicals induced by H_2O_2; decreased MMP; reduced levels of DNM1L and lessen SOD1 and GPx1 activities in rat PC-12 cells	(Zhang C. et al., 2017)
NOP-47	Bioactive peptides	Whey-derived	Improves vascular function in healthy subjects	(Ballard et al., 2009)
			Briefly improves endothelium-dependent vasodilation in aged subjects with low FMD	(Ballard et al., 2013)

Abbreviations: TAC, total antioxidant capacity; FMD, brachial artery flow-mediated dilation; MMP, mitochondrial membrane potential; DNM1L, dynamin-related protein 1; PC-12; transplantable rat pheochromocytoma cells; NO, nitric oxide; TransAC, transverse abdominal aortic constriction; NCX, Na^+/Ca^{2+} exchanger; eNOS, endothelial NO synthase; PTGS2/COX2, prostaglandin G/H synthase 2; HUVECs, human umbilical vein endothelial cells; TBARS, thiobarbituric acid reactive substances; p38-MAPK, p38 mitogen-activated protein kinase; JNK, c-Jun N-terminal kinase; PKB/Akt, protein kinase B; ERK1/2, extracellular signal-regulated kinase 1/2.

Table 7.7 Experimental therapeutic agents against oxidative stress species in CVD-related conditions

Agent	Outcome parameters	Associated condition	Outcome results	Study Designs	Trial ID	Reference
Glucosamine	FMD and intraerythrocyte glutathione levels	Vascular endothelial function, atherosclerosis	Improved FMD and intraerythrocyte GSH/GSSG ratios	Parallel, individual, randomized, single blind	UMIN000020586	Katoh et al., 2017
Resveratrol	Cardio-ankle vascular index (CAVI)	T2DM, atherosclerosis	Improved arterial stiffness and reduce oxidative stress	Double-blind, randomized, placebo-controlled		Imamura et al., 2017
Valsartan	Oxidative stress: nitrotyrosine levels; metabolic: serum glucose, insulin, lipid profile, and urine albumin and creatinine	T2DM, hypertension	Valsartan reduced the oxidative stress marker in T2DM with hypertension	Randomized	KCT0001498	Kim et al., 2017
Amlodipine		T2DM, hypertension	Amlodipine reduced the oxidative stress marker in T2DM with hypertension	Randomized	KCT0001498	Kim et al., 2017
palm-tocotrienols (TRF- 80)	Brachial artery FMD, lipid profile, glucose, insulin and inflammatory markers	T2DM	Increased bioavailability of palm-based carotenes and tocotrienols; no effects on vascular function or CVD risk factors	Randomized, placebo-controlled, double-blind, placebo (palm olein)	ACTRN12613001041741	Stonehouse et al., 2016

(*Continued*)

Table 7.7 (*Continued*)

Agent	Outcome parameters	Associated condition	Outcome results	Study Designs	Trial ID	Reference
palm-carotene (CC-60)	Brachial artery FMD, lipid profile, glucose, insulin and inflammatory markers	T2DM	Increased bioavailability of palm-based carotenes and tocotrienols; no effects on vascular function or CVD risk factors	Randomized, placebo-controlled, double-blind, placebo (palm olein)	ACTRN12613001041741	Stonehouse et al., 2016
Carvedilol	Enzymatic activity: PON and arylesterase; Lipid Peroxidation: MDA Assay; ELISA: sLOX-1 levels		Hypertension			Ayashi et al., 2016
Tauroursodeoxycholic acid (TUDCA)	Brachial artery FMD after 1 h and 2 h post-glucose challenge	Hyperglycaemia-induced endothelial dysfunction	Endothelial dysfunction mitigated by TUDCA	Healthy subjects submitted to acute glucose pulses. Tested on two independent visits in a single-blind randomized cross-over design		Walsh et al., 2016
Melatonin	EF%, HR, plasma CTnI, IL-1β, iNOS and caspase-3	MI, CABG	Ameliorates the degree of MI-reperfusion injury, dose-dependent effect	Randomized group allocation		Dwaich et al., 2016

Agent	Outcome parameters	Associated condition	Outcome results	Study Designs	Trial ID	Reference
Barnidipine + losartan	BP every month; plasma glucose, HbA1c, Hs-CRP, TNF-α, MMP-2, MMP-9, sVCAM-1, sICAM-1, isoprostanes and PON-1	Essential hypertension	Improvement of some parameters indicative of endothelial damage and oxidative stress	Randomized, double-blind, controlled study	NCT02064218	Derosa et al., 2016a
Olmesartan + amlodipine	BP monthly; baseline, and after 6 and 12 months: lipoprotein (a), MPO, isoprostanes, and PON-1	T2DM, hypertension		Comparison of single and combinations of drug, randomized, double-blind, controlled study	NCT02064218	Derosa et al., 2016b
Coenzyme Q10 (CQ10)		Acute viral myocarditis	Reduced inflammatory factors and oxidative stress	Randomized, blind		Shao et al., 2016
Coenzyme Q10		Hemodialysis	CVD prevention	Triple-masked, factorial assignment	NCT01408680	
Trimetazidine		Acute viral myocarditis	Reduced inflammatory factors and oxidative stress	Randomized, blind		Shao et al., 2016
CQ10 + trimetazidine	Heart function: chocardiographic examination; oxidative markers: SOD, MDA, GSH-px; inflammation markers: TNF-alpha, IL-8, MCP-1; CRP levels	Acute viral myocarditis	More powerful effect on anti-inflammation and anti-oxidative stress than CQ10 or trimetazidine alone	Randomized, blind		Shao et al., 2016

(*Continued*)

Table 7.7 (*Continued*)

Agent	Outcome parameters	Associated condition	Outcome results	Study Designs	Trial ID	Reference
Nebivolol	Inflammatory and oxidative stress biomarkers, microvascular function, endothelial function, and virtual histology intravascular ultrasound	Non-obstructive CAD	Both β-blockers had similar effects on oxidative stress, microvascular function, and endothelial function	Pilot double-blinded randomized controlled trial	NCT01230892	Hung et al., 2016
Atenolol	Inflammatory and oxidative stress markers, microvascular function, endothelial function, and virtual histology intravascular ultrasound	Non-obstructive Coronary Artery Disease	Both β-blockers had similar effects on oxidative stress, microvascular function, and endothelial function	Pilot double-blinded randomized controlled trial	NCT01230892	Hung et al., 2016
Liraglutide	First occurrence of death from cardiovascular causes, non-fatal (including silent) MI, or non-fatal stroke	Cardiovascular Outcomes in T2DM	Rates of non-fatal MI, non-fatal stroke, and hospitalization for HF were non-significantly lower in the liraglutide group than in the placebo group	Double-blind, randomized, placebo-controlled	NCT01179048	Marso et al., 2016

Agent	Outcome parameters	Associated condition	Outcome results	Study Designs	Trial ID	Reference
Irbesartan	Evaluate the hypouricemic effect in a clinical setting	Hyperuricemia	Beneficial effects on hyperuricemia and oxidative stress			Chida et al., 2015
Allopurinol	CCE; echocardiographic measures of LV remodeling; markers including UA, NT-pro-BNP, cystatin C, and MPO	Hyperuricemic, HF	High-risk HF patients with reduced EF and elevated uric acid levels, failed to improve clinical status, exercise capacity, quality of life, or LV EF	1:1 randomized, double-blind, placebo-controlled	NCT00987415	Givertz et al., 2015
***N*-acetyl cysteine and omega 6 Fish oil (PUFA)**		Type 1 Diabetes	Prevent birth defects in the offspring	Double-blinded with parallel assignment	NCT03056014	
Dietary Supplement: XXS OR non-XXS		Mild or moderate hyperlipidemia	Measurement of oxidative stress markers	Double-blinded with parallel assignment	NCT02826083	

Abbreviations: FMD, flow-mediated vasodilation; GSH/GSSG, glutathione/glutathione disulphide; CAVI, cardio-ankle vascular index; EF%, ejection fraction; HR, heart rate; CTnI, plasma levels of cardiac Troponin-I; IL-1β, interleukin-1beta; iNOS, inducible nitric oxide synthase; CABG, coronary artery bypass graft surgery; HbA1c, glycated haemoglobin; Hs-CRP, high-sensitivity C-reactive protein; TNF-α, tumour necrosis factor-α, MMP-2, metalloproteinase-2; MMP-9, metalloproteinase-9; sVCAM-1, soluble vascular adhesion protein-1; sICAM-1, soluble intercellular adhesion protein-1; PON-1, isoprostanes and paraoxonase-1; MPO, myeloperoxidase; CAD, coronary artery disease; HF, heart failure; CCE, clinical endpoint.

Curcumin is another drug that possesses antioxidant properties. It interacts with 1,1-diphenyl-2-picryl-hydrazyl (DPPH) to stabilize free radicals, making this compound a potent scavenger of superoxide radicals (Sreejayan and Rao, 1996). Curcumin is a natural active polyphenol found in turmeric with the potential to modulate the biological activity of signaling molecules and was shown to display anti-inflammatory and anti-cancer activity that targets ROS and RNS (Gupta et al., 2013). Its RNS modulatory role was demonstrated by inhibiting the induction of the nitric oxide synthase within macrophages (Sreejayan and Rao, 1997). Curcumin has the ability to induce expression of phase II antioxidant enzymes such as glutamate-cysteine ligase (GCL) (Dickinson et al., 2004) which affects glutathione synthesis by up-regulating GCL expression and activating the nuclear factor E2-related factor 2 (Nrf2)-dependent pathway (Dickinson et al., 2003).

Rhein is known to have hepatoprotective, nephroprotective, anti-cancer, anti-inflammatory, and other protective effects. It was shown that it protects endothelial cells from oxidative-induced injury via hydrogen peroxide which makes this agent a potential candidate for treatment of altherosclerosis. Rhein is an antioxidant compound that has two mechanisms, anti-oxidant and pro-oxidant (Zhong et al., 2012). The anti-oxidant mechanism is not well known and the pro-oxidant mechanism potentially involves inhibition of the mitochondrial respiratory complex I and subsequent facilitation of NADH- and NADPH-dependent lipid peroxidation (Glinn et al., 1997).

Lipoic acid, known as α-lipoic acid, is a potent antioxidant that holds an essential role in mitochondrial bioenergetics reactions and contains insulin-mimetic and anti-inflammatory activities (Rochette et al., 2015). It was shown that α-lipoic acid can regulate anti-oxidant and anti-inflammatory pathway-associated gene transcription (Park et al., 2014), and has the ability to scavenge and inactivate free radicals, protect from oxidative damage in diseases, and is a natural antioxidant (Tibullo et al., 2017). There is therapeutic potential for lipoic acid to help prevent vascular disease, hypertension, inflammation and to fight against diabetes.

Another source of experimental therapeutic agents against CVD are potential peptides. Therapeutic peptides are short amino acid monomer chains that are naturally occurring within the biological system (Recio et al., 2016). Potential peptides, such as biomimetic peptides (antioxidin-RL peptide), can protect cells from oxidative stress and are also more stable than vitamin C and can withstand light-induced degradation (Zhang C. et al., 2017). The antioxidin-RL peptide binds with residue tyrosine and catalyzes the cysteine residue, scavenging ROS at a high frequency in vitro. The protective nature of the antioxidin-RL peptide is due to the intracellular inhibition of hydrogen peroxide expression change of responsive genes under oxidative stress (Zhang C. et al., 2017). Another peptide for therapeutic experimental measure are food-derived bioactive peptides. Bioactive peptides are compounds that demonstrate anti-inflammatory, antioxidant, and anti-hypertensive properties (Erdmann et al., 2008; Sousa et al., 2012). Food-based bioactive peptides hold a wide range of beneficial effects on pathological conditions, especially towards inflammatory models (i.e. colitis, arthritis, atherosclerosis, and respiratory tract inflammation) (Chakrabarti et al., 2014). Bioactive peptides have gained a considerable amount of interest in applications of therapeutic techniques, but limitations such as proper dosage and administration frequency, wide variability, and biological effects remain hurdles still to overcome (Cam and de Mejia, 2012; Rutherfurd-Markwick, 2012). While bioactive peptides are considered safe, there is still uncertainty due to potential toxic side effects if a high dose is consumed (Bast and Haenen, 2002). Both bioactive peptides and peptide-rich hydrolysates show promise in anti-inflammatory properties and as antioxidant agents (Chakrabarti et al., 2014). Experimental CVD therapeutics using peptides is a possible strategy and is currently of high interest to improve the therapeutic potential that will provide prevention and treatment opportunities that could eventually surpass the current therapies of antioxidants and drug-inducement.

7.5.3 Antioxidants of Exogenous Source as Potential Therapeutic Targets

Another approach is to provide the full spectrum of vitamins on a recommend daily dose that the human body requires, hopefully thereby removing the need to identify the active ingredient. Some trials studies found reduced cardiovascular risk with this type of implementation (Holmquist et al., 2003; Watkins et al., 2000), others do not (Moyer, 2014; Muntwyler et al., 2002; Neuhouser et al., 2009; Park et al., 2011). Additionally, a recent systematic review and meta-analysis drew the conclusion that there is no supportive evidence to the use of antioxidants supplements with the aim of prevention of primary and secondary diseases, and that higher doses of beta-carotene, vitamin E and vitamin A may increase mortality (Bjelakovic et al., 2015). Moreover, dietary supplementation with antioxidant agents has shown little (Murer et al., 2014; Rosa et al., 2014) or no therapeutic effect (Myung et al., 2013; Ye et al., 2013), and even controversial outcomes in many clinical trials (Bjelakovic et al., 2012, 2015), though experimental animal studies have described than high doses of vitamin C (ascorbic acid) can increase H_2O_2 interstitial levels and impair NO bioavailability (Garry et al., 2009).

In part some of these results can be explained by an incorrect or inappropriate choice of study endpoints, advanced stage of disease, patients undergoing standard drug therapies (unknown interactions of some drug-supplements, or diminished effects), dosages used since not many animal studies are available to extrapolate an informed dose of the supplements to administer, chronic exposure to antioxidants, given the ROS nature as signalling molecules could arise to compensatory mechanisms, and ultimately lead to overproduction of ROS. On the other hand, antioxidant vitamins cannot be used as catalytic scavengers of H_2O_2 and hypochlorous acid (HOCl), both non-radical molecules with well-known key roles in the mechanism of vascular injury. The lack of studies regarding (bio)activity, (bio)accessibility and (bio)availability of these antioxidant compounds raises concern when their action is explored in human subjects. Additionally, the issue of rapid depletion of the antioxidant active forms rendering them inactive, along with accumulation of the both oxidized and pro-oxidant form of the parent molecule, still has to be addressed.

7.5.4 Antioxidants of Endogenous Source as Potential Therapeutic Targets

Direct delivery of antioxidants to the mitochondria has been demonstrated with the use of mitoquinone (MitoQ), a compound that mimics the functions of the endogenous coenzyme Q10

(Tables 7.4 and 7.7), which is capable to reduce radical production without damage to the mitochondria, and thus not impacting on the normal/basal electron transport chain process (Smith and Murphy, 2010). In many experimental animal models, it has proven to lessen myocardial ischemia/reperfusion injury (Dare et al., 2015), metabolic syndrome (Mercer et al., 2012), nitrate-associated endothelial injury (Gioscia-Ryan et al., 2014), hypertension (Graham et al., 2009), liver fibrosis (Rehman et al., 2016) and right ventricular remodelling (Pak et al., 2018). Although in clinical translation it has proven ineffective to ameliorate Parkinson's disease (Snow et al., 2010), nor in improving hepatitis C long-term infection (Gane et al., 2010) and failed in mitigating physiological adaptions due to endurance exercise (Shill et al., 2016), these trials aid to confirm drug oral safety when administered long-term.

Other mitochondria-targeting drugs, such as derivatives of triphenylphosphonium (TPP^+) (Leo et al., 2008) and resembling MitoQ's mode of action, include MitoE (conjugating α-tocopherol i.e. vitamin E with TPP^+) (26447154) and MitoTEMPOL, a piperidine nitroxides derivative(Trnka et al., 2009) that promotes recovery in sepsis induced by acute kidney injury (AKI) in rat models (Sims et al., 2014). SkQ1, a cationic plastoquinone derivative (Antonenko et al., 2008), also known as Visomitin, has shown effective protection to corneal damage in a multicentre clinical trial (Brzheskiy et al., 2015), and promising use against age-associated cardiomyopathy in mice (Manskikh et al., 2015).

Pathological sources of mitochondrial dysfunction could be targeted using peptides belonging to the Szeto-Schiller (SS) family such as the SS-31(also known as MTP-131 and Bendavia) (Szeto and Birk, 2014). They bind to cardiolipin and prevent its oxidation, thereby enhancing electron transfer within the electron transfer chain (ETC) and diminishing O_2^- production and ultimately restoring cellular bioenergetics (Birk et al., 2014). Treatment with SS-31 and the SS-20 analogues reduced lipid peroxidation and infarct size in ischemia reperfusion injury (IRI) murine models (Cho et al., 2007). Moreover, SS-31 mitigates release of cytokines in response to treatment with TNF-alpha in skeletal muscle cells using mouse myoblast cell lines such as C2C12 myotubes (Lightfoot et al., 2015).

Counter-balance of the deleterious effects of oxidative-stress can also be attainable by promoting activation of up-stream factors responsible for the regulation of cell-mediated response to oxidative stress (Gazaryan and Thomas, 2016). This is in part orchestrated by binding of the nuclear factor erythroid 2-related factor 2 (NFE2L2) to cis-regulatory elements, commonly known as antioxidant response elements (ARE) in the promotor region of target genes (Huang et al., 2000). Without external oxidoreductive stimuli, the Kelch-like ECH-associated protein 1(KEAP) promotes continuous ubiquitination and consequent proteasome degradation of NFE2L2, leading to suppression of its transcriptional activity (Lo et al., 2006). Along this line, a number of drugs are currently being developed to attempt to explore how to disrupt the NFE2L2-KEAP1 interface, like the tecfidera (dimethyl fumarate), bardoxolone, and BTB domain and CNC homolog 1 (Bach1) and consequently activating the NFE2L2/ARE pathway. This would then lead to an increased expression of ROS detoxifying enzymes and/or synthesis of pro-antioxidant molecules (Gazaryan and Thomas, 2016; Gesslbauer and Bochkov, 2017). Another approach encompasses mechanisms to cope with damage induced by oxidative stress without changing either basal activity of detoxifying enzymes or levels of pro-antioxidant molecules (Gesslbauer and Bochkov, 2017), thereby lessen the effect of inflammation and aiming to correct disrupted intracellular events due to ROS-induction damage (Gesslbauer and Bochkov, 2017). Attenuation of inflammation by pharmacological intervention has been addressed by altering or interfering with key players in the inflammatory process, such as acting on secretory mediators like inhibiting IL-18, chemokines, and TNF-alpha, using compounds such as Anakinra (Brown, 1989; Toldo et al., 2012), Etanercept (Gao et al., 2015), Infliximab (Gerlach et al., 2014), and Evasin-3 (Montecucco et al., 2010).

Interventions targeting cell-adhesion molecules (CAM) using antagonist like G31P, reparixin (Connell et al., 2015; Sousa et al., 2013), and TAK779 (Takami et al., 2002), which are known chemokine receptor (CCRs/CXCR) inhibitors and impair leukocyte migration, have also shown beneficial effects. Inhibition of many members of CAM, such as ICAM1 and PECAM1 with their respective counter-receptor antibodies, attenuates severity in many IRI experimental animal models (Merchant et al., 2003; Sun et al., 2001). Likewise, targeting selectins (P-, E-, and L-selectins) using pan-selectin drug antagonists like TB-1269, OC-229 (17454392), and anti-PSGL1 (Hayward et al., 1999) has shown to be effective in disrupting the recruiting of leukocytes from blood circulation to the lesion site in many experimental animal models of IRI. On the other hand, targeting the NF-kB transcriptional pathway with Bay-11-7082, pyrrolidine dithiocarbamate (PDTC), PS-519 have shown to be protective to the heart in various experimental rodent models of IRI (Kim et al., 2010; Rudolph et al., 2010; Stansfield et al., 2007), specifically showing potential to reduce infarction area (Kim et al., 2010; Rudolph et al., 2010) and ameliorate overall ventricular function (Stansfield et al., 2007). Nevertheless, any pharmacological intervention inhibiting activation of NF-kB should not neglect other functions that this gene and its targets are involved in, such as tissue regeneration and survival. Moreover, as explored in IRI experimental animal studies, treatment with NF-kB antagonists can, in the first stage of the disease, suppress anti-inflammatory response, while at later stages will most probably induce cell death and inhibit tissue regeneration (Gazaryan and Thomas, 2016). Therefore selecting the optimal time-frame to administer NF-kB inhibitors and as well as develop drugs with a half-life suitable for the selected treatment period is of utmost importance.

Modulation of the intracellular signaling by employing drug agents like sildenafil, vardenafil, tadalafil, and avanafil for the inhibition of cGMP-specific 3′,5′-cyclic phosphodiesterase (PDE5A) has been investigated in IRI experimental animal models and their usage, reduced myocardial fibrosis and hypertrophy (Korkmaz-Icoz et al., 2018) via activation of the NO/cGMP/PKG signalling pathway (Korkmaz-Icoz et al., 2018), and many more are currently in various stages of the drug development avenue.

7.6 Conclusions and Future Trends

Many cardiovascular complications arise from the build-up of modifiable metabolic risk factors, and those are themselves also well-described targets of ROS. Targeting ROS via pharmacological intervention seems a compelling strategy for both prevention and treatment of many associated-vascular diseases. Mono supplementation with vitamin C or E, investigated in many clinical trials, failed to provide a beneficial outcome for cardiovascular patients. Therefore it can be foreseen that compounds that inhibit the enzymatic sources of ROS over traditional supplement-based antioxidants will prove to be more useful in clinical practice. Although, it is broadly accepted and recommended intake of ROS scavenging molecules via consumption of fruits and vegetables as a way of promoting essential health benefits (Mozaffarian, 2016).

Following the current trend in the development of nano-delivery systems and novel sensors that dynamically probes the level of reactive oxygen and nitrogen species (ROS/RNS) within the intracellular microenvironment and that are able to properly discriminate physiologic, protective and harmful ROS/RNS levels. Then in any case of excess of ROS/RNS they are able to neutralize ROS/RNS derived oxidative-stress by modulating in-situ antioxidant release (Weissig and Guzman-Villanueva, 2015), which could lead to the rediscovery of antioxidant agents previously labelled as "ineffective" due to lack of bioavailability of its active form.

Several NADPH oxidase isoforms are dependent of the p47phox (NCF1) regulatory subunit for a complete enzymatic activity, therefore developing novel agonist that targets only this organizer element, and not elements required for enzymatic activation will still hold interest in pharmacological research. Many of these inhibitors can potentially be already available as therapeutics for other ailments; thus, drug repurposing could become a sustainable strategy to overcome many issues related with drug pharmacokinetics and pharmacological safety (Dao et al., 2015).

Based on the acquired knowledge from experimental animal studies, cell-based assays and cell-based systems, the development of novel approaches for the treatment of oxidative-stress should aim not only on the use of standard antagonists against catalysts that have superoxide radicals as end-products, but also mitigate ROS-induced by CVD-related pathologies, thereby therapeutically correcting disrupted intracellular molecular events and also promoting resolution of inflammation, and then ultimately further damage and progression of the disease.

Note: All text, tables and illustrations in this chapter are novel, have not been published before and were generated by the authors.

References

Al-Gubory K. H., *Reprod Biomed Online*, **29** (2014), 17–31.

Ali M., Pulli B., Courties G., Tricot B., Sebas M., Iwamoto Y., Hilgendorf I., Schob S., Dong A., Zheng W., Skoura A., Kalgukar A., Cortes C., Ruggeri R., Swirski F. K., Nahrendorf M., Buckbinder L., Chen J. W., *JACC: Basic to Trans Sci*, **1** (2016), 633–643.

Anatoliotakis N., Deftereos S., Bouras G., Giannopoulos G., Tsounis D., Angelidis C., Kaoukis A., Stefanadis C., *Curr Top Med Chem*, **13** (2013), 115–138.

Antonenko Y. N., Avetisyan A. V., Bakeeva L. E., Chernyak B. V., Chertkov V. A., Domnina L. V., Ivanova O. Y., Izyumov D. S., Khailova L. S., Klishin S. S., Korshunova G. A., Lyamzaev K. G., Muntyan M. S., Nepryakhina O. K., Pashkovskaya A. A., Pletjushkina O. Y., Pustovidko A. V., Roginsky V. A., Rokitskaya T. I., Ruuge E. K., Saprunova V. B., Severina, II, Simonyan R. A., Skulachev I. V., Skulachev M. V., Sumbatyan N. V., Sviryaeva I. V., Tashlitsky V. N., Vassiliev J. M., Vyssokikh M. Y., Yaguzhinsky L. S., Zamyatnin A. A. Jr., Skulachev V. P., *Biochemistry (Mosc)*, **73** (2008), 1273–1287.

Aoyama T., Paik Y. H., Watanabe S., Laleu B., Gaggini F., Fioraso-Cartier L., Molango S., Heitz F., Merlot C., Szyndralewiez C., Page P., Brenner D. A., *Hepatology*, **56** (2012), 2316–2327.

Aseervatham G. S., Sivasudha T., Jeyadevi R., Arul Ananth D., *Environ Sci Pollut Res Int*, **20** (2013), 4356–4369.

Ayashi S., Assareh A. R., Jalali M. T., Olapour S., Yaghooti H., *Indian J Pharmacol*, **48** (2016), 372–376.

Ayer A., Gourlay C. W., Dawes I. W., *FEMS Yeast Res*, **14** (2014), 60–72.

Back M., *Future Sci OA*, **3** (2017), Fso236.

Back M., Yin L., Nagy E., Ingelsson E., *Br J Clin Pharmacol*, **75** (2013), 280–281.

Bai X. J., Hao J. T., Wang J., Zhang W. F., Yan C. P., Zhao J. H., Zhao Z. Q., *Pharmacol Rep*, **70** (2018), 60–68.

Bain G., King C. D., Brittain J., Hartung J. P., Dearmond I., Stearns B., Truong Y. P., Hutchinson J. H., Evans J. F., Holme K., *J Clin Pharmacol*, **52** (2012), 1482–1493.

Bain G., King C. D., Schaab K., Rewolinski M., Norris V., Ambery C., Bentley J., Yamada M., Santini A. M., van de Wetering de Rooij J., Stock N., Zunic J., Hutchinson J. H., Evans J. F., *Br J Clin Pharmacol*, **75** (2013), 779–790.

Ballard K. D., Bruno R. S., Seip R. L., Quann E. E., Volk B. M., Freidenreich D. J., Kawiecki D. M., Kupchak B. R., Chung M. Y., Kraemer W. J., Volek J. S., *Nutr J*, **8** (2009), 34.

Ballard K. D., Kupchak B. R., Volk B. M., Mah E., Shkreta A., Liptak C., Ptolemy A. S., Kellogg M. S., Bruno R. S., Seip R. L., Maresh C. M., Kraemer W. J., Volek J. S., *Br J Nutr*, **109** (2013), 882–893.

Bast A., Haenen G. R., *Environ Toxicol Pharmacol*, **11** (2002), 251–258.

Baumer A. T., Kruger C. A., Falkenberg J., Freyhaus H. T., Rosen R., Fink K., Rosenkranz S., *Clin Exp Hypertens*, **29** (2007), 287–299.

Berry C. E., Hare J. M., *J Physiol*, **555** (2004), 589–606.

Bhandarkar S. S., Jaconi M., Fried L. E., Bonner M. Y., Lefkove B., Govindarajan B., Perry B. N., Parhar R., Mackelfresh J., Sohn A., Stouffs M., Knaus U., Yancopoulos G., Reiss Y., Benest A. V., Augustin H. G., Arbiser J. L., *J Clin Invest*, **119** (2009), 2359–2365.

Birben E., Sahiner U. M., Sackesen C., Erzurum S., Kalayci O., *World Allergy Organ J*, **5** (2012), 9–19.

Birk A. V., Chao W. M., Bracken C., Warren J. D., Szeto H. H., *Br J Pharmacol*, **171** (2014), 2017–2028.

Bjelakovic G., Nikolova D., Gluud L. L., Simonetti R. G., Gluud C., *Cochrane Database Syst Rev*, **3** (2012), Cd007176.

Bjelakovic G., Nikolova D., Gluud L. L., Simonetti R. G., Gluud C., *Sao Paulo Med J*, **133** (2015), 164–165.

Bolisetty S., Jaimes E. A., *Int J Mol Sci*, **14** (2013), 6306–6344.

Boonla O., Kukongviriyapan U., Pakdeechote P., Kukongviriyapan V., Pannangpetch P., Prachaney P., Greenwald S. E., *Nitric Oxide*, **42** (2014), 44–53.

Borbely G., Huszar M., Varga A., Futosi K., Mocsai A., Orfi L., Idei M., Mandl J., Keri G., Vantus T., *Med Chem*, **8** (2012), 174–181.

Borghi C., Verardi F. M., Pareo I., Bentivenga C., Cicero A. F., *Expert Rev Cardiovasc Ther*, **12** (2014), 1219–1225.

Bowler R. P., Crapo J. D., *J Allergy Clin Immunol*, **110** (2002), 349–356.

Brash A. R., *J Biol Chem*, **274** (1999), 23679–23682.

Brown W. V., *Curr Med Res Opin*, **11** (1989), 321–330.

Brown D. I., Griendling K. K., *Circ Res*, **116** (2015), 531–549.

Brzheskiy V. V., Efimova E. L., Vorontsova T. N., Alekseev V. N., Gusarevich O. G., Shaidurova K. N., Ryabtseva A. A., Andryukhina O. M., Kamenskikh T. G., Sumarokova E. S., Miljudin E. S., Egorov E. A., Lebedev O. I., Surov A. V., Korol A. R., Nasinnyk I. O., Bezditko P. A., Muzhychuk O.

P., Vygodin V. A., Yani E. V., Savchenko A. Y., Karger E. M., Fedorkin O. N., Mironov A. N., Ostapenko V., Popeko N. A., Skulachev V. P., Skulachev M. V., *Adv Ther*, **32** (2015), 1263–1279.

Cam A., de Mejia E. G., *Mol Nutr Food Res*, **56** (2012), 53–66.

Cayatte A. J., Rupin A., Oliver-Krasinski J., Maitland K., Sansilvestri-Morel P., Boussard M. F., Wierzbicki M., Verbeuren T. J., Cohen R. A., *Arterioscler Thromb Vasc Biol*, **21** (2001), 1577–1584.

Cervantes Gracia K., Llanas-Cornejo D., Husi H., *J Clin Med*, **6** (2017), 22.

Chakrabarti S., Jahandideh F., Wu J., *Biomed Res Int*, **2014** (2014), 608979.

Chakraborty J., Banerjee S., Ray P., Hossain D. M., Bhattacharyya S., Adhikary A., Chattopadhyay S., Das T., Sa G., *J Biol Chem*, **285** (2010), 33104–33112.

Chandrasekharan J. A., Sharma-Walia N., *J Inflamm Res*, **8** (2015), 181–192.

Chawengsub Y., Gauthier K. M., Campbell W. B., *Am J Physiol Heart Circ Physiol*, **297** (2009), H495–H507.

Chawla A., *Circ Res*, **106** (2010), 1559–1569.

Cheng J., Ou J. S., Singh H., Falck J. R., Narsimhaswamy D., Pritchard K. A., Jr., Schwartzman M. L., *Am J Physiol Heart Circ Physiol*, **294** (2008), H1018–H1026.

Chida R., Hisauchi I., Toyoda S., Kikuchi M., Komatsu T., Hori Y., Nakahara S., Sakai Y., Inoue T., Taguchi I., *Hypertens Res*, **38** (2015), 765–769.

Chistiakov D. A., Melnichenko A. A., Myasoedova V. A., Grechko A. V., Orekhov A. N., *J Mol Med (Berl)*, **95** (2017), 1153–1165.

Chistiakov D. A., Orekhov A. N., Bobryshev Y. V., *Front Physiol*, **6** (2015), 365.

Cho J., Won K., Wu D., Soong Y., Liu S., Szeto H. H., Hong M. K., *Coron Artery Dis*, **18** (2007), 215–220.

Cifuentes-Pagano E., Saha J., Csanyi G., Ghouleh I. A., Sahoo S., Rodriguez A., Wipf P., Pagano P. J., Skoda E. M., *Medchemcomm*, **4** (2013), 1085–1092.

Comte M., Maulet Y., Cox J. A., *Biochem J*, **209** (1983), 269–272.

Connell B. J., Gordon J. R., Saleh T. M., *Neurosci Lett*, **606** (2015), 117–122.

Conrad C., Lymp J., Thompson V., Dunn C., Davies Z., Chatfield B., Nichols D., Clancy J., Vender R., Egan M. E., Quittell L., Michelson P., Antony V., Spahr J., Rubenstein R. C., Moss R. B., Herzenberg L. A., Goss C. H., Tirouvanziam R., *J Cyst Fibros*, **14** (2015), 219–227.

Cosentino F., Eto M., De Paolis P., van der Loo B., Bachschmid M., Ullrich V., Kouroedov A., Delli Gatti C., Joch H., Volpe M., Luscher T. F., *Circulation*, **107** (2003), 1017–1023.

Cosentino F., Francia P., Camici G. G., Pelicci P. G., Luscher T. F., Volpe M., *Arterioscler Thromb Vasc Biol*, **28** (2008), 622–628.

Csanyi G., Cifuentes-Pagano E., Al Ghouleh I., Ranayhossaini D. J., Egana L., Lopes L. R., Jackson H. M., Kelley E. E., Pagano P. J., *Free Radic Biol Med*, **51** (2011), 1116–1125.

Cui H., Kong Y., Zhang H., *J Signal Transduct*, **2012** (2012), 646354.

Dahan I., Molshanski-Mor S., Pick E., *J Leukoc Biol*, **91** (2012), 501–515.

Dao V. T., Casas A. I., Maghzal G. J., Seredenina T., Kaludercic N., Robledinos-Anton N., Di Lisa F., Stocker R., Ghezzi P., Jaquet V., Cuadrado A., Schmidt H. H., *Antioxid Redox Signal*, **23** (2015), 1113–1129.

Dare A. J., Logan A., Prime T. A., Rogatti S., Goddard M., Bolton E. M., Bradley J. A., Pettigrew G. J., Murphy M. P., Saeb-Parsy K., *J Heart Lung Transplant*, **34** (2015), 1471–1480.

Daubney J., Bonner P. L., Hargreaves A. J., Dickenson J. M., *Basic Clin Pharmacol Toxicol*, **116** (2015), 96–109.

Dean O., Giorlando F., Berk M., *J Psychiatry Neurosci*, **36** (2011), 78–86.

Deavall D. G., Martin E. A., Horner J. M., Roberts R., *J Toxicol*, **2012** (2012), 645460.

Derosa G., Mugellini A., Pesce R. M., D'Angelo A., Maffioli P., *BMC Cardiovasc Disord*, **16** (2016a), 66.

Derosa G., Mugellini A., Pesce R. M., D'Angelo A., Maffioli P., *Medicine (Baltimore)*, **95** (2016b), e3084.

Diatchuk V., Lotan O., Koshkin V., Wikstroem P., Pick E., *J Biol Chem*, **272** (1997), 13292–13301.

Dickinson D. A., Iles K. E., Zhang H., Blank V., Forman H. J., *Faseb J*, **17** (2003), 473–475.

Dickinson D. A., Levonen A. L., Moellering D. R., Arnold E. K., Zhang H., Darley-Usmar V. M., Forman H. J., *Free Radic Biol Med*, **37** (2004), 1152–1159.

Dikalova A., Clempus R., Lassegue B., Cheng G., McCoy J., Dikalov S., San Martin A., Lyle A., Weber D. S., Weiss D., Taylor W. R., Schmidt H. H., Owens G. K., Lambeth J. D., Griendling K. K., *Circulation*, **112** (2005), 2668–2676.

Drummond G. R., Sobey C. G., *Trends Endocrinol Metab*, **25** (2014), 452–463.

Du K., Ramachandran A., Jaeschke H., *Redox Biol*, **10** (2016), 148–156.

Dwaich K. H., Al-Amran F. G., Al-Sheibani B. I., Al-Aubaidy H. A., *Int J Cardiol*, **221** (2016), 977–986.

Eckel R. H., Kahn S. E., Ferrannini E., Goldfine A. B., Nathan D. M., Schwartz M. W., Smith R. J., Smith S. R., *J Clin Endocrinol Metab*, **96** (2011), 1654–1663.

Elinder F., Liin S. I., *Front Physiol*, **8** (2017), 43.

Erdmann K., Cheung B. W., Schroder H., *J Nutr Biochem*, **19** (2008), 643–654.

Espinosa-Diez C., Miguel V., Mennerich D., Kietzmann T., Sanchez-Perez P., Cadenas S., Lamas S., *Redox Biol*, **6** (2015), 183–197.

Fearon I. M., Faux S. P., *J Mol Cell Cardiol*, **47** (2009), 372–381.

Fenyo I. M., Florea I. C., Raicu M., Manea A., *Vascul Pharmacol*, **54** (2011), 100–106.

Fortuno A., Bidegain J., Robador P. A., Hermida J., Lopez-Sagaseta J., Beloqui O., Diez J., Zalba G., *Hypertension*, **54** (2009), 744–750.

Fowdar K., Chen H., He Z., Zhang J., Zhong X., Zhang J., Li M., Bai J., *Heart Lung*, **46** (2017), 120–128.

Fu R., Wang Q., Guo Q. Xu J., Wu X., *Vascul Pharmacol*, **58** (2013), 78–86.

Gaggini F., Laleu B., Orchard M., Fioraso-Cartier L., Cagnon L., Houngninou-Molango S., Gradia A., Duboux G., Merlot C., Heitz F., Szyndralewiez C., Page P., *Bioorg Med Chem*, **19** (2011), 6989–6999.

Gane E. J., Weilert F., Orr D. W., Keogh G. F., Gibson M., Lockhart M. M., Frampton C. M., Taylor K. M., Smith R. A., Murphy M. P., *Liver Int*, **30** (2010), 1019–1026.

Gao C., Liu Y., Yu Q., Yang Q., Li B., Sun L., Yan W., Cai X., Gao E., Xiong L., Wang H., Tao L., *Am J Physiol Heart Circ Physiol*, **308** (2015), H1583–H1591.

Garrido-Urbani S., Jemelin S., Deffert C., Carnesecchi S., Basset O., Szyndralewiez C., Heitz F., Page P., Montet X., Michalik L., Arbiser J., Ruegg C., Krause K. H., Imhof B. A., *PLoS One*, **6** (2011), e14665.

Garry A., Edwards D. H., Fallis I. F., Jenkins R. L., Griffith T. M., *Cardiovasc Res*, **84** (2009), 218–226.

Gatto G. J., Jr., Ao Z., Kearse M. G., Zhou M., Morales C. R., Daniels E., Bradley B. T., Goserud M. T., Goodman K. B., Douglas S. A., Harpel M. R., Johns D. G., *J Enzyme Inhib Med Chem*, **28** (2013), 95–104.

Gazaryan I. G., Thomas B., *Neural Regen Res*, **11** (2016), 1708–1711.

Gerlach U. A., Atanasov G., Wallenta L., Polenz D., Reutzel-Selke A., Kloepfel M., Jurisch A., Marksteiner M., Loddenkemper C., Neuhaus P., Sawitzki B., Pascher A., *Transplantation*, **97** (2014), 732–739.

Gesslbauer B., Bochkov V., *Biochem Soc Trans*, **45** (2017), 1225–1252.

Gianni D., Nicolas N., Zhang H., Der Mardirossian C., Kister J., Martinez L., Ferguson J., Roush W. R., Brown S. J., Bokoch G. M., Hodder P., Rosen H., Optimization and characterization of an inhibitor for NADPH oxidase 1 (NOX-1), in *Probe Reports from the NIH Molecular Libraries Program* National Center for Biotechnology Information (US), Bethesda (MD) (2010).

Gianni D., Taulet N., Zhang H., DerMardirossian C., Kister J., Martinez L., Roush W. R., Brown S. J., Bokoch G. M., Rosen H., *ACS Chem Biol*, **5** (2010), 981–993.

Giardino G., Cicalese M. P., Delmonte O., Migliavacca M., Palterer B., Loffredo L., Cirillo E., Gallo V., Violi F., Pignata C., *Oxid Med Cell Longev*, **2017** (2017), 4590127.

Gioscia-Ryan R. A., LaRocca T. J., Sindler A. L., Zigler M. C., Murphy M. P., Seals D. R., *J Physiol*, **592** (2014), 2549–2561.

Givertz M. M., Anstrom K. J., Redfield M. M., Deswal A., Haddad H., Butler J., Tang W. H., Dunlap M. E., LeWinter M. M., Mann D. L., Felker G. M., O'Connor C. M., Goldsmith S. R., Ofili E. O., Saltzberg M. T., Margulies K. B., Cappola T. P., Konstam M. A., Semigran M. J., McNulty S. E., Lee K. L., Shah M. R., Hernandez A. F., *Circulation*, **131** (2015), 1763–1771.

Glinn M. A., Lee C. P., Ernster L., *Biochim Biophys Acta*, **1318** (1997), 246–254.

Graham D., Huynh N. N., Hamilton C. A., Beattie E., Smith R. A., Cocheme H. M., Murphy M. P., Dominiczak A. F., *Hypertension*, **54** (2009), 322–328.

Grant C. M., *J Biol*, **7** (2008), 1.

Gray S. P., Di Marco E., Okabe J., Szyndralewiez C., Heitz F., Montezano A. C., de Haan J. B., Koulis C., El-Osta A., Andrews K. L., Chin-Dusting J. P., Touyz R. M., Wingler K., Cooper M. E., Schmidt H. H., Jandeleit-Dahm K. A., *Circulation*, **127** (2013), 1888–1902.

Green D. E., Murphy T. C., Kang B. Y., Kleinhenz J. M., Szyndralewiez C., Page P., Sutliff R. L., Hart C. M., *Am J Respir Cell Mol Biol*, **47** (2012), 718–726.

Guo A. M., Arbab A. S., Falck J. R., Chen P., Edwards P. A., Roman R. J., Scicli A. G., *J Pharmacol Exp Ther*, **321** (2007), 18–27.

Gupta S. C., Kismali G., Aggarwal B. B., *Biofactors*, **39** (2013), 2–13.

Guzik T. J., Chen W., Gongora M. C., Guzik B., Lob H. E., Mangalat D., Hoch N., Dikalov S., Rudzinski P., Kapelak B., Sadowski J., Harrison D. G., *J Am Coll Cardiol*, **52** (2008), 1803–1809.

Guzik T. J., Sadowski J., Guzik B., Jopek A., Kapelak B., Przybylowski P., Wierzbicki K., Korbut R., Harrison D. G., Channon K. M., *Arterioscler Thromb Vasc Biol*, **26** (2006), 333–339.

Harrison R., *Free Radic Biol Med*, **33** (2002), 774–797.

Harvey A. P., Montezano A. C., Hood K. Y., Lopes R. A., Rios F., Ceravolo G., Graham D., Touyz R. M., *Life Sci*, **179** (2017), 110–119.

Hayward R., Campbell B., Shin Y. K., Scalia R., Lefer A. M., *Cardiovasc Res*, **41** (1999), 65–76.

He L., He T., Farrar S., Ji L., Liu T., Ma X., *Cell Physiol Biochem*, **44** (2017), 532–553.

Heumuller S., Wind S., Barbosa-Sicard E., Schmidt H. H., Busse R., Schroder K., Brandes R. P., *Hypertension*, **51** (2008), 211–217.

Hirano K., Chen W. S., Chueng A. L., Dunne A. A., Seredenina T., Filippova A., Ramachandran S., Bridges A., Chaudry L., Pettman G., Allan C., Duncan S., Lee K. C., Lim J., Ma M. T., Ong A. B., Ye N. Y., Nasir S., Mulyanidewi S., Aw C. C., Oon P. P., Liao S., Li D., Johns D. G., Miller N. D., Davies C. H., Browne E. R., Matsuoka Y., Chen D. W., Jaquet V., Rutter A. R., *Antioxid Redox Signal*, **23** (2015), 358–374.

Hoeschen R. J., *Can J Cardiol*, **13** (1997), 1021–1025.

Hohn T. J., Grune T., *Redox Biol*, **2** (2014), 388–394.

Holmquist C., Larsson S., Wolk A., de Faire U., *J Nutr*, **133** (2003), 2650–2654.

Huang H. C., Nguyen T., Pickett C. B., *Proc Natl Acad Sci U S A*, **97** (2000), 12475–12480.

Hulsmans M., Holvoet P., *J Cell Mol Med*, **14** (2010), 70–78.

Humayun Q., Mirza S., *Lancet*, **378** (2011), 565; author reply 565–566.

Hung O. Y., Molony D., Corban M. T., Rasoul-Arzrumly E., Maynard C., Eshtehardi P., Dhawan S., Timmins L. H., Piccinelli M., Ahn S. G., Gogas B. D., McDaniel M. C., Quyyumi A. A., Giddens D. P., Samady H., *J Am Heart Assoc*, **5** (2016), e002764.

Ikeda Y., Young L. H., Scalia R., Ross C. R., Lefer A. M., *Cardiovasc Res*, **49** (2001), 69–77.

Imamura H., Yamaguchi T., Nagayama D., Saiki A., Shirai K., Tatsuno I., *Int Heart J*, **58** (2017), 577–583.

Ingelsson E., Yin L., Back M., *J Allergy Clin Immunol*, **129** (2012), 702–707. e702.

Iso H., *J Atheroscler Thromb*, **18** (2011), 83–88.

Jaquet V., Marcoux J., Forest E., Leidal K. G., McCormick S., Westermaier Y., Perozzo R., Plastre O., Fioraso-Cartier L., Diebold B., Scapozza L., Nauseef W. M., Fieschi F., Krause K. H., Bedard K., *Br J Pharmacol*, **164** (2011), 507–520.

Joseph L. C., Kokkinaki D., Valenti M. C., Kim G. J., Barca E., Tomar D., Hoffman N. E., Subramanyam P., Colecraft H. M., Hirano M., Ratner A. J., Madesh M., Drosatos K., Morrow J. P., *JCI Insight*, **2** (2017), e94248.

Kang Y. E., Kim J. M., Joung K. H., Lee J. H., You B. R., Choi M. J., Ryu M. J., Ko Y. B., Lee M. A., Lee J., Ku B. J., Shong M., Lee K. H., Kim H. J., *PLoS One*, **11** (2016), e0154003.

Karra L., Haworth O., Priluck R., Levy B. D., Levi-Schaffer F., *Mucosal Immunol*, **8** (2015), 852–862.

Katoh A., Kai H., Harada H., Niiyama H., Ikeda H., *Int Heart J*, **58** (2017), 926–932.

Kaufman M., Guglin M., *Heart Fail Rev*, **18** (2013), 177–186.

Kayama Y., Minamino T., Toko H., Sakamoto M., Shimizu I., Takahashi H., Okada S., Tateno K., Moriya J., Yokoyama M., Nojima A., Yoshimura M., Egashira K., Aburatani H., Komuro I., *J Exp Med*, **206** (2009), 1565–1574.

Kennedy J. A., Beck-Oldach K., McFadden-Lewis K., Murphy G. A., Wong Y. W., Zhang Y., Horowitz J. D., *Eur J Pharmacol*, **531** (2006), 13–19.

Kim H. J., Han S. J., Kim D. J., Jang H. C., Lim S., Choi S. H., Kim Y. H., Shin D. H., Kim S. H., Kim T. H., Ahn Y. B., Ko S. H., Kim N. H., Seo J. A., Kim H. Y., Lee K. W., *Korean J Intern Med*, **32** (2017), 497–504.

Kim Y. S., Kim J. S., Kwon J. S., Jeong M. H., Cho J. G., Park J. C., Kang J. C., Ahn Y., *Int Heart J*, **51** (2010), 348–353.

Kofler P. A., Pircher H., von Grafenstein S., Diener T., Holl M., Liedl K. R., Siems K., Jansen-Durr P., *Planta Med*, **79** (2013), 244–252.

Konior A., Schramm A., Czesnikiewicz-Guzik M., Guzik T. J., *Antioxid Redox Signal*, **20** (2014), 2794–2814.

Korkmaz-Icoz S., Radovits T., Szabo G., *Br J Pharmacol*, **175** (2018), 223–231.

Krieglstein C. F., Granger D. N., *Am J Hypertens*, **14** (2001), 44s–54s.

Kuhn H., *Circ Res*, **94** (2004), 1527–1529.

Laleu B., Gaggini F., Orchard M., Fioraso-Cartier L., Cagnon L., Houngninou-Molango S., Gradia A., Duboux G., Merlot C., Heitz F., Szyndralewiez C., Page P., *J Med Chem*, **53** (2010), 7715–7730.

Lazarevic-Pasti T., Leskovac A., Vasic V., *Curr Drug Metab*, **16** (2015), 168–190.

Lee M. C., Velayutham M., Komatsu T., Hille R., Zweier J. L., *Biochemistry*, **53** (2014), 6615–6623.

Leo S., Szabadkai G., Rizzuto R., *Ann N Y Acad Sci*, **1147** (2008), 264–274.

Liemburg-Apers D. C., Willems P. H., Koopman W. J., Grefte S., *Arch Toxicol*, **89** (2015), 1209–1226.

Lightfoot A. P., Sakellariou G. K., Nye G. A., McArdle F., Jackson M. J., Griffiths R. D., McArdle A. *Redox Biol*, **6** (2015), 253–259.

Limon-Pacheco J., Gonsebatt M. E., *Mutat Res*, **674** (2009), 137–147.

Liu Y., Davidson B. P., Yue Q., Belcik T., Xie A., Inaba Y., McCarty O. J., Tormoen G. W., Zhao Y., Ruggeri Z. M., Kaufmann B. A., Lindner J. R., *Circ Cardiovasc Imaging*, **6** (2013), 74–82.

Liu G., Vogel S. M., Gao X., Javaid K., Hu G., Danilov S. M., Malik A. B., Minshall R. D., *Arterioscler Thromb Vasc Biol*, **31** (2011), 1342–1350.

Lo S. C., Li X., Henzl M. T., Beamer L. J., Hannink M., *Embo J*, **25** (2006), 3605–3617.

Lyle A. N., Deshpande N. N., Taniyama Y., Seidel-Rogol B., Pounkova L., Du P., Papaharalambus C., Lassègue B., Griendling K. K., *Circ. Res.*, **105** (2009), 249–259.

Madamanchi N. R., Hakim Z. S., Runge M. S., *J Thromb Haemost*, **3** (2005), 254–267.

Madrigal-Matute J., Fernandez-Garcia C. E., Gomez-Guerrero C., Lopez-Franco O., Munoz-Garcia B., Egido J., Blanco-Colio L. M., Martin-Ventura J. L., *Cardiovasc Res*, **95** (2012), 116–123.

Maffi P., Secchi A., *Dev Ophthalmol*, **60** (2017), 1–5.

Maia L. B., Pereira V., Mira L., Moura J. J., *Biochemistry*, **54** (2015), 685–710.

Manganas P., MacPherson L., Tokatlidis K., *Cell Tissue Res*, **367** (2017), 43–57.

Manskikh V. N., Gancharova O. S., Nikiforova A. I., Krasilshchikova M. S., Shabalina I. G., Egorov M. V., Karger E. M., Milanovsky G. E., Galkin, II, Skulachev V. P., Zinovkin R. A., *Histol Histopathol*, **30** (2015), 353–360.

Marso S. P., Daniels G. H., Brown-Frandsen K., Kristensen P., Mann J. F., Nauck M. A., Nissen S. E., Pocock S., Poulter N. R., Ravn L. S., Steinberg W. M., Stockner M., Zinman B., Bergenstal R. M., Buse J. B., *N Engl J Med*, **375** (2016), 311–322.

Mashima R., Okuyama T., *Redox Biol*, **6** (2015), 297–310.

Massart C., Giusti N., Beauwens R., Dumont J. E., Miot F., Sande J. V., *FEBS Open Bio*, **4** (2013), 55–59.

Mathers C. D., Loncar D., *PLoS Med*, **3** (2006), e442.

Matsuno K., Yamada H., Iwata K., Jin D., Katsuyama M., Matsuki M., Takai S., Yamanishi K., Miyazaki M., Matsubara H., Yabe-Nishimura C., *Circulation*, **112** (2005), 2677–2685.

Mattioni T. A., Zheutlin T. A., Sarmiento J. J., Parker M., Lesch M., Kehoe R. F., *Ann Intern Med*, **111** (1989), 574–580.

Medhora M., Chen Y., Gruenloh S., Harland D., Bodiga S., Zielonka J., Gebremedhin D., Gao Y., Falck J. R., Anjaiah S., Jacobs E. R., *Am J Physiol Lung Cell Mol Physiol*, **294** (2008), L902–L911.

Meijles D. N., Pagano P. J., *Hypertension*, **67** (2016), 14–19.

Meitzler J. L., Antony S., Wu Y., Juhasz A., Liu H., Jiang G., Lu J., Roy K., Doroshow J. H., *Antioxid Redox Signal*, **20** (2014), 2873–2889.

Mercer J. R., Yu E., Figg N., Cheng K. K., Prime T. A., Griffin J. L., Masoodi M., Vidal-Puig A., Murphy M. P., Bennett M. R., *Free Radic Biol Med*, **52** (2012), 841–849.

Merchant S. H., Gurule D. M., Larson R. S., *Am J Physiol Heart Circ Physiol*, **284** (2003), H1260–H1268.

Mironczuk-Chodakowska I., Witkowska A. M., Zujko M. E., *Adv Med Sci*, **63** (2018), 68–78.

Montecucco F., Lenglet S., Braunersreuther V., Pelli G., Pellieux C., Montessuit C., Lerch R., Deruaz M., Proudfoot A. E., Mach F., *Arterioscler Thromb Vasc Biol*, **30** (2010), 1371–1377.

Mousslim M., Pagano A., Andreotti N., Garrouste F., Thuault S., Peyrot V., Parat F., Luis J., Culcasi M., Thetiot-Laurent S., Pietri S., Sabatier J. M., Kovacic H., *Eur J Pharmacol*, **794** (2017), 162–172.

Moyer V. A., *Ann Intern Med*, **160** (2014), 558–564.

Mozaffarian D., *Circulation*, **133** (2016), 187–225.

Munson J. M., Fried L., Rowson S. A., Bonner M. Y., Karumbaiah L., Diaz B., Courtneidge S. A., Knaus U. G., Brat D. J., Arbiser J. L., Bellamkonda R. V., *Sci Transl Med*, **4** (2012), 127ra136.

Muntean D. M., Sturza A., Danila M. D., Borza C., Duicu O. M., Mornos C., *Oxid Med Cell Longev*, **2016** (2016), 8254942.

Muntwyler J., Hennekens C. H., Manson J. E., Buring J. E., Gaziano J. M., *Arch Intern Med*, **162** (2002), 1472–1476.

Murer S. B., Aeberli I., Braegger C. P., Gittermann M., Hersberger M., Leonard S. W., Taylor A. W., Traber M. G., Zimmermann M. B., *J Nutr*, **144** (2014), 193–201.

Murphy M. P., *Biochem J*, **417** (2009), 1–13.

Musset B., Clark R. A., DeCoursey T. E., Petheo G. L., Geiszt M., Chen Y., Cornell J. E., Eddy C. A., Brzyski R. G., El Jamali A., *J Biol Chem*, **287** (2012), 9376–9388.

Myung S. K., Ju W., Cho B., Oh S. W., Park S. M., Koo B. K., Park B. J., *BMJ*, **346** (2013), f10.

Neau D. B., Gilbert N. C., Bartlett S. G., Boeglin W., Brash A. R., Newcomer M. E., *Biochemistry*, **48** (2009), 7906–7915.

Neuhouser M. L., Wassertheil-Smoller S., Thomson C., Aragaki A., Anderson G. L., Manson J. E., Patterson R. E., Rohan T. E., van Horn L., Shikany J. M., Thomas A., LaCroix A., Prentice R. L., *Arch Intern Med*, **169** (2009), 294–304.

Nimse S. B., Pal D., *RSC Adv*, **5** (2015), 27986–28006.

Nita M., Grzybowski A., *Oxid Med Cell Longev*, **2016** (2016), 3164734.

O'Donnell V. B., Maskrey B., Taylor G. W., *Methods Mol Biol*, **462** (2009), 5–23.

O'Donnell B. V., Tew D. G., Jones O. T., England P. J., *Biochem J*, **290 (Pt 1)** (1993), 41–49.

Olukman M., Orhan C. E., Celenk F. G., Ulker S., *J Diabetes Complications*, **24** (2010), 415–423.

Ott M., Gogvadze V., Orrenius S., Zhivotovsky B., *Apoptosis*, **12** (2007), 913–922.

Ozcan A., Ogun M., Biochemistry of reactive oxygen and nitrogen species, basic principles and clinical significance of oxidative stress, in *Biochemistry, Genetics and Molecular Biology: Basic Principles and Clinical Significance of Oxidative Stress*, Gowder D. S. J. T., eds.; InTech (2015).

Pacher P., Beckman J. S., Liaudet L., *Physiol Rev*, **87** (2007), 315–424.

Pak O., Scheibe S., Esfandiary A., Gierhardt M., Sydykov A., Logan A., Fysikopoulos A., Veit F., Hecker M., Kroschel F., Quanz K., Erb A., Schafer K., Fassbinder M., Alebrahimdehkordi N., Ghofrani H. A., Schermuly R. T., Brandes R. P., Seeger W., Murphy M. P., Weissmann N., Sommer N., *Eur Respir J*, **51** (2018), 1701024.

Palomo L., Felix-Redondo F. J., Lozano-Mera L., Perez-Castan J. F., Fernandez-Berges D., Buitrago F., *Br J Gen Pract*, **64** (2014), e627–e633.

Panday A., Sahoo M. K., Osorio D., Batra S., *Cell Mol Immunol*, **12** (2015), 5–23.

Park S., Karunakaran U., Jeoung N. H., Jeon J. H., Lee I. K., *Curr Med Chem*, **21** (2014), 3636–3645.

Park S. Y., Murphy S. P., Wilkens L. R., Henderson B. E., Kolonel L. N., *Am J Epidemiol*, **173** (2011), 906–914.

Pashkow F. J., *Int J Inflam*, **2011** (2011), 514623.

Pasterkamp G., de Kleijn D. P., Borst C., *Cardiovasc Res*, **45** (2000), 843–852.

Patsouris D., Cao J. J., Vial G., Bravard A., Lefai E., Durand A., Durand C., Chauvin M. A., Laugerette F., Debard C., Michalski M. C., Laville M., Vidal H., Rieusset J., *PLoS One*, **9** (2014), e110653.

Petri M. H., Laguna-Fernandez A., Gonzalez-Diez M., Paulsson-Berne G., Hansson G. K., Back M., *Cardiovasc Res*, **105** (2015), 65–74.

Piomelli D., *Curr Opin Cell Biol*, **5** (1993), 274–280.

Poeckel D., Funk C. D., *Cardiovasc Res*, **86** (2010), 243–253.

Quinlan C. L., Orr A. L., Perevoshchikova I. V., Treberg J. R., Ackrell B. A., Brand M. D., *J Biol Chem*, **287** (2012), 27255–27264.

Radi R., *J Biol Chem*, **288** (2013), 26464–26472.

Radmark O., Werz O., Steinhilber D., Samuelsson B., *Trends Biochem Sci*, **32** (2007), 332–341.

Rahal A., Kumar A., Singh V., Yadav B., Tiwari R., Chakraborty S., Dhama K., *Biomed Res Int*, **2014** (2014), 761264.

Rahman K., *Clin Interv Aging*, **2** (2007), 219–236.

Ranayhossaini D. J., Rodriguez A. I., Sahoo S., Chen B. B., Mallampalli R. K., Kelley E. E., Csanyi G., Gladwin M. T., Romero G., Pagano P. J., *J Biol Chem*, **288** (2013), 36437–36450.

Recio C., Maione F., Iqbal A. J., Mascolo N., De Feo V., *Front Pharmacol*, **7** (2016), 526.

Rehman H., Liu Q., Krishnasamy Y., Shi Z., Ramshesh V. K., Haque K., Schnellmann R. G., Murphy M. P., Lemasters J. J., Rockey D. C., Zhong Z., *Int J Physiol Pathophysiol Pharmacol*, **8** (2016), 14–27.

Rey F. E., Cifuentes M. E., Kiarash A., Quinn M. T., Pagano P. J., *Circ Res*, **89** (2001), 408–414.

Rochette L., Ghibu S., Muresan A., Vergely C., *Can J Physiol Pharmacol*, **93** (2015), 1021–1027.

Rosa F. T., Freitas E. C., Deminice R., Jordao A. A., Marchini J. S., *Eur J Nutr*, **53** (2014), 823–830.

Roth G. A., Johnson C., Abajobir A., Abd-Allah F., Abera S. F., Abyu G., et al., *J Am Coll Cardiol*, **70** (2017), 1–25.

Rudolph V., Rudolph T. K., Schopfer F. J., Bonacci G., Woodcock S. R., Cole M. P., Baker P. R., Ramani R., Freeman B. A., *Cardiovasc Res*, **85** (2010), 155–166.

Rutherfurd-Markwick K. J., *Br J Nutr*, **108 Suppl 2** (2012), S149–S157.

Saltiel A. R., Kahn C. R., *Nature*, **414** (2001), 799–806.

Scharnagl H., Kleber M. E., Genser B., Kickmaier S., Renner W., Weihrauch G., Grammer T., Rossmann C., Winkelmann B. R., Boehm B. O., Sattler W., Marz W., Malle E., *Int J Cardiol*, **174** (2014), 96–105.

Schrijver I. T., Kemperman H., Roest M., Kesecioglu J., de Lange D. W., *Intensive Care Med Exp*, **5** (2017), 43.

Schuchardt M., Herrmann J., Tolle M., van der Giet M., *Curr Pharm Des*, **23** (2017), 3391–3404.

Schweigerer L., Fotsis T., *Klin Padiatr*, **203** (1991), 319–322.

Sedeek M., Hebert R. L., Kennedy C. R., Burns K. D., Touyz R. M., *Curr Opin Nephrol Hypertens*, **18** (2009), 122–127.

Shao L., Ma A., Figtree G., Zhang P., *J Cardiovasc Pharmacol*, **68** (2016), 150–154.

Shi J., Ross C. R., Leto T. L., Blecha F., *Proc Natl Acad Sci U S A*, **93** (1996), 6014–6018.

Shill D. D., Southern W. M., Willingham T. B., Lansford K. A., McCully K. K., Jenkins N. T., *J Physiol*, **594** (2016), 7005–7014.

Shinozaki K., Ayajiki K., Nishio Y., Sugaya T., Kashiwagi A., Okamura T., *Hypertension*, **43** (2004), 255–262.

Sims C. R., MacMillan-Crow L. A., Mayeux P. R., *Clin Pharmacol Ther*, **96** (2014), 662–664.

Smith S. M., Min J., Ganesh T., Diebold B., Kawahara T., Zhu Y., McCoy J., Sun A., Snyder J. P., Fu H , Du Y., Lewis I., Lambeth J. D., *Chem Biol*, **19** (2012), 752–763.

Smith R. A., Murphy M. P., *Ann N Y Acad Sci*, **1201** (2010), 96–103.

Snow B. J., Rolfe F. L , Lockhart M. M., Frampton C. M., O'Sullivan J. D., Fung V., Smith R. A., Murphy M. P., Taylor K. M., *Mov Disord*, **25** (2010), 1670–1674.

Son D. J., Ha S. J., Song H. S., Lim Y., Yun Y. P., Lee J. W., Moon D. C., Park Y. H., Park B. S., Song M. J., Hong J. T., *J Pharmacol Exp Ther*, **317** (2006), 627–634.

Song P., Xu J., Song Y., Jiang S., Yuan H., Zhang X., *Dis Markers*, **2015** (2015), 761939.

Sorescu G. P., Song H., Tressel S. L., Hwang J., Dikalov S., Smith D. A., Boyd N. L., Platt M. O., Lassegue B., Griendling K. K., Jo H., *Circ Res*, **95** (2004), 773–779.

Sorescu D., Weiss D., Lassegue B., Clempus R. E., Szocs K., Sorescu G. P., Valppu L., Quinn M. T., Lambeth J. D., Vega J. D., Taylor W. R., Griendling K. K., *Circulation*, **105** (2002), 1429–1435.

Soubhye J., Chikh Alard I., Aldib I., Prevost M., Gelbcke M., De Carvalho A., Furtmuller P. G., Obinger C., Flemmig J., Tadrent S., Meyer F., Rousseau A., Neve J., Mathieu V., Zouaoui Boudjeltia K., Dufrasne F., Van Antwerpen P., *J Med Chem*, **60** (2017), 6563–6586.

Sousa L. F., Coelho F. M., Rodrigues D. H., Campos A. C., Barcelos Lda S., Teixeira M. M., Rachid M. A., Teixeira A. L., *Clinics (Sao Paulo)*, **68** (2013), 391–394.

Sousa G. T., Lira F. S., Rosa J. C., de Oliveira E. P., Oyama L. M., Santos R. V., Pimentel G. D., *Lipids Health Dis*, **11** (2012), 67.

Sovari A. A., Dudley S. C., Jr., *Front Physiol*, **3** (2012), 311.

Sreejayan N., Rao M. N., *Arzneimittelforschung*, **46** (1996), 169–171.

Sreejayan, Rao M. N., *J Pharm Pharmacol*, **49** (1997), 105–107.

Stansfield W. E., Moss N. C., Willis M. S., Tang R., Selzman C. H., *Ann Thorac Surg*, **84** (2007), 120–125.

Stein L. R., Imai S., *Trends Endocrinol Metab*, **23** (2012), 420–428.

Stielow C., Catar R. A., Muller G., Wingler K., Scheurer P., Schmidt H. H., Morawietz H., *Biochem Biophys Res Commun*, **344** (2006), 200–205.

Stock N. S., Bain G., Zunic J., Li Y., Ziff J., Roppe J., Santini A., Darlington J., Prodanovich P., King C. D., Baccei C., Lee C., Rong H., Chapman C., Broadhead A., Lorrain D., Correa L., Hutchinson J. H., Evans J. F., Prasit P., *J Med Chem*, **54** (2011), 8013–8029.

Stolk J., Hiltermann T. J., Dijkman J. H., Verhoeven A. J., *Am J Respir Cell Mol Biol*, **11** (1994), 95–102.

Stonehouse W., Brinkworth G. D., Thompson C. H., Abeywardena M. Y., *Atherosclerosis*, **254** (2016), 205–214.

Sun Z., Wang X., Lasson A., Bojesson A., Annborn M., Andersson R., *Scand J Gastroenterol*, **36** (2001), 55–65.

Suzuki K., *Adv Drug Deliv Rev*, **61** (2009), 287–289.

Suzuki H., Kayama Y., Sakamoto M., Iuchi H., Shimizu I., Yoshino T., Katoh D., Nagoshi T., Tojo K., Minamino T., Yoshimura M., Utsunomiya K., *Diabetes*, **64** (2015), 618–630.

Szeto H. H., Birk A. V., *Clin Pharmacol Ther*, **96** (2014), 672–683.

Takahashi K., Okumura H., Guo R., Naruse K., *Int J Mol Sci*, **18** (2017).

Takai S., Jin D., Kawashima H., Kimura M., Shiraishi-Tateishi A., Tanaka T., Kakutani S., Tanaka K., Kiso Y., Miyazaki M., *J Atheroscler Thromb*, **16** (2009), 480–489.

Takami S., Minami M., Katayama T., Nagata I., Namura S., Satoh M., *J Cereb Blood Flow Metab*, **22** (2002), 780–784.

Tayebati S. K., Tomassoni D., Di Cesare Mannelli L., Amenta F., *Clin Exp Hypertens*, **38** (2016), 30–38.

ten Freyhaus H., Huntgeburth M., Wingler K., Schnitker J., Baumer A. T., Vantler M., Bekhite M. M., Wartenberg M., Sauer H., Rosenkranz S., *Cardiovasc Res*, **71** (2006), 331–341.

Tibullo D., Li Volti G., Giallongo C., Grasso S., Tomassoni D., Anfuso C. D., Lupo G., Amenta F., Avola R., Bramanti V., *Inflamm Res*, **66** (2017), 947–959.

Tirone F., Radu L., Craescu C. T., Cox J. A., *Biochemistry*, **49** (2010), 761–771.

Toldo S., Schatz A. M., Mezzaroma E., Chawla R., Stallard T. W., Stallard W. C., Jahangiri A., Van Tassell B. W., Abbate A., *Cardiovasc Drugs Ther*, **26** (2012), 273–276.

Trachootham D., Lu W., Ogasawara M. A., Nilsa R. D., Huang P., *Antioxid Redox Signal*, **10** (2008), 1343–1374.

Trnka J., Blaikie F. H., Logan A., Smith R. A., Murphy M. P., *Free Radic Res*, **43** (2009), 4–12.

Turrens J. F., *J Physiol*, **552** (2003), 335–344.

Ursini F., Maiorino M., Forman H. J., *Redox Biol*, **8** (2016), 205–215.

Uttara B., Singh A. V., Zamboni P., Mahajan R. T., *Curr Neuropharmacol*, **7** (2009), 65–74.

Valko M., Leibfritz D., Moncol J., Cronin M. T., Mazur M., Telser J., *Int J Biochem Cell Biol*, **39** (2007), 44–84.

Valko M., Rhodes C. J., Moncol J., Izakovic M., Mazur M., *Chem Biol Interact*, **160** (2006), 1–40.

Waldman M., Peterson S. J., Arad M., Hochhauser E., *Prostaglandins Other Lipid Mediat*, **125** (2016), 108–117.

Walsh L. K., Restaino R. M., Neuringer M., Manrique C., Padilla J., *Clin Sci (Lond)*, **130** (2016), 1881–1888.

Wang Y., Mo X., Piper M. G., Wang H., Parinandi N. L., Guttridge D., Marsh C. B., *PLoS One*, **6** (2011), e28081.

Wang Q., Zhou H., Gao H., Chen S. H., Chu C. H., Wilson B., Hong J. S., *J Neuroinflammation*, **9** (2012), 32.

Ward N. C., Puddey I. B., Hodgson J. M., Beilin L. J., Croft K. D., *Free Radic Biol Med*, **38** (2005), 1032–1036.

Watkins M. L., Erickson J. D., Thun M. J., Mulinare J., Heath C. W., Jr., *Am J Epidemiol*, **152** (2000), 149–162.

Waxman S. G., Ritchie J. M., *Science*, **228** (1985), 1502–1507.

Weissig V., Guzman-Villanueva D., *Expert Opin Drug Deliv*, **12** (2015), 1783–1790.

Wilkins E., Wilson L., Wickramasinghe K., Bhatnagar P., Leal J., Luengo-Fernandez R., Burns R., Rayner M., Townsend N., *European Heart Network: Brussels, Belgium* (2017).

Wind S., Beuerlein K., Eucker T., Muller H., Scheurer P., Armitage M. E., Ho H., Schmidt H. H., Wingler K., *Br J Pharmacol*, **161** (2010), 885–898.

Wu F., Han M., Wilson J. X., *Br J Pharmacol*, **157** (2009), 1014–1023.

Xiao Y. F., *J Geriatr Cardiol*, **8** (2011), 147–158.

Xu Q., Kulkarni A. A., Sajith A. M., Hussein D., Brown D., Guner O. F., Reddy M. D., Watkins E. B., Lassegue B., Griendling K. K., Bowen J. P., *Bioorg Med Chem*, **26** (2018), 989–998.

Ye Y., Li J., Yuan Z., *PLoS One*, **8** (2013), e56803.

Ying Z., Kampfrath T., Sun Q., Parthasarathy S., Rajagopalan S., *Inflamm Res*, **60** (2011), 219–225.

Zahedi M., Ghiasvand R., Feizi A., Asgari G., Darvish L., *Int J Prev Med*, **4** (2013), 777–785.

Zeng Q., Han Y., Bao Y., Li W., Li X., Shen X., Wang X., Yao F., O'Rourke S. T., Sun C., *Am J Physiol Heart Circ Physiol*, **299** (2010), H1109–H1117.

Zang M., Xu S., Maitland-Toolan K. A., Zuccollo A., Hou X., Jiang B., Wierzbicki M., Verbeuren T. J., Cohen R. A., *Diabetes*, **55** (2006), 2180–2191.

Zhang J., Dierckx R., Mohee K., Clark A. L., Cleland J. G., *ESC Heart Fail*, **4** (2017), 40–45.

Zhang Y., Shimizu H., Siu K. L., Mahajan A., Chen J. N., Cai H., *J Biol Chem*, **289** (2014), 23200–23208.

Zhang M., Swarts S. G., Yin L., Liu C., Tian Y., Cao Y., Swarts M., Yang S., Zhang S. B., Zhang K., Ju S., Olek D. J., Jr., Schwartz L., Keng P. C., Howell R., Zhang L., Okunieff P., *Adv Exp Med Biol*, **701** (2011), 283–289.

Zhang J., Wang X., Vikash V., Ye Q., Wu D., Liu Y., Dong W., *Oxid Med Cell Longev*, **2016** (2016), 4350965.

Zhang C., Zhou Y., Yang G. Y., Li S., *Am J Transl Res*, **9** (2017), 5518–5527.

Zhong X. F., Huang G. D., Luo T., Deng Z. Y., Hu J. N., *Mol Med Rep*, **5** (2012), 1261–1266.

Zhou M. S., Hernandez Schulman I., Pagano P. J., Jaimes E. A., Raij L., *Hypertension*, **47** (2006), 81–86.

Zhou Y. X., Xia W., Yue W., Peng C., Rahman K., Zhang H., *Evid Based Complement Alternat Med*, **2015** (2015), 578107.

Chapter 8

Enzymatic Amino Acid Deprivation Therapies Targeting Cancer

Carla S. S. Teixeira, Henrique S. Fernandes, Sérgio F. Sousa, and Nuno M. F. S. A. Cerqueira

UCIBIO@REQUIMTE, BioSIM, Departamento de Biomedicina, Faculdade de Medicina da Universidade do Porto, Alameda Professor Hernâni Monteiro 4200-319, Porto, Portugal

nunoscerqueira@med.up.pt

8.1 Introduction

Proteins are large and complex molecules that play critical roles in the cells. They are involved in virtually all cell functions, including their regulation and structure. Any alteration on the concentration or composition of the proteins will disturb the function of the cells and can compromise cellular viability. In severe cases, this can ultimately lead to the death of the organism.

Proteins are made up of hundreds or thousands of smaller units called amino acids, which are attached to one another through peptide bonds in long chains. There are 20 different types of

Pharmaceutical Biocatalysis: Fundamentals, Enzyme Inhibitors, and Enzymes in Health and Diseases
Edited by Peter Grunwald

ISBN 978-981-4800-61-7 (Hardcover), 978-0-429-29503-4 (eBook)
www.jennystanford.com

natural amino acids that can be combined and used by the cells to produce proteins. It is, therefore, important that these molecules are available in sufficient amount in the cells to maintain their normal function.

Most of the mammals, and particularly humans, have long ago ceased the capability to produce efficiently 10 of the 20 natural amino acids. This means that for the healthy development, growth, metabolic, and physiological functions of their cells, these compounds must be obtained from the diet. Otherwise, they would starve to death. These amino acids are termed essential and include isoleucine, leucine, lysine, methionine, phenylalanine, threonine, tryptophan, and valine. This class of amino acids can also include histidine and arginine. Both histidine and arginine are non-essential in healthy adults but become essential when their requirement increases and they need to be obtained from the diet. This usually occurs during the periods of growth, illness, or metabolic stress. For this reason, they are called semi-essential amino acids. All the other amino acids are termed non-essential and are efficiently synthesized by the enzymes present in the cells (alanine, asparagine, aspartic acid, cysteine, glutamic acid, glutamine, glycine, proline, serine, and tyrosine).

The metabolism of the amino acids has often been disregarded for clinical purposes. However, recent studies have shown that the amino acid can represent a great clinical opportunity, still poorly explored, for the treatment of several types of cancer that require high concentrations of one or more amino acid residues that are essential in their development, survival, and spread. In order to overcome these problems, tumours rely on the extracellular pool of these amino acids to satisfy protein biosynthesis demands and continue to grow without limitations. This means that if the concentration of those amino acids decreases, the development of tumour cells can be impaired or even annihilated. At the same time, healthy cells remain almost unchanged since they are less demanding (in the case on the essential amino acids) or can synthesize these compounds (non-essential or semi-essential amino acids) in sufficient amounts by alternative pathways.

The amino acid deprivation cancer therapy relies on these principles, and its main objective is to starve tumours to death while keeping normal cells unharmed (Fung and Chan, 2017). The optimal conditions for this therapy thus occur when the tumours are sensitive to one of the non-essential or semi-essential amino acid residues. However, a simple diet is not enough to achieve a therapeutically relevant amino acid deprivation for cancer therapy, and therefore, other strategies are used for this purpose.

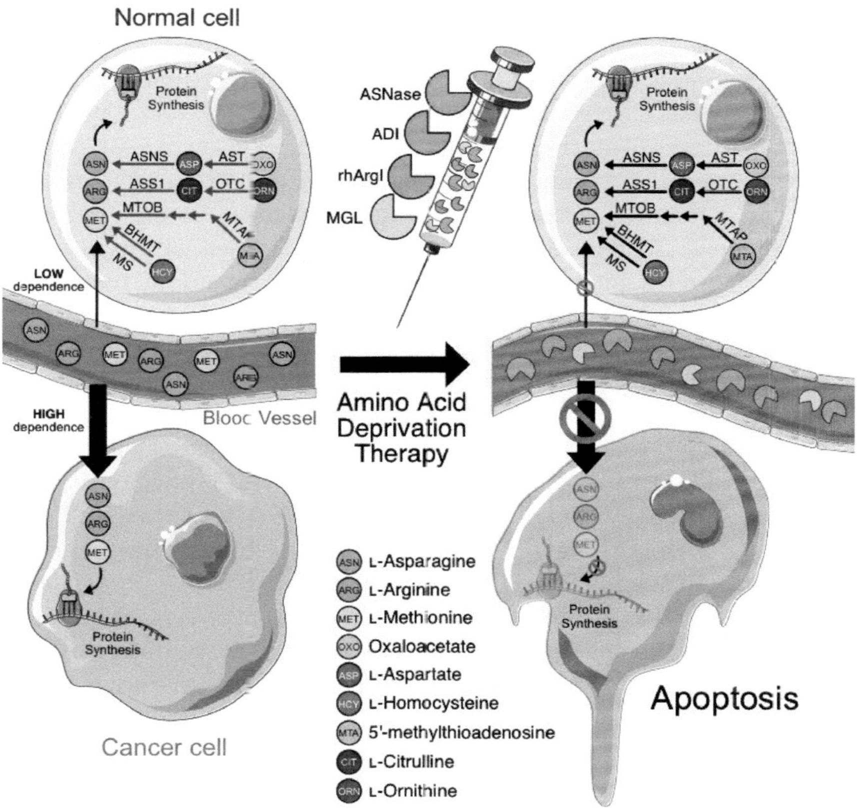

Figure 8.1 Schematic representation of the metabolic mechanisms involving the amino acid depletion therapies, using methionine-γ-lyase (METase), asparaginase (ASNase), arginase (rhARGI), and L-arginine deiminase (ADI).

During the last 50 years, enzymatic therapies have been found to be quite efficient for this purpose. These treatments involve the use of heterologous enzymes (from other organisms, such as bacteria), recombinant, or engineered human enzymes to decrease the concentration of a specific amino acid in the bloodstream that is essential for the development of a tumour (Pokrovsky et al., 2017).

There are successful examples where the enzymatic amino acid deprivation therapies are used to treat cancer. One example involves the use of arginine deiminase (ADI) in hepatocellular carcinoma (HCC), melanoma, and other urea cycle-deficient cancer cells that have a high demand for L-arginine (semi-essential amino acid). Another example, involves the use of the enzyme methionine-γ-lyase (MGL) in the treatment of several types of tumours, such as, colon, breast, prostate, ovary, lung, brain, kidney, stomach, and bladder cancers, as well as larynx melanoma, sarcoma, leukaemia and lymphomas that require a high concentration of L-methionine (essential amino acid) (Fernandes et al., 2017a). The non-essential amino acids can also be used in enzymatic amino acid depletion therapies. This only happens in tumours where the normal metabolism of one non-essential amino acid in the cells is disrupted or when there is some sort of defect in the ability to use a certain amino acid (DeBerardinis and Chandel, 2016). This is the case of childhood acute lymphoblastic leukaemia (ALL), and to a lesser extent, of non-Hodgkin's lymphoma, whose tumours require a high concentration of L-asparagine. In these cases, the use of L-asparaginase (ASNase) has been found to be remarkably successful in the treatment of these tumours and this method has already been approved by the FDA.

Taking into account the therapeutic potential of these enzymes in cancer therapies, they have been extensively studied in the last two decades. These therapies have been showing very promising results, albeit some challenges have been reported in clinical trials. In the following sections, the structure, function, and catalytic mechanism of therapeutic heterologous enzymes involved in L-ARG, L-MET, and L-ASN degradation will be revised. Particular attention will be given to the heterologous enzymes L-arginase (rhARGI), L-arginine deiminase (ADI) (Ni et al., 2008), L-methionine-γ-lyase (MGL) (Tan et al., 1996), and L-asparaginase (ASNase) (Rytting, 2010) taking into account their current therapeutic potential.

8.2 Amino Acid Deprivation Enzymes

Eukaryotic cells control the concentration of the amino acids in the cell through a tight regulation of their concentration. From this control results a delicate state of dynamic equilibrium that is known as proteome homeostasis. A crucial step in the maintenance of the proteome homeostasis is the synthesis of proteins that are required for cell growth and proliferation. Therefore, any perturbation in this equilibrium may affect permanently cell metabolism and even its survival. The amino acid deprivation therapies have, therefore, a very narrow window to work with, since the decrease in the concentrations of any amino acid in the bloodstream needs to be carefully addressed in order to prevent any catastrophic effect on the normal cells and in the full organism. For this reason, the amino acid deprivation therapy cannot be used in all types of cancers that are autotrophic for a certain amino acid, but only in the ones where a small decrease in that amino acid can lead to a therapeutic effect while keeping the other cells safe. The following sections describe cases where these assumptions have been fully addressed.

8.2.1 Arginine

Arginine (L-ARG) is in different ways, an important amino acid for all living organisms. Besides its common role as a building block for protein synthesis, in the case of mammals, it is also a key precursor/mediator of a series of metabolic, immunological, neurological and signalling pathways in the cells (Barbul et al., 1977, 1983; Billiar, 1995; Rogers and Visek, 1985).

For adult mammals, L-ARG is considered a semi or a conditionally essential amino acid. This means that although for healthy individuals arginine is not an essential dietary amino acid, the emergence of a catabolic stress, such as an inflammation or infection, raises the metabolic demands to levels that cannot be supplied by the endogenous synthesis, turning it into an essential amino acid (Daly et al., 1988; Morris Jr., 2007; Reeds, 2000).

A healthy individual obtains the metabolic required L-ARG from the diet, endogenous de novo synthesis and cellular breakdown of proteins (Morris Jr., 2006). When deprived of L-ARG, healthy cells undergo a cell cycle arrest at Go/G1 phase and become quiescent. This effect is reversible for the majority of the cells who recover to their normal proliferation status if reinstated with L-ARG [16].

In the case of tumour cells, its metabolism is so faulty that when deprived of arginine they continue the cell cycle, undergoing unbalanced growth, which eventually leads to the activation of apoptotic pathways (Patil et al., 2016). This particularity, common to all tumour cells, has been widely explored in the development of new anti-cancer therapies against some cancers with auxotrophy toward L-ARG (Dhankhar et al., 2018).

The susceptibility of some cancers to L-ARG deprivation was observed for more than 60 years (Bach and Swaine, 1965) although, by the time, the mechanism behind it was unknown. To date, there is evidence of two enzymes with proven ability to efficiently deprive L-ARG auxotrophic tumour cells from its essential amino acid: a Mycoplasma arginine deiminase (ADI) and a human recombinant arginase (rhArg). They have different catalytic mechanisms and are effective in cancer cells with different characteristics.

8.2.2 Arginine Deiminase

In adult mammals, 5 to 15% of the plasma arginine is obtained by de novo biosynthetic pathway, which involves the conversion of citrulline to L-ARG catalysed by argininosuccinate synthase (ASS1) and argininosuccinate lyase (ASL). ASS1 catalyzes the condensation of citrulline and aspartate to form argininosuccinate that is split into fumarate and L-ARG by ASL. ASS1 is a ubiquitous enzyme whose expression, localization, and regulation differ significantly depending on the tissue-specific needs for L-ARG (Haines et al., 2011).

Some cancers such as metastatic melanoma (MM), prostate carcinomas, hepatocellular carcinoma (HCC) (Dillon et al., 2004) non-Hodgkin's lymphoma, Hodgkin's lymphoma, pancreatic carcinoma, osteosarcoma, and malignant pleural mesothelioma and some breast tumours (Qiu et al., 2015) have an elevated requirement for L-ARG due to the deficient expression of ASS1 that results in L-ARG auxotrophy (Feun et al., 2008, 2012, 2015). Arginine deiminase (ADI,

EC 3.5.3.6) is an enzyme with high affinity to L-ARG that catalyses the conversion of L-ARG into citrulline and ammonia. The cancer cells who do not express ASS1 are unable to re-synthesize L-ARG from citrulline, which culminates in the full deprivation of an amino acid that is imperative for their survival (Shen et al., 2003).

ADI, together with ornithine transcarbamylase (OTC, EC 2.1.3.3) and carbamate kinase (CK, EC 2.7.2.2), belongs to the arginine deiminase pathway (ADI pathway), which is used by many L-ARG-dependent microorganisms to catabolize the amino acid–generating ATP. It has been identified, purified, and characterized in bacteria, archaea and some anaerobic eukaryotes (Horn, 1933; Knodler et al., 1994, 1995; Leopoldini et al., 2009), but there is no evidence of its existence in higher eukaryotes. It was demonstrated that ADI only possesses anti-tumour activity in ASS1 deficient tumours, being ineffective when ASS1 is restored (Ensor et al., 2002; Qiu et al., 2014). Further evidence suggests that ADI mediates the inhibition of tumour growth not only by exhausting the supplies of L-ARG, but also by its anti-angiogenic activity via suppression of NO generation (Beloussow et al., 2002; Park et al., 2003; Yoon et al., 2007).

The first evidence of the in vitro anti-tumour activity of ADI was reported in 1981 by Jones, who applied ADI isolated from *Pseudomonas putida* on murine leukaemia lymphoblast (Ni et al., 2008). A decade later, in 1990, Miyazaki and colleagues demonstrated that the addition of a small dose (5 ng/ml) of Mycoplasma ADI to the culture media depleted all the available L-ARG, which in turn lead to the inhibition of various human cancer cell lines (Miyazaki et al., 1990). From there onwards, several studies confirmed the potential use of Mycoplasma ADI as an anti-tumour agent in vitro and in vivo (Ashikaga et al., 1994; Sugimura et al., 1992; Takaku et al., 1992).

Although there are studies with other ADI sources (Zhang et al., 2015), the most commonly used ADI protein is derived from *M. arginini*. It is composed of two identical subunits with a molecular weight of 45 kDa each and an isoelectric point of 4.7. It is stable at neutral pH and exerted maximal enzyme activity at pH 6.0–7.5 and at 50°C (Takaku et al., 1992).

Since ADI from *M. arginini* has high immunogenicity and its half-life time is too short (approximately 5 h) to be directly used in vivo, it was conjugated with polyethylene glycol (PEG) via a succinimidyl succinate amide bond (Takaku et al., 1993). A formulated ADI-PEG20 (molecular weight: 20 kDa) showed significantly reduced antigenicity and an increased half-life time from 5 h to 7 days in mice and rabbit. In addition, it was also demonstrated that it efficiently degraded plasma arginine to undetectable levels (<5–7 μmol/L) (Ensor et al., 2002; Holtsberg et al., 2002) leading to the L-ARG deprivation in humans (Izzo et al., 2004).

Several studies showed that ADI is a potentially better therapeutic agent for the treatment of leukaemia than ASNase, which has been used for the treatment of acute leukaemia for over 20 years (Ensor et al., 2002; Durden and Distasio, 1980). ADI inhibited the growth of cultured leukaemia cells at concentrations about 20–100 times lower and does not have the serious side effects associated with ASNase (Gong et al., 2000; Muller and Boos, 1998).

The anti-tumour activity of ADI-PEG20 has been further demonstrated through laboratory studies in HCC (Dillon et al., 2004), pancreatic (Bowles et al., 2008), prostate (Kim et al., 2009), small cell lung (Kelly et al., 2012), head and neck (Huang et al., 2012), and breast cancers (Qiu et al., 2014), lymphoma (Delage et al., 2012), myxofibrosarcomas (Huang et al., 2013), melanoma (Dillon et al., 2004; Feun et al., 2012), and glioblastoma (Syed et al., 2013).

8.2.2.1 Therapeutic uses

By 2018, there are records of more than 20 clinical trials in different phases (phase I, II, and III) and different recruiting status whose general objective is to evaluate the safety and therapeutic efficacy of ADI-PEG20 in patients suffering from different cancers (ClinicalTrials.gov. Available at: https://clinicaltrials.gov).

The first clinical trials were completed 15 years ago and showed that the plasma L-ARG levels decrease from approximately 130 μmol/L to below the level of detection (<2 μmol/L) and remain low for more than 7 days after the first intramuscular injection of 160 IU/m^2 of ADI-PEG20 in both HCC (NCT00056992) and MM patients (NCT00029900). The response rates of these studies are in the range 25–47%, and they demonstrated good safety and tolerability (Izzo et al., 2004; Ascierto et al., 2005).

All the posterior ADI-PEG20 clinical trials completed confirmed that monotherapy with ADI-PEG 20 in patients with both MM and HCC is generally well tolerated.

In a HCC trial with patients with metastatic or inoperable histologically confirmed HCC, it was observed that the L-ARG levels remained below baseline for 50 days and had an overall mean survival of 15.8 months. The authors also observed a low rate of significant immunogenic events (2.5%) suggesting good tolerability of ADI (Glazer et al., 2010).

Another phase II study in Asian patients with HCC confirmed that weekly treatment with 160 IU/m^2 was sufficient to deplete circulating L-ARG for 7 days. Additionally, patients with depletion of circulating L-ARG for <4 weeks had a trend towards better survival than those with depletion of arginine for <4 weeks (Yang et al., 2010).

Based on these encouraging results, a phase III trial (NCT01287585) was conducted with HCC patients who had failed prior chemotherapy versus placebo. The results were not as encouraging as expected. ADI-PEG20 monotherapy at the dose of 18 mg/m^2 proved to be ineffective in prolonging overall survival in the selected HCC patients. However, it was also observed that the patients submitted to prolonged L-ARG depletion from ADI-PEG20 showed a relatively superior overall survival. Based on these results, the authors are now focused on the development of new strategies to enhance prolonged L-ARG depletion and synergize the effect of ADI-PEG20 (Abou-Alfa et al., 2018).

The results from recent clinical trials concerning MM are not so encouraging as the ones from HCC. Feun and colleagues performed a clinical trial with ocular and cutaneous melanoma patients (NCT00450372) whose results demonstrated that the response to L-ARG-depleting therapy was indeed correlated with lack of ASS expression. However, contrary to previous results, they also observed that it was necessary to raise the dose of ADI-PEG20 from 160 to 320 IU/m^2 per week to observe any patient response (Feun et al., 2012). In another contemporary clinical trial with unresectable stage III or stage IV melanoma patients (NCT00520299), it was also observed that ADI-PEG20 has limited clinical activity as a single agent at doses of 160 IU/m^2 or less.

This study also revealed an inverse correlation between mean levels of ADI-PEG20 and ADI-PEG20 antibody levels which is suggestive of some immunogenicity associated with ADI-PEG20. Unexpectedly, their results suggested that uveal melanoma might be more sensitive to L-ARG deprivation than cutaneous melanoma (Ott et al., 2013). A recently completed Phase II clinical trial with malignant pleural mesothelioma patients (NCT01279967) showed that ADI-PEG20 significantly improves progression-free survival, and possibly the overall survival (Szlosarek et al., 2017).

There are also several on-going phase I and II clinical trials using ADI-PEG20 monotherapy in patients with different tumours, such as relapsed sensitive or refractory small cell lung cancer (SCLC) (NCT01266018), prostate cancer and non-small cell lung cancer (NCT01497925), acute myeloid leukaemia (NCT01910012), advanced non-Hodgkin's lymphoma (NCT01910025).

Other phase-I and -II clinical trials testing ADI-PEG20 in combination with existing chemotherapy are completed or ongoing for advanced gastrointestinal cancer (NCT02102022), HER2 negative metastatic breast cancer (NCT01948843), advanced pancreatic cancer (NCT02101580), malignant pleural mesothelioma (NCT02709512), soft tissue sarcoma (NCT03449901), HCC (NCT02101593), and MM (NCT01665183), among others.

8.2.2.2 Structure and reaction mechanism

As already discussed, most studies performed to date used ADI isolated from *Mycoplasma arginini*. Based on the Protein Data Bank existing crystallographic structures and on their crystallographic structures, Das et al. (2004) proposed a catalytic mechanism for ADI. Some of its the most important structural features (Das et al., 2004; Galkin et al., 2005; Lu et al., 2004; Lu et al., 2006; Smith et al., 1978) are represented in Fig. 8.2 (PDB code 1S9R) (Das et al., 2004).

The analysis of existing data suggests that L-ARG enters into the enzyme's active site and is oriented to its more favourable catalytic position by a network of polar interactions between its guanidino moiety and the side chain carboxylates of ASP161 and ASP271 residues (Fig. 8.2).

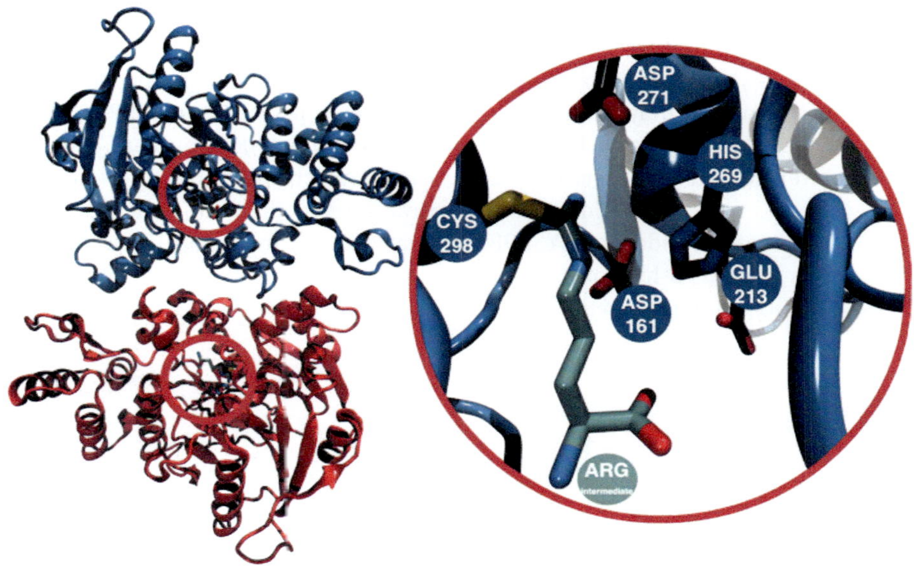

Figure 8.2 (Left) Quaternary structure of ADI homodimer (PDB code: 1S9R) coloured by subunit; (right) close-up view of the active site of ADI.

The proposed mechanism for the enzymes is represented in Fig. 8.3 and comprises six sequential steps.

The ADI mechanism starts with a nucleophilic attack from CYS398 to Cζ carbon of the substrate and a series of simultaneous proton transfers, from CYS398 to Nη2 nitrogen of the substrate, from Nη2 to HIS269 and from HIS269 to GLU213. At the end of the first step, a tetrahedral intermediate is generated with the Cζ carbon of the substrate covalently bonded to CYS398 side chain (Step 1).

In the second step, HIS269 gives a proton to nitrogen Nη2, generating an ammonia molecule. Simultaneously, HIS269 receives a proton from GLU213. At the end of this step, the ammonia molecule is released from the active site, and the reaction intermediate remains covalently bonded to CYS398 (Step 2).

The next step of the catalytic process requires the intervention of a water molecule that is in close proximity to the ASP271 and HIS269 residues. This water molecule is involved in a nucleophilic attack to carbon Cζ of the reaction intermediate and simultaneously donates a proton to HIS269 that, by its turn, donates a proton to GLU21. The structural data suggests that this reaction may be endorsed by the close proximity of ASP161 that establishes two hydrogen bonds with the –NH groups of the intermediate that turns Cζ carbon more electrophilic. At the end of this reaction, another tetrahedral intermediate is generated covalently bonded to CYS398. Contrary to the first step (Step 1), this one has a hydroxyl group attached to Cζ carbon instead of an amino group (Step 4).

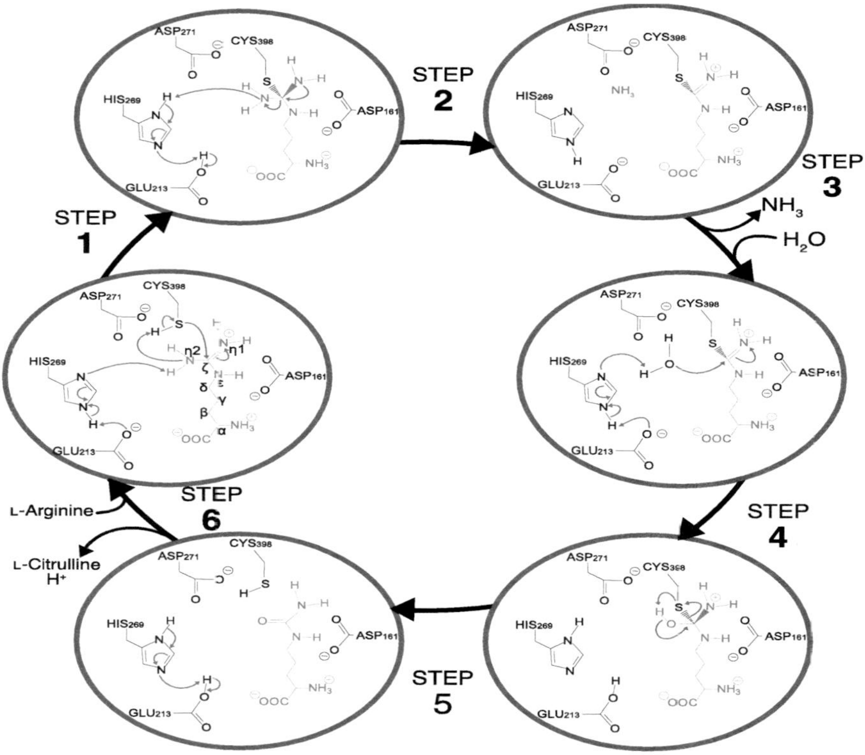

Figure 8.3 Schematic representation of the current proposal of the catalytic mechanism of ADI.

In the following step, a proton is transferred from the hydroxyl group of the reaction intermediate to CYS398, with the consequent cleavage of the covalent bond between CYS398 and the substrate. Simultaneously, there is a proton transfer from GLU213 to HIS269. This reaction generates a molecule of L-citrulline that is ready to leave the active site (Step 5).

The last step is the enzymatic turnover in which HIS398 loses a proton to the solvent or an unknown neighbouring active site residue and the enzyme becomes ready for a new catalytic cycle (Step 6).

8.2.3 Arginase

Arginase (EC 3.5.3.1) was the first enzyme used to deprive cancer cells of L-ARG. It is found in organisms from the five kingdoms and is one of the very few enzymes that specifically requires a spin-coupled Mn^{2+}–Mn^{2+} cluster for catalytic activity in vitro and in vivo (Maharem et al., 2018; Reczkowski and Ash, 1992).

There are two genetically distinct isozymes named types I and II. They share approximately 60% amino acid sequence homology and differ in their expression, regulation, tissue distribution, subcellular locations in mammals, immunologic reactivity, physiologic function, and certain enzymatic properties (Dizikes et al., 1986; Herzfeld and Raper, 1976; Jenkinson et al., 1996; Spector et al., 1983; Vockley et al., 1996).

Arginase I (ArgI) is the only of the two isozyme types with demonstrated anti-cancer therapeutic potential. It is a cytosolic enzyme found predominantly in liver hepatocytes, where it catalyses the hydrolysis of L-ARG to L-ornithine and urea. This is the final cytosolic step of the urea cycle, which detoxifies ammonia in mammals (Krebs, 1973). L-ornithine can be further metabolized to polyamines via ornithine decarboxylase (ODC). Then, it can generate L-proline via ornithine aminotransferase (OAT), or it can be converted to citrulline by ornithine transcarbamylase (OTC) (Morris Jr., 2002).

Recent studies show that tumour inhibition mediated by ArgI is only effective in OTC deficient cells and is independent of ASS expression (Cheng et al., 2007). OTC is an enzyme highly expressed in liver and small intestine, but its expression in cancer and other normal tissues is mostly down-regulated (Delers et al., 1984).

A metabolic functional cell possessing OTC will avoid the intracellular depletion of arginine by converting the ArgI product ornithine into citrulline, which can be converted back to L-ARG via ASS. An OCT-deficient cell when deprived of L-ARG is not capable to generate the intermediate metabolite citrulline, which is the substrate of ASS, culminating in the complete depletion of intracellular L-ARG. In the absence of its substrate, the ASS presence in the OTC-deficient cells is irrelevant. A recent study showed that this simple explanation might not be fully correct and applicable to all cancer cells. In particular, the results showed that some OTC and, partially ASS, deficient cancer cells exhibit high resistance to L-ARG deprivation in vitro (Bobak et al., 2010). This means that the inhibitory effects of L-ARG depleting agents on tumour growth require further investigation in order to be fully understood.

The individual evaluation of the expression of both OTC and ASS in each cancer seems to be a useful request before deciding which of the two enzymes should be used as in the l-ARG depriving therapy.

The administration of ArgI for the treatment of tumours has been reported since the 1950s (Simon-Reuss, 1953; Wiswell, 1951). Although the in vitro experiments were very promising at the time, the scale-up to in vivo tests was discouraging due to ArgI low affinity for L-ARG (K_M: 6 mmol/L at physiologic pH), its alkaline optimum pH (pH 9.6), and its very short circulatory half-life time (approximately 30 min) (Savoca et al., 1979).

The construction of a recombinant human arginase I (rhArgI), which was only possible after the advent of recombinant DNA technology, brought a renewed interest in the use of arginase as an anti-tumour agent.

In the last years a pegylated cobalt-modified enzyme was developed. The substitution of the Mn^{2+} metal centre in human ArgI with Co^{2+} (Co-rhArgI) results in an enzyme that displays 10-fold higher catalytic efficiency for l-ARG hydrolysis, 12–15-fold reduction in the IC_{50} towards a variety of malignant cell lines and a half-life time of 22 h in serum. The enhanced pharmacological properties of Co-hArgI translate into a 12–15-fold improvement in the killing of melanomas and hepatocellular carcinomas in vitro (Stone et al., 2010). Later, the same group reported that the modification of Co-hArgI with PEG-5K NHS esters increased the retention of the enzyme in circulation by about 2 orders of magnitude (Glazer et al., 2011).

This modified version of bioengineered Co-rhArgI–PEG was proven to be cytotoxic in HCC (Cheng et al., 2007), melanoma (Lam et al., 2011), pancreatic carcinoma tumour xenografts (Glazer et al., 2011), acute myeloid leukaemia (Tanios et al., 2013), and glioblastoma cells (Khoury et al., 2015).

Beyond the already discussed particularities between rhArgI and ADI sensitive cancers, there are other differences that must be mentioned. One is the fact that contrary to ADI, rhArgI is a human enzyme and consequently non-immunogenic. The other is the greater toxicity of the rhArgI relative to ADI, which according to some authors may be due the lack OTC in normal tissues and the consequent impossibility to rescue L-ARG from ornithine. The authors proposed that providing citrulline supplementation to patients undergoing rhArgI mediated L-ARG deprivation could lower the treatment toxicity and eventually improve the therapeutic index, particularly in patients with ASS-negative metastatic cancers (Mauldin et al., 2012).

8.2.3.1 Therapeutic uses

By 2018, the number of clinical trials using rhArgI is quite lower than the number of clinical trials with ADI. Also, to our knowledge, to date there is no phase III clinical trial completed. There are at least four clinical trials completed, involving phase I and II. One of them aimed to determine whether rhArgI-PEG was safe and effective in the treatment of advanced HCC (NCT01092091). It was concluded that rhArgI-PEG has a manageable safety profile and is potentially a superior L-ARG depleting agent than ADI in the treatment of HCC due to its low toxicity profile, apparently inexistence of immunogenicity and sustainable l-ARG depletion. The optimal therapeutic dose of rhArgI-PEG was defined as 1600 IU/kg in advanced HCC patients. In contrast to ADI-PEG20, rhArgI-PEG was administrated intravenously (Yau et al., 2013). This clinical study followed a first one performed by the same research group, where pegylated recombinant human ArgI was administrated in 8 escalation doses alone and in combination with standard doses of doxorubicin (NCT00988195).

Two other completed studies evaluated the safety and efficacy of rhArgI-PEG in combination therapy in patients with advanced HCC (NCT02089633) and the safety of increasing dose levels of Co-rhArgI-PEG in patients with relapsed or refractory acute myeloid leukaemia or myelodysplastic syndrome (NCT02732184).

There are at least two clinical trials recruiting patients: One of them intends to evaluate the safety of rhArgI-PEG to treat patients with L-ARG auxotrophic advanced solid tumours (melanoma, renal cell carcinoma, prostate cancer, and hepatocellular carcinoma), and that have progressed after receiving approved or established therapies (NTC02285101). The other one aims to study the safety of increasing dose levels of Co-rhArgI-PEG in patients with advancer cancers (NCT02561234). One phase II clinical trial is active but not recruiting. Its objective is to evaluate the efficacy of rhArgI-PEG in patients with relapsed or refractory acute myeloid leukaemia in terms of remission rate (NCT02899286). Finally, there is an international phase I/II trial that is not yet recruiting and whose objective is to evaluate the safety and activity of rhArgI-PEG in children and young people with relapsed/refractory leukaemia, neuroblastoma, sarcoma, and high-grade gliomas (NCT03455140).

8.2.3.2 Structure and reaction mechanism

hArgI is an oligomer with 107 kDa with optimum pH at 9.3 and a K_M for l-ARG of about 10.5 mM (Beruter et al., 1978). The enzyme has some particular structural features that includes spin-coupled Mn^{2+}–Mn^{2+} cluster and an S-shaped tail, 19 amino acids long, located at the subunit-subunit interface. This S-shaped tail-mediated 54% of all the intersubunit interactions (Lavulo et al., 2001).

There are several proposals for the catalytic mechanism of hArgI, (Leopoldini et al., 2009; Christianson and Cox, 1999; Cox et al., 2001; Khangulov et al., 1998), although many aspects remain elusive. The analysis of the crystallographic structure of the human ARGase in complex with L-ornithine (PDB code: 3GMZ) (Fig. 8.4) (Ilies et al., 2011) reveals some clues about the important catalytic residues, and their possible role in the catalytic mechanism. GLU277 is one of the residues that stands out. It is an active site residue that is apparently implicated in the substrate binding through the formation of a strong electrostatic interaction with the guanidium group. This interaction places the substrate in a favourable position with respect to the metal-bound hydroxide anion with whom it must react (Cavalli et al., 1994; Kanyo et al., 1996; Scolnick et al., 1997). Additionally, GLU277, along with a hydroxide anion and ASP232 and ASP128 are coordinated with the Mn^{2+} ions.

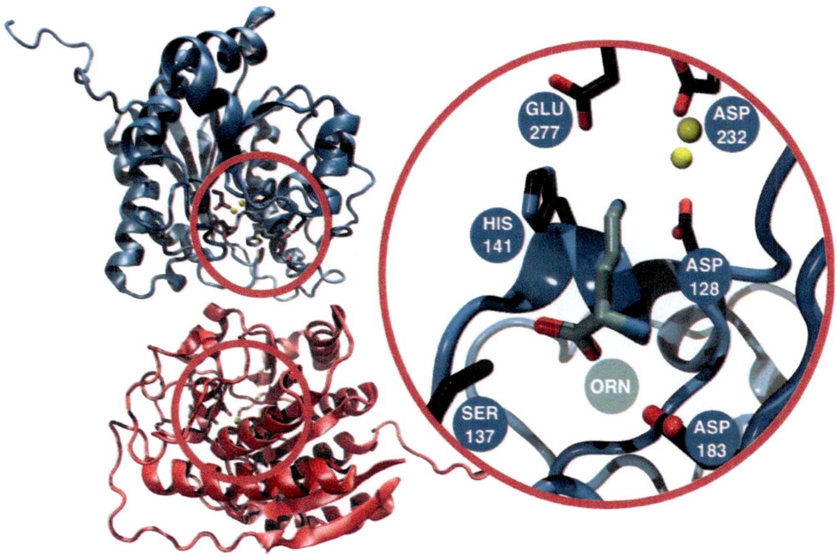

Figure 8.4 (Left) Quaternary structure of hArgl homodimer (PDB code: 3GMZ) coloured by subunit; (right) close-up view of the active site of the enzyme.

The proposed mechanism for this reaction is represented in Fig. 8.5 and comprises six sequential steps.

The reaction starts with the binding of the L-ARG's guanidine group to Mn^{2+} ions. After its correct positioning, L-ARG undergoes a nucleophilic attack from the bridging hydroxy anion to its carbon Cζ (Step 1) originating a tetrahedral intermediate. In the second step, ASP128, which is thought to be deprotonated (Leopoldini et al., 2009), acts as an acid/base catalyst, abstracting a proton from the hydroxyl group and shuttling it to nitrogen Nε of the substrate (Step 2). In the third step, the bond between nitrogen Nε and Carbon Cζ from the substrate breaks, generating the reaction products: L-ornithine and urea (Step 3). The next steps will promote product dissociation from the enzyme. It requires the entry of a water molecule that will bridge the binuclear Mn^{2+} cluster, causing the urea product to move to a terminal coordination site on Mn^{2+} (Step 4). Then, HIS141 residue acts as a proton shuttle from the bulk solvent to nitrogen Nε of ornithine. This proton transfer facilitates product release and regenerates the nucleophilic metal-bridging hydroxide anion (Step 5). Finally, ornithine dissociates and the enzymatic turnover takes place preparing the enzyme for a new catalytic cycle (Step 6).

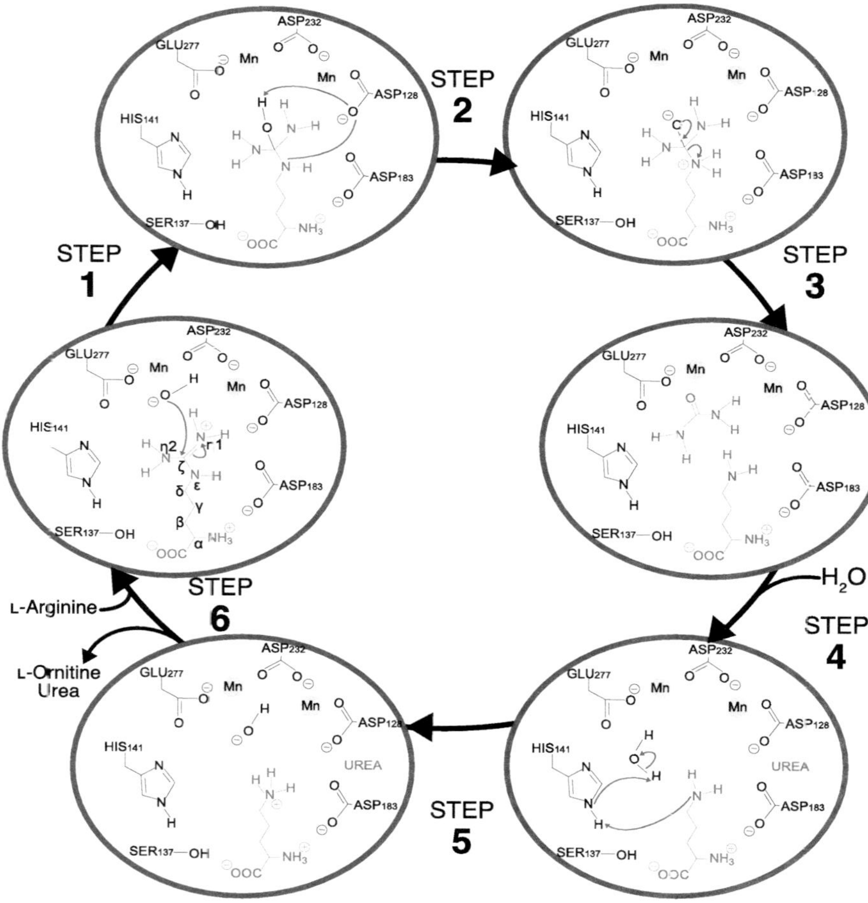

Figure 8.5 Schematic representation of the Current proposal of the catalytic mechanism of hArgI.

8.2.4 Asparagine

L-Asparagine (L-ASN) is a non-essential amino acid that is synthesized in humans by the asparagine synthetase enzyme (Lomelino et al., 2017). However, L-ASN is indispensable for the synthesis of several proteins, in particular, the ones that are glycosylated through N-glycosylation (Cherepanova and Gilmore, 2016) Moreover, this amino acid is involved in the control of some cell functions in the nervous system, and also in the metabolism of ammonia to enable the elimination of this toxic metabolic by-product (Erecinska et al., 1991; Chen and Chen, 1992).

In humans, the asparagine synthetase catalyses the production of L-ASN through an ATP-dependent reaction, where L-aspartate and L-glutamine are converted to L-ASN and L-glutamate (Lomelino et al., 2017; Richards and Kilberg, 2006).

Although cells can produce their own supply of L-ASN, some types of cancer, such as leukaemia (Shahriari et al., 2017; Wetzler et al., 2007), breast (Knott et al., 2018; Yang et al., 2014), gastric (Yu et al., 2016), and lung (Xu et al., 2016) cancers, are impaired in the pathways related to the synthesis of L-ASN. Consequently, these tumours become dependent on an external supply of L-ASN, which offers a specific opportunity window to target cancer using L-ASN deprivation therapies.

L-Asparaginase (ASNase) is an enzyme available in bacteria, plants, and fungi that catalyses the hydrolysis of L-ASN to L-aspartate and ammonia (Gesto et al., 2013). The adequate administration of ASNase in the bloodstream can reduce the plasma levels of L-ASN and promote tumour growth slowdown by starving the cancer cells.

Kidd observed, for the first time, the therapeutic potential of ASNase as an anti-tumour approach in two different types of lymphoma. (Kidd, 1953, 1953a) However, at the time, the role of ASNase was unknown because Kidd only reported that the gynae pig serum was able to regress the lymphomas in murine. Later on, Broome and Dolowy were able to identify and isolate ASNase from the gynae pig serum, and then, they understood that the anti-tumour effect was due to the ASNase (Broome, 1961, 1963; Dolowy, 1966). These studies were pioneer to point out ASNase as an enzyme with great therapeutic potential against cancer.

ASNase (EC 3.5.1.1) is an enzyme available in several bacteria genus, namely *Bacillus*, *Pseudomonas* (Saeed et al., 2018), *Serratia*, *Aerobacter*, *Photobacterium*, *Xanthomonas*, (Peterson and Ciegler, 1969) *Proteus* (Tosa et al., 1971), *Streptomyces* (DeJong, 1972), *Vibrio* (Kafkewitz and Goodman, 1974), and *Aspergillus* (Sarquis et al., 2004). Besides the broad spectrum of microorganism that has ASNase in their proteome, ASNases derived from *Escherichia coli*, and *Erwinia chrysanthemi* showed the most promising results in anti-cancer therapy. The *E. coli* form is the main source of ASNase because of its capability to produce the enzyme in great quantities. The *E. chrysanthemi* form is an alternative when hypersensitivity reactions occur in patients treated with the *E. coli* ASNase. In fact, the immunogenicity is the biggest drawback of the use of these non-human enzymes in anti-cancer therapy.

ASNase from *E. coli* presents two different forms, ASNase-I and ASNase-II, which are different regarding its cellular location and affinity for the substrate. ASNase-I is present in the cytosol, and it has a lower affinity for the substrate (L-ASN) when compared with ASNase-II that is located in the periplasmic space (region between the inner cytoplasmic membrane and the bacterial outer membrane) (Yao et al., 2005). Due to the higher affinity of ASNase-II for the substrate, this enzyme has been extensively studied for application in anti-cancer therapy.

8.2.4.1 Therapeutic uses

ASNase-II has been used as a general therapy agent for the treatment of acute lymphoblastic leukaemia, and other related blood cancers (Shrivastava et al., 2016). The administration of ASNase causes a decrease in the circulating pool of L-ASN, starving the susceptible cancer cells whose cell cycle becomes arrested in the G1-phase, which causes apoptosis (Lee et al., 1989; Keating et al., 1993).

Besides the excellent therapeutic efficiency, ASNase has a short half-life time in the bloodstream, causing a higher demand for injections to keep the therapeutic level. Consequently, this procedure leads to some side effects, namely allergies, exacerbated immune responses, and even anaphylactic shock (Mashburn and Wriston, 1964; Moola et al., 1964; Soares et al., 2002; Henriksen et al., 2015).

Similarly to what happens with other enzymes used in therapy, pegylation is one of the most appropriate manners to improve de plasma half-life and reduce the immunogenicity of the ASNase. Pegylation is a process in which the enzyme is modified with 5 kDa units of monomethoxypolyethylene glycol (PEG) to reduce de exposition of immunogenic patterns in the enzyme's surface (Rytting, 2010). The pegylated ASNase keeps its enzymatic activity, but the half-life time in the bloodstream increases about five times, and the immunogenicity is also lower (Zeidan et al., 2008).

Currently, pegylated ASNase is being used as first-line therapy in the USA, UK, Australia, and New Zealand, and as second-line therapy in Europe, where it is used when the patients are known to have an allergy to the free ASNase (Fu and Sakamoto, 2007). The application of ASNase was approved by the FDA in 2006 and it is commercialized under the name Oncaspar®. Ten years later, in 2016, EMA approved Oncaspar® to be used in European Union.

Although all these formulations make possible the use of ASNase as an anticancer therapeutic target, the formulations currently available are far from perfect. The major side effect continues to be the immunogenicity derived from the bacterial source of the ASNase. There are well-known side effects caused by the administration of ASNase in patients with cancer, namely hypersensitivity reactions (Avramis and Tiwari, 2006), coagulation disorders (Hernandez-Espinosa et al., 2006), pancreatitis, hyperglycaemia (Raja et al., 2012), and hepatotoxicity (Bodmer et al., 2006). These side effects give rise to an intense search for alternatives in order to identify other isoforms, and enzymes from other organisms, that show, at least, the same efficacy, and attenuated side effects (Thomas et al., 2010; Shrivastava et al., 2012; Mendes et al., 2012).

8.2.4.2 Structure and reaction mechanism

ASNase-II from *E. coli* is a homotetramer whose subunits have 326 amino acids each (Fig. 8.6). The pocket where the enzyme binds the substrate and actively catalyses the reaction is composed of amino acid residues from two adjacent subunits: THR12, TYR25, SER58, GLN59, THR89, ASP90, and LYS162 from one subunit, and ASN248 and GLU283 from the adjacent subunit. These residues establish a strong set of hydrogen bond interactions that includes a conserved water molecule (Fig. 8.6). There were several plausible proposals for the catalytic mechanism of ASNase-II, but the symmetry of the location of THR12 and THR89 residues, which enclose the C2 carbon of the substrate, suggests a direct involvement in the catalytic mechanism (Swain et al., 1993).

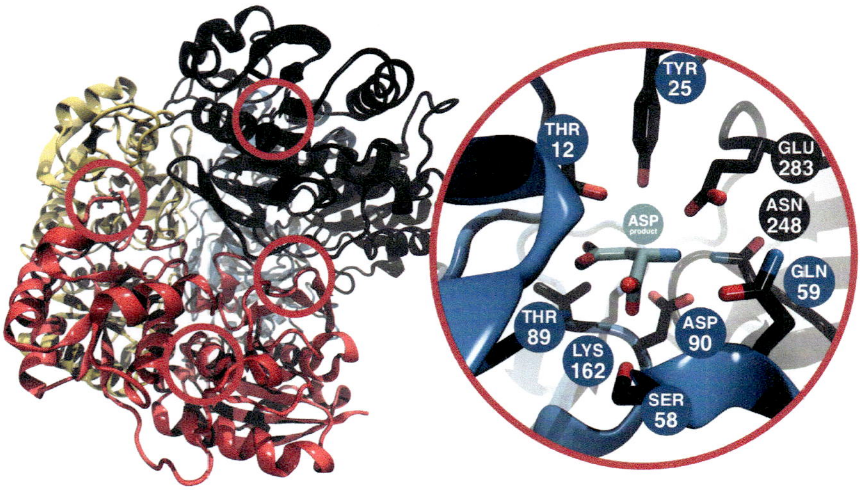

Figure 8.6 (Left) Quaternary structure of the ASNase (PDB code: 3ECA) coloured by subunit. (Right) Representation of the important amino acid residue, and the product in the active site of the ASNase. The molecules are tagged by residue name and residue index.

Currently, the catalytic mechanism of ASNase-II is known and supported by several computational and experimental studies (Gesto et al., 2013; Borek et al., 2014; Schalk et al., 2016).

The catalytic process starts with the formation of a tetrahedral intermediate after the nucleophilic attack of the water molecule to carbon Cα of the L-ASN (substrate) (Fig. 8.7, Step 1). Since the water molecule establishes a new covalent bond with the substrate, one of its protons is passed to the amino group of LYS162 that becomes positively charged.

In the second step, LYS162 residue protonates the amino group of the substrate in a proton transfer mediated by the THR89 residue (Fig. 8.7, Step 2). Afterwards, the amino group is eliminated as ammonia, and the aspartic acid is formed (Fig. 8.7,

Step 3). Finally, both products are released, and the enzyme becomes available to bind a new substrate molecule to initiate another catalytic cycle (Fig. 8.7, Step 4) (Gesto et al., 2013).

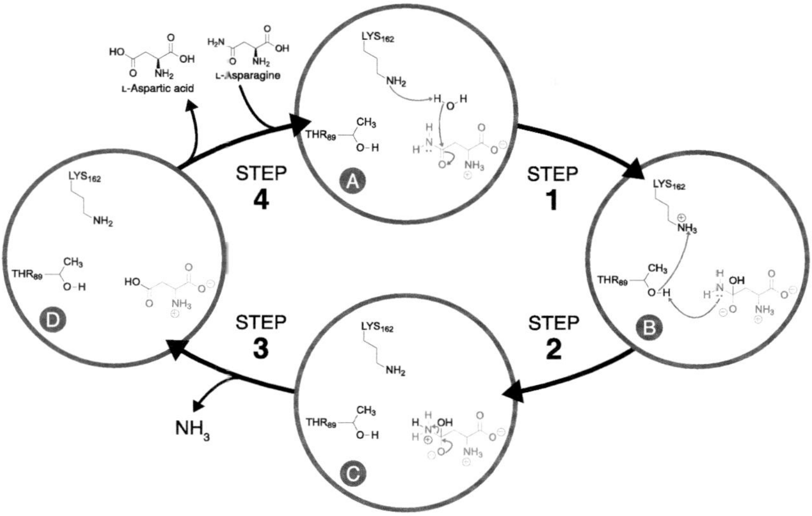

Figure 8.7 A schematic representation showing a proposal for the catalytic mechanism of the ASNase. A–D are reaction intermediates.

8.2.5 Methionine

L-Methionine (L-MET) is one of two sulphur-containing proteinogenic amino acids. This essential amino acid for humans (Young, 1994) acts as an essential precursor for the synthesis of cysteine (through the production of S-adenosylmethionine), taurine, glutathione (a vital tripeptide that reduces reactive oxygen species (ROS), protecting cells from oxidative stress), and polyamines spermidine and spermine (crucial for nuclear and cell division (Cellarier et al., 2003; Davis and Uthus, 2004; Ho et al., 2011; Thivat et al., 2007). Although all these roles are essential for the cell wellbeing, the most important role of L-MET is as a source of methyl units for the DNA methylation through its intermediate S-adenosylmethionine.

It is well known that the DNA methylation is one of the most significant epigenetic signalling markers, controlling the gene expression in several processes such as X-chromosome inactivation, genomic imprinting, repression of repetitive elements, and carcinogenesis (Geiman and Muegge, 2010; Ramani et al., 2010). Regarding the carcinogenesis, in some types of cancer, e.g. endometrial (Fiolka et al., 2013), breast (Chimonidou et al., 2011), lung (Powrózek and Małecka-Massalska, 2016), and myeloma (Wong and Chim, 2015) cancers, the tumourigenic process has been associated with the hypermethylation of the promoter regions of tumour-suppressor genes (Kulis and Esteller, 2010). Considering the dependence of cells of L-MET for the DNA methylation event, it was suggested that strategies leading to L-MET deprivation could be used as therapy against cancer. Since, L-MET is a crucial amino acid in the synthesis of almost all proteins of the cell, this means that the enzyme cannot be applied in long treatments and must be inactivated.

Since L-MET is an essential amino acid, the L-MET deprivation could be harmful for the human body, but it was proved that long-term nutritional deprivation of L-MET does not compromise the human life because cells have several pathways to ensure the L-MET recycle. L-MET can be recovered through the re-methylation of homocysteine by the L-methionine synthase (MS) or, in the liver, by the betaine-homocysteine methyltransferase (BHMT) (Cellarier et al. 2003; Hoffman, 1984; Guo et al., 1993). Additionally, methylthioadenosine phosphorylase (MTAP) is also able to catalyse the production of L-MET from 5'-methylthioadenosine.

Although all of these salvage pathways can restore the required levels of L-MET, several malignant cell lines, including breast, lung, colon, kidney and bladder cancers, glioblastoma and melanoma, have suppressed these salvage pathways (Cavuoto and Fenech, 2012). Consequently, the cancer cells are dependent on an exogenous supply of L-MET, manifesting a growth inhibition when this supply is suppressed (Jeon et al., 2016; Chaturvedi et al., 2018).

The high demand for L-MET in the tumour cells as a result of their great protein synthesis and the lack of L-MET salvage pathways make L-MET essential for the tumour growth and proliferation. Therefore, the L-MET restriction can be exploited as a cancer therapy, in particular for these L-MET-dependent tumours (Chaturvedi et al., 2018).

In the literature, two different approaches have been documented with therapeutic promising results using the dietary deprivation of L-MET and the use of L-MET depletion enzymes. The nutritional restriction of L-MET showed promising results in animal studies, but this diet showed a limited efficiency in clinical trials (Chaturvedi et al., 2018; Thivat et al., 2009). For example, a phase I clinical trial reported a poor reduction of L-MET in plasma (58%) after 17.3 weeks of L-MET dietary restriction (Epner et al., 2002)

Due to the lack of effectiveness in reducing the serum levels of L-MET through nutritional restriction, methioninase or methionine γ-lyase (MGL) has been studied to achieve the desired level of reduction. Several studies reported a fast and effective depletion of L-MET from serum, showing a vast efficiency against human tumours graphs implanted in mice. A pilot study was conducted in three patients showing a drastic reduction in plasma L-MET from 23.1 μM to 0.1 μM (reduction of approximately 99.6%) when the enzyme was administrated by a 10 h intravenous infusion (Tan et al., 1996a). Based on this report, MGL corroborated its efficiency in reducing serum L-MET levels. Moreover, a slowdown in the tumour growth and spread necrosis in tumour regions were observed in either the MHL administered alone or in combination with other drugs (Yoshioka et al., 1998; Tan et al., 1999; Murakami et al., 2017, Kawaguchi et al., 2017, Kokkinakis et al., 2001).

Beyond the depletion effect of MGL, this enzyme can also be administered in combination with the prodrug selenomethionine. The MGL can produce methyl-selenol from the selenomethionine, which causes oxidation of thiol groups generating toxic superoxide molecules that cause cell death. By this way, MGL can be fused with a specific antigen to target the tumour cell and, then, the selenomethionine is administrated, being active only in the area where the MGL is located (Zhao et al., 2006; Guillen et al., 2014).

8.2.5.1 Therapeutic uses

MGL has a broad range of application in cancer therapies. Since this enzyme is absent in mammals, MGL was initially isolated from *Clostridium sporogene* (Sharma et al., 2014). Later on, the enzyme from *Pseudomonas putida* was chosen for clinical purposes, due to its greater stability and affinity for the substrate (low K_M) (Esaki and, Soda, 1987). Therefore, the MGL gene from *P. putida* was isolated and expressed in *Escherichia coli* to produce the recombinant enzyme in great quantities for administration in humans (Hori et al., 1996).

The use of an enzyme from another organism raises complications related to the immunogenicity of the recombinant enzyme and its low stability in the bloodstream. In order to overcome this problem, the recombinant MGL was modified with polyethylene glycol (PEG). The pegylated enzyme presented low immunogenicity, a longer half-life time, and greater resistance against degradation by other enzymes in the bloodstream (Tan et al., 1999; Kudou et al., 2007, Takakura et al., 2006). Pegylation can overcome these issues but causes a decrease in the affinity for the PLP cofactor which decreases the efficiency and specificity of the MGL. Fortunately, this issue was easily solved through the supplementation of the patients with PLP and oleic acid (Takakura et al., 2006).

Although all these promising results, there are only one phase I clinical trials addressing the viability to administered intravenously MGL to deplete plasma L-MET. The pegylated MGL was tested in three healthy people reporting no significant toxicity and a considerable capacity to decrease serum L-MET level. In sum, this pilot phase I clinical trial reported that the administration of MGL is safe and able to effectively reduce L-MET level in the bloodstream (Tan et al., 1996; Tan et al., 1997). However, the clinical trials must continue to corroborate the effectiveness of MGL to treat cancer.

Since the recombinant MGL presents strong applicability for the cancer treatment, the molecular details, in particular, the catalytic mechanism can provide valuable insights to enhance its efficiency.

8.2.5.2 Structure and reaction mechanism

MGL (EC 4.4.1.11) is a pyridoxal-5′-phosphate (PLP)-dependent enzyme that catalyses the reversible conversion of L-MET into methanethiol, ammonia, and 2-oxobutanoate (Kreis and Hession, 1973). The enzyme is a homotetramer with 398 amino acid residues in each subunit, and it has four active sites placed in each interface of two adjacent subunits (Fukumoto et al., 2012). Each one of the four active sites binds a PLP cofactor molecule, which is required for binding the substrate (L-MET) and triggering the reaction (Kudou et al., 2007, Motoshima et al. 2000; Tone, 2005).

Although MGL belongs to the γ-family of PLP-dependent enzymes, there is an important cysteine residue (CYS116) in the active site that is absent in the other enzymes of the same family. Mutagenic studies reported that the CYS116 residue is crucial for the enzymatic activity (Kudou et al., 2008) and substrate specificity (Kudou et al. 2007, Inoue et al., 2000). Beyond the CYS116 residue, there are five more residues with a key role in the mechanism based on structural, enzymatic, and mutagenesis studies: TYR59, ARG61, TYR114, ASP186, LYS211, LYS240, and ASP241 (Fig. 8.8).

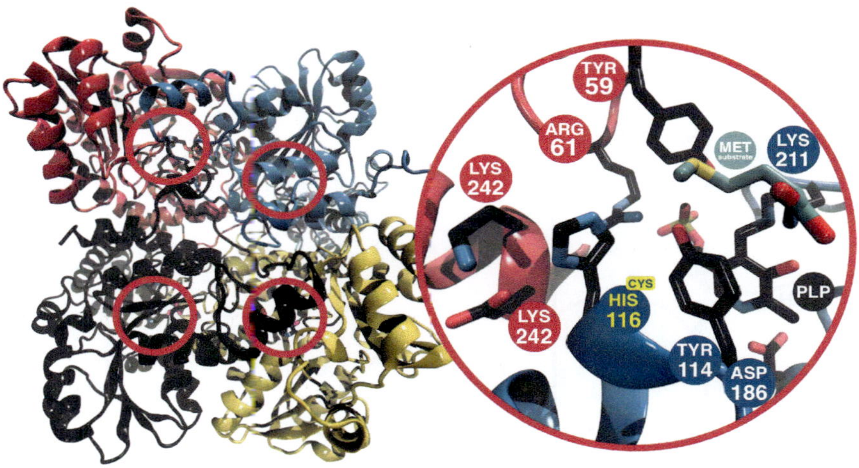

Figure 8.8 (Left) Quaternary structure of the MGL (PDB code: 3VK3) coloured by subunit. (Right) Representation of the important amino acid residues, the PLP cofactor, and the substrate in the active site of the MGL. The molecules are tagged by residue name and residue index. The CYS116 was mutated by a HIS residue to allow the co-crystallization of the MGL with the natural substrate (L-MET).

The proposed catalytic mechanism for MGL is based on these studies and the available knowledge for other PLP-dependent enzymes (Kudou et al. 2007, Cerqueira et al., 2011; Oliveira et al., 2011; Fernandes et al., 2017b, 2018).

All PLP-dependent enzymes wait for their substrate in a steady state where the PLP is covalently bonded to a lysine residue (LYS211) forming an internal aldimine (Fig. 8.9A). When a molecule of substrate enters the active site, its amino group makes a nucleophilic attack (Fig. 8.9, Step 1) to the C8 of the PLP cofactor generating a gem-diamine intermediate (Fig. 8.9B). Immediately after, the bond between the LYS211 residue and the PLP cofactor is broken, and the substrate (L-MET) becomes covalently bonded to the PLP cofactor (Fig. 8.9, Step 2), originating an intermediate commonly called external aldimine (Fig. 8.9C).

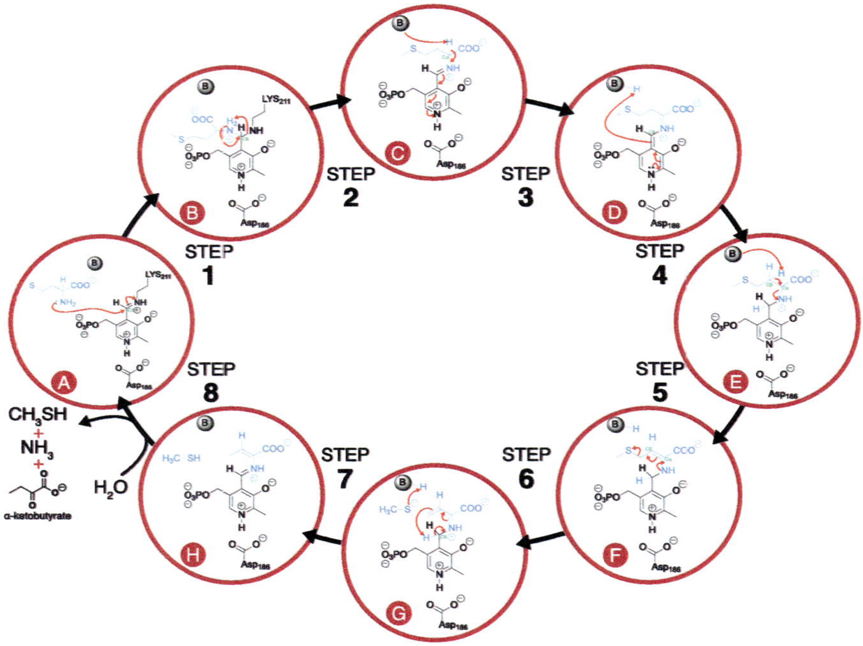

Figure 8.9 A schematic representation showing a proposal for the catalytic mechanism of the MGL. A–H are reaction intermediates, and the "B" sphere represents a generic molecule in the active that can behave as a base.

In the next step, the proton bonded to the Cα of the L-MET is abstracted from an unknown base from the active site (Fig. 8.9, Step 3). Based on the crystallographic structure and the catalytic mechanism of other PLP-dependent enzymes, the LYS211 residue could be the base involved in this reaction. The resulting electron pair is stabilized across the PLP conjugated system forming a quinonoid intermediate (Fig. 8.9D).

In the fourth step, the proton abstracted in the previous step is transferred to carbon C8 of the PLP cofactor, which increases the acidity of the protons bonded to the Cβ of the L-MET. Consequently, in the fifth step, the base can get one of these protons.

The abstraction of the proton bonded to carbon Cβ atom triggers the elimination of the methanethiol group (γ-elimination) (Fig. 8.9, Step 6), which is rapidly protonated by the base in the active site. In the meantime, the obtained amino-crotonate intermediate (Fig. 8.9H) undergoes hydrolysis of its amino group. In this eighth step, the amino group is eliminated as an ammonia molecule, and a water molecule replaces that position as a carbonyl group.

In the end, the LYS211 becomes covalently bonded to the PLP cofactor, and the enzyme restores its initial configuration where the internal aldimine (Fig. 8.9A) is ready to catalyse a new reaction as soon as the substrate becomes available.

8.3 Conclusions

Amino acids are no longer regarded solely as building materials but also as sensing markers that regulate important signalling pathways that can be used for therapeutic purposes.

In the last decades, research on amino acid depletion in cancer cells using enzymes has provided valuable insights on the potential impact of metabolic control and regulation in the tumour microenvironment These studies have shown that decreasing the concentration of specific amino acids in the bloodstream, which are required in high demands by tumour, can be an effectiveness way to treat the tumour.

The arginine dependence of some tumour cells, although known for more than 60 years, has failed to reunite by 2018 all the safety and efficacy conditions to become approved for use in therapy.

Although arginase was initially pointed as a potential new medicine against cancer cells, its inherent limitations, insurmountable at the time, turned the attentions to ADI. In 2018, there are more than 20 clinical trials ongoing or completed whose aim is to evaluate the potential of ADI as a safe and effective anti-cancer treatment target. The results from a recently finished phase III clinical trial (NCT01287585) were not as encouraging as expected but did not withdraw the researcher's attention from ADI.

In the meanwhile, new strategies to improve arginase potential have arouse and at the moment there is a completed clinical trial (NCT01092091) were it was shown that rhArgI-PEG has a manageable safety profile and is potentially a superior arginine depleting agent than ADI in the treatment of HCC. At the moment there are about 10 clinical trials in different stages evaluating the effectiveness and safety of arginase as a potential anti-cancer therapeutic targets. The combination of the different arginine depleting enzymes with existing therapy is also being tested. The observation that patients submitted to prolonged arginine depletion showed a relative superior overall survival is a valid reason to keep the exploitation of arginine deprivation as a potential therapy against some cancers.

The L-ASN depletion has already been used in therapy after some modifications to overcome some of the most common side effects as immunogenicity. Therefore, the pegylated ASNase is used as first-line therapy in the USA, UK, Australia, and New Zealand, and as second-line therapy in Europe against acute lymphoblastic leukaemia and other related blood cancers. Although the evolution of these formulations, the research on improvements that can increase the efficacy and suppress the side effects goes on, in particular, to overcome some cases of hypersensitivity reactions, pancreatitis, hyperglycaemia, and hepatotoxicity. Contrary to he L-ASN, the L-MET depletion therapies against cancer are taking the first steps. The bacterial origin of the MGL enzyme is the source of major complications in the use of this enzyme in therapy. Several cases of immunogenicity reactions have been reported, and although the pegylation strategy suppressed the immunogenicity and low half-life time of the enzyme in the bloodstream, this modification causes a decrease in the affinity of the enzyme for the substrate. Thus, the amino acid deprivation strategy to treat cancer targeting the serum L-MET supply is currently in an early stage of development comparatively to L-ASN, for example.

As any new therapy targeting cancer, amino acid deprivation therapies must be explored with caution in order to eliminate potential negative impacts that can be caused by the side effects observed when the metabolism of normal cells is also affected. It is thus important to understand at the atomic-level the underlying mechanisms of metabolic function of the enzymes that are used in these therapies. At the same time, this knowledge can be used to guide the development of new and more specific therapies alone or in combination with other drugs that can decrease the number of side effects, and turn these treatments more effective.

References

Abou-Alfa, G. K., et al., *Ann. Oncol.*, **29** (2018), 1402–1408.

Ascierto P. A., et al., *J. Clin. Oncol.*, **23** (2005), 7660–7668.

Ashikaga T., et al., *Biosci. Biotechnol. Biochem.*, **58** (1994), 839–842.

Avramis V. I., Tiwari P. N., *Int. J. Nanomedicine*, **1** (2006), 241–254.

Bach S. J., Swaine D., *Br. J. Cancer*, **19** (1965), 379–386.

Barbul A., et al., *Surgical Forum*, 28 (1977), 101–103.

Barbul, A., et al., *Am. J. Clin. Nutr.*, 37 (1983), 786–794.

Beloussow K., et al., *Cancer Lett.*, 183 (2002), 155–162.

Beruter J., Colombo J. P., Bachmann C., *Biochem. J.*, 175 (1978), 449–454.

Billiar T. R., *Ann. Surgery*, **221** (1995), 339–349.

Bobak Y. P., et al., *Cell. Biol. Int.*, **34** (2010), 1085–1089.

Bodmer M., et al., *Digestion*, **74** (2006), 28–32.

Borek D., et al., *FEBS J.*, **281** (2014), 4097–4111.

Bowles T. L., et al., *Int. J. Cancer*, **123** (2008), 1950–1955.

Broome J. D., *Nature*, **191** (1961), 1114–1115.

Broome J. D., *J. Experim. Med.*, **118** (1963), 99–120.

Cavalli R. C., et al., *Biochemistry*, **33** (1994), 10652–10657.

Cavuoto P., Fenech M. F., *Cancer Treat. Rev.*, **38** (2012), 726–736.

Cellarier E., et al., *Cancer Treat. Rev.*, **29** (2003), 489–499.

Cerqueira N. M., Fernandes P. A., Ramos M. J., *J. Chem. Theory Comput.*, **7** (2011), 1356–1368.

Chaturvedi S., Hoffman R. M., Bertino J. R., *Biochem. Pharmacol.*, **154** (2018), 170–173.

Chen Z. P., Chen K. Y., *J. Biol. Chem.*, **267** (1992), 6946–6951.

Cheng P. N., et al., *Cancer Res.*, **67** (2007), 309–317.

Cherepanova, N. A., Gilmore R., *Sci. Rep.*, **6** (2016), 20946.

Chimonidou M., et al., *Clin. Chem.*, **57** (2011), 1169–1177.

Christianson D. W., Cox J. D., *Annu. Rev. Biochem.*, 68 (1999), 33–57.

Cox J. D., et al., *Biochemistry*, **40** (2001), 2689–2701.

Daly J. M., et al., *Ann. Surgery*, **208** (1988), 512–523.

Das K., et al., *Structure*, 12 (2004), 657–667.

Davis C. D., Uthus E. O., *Exp. Biol. Med.* (Maywood), **229** (2004), 988–995.

DeBerardinis R. J., Chandel N. S., *Sci. Adv.*, **2** (2016), e1600200.

DeJong P. J., *Appl. Microbiol.*, **23** (1972), 1163–1164.

Delage B., et al., *Cell Death Dis.*, **3** (2012), e342.

Delers A., et al., *Mol. Cell. Biol.*, **4** (1984), 809–812.

Dhankhar R., et al., *Expert Opin. Ther. Pat.*, **28** (2018), 603–614.

Dillon B. J., et al., *Cancer*, **100** (2004), 826–833.

Dizikes G. J., et al., *Biochem. Biophys, Res. Commun.*, **141** (1986), 53–59.

Dolowy W. C., et al., *Cancer*, **19** (1966), 1813–1819.

Durden D. L., Distasio J. A., *Cancer Res.*, **40** (1980), 1125–1129.

Ensor C. M., et al., *Cancer Res.*, **62** (2002), 5443–5450.

Epner D. E., et al., *Nutr. Cancer*, **42** (2002), 158–166.

Erecinska M., et al., *J. Neurochem.*, **57** (1991), 491–498.

Esaki N., Soda K., *Methods Enzymol.*, **143** (1987), 459–465.

Fernandes H. S., et al., *Expert. Opin. Ther. Pat.*, **27** (2017a), 283–297.

Fernandes H. S., Ramos M. J., Cerqueira N., *Chemistry*, **23** (2017b), 9162–9173.

Fernandes H. S., Ramos M. J., Cerqueira N. M. F. S. A., *ACS Catalysis*, **8** (2018), 10096–10110.

Feun L., et al., *Curr. Pharm. Des.*, **14** (2008), 1049–1057.

Feun L. G., et al., *Br. J. Cancer*, **106** (2012), 1481–1485.

Feun L. G., Kuo M. T., Savaraj N., *Curr. Opin. Clin. Nutr. Metabol. Care*, 18 (2015), 78–82.

Fiolka R., et al., *Oncol. Rep.*, **30** (2013), 2878–2886.

Fu C. H., Sakamoto K. M., *Expert. Opin. Pharmacother.*, **8** (2007), 1977–1984.

Fukumoto M., et al., *Biosci. Biotechnol. Biochem.*, **76** (2012), 1275–1284.

Fung M. K. L., Chan G. C., *J. Hematol. Oncol,* **10** (2017), 144.

Galkin A., et al., *J. Biol. Chem.,* **280** (2005), 34080–34087.

Geiman T. M., Muegge K., *Mol. Reprod. Dev.,* **77** (2010), 105–113.

Gesto D. S., et al., *J. Am. Chem. Soc.,* **135** (2013), 7146–7158.

Glazer E. S., et al., *J. Clin. Oncol.,* **28** (2010), 2220–2226.

Glazer E. S., et al., *Translat. Oncol.,* **4** (2011), 138–146.

Gong H., et al., *Leukemia,* **14** (2000), 826–829.

Guillen K. P., Kurkjian C., Harrison R. G., *J. Biomed. Sci.,* **21** (2014), 65.

Guo H., et al., *Cancer Res.,* 53 (1993), 5676–5679.

Haines R. J., L. C. Pendleton L. C., Eichler D. C., *Int. J. Biochem. Mol. Biol.,* **2** (2011), 8–23.

Henriksen L. T., et al., *Pediatr. Blood Cancer,* **62** (2015), 427–433.

Hernandez-Espinosa D., et al., *Am. J. Pathol.,* **169** (2006), 142–153.

Herzfeld A., Raper S. M., *Biochem. J.,* **153** (1976), 469–478.

Ho E., et al., *Adv. Nutr.,* **2** (2011), 497–510.

Hoffman R. M., *Biochim. Biophys. Acta,* **738** (1984), 49–87.

Holtsberg F. W., et al., *J. Control. Release,* **80** (2002), 259–271.

Hori H., et al., *Cancer Res.,* **56** (1996), 2116–2122.

Horn F., *Hoppe Seylers Z. Physiol. Chem.,* **216** (1933), 244–247.

Huang C. C., et al., *Oral. Oncol.,* **48** (2012), 1227–1235.

Huang H. Y., et al., *Clin. Cancer Res.,* **19** (2013), 2861–2872.

Ilies M., et al., *J. Med. Chem.,* **54** (2011), 5432–5443.

Inoue H., et al., *Biosci. Biotechnol. Biochem.,* **64** (2000), 2336–2343.

Izzo F., et al., *J. Clin. Oncol.,* **22** (2004), 1815–1822.

Jenkinson C. P., Grody W. W., Cederbaum S. D., *Comp. Biochem. Physiol. B Biochem. Mol. Biol.,* **114** (1996) 107–132.

Jeon H., et al., *Oncotarget,* **7** (2016), 67223–67234.

Kafkewitz D., Goodman D., *Appl. Microbiol.,* **27** (1974), 206–209.

Kanyo Z. F., et al., *Nature,* **383** (1996), 554–547.

Kawaguchi K., et al., *Oncotarget,* **8** (2017), 85516–85525.

Keating M. J., et al., *Leuk. Lymphoma,* **10** Suppl (1993), 153–157.

Kelly M. P., et al., *Br. J. Cancer* **106** (2012), 324–332.

Khangulov S. V., et al., *Biochemistry,* **37** (1998), 8539–8550.

Khoury O., et al., *J. Neurooncol.,* **122** (2015), 75–85.

Kidd J. G., *J. Exp. Med.*, **98** (1953), 583–606.

Kidd J. G., *J. Exp. Med.*, **98** (1953a), 565–582.

Kim R. H., et al., *Cancer Res.*, **69** (2009), 700–708.

Knodler L. A., Edwards M. R., Schofield P. J., *Exp. Parasitol.*, **79** (1994), 117–125.

Knodler L. A., Schofield P. J., Edwards M. R., *Microbiology*, **141** (1995), 2063–2070.

Knott S. R. V., et al., *Nature*, **554** (2018), 378–381.

Kokkinakis D. M., et al., *Cancer Res.*, **61** (2001), 4017–4023.

Krebs H., *Biochem. Education*, **1** (1973), 19–23.

Kreis W., Hession C., *Cancer Res.*, **33** (1973), 1866–1869.

Kudou D., et al., *J. Biochem.*, **141** (2007), 535–544.

Kudou D., et al., *Biosci. Biotechnol. Biochem.*, **72** (2008), 1722–1730.

Kulis M., Esteller M., *Adv. Genet.*, **70** (2010), 27–56.

Lam T. L., et al., *Pigment Cell Melanoma Res.*, **24** (2011) 366–376.

Lavulo L. T., Sossong T. M. Jr., Brigham-Burke M. R., et al., *J. Biol. Chem.*, **276** (2001), 14242–14248.

Lee, S.-M., Wroble M. H., Ross J. T., *Appl. Biochem. Biotechnol.*, **22** (1989), 1–11.

Leopoldini M., N. Russo N., and M. Toscano M., *Chemistry* (Weinheim an der Bergstrasse, Germany), **15** (2009), 8026–8036.

Lomelino C. L., et al., *J. Biol. Chem.*, **292** (2017), 19952–19958.

Lu X., et al., *J. Am. Chem. Soc.*, **126** (2004), 5374–5375.

Lu X., et al., *Biochemistry*, **45** (2006), 1162–1172.

Maharem T. M., et al., *Int. J. Biol. Macromol.*, **108** (2018), 88–97.

Mashburn L. T., Wriston J. C., *Arch. Biochem. Biophys.*, 105 (1964), 450–453.

Mauldin J. P., et al., *Translat. Oncol.*, **5** (2012), 26.

MendesA. V., et al., *Perspect. Psychiatr. Care*, **48** (2012), 138–148.

Miyazaki, K., et al., *Cancer Res.*, **50** (1990), 4522–4527.

Moola Z. B., et al., *Biochem. J.*, **302** (1964), 921–927.

Morris S. M., Jr., *Annu. Rev. Nutr.*, **22** (2002) 87–105.

Morris S. M., Jr., *Am. J. Clin. Nutr.*, **83** (2006), 508S–512S.

Morris, S. M., Jr., *J. Nutr.*, **137** (2007), 1602S–1609S.

Motoshima H., et al., *J. Biochem.*, **128** (2000), 349–354.

Muller H. J., Boos J., *Crit. Rev. Oncol./Hematol.*, **28** (1998), 97–113.

Murakami T., et al., *Oncotarget*, **8** (2017), 35630–35638.

Ni Y., Schwaneberg U., Sun Z. H., *Cancer Lett.*, **261** (2008), 1–11.

Oliveira E. F., et al., *J. Am. Chem. Soc.*, **133** (2011), 15496–15505.

Ott P. A., et al., *Invest. New Drugs*, **31** (2013), 425–434.

Park, I. S., et al., *Br. J. Cancer*, **89** (2003), 907–914.

Patil M. D., et al., *Oncogene*, **35** (2016), 4957–4972.

Peterson R. E., Ciegler A., *Appl. Microbiol.*, **17** (1969), 929–930.

Pokrovsky V. S., et al., *Curr. Med. Chem.*, (2017); doi: 10.2174/0929867324666171006132729.

Powrózek T., Małecka-Massalska T., *Transl. Cancer Res.*, **5** (2016), S1531–S1538.

Qiu F., et al., *Sci. Signal.*, **7**(319) (2014), ra31.

Qiu F., Huang J., Sui M., *Cancer Lett.*, **364** (2015), 1–7.

Raja R. A., Schmiegelow K., Frandsen T. L., *Br. J. Haematol.*, **159** (2012), 18–27.

Ramani K., et al., *Hepatology*, **51** (2010), 986–995.

Reczkowski R. S., Ash D. E., *J. Am. Chem. Soc.*, **114** (1992), 10992–10994

Reeds P. J., *J. Nutr.*, **130** (2000), 1835S–1840S.

Richards N. G., Kilberg M. S., *Annu. Rev. Biochem.*, **75** (2006), 629–654.

Rogers Q. R., Visek W. J., *J. Nutr.*, **115** (1985), 505–508.

Rytting M., *Expert. Opin. Biol. Ther.*, **10** (2010), 833–839.

Saeed H., et al., *Protein J.* **37** (2018), 461–471.

Sarquis M. I., et al., *Mem. Inst. Oswaldo Cruz*, **99** (2004), 489–492.

Savoca K. V., et al., *Biochim. Biophys. Acta*, **578** (1979) 47–53.

Schalk A. M., et al., *J. Biol. Chem.*, **291** (2016), 5088–5100.

Scolnick L. R., et al., *Biochemistry*, **36** (1997), 10558–10565.

Shahriari S., et al., *Iran. J. Public Health*, **46** (2017), 357–367.

Sharma B., Singh S., Kanwar S. S., *Biomed. Res. Int.*, **2014**, Article ID 506287.

Shen L.-J., et al., *Cancer Lett*, **191** (2003), 165–170.

Shrivastava A., et al., *Prep. Biochem. Biotechnol.*, **42** (2012), 574–581.

Shrivastava A., et al., *Crit. Rev. Oncol. Hematol.*, **100** (2016), 1–10.

Simon-Reuss I., *Biochim. Biophys. Acta*, **11** (1953), 396–402.

Smith D. W., Ganaway R. L., Fahrney D. E., *J. Biol. Chem.*, **253** (1978), 6016–6020.

Soares A. L., et al., *Int. J. Pharmaceutics*, **237** (2002), 163–170.

Spector E. B., Rice S. C., Cederbaum S. D., *Pediatric Res.*, **17** (1983), 941–944.

Stone E. M., et al., *ACS Chem. Biol.*, **5** (2010), 333–342.

Sugimura K., et al., *Microbiol. Immunol.*, **36** (1992), 667–670.

Swain A. L., et al., *Proc. Natl. Acad. Sci. USA*, **90** (1993), 1474–1478.

Syed N., et al., *Cell Death Dis.*, **4** (2013), e458.

Szlosarek P. W., et al., *JAMA Oncol.*, **3** (2017), 58–66.

Takaku H., et al., *Int. J. Cancer*, **51** (1992), 244–249.

Takaku H., et al., *Jpn. J. Cancer Res.*, **84** (1993), 1195–1200.

Takakura T., et al., *Cancer Res.*, **66** (2006), 2807–2814.

Tan Y., et al., *Anticancer Res*, **16** (1996), 3931–3936.

Tan Y., et al., *Anticancer Res.*, **16** (1996a), 3937–3942.

Tan Y., et al., *Anticancer Res.*, **17** (1997), 3857–3860.

Tan Y., et al., *Clin. Cancer Res.*, **5** (1999), 2157–2163.

Tanios R., et al., *Leukemia Res.*, **37** (2013) 1565–1571.

Thivat E., et al., *Anticancer Res.*, **27** (2007), 2779–2783.

Thivat E., et al., *Anticancer Res.*, **29** (2009), 5235–5240.

Thomas X., et al., *Bull. Cancer*, **97** (2010), 1105–1117.

Tone, M. D., Arch. *Biochem. Biophys.*, **433** (2005), 279–287.

Tosa T., et al., *Appl. Microbiol.*, **22** (1971), 387–392.

Vockley J. G., et al., *Genomics*, **38** (1996), 118–123.

Wetzler M., et al., *Blood*, **109** (2007), 4164–4167.

Wiswell O. B., *Proc. Soc. Experim. Biol. Med.* (New York, N Y), **76** (1951), 588–589.

Wong K .Y., Chim C. S., *Epigenomics*, **7** (2015), 985–1001.

Xu Y., et al., *Cancer Gene Ther.*, **23** (2016), 287–294.

Yang, T. S., et al., *Br. J. Cancer*, **103** (2010), 954–960.

Yang H., et al., *Chem. Biol. Drug Des.*, **84** (2014), 578–584.

Yao M., et al., *Acta Crystallogr. D Biol. Crystallogr.*, **61** (2005), 294–301.

Yau T., et al., *Invest. New Drugs*, **31** (2013), 99–107.

Yoon, C. Y., et al., *Int. J. Cancer*, **120** (2007), 897–905.

Yoshioka T., et al., *Cancer Res.*, **58** (1998), 2583–2587.

Young V. R., *J. Nutr.*, **124** (1994), 1517S–1523S.

Yu Q., et al., *Scand. J. Gastroenterol.*, **51** (2016), 1220–1226.

Zeidan A., Wang E. S., Wetzler M., *Expert Opin. Biol. Ther.*, **9** (2008), 111–119.

Zhang L., et al., *Cancer Lett.*, **357** (2015), 346–354.

Zhao R., Domann F. E., Zhong W., *Mol. Cancer. Ther.*, **5** (2006), 3275–3284.

Chapter 9

Carboxylesterase Inhibitors: Relevance for Pharmaceutical Applications

Li-Wei Zou and Guang-Bo Ge

Institute of Interdisciplinary Integrative Medicine Research, Shanghai University of Traditional Chinese Medicine, Shanghai, 201203, China

geguangbo@dicp.ac.cn

9.1 Introduction

Mammalian carboxylesterases (CES, E.C. 3.1.1.1) are key members of the serine hydrolase superfamily, which are localized within the lumen of the endoplasmic reticulum (ER) in various tissues (Satoh and Hosokawa, 1998; 2006; Wang et al., 2018). As their name implies, CES catalyze the ester cleavage of a large number of structurally diverse ester- or amide-containing substrates into the corresponding alcohol/amine and carboxylic acid. Actually, CES can hydrolyze ester, thioester, amide, and carbamate linkages in a wide variety of endo- and xenobiotic compounds, thus play key roles in endobiotic metabolism, as well as in activation and

Pharmaceutical Biocatalysis: Fundamentals, Enzyme Inhibitors, and Enzymes in Health and Diseases
Edited by Peter Grunwald

ISBN 978-981-4800-61-7 (Hardcover), 978-0-429-29503-4 (eBook)
www.jennystanford.com

detoxification of xenobiotics (Satoh and Hosokawa, 1998; Wang et al., 2018). In the human body, human carboxylesterase 1 (CES1) and human carboxylesterase 2 (CES2) are two predominant isoenzymes involved in both endogenous and xenobiotic metabolism, which have been extensively studied over the past decade (Satoh et al., 2002; Redinbo and Potter, 2005; Imai, 2006; Hosokawa, 2008; Wang et al., 2018; Zou et al., 2018). Both CES1 and CES2 play crucial roles in the hydrolytic metabolism of various ester xenobiotics including many ester drugs (such as oseltamivir, clopidogrel, irinotecan, and capecitabine) and environmental toxicants (such as pyrethroids) (Potter et al., 1998; Nishi et al., 2006; Zhu and Markowitz, 2009; Imai and Ohura, 2010; Zhu et al., 2013). Meanwhile, these two enzymes (CES1 and CES2) are also responsible for the hydrolysis of endogenous esters, including cholesteryl esters, triacylglycerols, and other endogenous lipids, and thus play vital physiological functions in lipid homeostasis (Alam et al., 2002; Crow et al., 2010; Wang et al., 2013; Ruby et al., 2017; Lian et al., 2018).

Over the past 20 years, the relevance of two predominant CES to human diseases and the roles of CES in xenobiotic metabolism have been extensively investigated (Wang et al., 2018; Ruby et al., 2017). The key roles of CES in both endogenous and xenobiotic metabolism have attract great interest in the discovery of CES modulators to regulate lipid metabolism or to enhance the outcomes of ester drugs (Yoon et al., 2004; Crow et al., 2008; Hicks et al., 2009; Dominguez et al., 2014; Xu et al., 2016; Wang et al., 2017). This review covers the structural features and catalytic properties of CES, tissue distribution, biological functions, substrate specificities, and inhibitor spectra of two predominant CES, as well as the research significance and recent progress on the discovery of CES inhibitors. All information and knowledge presented in this review will be very helpful for pharmacologists to explore the relevance of CES to human diseases or to investigate the contribution of CES in xenobiotic metabolism, as well as for medicinal chemists to design and develop more CES inhibitors with improved potency and specificity for potential pharmaceutical applications.

9.2 Structural Features and Catalytic Properties of CES

9.2.1 Structural Features of CES

CES belong to the α/β-hydrolase fold superfamily of proteins. The majority of mammalian carboxylesterases are intracellular proteins found predominantly in the microsomal fraction that encompass the ER (Satoh and Hosokawa, 1998; Furihata et al., 2004). It has been revealed that microsomal CES from human, rabbit, and mouse carry the HXEL motifs of the KDEL consensus ER retrieval sequence at their C-terminal (such as HIEL and HTEL for CES1 and CES2, respectively), which is very essential for the localization of these enzymes to the ER lumen in mammalian cells (Robbi and Beaufay, 1991)·. Following cleavage of the C-terminal signal peptide, microsomal carboxylesterases can be released from their membrane-associated state, suggesting that these enzymes are not transmembrane proteins but soluble proteins that reside in the ER lumen (Potter et al., 1998).

The three-dimensional structures of several mammalian CES, including human carboxylesterase 1 (CES1), have been solved by X-ray crystallography with several ligands (Potter et al., 1998; Bencharit et al., 2002; Bencharit et al., 2003a; Bencharit et al., 2003b; Furihata et al., 2004; Fleming et al., 2007). As depicted in Fig. 9.1, CES1 is composed by a central catalytic domain, an αβ domain, and an adjacent regulatory domain which containing the low-affinity surface ligand-binding Z-site. The X-ray crystal structure of CES1 demonstrated its existence as monomer, trimer, or hexamer, with substrate dependent equilibrium of homooligomer formation. In contrast, CES2 exist as monomers. Both sequence alignments and secondary sequence predictions have suggested that these three CES are members of α/β hydrolase family (Holmes et al., 2009). Although the three-dimensional structures of CES2 have not been reported, the 3D structure modeling of CES2 could be downloaded from the SWISS-MODEL repository (a database of annotated 3D protein structure models generated by the SWISS-MODEL homology-modeling pipeline).

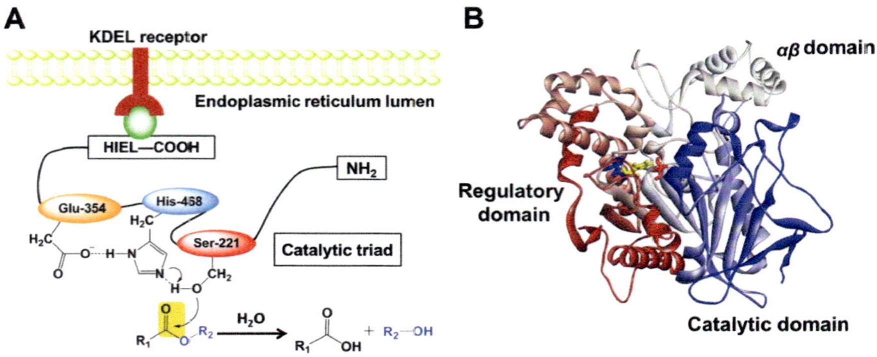

Figure 9.1 The structural features of CES1. Reproduced from Wang et al. (2018).

The molecular properties of CES1 and CES2 have been listed in Table 9.1 (Bencharit et al., 2002; Bencharit et al., 2003b; Furihata et al., 2004; Wang et al., 2017). Similar to all known serine hydrolases, the catalytic domain of human CES1 or CES2 contains a catalytic triad (such as Ser^{221}, Glu^{354}, and His^{468} in CES1) at the interface of the three domains, which is highly conserved among all mammalian carboxylesterases and is very crucial for CES-mediated catalysis (Fig. 9.1B) (Bencharit et al., 2003a; Yoon et al., 2004). Mutation of any residue of the catalytic triad will lead to the loss of carboxylesterase activity (Robbi and Beaufay, 1991). Furthermore, the oxyanion hole formed by Gly^{142} and Gly^{143} in the HGGG motif is also highly conserved among all mammalian carboxylesterases and is very crucial for carboxylesterase activity. The catalytic cavity of human CES1 is quite large (~1,300 Å^3) and is lined mainly by hydrophobic amino acids, except the residues (such as Ser^{221}) of the catalytic triad. The residue Ser^{221} divides the whole ligand-binding pocket of CES1 into two pockets, one is a rigid pocket which makes CES1 selective for those substrates with small acyl group, and another is a flexible pocket which makes CES1 promiscuous towards a wide range of esters with various acyl groups (Bencharit et al., 2003b). These features make CES1 capable of interacting with a wide variety of chemically diverse ligands.

Table 9.1 Molecular properties of CES1 and CES2

Properties	CES1	CES2
Molecular weight	60 kD (monomer) 180 kD (trimer)	60 kD (monomer)
Isoelectric point	5.6–5.8	4.8–5.0
Optimal pH	6.5	7.5–8.0
C-terminal signal peptide	HIEL	HTEL
Catalytic triad	Ser^{221}, Glu^{354} and His^{468}	Ser^{228}, Glu^{345} and His^{457}
Glycosylation site	Asn–X–Thr Asn^{79}	Asn–X–Ser/Thr Asn^{103}, Asn^{267}

Note: The information is taken from Potter et al. (1988), Bencharit et al. (2002), Bencharit et al. (2003a), and Furihata, et al. (2004).

9.2.2 The Catalytic Properties of CES

The CES hydrolyze substrates using a classic base-catalyzed mechanism via a two-step reaction which is conserved in all serine hydrolases, including lipases, proteases, and peptidases (Fig. 9.2). This process is dependent on an essential catalytic triad which is generally composed by three residues (Ser, His, and Glu) within the active cavity of mammalian carboxylesterases (Satoh and Hosokawa, 1998; Imai, 2006; Hosokawa, 2008). First, a nucleophilic attack by the base-activated serine oxygen atom (such as Ser^{221} in human CES1) on the carbonyl carbon of the substrate lead to the formation of an acyl-enzyme intermediate and the release of an alcohol, thiol, or amine product (Fig. 9.2). Second, the acyl-enzyme intermediate is attacked in an identical fashion with water acting as the nucleophile, leading to the release of the carboxylic acid metabolite, which regenerates the carboxylesterase to its original state with the free serine residue (Imai, 2006; Hosokawa, 2008).

Notably, several mammalian CES, especially CES1, display transesterification activities. When alcohol presents in the CES1 reaction system, alcohol replaces water to attack the acyl-enzyme intermediate to generate an ethyl ester product. One of the well-studied examples of this reaction is the formation of cocaethylene in individuals abusing both cocaine and alcohol (Brzezinski et al., 1994; Beckemeier and Bora, 1998; Parker and Laizure, 2010). Under these conditions, the ethyl group from ethanol replaces the methyl group of cocaine to produce a more toxic metabolite cocaethylene. Furthermore, CES1 can process the creation of cholesteryl esters from cholesterol and fatty acids, as well as to generate fatty acid ethyl esters (FAEEs) from fatty acyl-Coenzyme A (CoA) and ethanol, using transesterification reaction (Brzezinski et al., 1994; Bora et al., 1996; Beckemeier and Bora, 1998; Tang et al., 2006). Since CES have cholesteryl esters hydrolysis and FAEE hydrolysis activity, the formation of these endogenous esters is the result of the balance between classic hydrolysis reaction and transesterification reaction.

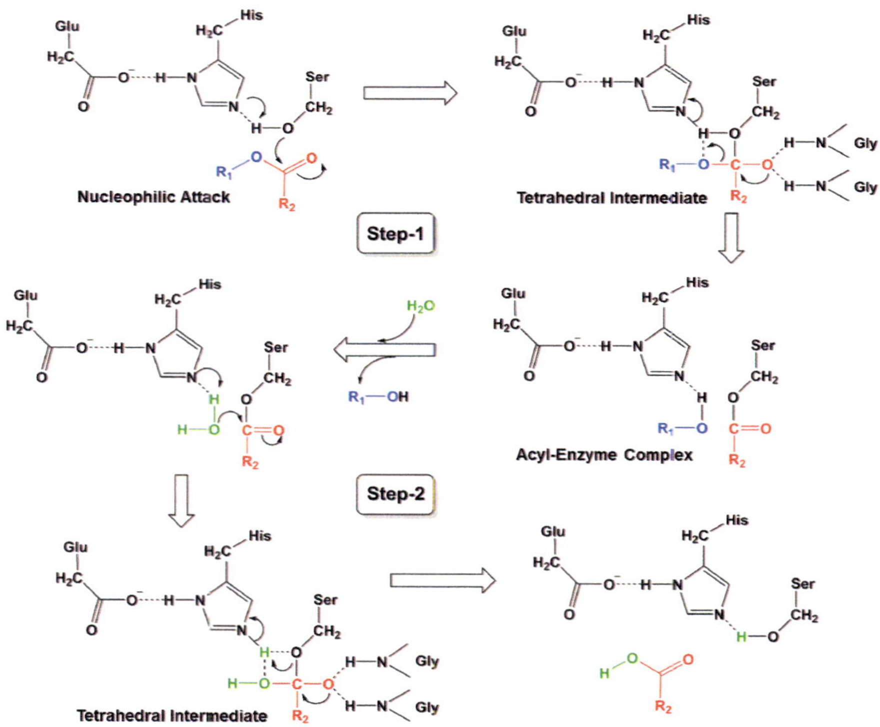

Figure 9.2 The two-step catalytic mechanism of carboxylesterases. Reproduced from Wang et al. (2018).

9.3 Tissue Distribution and Substrate Preference of Human CES

9.3.1 Tissue Distribution of Human CES

Although CES1 and CES2 share 47% amino acid sequence identity, these two enzymes exhibit extremely different substrate distribution and specificity (Imai, 2006; Hosokawa, 2008). Generally, CES1 and CES2 are highly expressed in the epithelia of most metabolic organs, including liver, intestine, and kidney, indicating that these two enzymes play a protective role against xenobiotics. CES1 is abundant expressed in the liver and adipocytes, with lesser amounts in the kidney, monocytes, lung, intestine, testis, heart, and macrophages (Diczfalusy et al., 2001; Satoh et al., 2002; Taketani et al., 2007; Williams et al., 2010; Sato et al., 2012; Lv et al., 2016). In contrast, CES2 is expressed mainly in the small intestine and colon, and also observed in kidney, liver, heart, brain, and testis (Wang et al., 2018). Recently, the quantitative data of CES abundance in the human liver microsomes (HLM) and liver cytosol (HLC) has been reported. The average protein levels of CES1 and CES2 in HLM were 402 and 29.8 pmol/mg, respectively, while in HLC were 54.5 and 2.76 pmol/mg, respectively (Sato et al., 2012). Furthermore, the secreted forms of CES with significant CES activities could be detected in the blood from rodent animals, but in the blood from human beings, the CES and CES activities are hardly detected (Li et al., 2005; Lv et al., 2016). Notably, the expression of CES1 and CES2 in tumor tissues or cancer cell lines is much different from non-cancerous healthy cells. It has been proved that Caco-2 cells mainly express the CES1 isoenzyme, but the protein levels of CES1 in the human intestine is extremely low (Taketani et al., 2007). In addition, CES2 is overexpressed in several types of cancer or cancer cell lines, including multiple myeloma, thyroid papillary carcinoma, esophagous squamous carcinoma, and kidney adenocarcinoma (Yano et al., 2008). Notably, CES2 expression in cancer cell lines and tumor tissue significantly correlate with bioactivation of several cancer prodrugs (such as irinotecan and LY2334737, the prodrug of gemcitabine), which agrees well with their anticancer effects (Xu et al., 2002; Yano et al., 2008; Pratt et al., 2013). These results suggested that the medicinal chemists can rational design CES2-bioactivated prodrugs on the basis of CES2 expression in cancer cell lines, which will be very helpful for cancer therapy.

9.3.2 Substrate Preferences of Human CES

Human carboxylesterases have broad substrate spectra, which can hydrolyze a vast number of endo- and xenobiotic substrates with ester, thioester, carbamate, and amide bonds (Imai et al., 2006; Shi et al., 2008; Takahashi et al., 2008; Williams et al., 2008; Wang et al., 2011; Thomsen et al., 2014; Wang et al., 2016; Yoshida et al., 2018). Over the past decade, many studies have reported that human CES1 and human CES2 exhibit distinct substrate preferences. The representative substrates for CES1 and CES2 have been depicted in Fig. 9.3 (Imai, 2006; Hosokawa, 2008; Wang et al., 2016), and Fig. 9.4 (Potter et al., 1998; Sanghani et al., 2004; Imai et al., 2006; Wang et al., 2011; Yoshida et al., 2018), respectively. In general, CES1 prefers to metabolize the ester substrates that contain a small alcohol group and a bulky acyl group, such as enalapril, oseltamivir, imidapril, clopidogrel, meperidine, D-luciferin methyl ester (DME), and the illegal drugs heroin and cocaine (Imai, 2006; Hosokawa, 2008; Wang et al., 2016). In contrast, CES2 prefers to hydrolyze esters with a relatively large alcohol group and a small acyl group, such as irinotecan, prasugrel, capecitabine, flutamide and fluorescein diacetate (Potter et al., 1998; Sanghani et al., 2004; Imai et al., 2006; Wang et al., 2011; Yoshida et al., 2018). However, several substrates with a small acyl group, such as *R*-propionyl propranolol, can also be hydrolyzed by CES1. As mentioned above, CES1 has two ligand-binding pockets, one is a rigid pocket and another is a flexible pocket, which makes CES1 promiscuous towards a vast number of substrates Holmes et al. (2009). In recent years, based on the substrate preferences of both CES1 and CES2, various optical probe substrates have been developed for highly selective sensing the real activities of CES1 or CES2 in complex biological systems (Feng et al., 2014; Liu et al., 2014a; Liu et al., 2014b; Feng et al., 2015; Jin et al., 2015; Jin et al., 2016; Ding et al., 2017). These optical probes provide efficient tools for high-throughput screening (HTS) of CES modulators in cell/tissue preparations or even in living cells, due to the inherent advantages including non-destructive, highly sensitive, easily managed, and applicable to HTS assay.

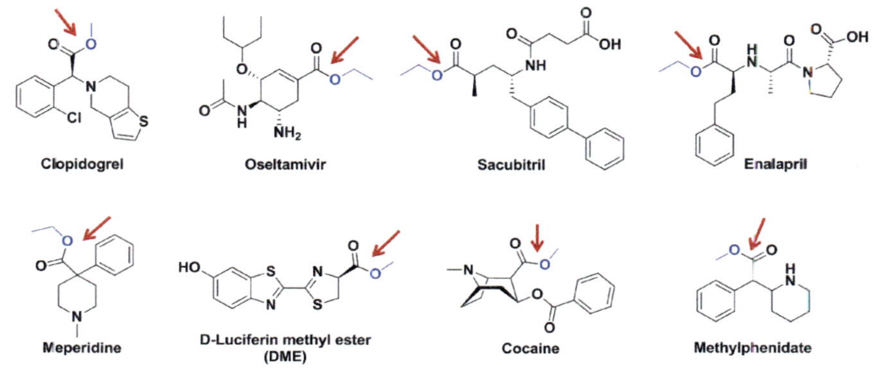

Figure 9.3 Substrate preference of human CES1. All representative substrates for CES1 contain a small alcoholic group and a bulky acyl group.

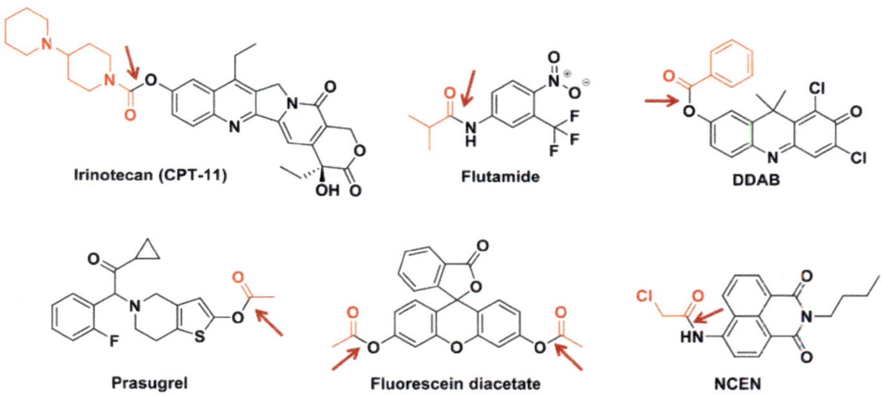

Figure 9.4 Substrate preference of human CES2. CES2 prefers to hydrolyze the esters with a relatively large alcohol group and a small acyl group.

9.4 Biological Functions of CES and Applications of CES Inhibitors

The primary biological function of CES is xenobiotic metabolism. Over the past 20 years, CES have been regarded as the classic xenobiotic-metabolizing enzymes which are responsible for the hydrolysis of a variety of ester-containing drugs, prodrugs, and environmental toxins. Many clinical drugs with ester moieties could be readily hydrolyzed by CES, including the anticancer prodrugs (such as irinotecan, capecitabine), narcotics (cocaine, heroin, and meperidine), angiotensin-converting enzyme inhibitors (enalapril, temocapril, imidapril, and quinapril), and other drugs with ester moieties (oseltamivir, clopidogrel, flumazenil, procaine, oxybutynin, delapril, flutamide, and prasugrel) (Satoh and Hosokawa, 1998; Imai, 2006; Hosokawa, 2008; Zou et al., 2018). In addition to drug metabolism, CES also participate in the detoxification and metabolic clearance of many ester-containing toxins, such as the pyrethroid insecticides (deltamethrin and permethrin) (Nishi et al., 2006; Merali et al., 2014). Notably, strong inhibition on CES may slow down the hydrolysis of CES substrates, which may affect their pharmacokinetic properties and thus modulate their outcomes in vivo (Bora et al., 1996; Imai and Ohura, 2010). For example, CPT-11 (a CES2 substrate drug) could trigger severe delayed diarrhea due to the overproduction of SN-38 (the hydrolytic product of CPT-11) in the small intestine, co-administration with potent CES2 inhibitors may ameliorate CPT-11 caused severe diarrhea in patients and thus improve the therapeutic effect (Alimonti et al., 2004; Quinney et al., 2005).

In addition to the well-known roles of CES in xenobiotic metabolism, these enzymes also participate in endogenous metabolism. As one of the most abundant serine hydrolases distributed in human hepatocytes and adipocytes, CES1 is responsible for the hydrolysis of a vast number of endogenous esters (such as cholesteryl esters and triacylglycerols) and thus participates in physiological and pathological processes, such as lipid metabolism, cholesterol homeostasis, and fatty liver disease (Alberti et al., 2009; Faulds and Dahlman-Wright, 2012; Bie et al., 2013; Lian et al., 2018). Notably, it has been reported that both expression and enzymatic activities of CES1 in the adipose tissues from obese and type 2 diabetic patients are markedly elevated compared to lean subjects, while treatment with CES1 inhibitors brings multiple beneficial effects on both glucose and lipid homeostasis in mice (Dominguez et al., 2014). More recently, CES2 has been reported with triacylglycerol and diacylglycerol hydrolase activity; its expression and function in liver are strongly related with several metabolic diseases, such as obesity and non-alcoholic steatohepatitis (Dominguez et al., 2014; Li et al., 2016; Ruby et al., 2017).

Besides the key roles in endo- and xenobiotic metabolism, CES also have other biological roles, such as trafficking and retention of proteins in the ER. In the ER, CES1 and CES2 appear to regulate protein trafficking, including release of proteins. For instance, CES1 can directly bind to the C-reactive protein (CRP) and holds this small protein before its release into the plasma (Yue et al., 1996; Li et al., 2016). CES use a region of amino acid sequence adjacent to the "side door," which comprises the loop between α15 and β18, to contact CRP. These CES could hold a small reservoir of CRP within the ER, and then release it during the stage of tissue injury. It has also been reported that CES can directly interact with β-glucuronidases, the enzymes responsible for the removal of glucuronic acid moieties which are typically conjugated to drugs and endobiotics by the UDP-glucuronosyltransferase (UGT) enzymes, in the ER (Zhen et al., 1993). Although the interactions between β-glucuronidases with CES have not been extensively investigated, several studies have demonstrated that some compounds (such as organophosphate) are capable of inducing the release of β-glucuronidases from the ER by disrupting the β-glucuronidase-CES1 complex (Zhen et al., 1993).

9.5 Recent Advances on Discovery of CES Inhibitors

The key roles of CES in both human health and xenobiotic metabolism arouse great interest in the discovery of CES inhibitors to modulate endogenous metabolism or to improve the outcomes of patients administrated ester drugs. With these goals in mind, many small molecule inhibitors of CES have been identified with the specific intention of altering enzyme activity for therapeutic purposes. Over the past decade, a variety of compounds with diverse scaffolds have been found with potent inhibitory effects against CES, which arouse great interest in medicinal chemists to develop safer and more effective lead compounds or drug candidates. This section has focused on the recent advances in the discovery and development of chemical inhibitors against two major human CES, including the reported CES inhibitors and their inhibition potency, as well as the structural features of CES inhibitors and the SARs.

9.5.1 Reversible Inhibitors of Human CES

9.5.1.1 1,2-Diones

1,2-Diones have been identified as the most potency inhibitors of human CES (Hyatt et al., 2007b; Hatfield and Potter, 2011; Hatfield et al., 2013; Binder et al., 2018), such compounds include benzils (Wadkins et al., 2005), alkyl-1,2-diones (Parkinson et al., 2011), isatins (Hyatt et al., 2007a), and 1,2-quinones (Hatfield et al., 2017). These compounds demonstrate high potency and relative high selectivity toward CES1 or CES2, due to they do not inhibit human acetylcholinesterase (AchE) or butyrylcholinesterase (BchE). SAR studies suggest that the 1,2-dione moiety is very essential for CES inhibition (Hatfield et al., 2013). If the dione oxygen atoms are *trans*-coplanar or not planar, the compounds are more potent in CES2 inhibition, such as benzils (Table 9.2). In contrast, replacement of the benzene rings with alkyl substituents results in an increase of selectivity towards CES1 (Wadkins et al., 2005). Furthermore, introduction of different atoms (N, O, S) vicinal to the benzene ring and the carbonyl carbon atom results in the loss of inhibitory effects on these two CES. Potter et al. suggest that the benzils show the excellent inhibitory activity against CES, partially because the 1,2-dione moiety mimics the ester chemical character allowing for the initiation of the nucleophilic addition-elimination reactions with serine in the active site to one of the carbonyl groups (Fig. 9.5) (Wadkins et al., 2005). However, abortive nucleophilic attack result in an intermediate that will be unlikely to generate the C–C cleavage, therefore, leading to enzyme inhibition for the repetitious attack and release by the serine oxygen atom (such as Ser^{221} of CES1) on the carbonyl groups. This supposition suggests that decreasing the electron density around the carbonyl carbon atom, such as substituted ethane-1,2-diones, will make the carbonyl carbon atom more susceptible to react with the serine oxygen.

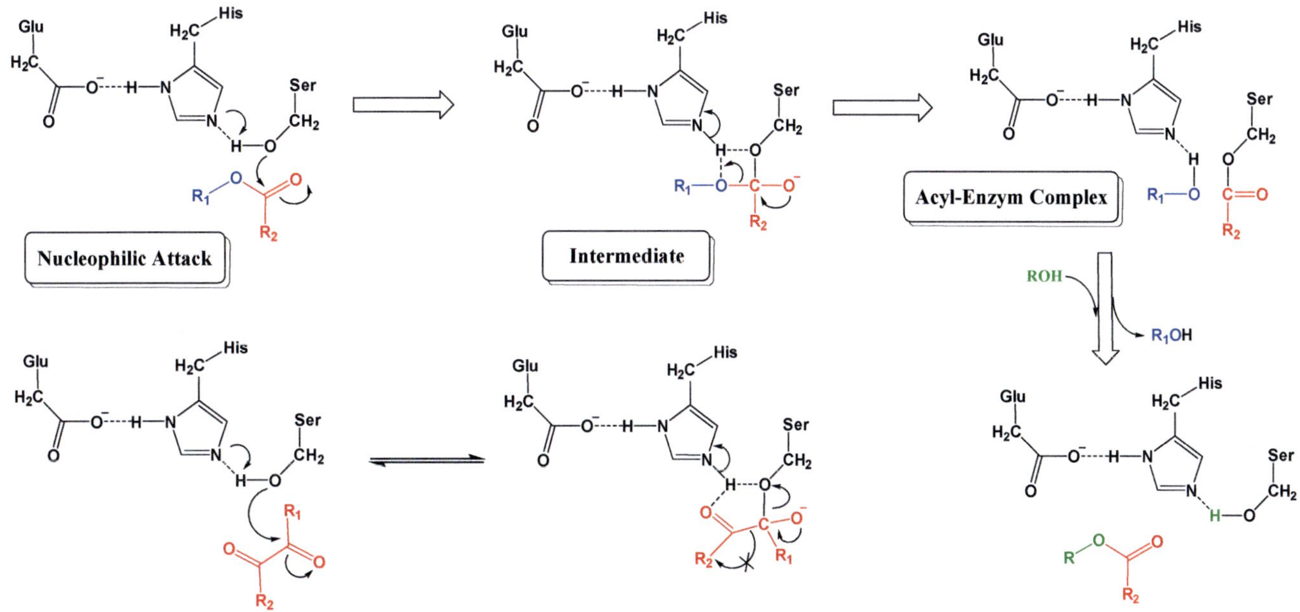

Figure 9.5 Proposed mechanism of CES inhibition by benzils. Reproduced from Zou et al. (2018).

Taking into account that the catalytic residues of CES are buried at the bottom of deep hydrophobic gorges in these enzymes (Wadkins et al., 2001; Redinbo and Potter, 2005; Imai and Hosokawa, 2010), it is easily conceivable that this area favors the lipophilic compounds. Potter et al. assume that the inhibitory potency of 1,2-diones depends primarily on the electrophilicity of the carbonyl groups, but also on the hydrophobicity of the substituent groups (Hyatt et al., 2007a; Hyatt et al., 2007b).

When the two carbonyl oxygen atoms of 1,2-dione moiety are *cis*-coplanar or fused within a ring structure, such as 1,2-quinones and isatins, demonstrate high specificity towards CES1 over CES2 (Table 9.2). Tanshinones, the natural phenanthrene-quinone compounds isolated from Chinese herb Danshen (De Palma et al., 2008), are found to be potent inhibitors of both CES1 and CES2 (Hatfield et al., 2013). The entrance to the active site gorge in CES1 is considerably smaller, and any substitution introduced to the benzene ring of the 1,2-quinone or isatin should be small enough to ensure the access of the inhibitors to the active cavity (Hyatt et al., 2007a). β-Lapachone with 1,2-dione moiety in *cis*-coplanar is a potent and reversible CES inhibitor, and demonstrates relatively high specificity towards CES2 over CES1 (Hatfield et al., 2017). 4-Phenoxynaphthalene-1,2-diones demonstrated high selectivity toward CES1 as compared with β-lapachone. The amino analogues (4-phenyl- aminonaphthalene-1,2-diones) with the imino forms were inactive toward human CES when compared with phenoxy analogues, suggesting that the imino group at the C-4 site was unbeneficial for CES inhibition. The potency of *N*-methylated amino analogues (4-phenyl(methyl)aminonaphthalene-1,2-diones) were equal to the phenoxy analogues and demonstrated selectivity towards CES1. Further investigation on inhibition kinetic analyses and docking simulations showed that the tautomerism of the amino derivatives to the imino forms likely accounts for their loss in biological activity.

The SAR analysis of these 1,2-diones revealed that the 1,2-dione moiety, aromatic rings or increasing hydrophobicity, and the substitution which did not hinder the access of the inhibitor to the active cavity of the enzyme, were indispensable for CES inhibition. Meanwhile, the specificity for CES1 could be achieved by allowing the 1,2-dione in a *cis*-coplanar, the specificity towards

CES2 could be accomplished by constraining the 1,2-dione in a *trans*-coplanar or non-planar configuration.

Table 9.2 The inhibition potency of 1,2-diones towards human carboxylesterases

Inhibitors	**Structures**	**K_i (CES1)* μM**	**K_i (CES2)* μM**	**Refs**
Benzil (diphenylethane-1,2- dione)	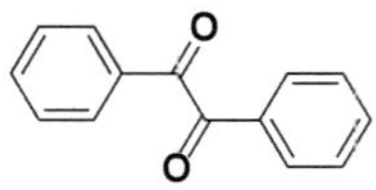	0.0451	0.0147	Wadkins et al. (2005)
1,2-Dicyclohexylethane-1,2-dione	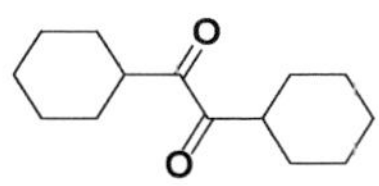	0.072	0.005	Hyatt et al. (2007b)
1-Phenylpropane-1,2- dione	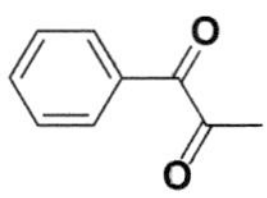	1.84	5.27	Wadkins et al. (2005)
Butan-2,3-dione	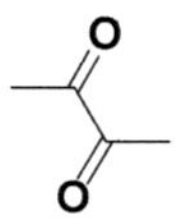	>100	>100	Wadkins et al. (2005)
Phenyl-1,2-octanedione	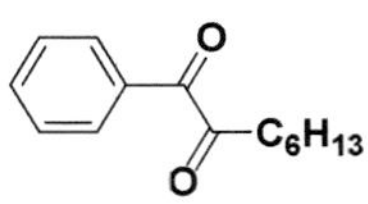	0.0022	0.0287	Parkinson et al. (2011)
Tetradecane-7,8-dione	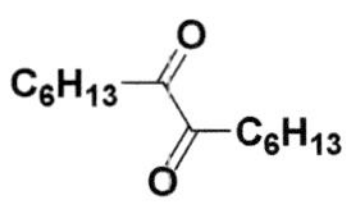	0.0048	0.0764	Parkinson et al. (2011)
Octadecane-9,10-dione	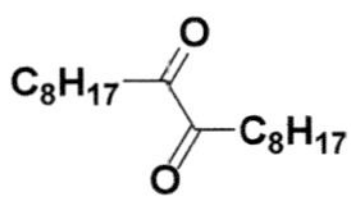	0.00084	0.0057	Parkinson et al. (2011)
Substituted ethane-1,2-diones	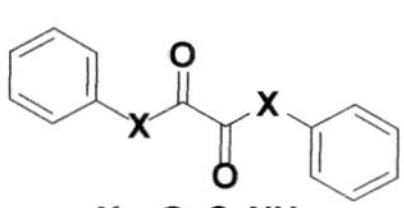 X = O, S, NH	>100	>100	Parkinson et al. (2011)

(*Continued*)

Table 9.2 (*Continued*)

Inhibitors	Structures	K_i (CES1)* μM	K_i (CES2)* μM	Refs
2,2′-Thenil (1,2-dithien-2-ylethane-1,2-dione)	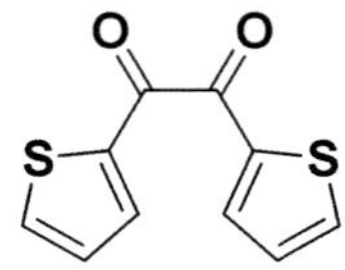	0.212	1.04	Hyatt et al. (2005)
Phenanthrene-9,2-dione	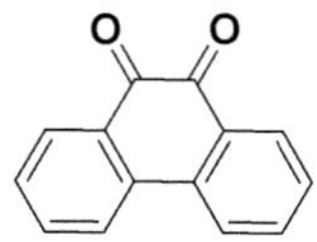	0.0025	0.052	Hatfield et al. (2013)
1,2-Naphthoquinone	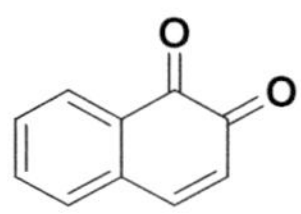	0.93	2.40	Hatfield et al. (2013)
Cyclohexane-1,2-dione	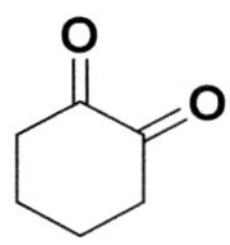	>100	>100	Wadkins et al. (2005)
1-Phenyl-1H-indole-2,3-dione	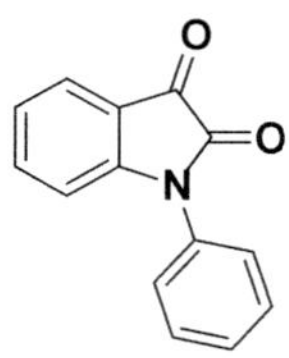	0.023	0.95	Hyatt et al. (2007a)
β-Lapachone	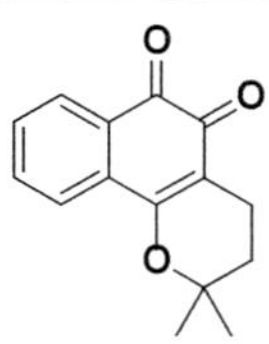	1.220	0.109	Hatfield et al. (2017)
4-Phenoxynaphthalene-1,2-dione	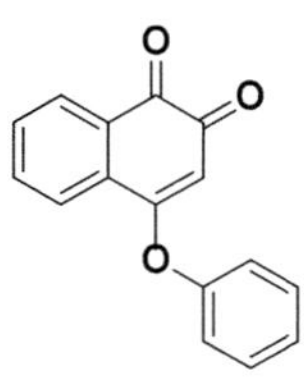	0.0238	0.1998	Hatfield et al. (2017)

Inhibitors	Structures	K_i (CES1)* μM	K_i (CES2)* μM	Refs
4-(Methyl(phenyl)amino)naphthalene-1,2-dione	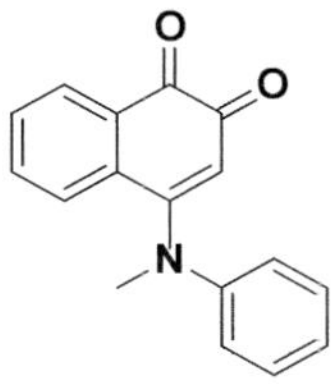	0.032	>10	Hatfield et al. (2017)
Tanshinone IIA	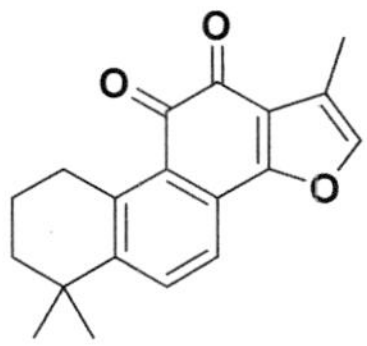	6.89	2.45	Hatfield et al. (2013)
Dihydrotanshinone	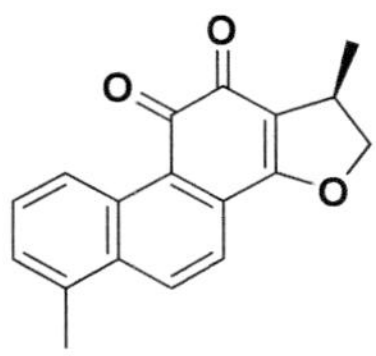	0.40	0.12	Hatfield et al. (2013)

*O-Nitrophenyl acetate as a substrate for both CES1 and CES2.

9.5.1.2 Benzene sulfonamides

Benzene sulfonamides displayed potent inhibitory effects against CES2 with K_i values at micromolar to nanomolar level and relative high selectivity over CES1 (Table 9.3), and had no inhibitory effects on either human AchE or BchE (Wadkins et al., 2004; Hicks et al., 2009). Inhibition kinetic analyses demonstrated that these compounds were partially competitive inhibitors against human CES. Quantitative SAR (QSAR) analyses and the clogP correlations indicated that the inclusion of an aromatic, planar, hydrophobic domain within the center of the compound greatly favored for CES2 inhibition (Wu et al., 2003; Hicks et al., 2009; Lamego et al., 2013). Introduction of halogen in terminal benzene rings was beneficial for CES inhibition. However, these compounds containing large bulky substituent within either the central domain or the terminal benzene rings of the benzene sulfonamide analogues demonstrated less potency on CES2. These findings suggested that the hydrophobicity of these sulfonamides was a key factor affecting the inhibitory effects on CES2.

Table 9.3 The inhibition potency of sulfonamides towards carboxylesterases

Inhibitors	Structures	K_i (CES1)* μM	K_i (CES2)* μM	Refs
N,N'-(1,4-phenylene) dibenzene sulfonamide	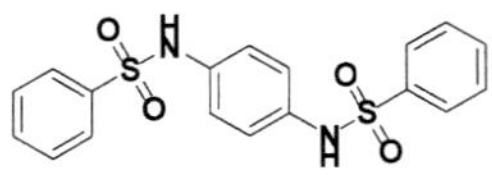	>100	1.06	Wadkins et al. (2004)
N,N'-(1,4-phenylene)bis(4-chloro-benzene sulfonamide)	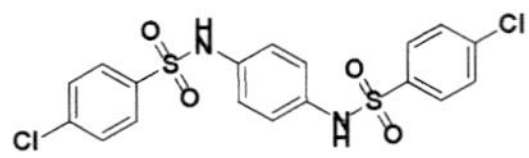	13.7	0.053	Wadkins et al. (2004)
N,N'-(1,4-phenylene)bis(2,4,5-trichloro-benzene sulfonamide)	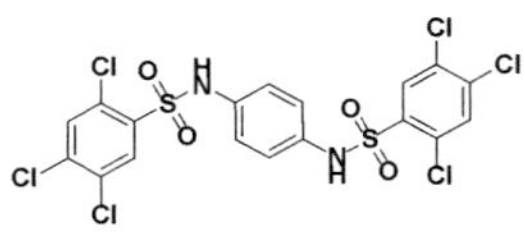	>100	>100	Hicks et al. (2009)
N,N'-(2,5-dichloro-1,4-phenylene)bis(4-chloro-benzene sulfonamide)	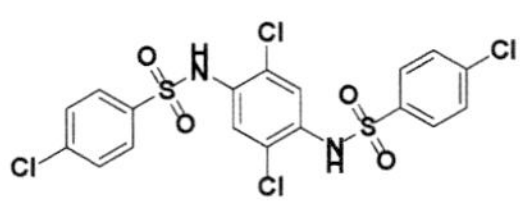	>100	0.45	Hicks et al. (2009)
N,N'-(2,5-dibromo-1,4-phenylene)bis(4-chloro-benzene sulfonamide)	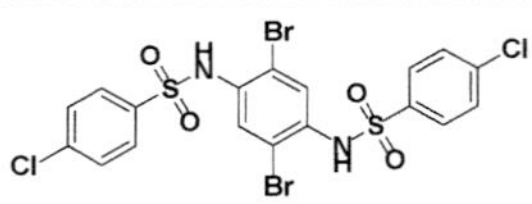	>100	>100	Hicks et al. (2009)
N,N'-(naphthalene-1,4-diyl)bis(4-chloro-benzene sulfonamide)	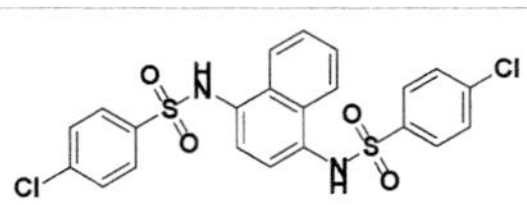	>100	0.20	Wadkins et al. (2004)
N,N'-(perfluoro-1,4-phenylene)bis(4-bromo-benzenesulfonamide)	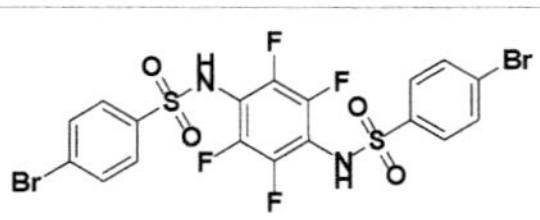	>100	0.023	Hicks et al. (2009)
N,N'-(2,3,5,6-tetramethyl-1,4-phenylene)bis(4-bromo-benzenesulfonamide)	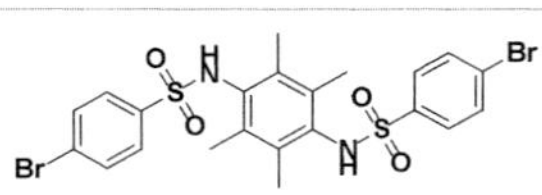	>100	3.24	Hicks et al. (2009)
2,7-Bis(phenyl-sulfonyl)-9*H*-fluorene	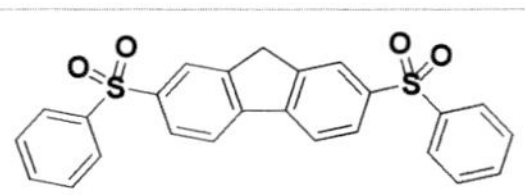	>100	0.091	Hicks et al. (2009)

*O-Nitrophenyl acetate as a substrate for both CES1 and CES2.

9.5.1.3 Trifluoroketones

Trifluoroketones are potent inhibitors of a variety of hydrolases, such as juvenile hormone esterase (Prestwich et al., 1984), AchE (Gelb et al., 2002; Wadkins et al., 2007), chymotrypsin (Brady and Abeles, 1990), angiotensin converting enzyme (Ojima et al., 1994) and fatty acid amidohydrolase (Boger et al., 1999). Trifluoroketones also displayed strong inhibitory effects on human CES with K_i values at nanomolar level, and mostly demonstrated poor specificity toward CES1 or CES2 (Table 9.4) (Wadkins et al., 2007). The trifluoro moiety of these compounds could be easily hydrated to form the *gem*-diol (Gelb et al., 2002; Wadkins et al., 2007). However, the equilibrium between the ketone and *gem*-diol forms must be sufficiently dynamic such that adequate concentrations of the ketone are available for enzyme inhibition (Rosell et al., 1996). Thus the rates of the ketone/*gem*-diol equilibrium would affect the interactions between trifluoroketones and the amino acid residues within the CES active site, and thus affect their inhibition potency. However, this process was poorly understood and would require further investigation to explore the mechanism of trifluoroketones as CES inhibitors. In most cases, thioether-containing compounds were more potent inhibitors than sulfinyl or sulfonyl analogs, and these compounds containing longer, more hydrophobic alkyl chains were more potent inhibitors of CES (Table 9.4).

Table 9.4 The inhibition potency of trifluoroketones towards carboxylesterases

Inhibitors	Structures	K_i (CES1) μM	K_i (CES2) μM	Refs.
3-(Dodecylthio)-1,1,1-trifluoropropan-2-one	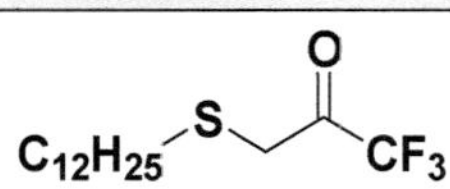	0.071	0.015	Wadkins et al. (2007)
3-(Dodecylsulfinyl)-1,1,1-trifluoropropane-2,2-diol	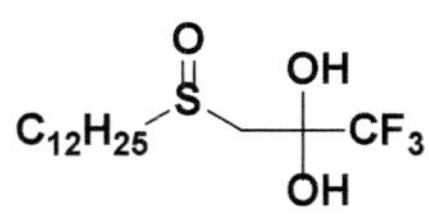	1.90	0.054	Wadkins et al. (2007)

(*Continued*)

Table 9.4 (*Continued*)

Inhibitors	Structures	K_i (CES1) μM	K_i (CES2) μM	Refs.
3-(Dodecylsulfonyl)-1,1,1-trifluoropropane-2,2-diol	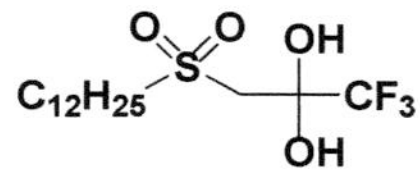	0.74	0.032	Wadkins et al. (2007)
3-(Decylthio)-1,1,1-trifluoropropan-2-one	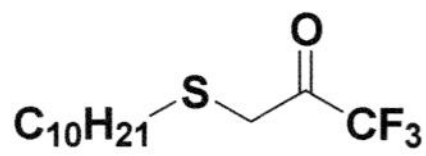	0.13	0.15	Wadkins et al. (2007)
1,1,1-Trifluoro-3-(octylthio)propan-2-one	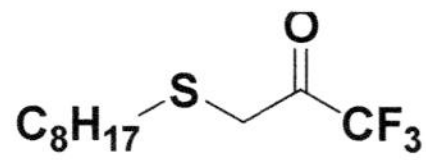	0.19	0.22	Wadkins et al. (2007)
1,1,1-Trifluoro-3-(hexylthio)propan-2-one	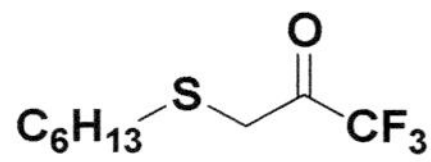	0.33	0.56	Wadkins et al. (2007)
3-(Butylthio)-1,1,1-trifluoropropan-2-one	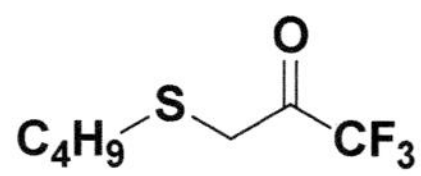	0.80	1.70	Wadkins et al. (2007)

*O-Nitrophenyl acetate as a substrate for both CES1 and CES2.

9.5.1.4 Acyl glucuronides

Acyl glucuronides are a class of phase II metabolites where carboxylic acids are conjugated with a glucuronic acid via an ester bond (Buchheit et al., 2011; Di Meo et al., 2013). Generally, the acyl glucuronides were anticipated to be rapidly excreted into urine or bile without exhibiting significant pharmacological activities. However, Williams et al. found that some acyl glucuronides are potential CES inhibitors, including diclofenac-β-D-glucuronide, and clopidogrel-β-D-glucuronide (Table 9.5) (Inoue et al., 2013). These two glucuronides could inhibit CES1 mediated 4-nitrophenyl acetate hydrolysis, with the K_i values of 4.32 and 24.8 μM, respectively. In contrast, these two glucuronides did not significantly inhibit CES2, indicating that the acyl glucuronides was not beneficial for CES2 inhibition.

Table 9.5 The inhibition potency of acyl glucuronides towards carboxylesterases

Inhibitors	Structures	K_i (CES1)* μM	References
Diclofenac-β-D-glucuronide	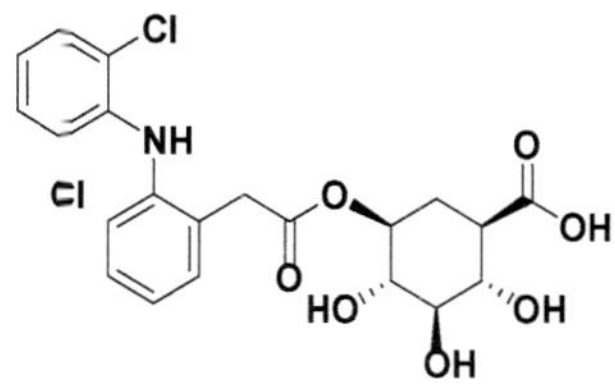	4.32	Inoue et al. (2013)
Clopidogrel-β-D-glucuronide	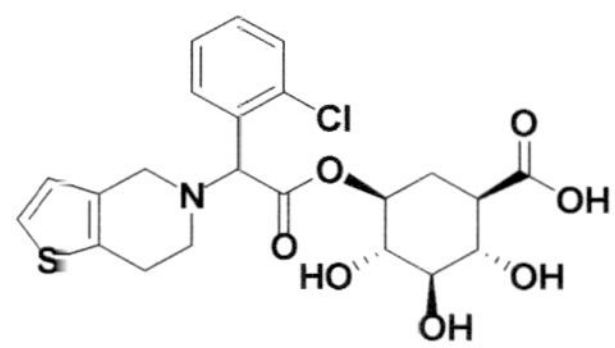	24.8	Inoue et al. (2013)
Ibuprofen-β-D-glucuronide	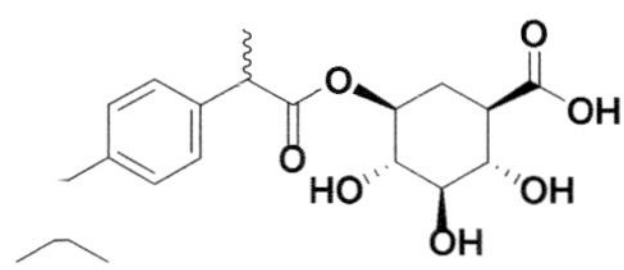	355	Inoue et al. (2013)

*4-Nitrophenyl acetate as a substrate for both CES1 and CES2.

9.5.1.5 Triterpenoids

Triterpenoids are a diverse group of natural products with wide distribution, high chemical diversity and important pharmacological properties (Zebiri et al., 2017; Zhou et al., 2017). Over the past decades, many groups in phytochemistry and medicinal chemistry focused on the discovery and development of bioactive compounds from triterpenoids as specific and potent inhibitors to a given enzyme via efficient chemical modifications (Kazakova et al., 2009; Callies et al., 2015). Zou et al. collected a series of natural triterpenoids and tested their inhibitory effects against CES using specific optical substrate for CES1 and CES2, respectively (Zou et al., 2017). Two pentacyclic triterpenoids including oleanolic acid (OA) and ursolic acid (UA), displayed strong inhibitory effects on CES1, were selected as the lead compounds for the development of potent and selective inhibitors against CES1 (Table 9.6). The SAR analysis revealed that the carboxyl group at the C-28 site of OA and UA was essential for CES1 inhibition, and any modifications on this group with amide, ester or alcohol are unbeneficial for CES1 inhibition. In contrast, the replacement of C-3 hydroxyl group with a ketone or ester could improve the inhibitory effects on CES1 and high selectivity over CES2. Guided by these SAR, the C-3 hydroxyl group of OA or UA was modified to 3-*O*-β-carboxypropionyl, the two derivatives including 3β-*O*-(β-carboxypropionyl)-olean-12-en-28-oic acid and 3β-*O*-(β-carboxy-propionyl)-urs-12-en-28-oic acid displayed excellent inhibition potency on CES1 and high selectivity over CES2 (3296-fold and 6919-fold over CES2, respectively (Zou et al., 2017). Furthermore, inhibition kinetic analyses and docking simulations demonstrated that 3β-*O*-(β-carboxypropionyl)-urs-12-en-28-oic acid functioned as a competitive inhibitor against CES1-mediated DME hydrolysis, with the K_i value of 12.6 nM. To the best of our knowledge, this compound is the most potent and selective inhibitor of CES1, which can be used as a promising lead compound for the development of CES1 inhibitor as drug candidate.

Other natural pentacyclic triterpenoids, such as β-boswellic acid with carboxyl group at the C-23 site, demonstrated strong inhibitory effects on CES2 and high selectivity over CES1 (Zou et al., 2017). Glycyrrhetinic acid (GA), the major bioactive ingredient of the roots and rhizomes of licorice (*Glycyrrhiza* species), which displays satisfying safety during long history of medicinal use, was selected as a reference compound for the development of potent and specific inhibitors against CES2 (Zou et al., 2016). Guided by the structure-CES2 inhibition relationships of a series of GA derivatives, Ge et al. designed and developed a more potent compound 3-*O*-(β-carboxypropionyl)-11-deoxo- glycyrrhetinic acid-30-ethyl ester as a novel and highly selective inhibitor against CES2, with the IC_{50} value of 20 nM and very high selectivity over CES1 (1000-fold), which was 3463-fold more potent than the parent compound GA. The SARs of these pentacyclic triterpenoids as CES1 or CES2 inhibitors were summarized in Fig. 9.6, which are very helpful for medical chemists to design and develop more potent and highly selective CES1 or CES2 inhibitors for biomedical applications.

Table 9.6 The inhibition potency of pentacyclic triterpenoids towards carboxylesterases

Inhibitors	Structures	IC_{50} (CES1)* μM	IC_{50} (CES2)* μM	References
Oleanolic acid (OA)	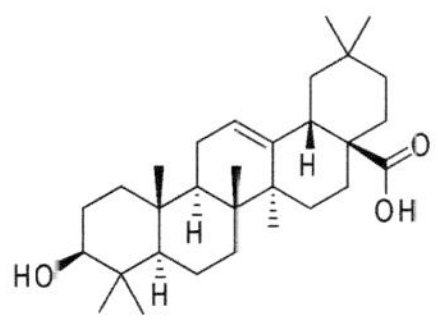	0.28	5.49	Zou et al. (2017)
3β-O-(β-Carboxypropionyl)-olean-12-en-28-oic acid	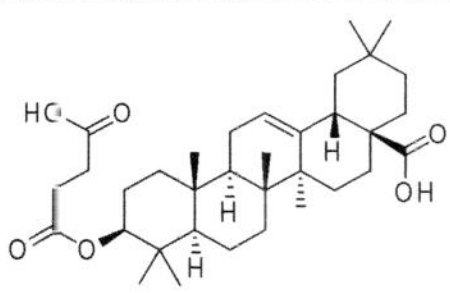	0.017	56.04	Zou et al. (2017)
Ursolic acid (UA)	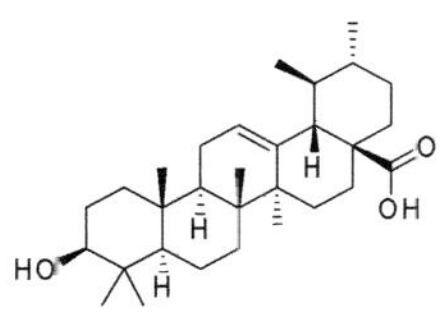	0.24	6.05	Zou et al. (2017)
3β-O-(β-Carboxypropionyl)-urs-12-en-28-oic acid	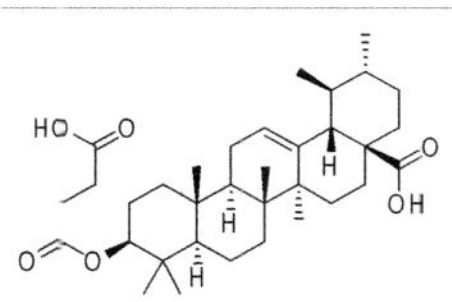	0.012	83.03	Zou et al. (2017)
β-Boswellic acid	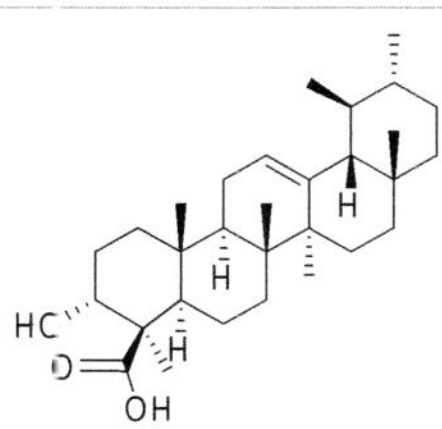	>400	2.12	Zou et al. (2017)
Glycyrrhetinic acid	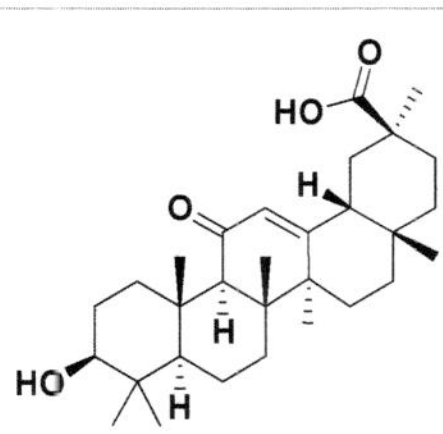	20.97	69.26	Zou et al. (2016)

(*Continued*)

Table 9.6 (*Continued*)

Inhibitors	Structures	IC_{50} (CES1)* μM	IC_{50} (CES2)* μM	References
11-Deoxo-glycyrrhetinic acid	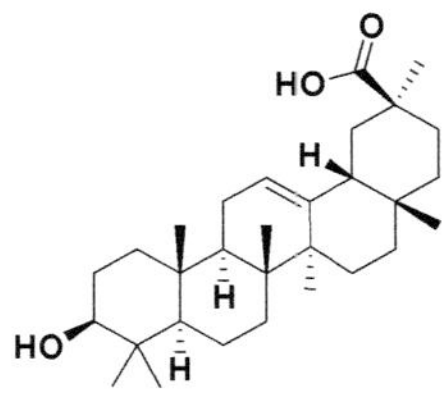	10.53	6.95	Zou et al. (2016)
3-O-(β-Carboxypropionyl)-11-deoxo-glycyrrhetinic acid-30-ethyl ester	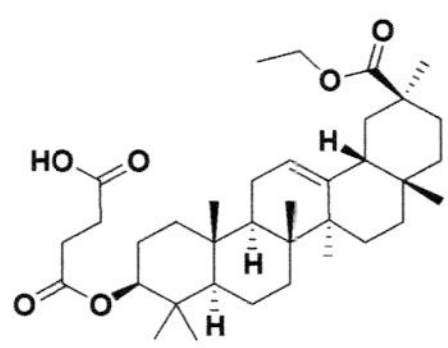	20.41	0.02	Zou et al. (2016)

*D-luciferin methyl ester (DME) and fluorescein diacetate (FD) were used as the highly specific probe substrate for hCE1, and hCE2, respectively.

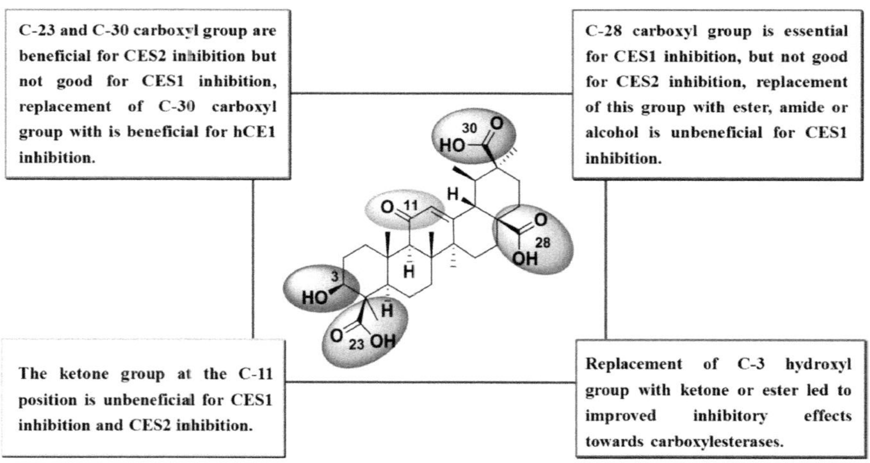

Figure 9.6 SAR summaries of pentacyclic triterpenoid derivatives as CES inhibitors.

Natural protostane triterpenoids, such as protopanaxadiol and protopanaxatriol, exhibit less potency and poor selectivity on CES1, but display strong inhibitory effects on CES2, suggesting that the long alkyl chain at the C-20 site is unbeneficial for CES1 inhibition (Zou et al., 2017). Recently, 22 protostane triterpenoids have been isolated from the rhizome of *Alismaorientale* (Mai et al., 2015b), five of which, including alismanol B, 25-O-ethylalisol A, alismanol D, alismanol F, and 11-deoxyalisol A, strongly inhibited CES2 with the IC_{50} values between 2.02 and 8.68 μM (Table 9.7) (Mai et al., 2015b). Alisol G (25-anhydroalisol A) is a major protostane triterpene obtained from dried rhizomes of *Alismaorientalis.* Following the biotransformation mediated by *P. janthinellumAS 3.510*, the metabolites of alisol G, including 25*S*-25,26-epoxy-alisol G, alisol G (23,24)-diol acetonide and 25,26-dihydroxy-alisol G (23,24)-diol acetonide were obtained, which showed significant CES2 inhibitory effects, with the IC_{50} values of 6.81, 3.38, and 6.33 μM, respectively (Mai et al., 2015a).

Table 9.7 The inhibition potency of protostane triterpenoids towards carboxylesterases

Inhibitors	Structures	IC_{50} (CES1) μM	IC_{50} (CES2)* μM	References
Protopanaxadiol	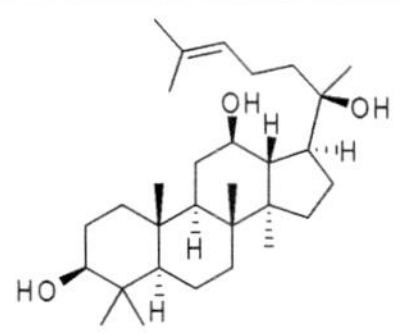	>500	2.658	Zou et al. (2017)
Protopanaxatriol	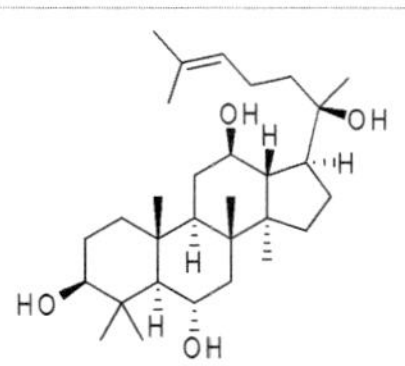	>500	9.800	Zou et al. (2017)
Alismanol B	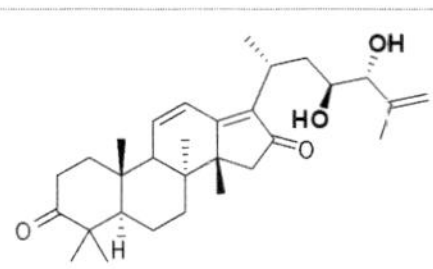	—	8.68	Mai et al. (2015b)
25-O-Ethylalisol A	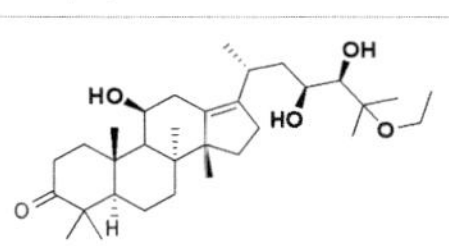	—	4.72	Mai et al. (2015b)

(*Continued*)

Table 9.7 (*Continued*)

Inhibitors	Structures	IC_{50} (CES1) μM	IC_{50} (CES2)* μM	References
Alismanol D	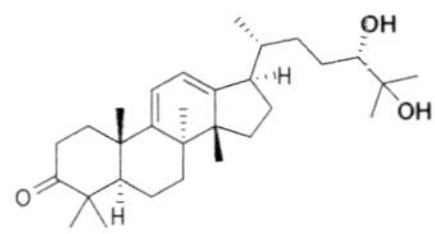	—	4.58	Mai et al. (2015b)
Alismanol F	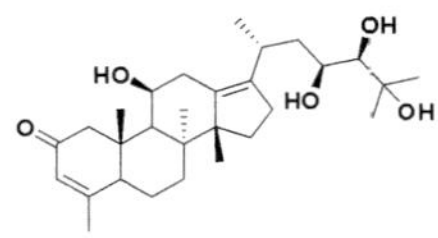	—	2.02	Mai et al. (2015b)
11-Deoxyalisol A	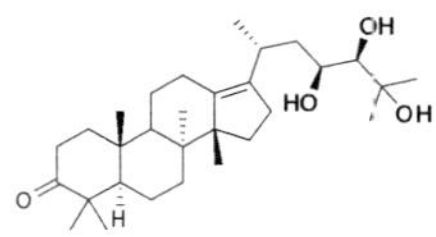	—	4.63	Mai et al. (2015b)
25-Anhydroalisol A (alisol G)	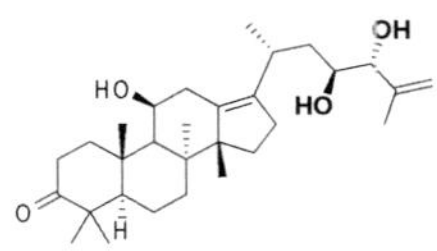	—	3.85	Mai et al. (2015a)
25*S*-25,26-Epoxy-alisol G	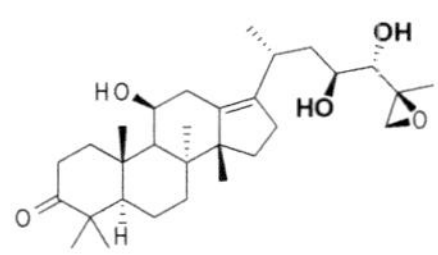	—	6.81	Mai et al. (2015a)
Alisol G (23,24)-diol acetonide	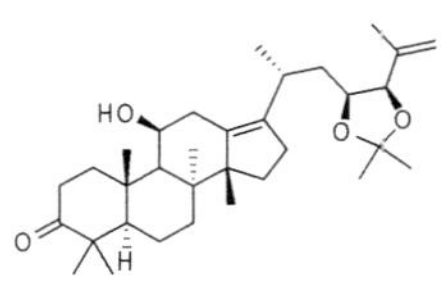	—	3.38	Mai et al. (2015a)
25,26-Dihydroxy-alisol G (23,24)-diol acetonide	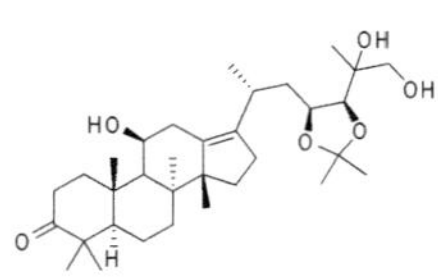	—	6.33	Mai et al. (2015a)
Alismanol H	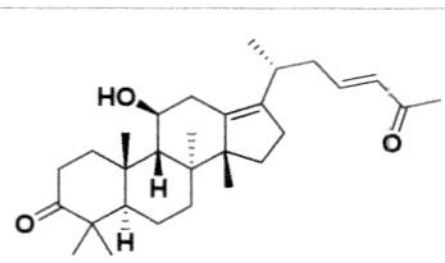	—	2.67	Zhang et al. (2017)

Inhibitors	Structures	IC$_{50}$ (CES1) μM	IC$_{50}$ (CES2)* μM	References
25-O-butyl alisol A	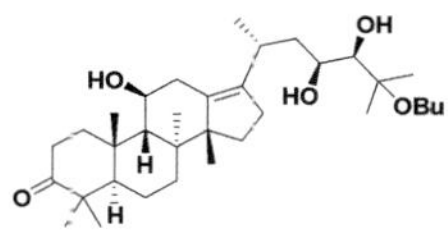	—	0.51	Zhang et al. (2017)
Alisol A 23,24-acetonide	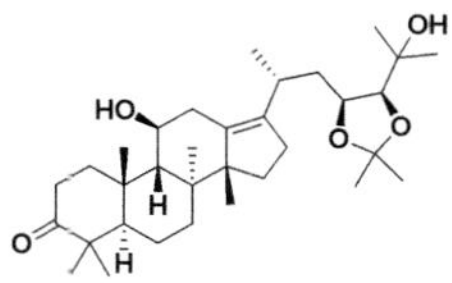	—	2.99	Zhang et al. (2017)
24-Deacetyl alisol O	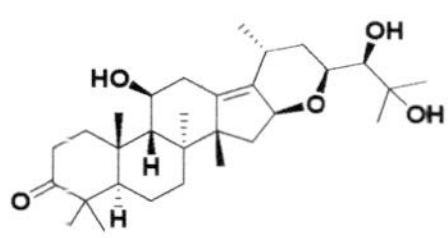	—	2.58	Zhang et al. (2017)
16,23-Oxido alisol B	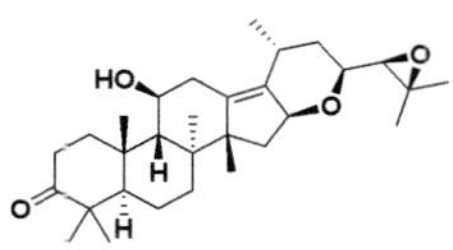	—	2.66	Zhang et al. (2017)

*4-Benzoyl-*N*-butyl-1,8-naphthalimide (MPN) as probe substrate for hCE2.

Following the phytochemical investigation on *A. orientalis* and the biotransformation of two major triterpenoids (alisol A and alisol B 23-acetate) of *A. orientalis*, 22 protostane triterpenoids including 10 new protostane-type triterpenoids were obtained (Zhang et al., 2017). Among them, 14 compounds are potent inhibiors of CES2, with the IC$_{50}$ values lower than 10 μM. The inhibition kinetics demonstrated that 25-O-butyl alisol A could inhibit the CES2-catalyzed 4-benzoyl-*N*-butyl-1,8-naphthalimide (MPN) hydrolysis, with the K_i value of 0.57 μM.

9.5.1.6 Flavonoids

Flavonoids are polyphenolic products widely distributed in vegetables, fruits, and beverages such as tea and wine (Santos-Buelga and Feliciano, 2017; Wen et al., 2017). Recent studies have demonstrated that some natural flavonoids, including bavachinin, coryfolin, corylin, neobavaisoflavone, corylifol A, and corylifolinin, potently inhibit both CES1 and CES2 (Table 9.8) (Li et al., 2015;

Table 9.8 The inhibition potency of flavonoids towards carboxylesterases

Inhibitors	Structures	K_i (CES1)* µM	K_i (CES2)* µM	References
Bavachinin	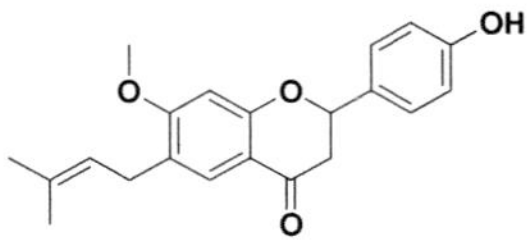	0.5[a]	1.12[b]	(Li et al., 2015; Sun et al., 2016)
Coryfolin	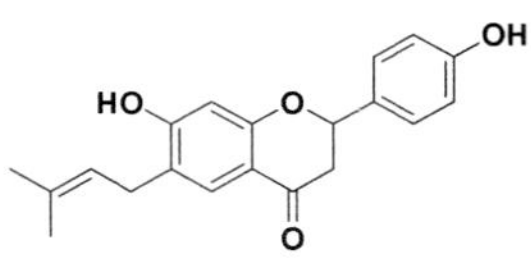	9[a]	—	Sun et al. (2016)
Corylin	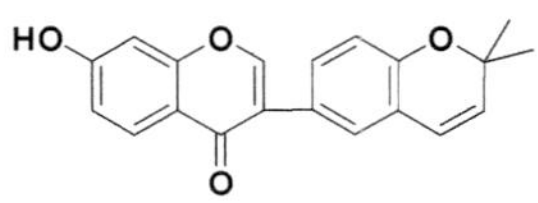	0.7[a]	—	Sun et al. (2016)
Neobavaisoflavone	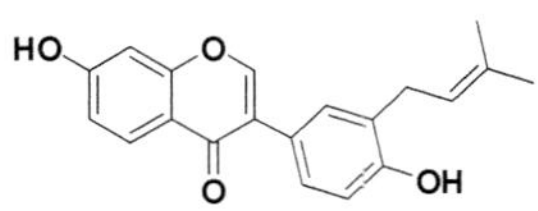	5.3[a]	3.89[a]	Li et al. (2015), Sun et al. (2016)
Corylifol A	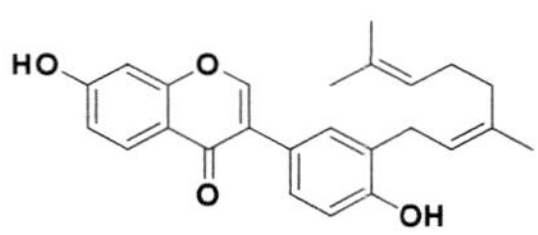	—	0.62[a]	Li et al. (2015)
Corylifolinin	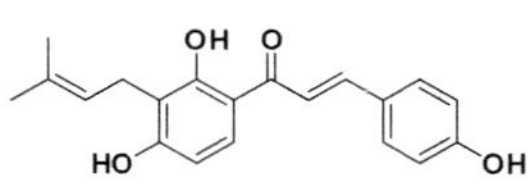	9.4[a]	1.61[a]	Sun et al. (2016)
5,6-Dihydroxyflavone	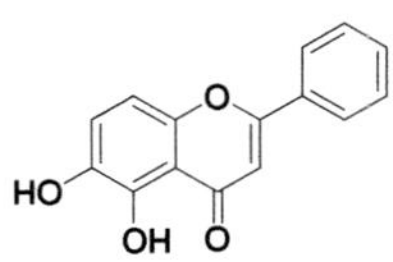	>200	3.50	Weng et al. (2018)
Isorhamnetin	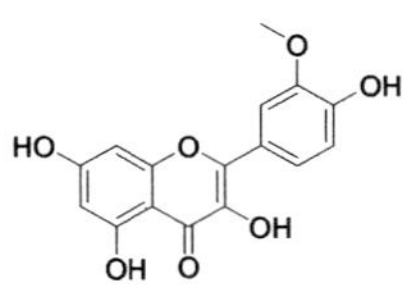	>200	3.63	Weng et al. (2018)

Inhibitors	Structures	K_i (CES1)* μM	K_i (CES2)* μM	References
Apigenin 7-O-methyl ether	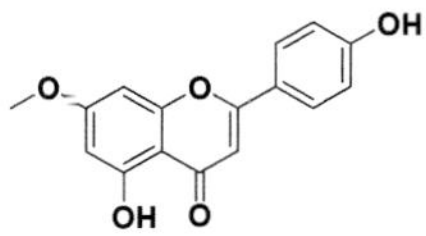	>200	4.73	Weng et al. (2018)
3,6-Dihydroxyflavone	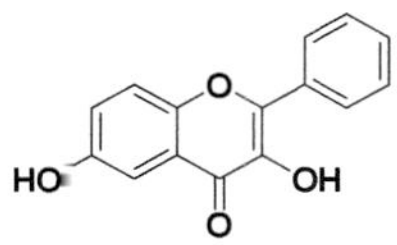	>200	5.06	Weng et al. (2018)
Eupatilin	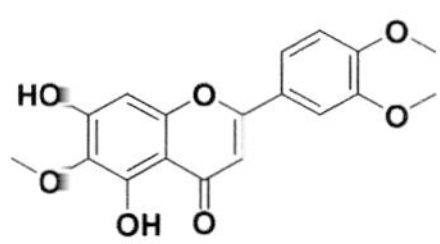	>200	6.85	Weng et al. (2018)
Hispidulin	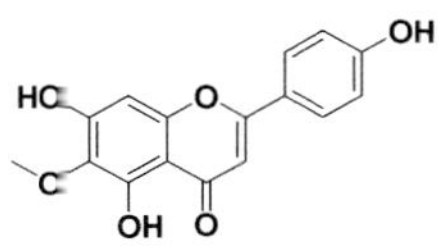	>200	7.64	Weng et al. (2018)
Sanggenone C	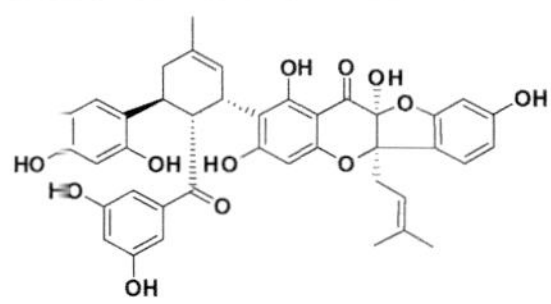	—	0.76[a]	Liu et al. (2016)
Sanggenone D	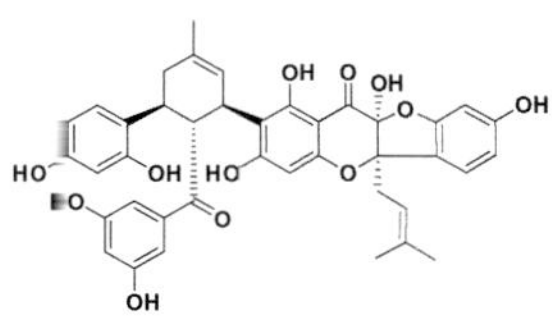	—	0.97[a]	Liu et al. (2016)
Kuwanon G	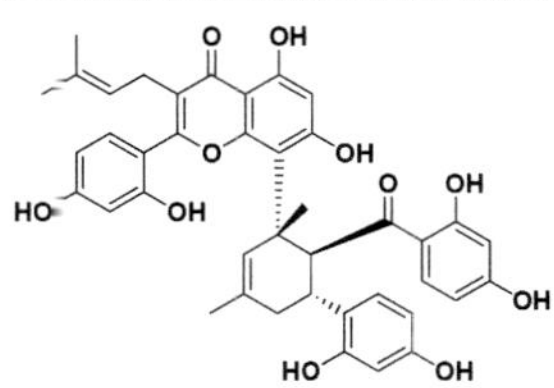	—	1.09[a]	Liu et al. (2016)

*2-(2-Benzoyloxy-3-methoxyphenyl) benzothiazole (BMBT) and fluorescein diacetate (FD) were used as the highly specific probe substrate for hCE1, and hCE2, respectively.

Sun et al., 2016). Coryfolin, corylin and neobavaisoflavone are noncompetitive inhibitors against CES1-mediated BMBT hydrolysis, while bavachinin is a competitive inhibitor against CES1-mediated BMBT hydrolysis (Sun et al., 2016). Neobavaiso-flavone, isobavachalcone, bavachinin and corylifol A, also found in *Fructus Psoraleae* (a famous Chinese medicinal herb), are potent inhibitors of CES2 (Li et al., 2015), with the K_i values ranging from 0.62 μM to 3.89 μM. Weng et al. (2018) collected more than 50 flavonoids to assay their inhibitory effects against CES2 using a fluorescence-based method. Among all tested flavonoids, 5,6-dihydroxyflavone, isorhamnetin, apigenin 7-*O*-methyl ether, 3,6-dihydroxyflavone, eupatilin, and hispidulin, displayed relatively strong inhibitory effects against CES2, with the IC_{50} values lower than 10 μM. Meanwhile, the inhibitory effects of these six flavonoids on CES1 were assayed by using DME as a specific optical substrate, with the IC_{50} values larger than 200 μM. The SARs of these tested flavonoids demonstrated that the phenolic groups at the C3 and the C6 sites of flavones were essential for CES2 inhibition, while introduction of a hydrophilic group at C-8 site or a bulky hydrophilic group at B ring might be beneficial for CES2 inhibition. The ethanolic extract of the root-bark of white mulberry strongly inhibited CES2, and its three major constituents, including sanggenone D, kuwanon G, andsanggenone C, could non-competitively inhibit CES2-mediated FD hydrolysis in HLM (Table 9.8) (Liu et al., 2016). These groundbreaking studies suggest that the flavonoids and related phenylpropanoids represent a gold mine, from which carboxylesterase modulators could be developed.

9.5.1.7 Other CES inhibitors

Besides the above-mentioned compounds, other compounds, including fatty acids, sterols, pyrethroids, and therapeutic drugs, also displayed strong inhibitory effects against carboxylesterase (Xu et al., 2016; Lei et al., 2017; Wang et al., 2017). Crow et al. found that most naturally occurring fatty acids strongly inhibited the hydrolytic activities of recombinant CES1, with the IC_{50} values at micromolar range Crow et al. (2010). Unsaturated fatty acids displayed potent inhibitory effects on CES1 than saturated ones, while they also display high specificity towards CES1 over CES2. 27-Hydroxycholesterol (27-HC), an oxidized form of cholesterol, also showed promising inhibitory activity against recombinant CES1 (IC_{50} 46 nM) and high selectivity over CES2. 27-HC functioned as noncompetitive inhibitor against CES1, with the very low K_i value (10 nM) (Crow et al., 2010). Bakuchiol, a natural phenolic compound isolated from *Fructus Psoraleae* (Bu-gu-zhi in Chinese), strongly inhibited CES2 (Li et al., 2015). Pyrethroids are a class of organic compounds similar to the natural pyrethrins produced by the flowers of pyrethrums (*Chrysanthemum cinerariaefolium*). Pyrethroids are popular household insecticides for their relatively low toxicity to mammals in contrast to organophosphorus insecticides. Recently, Ge et al. found that six commonly used pyrethroids including deltamethrin showed moderate inhibitory effects against both CES1 and CES2 (Table 9.9) (Lei et al., 2017). Deltamethrin strongly inhibited CES1-mediated DME hydrolysis in HLM, with the IC_{50} value of 2.39 μM. Deltamethrin was a competitive inhibitor against CES1-mediated BMBT hydrolysis, but it functioned as a noncompetitive inhibitor against CES1-mediated DME or DMCB hydrolysis in HLM. Further investigations on inhibition kinetic analyses and docking simulations suggested that CES1 had at least two ligand-binding sites, and deltamethrin (DMT) could bind with the same ligand binding site as BMBT. Physostigmine, a natural alkaloid, was a highly specific CES2 inhibitor with the K_i value of 0.358 μM (Umehara et al., 2016). Physostigmine potently inhibited the hydrolysis of irinotecan but did not affect the hydrolysis of clopidogrel in human liver S9. By the catalytic asymmetric [3+2] cyclization of novel 4-isothiocyanato pyrazolones and isatin-derived ketimines, Wang and Zou et al. systhsized a wide range of intriguing dispirotriheterocyclic products in high yield with excellent diastereoselectivity and enantioselectivity (Bao et al., 2018). In addition, a chiral sulfoxide derivative (*tert*-butyl (3*R*,4′*R*)-1-benzyl-2′-((*S*)-methylsulfinyl)-2,5″-dioxo-1″-phenyl-3″-(*p*-tolyl)-1″,5″-dihydro-1′H-dispiro[indoline-3,5′-imidazole-4′,4″-pyrazole]-1′-carboxylate) of this dispirocyclic product was identified to be a promising CES1 inhibitor. The IC_{50} of this chiral sulfoxide derivative was 0.39 μM, which activity was 20-fold stronger than its enantiomer.

Table 9.9 The inhibition potency of natural compounds towards carboxylesterases

Inhibitors	Structures	IC_{50} or K_i (CES1) μM	IC_{50} or K_i (CES2) μM	References
Myristic acid	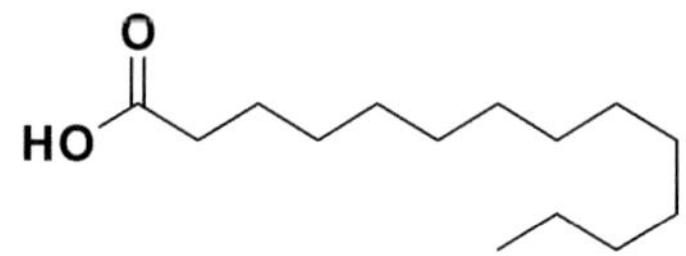	9	>100	Crow et al. (2010)
27-Hydroxy cholesterol	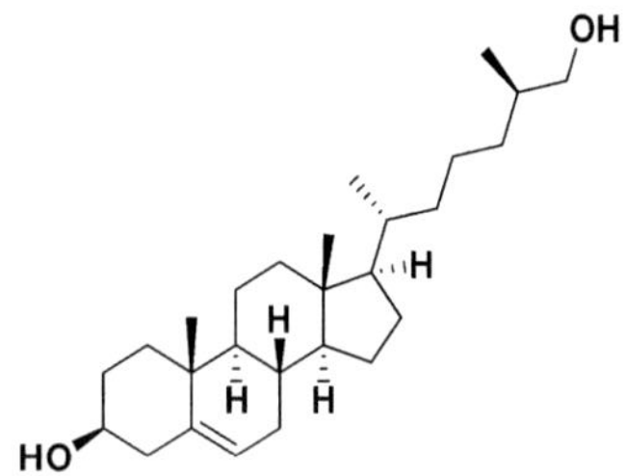	0.046	>100	Crow et al. (2010)
Bakuchiol	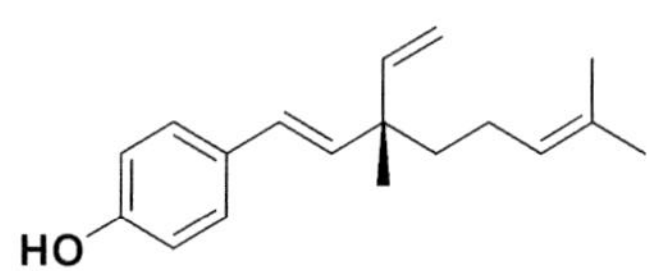	NA*	2.12	Li et al. (2015)

Inhibitors	Structures	IC_{50} or K_i (CES1) μM	IC_{50} or K_i (CES2) μM	References
Deltamethrin	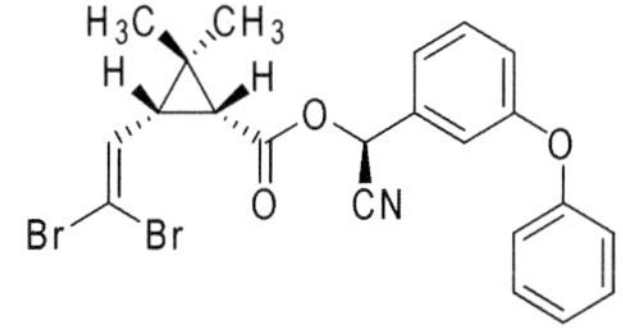	2.39	25.16	Lei et al. (2017)
Physostigmine	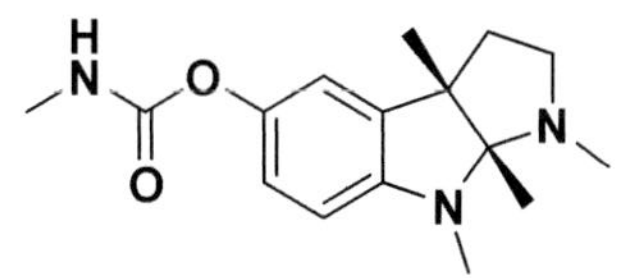	1198	0.358	Umehara et al. (2016)
tert-butyl (3*R*,4′*R*)- 1-benzyl-2′-((*S*)-methylsulfinyl)-2,5″-dioxo-1″-phenyl-3″-(*p*-tolyl)-1″,5″-dihydro-1′H-dispiro[indoline-3,5′-imidazole-4′,4″-pyrazole]-1′-carboxylate	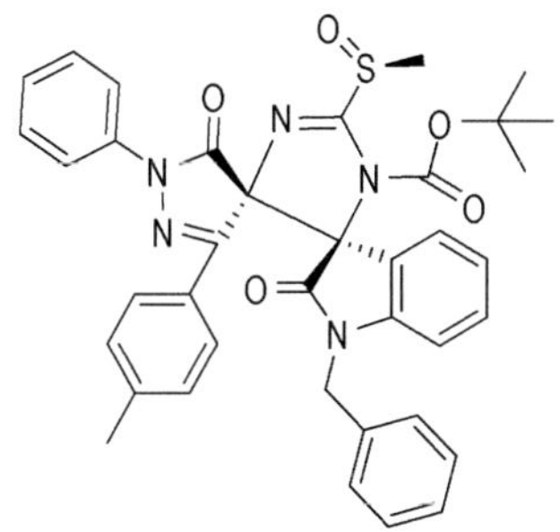	0.39	—	Bao et al. (2018)

*NA means not assayed.

Table 9.10 The inhibition potency of therapeutic drugs towards carboxylesterases

Inhibitors	Structures	K_i (CES1) µM	K_i (CES2) µM	References
Simvastatin	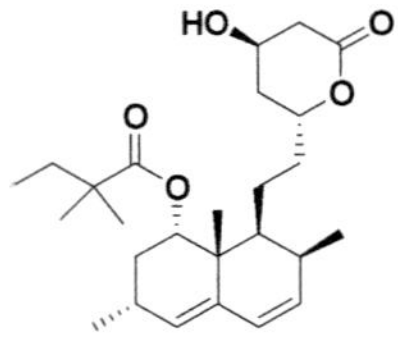	0.76	1.85	Fukami et al. (2010)
Troglitazone	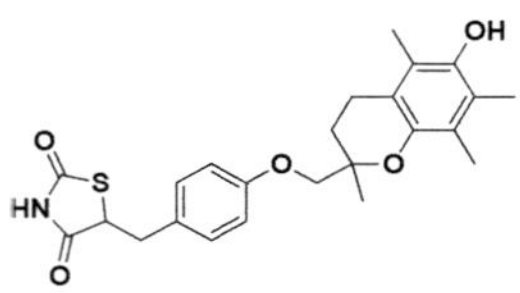	5.64	—	Fukami et al. (2010)
Fenofibrate	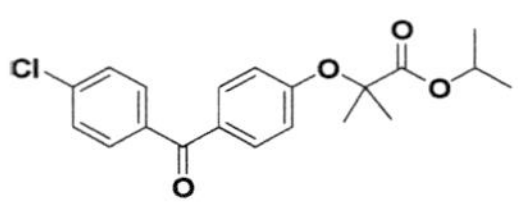	—	0.04	Fukami et al. (2010)
Nitrendipine	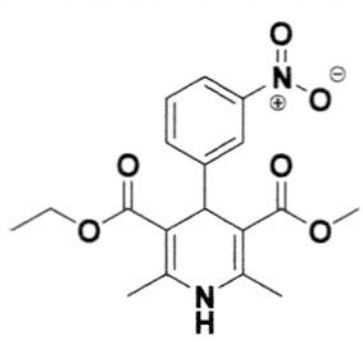	1.24	—	Xu et al. (2013)
Telmisartan	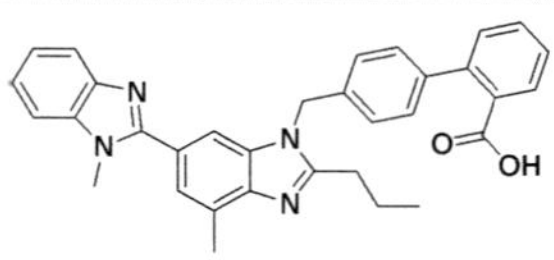	1.69	—	Xu et al. (2013)
Diltiazem	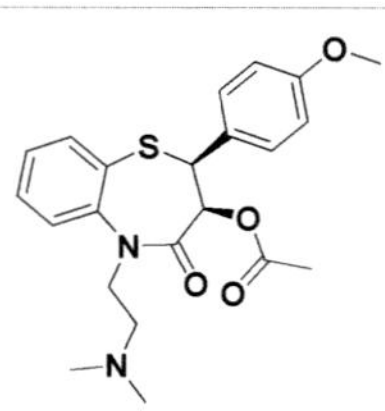	—	0.25	Xu et al. (2013)
Verapamil	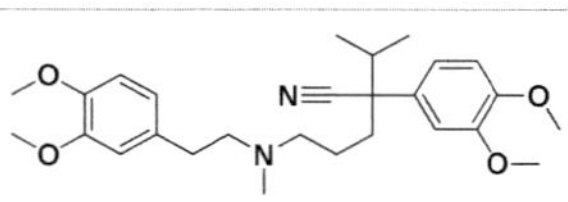	—	3.84	Xu et al. (2013)

Inhibitors	Structures	K_i (CES1) μM	K_i (CES2) μM	References
Nelfinavir	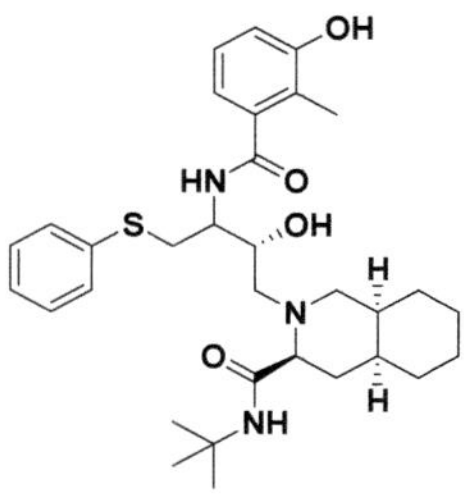	3.7	—	Rhoades et al. (2012)
Loperamide	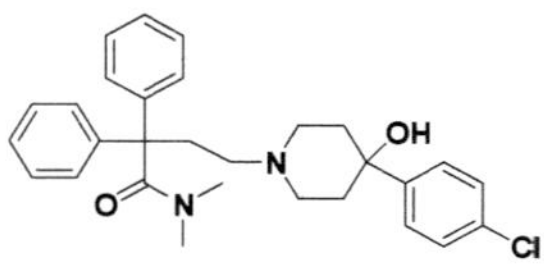	—	1.5 (IC_{50})	Quinney et al. (2005)
Sodium lauryl sulfate	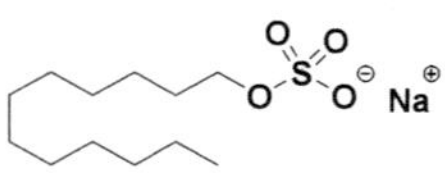	0.12 (μg/mL)	—	Zhang et al. (2014)
Polyoxyl 40 hydrogenated castor oil	—	0.76 (μg/mL)	—	Zhang et al. (2014)
Tween 20	—	—	0.93 (μg/mL)	Zhang et al. (2014)
Polyoxyl 35 castor oil	—	—	4.40 (μg/mL)	Zhang et al. (2014)

As shown in Table 9.10, some antihyperlipidemic and antidiabetic drugs, such as simvastatin and troglitazone, significantly inhibited the activity of CES (Fukami et al., 2010). Simvastatin and troglitazone strongly inhibited CES1, with the K_i values of 0.76 and 5.64 μM, respectively. Simvastatin and fenofibrate showed excellent inhibitory effects against CES2 with lower K_i value of 1.85 and 0.04 μM, respectively. Antihypertensive drugs including nitrendipine and telmisartan could also inhibit CES1-mediated imidapril hydrolysis in HLM (Table 9.10) (Xu et al., 2013). Diltiazem and verapamil showed excellent inhibitory effect against CES2 with lower K_i value as 0.25 and 3.84 μM, respectively (Xu et al.,

2013). Antiviral drug nelfinavir exhibited strong inhibition on CES1 via a competitive mechanism with the K_i value of 3.7 μM (Rhoades et al., 2012). Other protease inhibitors, e.g., atazanavir, amprenavir, ritonavir, and saquinavir, displayed less potent inhibition when compared with nelfinavir. Loperamide was often used to treat chemotherapy-associated diarrhea, and it was a potent and selective CES2 inhibitor with the IC_{50} value of 1.5 μM (Quinney et al., 2005). Among 25 pharmaceutical excipients, sodium lauryl sulfate (SLS) and polyoxyl 40 hydrogenated castor oil (RH40) could inhibit CES1-mediated imidapril hydrolysis in recombinant CES1A1 and HLM, and Tween 20 and Polyoxyl 35 castor oil (EL35) could inhibit CES2-mediated CPT-11 hydrolysis in recombinant CES1 and HLM (Table 9.11) (Zhang et al., 2014). These results demonstrate that surfactants such as SLS, RH40, Tween 20 and EL35 may attenuate CES activity, therefore special care should be taken when these pharmaceutical excipients were co-administrated with CES substrate drugs.

9.5.2 Irreversible Inhibitors of Human CES

9.5.2.1 Carbamates

Carbamate compounds were developed to inhibit almost all members of the serine hydrolase family via covalent modification of serine at the active site (Kathuria et al., 2003; Long et al., 2009; Chang et al., 2013). Carbamates have been identified as potent inhibitors of AchE and these compounds have been widely used for the pest control in several crops, as well as in poultry and domestic animals (Roy et al., 2012; Casida and Durkin, 2013). Recently, several cholinesterase inhibitors containing the carbamate moiety, such as JZL184 and phenethylcymserine, were found to be CES inhibitors (Table 9.11) (Crow et al., 2012; Tsurkan et al., 2013). But all these compounds displayed poor isoform selectivity towards mammalian CES. Strong inhibition of carbamates on CES could be attributed to the very slow hydrolysis rates of the carbamoyl-enzyme intermediate, which was much slower than that of the acetyl-enzyme intermediate (Fig. 9.7) (Scaloni et al., 1994).

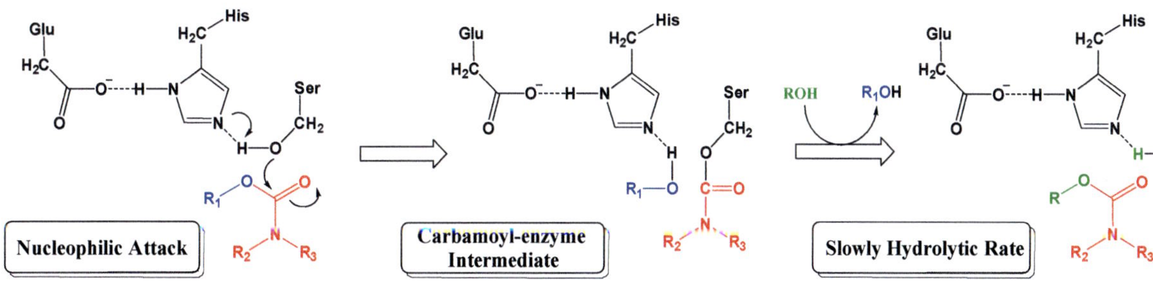

Figure 9.7 Proposed mechanism of carboxylesterases inhibition by carbamates

Table 9.11 The inhibition potency of carbamates towards carboxylesterases

Inhibitors	Structures	IC_{50} or K_i (CES1) μM	IC_{50} or K_i (CES2) μM	References
JZL184	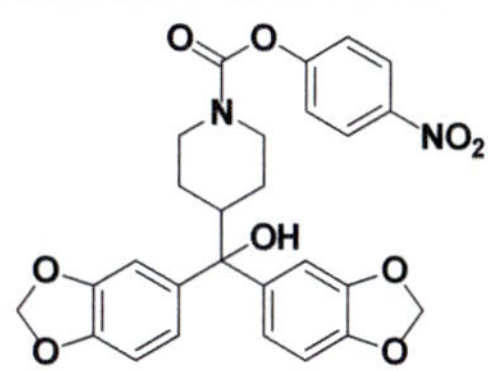	0.48	0.63	Crow et al. (2012)
Phenethylcymserine	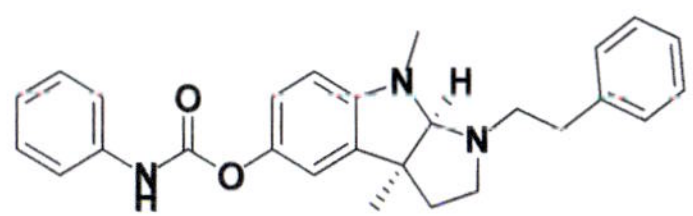	20.5	5.6	Tsurkan et al. (2013)

9.5.2.2 Organophosphates

Organophosphates (OPs) exert their toxicity via inhibition on synaptic AchE (Scholz et al., 2006; Lee et al., 2015). OPs have been utilized extensively as insecticides since 1940s and also been developed as nerve gas agents (Qadri and Kashi, 1971; Tong et al., 2013). A number of serine hydrolases including cholinesterases and CES including CES1 and CES2 have been reported to react with OPs and their catalytic activities could be significantly inhibited following exposure to OPs (Kovach, 1988; Nomura et al., 2006). As shown in Table 9.12, OPs could react with CES and generate a stable phosphate ester that was covalently linked to the catalytic residue (such as Ser221 of CES1) of CES. Several OPs, including bis(4-nitrophenyl)-phosphate (BNPP), chlorpyrifos oxon, and paraoxon, are potent irreversible inhibitors of human CES with the IC_{50} values at the nM level (Heymann et al., 1979; Casida and Quistad, 2005; Crow et al., 2012; Zou et al., 2016). SAR analysis revealed that the phosphate group of these compounds was essential for CES inhibition, while introduction of phosphate group to target compounds were likely beneficial for CES inhibition.

Table 9.12 The inhibition potency of organophosphates towards carboxylesterases

Inhibitors	Structures	IC_{50} (CES1) nM	IC_{50} (CES2) nM	References
Bis(4-nitrophenyl) phosphate (BNPP)	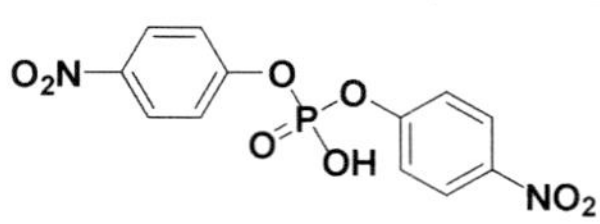	140	6110	Zou et al. (2016)
Chlorpyrifosoxon	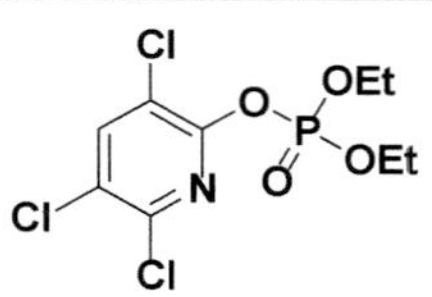	0.15	0.33	Crow et al. (2012)
Paraoxon	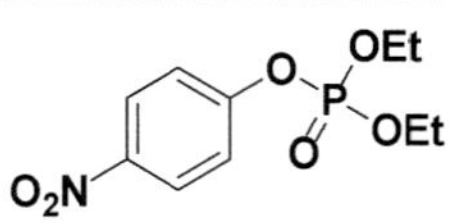	0.38	6.2	Crow et al. (2012)

A series of synthesized phosphorylated flavonoids exhibited significantly improved inhibition potency towards porcine CES when compared with their parent compounds (Wei et al., 2013; Wei et al., 2014), while their IC_{50} values were as low as 1.79 nM. These phosphorylated flavonoids were irreversible competitive inhibitors against mammalian CES.

9.6 Conclusion and Future Perspectives

Over the past decade, the medicinal chemists have made significant breakthrough on the discovery and development of highly selective and potent inhibitors against mammalian CES. Thanks to the newly developed optical substrates for CES1 and CES2, the present screening of CES inhibitors is more convenient and efficient. Until now, a wide range of potent CES inhibitors with structurally diverse scaffolds (such as 1,2-diones and triterpenoid) have been reported, some of which displayed very high efficacy with the IC_{50} values at nanomolar level. However, the potent and highly selective CES inhibitors are rarely reported, while the key structural features and the SAR of CES1 or CES2 inhibitors have not been well-characterized. Thus, it is necessary to explore the key structural determinant and SAR for CES1 or CES2 inhibition, by using a series of compounds with good structural connectivity. Apart from the optimization of potency and selectivity, the ADME/Tox properties of CES inhibitors should also be evaluated and optimized, which will be very helpful for the development of CES inhibitors as drug candidates. Currently, most of reported CES inhibitors were assayed in vitro, and their ability targeting intracellular CES and their potency against target CES in living systems have not been well investigated. Therefore, it is highly desirable to find more efficacious CES inhibitors and to test their safety and efficacy in vivo. Prior to the whole body test in animals, it is necessary to characterize the species differences in the function and the responses towards inhibitor(s) of mammalian CES among different species (Ma et al., 2017). Additionally, given that CES have multiple ligand-binding sites, it is necessary to characterize the binding site(s) and the inhibition types for the CES inhibitors of interest, which will be very helpful for the deeper understanding of the interactions between ligands and CES. Taken together, current findings call for more in-depth studies including the biological functions of CES, the relevance of CES to human diseases, the species differences of mammalian CES, as well as the interactions between CES and ligands, which will facilitate the discovery and development of CES inhibitors for potential pharmaceutical applications.

Acknowledgments

We thank the financial support of the National Key Research and Development Program of China (2016YFC1303900, 2017YFC1700200, 2017YFC1702000), the National S&T Major Projects of China (2017ZX09101004), the NSF of China (81703604, 81773687, 21602219, 81573501 and 81473181), Program of Shanghai Academic/Technology Research Leader (18XD1403600), and the Innovative Entrepreneurship Program of High-level Talents in Dalian (2016RQ025 & 2017RQ121).

Conflict of Interest

The authors declare no conflict of interest, financial or otherwise.

List of Abbreviations

AchE: acetylcholinesterase

ADME/Tox: absorption, distribution, metabolism and excretion/toxicity

BchE: butyrylcholinesterase

BMBT: 2-(2-Hydroxy-3-methoxyphenyl) benzothiazole

BNPP: bis(4-nitrophenyl)phosphate

CES: carboxylesterases

CES1: human carboxylesterase 1

CES2: human carboxylesterase 2

CoA: acyl-coenzyme A

CPT-11: irinotecan

CRP: C-reactive protein

DMCB: 3,5-dimethyl-8-methoxycarbonyl-BODIPY
DME: D-luciferin methyl ester
DMT: deltamethrin
EL35: polyoxyl 35 castor oil
ER: endoplasmic reticulum
FAEEs: fatty acid ethyl esters
FD: fluorescein diacetate
GA: glycyrrhetinic acid
27-HC: 27-hydroxycholesterol
HLC: human liver cytosol
HLM: human liver microsomes
HTS: high throughput screening
MPN: 4-benzoyl-N-butyl-1,8-naphthalimide
OA: oleanolic acid
OPs: organophosphates
QSAR: quantitative structure-activity relationship
RH40: polyoxyl 40 hydrogenated castor oil
SARs: structure-activity relationships
SLS: sodium lauryl sulfate
SN38: 7-Ethyl-10-hydroxy-camptothecin
UA: ursolic acid
UGT: uridine diphosphateglucuronosyltransferase

References

Alam M., Ho S., Vance D. E., Lehner R., *Protein Expres. Purif.*, **24** (2002), 33–42.

Alberti K. G. M. M., Eckel R. H., Grundy S. M., Zimmet P. Z., Cleeman J. I., Donato K. A., Fruchart J. C., James W. P. T., Loria C. M., Smith, S. C., *Circulation*, **120** (2009), 1640–1645.

Alimonti A., Gelibter A., Pavese I., Satta F., Cognetti F., Ferretti G., Rasio D., Vecchione A., Di Palma M., *Cancer Treat. Rev.*, **30** (2004), 555–562.

Bao X. Z., Wei S. Q., Qian X. K., Qu J. P., Wang B. M., Zou L. W., Ge G. B. *Org. Lett.*, **20** (2018), 3394–3398.

Beckemeier M. E., Bora P. S., *J. Mol. Cell Cardiol.*, **30** (1998), 2487–2494.

Bencharit S., Morton C. L., Howard-Williams E. L., Danks M. K., Potter P. M., Redinbo M. R., *Nat. Struct. Biol.*, **9** (2002), 337–342.

Bencharit S., Morton C. L., Hyatt J. L., Kuhn P., Danks M. K., Potter P. M., Redinbo M. R., *Chem. Biol.*, **10** (2003a), 341–349.

Bencharit S., Morton C. L., Xue Y., Potter P. M., Redinbo M. R., *Nat. Struct. Biol.*, **10** (2003b), 577–577.

Bie J. H., Wang J., Marqueen K. E., Osborne R., Kakiyama G., Korzun W., Ghosh S. S., Ghosh S., *Arterioscl. Throm. Vas.*, **33** (2013), 1795–1802.

Binder R. J., Hatfield M. J., Chi L. Y., Potter, P. M., *Eur. J. Med. Chem.*, **149** (2018), 79–89.

Boger D. L., Sato H., Lerner A. E., Austin B. J., Patterson J. E., Patricelli M. P., Cravatt B. F., *Bioorg. Med. Chem. Lett.*, **9** (1999), 265–270.

Bora P. S., Guruge B. L., Miller D. D., Chaitman B. R., Ruyle M. S., *J. Mol. Cell Cardiol.*, **28** (1996), 2027–2032.

Brady K., Abeles R. H., *Biochemistry*, **29**(1990), 7608–7617.

Brzezinski M. R., Abraham T. L., Stone C. L., Dean R. A., Bosron W. F., *Biochem. Pharmacol.*, **48**(1994), 1747–1755.

Buchheit D., Dragan C. A., Schmitt E. I., Bureik, M., *Drug Metab. Dispos.*, **39** (2011), 2174–2181.

Callies O., Bedoya L. M., Beltran M., Munoz A., Calderon P. O., Osorio A. A., Jimenez I. A., Alcami J., Bazzocchi I. L., *J. Nat. Prod.*, **78** (2015), 1045–1055.

Casida J. E., Durkin. K. A., *Chem-Biol Interact.*, **203** (2013), 221–225.

Casida J. E., Quistad G. B., *Chem-Biol Interact.*, **157** (2005), 277–283.

Chang J. W., Cognetta A. B., Niphakis M. J., Cravatt B. F., *ACS Chem. Biol.*, **8** (2013), 1590–1599.

Crow J. A., Bittles V., Borazjani A., Potter P. M., Ross M. K., *Biochem. Pharmacol.*, **84** (2012), 1215–1222.

Crow J. A., Herring K. L., Xie S. Q., Borazjani A., Potter P. M., Ross M. K., *BBA-Mol. Cell Biol. L.*, **1801** (2010), 31–41.

Crow J. A., Middleton B. L., Borazjani A., Hatfield M. J., Potter P. M., Ross M. K., *BBA-Mol. Cell Biol. L.*, **1781** (2008), 643–654.

De Palma A., Rossi R., Carai M., Cabras C., Colombo G., Arnoldi L., Fuzzati N., Riva A., Morazzoni P., Mauri P. L., *Curr. Pharm. Anal.*, **4** (2008), 249–257.

Di Meo F., Steel M., Nicolas P., Marquet P., Duroux J. L., Trouillas P., *J. Mol. Model.*, **19** (2013), 2423–2432.

Diczfalusy M. A., Bjorkhem I., Einarsson C., Hillebrant C. G., Alexson S. E. H., *J. Lipid Res.*, **42** (2001), 1025–1032.

Ding L. L., Tian Z. H., Hou J., Weng Z. M., Cui J. N., Yang L., Ge G. B., *Acta Pharm. Sin.*, **52** (2017), 58–65.

Dominguez E., Galmozzi A., Chang J. W., Hsu K. L., Pawlak J., Li W. W., Godio C., Thomas J., Partida D., Niessen S., O'brien P. E., Russell A. P., Watt M. J., Nomura D. K., Cravatt B. F., Saez E., *Nat. Struct. Biol.*, **10** (2014), 113–121.

Faulds M. H., Dahlman-Wright K., *Curr. Opin. Oncol.*, **24** (2012), 58–61.

Feng L., Liu Z. M., Hou J., Lv X., Ning J., Ge G. B., Cui J. N., Yang L., *Biosens. Bioelectron.*, **65** (2015), 9–15.

Feng L., Liu Z. M., Xu L., Lv X., Ning J., Hou J., Ge G. B., Cui J. N., Yang L., *Chem. Commun.*, **50** (2014), 14519–14522.

Fleming C. D., Edwards C. C., Kirby S. D., Maxwell D. M., Potter P. M., Cerasoli D. M., Redinbo M. R., *Biochemistry*, **46** (2007), 5063–5071.

Fukami T., Takahashi S., Nakagawa N., Maruichi T., Nakajima M., Yokoi T., *Drug Metab. Dispos.*, **38** (2010), 2173–2178.

Furihata T., Hosokawa M., Koyano N., Nakamura T., Satoh T., Chiba K., *Drug Metab. Dispos.*, 32(2004), 1170–1177.

Gelb M. H., Svaren J. P., Abeles R. H., *Biochemistry*, **24** (2002), 1813–1817.

Hatfield M. J., Chen J. W., Fratt E. M., Chi L. Y., Bollinger J. C., Binder R. J., Bowling J., Hyatt J. L., Scarborough J., Jeffries C., Potter P. M., *J. Med. Chem.*, **60** (2017), 1568–1579.

Hatfield M. J., Potter P. M., *Expert Opin. Ther. Pat.*, **21** (2011), 1159–1171.

Hatfield M. J., Tsurkan L. G., Hyatt J. L., Edwards C. C., Lemoff A., Jeffries C., Yan B., Potter P. M., *J. Nat. Prod.*, **76** (2013), 36–44.

Heymann E., Mentlein R., Schmalz R., Schwabe C., Wagenmann F., *Eur. J. Biochem.*, **102** (1979), 509–519.

Hicks L. D., Hyatt J. L., Stoddard S., Tsurkan L., Edwards C. C., Wadkins R. M., Potter P. M., *J. Med. Chem.*, **52** (2009), 3742–3752.

Holmes R. S., Glenn J. P., Vandeberg J. L., Cox L. A., *J. Med. Primatol.*, **38** (2009), 27–38.

Hosokawa M., *Molecules,* **13** (2008), 412–431.

Hyatt J. L., Moak T., Hatfield M. J., Tsurkan L., Edwards C. C., Wierdl M., Danks M. K., Wadkins R. M., Potter P. M., *J. Med. Chem.*, **50** (2007a), 1876–1885.

Hyatt J. L., Stacy,V., Wadkins R. M., Yoon K. J. P., Wierdl M., Edwards C. C., Zeller M., Hunter A. D., Danks M. K., Crundwell G., Potter P. M., *J. Med. Chem.*, **48** (2005), 5543–5550.

Hyatt J. L., Wadkins R. M., Tsurkan L., Hicks L. D., Hatfield M. J., Edwards C. C., Ross C. R., Cantalupo S. A., Crundwell G., Danks M. K., Guy R. K., Potter P. M., *J. Med. Chem.*, **50** (2007b), 5727–5734.

Imai T., *Drug Metab. Pharmacok.*, **21** (2006), 173–185.

Imai T., Hosokawa M., *J. Pestic. Sci.*, **35** (2010), 229–239.

Imai T., Ohura K., *Curr. Drug Metab.*, **11** (2010), 793–805.

Imai T., Taketani M., Shii M., Hosokawa M., Chiba K., *Drug Metab. Dispos.*, **34** (2006), 1734–1741.

Inoue N. R., Hall A., Lai W. G., Williams E. T., *Drug Metab. Dispos.*, **41** (2013), 698–703.

Jin Q., Feng L., Wang D. D., Dai Z. R., Wang P., Zou L. W., Liu Z. H., Wang J. Y., Yu Y., Ge G. B., Cui J. N., Yang L., *ACS Appl. Mater. Inter.*, **7** (2015), 28474–28481.

Jin Q., Feng L., Wang D. D., Wu J. J., Hou J., Dai Z. R., Sun S. G., Wang J. Y., Ge G. B., Cui J. N., Yang L., *Biosens Bioelectron.* **83** (2016), 193–199.

Kathuria S., Gaetani S., Fegley D., Valino F., Duranti A., Tontini A., Mor M., Tarzia G., La Rana G., Calignano A., Giustino A., Tattoli M., Palmery M., Cuomo V., Piomelli D., *Nat. Med.*, **9** (2003), 76–81.

Kazakova O. B., Giniyatullina G. V., Tolstikov G. A., Kataev V. E., Musin R. Z., *Russ. J. Bioorg. Chem.*, **35** (2009), 645–650.

Kovach I. M., *J. Enzyme Inhib.*, **2** (1988), 199–208.

Lamego J., Cunha B., Peixoto C., Sousa M. F., Alves P. M., Simplicio A. L., Coroadinha A. S., *Appl. Microbiol. Biotechnol.*, **97** (2013), 1161–1173.

Lee I., Eriksson P., Fredriksson A., Buratovic S., Viberg H., *Toxicol. Appl. Pharm.*, **288** (2015), 429–438.

Lei W., Wang D. D., Dou T. Y., Hou J. E., Feng L., Yin H., Luo Q., Sun J. E., Ge G. B., Yang L., *Toxicol. Appl. Pharm.*, **321**(2017), 48–56.

Li Y. G., Hou J., Li S. Y., Lv X., Ning J., Wang P., Liu Z. M., Ge G. B., Ren J. Y., Yang L., *Fitoterapia*, **101** (2015), 99–106.

Li B., Sedlacek M., Manoharan I., Boopathy R., Duysen E. G., Masson P., Lockridge O., *Biochem. Pharmacol.*, **70** (2005), 1673–1684.

Li Y. Y., Zalzala M., Jadhav K., Xu Y., Kasumov T., Yin L. Y., Zhang Y. Q., *Hepatology*, **63** (2016), 1860–1874.

Lian J. H., Nelson R., Lehner R., *Protein Cell*, **9** (2018), 178–195.

Liu Z. M., Feng L., Ge G. B., Lv X., Hou J., Cao Y. F., Cui J. N., Yang L., *Biosens Bioelectron*, **57** (2014b), 30–35.

Liu Z. M., Feng L., Hou J., Lv X., Ning J., Ge G. B., Wang K. W., Cui J. N., Yang L., *Sens. Act. B: Chem.*, **205** (2014a), 151–157.

Liu Y. J., Li S. Y., Hou J., Liu Y. F., Wang D. D., Jiang Y. S., Ge G. B., Liang X. M., Yang L., *Fitoterapia*, **115** (2016), 57–63.

Long J. Z., Li W. W., Booker L., Burston J. J., Kinsey S. G., Schlosburg J. E., Pavon F. J., Serrano A. M., Selley D. E., Parsons L. H., Lichtman A. H., Cravatt B. F., *Nat. Chem. Biol.*, **5** (2009), 37–44.

Lv X., Wang D. D., Feng L., Wang P., Zou L. W., Hao D. C., Hou J., Cui J. N., Ge G. B., Yang, L., *RSC Adv.*, **6** (2016), 4302–4309.

Ma H. Y., Yang J. D., Hou J., Zou L. W., Jin Q., Hao D. C., Ning J., Ge G. B., Yang L., *Toxicol. in Vitro*, **44** (2017), 280–286.

Mai Z. P., Xin X. L., Sun Z., Zhang N., Huang S. S., Wang C., Chen L., Li Y., Huo X. K., Fan G. J., *Phytochem. Lett.*, **13** (2015a), 228–233.

Mai Z. P., Zhou K., Ge G. B., Wang C., Huo X. K., Dong P. P., Deng S., Zhang B. J., Zhang H. L., Huang S. S., Ma X. C., *J. Nat. Prod.*, **78** (2015b), 2372–2380.

Merali Z., Ross S., Pare G., *Drug Metabol. Drug Interact.*, **29** (2014), 143–151.

Nishi K., Huang H. Z., Kamita S. G., Kim I. H., Morisseau C., Hammock B. D., *Arch. Biochem. Biophys.*, **445** (2006), 115–123.

Nomura D. K., Durkin K. A., Chiang K. P., Quistad G. B., Cravatt B. F., Casida J. E., *Chem. Res. Toxicol.*, **19** (2006), 1142–1150.

Ojima I., Jameison F. A., Pete B., Radunz H., Schittenhelm C., Lindner H. J., Emith A. E., *Drug Des. Discov.*, **11** (1994), 91–113.

Parker R. B., Laizure S. C., *Drug Metab. Dispos.*, **38** (2010), 317–322.

Parkinson E. I., Hatfield M. J., Tsurkan L., Hyatt J. L., Edwards C. C., Hicks L. D., Yan B., Potter P. M., *Bioorg. Med. Chem.*, **19** (2011), 4635–4643.

Potter P. M., Wolverton J. S., Morton C. L., Wierdl M., Danks M. K., *Cancer Res.*, **58** (1998), 3627–3632.

Pratt S. E., Durland-Busbice S., Shepard R. L., Heinz-Taheny K., Iversen P. W., Dantzig A. H., *Clin. Cancer Res.*, **19** (2013), 1159–1168.

Prestwich G. D., Eng W. S., Roe R. M., Hammock B. D., *Arch. Biochem. Biophys.*, **228** (1984), 639–645.

Qadri S. S., Kashi K. P., *J. Econ. Entomol.*, **64** (1971), 1015–1018.

Quinney S. K., Sanghani S. P., Davis W. I., Hurley T. D., Sun Z., Murry D. J., Bosron W. F., *J. Pharmacol. Exp. Ther.*, **313** (2005), 1011–1016.

Redinbo M. R., Potter P. M., *Drug Discov. Today*, **10** (2005), 313–325.

Rhoades J. A., Peterson Y. K., Zhu H. J., Appel D. I., Peloquin C. A., Markowitz J. S., *Pharm. Res. Dordr.*, **29** (2012), 972–982.

Robbi M., Beaufay H., *J. Biol. Chem.*, **266** (1991), 20498–20503.

Rosell G., Herrero S., Guerrero A., *Biochem. Bioph. Res. Co.*, **226** (1996), 287–292.

Roy K. K., Tota S., Tripathi T., Chander S., Nath C., Saxena A. K., *Bioorg. Med. Chem.*, **20** (2012), 6313–6320.

Ruby M. A., Massart J., Hunerdosse D. M., Schonke M., Correia J. C., Louie S. M., Ruas J. L., Naslund E., Nomura D. K., Zierath J. R., *Cell Rep.*, **18** (2017), 636–646.

Sanghani S. P., Quinney S. K., Fredenburg T. B., Davis W. I., Murry D. J., Bosron W. F., *Drug Metab. Dispos.*, **32** (2004), 505–511.

Santos-Buelga C., Feliciano A. S., *Molecules*, **22** (2017), 477.

Sato Y., Miyashita A., Iwatsubo T., Usui T., *Drug Metab. Dispos.*, **40** (2012), 1389–1396.

Satoh T., Hosokawa M., *Ann. Rev. Pharmacol.*, **38** (1998), 257–288.

Satoh T., Hosokawa M., *Chem. Biol. Interact.*, **162** (2006), 195–211.

Satoh T., Taylor P., Bosron W. F., Sanghani S. P., Hosokawa M., La Du B. N., *Drug Metab. Dispos.*, **30** (2002), 488–493.

Scaloni A., Barra D., Jones W. M., Manning J. M., *J. Biol. Chem.*, **269** (1994), 15076–15084.

Scholz N. L., Truelove N. K., Labenia J. S., Baldwin D. H., Collier T. K., *Environ. Toxicol. Chem.*, **25** (2006), 1200–1207.

Shi D., Yang J., Yang D. F., Yan B., *Toxicology*, **254** (2008), 97–105.

Sun D. X., Ge G. B., Dong P. P., Cao Y. F., Fu Z. W., Ran R. X., Wu X., Zhang Y. Y., Hua H. M., Zhao Z. Y., Fang Z. Z., *Xenobiotica*, **46** (2016), 503–510.

Takahashi S., Katoh M., Saitoh T., Nakajima M., Yokoi T., *J. Pharm. Sci. US*, **97** (2008), 5434–5445.

Taketani M., Shii M., Ohura K., Ninomiya S., Imai T., *Life Sci.*, **81** (2007), 924–932.

Tang M., Mukundan M., Yang J., Charpentier N., Lecluyse E. L., Black C., Yang D. F., Shi D. S., Yan B. F., *J. Pharmacol. Exp. Ther.*, **319** (2006), 1467–1476.

Thomsen R., Rasmussen H. B., Linnet K., Consortium I., *Drug Metab. Dispos.*, **42** (2014), 126–133.

Tong H., Su Q., Zhou X., Bai L., *J. Pest. Sci.*, **86** (2013), 599–609.

Tsurkan L. G., Hatfield M. J., Edwards C. C., Hyatt J. L., Potter P. M., *Chem. Biol. Interact.*, **203** (2013), 226–230.

Umehara K. I., Zollinger M., Kigondu E., Witschi M., Juif C., Huth F., Schiller H., Chibale K., Camenisch G., *Xenobiotica*, **46** (2016), 862–867.

Wadkins R. M., Hyatt J. L., Edwards C. C., Tsurkan L., Redinbo M. R., Wheelock C. E., Jones P. D., Hammock B. D., Potter P. M., *Mol. Pharmacol.*, **71** (2007), 713–723.

Wadkins R. M., Hyatt J. L., Wei X., Yoon K. J. P., Wierdl M., Edwards C. C., Morton C. L., Obenauer J. C., Damodaran K., Beroza P., Danks M. K., Potter P. M., *J. Med. Chem.*, **48** (2005), 2906–2915.

Wadkins R. M., Hyatt J. L., Yoon K. J. P., Morton C. L., Lee R. E., Damodaran K., Beroza P., Danks M. K., Potter P. M., *Mol. Pharmacol.*, **65** (2004), 1336–1343.

Wadkins R. M., Morton C. L., Weeks J. K., Oliver L., Wierdl M., Danks M. K., Potter P. M., *Mol. Pharmacol.*, **60** (2001), 355–362.

Wang R., Borazjani A., Matthews A. T., Mangum L. C., Edelmann M. J., Ross M. K., *Biochemistry*, **52** (2013), 7559–7574.

Wang D. D., Jin Q., Zou L. W., Hou J., Lv X., Lei W., Cheng H. L., Ge G. B., Yang L., *Chem. Commun.*, **52** (2016), 3183–3186.

Wang J., Williams E. T., Bourgea J., Wong Y. N., Patten C. J., *Drug Metab. Dispos.*, **39** (2011), 1329–1333.

Wang D. D., Zou L. W., Jin Q., Hou J., Ge G. B., Yang L., *Fitoterapia*, **117** (2017), 84–95.

Wang D. D., Zou L. W., Jin Q., Hou J., Ge G. B., Yang L., *Acta Pharm. Sin. B*, **8** (2018), 699–712.

Wei Y. L., Peng A. Y., Huang J., *Chem. Biol. Interact.*, **204** (2013), 75–79.

Wei Y. L., Peng A. Y., Wang B., Ma L., Peng G. P., Du Y. D., Tang J. M., *Eur. J. Med. Chem.*, **74** (2014), 751–758.

Wen L. R., Jiang Y. M., Yang J. L., Zhao Y. P., Tian M. M., Yang B., *Ann. N. Y. Acad. Sci.*, **1398** (2017), 120–129.

Weng Z. M., Ge G. B., Dou T. Y., Wang P., Liu P. K., Tian X. H., Qiao N., Yu Y., Zou L. W., Zhou Q., Zhang W. D., Hou J., *Bioorg. Chem.*, **77** (2018), 320–329.

Williams E. T., Jones K. O., Ponsler G. D., Lowery S. M., Perkins E. J., Wrighton S. A., Ruterbories K. J., Kazui M., Farid N. A., *Drug Metab. Dispos.*, **36** (2008), 1227–1232.

Williams E. T., Wang H., Wrighton S. A., Qian Y. W., Perkins E. J., *Mol. Phylogenet. Evol.*, **57**(2010), 23–34.

Wu M. H., Chen P. X., Remo B. F., Cook E. H., Das S., Dolan M. E., *Pharmacogenetics*, **13** (2003), 425–435.

Xu Y. J., Zhang C. L., He W. X., Liu D., *Eur. J. Drug Metab. Ph.*, **41** (2016), 321–330.

Xu Y. J., Zhang C. L., Li X. P., Wu T., Ren X. H., Liu D., *Drug Metab. Pharmacok.*, **28** (2013), 468–474.

Xu G., Zhang W. H., Ma M. K., Mcleod H. L., *Clin. Cancer Res.*, **8** (2002), 2605–2611.

Yano H., Kayukawa S., Iida S., Nakagawa C., Oguri T., Sanda T., Ding J., Mori F., Ito A., Ri M., Inagaki A., Kusumoto S., Ishida T., Komatsu H., Inagaki H., Suzuki A., Ueda R., *Cancer Sci.*, **99** (2008), 2309–2314.

Yoon K. J. P., Hyatt J. L., Morton C. L., Lee R. E., Potter P. M., Danks M. K., *Mol. Cancer Ther.*, **3** (2004), 903–909.

Yoshida T., Fukami T., Kurokawa T., Gotoh S., Oda A., Nakajima M., *Eur. J. Pharm. Sci.*, **111** (2018), 167–176.

Yue C. C., Mullergreven J., Dailey P., Lozanski G., Anderson V., Macintyre S., *J. Biol. Chem.*, **271** (1996), 22245–22250.

Zebiri I., Haddad M., Duca L., Sauvain M., Paloque L., Cabanillas B., Rengifo E., Behr J. B., Voutquenne-Nazabadioko L., *Nat. Prod. Res.*, **31** (2017), 1333–1338.

Zhang Z. J., Huo X. K., Tian X. G., Feng L., Ning J., Zhao X. Y., Sun C. P., Wang C., Deng S., Zhang B. J., Zhang H. L., Liu Y., *Rsc Adv.*, **7** (2017), 28702–28710.

Zhang C. L., Xu Y. J., Zhong Q. N., Li X. P., Gao P., Feng C. Y., Chu Q., Chen Y., Liu D., *Plos One*, **9** (2014), c93819.

Zhen L. D., Baumann H., Novak E. K., Swank R. T., *Arch. Biochem. Biophys.*, **304** (1993), 402–414.

Zhou M., Zhang R. H., Wang M., Xu G. B., Liao S. G. *Eur. J. Med. Chem.*, **131** (2017), 222–236.

Zhu H. J., Markowitz J. S., *Drug Metab. Dispos.*, **37** (2009), 264–267.

Zhu H. J., Wang X. W., Gawronski B. E., Brinda B. J., Angiolillo D. J., Markowitz J. S., *J. Pharmacol. Exp. Ther.*, **344** (2013), 665–672.

Zou L. W., Dou T. Y., Wang P., Lei W., Weng Z. M., Hou J., Wang D. D., Fan Y. M., Zhang W. D., Ge G. B., Yang L., *Front. Pharmacol.*, **8** (2017), 435.

Zou L. W., Jin Q., Wang D. D., Qian Q. K., Hao D. C., Ge G. B., Yang L., *Curr. Med. Chem.*, **25** (2018), 1627–1649.

Zou L. W., Li Y. G., Wang P., Zhou K., Hou J., Jin Q., Hao D. C., Ge G. B., Yang L., *Eur. J. Med. Chem.*, **112** (2016), 280–288.

Chapter 10

Molecular Aspects of the Activity and Inhibition of the FAD-Containing Monoamine Oxidases

Rona R. Ramsay

University of St Andrews,
Biomolecular Sciences Building,
North Haugh, St Andrews KY16 9ST, UK

rrr@st-andrews.ac.uk

10.1 Introduction

The flavin-containing monoamine oxidases (MAO, E.C. 1.4.3.4) are redox enzymes found on the mitochondrial outer membrane in every animal cell. Their pharmaceutical importance is due to the fact that MAOs oxidize neurotransmitters such as serotonin and dopamine, changed levels of which are associated with behavioural changes, depression, addiction, aggression and neurodegeneration. High or low MAO activity is associated with pathology. For example, elevated MAO B is observed in neurodegenerative diseases such as Alzheimer's disease. Above

Pharmaceutical Biocatalysis: Fundamentals, Enzyme Inhibitors, and Enzymes in Health and Diseases
Edited by Peter Grunwald

ISBN 978-981-4800-61-7 (Hardcover), 978-0-429-29503-4 (eBook)
www.jennystanford.com

average activity of MAO A has been associated with major depression whereas low MAO A is associated with aggression.

MAO inhibitors (MAOI) are widely used as antidepressants, mood stabilizers, and to offset the effects of brain degenerative diseases such as Parkinson's disease. The social and economic impact of neurodegenerative diseases in society remains a good reason for exploring new inhibitors of MAO to elevate remaining neurotransmitters, particularly in multi-target drugs for these complex pathologies.

MAO catalysis was first characterized in 1928, but it was the observation of the antidepressant effect of MAOI that triggered the rise of interest in the enzymes in the 1950s (reviewed in Tipton (2018). Figure 10.1 illustrates the scientific interest in monoamine oxidases expanding during the pharmacological exploration of MAO inhibitors (MAOI) as drugs, mainly used as antidepressants, then beginning to decline in the late 1980s as pharmaceutical interest waned. The publication of the crystal structure of MAO B in 2002 and MAO A in 2005 renewed interest and enabled the modern era of drug design, engaging chemists in particular in discovery of new compounds using computational and high throughput methods (23% of publications from 2005).

The two monoamine oxidase isoenzymes, MAO A and MAO B, are encoded by genes on the X-chromosome (MAOA and MAOB) and are expressed in all tissues. In addition to their well-known role in the brain, peripheral MAOs especially in gut (MAO A), liver (both), platelet (MAO B) and placenta (MAO A) play significant roles in protection against biogenic amines and contribute to Phase 1 drug metabolism. In heart, hydrogen peroxide and aldehydes produced by MAO oxidation of amines may contribute to heart failure. These flavoprotein oxidases (E.C. 1.4.3.4) are located on the mitochondrial outer membrane, and therefore metabolize intra-cellular amines. In contrast, the copper-dependent primary amine oxidases (the soluble semicarbazide-sensitive amine oxidase (SSAO) and its membrane form, Vascular Adhesion Protein-1, E.C. 1.4.3.21) act on extra-cellular amines and function in inflammation (Becchi et al., 2017). SSAO, lysyl oxidase (1.4.3.13) and diamine oxidase (E.C. 1.4.3.22 which metabolizes histamine) contain 2,4,5-trihydroxyphenylalaninequinone as the cofactor (www.brenda-enzymes.org). These non-FAD enzymes will not be considered in this article. Other FAD–containing oxidases include the related amino acid oxidases, L-amino acid oxidase (LAAO, E.C. 1.4.3.2) and D-amino acid oxidase (DAAO, E.C. 1.4.3.3), and the spermine metabolizing enzyme, polyamine oxidase (E.C. 1.5.3.13), all of which have covalently bound FAD. A possible evolutionary precursor of MAO is found in the peroxisomes of Aspergillus niger and has a non-covalently bound FAD co-factor (Sablin et al., 1998) making it interesting for chemical applications (Bailey et al., 2007).

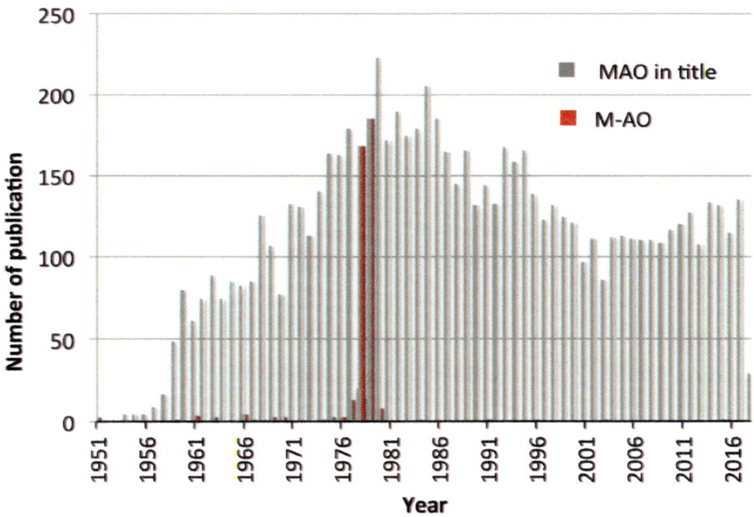

Figure 10.1 Publications with monoamine oxidase in the title. Note the use of mono-amine oxidase in 1978–79 (compiled from Web of Science, May 2018).

The reaction catalysed by MAO removes two hydrogen atoms from the amine forming an imine, and reduces FAD. The reduced FAD is then reoxidized by oxygen to form hydrogen peroxide. The imine product is hydrolysed by water giving the corresponding aldehyde and ammonia (Fig. 10.2). This chapter will consider the structure of the enzyme and the redox reaction with the cofactor and before turning to the substrate specificity, kinetics and mechanism that are important for drug design. MAO A and MAO B show broad specificity for both substrates and inhibitors, yet can display exquisite selectivity as will be discussed in the last section on inhibitors.

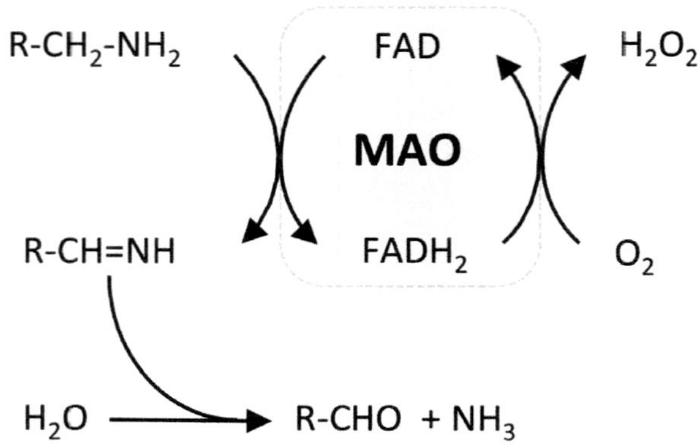

Figure 10.2 The redox reaction catalysed by MAO is followed by spontaneous hydrolysis of the imine product to give the aldehyde and ammonia. The $FADH_2$ is re-oxidized by oxygen to give the second product, hydrogen peroxide.

10.2 FAD: The Catalytic Cofactor

10.2.1 FAD Is Covalently Attached to MAO

FAD, a redox cofactor important for catalysis in many enzymes, is derived from the vitamin riboflavin. As in most FAD-containing enzymes, the adenine part of the molecule binds to a Rossman fold in the protein. The isoalloxazine ring part of FAD is held in position by a covalently bond to a cysteine residue in MAO (Fig. 10.3). In all crystal structures of MAO B, the isoalloxazine ring is bent by about 30° (Binda et al., 2003) with about 0.3° difference between the oxidized and reduced forms. However, molecular dynamics has now enabled a more realistic picture of the shape of the flavin in the flexible protein active site. Two different studies have shown that it is almost planar in the oxidized form (FAD) (Vianello et al., 2012; Zapata-Torres et al., 2015), but when it is reduced either by a hydride ion from the substrate transferred to N5 in the first step of the catalytic mechanism or after the propargylamine adduct has formed at N5, the $FADH_2$ is bent by almost 30 degrees (Borstnar et al., 2011) in agreement with the crystal structure.

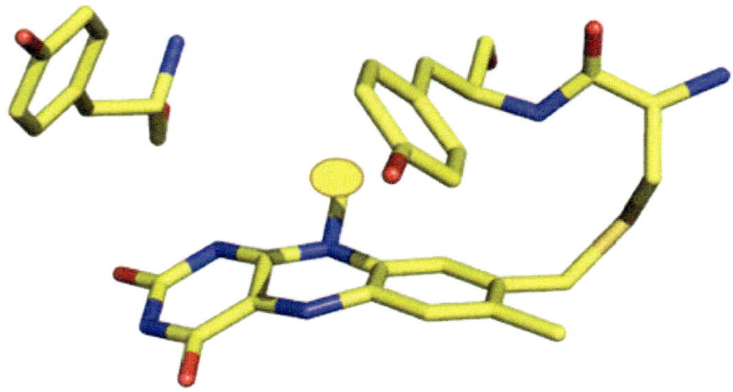

Figure 10.3 The 8-α-*S*-cysteinyl-isoalloxazine ring of FAD in the active site of MAO. The yellow circle at N10 indicates the rest of the FAD molecule. In MAO B (pdb: 1GOS), the flavin is bonded to Cys397 and the two tyrosines that are key to the aromatic cage for the substrate are Tyr398 and Tyr 435. The N5 (labelled) is involved in catalysis of the redox reaction.

Covalent attachment (at the C8 position of FAD in MAO), bending, and the surrounding protein groups alter the redox potential in flavoproteins. The two-electron redox potential for FAD in solution is −0.23 V and is the similar for the cysteinyl-FAD in MAO (about −0.2 V). The redox potential for amine oxidation is greater than +1 V, so for amine to reduce MAO requires a large adjustment from binding to the protein. $FADH_2$ readily reacts with oxygen producing hydrogen peroxide (H_2O_2) (+0.3 V). In amine oxidases, the reoxidation is thought to proceed via a C4a hydroperoxy intermediate (Mattevi, 2006). Although the catalytic reaction is a two-electron reduction, with hydride transfer the most likely mechanism, FAD is a versatile one or two electron acceptor. When MAO is reduced by dithionite in the presence of mediator dyes, inhibitors stabilize the anionic semiquinone form (Hynson et al., 2004).

10.2.2 FAD Is Modified by Irreversible Inhibitors

The major pharmaceutical drugs used to inhibit MAO modify the FAD irreversibly. Phenelzine (Nardil), tranylcypromine, and deprenyl (Selegiline) represent three classes of MAO inactivators, all of which form a covalent bond to the FAD as shown in Fig. 10.4. Phenelzine and tranylcypromine are successful antidepressants achieving about 50% response rates, slightly better than the tricyclic antidepressants (Nutt, 2006). Deprenyl is now used in both oral and patch formulations both to delay the need for L-DOPA and as adjunct therapy in Parkinson's disease. Although the structures of these MAO-inhibitor adducts were characterized chemically and by the crystal structures, the mechanisms of adduct formation remain incompletely understood. Inhibition of MAO will be discussed in more detail below.

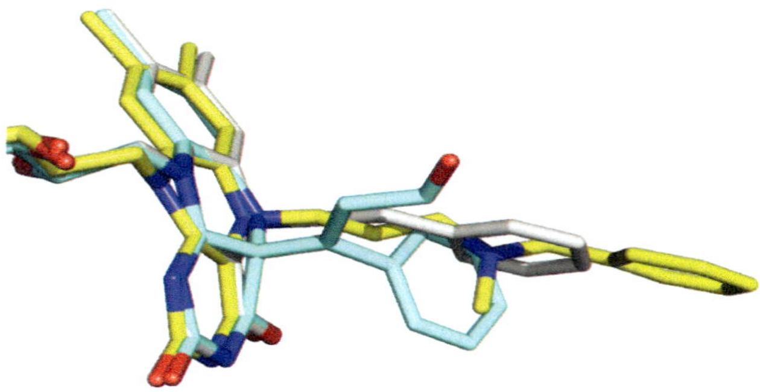

Figure 10.4 Irreversible MAO inhibitor drugs form flavin adducts. The figure was prepared in MAcPyMol with the protein main chains aligned. The adducts shown are MAO B with phenelzine (grey, 2VRM) and deprenyl (1GOS) bonded to N5, but tranylcypromine (cyan, 2XFU) bonded at C4a of the FAD.

10.3 MAO Proteins

10.3.1 MAO Protein Expression and Activity

Abnormal MAO activity has long been associated with neuropathology and behavioural changes (Finberg and Rabey, 2016). Humans with low MAO A activity and MAO A knockout mice display aggression (Bortolato and Shih, 2011). Low platelet MAO B is associated with alcoholism and behavioural disinhibition (van Amsterdam et al., 2006; Bortolato and Shih, 2011).

Expression of MAO A and B differs by cell type and they are regulated differently. The upstream 30 base pair VTNR polymorphism in the MAO A promotor with 3 or 5 repeats results in as much as a 10-fold lower expression compared to the normal 4 repeats (Sabol et al., 1998), and a downstream VTNR has been reported to decrease expression of a novel isoform (Manca et al., 2018). Both MAO A and B have similar GC-rich promoter regions activated via Sp1 sites. Both promotors are down-regulated by the transcription factor R1 but MAO A expression is increased via the Glucagon Receptor Element and MAO B expression is increased by retinoic acid (Shih et al., 2011).

Epigenetic regulation also plays a role and may provide the link with environmental influence on behaviour. One study demonstrated that MAOA promoter hypermethylation was associated with antisocial personality disorder and with high serotonin levels via down-regulation of MAO A gene expression (Checknita et al., 2015). Epigenetic influence was explored using allelic expression imbalance (AEI) of two single nucleotide polymorphisms in female human brain tissue. AEI ratios varied from 0.3 to 4 in prefrontal cortex samples indicating transcription-level regulation of mRNA production. CpG methylation in the MAOA promoter region was high in females and the extent of allelic MAOA methylation correlated with AEI ratio, showing that CpG methylation regulates gene expression (Pinsonneault et al., 2006). Using molecular biology combined with Positron Emission Tomography (PET) to measure regional MAO activity, a robust association of CpG site-specific methylation of the MAOA promoter with regional brain MAO A levels was demonstrated, strongly supporting the notion that epigenetic status (which can be influenced by environment, such as smoking) contributes to inter-individual differences in behaviour (Shumay et al., 2012). The activity of MAO in human brain studied by PET has confirmed earlier studies that recovery of MAO activity after inhibition by irreversible inhibitors such as deprenyl is slow, with a half-life of 40 days in the human brain, because it depends on the replacement of the inactive adduct enzyme with newly synthesized protein. PET has also proved useful for dosing studies, indicating that more than 75% occupancy of MAO active sites is required for clinical efficacy (reviewed in Fowler et al. (2005)).

10.3.2 Structures and Active Sites

The first purified MAO B was extracted from beef liver (Salach, 1979) and MAO A from human placenta. The first large scale heterologous expression of MAO A in *S. cerevisiae* provided pure enzyme for kinetic studies (Weyler et al., 1990), but it was high-level expression in *Pichia pastoris* (Newton-Vinson et al., 2000; Li et al., 2002) that enabled crystallization first of human MAO B protein, and then of rat and human MAO A (Binda et al., 2002; Ma et al., 2004; De Colibus et al., 2005; Son et al., 2008). The terminal hydrophobic helix of each of the two subunits of MAO is monotopically inserted in the mitochondrial outer membrane and further residues associate with the surface of the membrane, so that high amounts of detergent are required to solubilize the protein. The active site details from the crystal structures opened the way for structure-based drug design.

Comparison of the rat and human MAO structures revealed minor differences reflecting the reported species differences in substrate and inhibitor specificity (Krueger et al., 1995; Hubalek et al., 2005; Ramadan et al., 2007). In all structures, the catalytic FAD is at the end of a tunnel leading from the surface of the protein. The tunnel is generally hydrophobic, ending in an aromatic cage near the flavin where tyrosines (Tyr398 and Tyr435 in MAO B) align the neutral amine substrate towards the C4-N5 region of the flavin. The flavin is bent at about 30° along the N5-N10 axis favouring reduction as shown in Fig. 10.3, and there is a lysine-water-flavin(N5) motif important for the redox reactions of the flavin (Binda et al., 2002).

The two MAOs share 70% sequence identity but the active sites have subtle differences. In particular, MAO B has a constriction in the middle of the cavity due to residues isoleucine 199 and tyrosine 326 and MAO B has one cysteine (Cys172) not present in the MAO A active site (Edmondson et al., 2007; Milczek et al., 2011; Orru et al., 2013). The differences allow discovery or design of compounds for selective inhibition, as will be discussed in the inhibition section.

10.3.3 MAO Chemical Mechanism

Supported by extensive molecular model chemistry and recent thermodynamic calculations, three mechanisms have been considered for MAO catalysis: the polar nucleophilic mechanism (Orru et al., 2013), the radical (Silverman, 1995), and the hydride transfer mechanism (Kay et al., 2007). The hydride transfer mechanism (Fig. 10.5) is found in other oxidases and is supported by theoretical studies based on transition state theory (Vianello et al., 2016). However, the versatility of flavin catalysis could allow alternative mechanisms with specific chemicals, for example with the cyclopropylamines.

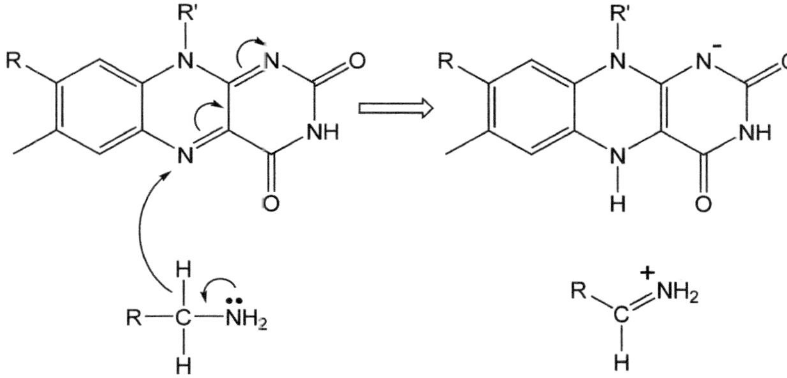

Figure 10.5 Hydride transfer is a probable mechanism for amine oxidation by MAO.

10.3.4 Two Substrate Kinetics and Consequences for Amine Turnover

MAO catalyses the oxidation of amines by molecular oxygen. Depending on the conditions used, this two-substrate enzyme can reveal either ping-pong kinetics or a ternary complex kinetic mechanism in which a new amine substrate molecule binds to the reduced MAO before reaction with oxygen (Tan and Ramsay, 1993). Fig. 10.6 illustrates the three possible re-oxidation pathways.

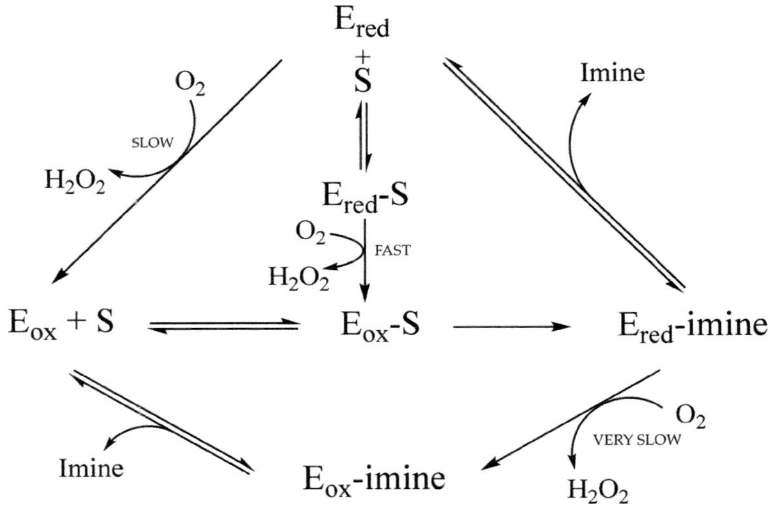

Figure 10.6 The catalytic cycle in MAO showing the alternative pathways for re-oxidation of FAD.

Both steady-state and stopped flow kinetics have been used to establish that ligands (substrate or inhibitor) can bind to either the oxidized or reduced forms of the enzyme, albeit with very different affinities (Tan and Ramsay, 1993; Ramsay et al., 2011). The proportion of MAO in the reduced state will vary depending on the amine concentration, the rates of reduction and oxidation of the flavin, and on the oxygen levels. Thus, the turnover of amines in vivo will be influenced by the local concentration of oxygen and by the redox state of the enzyme in the presence of the predominant substrate (Ramsay, 1998; Sablin and Ramsay, 2001). For MAO A, the rates of the reductive half reaction (k_{red}) are all similar to the steady-state rate (k_{cat}), and the oxidative half reaction in the presence of amine is faster (k_{ox}). For MAO B, the rate of amine oxidation can be faster than k_{cat}. For some substrates, notably phenylethylamine (PEA), the oxidative half-reaction determines the rate of turnover (Table 10.1). These rates mean that MAO A oxidizing PEA will be predominantly oxidized but MAO B will be fully reduced in the steady state. Another important difference between MAO A and B is the K_M for oxygen. MAO A continues working at low oxygen concentration (K_M = 0.06 mM), whereas

MAO B has a K_M close the concentration of dissolved oxygen (0.28 mM), so that the rate of MAO B catalysis will decrease with the oxygen concentration in the cell.

Table 10.1 Rates for turnover (k_{cat}), reductive half-reaction (k_{red}), and the oxidative half reaction in the presence of amine (k_{ox}) in purified MAO A and MAO B

	MAO A rates (min^{-1})			MAO B rates (min^{-1})		
Amine	k_{cat}	k_{red}	k_{ox}	k_{cat}	k_{red}	k_{ox}
5HT	2.8	2.1	6	0.08	0.01	2
PEA	0.75	1.10	12	3.6	572	2
MPTB	0.20	0.20	40	0.16	0.12	6
No amine			1			1

Note: Data from Tan and Ramsay (1993).

10.4 Substrate Specificity of these Promiscuous Enzymes

10.4.1 Neurotransmitter Metabolism

Substrate specificity depends not only on the k_{cat} discussed above but also on the recognition of the substrate by the enzyme, measured in kinetic terms by K_M. For example, the K_M for serotonin in rat brain homogenates is 0.18 mM with MAO A but is 1.2 mM with MAO B so clearly serotonin concentrations would have to be very high before MAO B contributed much to its metabolism. Comparison of specificity is best informed by k_{cat}/K_M alongside the expected or measured concentrations in the cell. Some values for k_{cat}/K_M in rat brain are given in Table 10.2. Note that the location of MAO on the mitochondrial surface means that it is the intracellular concentration of amine that is important rather than the synaptic concentration. Reuptake systems bring the extracellular amine into the presynaptic nerve terminal where mitochondria are clustered and the storage vesicles are also located. The vesicular reuptake system (VMAT2) with affinity for the monoamine neurotransmitters of 1.4 μM for dopamine and 0.9 μM for serotonin (Erickson et al.,

1996) efficiently clears dopamine to leave cytosolic levels below 50 nM (Mosharov et al., 2006; Mosharov et al., 2009). The MAO that predominates in dopaminergic neurons is MAO A with a K_M for dopamine of 212 μM. In serotonergic cells, MAO B predominates with a K_M for serotonin of 1 mM (Youdim et al., 2006). Thus most of the neurotransmitters will be returned to storage vesicles and only a very little metabolized. However, inhibition of MAO B in particular raises the overall content of amines in the brain. MAO B is the major form in all the non-neuronal cells and has been reported to increase with aging and in Alzheimer's disease, making it an attractive pharmaceutical target to help maintain global levels of amines outside the synapses in deteriorating brain. The importance of MAO for metabolism of neurotransmitter is apparent in serotonin toxicity where excessive levels of extracellular serotonin accumulate. It occurs only when a serotonin reuptake inhibitor is co-administered with a non-selective irreversible MAOI (Gillman, 2011).

Table 10.2 Relative rates of amine oxidation by rat MAO A and MAO B, calculated from Fowler and Tipton (1984)

Substrate	Catalytic constant (V/K_M) $\mu mol \cdot min^{-1} \cdot mg^{-1} \cdot mM^{-1}$	
	MAO A	MAO B
Serotonin	540	7
Tyramine	309	113
Phenylethylamine	46	4350
Dopamine	231	232
Noradrenaline	232	143

10.4.2 Metabolism of Biogenic Amines

MAO enzymes in the periphery are important for metabolism of exogenous amine compounds. Ingested amines have variable proportions present as the neutral compound that can diffuse cross membranes. Where there is a pH difference, such as between cytosol and the mitochondrial matrix, reprotonation results in accumulation of the charged species in the new compartment. The mitochondrial transmembrane potential also favours retention of positively charged amines. Biogenic amines from the diet, principally tyramine, are extensively metabolized in the intestinal wall by MAO A (Anderson et al., 1993). If tyramine increases in the circulation, complex sympathetic nervous system perturbations ensue, with consequent vasoconstriction and bradycardia (reviewed in Finberg and Gillman (2011)). Tyramine accumulation results in this "cheese effect" when MAO A in the gut and MAO A and MAO B in the liver are inhibited by non-selective drugs such as tranylcypromine or phenelzine. The tyramine-induced cheese effect is avoided by using selective MAO B inhibitors such as selegiline in adjunct therapy of Parkinson's disease or reversible inhibitors of MAO A in depression (Lum and Stahl, 2012).

A wide variety of amine-containing drugs are also metabolized by MAO, contributing 1% of the enzymatic metabolism of marketed drugs (compared to 95% by the P450 family) (Rendic and Guengerich, 2015). The biotransformation of drugs and other xenobiotics by the amine oxidases in contrast to the P450 family has been reviewed (Benedetti et al., 2001). Metabolism by MAO can also transform a compound into a toxic product. A well-known example is the four electron oxidation of l-methyl-4-phenyl-l,2,3,6-tetrahydropyridine (MPTP) by MAO to 1-methyl-4-phenylpyridinium (MPP^+). The toxic MPP^+ is taken up into dopaminergic cells in the substantia nigra and is accumulated into mitochondria where it inhibits NADH oxidation by the respiratory chain, hence preventing ATP production and consequent dopaminergic cell death leading to Parkinson's disease (Singer and Ramsay, 1990). To avoid such adverse side effects in development of new compounds, computational methods for the prediction of probable metabolic products has been developed based on databases of approved drugs with well-characterized metabolic profiles (Kar and Leszczynski, 2017).

10.4.3 Products from MAO Catalysis

All amine substrates are oxidized by MAO to the imine which is hydrolysed non-enzymically to the corresponding aldehyde (Fig. 10.1). Aldehydes are somewhat reactive molecules that can damage biological molecules by nucleophilic addition. In the cell, aldehyde dehydrogenase metabolizes the aldehyde to the carboxylic acid. The second product of MAO catalysis is hydrogen peroxide (H_2O_2). The toxic effect of MAO-generated H_2O_2 was clearly demonstrated for serotonin metabolism in heart (Pena-Silva et al., 2009). Decreased generation of H_2O_2 in the brain may contribute to the beneficial effects of selegiline.

10.5 Inhibition of MAO

The pharmacological exploitation of MAO inhibitors as drugs began with the serendipitous discovery that the tubercular drug iproniazid had antidepressant effects (reviewed in Tipton, 2018). Despite the cheese effect of the non-selective irreversible inhibitors, the most effective MAO inhibition is still achieved by irreversible inhibition. Drugs such as pargyline and tranylcypromine are underused in psychiatry, yet the reversible MAO A-selective antidepressant drug, moclobemide, works only in some cases of depression. MAO B-selective reversible inhibitors have been slow to appear but safinamide (K_i for MAO B 450 nM) has now been approved by the FDA for use as adjunct therapy in Parkinson's disease. The therapeutic potential of MAOI is clearly recognized (Youdim et al., 2006), with recent efforts yielding patents for new reversible inhibitors or for inhibitors that also target other features of degenerative disease (multi-target compounds) as will be discussed below.

The definitions used for assessment of reversible and irreversible inhibitors discussed here are given in Table 10.3. The best pharmacological approach for MAO inhibition is not clear. Proven history of over 50 years has established the effectiveness of irreversible drugs such as selegiline. Extensive human studies in vivo over the last 20 years using PET have established that 75–80% inhibition of MAO is required before beneficial effects are seen, emphasizing the spare capacity in this enzyme (Fowler et al., 2015a). After irreversible adduct formation the in vivo inhibition is long lasting, with slow re-synthesis of MAO required to replace the inactivated enzyme. Selective irreversible inhibitors of MAO B (for example, selegiline) now have a proven safety record, improved by delivery in transdermal patches to avoid major inhibition of MAO B in the liver. Reversible inhibitors require constant dosing, and even they are not completely free of the cheese effect (Lum and Stahl, 2012). The inhibition of MAO increases brain neuroamines, but the subsequent downstream changes in receptors and heterologous effects on other neurotransmitter systems are slow. Thus, the prolonged effect of irreversible inhibition may be more suited to this set of events. The sections below consider key reversible inhibitors, effective tight-binding inhibitors and irreversible inhibitors in turn before looking briefly at new computational approaches to drug discovery.

Table 10.3 Reversible and irreversible inhibition of MAO

Type	Reaction	Measure	Comment
Reversible	$E + I \underset{k_{-1}}{\overset{k_{+1}}{\rightleftharpoons}} E{\cdot}I$	$K_i = k_{-1}/k_{+1}$	Reversible binding can be at equilibrium, or slow, or tight, depending on rates
Irreversible (mechanism-based)	$E + I \underset{k_{-1}}{\overset{k_{+1}}{\rightleftharpoons}} E{\cdot}I \xrightarrow{k_{+3}} E\text{-}I$	K_I and k_{inact}	Depends on binding, catalytic conversion of I, and then chemical reaction

10.5.1 Reversible Inhibitors of MAO

The reluctance to prescribe irreversible inhibitors for neuoro-pathology because of the cheese effect has supported extensive exploration of new scaffolds by medicinal chemists. Aspects of the reversible inhibition from structure-activity studies were reviewed before the publication of the crystal structure of the enzymes (Wouters, 1998). More recently, many compounds, mostly with common features of at least one aromatic ring and a nitrogen atom, and having a relatively linear structure, have been discovered through data mining or by structure-based design. A recent review pulls together the many chemical entities that can inhibit reversibly and selectively MAO A or MAO B (Tripathi et al., 2018). Of particular interest are the various derivatives based on the coumarin scaffold, for use as multi-target compounds against neurodegeneration (reviewed in (Stefanachi et al., 2018)). Many medicinal chemistry groups are aiming to develop other MAO-A and MAO-B reversible inhibitors, only some of which reach the stage of pharmacology (Finberg and Rabey, 2016). In light of the growing scale of neurodegenerative disease, promising compounds are promptly patented, and patent updates appear regularly, such as one surveying patents of clinical interest filed in 2015–2017 (Carradori et al., 2018). These patents include 108 compounds, both synthetic and from natural sources, including coumarins, pyrazoles, and arylamides.

The initial experimental assessment of novel reversible inhibitors is usually done by determining the IC_{50} values of a series of chemicals. The two-substrate, alternative pathway kinetic behaviour described above means that great care is required to set conditions that allow measurement of competition with the amine substrate: i.e. that the amine substrate is rate limiting. For comparison of inhibition of MAO A and MAO B, the fractional saturation with amine substrate should be the same (usually 2 or 3 times K_M). Further details can be found elsewhere (Ramsay et al., 2011; Ramsay and Tipton, 2017). Compounds of interest are then characterized to confirm reversibility and identify mode of inhibition.

Most inhibitors are competitive with the amine substrate, as can be understood by inspection of the active site: both substrate and inhibitor bind in the same place. However, apparent deviation from competitive inhibition is sometimes found for MAO B. The MAO B K_M for oxygen of 0.23 mM is close to the concentration of oxygen in air-saturated buffer at 37°C. The deviations can arise when the conditions chosen result in both oxidized and reduced MAO B being present in the steady state, so that the affinity of the compound for both forms influences the shape and concentration-dependence of the dose-response curve (Ramsay et al., 2011).

The one clear example of mixed inhibition that has been fully established is for the imidazoline compound 2-(2-benzofuranyl)-2-imidazoline (2-BFI). The 2-BFI binds in the entrance cavity of MAO B when the cavity is in a particular conformation that can be stabilized by tranylcypromine bound in the catalytic site (Bonivento et al., 2010; McDonald et al., 2010). The I2-type imidazoline compounds bind to a subset of MAO molecules with nanomolar affinity but inhibition of the bulk population of MAO requires micromolar concentrations (Jones et al., 2007; McDonald et al., 2010). The significance of imidazoline binding to MAO in the cell is not clear.

10.5.2 Examples of Tight Binding Reversible Inhibitors of MAO

The rate of binding of harmine to MAO measured by rapid scan stopped-flow spectrometry was 2×10^8 M^{-1} s^{-1}, close to diffusion-limited rate. Yet the onset of inhibition in a steady-state assay required almost 2 min. Many inhibition studies preincubate the ligand and enzyme for 5–30 min before adding substrate to start the reaction. This allows equilibrium of the inhibitor in the absence of substrate (binding affinity) but if the off-rate is slow, curvature of the time course in the presence of substrate will be observed. Thus, continuous assays with or without preincubation can give additional information. The harmine on-rate is fast, so its off-rate must be slower to give the tight-binding K_i of 5 nM (Kim et al., 1997).

The thionine dye, methylene blue (MB), also binds tightly to MAO A. Clinical uses (surgical, anti-malarial) and preclinical information on MB have recently been reviewed (Delport et al., 2017). MB inhibits nitric oxide synthase, guanyl cyclase and MAO A, the latter with a K_i of 28 nM (Table 10.4). Its potency as an MAO inhibitor in vivo became evident as a result of serotonin syndrome, sometime fatal when it was administered during surgery to patients taking serotonin reuptake inhibitor antidepressants (Gillman, 2006). MB also acts as an electron donor and acceptor with MAO A (Ramsay et al., 2007) as it does in mitochondria (Delport et al., 2017).

Safinamide binds tightly to MAO B but also blocks voltage-gated sodium and calcium channels, and inhibits glutamate release in rat hippocampal synaptosomes (Caccia et al., 2006; Stocchi et al., 2006). Safinamide, a selective reversible inhibitor of MAO B (Table 10.4; Binda et al., 2007), was used to explore the structural effect of the two residues specific to the active site of MAO B. Isoleucine 119 and tyrosine 326 constrict the cavity of MAO B into entrance and catalytic parts. Long ligands such as safinamide span this gate, pushing it open and flexing around the constriction. MAO A has a wider active site without this constriction so that safinamide binds poorly (K_i = 365 uM, Table 10.4). The strong inhibition of MAO B by safinamide was increased from 450 nM to 21 nM after the mutation to alanine of Ile199, one of the gating residues between the entrance and catalytic cavities of MAO B. However, in the double mutant, Ile199Ala-Tyr326Ala, affinity was much less (K_i = 4 μM (Milczek et al., 2011)).

Table 10.4 Examples of drugs that inhibit human MAO A and MAO B

Drug	Indication	MAO A *Ki* (μM)	MAO B *Ki* (μM)	Reference
Reversible				
D-Amphetamine	Controlled	18	250	Ramsay and Hunter, 2002
Harmine	Endogenous	0.005		Kim et al., 1997
			121[a]	Reniers et al., 2011
Methylene Blue	Trial AD	0.028	5.5	Ramsay et al., 2007
Moclobemide	Depression	11.5	>100	Da Prada et al., 1989; Curet et al., 1996
Pirlindole	Depression	0.045	>100	Medvedev et al., 1999
Safinamide	PD adjunct	365	0.45	Binda et al., 2007
Irreversible				
		IC50[b] (μM)	IC50[b] (μM)	
Clorgyline	Imaging	0.0004	3.57	Malcomson et al., 2015
Deprenyl (Selegiline)	PD adjunct	16.9	0.004	Unpublished[c]
Pargyline	Depression	2.47	0.0077	Unpublished[c]
Tranylcypromine	Depression	0.237	0.074	Malcomson et al., 2015
Phenelzine	Depression	4.5	9	Binda et al., 2008
Isocarboxid[d]	Depression	0.3	0.18	Da Prada et al., 1989

[a]Estimate from invalid use of Cheng–Prusoff equation.

[b]IC_{50} values from different labs show considerable variation.

[c]In the unpublished data, activity was measured, using 1 mM tyramine as substrate, after 30 min pre-incubation of MAO with the inhibitor (P. L. Joffrin (2016), Honours Thesis, University of St Andrews).

[d]Rat brain—the reference also includes values for some of the other irreversible inhibitors in rat brain homogenates.

10.5.3 Examples of Irreversible Inhibitors of MAO

Table 10.4 lists the irreversible inhibitors of MAO in clinical use. The three major classes of drugs that form covalent adducts with MAO are the older hydrazines (phenelzine, isocarboxazid), a cyclopropylamine (tranylcypromine), and the propargylamines (pargyline, clorgyline, deprenyl). The general scheme in Fig. 10.7 shows initial enzyme-catalysed oxidation of the inhibitor to an activated product that either decays to product or reacts with the enzyme (*N*5 of the FAD for propargylamines) to form a covalent adduct. This section will discuss the molecular interactions of the three classes of compounds rather than the extensive pharmacological data from rat studies or from the clinic.

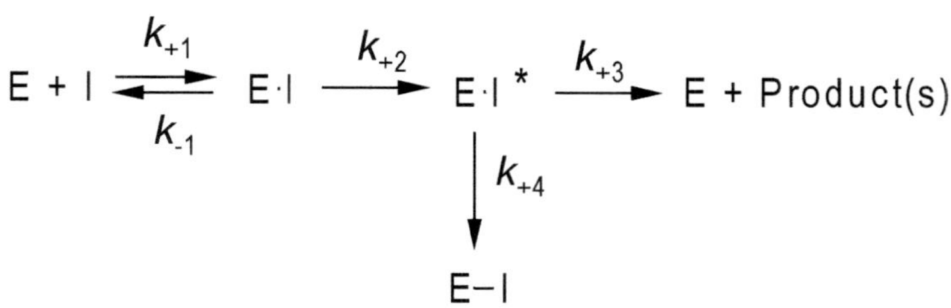

Figure 10.7 Mechanism-based inhibition of MAO: the reactive product can either dissociate or react to form a covalent adduct.

The original discovery of the antidepressant effect of hydrazines came from the observation of mood elevation in tuberculosis patients treated with isoniazid. Hydrazines were then characterized as effective MAO inhibitors (Zeller et al., 1955). Both phenelzine and isocarboxazid carry the risk of liver toxicity but remain useful for treatment-resistant depression. The reversible inhibition of MAO by phenelzine is poor (K_i is 47 μM for MAO A, 15 μM for MAO B (Binda et al., 2008), as it is in rat and pig brain (Tipton and Spires, 1971). Inactivation requires that the hydrazine act first as a substrate. The normal product from the oxidation of the hydrazine is an imine that is hydrolysed to the aldehyde and released as usual with concomitant H_2O_2 production. Alternatively, the product could be a diazene that reacts with oxygen giving a radical that could alkylate the flavin. Inactivation of MAO A and of MAO B follows the same O_2-dependent mechanism with some turnover for each inactivation event (the partition ratio). The kinetics of inactivation give k_{inact}/K_{inact} equal to 18 $min^{-1}M^{-1}$ for MAO A and 3 $min^{-1}M^{-1}$ for MAO, B both with a partition ratio of turnover to inactivation of about 40 (Binda et al., 2008). The crystal structure shows that the adduct formed is at *N*5 of the FAD (Fig. 10.4).

Tranylcypromine (TCP) inactivation of MAO has been studied in depth using a variety of cyclopropylamine derivatives. TCP itself modifies the flavin at the *C*4a position (Fig. 10.4) but *N*-cyclo-α-methylbenzylamine modifies a cysteine. Chemical and mechanistic studies of the inactivation support a radical mechanism in which one electron passes from the substrate amino group to the oxidized flavin to give the amine radical cation and the flavosemiquinone, followed by opening of the cyclopropyl ring and adduct formation (Silverman, 1995). Where adduct formation does not occur, further oxidation gives the iminium ion which can either leave the active site or react with nearby cysteines. Reversible inhibition by TCP is weak at around 100 μM for both MAO A and B (Table 10.5) (Bonivento et al., 2010; Malcomson et al., 2015). Inactivation with low partition ratios in both forms, is marginally better in MAO B (k_{inact}/K_{inact} is 0.2 compared to 0.1 $min^{-1}\cdot\mu M^{-1}$ in MAO A). Interestingly, in MAO B the covalently bound TCP stabilizes the entrance cavity residues to favour binding of 2-BFI, an imidazoline site ligand (Bonivento et al., 2010; McDonald et al., 2010). The clinical pharmacology for tranylcypromine use in depression has recently been reviewed in a meta-analysis which found that, when administered with dietary restriction of tyramine, it is a safe and effective antidepressant (Ricken et al., 2017).

Table 10.5 Kinetic parameters for initial binding and for inactivation of MAO by irreversible inhibitors

	MAO A			MAO B		
Compound	**K_i (μM)**	**K_{inact} (μM)**	**k_{inact} (min^{-1})**	**K_i (μM)**	**K_{inact} (μM)**	**k_{inact} (min^{-1})**
Phenelzine	47	3.1	0.12	15	0.9	0.9
Tranylcypromine	91	7.7	0.77	105	o.16	0.16
Pargyline	15	22	0.65	1.8	0.35	0.35
Deprenyl	75	193	0.25	0.97	103	0.53

The propargylamine class of inactivators provides clear examples of the importance of initial ligand binding in selectivity, as seen in the 100-fold better K*i* for deprenyl on MAO B than for MAO A (Table 10.5). As with both the hydrazines and cyclopropylamines, the first step in inactivation by propargylamines is the mechanism-based oxidation to the propargylimine. With the product retained in the active site by the positive charge, the allenyl resonance form undergoes nucleophilic attack by the FADH–forming a substituted 1,3-diaminoallene which, mediated by a water molecule, rearranges to the final delocalized product. The stable adduct to *N*5 (Fig. 10.4) is the same for all propargylamines studied (Binda et al., 2005; Albreht et al., 2018). The rate constants for inactivation (k_{inact}) are similar in MAO A and MAO B (Table 10.5) but the K_{inact} values reflect the selectivity of pargyline and deprenyl for MAO B inactivation. An even bigger selectivity difference is seen in the partition ratios for deprenyl: 482 in MAO A but only 1 for MAO B, indicating that every oxidation results in adduct formation in MAO B but that 482 product molecules are released from MAO A for each inactivation event. Thus, design of the optimum propargylamine inactivator must consider initial binding, oxidation rate and product retention and reactivity with the reduced flavin.

The irreversible inhibitors are clinically important and the propargylamine moiety is of particular interest for combination into multi-target drugs (see below). In the propargylamine class, pargyline has been in clinical use longest but deprenyl (selegiline) now leads the market as adjunct therapy for Parkinson's disease. The highly MAO A selective clorgyline and MAO B selective deprenyl are used in PET scans (Fowler et al., 2015a). It must be noted that under chronic administration of higher doses, each of these selective inactivators will inactive the other form of the enzyme (Fowler et al., 2015b).

10.6 Computational Innovation

Advance drug design is facilitated by good crystal structures, by mechanistic understanding and by large datasets of tested compounds as well as libraries of fragments to allow discovery of novel compounds to inhibit the enzyme. The rapid expansion of computational power to analyse large data and development of algorithms and programs techniques to explore the information has provided major advances in the last 10 years.

10.6.1 Theoretical Elucidation of Mechanism

Insight into the mechanism has come from transition state theory using multiscale simulations to calculate the free energy barriers for the catalytic step in the oxidation of amines by MAO (reviewed in Vianello et al. (2016)). The lower the free energy barriers between substrate, transition state intermediate, and product, the faster the reaction. The highest resolution structure of MAO B was optimized, fitted to the electrostatic potential, and the solvent reaction field included, then EVB calculations applied to the oxidation of a substrate such as dopamine by lumiflavin in the gas phase compared to the reaction in the enzyme. The work revealed that dopamine is stripped of its proton (pK_a 8.3) for the catalytic reaction that requires the neutral amine before oxidation via a two-step hydride transfer in which H^- (one proton and two electrons) is transferred to the *N*5 of the flavin and the other H^+ is removed from the substrate nitrogen and transferred via a water relay to *N*1 of the flavin to yield the neutral product required for dissociation. The calculated energy for the reaction was 16.1 kcal/mol in good agreement with the experimentally determined 16.5 kcal/mol. The method not only provides insight into the mechanism but also provides information relevant to design of the propargylamine inactivators of MAO. As described above, these mechanism-based inhibitors are first oxidized as substrates generating a positively charged product that could be retained close to the reduced flavin by electrostatic attraction and by favourable cation-π interactions in the active site "aromatic cage", where it can react with the reduced flavin to form a covalent adduct (Albreht et al., 2018).

10.6.2 Data-Mining and Tools for Drug Discovery

By far the most important computational advance for drug discovery has been the ability to search databases of millions of compounds for "hits"—compounds with a pharmacophore that will have good affinity for a target-binding site. With a clinically and biologically validated target identified, crystallized proteins and proven ligands enable the development of the descriptors for key interactions with the target binding site. The pharmacophore derived from effective, selective molecules allows chemi-informatics to identify similarity-based compounds and, with advanced data-mining, to exclude binding to undesirable off-targets (Nikolic et al., 2015). The dual resource of ligand and target information provides the basis for virtual screening that is now routine in drug discovery. Objective ranking of the hits can then be achieved by docking and model refinement. Molecular dynamics to check the ligand to protein fit is also important, allowing for the target flexibility, and subsequent optimization of the ligand structure. A much shorter list must then be synthesized and experimentally verified. The process has recently been summarized with examples across target fields (Ramsay et al., 2018).

To address the failure rate between ligand and drug, many computational approaches have been developed for the optimization of druggability, blood-brain barrier penetration, lipophilicity, and metabolism. All the computational approaches contribute to more cost-effective discovery and optimization of compounds, accelerating progress toward the clinic. Such high-throughput methods with the power to consider multiple targets at the same time are particularly effective and useful for designing multi-target drugs as will be considered in the next section.

10.7 Conclusion: The Future for MAO Inhibition in Multi-Target Drugs

The concept of multi-target drugs (MTD) to treat complex, multifactorial diseases such as neurodegeneration and cancer emerging at the beginning of this century. By 2017, 21% of the new compounds approved by the US Food and Drug Administration (FDA) for clinical use were aimed at more than one target, a testament to the rapid progress facilitated by computational approaches and high through-put technology as well as the progress in target identification and assay development at the experimental level. Drugs against cancer and neuropathologies were the largest classes of these MTD (Ramsay et al., 2018). MAOI with their proven effects of slowing the breakdown of neurotransmitters are a key part of the anti-neurodegenerative MTD.

Examples of MTD already on market include inhibitors of neurotransmitter reuptake systems (SNRIs) and antagonists for combinations of receptors such the dopamine D2, serotonin 5-HT2A and α1-adrenergic receptors that can treat schizophrenia. For the brain, the main enzyme targets for recent compound discovery have been the cholinesterases (ChE) that modulate release of other neurotransmitters, the monoamine oxidases (MAO) to elevate levels of monoamines, and betasecretase 1 (BACE1) to decrease the generation of the toxic amyloid beta (Aβ) peptide. One compound that has been studied in phase 2 clinical trials for cognitive impairment is ladostigil which combines two pharmacophores taken from rivastigmine (a ChE inhibitor) and rasagiline (a propargylamine inactivator of MAO). Ladostigil inhibits acetylcholinesterase, butyrylcholinesterase, and monoamine oxidases A and B, and has neuroprotective effects (reviewed in Finberg and Rabey, 2016). The main dilemma in the design of multi-target drugs is choosing the correct targets, and finding the correct balance of potency for all targets in one compound in order to have the desired effect in vivo.

Although inhibition of enzyme catalysts decreases turnover, inhibition of receptors stops cascading of signals. The potential inherent in tackling both enzyme and receptor in one compound led to the design of a tripotent compound to inhibit known enzyme targets (ChE and MAO) that also blocks the histamine receptor H3, inhibition of which is known to elevate acetylcholine, serotonin, dopamine and norepinephrine in the central nervous system. The resulting tripotent compound, contilisant, may be a useful MTD in neurodegenerative diseases (Bautista-Aguilera et al., 2017).

Medicinal chemical approaches, facilitated by computation, will provide many more similar candidates in the future. However, pharmacology to identify and validate the correct targets remains the key step in getting the correct compounds into clinical trials and on towards the market.

References

Albreht A., Vovk I., Mavri J., Marco-Contelles J., Ramsay R. R., *Front. Chem.*, **6** (2018), 169.

Anderson M. C., Hasan F., McCrodden J. M., Tipton K. F., *Neurochem. Res.*, **18** (1993), 1145–1149.

Bailey K. R., Ellis A. J., Reiss R., Snape T. J., Turner N. J., *Chem. Commun.*, (2007), 3640–3642.

Bautista-Aguilera O. M., Hagenow S., Palomino-Antolin A., Farré-Alins V., Ismaili L., Joffrin P.-L., et al., *Angew. Chem. Int. Ed.*, **56** (2017), 12765–12769

Becchi S., Buson A., Foot J., Jarolimek W., Balleine B. W., *Brit. J. Pharmacol.*, **174** (2017), 2302–2317.

Benedetti M. S., Tipton K. F., Whomsley R., *Fundam. Clin. Pharmacol.*, **15** (2001), 75–84.

Binda C., Hubalek F., Li M., Herzig Y., Sterling J., Edmondson D. E., et al., *J. Med. Chem.*, **48** (2005), 8148–8154.

Binda C., Li M., Hubalek F., Restelli N., Edmondson D. E., Mattevi A., *Proc. Natl. Acad. Sci. U.S.A.*, **100** (2003), 9750–9755.

Binda C., Newton-Vinson P., Hubalek F., Edmondson D. E., Mattevi A., *Nat. Struct. Biol.*, **9** (2002), 22–26.

Binda C., Wang J., Li M., Hubalek F., Mattevi A., Edmondson D. E., *Biochemistry*, **47** (2008), 5616–5625.

Binda C., Wang J., Pisani L., Caccia C., Carotti A., Salvati P., et al., *J. Med. Chem.*, **50** (2007), 5848–5852.

Bonivento D., Milczek E. M., McDonald G. R., Binda C., Holt A., Edmondson D. E., et al., *J. Biol. Chem.*, **285** (2010), 36849–36856.

Borstnar R., Repic M., Krzan M., Mavri J., Vianello R., *Eur. J. Org. Chem.*, (2011), 6419–6433.

Bortolato M., Shih J. C., *Intl. Rev. Neurobiol.*, **100** (2011), 13–42.

Caccia C., Maj R., Calabresi M., Maestroni S., Faravelli L., Curatolo L., et al., *Neurology*, **67** (2006), S18-S23.

Carradori S., Secci D., Petzer J. P., *Expert Opin. Ther. Pat.*, **28** (2018), 211–226.

Checknita D., Maussion G., Labonte B., Comai S., Tremblay R. E., Vitaro F., et al., *Brit. J. Psychiatry*, **206** (2015), 216–222.

Curet O., Damoiseau G., Aubin N., Sontag N., Rovei V., Jarreau F. X., *J. Pharmacol. Exp. Ther.*, **277** (1996), 253–264.

Da Prada M., Kettler R., Keller H. H., Burkard W. P., Mugglimaniglio D., Haefely W. E., *J. Pharmacol. Exp. Ther.*, **248** (1989), 400–414.

De Colibus L., Li M., Binda C., Lustig A., Edmondson D. E., Mattevi A., *Proc. Natl. Acad. Sci. U.S.A.*, **102** (2005), 12684–12689.

Delport A., Harvey B. H., Petzer A., Petzer J. P., *Toxicol. App. Pharmacol.*, **325** (2017), 1–8.

Edmondson D. E., Binda C., Mattevi A., *Arch. Biochem. Biophys.*, **464** (2007), 269–276.

Erickson J. D., Schafer M. K., Bonner T. I., Eiden L. E., Weihe E., *Proc. Natl. Acad. Sci. U.S.A.*, **93** (1996), 5166–5171.

Finberg J. P. M., Gillman K., *Intl. Rev. Neurobiol.*, **100** (2011), 169–190.

Finberg J., Rabey J., *Front. Pharmacol.*, **7** (2016), 340.

Fowler C. J., Tipton K. F., *J. Pharm. Pharmacol.*, **36** (1984), 111–115.

Fowler J. S., Logan J., Shumay E., Alia-Klein N., Wang G.-J., Volkow N. D., *J Labelled Compounds Radiopharm.*, **58** (2015a), 51–64.

Fowler J. S., Logan J., Volkow N. D., Shumay E., McCall-Perez F., Jayne M., et al., *Neuropsychopharmacology*, **40** (2015b), 650–657.

Fowler J. S., Logan J., Volkow N. D., Wang G. J., *Mol. Imaging Biol.*, **7** (2005), 377–387.

Gillman P. K., *Anaesthesia*, **61** (2006), 1013–1014.

Gillman P. K., *J. Psychopharmacol.*, **25** (2011), 429–436.

Hubalek F., Binda C., Khalil A., Li M., Mattevi A., Castagnoli N., et al., *J. Biol. Chem.*, **280** (2005), 15761–15766.

Hynson R., Kelly S., Price N., Ramsay R., *Biochim. Biophys. Acta*, **1672** (2004), 60–66.

Jones T. Z. E., Giurato L., Guccione S., Ramsay R. R., *FEBS J.*, **274** (2007), 1567–1575.

Kar S., Leszczynski J., *Curr. Drug Metab.*, **18** (2017), 1106–1122.

Kay C. W. M., El Mkami H., Molla G., Pollegioni L., Ramsay R. R., *J. Am. Chem. Soc.*, **129** (2007), 16091–16097.

Kim H., Sablin S., Ramsay R., *Arch. Biochem. Biophys.*, **337** (1997), 137–142.

Krueger M. J., Mazouz F., Ramsay R. R., Milcent R., Singer T. P., *Biochem. Biophys. Res. Commun.*, **206** (1995), 556–562.

Li M., Hubalek F., Newton-Vinson P., Edmondson D. E., *Protein Expr. Purif.*, **24** (2002), 152–162.

Lum C. T., Stahl S. M., *CNS Spectr.*, **17** (2012), 107–120.

Ma J. C., Yoshimura M., Yamashita E., Nakagawa A., Ito A., Tsukihara T., *J. Mol. Biol.*, **338** (2004), 103–114.

Malcomson T., Yelekci K., Borrello M. T., Ganesan A., Semina E., De Kimpe N., et al., *FEBS J.*, **282** (2015), 3190–3198.

Manca M., Pessoa V., Lopez A. I., Harrison P. T., Miyajima F., Sharp H., et al., *J. Mol. Neurosci.*, **64** (2018), 459–470.

Mattevi A., *TIBS*, **31** (2006), 276–283.

McDonald G. R., Olivieri A., Ramsay R. R., Holt A., *Pharmacol. Res.*, **62** (2010), 475–488.

Medvedev A. E., Ramsay R. R., Ivanov A. S., Veselovsky A. V., Shvedov V. I., Tikhonova O. V., et al., *Neurobiology*, **7** (1999), 151–158.

Milczek E. M., Binda C., Rovida S., Mattevi A., Edmondson D. E., *FEBS J.*, **278** (2011), 4860–4869.

Mosharov E. V., Larsen K. E., Kanter E., Phillips K. A., Wilson K., Schmitz Y., et al., *Neuron*, **62** (2009), 218–29.

Mosharov E. V., Staal R. G., Bove J., Prou D., Hananiya A., Markov D., et al., *J. Neurosci.*, **26** (2006), 9304–9311.

Newton-Vinson P., Hubalek F., Edmondson D. E., *Protein Expr. Purif.*, **20** (2000), 334–345.

Nikolic K., Mavridis L., Bautista-Aguilera O. M., Marco-Contelles J., Stark H., Carreiras M. D., et al., *J. Comput. Aid Mol. Des.*, **29** (2015), 183–198.

Nutt D. J., *J. Clin. Psychiatry*, **67 Suppl 6** (2006), 3–8.

Orru R., Aldeco M., Edmondson D. E., *J. Neural Transm.*, **120** (2013), 847–851.

Pena-Silva R. A., Miller J. D., Chu Y., Heistad D. D., *Am. J. Physiol: Heart Circ. Pysiol.*, **297** (2009), H1354-H1360.

Pinsonneault J. K., Papp A. C., Sadee W., *Human Mol. Genet.*, **15** (2006), 2636–2649.

Ramadan Z. B., Wrang M. L., Tipton K. F., *Neurochem. Res.*, **32** (2007), 1783–1790.

Ramsay R. R., Dunford C., Gillman P. K., *Brit. J. Pharmacol.*, **152** (2007), 946–951.

Ramsay R. R., Hunter D. J. B., *Biochim. Biophys. Acta Proteins Proteom.*, **1601** (2002), 178–184.

Ramsay R. R., *J. Neural Transm.*, **52** (1998), 139–147.

Ramsay R. R., Olivieri A., Holt A., *J. Neural Transm.*, **118** (2011), 1003–1019.

Ramsay R. R., Popovic-Nikolic M. R., Nikolic K., Uliassi E., Bolognesi M. L., *Clin. Transl. Med.*, **7** (2018), 3.

Ramsay R. R., Tipton K. F., *Molecules (Basel, Switzerland)*, **22** (2017).

Rendic S., Guengerich F. P., *Chem. Res. Toxicol.*, **28** (2015), 38–42.

Reniers J., Meinguet C., Moineaux L., Masereel B., Vincent S. P., Frederick R., et al., *Eur. J. Med. Chem.*, **46** (2011), 6104–6111.

Ricken R., Ulrich S., Schlattmann P., Adli M., *Eur. Neuropsychopharmacol. J. Eur. Coll. Neuropsychopharmacol.*, **27** (2017), 714–731.

Sablin S. O., Ramsay R. R., *Antiox. Redox Sign.*, **3** (2001), 723–729.

Sablin S. O., Yankovskaya V., Bernard S., Cronin C. N., Singer T. P., *Eur. J. Biochem.*, **253** (1998), 270–279.

Sabol S. Z., Hu S., Hamer D., *Human Genet.*, **103** (1998), 273–279.

Salach J. I., *Arch. Biochem. Biophys.*, **192** (1979), 128–137.

Shih J. C., Wu J. B., Chen K., *J. Neural Transm.*, **118** (2011), 979–986.

Shumay E., Logan J., Volkow N. D., Fowler J. S., *Epigenetics*, **7** (2012), 1151–1160.

Silverman R. B., *Acc. Chem. Res.*, **28** (1995), 335–342.

Singer T. P., Ramsay R. R., *FEBS Lett.*, **274** (1990), 1–8.

Son S. Y., Ma A., Kondou Y., Yoshimura M., Yamashita E., Tsukihara T., *Proc. Natl. Acad. Sci. U.S.A.*, **105** (2008), 5739–5744.

Stefanachi A., Leonetti F., Pisani L., Catto M., Carotti A., *Molecules (Basel, Switzerland)*, **23** (2018), 250.

Stocchi F., Vacca L., Grassini P., De Pandis M. F., Battaglia G., Cattaneo C., et al., *Neurology*, **67** (2006), S24.

Tan A. K., Ramsay R. R., *Biochemistry*, **32** (1993), 2137–2143.

Tipton K. F., *J. Neural Transm.*, (2018).

Tipton K. F., Spires I. P. C., *Biochem. J.*, **125** (1971), 521–524.

Tripathi A. C., Upadhyay S., Paliwal S., Saraf S. K., *Eur. J. Med. Chem.*, **145** (2018), 445–497.

van Amsterdam J., Talhout R., Vleeming W., Opperhuizen A., *Life Sci.*, **79** (2006), 1969–1973.

Vianello R., Domene C., Mavri J., *Front. Neurosci.*, **10** 327 (2016).

Vianello R., Repic M., Mavri J., *Eur. J. Org. Chem.*, (2012), 7057–7065.

Weyler W., Titlow C. C., Salach J. I., *Biochem. Biophys. Res. Commun.*, **173** (1990), 1205–1211.

Wouters J., *Curr. Med. Chem.*, **5** (1998), 137–162.

Youdim M. B. H., Edmondson D., Tipton K. F., *Nat. Rev. Neurosci.*, **7** (2006), 295–309.

Zapata-Torres G., Fierro A., Barriga-Gonzalez G., Cristian Salgado J., Celis-Barros C., *J. Chem. Inf. Model.*, **55** (2015), 1349–1360.

Zeller E. A., Barsky J., Berman E. R., *J. Biol. Chem.*, **214** (1955), 267–274.

Chapter 11

Multi-Functional Monoamine Oxidase and Cholinesterase Inhibitors for the Treatment of Alzheimer's Disease

Ireen Denya, Sarel F. Malan, and Jacques Joubert

Pharmaceutical Chemistry, School of Pharmacy,
University of the Western Cape, Private Bag X17,
Bellville 7535, South Africa

jjoubert@uwc.ac.za

11.1 Introduction: Alzheimer's Disease

Alzheimer's disease (AD) is a progressive neurodegenerative disorder of subtle onset, characterised by memory loss and cognitive impairment (Khachaturian, 1985). Only 1 in 4 of the almost 47 million cases of AD worldwide has been diagnosed. Although highly prevalent in Western Europe, the disease is essentially a global problem incurring an estimated cost of $818 billion in 2015 and is expected to reach $1 trillion in 2018 (Prince et al., 2015).

Pharmaceutical Biocatalysis: Fundamentals, Enzyme Inhibitors, and Enzymes in Health and Diseases
Edited by Peter Grunwald

ISBN 978-981-4800-61-7 (Hardcover), 978-0-429-29503-4 (eBook)
www.jennystanford.com

The symptoms of AD can be divided into four stages, with a progressive pattern of functional and cognitive impairment. It starts with the mild effects of ageing on memory such as an occasional forgetfulness of exact details and minor short-term memory loss (Backman et al., 2004). In early stage AD an increasing impairment of cognition and memory, including not remembering the episodes of forgetfulness, is observed. At this point, one may forget the names of friends and family and have some confusion in situations outside the familiar. These changes may only be noticed by those close to the patient (Black, 2012). In the moderate stage of the disease, the subjects are not able to perform the common activities of daily living thus hindering their independence. Motor activities become less coordinated, the memory worsens and insomnia sets in (Frank, 1994). At this point, it becomes very difficult for the patient to remember recently acquired information or to know where they are at a given moment. The late stage, which is the last and dementia stage of the disease, involves the complete dependence and mental incapacity of the patient (Förstl and Kurz, 1999). Language is reduced to simple phrases and single words, eventually leading to loss of speech (Frank, 1994). People with severe AD will ultimately not be able to perform any task on their own and will get to a point where they are bedridden and unable to feed themselves. Death eventually occurs due to infection or pneumonia and not the disease itself (Förstl and Kurz, 1999; Black, 2012).

11.2 Cholinesterases: Isoforms and Function

Cholinesterases are widespread enzymes found in both cholinergic and non-cholinergic tissues as well as in plasma and other bodily fluids (Massoulié et al., 1993). They are classified into two forms according to their substrate specificity, behaviour in excess substrate and susceptibility to inhibitors (Tarihi et al., 2003). Acetylcholinesterase (AChE) also referred to as 'true cholinesterase' is encoded by a gene on chromosome 7 (Aldridge, 1953). It is the more prominent of the cholinesterases and the only one consistently associated with cholinergic pathways. It was found that cholinergic neurons display high AChE enzyme activity in the perikaryon, proximal dendrite and axon (Mesulam and

Geula, 1991). AChE is known to be abundantly available in the brain, muscle and erythrocyte membrane (Dave et al., 2000). The second cholinesterase, found mainly in the liver, heart and lungs is called butyrylcholinesterase (BuChE). It is also known as pseudocholinesterase or non-specific cholinesterase and is present in much lower concentrations than AChE with limited distribution in the central nervous system (Dave et al., 2000; Mesulam, 2000). The two cholinesterases present similar soluble molecular forms in tissues and body fluids with differing tissue distribution.

AChE and BuChE both have the capacity to hydrolyse acetylcholine (ACh), be it in different ratios. Excess ACh levels in the brain cause saturation of AChE and in turn increase the activity of BuChE towards the neurotransmitter (Giacobini, 2004). Several lines of evidence have demonstrated that in an AD brain, with progression of the disease the activity of AChE gradually decreases while that of BuChE increases significantly in the hippocampus and temporal cortex of the brain (Alipour et al., 2012). This implies that in the case of AD, although the AChE activity is inhibited there is still significant ACh hydrolysis by BuChE. Therefore, if the activity of BuChE is also inhibited, the ACh concentration could be considerably increased in the brain. In as much as AChE hydrolyses more ACh than BuChE, the latter contributes more to AD because of the decreased levels of the true cholinesterase (Mesulam et al., 2002).

AChE and BuChE share a 65% amino acid sequence homology and have similar molecular forms and active site structure (Allderdice et al., 1991). Early kinetic studies indicate that the active site of AChE contains two subsites, the esteratic and anionic subsites, corresponding to the catalytic and choline–binding pockets, respectively. The anionic active subsite interacts with the charged quaternary group of the choline moiety of acetylcholine (Augustinsson et al., 1950; Rosenberry, 2006). It contains the amino acids Trp 86, Tyr 133, Tyr 337 and Phe 338 and these bind the quaternary trimethylammonium choline moiety of the substrate mainly through π-cation interactions (Harel et al., 1993) positioning the ester optimally at the acylation site. The acyl pocket, responsible for substrate selectivity by preventing access of the larger choline esters is composed of Phe 295 and Phe 297.

The oxyanion hole, Gly 121, Gly 122 and Ala 204, provides hydrogen bond donors that stabilise the tetrahedral transition state of the substrate (Ordentlich et al., 1998). Cationic substrates are not bound by a negatively charged amino acid in the anionic site, but by interaction of 14 aromatic residues that line the gorge leading to the active site (Colovic et al., 2013). Among the aromatic amino acids, Trp 84 is reported to be critical and not substitutable (Tougu, 2001). The esteratic subsite or cationic active site (CAS) where ACh is hydrolysed contains a catalytic triad of three amino acids, namely Ser 203, His 447 and Glu 334 (Colovic et al., 2013).

In addition to these two subsites of the catalytic centre, AChE also possesses one or more additional binding sites for its substrates. This 'peripheral' anionic site (PAS), very different from the choline-binding pocket of the active site, has been implicated in ligand-binding activity of AChE (Mooser and Sigman, 1974; Reiner et al., 1991). The PAS lies essentially on the surface of AChE, and it binds acetylcholine as the first step in the catalytic pathway (Sussman et al., 1991; Hosea et al., 1996). It is also responsible for modulating the catalysis as well as the binding of specific inhibitory compounds (Johnson and Moore, 2006). The PAS has also been identified as a site promoting non-cholinergic functions such as cell adhesion and neurite outgrowth in developing and transformed neural cells and amyloiodosis through an interaction with the amyloid β (Aβ) protein in AD (Hosea et al., 1996; Inestrosa et al., 1996; Inestrosa and Alarcón, 1998). The PAS consists of 5 residues, namely Tyr 72, Asp 74, Tyr 124, Trp 286 and Tyr 341 clustered around the entrance of the active site gorge (Harel et al., 1993). The site has also been shown to have interaction with a large omega loop that incorporates Tyr 72 and Asp 74. It has been proposed that this loop may be involved in the accessibility of small molecules to the active site as well as allosterism (Shi et al., 2003). There are 10 acidic residues in the area surrounding the PAS. This concentration of negatively charged residues known as the 'annular electrostatic motif' is shared by AChE and other homologous signalling molecules but not by BuChE (Botti et al., 1998). The aromatic PAS residues, with Trp 286 at their core, have been shown to act synergistically due to the very large increases in inhibition constants induced by multiple mutants (Bourne et al., 1999). The aromatic rings of Tyr 72 and Tyr 124 flank the indole ring of Trp 286 and together interact with charged groups of ligands (Johnson and Moore, 2006). The indole ring of Trp 286 exhibits a variety of potential interactions including aromatic-aromatic-, stacking- and π-cation interactions (Kreienkamp et al., 1991) depending on the nature of the ligand. Significant shifting of the indole is also observed when comparing crystal structures of apo- and complexed AChE (Bourne et al., 1995) indicating its importance in ligand binding. Asp 74, like Trp 286, is also able to utilise several different interaction modes, operating by ionic or hydrogen bond interactions mediated by its carboxylated moiety that uniquely act as a trap for charged substrate (Mallender, et al., 2000; Hosea, et al., 1996).

In the structure of BuChE, each monomer has a narrow 20 Å deep active site gorge lined by approximately 55 amino acid residues. The PAS is found at the mouth of the gorge with Asp 70 and Tyr 332 involved in the initial binding of positively charged substrates and in activation control. BuChE contains a hydrogen bond between these two residues which controls the functional design of the active site gorge. When a positively charged substrate is bound to the enzyme by a cation-π interaction with the aromatic ring of Tyr 332, the substrate interacts with the negatively charged Asp 70 and this process triggers a change in the conformation of the monomer. Thereafter the two flexible arms of the omega loop come close to each other and the substrate moves down to the Trp 82 residue of the choline-binding site. Trp 82 also forms a cationic-π complex with the substrate (Masson et al., 1997; Masson et al., 1999; Masson et al., 2001). The oxyanion hole, found near the choline-binding site, includes Gly 116, Gly 117 and Ala 199, helps to rotate the substrate to a position where it can be hydrolysed by Ser 198 (Ekholm and Konschin, 1999). When the substrate rotates horizontally, it binds to the acyl-binding pocket of the enzyme active site. Leu 286 and Val 288 are found in the acyl-binding pocket of BuChE (Phe 295 and Phe 297 in AChE) and allow for catalysis of larger acyl group containing substrates such as butyrylcholine (Radic et al., 1993). The stabilised substrate is hydrolysed by the catalytic triad, composed of Ser 198, His 438 and Glu 325, in the esteratic site of the active centre (Masson et al., 1997). It is important to note that six of the 14 aromatic amino acid residues lining the active site gorge of

AChE are replaced by aliphatic residues in BuChE. This explains the larger volume of the BuChE active site gorge in comparison. The replacement is also responsible for selective sensitivity against different inhibitors of the enzymes (Saxena et al., 1999).

11.3 Cholinergic Hypothesis of AD

Choline acetyltransferase (CAT) is an enzyme that catalyses the acetylation of choline by acetyl coenzyme A (CoA) to produce acetylcholine and coenzyme A (Fig. 11.1). The original building blocks, acetate and choline, enter the nerve cell via specific transporters. After CAT intervention, the resultant ACh is stored in presynaptic neuronal vesicles until it is released into the synapse during neurotransmission. In the synapse acetylcholine binds to nicotinic and muscarinic (cholinergic) receptors of the postsynaptic neuron resulting in a transfer between neurons (Fig. 11.2). CAT is produced mainly by the nucleus basalis of Meynert in the basal forebrain (Hebb and Whittaker, 1958).

Memory impairment and dementia are primary symptoms of AD and this was linked to the fact that aged cholinergic neurons were unable to synthesise acetylcholine without potassium stimulation Gibson and Peterson (1981). This impairment results in a consistent reduction in CAT and thus acetylcholine levels. Excess acetylcholine post synapse is hydrolysed by the cholinesterase enzyme to recyclable choline and acetate and terminates transmission of the impulse. In the brain of an AD patient production of acetylcholine is lower than its hydrolysis resulting in idle nerve cells that eventually die and waste away (Francis et al., 1999).

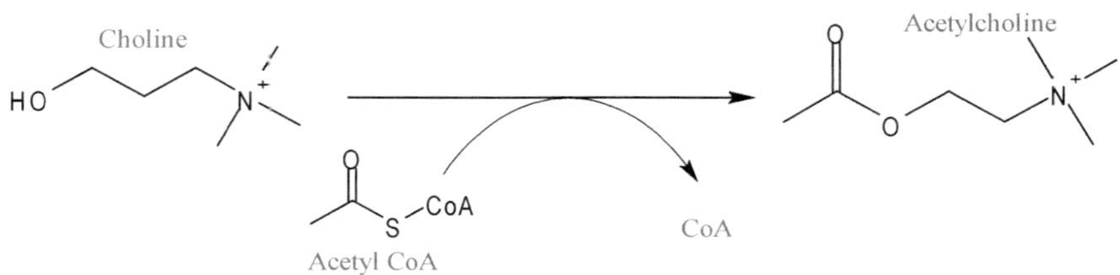

Figure 11.1 Synthesis of acetylcholine.

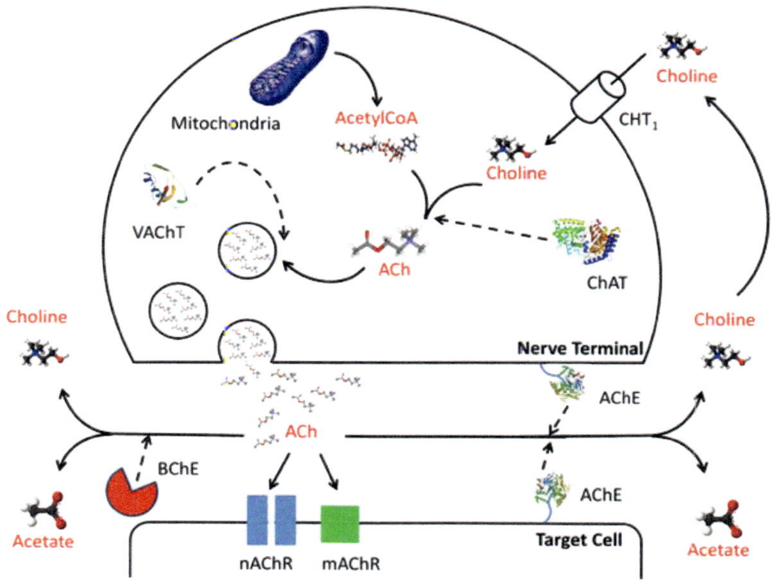

Figure 11.2 The cycle of events of Ach synthesis, release, action and breakdown at a cholinergic nerve terminal. ChAT, choline acetyltransferase; CHT 1, high-affinity choline transporter-1; VAChT, vesicular ACh transporter; mAChR, muscarinic receptors and nAChR nicotinic receptors. Adapted from Zoheir et al. (2012).

Multiple human studies have indicated that forebrain cholinergic pathways to the thalamus serve important functional roles in conscious awareness, attention, working memory and other cognitive processes (Perry et al., 1999). The activity of CAT, the enzyme responsible for the synthesis of Ach was found to be significantly reduced in pathological samples from the cortex and hippocampus in AD patients (Davies and Maloney, 1976; Perry et al., 1977). Figure 11.2 shows the cycle of events of acetylcholine synthesis and breakdown in the CNS.

A study by Bartus (2000) stated the cholinergic hypothesis to be essentially a loss of cholinergic function in the CNS that contributes to the cognitive decline associated with advanced age and AD. The cortex and hippocampus have also exhibited a depolarisation-induced acetylcholine release and choline uptake in nerve terminals to replenish the acetylcholine (Rylett, Ball, and Colhoun, 1983; Nilsson et al., 1986). At the same time, a reduced number of cholinergic projection neurons in the nucleus basalis of Meynert led to the conclusion that the pathogenesis of AD was mostly represented by the degeneration of the cholinergic connection from the nucleus basalis of Meynert to the cortex and hippocampus. There thus exists ample evidence to conclude that assault to the forebrain cholinergic neurons could result in interference with ACh release for normal brain function (Burk et al., 2002).

11.4 Monoamine Oxidase: Isoforms and Function

Monoamine oxidase (MAO) is an outer mitochondrial membrane bound flavin containing enzyme with a molecular weight of about 60 kDa, it plays an important role in the metabolism of endogenous and exogenous amines (Weyler et al., 1990; Shih et al., 1999). Intraneuronal MAO terminates the action of endogenous amine neurotransmitters, regulates intraneuronal amine stores and protects cells from dietary amines (false transmitters) (Youdim et al., 2006).

The enzyme exists in two isoforms, namely MAO-A and MAO-B; these isoforms have different substrates specificities in that MAO-A catalyses the oxidation of serotonin, whilst MAO-B catalyses the oxidation of both benzylamine and 2-phenylethylamine. Other neurotransmitters like dopamine, norepinephrine, epinephrine, tryptamine and tyramine are oxidised by both isoforms of the enzyme (Youdim et al., 1988). MAO enzyme activity is different in various regions of the brain, where the hypothalamus and basal ganglia show the highest activity while the cerebellum and neocortex show the lowest activity (O'Carroll et al., 1983). Fowler and company mapped the human brain MAO in 1987 to identify the distribution of the MAO isozymes throughout the brain and it was found that basal ganglia contain mainly MAO-B.

Although the overall chain-folds of the two isozymes are quite similar, there are similarities and differences in the respective active sites. The structure of the covalent flavin adenine dinucleotide (FAD) coenzymes and the two tyrosines constituting the aromatic cage in the active sites are identical (De Colibus et al., 2005). Since this region is directly involved with substrate oxidation, this close similarity supports the notion that both enzymes follow the same catalytic mechanisms (Edmondson et al., 2004). However, differences in the active site residues constituting the active site of MAO-A (7 residues in total) and the conformation of loop 20–216 cause the shape and sizes of the active sites to differ significantly and are the basis for the specificity of the isozymes (De Colibus et al., 2005). The entrances to the active sites of either enzyme have been proposed to be at the surface of the negatively charged outer membrane which could function to increase the effective substrate concentration by electrostatic attraction (Edmondson et al., 2007). Qualitative structure activity relationship (SAR) comparison on the catalytic properties of the two enzymes showed that the substrate has less freedom for rotation in the MAO-B site than the MAO-A site (De Colibus et al., 2005; Edmondson et al., 2007).

11.4.1 Monoamine Oxidase A

Human MAO-A contains 527 residues and is specific to serotonin, dopamine and norepinephrine. Its active site is formed by a single cavity that extends from the flavin ring to the cavity shaping loop of residues 210–216. The volume of this cavity is estimated to be approximately 550 Å^3 and is lined by 11 aliphatic and 5 aromatic residues, which demonstrate that it is hydrophobic (De Colibus et al., 2005).

Two cysteine residues (Cys 321 and Cys 323) are located near the entry of the catalytic site. The 'face' of the covalent FAD is situated at one end of the active site opposite to the entrance. It also has two parallel tyrosyl 407 and 444 residues forming an aromatic cage in front of the FAD (Edmondson et al., 2007).

11.4.2 Monoamine Oxidase B

Monoamine oxidase B (MAO-B) is made up of 520 amino acids and FAD cofactor. It has a bipartite elongated cavity that occupies a volume of approximately 700 Å when the side chain of Ileu199 is in the 'open' conformation The monomer involves a membrane-binding domain, a substrate domain and a flavin-binding domain (Kearney et al., 1971; Binda et al., 2001) MAO-B undergoes modification post translation i.e. acetylation where the amino terminal methionine is cleaved and the resultant *N*-terminal serine is acetylated (Li et al., 2002; Newton-Vinson et al., 2000). It is tightly bound to the outer membrane of the mitochondria by the C-terminal amino acids 461–520. Analysis of the enzyme residue sequence revealed that residues 489–515 form a transmembrane helix 27 amino acids long (Binda et al., 2001). The substrate-binding site is formed by a flat cavity of 420 Å. This cavity is lined by a number of aromatic and aliphatic amino acids, providing a highly hydrophobic environment (Walker and Edmondson, 1994). Adjoining the substrate cavity is a separate, smaller hydrophobic cavity lined by residues Phe 103, Pro 104, Trp 119, Leu 164, Leu 167, Phe 168, Leu 171, Ile 199, Ile 316 and Tyr 326. This cavity is situated between the active site and the protein surface, and is shielded from solvent by loop 99–112 to open access to the smaller entrance cavity. The substrate and hydrophobic entrance cavity are separated by residues Trp 119, Leu 171, Ile 199 and Tyr 326. When a substrate reaches the entrance cavity, a brief movement of the four separatory residues must occur to allow its diffusion into the active site. Loop 99–112 functions as a sort of gating switch to the entrance cavity (Walker and Edmondson, 1994; Binda et al., 2001).

In order to covalently bind with the FAD cofactor and undergo metabolism the substrate must pass through two hydrophobic cavities, an entrance cavity and a larger substrate cavity (Binda et al., 2001). The outer mitochondrial membrane is covered with negative charges and these attract amine substrates that are positively charged into the enzyme (Edmondson et al., 2004). The FAD cofactor is located at the end of the substrate cavity, covalently bound to the apo-enzyme and considered an important component for MAO catalytic activity (Edmondson, 2004). It is of significance to note that the amino acids that interact with FAD in the MAO-B enzyme are conserved in MAO-A suggesting that FAD is identical in both enzymes (Edmondson, 2004; Kirksey et al., 1998).

11.4.3 Monoamine Oxidase in AD

There are many proteins involved in oxidative stress and MAO is one of them. Research has shown that MAO-B inhibitors, such as iproniazid, reduce the formation of hydrogen peroxide (H_2O_2) and reactive oxygen species, the neurotoxic compounds that induce oxidative stress and ultimately contribute to increase neuronal damage (Foley et al., 2000). This stems from the MAO catalysed dopamine oxidation in the brain that forms the H_2O_2 (Bhat et al., 2015). Several studies have reported increased activities of MAO-B in the brain and blood platelets of patients suffering from AD, resulting in an elevation of brain levels of neurotoxic free radicals and thus contributing to oxidative stress. The H_2O_2 in normal circumstances is an anodyne cellular metabolite but once their rate of synthesis exceeds the rate of endogenous breakdown capacity oxidative stress results (Christen, 2000; Pratico, and Sung, 2004). When these levels increase, more H_2O_2 gets into contact with iron (II), which promotes the conversion of the peroxide into a highly reactive hydroxyl radical by means of a Fenton reaction (Foley et al., 2000). Studies demonstrate a general increase in MAO-B activity and slight variation in MAO-A activity with aging (Benedetti and Doster, 1989). In addition to this, a substantial increase in MAO-B levels was seen in the brain of individuals older than 50 to 60 years (Nicotra et al., 2004; Fowler, 1980). The surge in brain MAO-B was hypothesised to be due to enhanced transcription of the MAO B protein (Riederer, 2004). This thus implies that inhibition of MAO-B could be beneficial in AD treatment. The MAO catalytic reaction is summarised in Fig. 11.3.

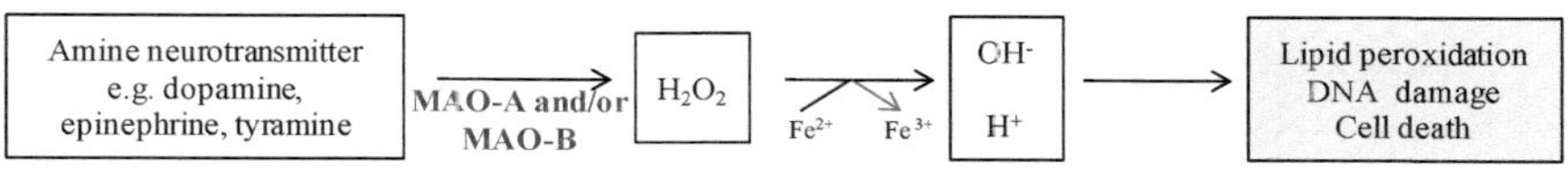

Figure 11.3 Involvement of MAO in AD.

The loss of MAO-A-immunoreactive cells is exacerbated in brainstem monoaminergic nuclei and other regions in late-stage cognitive decline (Chan-Palay et al., 1993). The development of dementia presents an imbalance between MAO-A-immunoreactive neurons and the MAO-A that impacts cognitive dysfunction. This reveals that activated MAO-A in neurons is involved in the pathology of AD and that therapeutic strategies based upon correction of MAO-A activities (by inhibition) may be important to ameliorating symptoms observed in AD (Chan-Palay et al., 1993).

11.5 Treatment of AD

The goals of treatment in patients with AD are to improve or slow down the progression of the disease, i.e. loss of memory and cognition and to maintain independent function (Mayeux and Sano, 1999; Katzman and Saitoh, 1991). Current drug treatments for AD target cholinergic and glutamatergic transmission to improve disease symptoms (Table 11.1). AChE inhibitors decrease the breakdown of the neurotransmitter acetylcholine to maintain its levels. They, therefore, boost cholinergic transmission in the basal forebrain and compensate for the loss of functional brain cells (Stahl, 2000). Medications currently approved by regulatory agencies to treat the cognitive manifestations of AD and improve quality of life are donepezil (Aricept®), rivastigmine (Exelon®) and galantamine (Razadyne®, Nivalin®). These drugs are reversible inhibitors of the enzyme. Tacrine, the first inhibitor to be approved in 1993 was discontinued due to high incidence of side effects, including hepatotoxicity (Bond et al., 2012; Watkins et al., 1994). Memantine is a moderate-affinity, voltage dependant, uncompetitive NMDA receptor antagonist approved for moderate to severe dementia of the Alzheimer's type. It inhibits excessive calcium influx induced by overexcitation of the NMDA receptor. Memantine, marketed as Ebixa®, Axura®, Namenda® and Memor®, is associated with less serious side-effects such as appetite loss and nausea (Robinson and Keating, 2006).

Table 11.1 Approved treatment options for AD

Drug	Donepezil	Galantamine	Rivastigmine	Memantine
Class	Piperidine derivative	Alkaloid	Carbamate	
FDA Approval	1996	2001	2000	2003
Use	Palliative treatment mild-to moderate AD	Palliative treatment mild-to moderate AD	Palliative treatment mild-to moderate AD	Palliative treatment mild-to moderate AD
Targets	AChE	AChE	AChE and BuChE	NMDAR

MoA	Inhibits the activity of AChE by binding to the CAS and PAS	Inhibits the activity of AChE by binding to the CAS	Inhibits the activity of AChE and BuChE by binding to the CAS and PAS	Uncompetitive, voltage-dependent NMDA receptor antagonist
Benefits	Improve cognition, daily function and speech	Improve cognition, daily function and speech	Improve cognition, daily function and speech	Decrease emergence of behavioural symptoms such as cognition
Half-life	70 h	7 h	10 h	60–80 h
Reference	Mehta et al., 2012			Parsons et al., 2013

MoA: Mechanism of action.

11.5.1 Multi-Functional MAO and ChE Agents

The drugs currently registered for the treatment AD (Table 11.1) provide only a palliative solution in the management of the disease. However, AD is a multi-factorial disease caused by various factors including protein misfolding, aggregation, oxidative stress, free radical formation and neuro-inflammatory processes. These insights together with ongoing research have provided the rationale for therapies directly targeting AD molecular causes (Cavalli et al., 2008; Munoz-Torrero and Camps, 2006). In the context of such a complex disease it is not insignificant to state that these hypotheses are not mutually exclusive but rather complimentary.

Multiple medication therapy or polypharmacy is a successful concept for diseases like cancer where it achieves maximum efficacy by attacking several targets simultaneously, exploiting synergy and minimising individual toxicity. Clinically, memantine and an acetylcholinesterase inhibitor (rivastigmine or donepezil) appear to produce an additional effect in the treatment of AD (Bolognesi et al., 2006; Grossberg et al., 2006). However, polypharmacy comes with issues of patient adherence and high cost. Natural evolution of polypharmacy takes us to the emergence of the multi-target drug ligand (MTDL) approach, a relatively novel drug design strategy (Zhang, 2005).

The MTDL strategy is an approach used to combine distinct pharmacophores of different drugs in the same structure to obtain hybrid molecules. Principally each pharmacophore should retain the ability to interact with its specific sites on the target and, consequently, produce specific pharmacological responses that slow down or block a neurodegenerative process (Pisani et al., 2011; Knez et al., 2017). These combinations occur in various ways but the most widely adopted, and the focus for this chapter, is the combination of ChE and MAO inhibitors to attain potential derivatives for the treatment of AD.

11.5.1.1 Donepezil derivatives

Donepezil also known as E2020 or by its full chemical name ((*R,S*)-1-benzyl-4-[(5,6-dimethoxy-1-indanon)-2-yl])methylpiperidine, was one the most potent AChE inhibitors approved by the Food and Drug Administration (FDA) in 1996 (Kryger et al., 1999). It is used in the palliative treatment of mild to severe AD. Donepezil (Fig. 11.4) is devoid of the side effects observed with the use of a previously registered AChE inhibitor tacrine, including; hepatotoxicity, slow pharmacokinetics and high incidence of side effects (Kryger et al., 1999).This was noted to be due to the presence of the benzylpiperidine moiety in the structure of donepezil (as compared to tacrine and physostigmine) (Kawakami, et al., 1996). Since the discovery of donepezil, it has been used as a lead compound in the design of novel drugs aimed at treating or even potentially curing AD. Its high inhibitory capacity is not the only reason for its widespread use as a lead compound, but also the fact that it is possible to do some functionalisation on its structure without affecting its activity.

Donepezil is an AChE inhibitor consisting of an *N*-benzylpiperidine and an indanone moiety (Fig. 11.4, Sugimoto et al., 1995) which shows higher affinity and selectivity for AChE than it does for BuChE, with an IC_{50} of 5.7 nM for AChE and 7000 nM for BuChE (Kawakami et al., 1996). Sugimoto et al. (1995) and Sussman et al. (1991) have done extensive studies regarding the structure activity relationship (SAR) of donepezil. They were able to show that donepezil can be divided into four main parts that play a role in its AChE inhibitory activity. These include the indanone moiety, the linkage moiety, the piperidine moiety and the benzyl moiety. Sussman et al. (1991) have demonstrated that the *N*-benzylpiperidine of donepezil interacts with the active CAS of AChE whereas the indanone moiety interacts partially with the outer PAS. Further research done by Sugimoto et al. (1995) demonstrated that the binding activity of donepezil is strongly dependent on its interaction with Trp-279 (the amino acid that interacts with the indanone at the PAS) and Phe-330 (the amino acid that interacts with the nitrogen in the piperidine moiety of donepezil in the active-site gorge). These amino acids are absent in BuChE (Sussman et al., 1991), and as such reduces the chances for interaction between the enzyme and donepezil. The ideal chain length for proper AChE interaction is 3-carbons between the piperidine moiety and the indanone moiety. A longer chain will dramatically reduce activity while the absence of a linkage abolishes activity. Introduction of rotatable bonds in the linkage moiety, according to Kryger et al. (1999), has the potential to improve the affinity of the compound. Any substitution on the piperidine moiety leads to a significant decrease in activity and replacing the piperidine with piperazine reduces the activity but does not cancel inhibitory capacity of the compound. Donepezil is sensitive to large molecule substitution on the benzyl moiety. This may be due to the limited space available between the benzyl ring and the gorge surface (Kryger et al., 1999). Small size substitutions on the benzyl ring are thus desirable on the meta position. Substitutions on the indanone moiety have lesser effect than on the benzyl ring; however, removal of the carbonyl group will abolish the activity of the compound (Kryger et al., 1999).

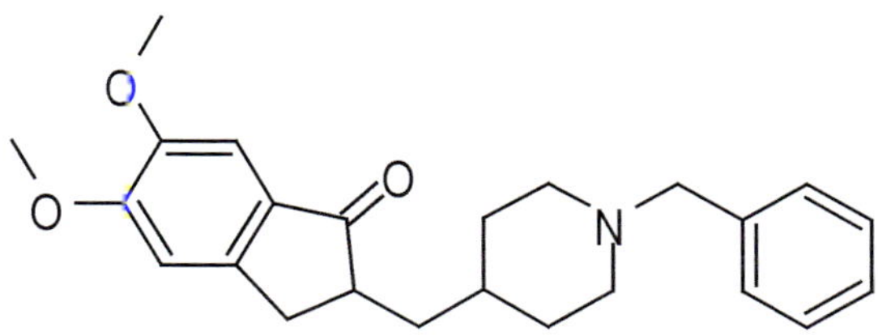

Figure 11.4 Structure of donepezil.

Given the major role of the *N*-benzylpiperidine moiety on the activity of donepezil, several studies have made use of this moiety in order to design novel multi-functional compounds for the treatment of AD (Bolea et al., 2011, Rodríguez-Franco et al., 2005).

In a study by Li et al. (2016), a series of compounds were synthesised with a strategy to retain the benzylpiperidine fragment with ChE inhibitory properties and introduce the benzamide function of the 2-picolinamide moiety from lazabemide with MAO inhibitory capacity. The team also introduced a 2-thiophene-carboxamide moiety because of its structural similarity to benzamide and compared these to pyridine, quinoline and/or substituted phenyl moieties. SAR analysis showed that compounds with nitrophenyl- or 2-quinoline amide substitutions together with a 2C linker linked to the benzylpiperidine moiety were the most potent ChE inhibitors. The 6-methyl-2-pyridine amide with a 2C linker portrayed the best MAO IC_{50} values. The representative compound **1** (Fig. 11.5) possessed the best-balanced MAO and ChE activity across all the tested compounds.

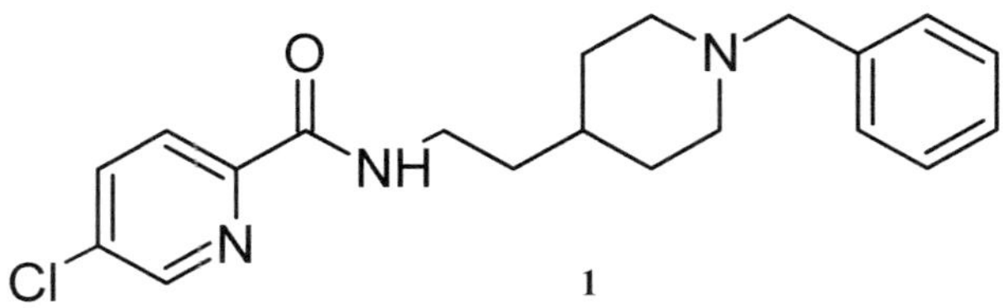

1

eeAChE: IC_{50} = 0.322 μM eqBuChE: IC_{50} = 1.23 μM

hMAO-A: IC_{50} = 3.14 μM hMAO-B: IC_{50} = 13.4 μM

Figure 11.5 Donepezil-lazabemide derived hybrid molecule.

Compounds **2a-c** and **2d-f** (Fig. 11.6) is a combination of the benzylpiperidine group with a metal chelating moiety represented by the 8-hydroxyquinoline system functionalised at position 2. MAO inhibitory properties were introduced by the addition of the propargylamine moiety (Zeller, 1963; Zindo et al., 2015). This approach yielded potent ChE inhibitors in the nanomolar range combined with micromolar MAO-A/B inhibition, metal chelating and antioxidant properties. Among the series of chemical entities assayed, compound **2f** exhibited the most interesting inhibitory profile as a mixed type ChE inhibitor of nanomolar potency and irreversible MAO inhibitor (Wu et al., 2016).

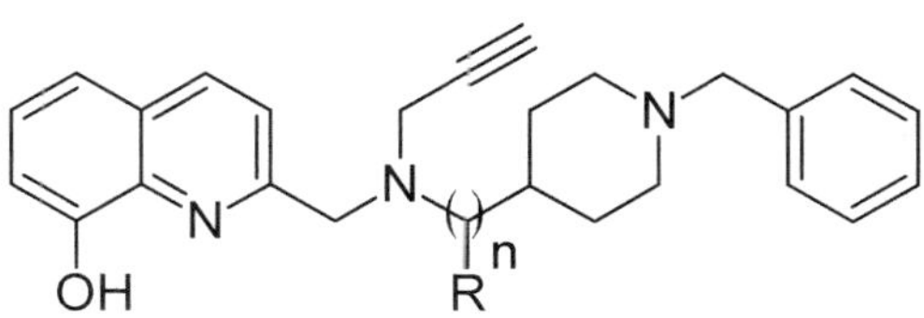

2a - 2c: n = 0 - 2, R = H
2d - 2f: n = 0 - 2, R = CN

2f eeAChE: IC_{50} = 0.322 μM
BuChE: IC_{50} = 1.23 μM
MAO-A: IC_{50} = 3.14 μM
MAO-B: IC_{50} = 13.4 μM

Figure 11.6 General structure of benzylpiperidine-propargylamine-8-hydroxyquinoline hybrid compounds.

Replacing the indanone fragment of donepezil with additional bioactive molecules to produce multi-targeted inhibitors is another exploratory strategy. For example, donepezil and *N*-[(5-(benzyloxy)-1-methyl-1H-indol-2-yl)methyl]-*N*-methylprop-2-yn-1-amine hybrids with a methylene linker have been designed as MTDLs capable of inhibiting ChEs and MAOs (Bolea et al., 2011). The length of the tether that connects the structural fragments had a major effect on MAO-binding activity. **ASS234** is a promising drug candidate with potential impact in AD therapy (Fig. 11.7). It is a very potent, irreversible human MAO-A and MAO-B inhibitor with IC_{50} values of 5.44 and 177 nM, respectively and also a reversible inhibitor of both ChEs (hAChE IC_{50} = 0.81 μM; hBuChE IC_{50} = 1.82 μM) (Esteban et al., 2014; Marco-Contelles et al., 2016).

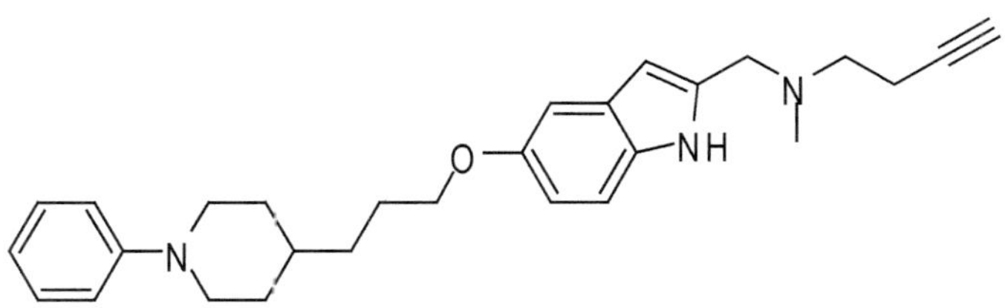

Figure 11.7 Structure of **ASS234**.

A new family of multi-targeted inhibitors was described by Estrada et al. (2016) which was obtained by linking cinnamic related structures (umbellic acid, caffeic acid and ferulic acid) with *N*-benzylpiperidine or *N*,*N*-dibenzyl(*N*-methyl)amine fragments (Fig. 11.8). The derived hybrids **3** (umbellic), **4** (caffeic) and **5** (ferulic) displayed balanced biological profiles with IC_{50} values in the low-to submicromolar range for human ChE and human MAOs. The compounds displayed similar SARs to donepezil (Estrada et al., 2016).

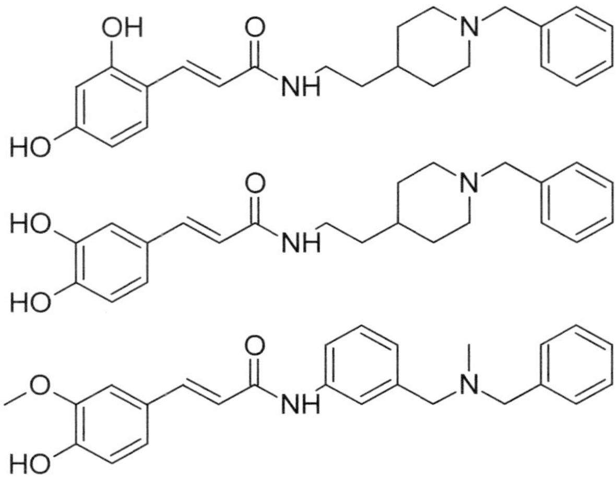

3
hAChE: $IC_{50} = 0.99\ \mu M$
hBuChE: $IC_{50} = 0.26\ \mu M$
hMAO-A: $IC_{50} = 5.5\ \mu M$
hMAO-B: $IC_{50} = 8.3\ \mu M$

4
hAChE: $IC_{50} = 1.75\ \mu M$
hBuChE: $IC_{50} = 0.69\ \mu M$
hMAO-A: $IC_{50} = 3.5\ \mu M$
hMAO-B: $IC_{50} = 6.0\ \mu M$

5
hAChE: $IC_{50} = 5.99\ \mu M$
hBuChE: $IC_{50} = 5.89\ \mu M$
hMAO-A: $IC_{50} = 4.5\ \mu M$
hMAO-B: $IC_{50} = 7.7\ \mu M$

Figure 11.8 Multi-targeted inhibitors described by Estrada et al. (2016).

A three-dimensional quantitative SAR study was used by Marco-Contelles and colleagues (2014) to define pharmacophores for multi-modal inhibition of relevant molecular targets and to design donepezil-pyridil hybrids with potential impact in the therapy of AD. Representative compound **6** was shown to be 318-fold more potent for the inhibition of AChE over BuChE with balanced MAO-A and MAO–B activities (Fig. 11.9, Marco-Contelles et al., 2014). The donepezil-pyridil hybrids with the most potent MAO-B inhibitory profiles interestingly had the longer linker lengths (5C-6C). Molecular modelling also revealed that compound **6** should bind with both the CAS and PAS of eeAChE and indicated that the cyano groups were important features in order to exhibit both AChE and BuChE inhibitory activities (Marco-Contelles et al., 2014).

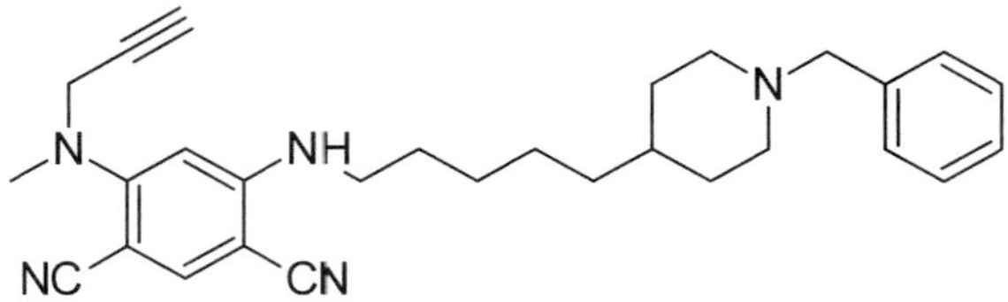

6
eeAChE: $IC_{50} = 1.30$ nM
eqBuChE: $IC_{50} = 530$ nM
hMAO-A: $IC_{50} = >50$ nM
hMAO-B: $IC_{50} = >50$ nM

Figure 11.9 A multifunctional donepezil-pyridil hybrid.

11.5.1.2 Coumarin derivatives

The coumarin nucleus (Fig. 11.10) has been acknowledged as a promising scaffold for the design of hybrid molecules that can concurrently impede AChE catalytic activity and AChE induced amyloid β aggregation (Fallarero et al., 2008) and act as MAO inhibitors. Several studies have shown that in order to modulate MAO and cholinesterase activity and selectivity, varying substitutions on positions 3 and 4 of the coumarin nucleus can be done (Soto-Ortega et al., 2011). Substitution at position 7 of the coumarin nucleus with small groups significantly reduces the cholinesterase inhibitory capacity of the coumarin nucleus, compared to when substituting with larger functional moieties (Fallarero et al., 2008). Many coumarin scaffold containing compounds such as AP2238 and ensaculin (Fig 11.10) have been successfully used as AChE inhibitors (Asadipour et al., 2013). Given the coumarins' capacity to bind at the PAS site of the AChE, and its relatively easy functionalisation, it becomes clear that by attaching a moiety to the coumarin function capable of binding to the catalytic site of AChE, a dual binding site compound could be designed. Patil et al. (2013) reported a comprehensive review on the establishment of coumarins for MAO inhibitory activity. A large number of research groups have effectively designed coumarin derivatives with significant MAO inhibitory activities (Piazzi et al., 2008; Patil et al., 2013). Coumarins are thus privileged molecules that can be used to design multi-potent compounds that could potentially be used in AD treatment (Samadi et al., 2012).

Driven by the study on ensaculin, Bruhlmann and colleagues (2001) undertook an exploratory study of AChE inhibition using some coumarinic MAO-B inhibitors in their library. SAR evaluation indicated that substitutions at position 3 of the coumarin nucleus modulated AChE (non-competitively) as well as MAO-B inhibitory activity. Substitution with methyl groups at both positions 3 and 4 led to more active compounds toward both AChE and MAO enzymes. Substitutions on the aromatic group of the benzyloxy moiety were also explored. Methyl groups in the ortho, meta, or para positions slightly decreased AChE activity, ortho substitution being the least favourable. Substitution in the ortho position also produced a decrease in MAO inhibition. Introducing electron-donating substituents in the meta position, i.e., OH, OCH_3, or NH_2, produced a decrease in AChE activity. With the exception of compound **7** (Fig. 11.10), which was the best AChE inhibitor (pK_{inc} = 5.47), substitution in the meta and/or para position by a halogen also produced a loss of activity toward AChE, while it led to a higher activity toward MAOs. Compound **7** is particularly interesting because it is among the best MAO-B inhibitors (pIC_{50} = 8.48), with a good MAO-B/A selectivity (Brühlmann et al., 2001).

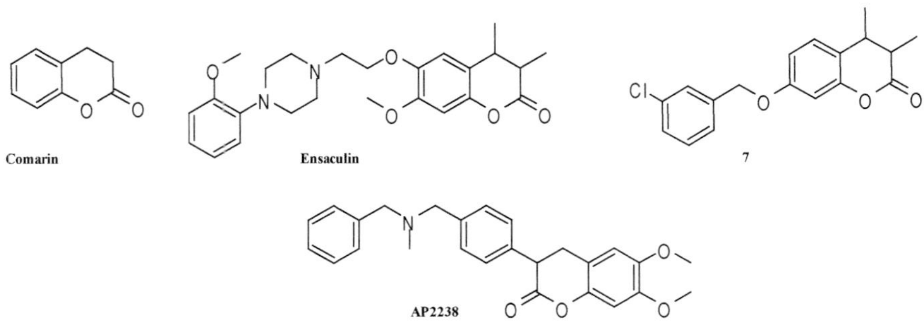

Figure 11.10 Structure of coumarin and its derivatives.

Tacrine-coumarin hybrids connected via piperazine alkane spacers with potent inhibitory activity towards AChE and BuChE and selective inhibition for MAO-B were developed by Xie and colleagues (2015) (Fig. 11.11).

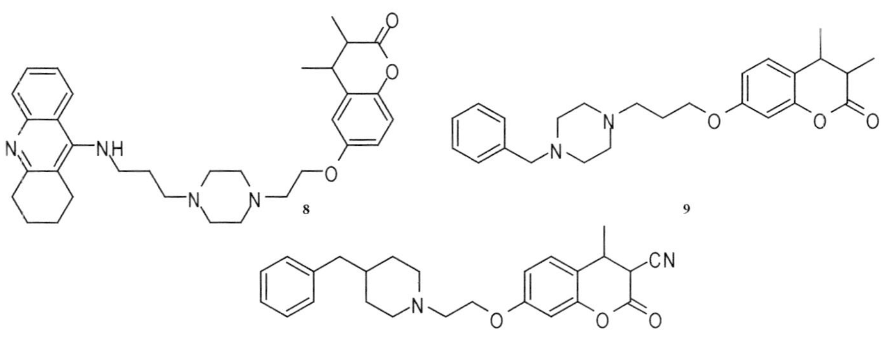

Figure 11.11 Coumarin-tacrine and coumarin–piperazine hybrid molecules.

These compounds were able to simultaneously bind to the PAS and CAS of AChE and because of the inclusion of the piperazine alkane spacer give additional contribution to AChE inhibition by binding to the mid-gorge site of the enzyme. From this series compound **8** was identified as a mixed-type AChE inhibitor (IC_{50} = 33.63 nM for eeAChE and 80.72 nM for eqBuChE) that was able to binding simultaneously to the CAS, PAS and mid-gorge of AChE. It was also able to competitively and selectively inhibit MAO-B with an IC_{50} of 45.07 μM for hMAO-A and 0.24 μM for hMAO-B (Xie et al., 2015). This group, went on to disclose another derivative of similar design to compound **8** (Xie et al., 2016). This time they combined the coumarin scaffold and the *N*-benzylpiperidine moiety of donepezil into one molecule. The biological results indicated that most of compounds displayed potent inhibitory activity for AChE and BuChE, and selective inhibition toward MAO-B over MAO-A. Of these compounds, **9** was the most potent inhibitor of eeAChE and eqBuChE (0.87 and 0.93 μM, respectively), and it was also a good and balanced inhibitor of hChEs and hMAO-B (IC_{50}: 1.37 μM for hAChE; 1.98 μM for hBuChE; 2.62 μM for hMAO-B). Molecular modelling and kinetic studies revealed that **9** was also a mixed-type inhibitor that binds simultaneously to the CAS, PAS and mid-gorge site of AChE. In addition, it was also found to be a competitive inhibitor of MAO-B (Xie et al., 2016).

Four series of coumarin-based derivatives were synthesised by Joubert et al. (2017) with a benzyl-, piperidine-, *N*-benzylpiperidine- or *p*-bromo-*N*-benzylpiperizine moiety, resembling the *N*-benzylpiperidine function of donepezil connected via an alkyl ether linkage. All the compounds displayed selective MAO-B inhibition with the benzyloxy series showing the most potent MAO-B activities. This series, however, possessed weak ChE activity. The most balanced ChE and MAO-B inhibitor was compound **10** (Fig. 11.11) from the benzylpiperazine series showing good eeAChE (IC_{50} = 9.10 μM) and eqBuChE activities (IC_{50} = 5.90 μM) and potent and selective hMAO-B inhibition (IC_{50} = 0.30 μM, selectivity index = >33) (Joubert et al., 2017).

Repsold et al. (2018) reported coumarin conjugates of morpholine, piperidine, thiophene and erucic acid (Fig. 11.12). The compounds portrayed varying degrees of inhibitory capacity towards the AChE and MAO-B enzymes and serve as new lead structures in the development of more active derivatives. Coumarin derivatives **11** and **12** stood out as the most promising compounds with IC_{50} values of 0.37 and 3.09 μM for hMAO-B and 30.9% and 21.56% inhibition of AChE at 1 μM, respectively. Docking studies revealed that the morpholine–coumarin compound **11** was able to occupy both the entrance and substrate cavities of the MAO-B active site, with the coumarin moiety residing in the substrate cavity, while the morpholine moiety resides in the entrance cavity. Moreover, compound **11** shows π-interactions with residues of Cys172, Leu171 and Ile198, and a relatively strong H-bond is present between the pyrone ring and Cys172. The coumarin entity of this compound was well positioned in the 'aromatic cage' of the MAO-B substrate cavity.

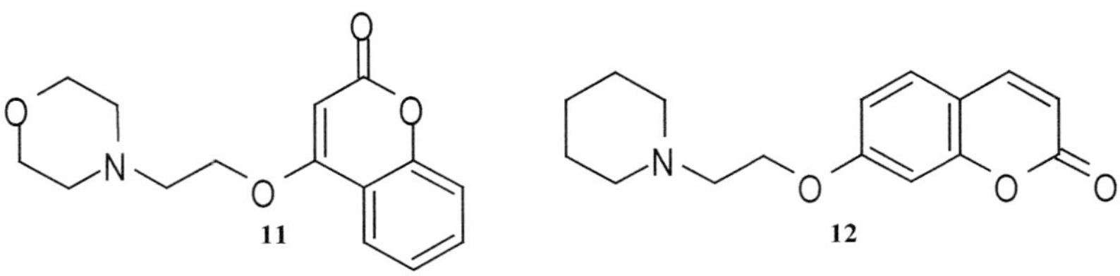

Figure 11.12 Coumarin conjugates by Repsold et al., 2018.

Further exploration of the effect of substitutions at position 3 of the coumarin nucleus on MAO-B inhibition was carried out by Matos et al. (2014) where they synthesised compounds with methoxy, hydroxyl and/or methyl substitutions at various positions of a fused benzocoumarin (compound **13**, Fig. 11.13). The highest studied compound inhibited AChE by up to 50% at a 100 μM concentration. The presence of a hydroxyl group at the ortho position of the phenyl group resulted in the most potent BuChE inhibitor (IC_{50} = 2.78 μM). When the group occupied the meta position, a 40 times decrease in activity was observed. The presence of a second hydroxyl group on the phenyl rings showed complete loss of activity. Replacement of the meta hydroxyl by a methoxy group diminished almost all activity. In this study, compounds with smaller molecular volume and higher molecular weight in general displayed better activity against BuChE. The MAO assay revealed that the compounds with a hydroxyl substitution were selective MAO-B inhibitors in the nanomolar to micromolar range. It was observed that a single methoxy group, whether meta, ortho or para, on the phenyl ring resulted in compounds with lower activity. Replacing it with a methyl group decreased the potency 1000-fold.

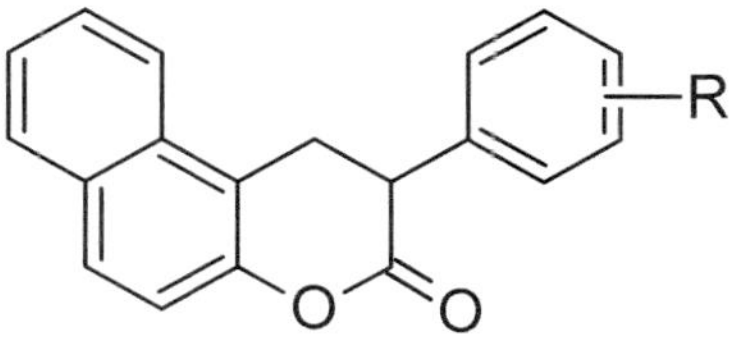

13
R = 2' -OH
hAChE: IC_{50} = 4.9 % at 100 μM
hBuChE: IC_{50} = 1.13 μM
hMAO-B: IC_{50} = 3.19 μM

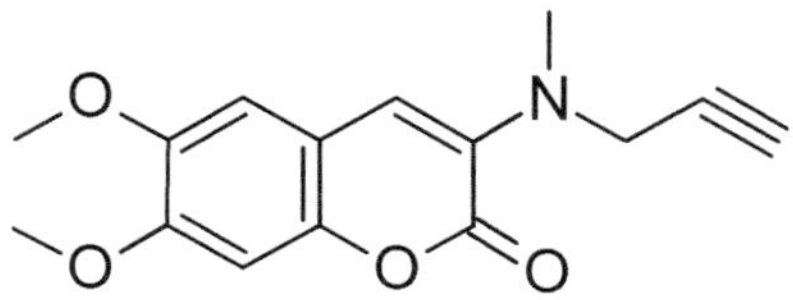

14
eeAChE: IC_{50} = 8.99 μM
eqBuChE: IC_{50} = < 5 % at 10 μM
rMAO-A: IC_{50} = < 5 % at 10 μM
MAO-B: IC_{50} = 2.8 μM

Figure 11.13 Substituted coumarin hybrids inhibiting ChE and MAO.

Various substituted protonatable basic moieties were introduced at position 3 of the coumarin to observe more SARs and positions 6 and 7 were used to study the effect of methoxy groups (Pisani et al., 2016). In vitro enzymatic screening towards ChEs and MAOs led to the identification a dual inhibitor **14** (Fig. 11.13) endowed with potent and selective MAO-B inhibitory activity in the low nanomolar range (IC_{50} = 2.8 nM) together with a moderate and selective AChE inhibitory activity (IC_{50} = 8.99 μM). The derivative had 2 methoxy groups at positions 6 and 7 and a propargylamine moiety at position 3. Compound **14** was studied further using the Madin-Darby canine kidney II permeability model in order to evaluate its ability to cross the blood-brain barrier (BBB) and penetrate into CNS. Furthermore, cell-based assays were used to assess the neuroprotective effect against different oxidative insults in human neuroblastoma SH-SY5Y cell lines. The results demonstrated a non-cytotoxic, neuroprotective and CNS active agent able to rapidly permeate BBB without interacting with efflux systems (Pisani et al., 2016).

11.5.1.3 Propargylated compounds

Pioneering work in the field of hybrid multi-target anti-AD drugs was carried out by Youdim and colleagues (Sterling et al., 2002).

They synthesised a large series of *N*-propargylaminoindans and *N*-propargyl-phenylethylamines as dual MAO/AChE inhibitors. The successful combination of the carbamate moiety of rivastigmine with rasagiline led to the compound ladostigil (**15**, Fig. 11.14). Ladostigil was shown to inhibit both cholinesterases (AChE = 31.80 μM and BuChE = 1.98 μM), and exhibited cognitive effects in animal models comparable to those of rivastigmine (Sterling, et al., 2002; Weinreb, et al., 2012). Once daily administration of ladostigil to rats for 2 weeks inhibited MAO-A and MAO–B by more than 70% with very little to no effect on intestinal enzymes (Weinstock, et al., 2000). The drug is currently in phase IIb clinical trials for mild cognitive impairment. In preclinical studies, ladostigil was shown to have broad neuroprotective activities against a variety of neurotoxins using both in vitro and in vivo models of neurodegeneration. This study itself was inspired by a study by Fink et al. (1996), who combined a tricyclic indole carbamate moiety with the propargylamine group. This approach was not pursued further due to low oral activity and poor brain penetration but it provided the framework for subsequent multi-functional inhibitors as seen in the case of ladostigil (Bolea et al., 2012).

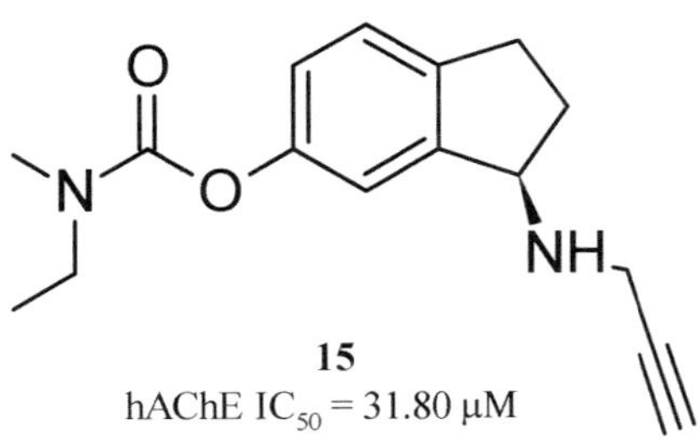

15
hAChE IC_{50} = 31.80 μM
eqBuChE IC_{50} = 1.98 μM
rMAO-B IC_{50} = 36.20 μM
rMAO-A IC_{50} = 300 μM

Figure 11.14 The structure of ladostigil.

In the design of new compounds structurally related ladostigil, studies have focused on combining the propargylamine and carbamoyl moiety utilising various different scaffolds and design strategies. One such study focused on combining the propargylamine and carbamoyl moiety connected through a phenyl scaffold. These compounds represented by **16** and **17** (Fig. 11.15) were good inhibitors of both MAO-A and MAO-B. In this study, the R-isomers, showed higher potency compared to their S-isomers. These results may indicate a common characteristic of chiral derivatives with the propargyl moiety for future drug design endeavours (Liu et al., 2016). The compounds were also found to have moderate ChE inhibitory activities with no significant difference between enantiomers. Docking studies of these compounds on the AChE, MAO–A and MAO–B enzymes were performed to elucidate possible binding interactions in the active sites. The AChE modelling results showed that the compounds possessed vital interactions with the cholinesterase active site. Both series of compounds showed interaction with Trp84, a key residue in the CAS. The propargylamine moieties of the structures were oriented towards the FAD cofactor of the MAO isozymes.

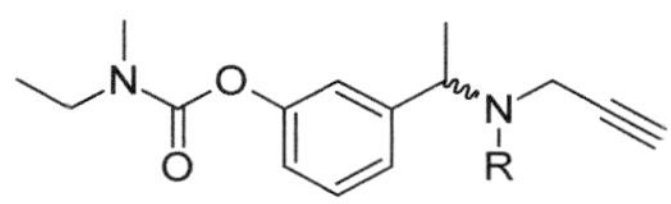

(16)(*R*)	(16)(*S*)	(17)(*R*)	(17)(*S*)
R = H	R = H	R = Me	R = Me
AChE IC_{50} = 7.7 μM	AChE IC_{50} = 11.5 μM	AChE IC_{50} = 48.2 μM	AChE IC_{50} = 58.3 μM
BuChE IC_{50} = 10.3 μM	BuChE IC_{50} = 12.8 μM	BuChE IC_{50} = 27.7 μM	BuChE IC_{50} = 34.6 μM
MAO-B IC_{50} = >1000 μM	MAO-B IC_{50} = >1000 μM	MAO-B IC_{50} = 4.57 μM	MAO-B IC_{50} = >1000 μM
MAO-A IC_{50} = 1.04 μM	MAO-A IC_{50} = 30.2 μM	MAO-A IC_{50} = 0.013 μM	MAO-A IC_{50} = 0.71 μM

Figure 11.15 Prorpargylamine-carbamoyl derived hybrid molecules with dual ChE and MAO inhibitory activities.

In a study by Denya et al. (2018) in vitro testing of indole-based compounds similar to ladostigil for ChE and MAO inhibitory activity was carried out using eeAChE, eqBuChE and hMAO enzymes. The assay results showed that the compounds were dual and non-selective inhibitors of the ChE enzymes. They all possessed IC_{50} values lower than 5 μM in the presence or absence of the propargylamine substitution on *N*1 of the indole. The compounds possessed good activity towards MAO–A and B and the presence of the urea moiety versus the carbamate moiety resulted in compounds that have comparably better inhibitory capacities than the carbamate linked compounds. In molecular modelling studies some interactions were observed between the indole ring and Trp 279 in AChE and Ille 199 in MAO, which showed that it probably plays a role in the inhibitory capacity of the compounds. Compound **18** had the best overall activity of all the tested compounds (Fig. 11.16). When compared to ladostigil, compound **18** demonstrated an AChE inhibitory capacity 9 times that of ladostigil and a 13-fold increase in MAO–B inhibition. Moreover, compound **18** also possesses superior MAO–A inhibitory capacity in vitro. It also proved to be more chemically stable compared to the carbamate-linked compound based on a 7-day forced degradation study (Denya et al., 2018).

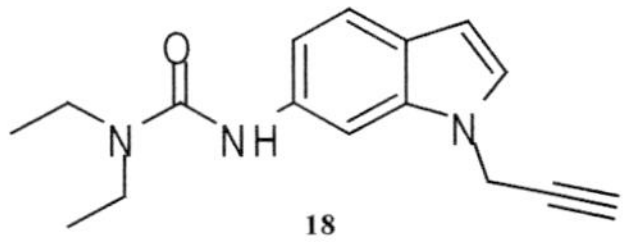

18

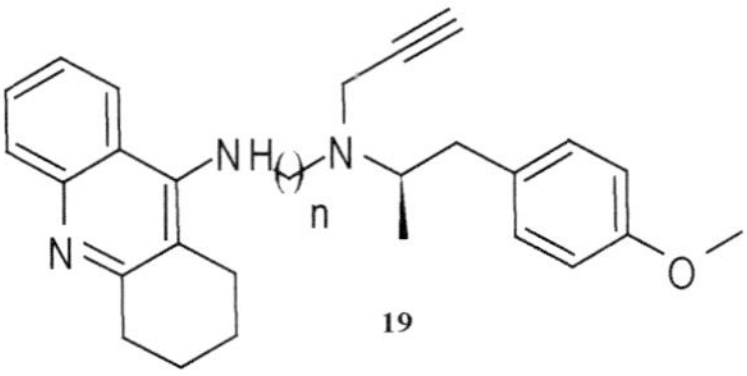

19

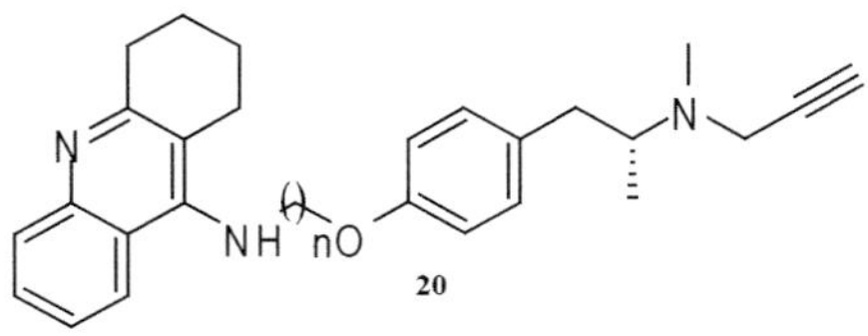

20

Figure 11.16 Various propargylamine incorporated hybrid molecules as MTDLs.

Lu et al. (2013a) combined a tacrine moiety (ChE inhibition) and the selegiline core (MAO inhibition) using carbon spacers of different lengths. Compounds **19** and **20** are representative structures for this series (Fig. 11.16). The results showed that all the target compounds were potent inhibitors of AChE and BuChE with IC_{50} values in the sub-micromolar range. In addition, the length of the alkylene linkage affected the inhibition of both AChE and BuChE. Compounds containing 2–5 carbon spacers between the tacrine and selegiline moiety provided stronger AChE inhibitory activities than tacrine. However, compounds with six and eight-carbon spacers exhibited a significant decrease in IC_{50}. It is interesting that the inhibitory activities significantly increased when the compounds possess longer (9 and 10) methylene spacers. The compound which had a 10-carbon spacer, exhibited the best IC_{50} value (2.03 nM) against AChE. Generally, the BuChE and MAO inhibitory activity was directly proportional to the number of methylene groups. Among the synthesised compounds, an eight-carbon spacer between the tacrine and selegiline moieties, gave the best results for both MAO-A (IC_{50} = 0.19 μM) and MAO-B (IC_{50} = 0.13 μM). Compounds with 3-6C chain are moderate MAO-A/B inhibitors in the micromolar range (Lu et al., 2013a).

Wang et al., 2014, reported on the synthesis, biochemical evaluation, ADME, toxicity and molecular modelling of novel multi-target directed hybrids of donepezil, selegiline and 8-hydroxyyquinoline for the potential prevention and treatment of AD. The most interesting derivative was racemic compound **21** (Fig. 11.17), an irreversible MAO-A/B inhibitor and mixed-type ChE inhibitor with metal-chelating properties. According to docking studies, both the enantiomers of **21** interact simultaneously with the CAS and PAS of eeAChE through a linker of appropriate length, supporting the observed mixed-type AChE inhibition. Both enantiomers exhibited a relatively similar position of both hydroxyquinoline and benzyl moieties with the rest of the molecule easily accommodated in the relatively large cavity of MAO-A. For MAO-B, the quinoline system was hosted at the entrance cavity whereas for MAO-A this system occupied the substrate cavity. In this disposition, the quinoline moiety interacted directly with the FAD aromatic ring. Very similar binding affinity values were also observed for both enantiomers with ChE and MAO enzymes. In addition, they established an essential role for the cyano group in properly positioning the ligand by hydrogen-bond formation with the amino acid residues of the binding sites of AChE, BuChE and MAO-A (Wang et al., 2014).

A series of novel propargylamine-modified pyrimidinylthiourea derivatives was developed as MTDLs for AD therapy. Among the derivatives **22** (Fig. 11.17) displayed a good and selective inhibitory profile against the AChE and MAO-B enzymes. Molecular docking studies showed that the pyrimidinylthiourea moiety of **22** could bind to the CAS of AChE, and the propargylamine moiety interacted directly with the FAD of MAO-B. Compound **22** does not stand out as a highly potent inhibitor compared to published literature, but it does introduce a new pyrimidinylthioureal scaffold into the pool of functional moieties (Xu et al., 2018).

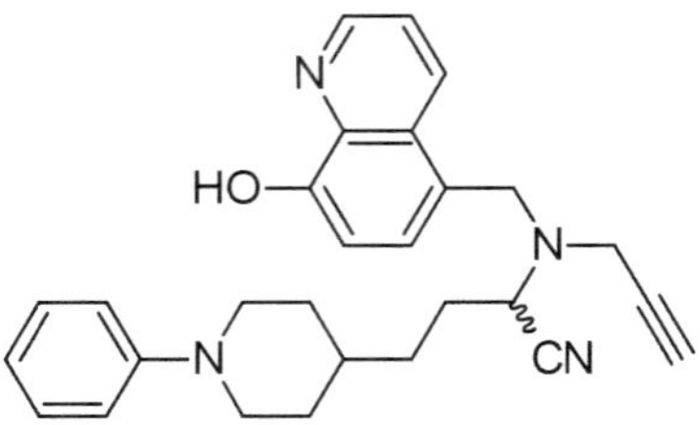

21

eeAChE: IC_{50} = 1.80 µM
BuChE: IC_{50} = 1.60 µM
MAO-A: IC_{50} = 6.20 µM
MAO-B: IC_{50} = 10.20 µM

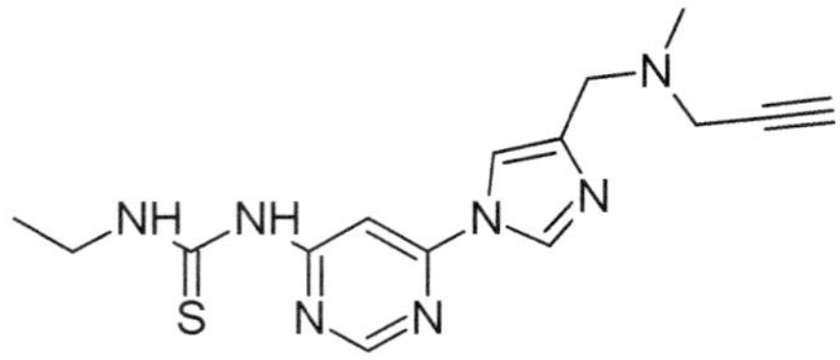

22

rAChE: IC_{50} = 0.32 µM
rBuChE: IC_{50} = 12.59 % at 40 µM
MAO-A: IC_{50} = 20.22 % at 50 µM
MAO-B: IC_{50} = 1.43 µM

Figure 11.17 Chemical structures of MTDLs **21** and **22** developed for AD therapy.

Another study was carried out by Košak et al., 2017 on *N*-propargylpiperidines conjugated tonaphthalene-2-carboxamide or naphthalene-2-sulfonamide moieties (Fig. 11.18). The *N*-propargylpiperidine sulfonamides were more potent hBuChE inhibitors compared to their analogous carboxamides. Removal of the *N*-alkyl chain from the carboxamide and sulfonamide nitrogen reduced the inhibitory potency, as secondary carboxamides and sulfonamides were significantly weaker inhibitors than their tertiary counterparts. Elongation of the *N*-alkyl chain with an additional methylene group did not affect the BuChE inhibitory potency significantly. Compounds bearing the propargyl on the carboxamide and sulfonamide nitrogen showed that this moiety is imperative for MAO-B inhibition. Compound **23** selectively inhibited hBuChE and MAO-B in a non-cytotoxic and neuroprotective manner.

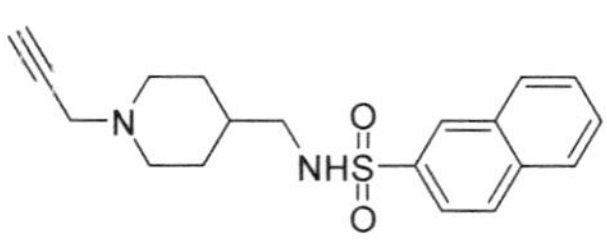

23

mAChE: IC_{50} = >100 µM
hBuChE: IC_{50} = 2.60 µM
hMAO-A: IC_{50} = >100 µM
hMAO-B: IC_{50} = 53.90 µM

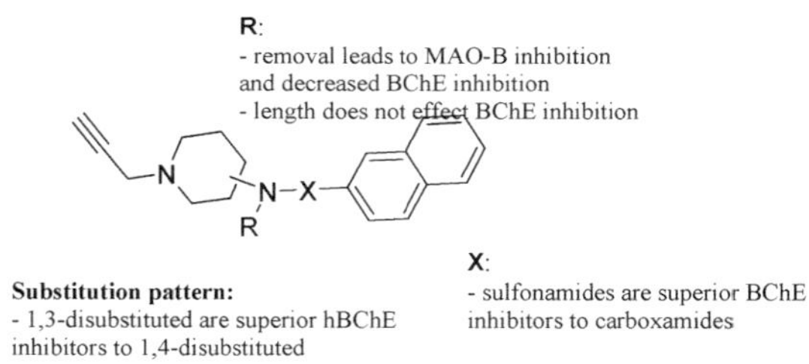

Figure 11.18 Structure of **23** and a summary of SAR findings on this class of compounds.

Compounds **24** and **25**, Fig. 11.19, were built from the *N*-benzylpiperidine and propargylamine moieties and connected through an appropriate linker to a central pyridine (**24**) or naphthyridine ring (**25**). The compounds were shown to be equipotent with donepezil and more potent than tacrine for eeAChE inhibition. All compounds were less potent that the reference compound for BuChE inhibition with IC_{50} values over 100 μM. For the naphthalene compounds inhibition of both ChEs increased with increased chain length. The latter also applied to MAO inhibition. This article concluded that in as much as this specific combination of moieties is not the best for multi-potent compounds, they established the importance of spacer length in interaction with both the CAS and PAS of AChE (Samadi et al., 2011).

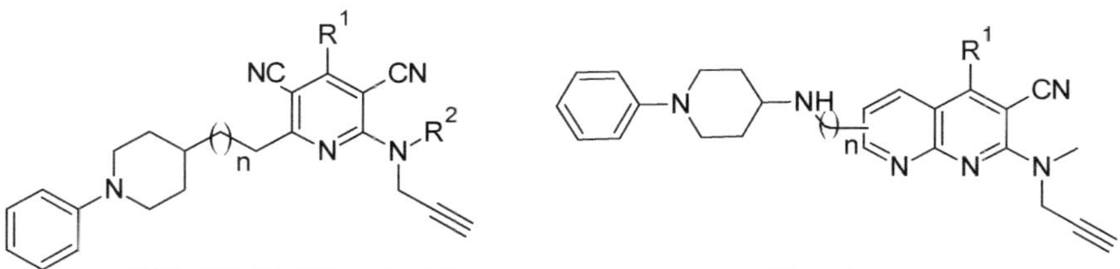

24 (R^1 = H, Ph; R^2 = H, Me; n = 0, 1, 2, 3) **25** (R^1 = H, Ph; n = 0, 1, 2, 3)

Figure 11.19 General formulas of **24** and **25**.

Multi-functional trihybrids were designed by Hirемathad et al. (2016) based on the combination of a benzothiazole moiety to tacrine linked via either a *S*-allylcysteine or *S*-propargylcysteine derivative (**26**, Fig. 11.20). These series of compounds showed AChE inhibitory activity in the submicromolar range, slightly better than tacrine. The compounds with the chloro-substituted tacrine moiety showed higher AChE inhibition (IC_{50}: 0.25–0.28 μM), with no significant difference between the propargyl/allyl cysteine substituents. Inclusion of the third benzothiazole moiety resulted in a considerable enhancement of AChE inhibitory activity. This may be attributed to stronger dual-binding mode of the trihybrid with both the CAS and PAS (Hiremathad et al., 2016). A very small difference in IC_{50} was observed after 30 min preincubation with MAO for both the non-propargyl and propargyl compounds suggesting that the new propargyl derivatives do not inactivate MAO to any notable difference. The high IC_{50} values obtained when inhibitor and substrate are added at the same time indicate low binding affinity, so it is likely that the propargyl group is not close enough to the *N*5 of the FAD in the active site of MAO to stop enzymatic activation (Hiremathad et al., 2016).

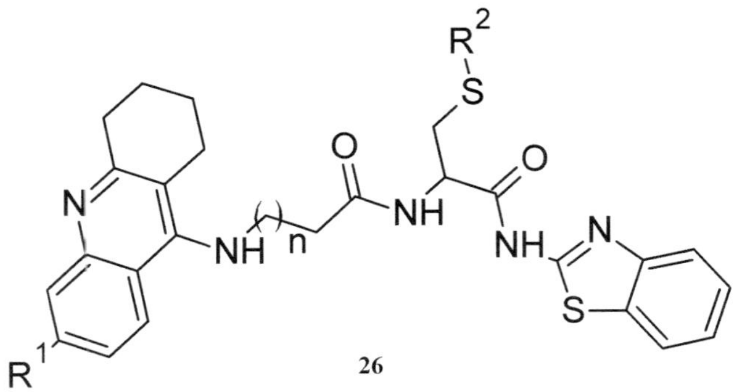

26

Figure 11.20 Multifunctional trihybrids designed by Hiremathad et al. (2016).

11.5.2 Miscellaneous Inhibitors

A potent MAO-B inhibiting analog of 2-acetyl-5-alkoxyphenol (IC_{50} = 2.9 nM) has recently been shown to positively affect learning and memory symptoms in AD animal models (Legoabe et al., 2015). Benzylamine side chains such as benzylpiperazine and *N*-methylbenzylamine derivatives has been shown to be active pharmacophoric groups for AChE inhibition (Sang et al., 2015). It is in light of this that Sang et al. (2017a) combined 2-acetylphenol with different length and functionalised benzylamine fragments to design a series of MTDLs. All the target compounds were selective AChE inhibitors with IC_{50} values ranging from 0.96 to 57.0 μM. Notably a longer methylene chain with piperidine and pyrrolidine terminal groups resulted in good activity. The opposite, however, was observed with regard to MAO inhibition. A satisfactory balance for the two enzymes was achieved with compound **27** (Fig. 11.21), a mixed type inhibitor that binds to both the CAS and PAS of AChE and acts as selective inhibitor of MAO-B (IC_{50} = 6.8 μM). Moreover, **27** showed the ability to act as an antioxidant, neuroprotectant and selective metal chelating agent in relevant assays (Sang et al., 2017a).

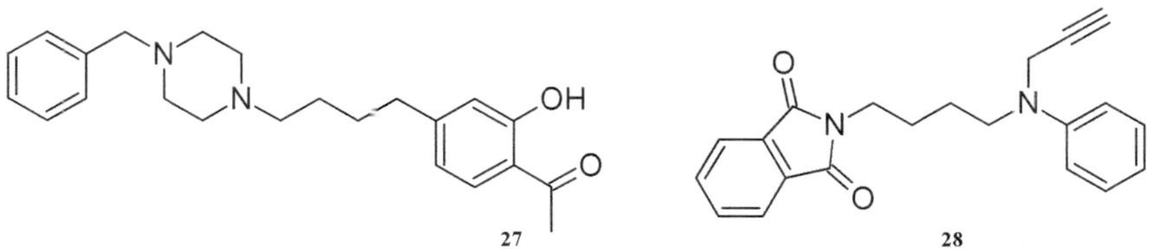

27
eeAChE: IC_{50} = 0.96 μM
eqBuChE: IC_{50} = 31.4 μM
hMAO-A: IC_{50} = 6.80 μM
hMAO-B: IC_{50} = 1.50 μM

28
eeAChE: IC_{50} = 0.84 μM
rBuChE: IC_{50} = 0.27 μM
hMAO-A: IC_{50} = < 5 % at 20 μM
hMAO-B: IC_{50} = 2.6 μM

Figure 11.21 Representative compounds **27** and **28** as MTDLs.

The researchers went on to apply a similar approach to Sang et al., 2017a, while replacing 2-acetylphenol with phthalimide fragment for MAO-B inhibition (Fig 11.21, Sang, et al., 2017b). The ChE assay in comparison with donepezil showed the compounds to be selective AChE inhibitors in the micromolar range. The compound with the *N*-(2-methoxybenzyl)ethanamine moiety connected via a linker to the phthalimide moiety was a highly selective eeAChE inhibitor with an IC_{50} value of 0.13 μM and selectivity index of 238.5. However, when BuChE was taken into consideration compound **28** had more relevant IC_{50} values (eeAChE IC_{50} = 0.84 μM and BuChE IC_{50} = 2.5 μM; SI = 3.0). The most active ChE inhibitors, including compound **28** were assayed for MAO. All of the compounds showed less than 5% inhibition at 20 μM for MAO-A (Sang et al., 2017b). Compound **28** exhibited the most potent MAO-B IC_{50} (2.6 μM). The presence of a benzylpropargylamine promoted good MAO-B activity as reported in literature (Sang et al., 2017b).

Sun et al., 2013 demonstrated that tacrine-homoisoflavonoid hybrids could yield better ChE and MAO-B inhibitory activity than either fragment alone (Fig. 11.22). A 1.8-fold increase in potency was observed for derivative **29** against hAChE compared to tacrine. The compound, featuring a methoxy group on the para-position of the homoisoflavonoid and a six-carbon linker displayed the most potent MAO-B activity. When the methoxy group was substituted with other groups or was at other positions of the homoisoflavonoid, weaker analogues were produced. In a similar study, Wang et al., 2015 synthesised compound **30** where the homoisoflavonoid moiety was connected via an appropriate carbon linker to a piperidine moiety (Fig. 11.22). This compound portrayed weaker ChE and similar MAO-B inhibitory potencies compared to **29**. Both **29** and **30** were shown to simultaneously interact with the CAS and the PAS of AChE. Near the bottom of the gorge, the tacrine moiety of **29** and the piperidine moiety of **30** interacted with important amino acids within the CAS. The homoisoflavonoid part of **29** and **30** reached the PAS and interacted with aromatic residues through π-π stacking. These studies have demonstrated the potential of the homoisoflavonoid as a promising scaffold in the design of MTDL.

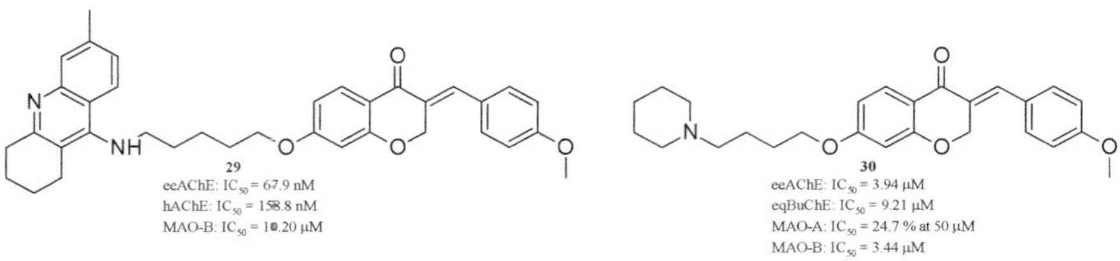

Figure 11.22 Homoisoflavonoid scaffolds used in the design of MTDLs.

A library of 3,4-(methylenedioxy)aniline-derived semi-carbazones were synthesised and evaluated for MAO and AChE inhibition for the treatment of AD. Compound **31** (Fig. 11.23) presented a balanced multi-functional profile with selective MAO-B (over MAO-A, selectivity index = 77) and AChE inhibition. In general the tested compounds were not neurotoxic. The presence of the semicarbazino (-NH-CO-NH-N=C-) linker was likely an important feature for these derivatives to exhibit both MAO-B and AChE inhibitory activities. The flexible semicarbazino linker, possessing functional moieties for hydrogen bonding, is critical for guiding the optimal orientation of the hydrophobic aryl/heteroaryl moieties of both terminals of **31** into their respective binding pockets within the active sites of MAO-B and AChE (Tripathi et al., 2016, 2017).

The xanthone moiety has been used by a number of research groups as a promising scaffold in the design of MTDL with MAO and ChE inhibitory activities (Cruz et al., 2017; Luo et al., 2017; Pinto et al., 2005; Azevedo et al., 2012). Xanthones are a class of oxygenated heterocyclic compounds with a dibenzo-γ-pyrone scaffold (Pinto et al., 2005). In addition to natural xanthones, there are a large variety of xanthones of synthetic origin. Xanthones obtained by synthesis may hold simple substituents such as hydroxy, methoxy, methyl and carboxy moieties to more complex substitution patterns such as epoxide, azole, methylidenebutyrolactone, aminoalcohol, sulfamoyl, methylthiocarboxylic acid and dihydropyridine functional groups. Cruz et al. (2017) went on to summarise the SAR findings from numerous studies on xanthone scaffolds used to develop compounds with dual AChE and MAO inhibitory activities (Fig. 11.24).

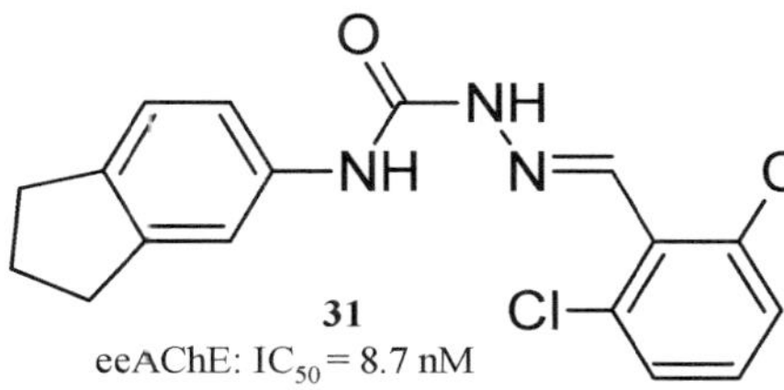

31

eeAChE: IC_{50} = 8.7 nM
MAO-A: IC_{50} = 4.52 μM
MAO-B: IC_{50} = 59.0 nM

Figure 11.23 Semicarbazone derived MTDL.

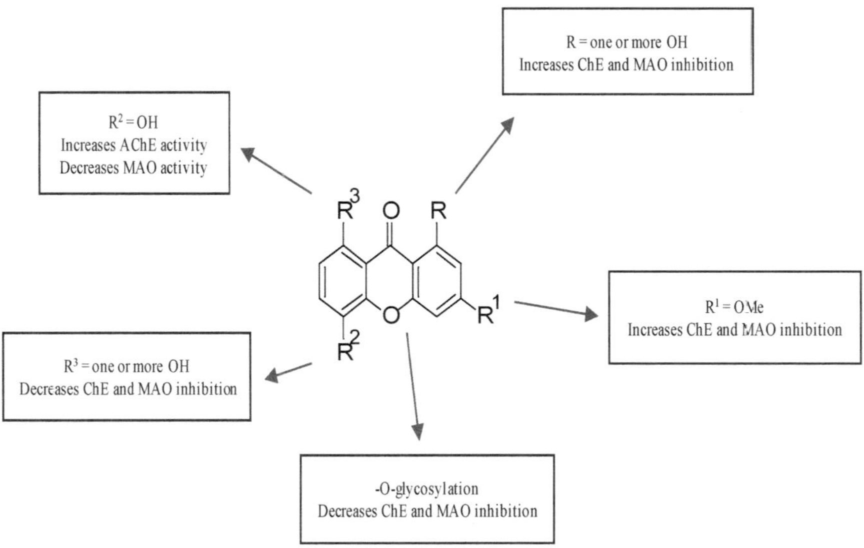

Figure 11.24 Applicable SARs for dual MAO and ChE inhibitory activities in xanthones. Adapted from Cruz et al., 2017).

In continuation on the research of xanthone derivatives as MTDLs a recent study by Luo et al., 2017 described the synthesis and evaluation of thioxanthone derivatives connected by an appropriate spacer to a benzylamine moiety represented by compound **32** (Fig. 11.25). Within this series the compounds possessing an *N*-methyl group usually displayed better AChE inhibitory activity than those with an *N*-ethyl group. Furthermore, the inclusion of a 2-dimethylamino group on the benzylamine moiety dramatically improved AChE inhibitory activity. Almost all synthesised compounds exhibited weak activity against BuChE. This result showed that these derivatives were potent and highly selective AChE inhibitors. These compounds also showed promising MAO inhibitory activities. Molecular modelling revealed that the thioxanthone ring was not located in close proximity to the FAD in MAO-B. However, the xanthine moiety was able to form two hydrogen bonds between the hydroxyl and the active site residues Ile199 and Tyr326. In addition, the benzylamine moiety of **30** interacted with Tyr398 and Tyr435 near the FAD cofactor via parallel π-π interactions. The docking studies, therefore explained the high MAO-B inhibitory activity of these derivatives. These compounds were also able to inhibit amyloid-β aggregation with good metal chelating properties (Luo et al., 2017).

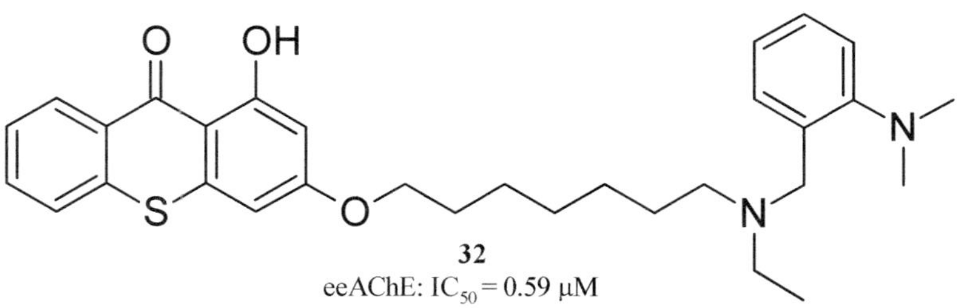

32

eeAChE: IC_{50} = 0.59 μM

rBuChE: IC_{50} = 28.7 μM

hMAO-A: IC_{50} = 1.01 μM

hMAO-B: IC_{50} = 0.90 μM

Figure 11.25 Thioxanthone derivatives as MTDLs represented by **32**.

Indolotacrine analogues that resemble the MAO inhibiting β-carboline alkaloid, harmine, were designed as multi-functional agents for AD (Benek et al., 2015). The rationale behind the hybridisation of the indole and tacrine moieties was because of the reported potency of tacrine as AChE inhibitor as well as the potential of indole derivatives to act as MAO inhibitors (Pérez et al., 1999). The most promising derivative, **33**, was able to inhibit ChE and MAO enzymes in the low micromolar range in vitro (Fig. 11.26). The compounds, compared to the parent compounds, displayed a slightly higher hepatotoxic and cytotoxic profile, but serve as a promising lead compound for further optimisation.

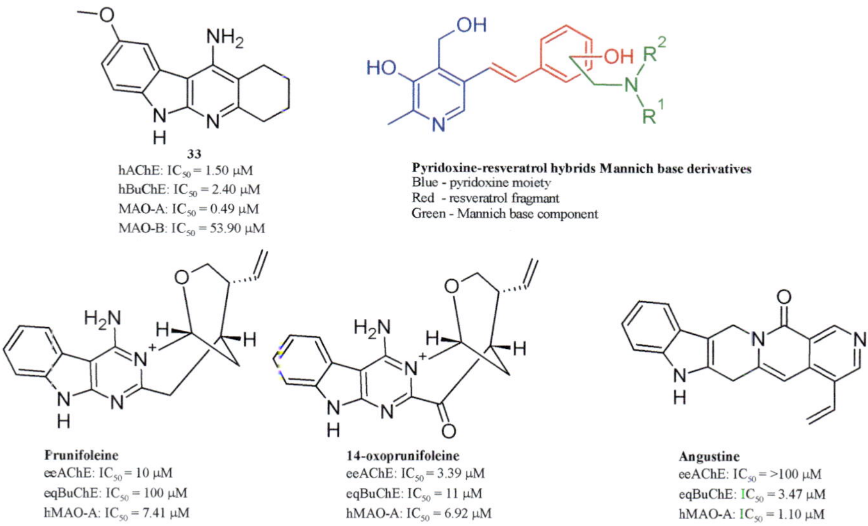

Figure 11.26 Various MTDLs as dual MAO and ChE inhibitors for the treatment of AD.

Resveratrol, a natural product with a stilbene structure, has been widely studied for its wide range of biological activities. Recent research has suggested that resveratrol is an anti-AD agent which possesses antioxidant, anti-inflammatory, MAO inhibitory and neuroprotective properties (Baur and Sinclair, 2006; Lu et al., 2013b). Moreover, pyridoxine (vitamin B6) has been shown to be a potent antioxidant that may have some benefit in AD models (Yang et al., 2017). Mannich base derivatives (another good antioxidant) were introduced to the pyridoxine-resveratrol hybrids (Fig. 11.26) and evaluated for ChE, MAO and antioxidant activity. The results confirmed what other studies have described and established SARs such as the necessity of a pyrollidine or piperidine moiety for AChE activity. Introduction of the Mannich base to the hybrids, however, decreases both ChE and MAO activity. Overall the compounds displayed poor inhibitory profiles, but these preliminary findings could serve as a starting point for the design of new MTDLs (Yang et al., 2017).

Alkaloids have also played a part in the field of MTDLs. In an initial exploratory study psychotria alkaloids were evaluated for activity as dual inhibitors of ChEs and MAOs. The quaternary β-carboline alkaloids prunifoleine and 14-oxoprunifoleine shown in Fig. 11.26 inhibited AChE (non-competitive), BuChE and MAO-A (time-dependent). The monoterpene indole alkaloid angustine was also able to inhibit BuChE and MAO-A reversibly and competitively. It bound to the MAO-A active site in a similar mannerto harmine, a known MAO inhibitor (Passos et al., 2013). All the alkaloids proved to be poor inhibitors of MAO-B with the most potent being prunifoleine having an IC_{50} of 340 μM. Taken together, these findings highlight the potential of natural metabolites as scaffolds for MTDL development.

11.6 Conclusion

Our understanding of AD has advanced enormously over the past few years, with much effort being devoted into research ligands that portray excellent in vitro activity. Unravelling the complex pathogenesis of AD has led to the synthesis of novel compounds capable of acting on different targets involved in AD development. With all the hypotheses regarding the pathophysiology of the disease, the cholinesterase and MAOs seem to always play a role either directly as is seen with AChE and BuChE in the cholinergic hypothesis, or indirectly as is the case with MAO-B and oxidative stress and AChE and amyloid plaque formation.

Literature has reported highly selective ligands that do not necessarily result into clinically effective drugs due to the multifactorial nature of the disease (Cavalli et al., 2008). Recent research has introduced the concept of MTDLs which involves the design of a molecule that can target multiple pathways of AD that contribute to disease pathogenesis (Bolognesi et al., 2006, Zhang, 2005). The rational design of compounds that simultaneously modulate different protein targets, either positively or negatively, at a comparable concentration for each target remains a challenging task. As shown by the reported examples, the most frequently adopted MTDLs design strategy has been based mainly on linking two properly selected pharmacophores through a spacer. During the subsequent optimisation phase, analogues are made to appropriately balance the desired activities for the two targets.

Various studies have capitalised on the structural relevance of single target molecules to design series' of compounds where pharmacophoric moieties such as donepezil, selegiline, propargylamine, tacrine and coumarins are employed to obtain compounds proposed to inhibit both cholinesterase and MAO enzymes. Evidence, as discussed in this chapter, suggests that the pool of possible multi-functional derivatives is far from exhausted and with further understanding of the disease more effective MTDL ligands will be developed and may lead to new treatment options for patients suffering from AD.

References

Aldridge, W. N., *Biochem. J.*, **53** (1953), 62–67.

Alipour, M., Khoobi, M., Foroumadi, A., Nadri, H., Moradi, A., Sakhteman, A., Ghandi, M., and Shafiee, A. *Bioorg. Med. Chem.*, **20** (2012), 7214–7222.

Allderdice, P. W., Gardner, H. A. R., Galutira, D., Lockridge, O., Ladu, B. N., and McAlpine, P. J., *Genomics*, **11** (1991), 452–454.

Asadipour, A., Alipour, M., Jafari, M., Khoobi, M., Emami, S., Nadri, H., Sakhteman, A., Moradi, A., Sheibani, V., Moghadam, F. H., and Shafiee, A., *Eur. J. Med. Chem.*, **70** (2013), 623–630.

Augustinsson, K.-B., Virtanen, A. I., Åqvist, S., and Rottenberg, M., *Acta. Chem. Scand.*, **4** (1950), 1149–1150.

Azevedo, C., Afonso, C., and Pinto, M. *Curr. Org. Chem.*, **16** (2012), 2818–2867.

Backman, L., Jones, S., Berger, A., Laukka, E. J., and Small, B. J., *J. Intern. Med.*, **256** (2004), 195–204.

Bartus, R. T., *Exp. Neurol.*, **163** (2000), 495–529.

Baur, J., and Sinclair, D. *Nat. Rev. Drug Discov.*, **5** (2006), 493–506.

Benedetti, M. S., and Doster, P., *Biochem. Pharmacol.*, **38** (1989), 20159.

Benek, O., Soukup, O., Pasdiorova, M., Hroch, L., Sepsova, V., Jost, P., Hrabinova, M., Jun, D., Kuca, K., Zala, D., Ramsay, R., Marco-Contelles, J., and Musilek, K., *ChemMedChem*, **11** (2015), 1264–1269.

Bhat, A. H., Dar, K. B., Anees, S., Zargar, M. A., Masood, A., Sofi, M. A., Ganie, S. A., *Biomed. Pharmacother.*, **74** (2015), 101–10.

Binda, C., Li, M., Hubalek, F., Restelli, N., Edmondson, D. E., and Mattevi, A., *Proc. Natl. Acad. Sci. U. S. A.*, **100** (2003), 9750–9755.

Binda, C., Newton-Vinson, P., Hubálek, F., Edmondson, D. E., and Mattevi, A., *Nat. Struct. Mol. Biol.*, **9** (2001), 22–26.

Black, R., *Evaluating Prescription Drugs Used to Treat Alzheimer's Disease: Comparing Effectiveness, Safety, and Price.* Available from: http://www.consumerreports.org/health/resources/pdf/best-buy-drugs/AlzheimersFINAL.pdf (2012) (accessed: 14 June 2016).

Bolea, I., Gella, A., and Unzeta, M., *J. Neural. Transm.*, **120** (2012), 893–902.

Bolea, I., Ju rez-Jim nez, J., de los os, C., Chioua, M., Pouplana, R., Luque, F. J., Unzeta, M., Marco-Contelles, J., and Samadi, A., *J. Med. Chem.*, **54** (2011), 8251–8270.

Bolognesi, M. L., Minarini, A., Tumiatti, V., and Melchiorre, C., *Expert Opin. Ther. Pat.*, **16** (2006), 811–823.

Bond, M., Rogers, G., Peters, J., Anderson, R., Hoyle, M., Miners, A., Moxham, T., Davis, S., Thokala, P., Wailoo, A., Jeffreys, M., and Hyde, C., *Health Technol. Assess.*, **16** (2012), 1–470.

Botti, S. A., Felder, C. E., Sussman, J. L., and Silman, I., *PEDS*, **11** (1998), 415–420.

Bourne, Y., Taylor, P., Bougis, P. E., and Marchot, P., *J. Biol. Chem.*, **274** (1999), 2963–2970.

Bourne, Y., Taylor, P., and Marchot, P., *Cell*, **83** (1995), 503–512.

Brühlmann, C., Ooms, F., Carrupt, P., Testa, B., Catto, M., Leonetti, F., Altomare, C., and Carotti, A., *J. Med. Chem.*, **44** (2001), 3195–3198.

Burk, J., Herzog, C., Porter, M., and Sarter, M., *Neurobiol. Aging*, **23** (2002), 467–77.

Cavalli, A., Bolognesi, M. L., Minarini, A., Rosini, M., Tumiatti, V., Recanatini, M., and Melchiorre, C. *J. Med. Chem.*, **51** (2008), 347–372.

Chan-Palay, V., Höchli, M., Savaskan, E., and Hungerecker, G., *Dement. Geriatr. Cogn.*, **4** (1993).

Christen, Y., *Am. J. Clin. Nutr.*, **71** (2000), 621S–629S.

Colovic, M. B., Krstic, D. Z., Lazarevic-Pasti, T. D., Bondzic, A. M., and Vasic, V. M., *Curr. Neuropharmacol.*, **11** (2013), 315–335.

Cruz, M., Cidade, H., and Pinto, M., *Future Med. Chem.*, **9** (2017), 1611–1630.

Dave, K. R., Syal, A. R., and Katyare, S. S., *Z. Naturforsch C Bio. Sci.*, **55** (2000).

Davies, P., and Maloney, A., *Lancet*, **308** (1976), 1403.

De Colibus, L., Li, M., Binda, C., Lustig, A., Edmondson, D. E., and Mattevi, A., *Proc. Natl. Acad. Sci. U. S. A.*, **102** (2005), 12684–12689.

Denya, I., Malan, S., Enogieru, A., Omoruyi, S., Ekpo, O., Kapp, E., Zindo, F., and Joubert, J., *MedChemComm*. **00** (2018), 1–14.

Edmondson, D., *Neurotoxicology*, **25** (2004), 63–72.

Edmondson, D. E., Binda, C., and Mattevi, A., *Arch. Biochem. Biophys.*, **464** (2007), 269–276.

Edmondson, D. E., Mattevi, A., Binda, C., Li, M., and Hubalek, F., *Curr. Med. Chem.*, **11** (2004), 1983–1993.

Ekholm, M., and Konschin, H., *Comput. Theor. Chem.*, **467** (1999), 161–172.

Esteban, G., Allan, J., Samadi, A., Mattevi, A., Unzeta, M., Marco-Contelles, J., Binda, C., and Ramsay, R., *BBA-Proteins Proteom.*, **1844** (2014), 1104–1110.

Estrada, M., Herrera-Arozamena, C., Pérez, C., Viña, D., Romero, A., Morales-García, J., Pérez-Castillo, A., and Rodríguez-Franco, M., *Eur. J. Med. Chem.*, **121** (2016), 376–386.

Fallarero, A., Oinonen, P., Gupta, S., Blom, P., Galkin, A., Mohan, C. G., and Vuorela, P. M., *Pharmacol. Res.*, **58** (2008), 215–221.

Fink, D., Palermo, M., Bores, G., Huger, F., Kurys, B., Merriman, M., Olsen, G., Petko, W., and O'Malley, G. *Bioorg. Med. Chem Lett.* **6** (1996), 625–630.

Foley, P., Gerlach, M., Youdim, M. B. H., and Riederer, P., *Parkinsonism Relat. D*, **6** (2000), 25–47.

Förstl, H., and Kurz, A., *Eur. Arch. Psychiatry Clin. Neurosci.*, **249** (1999), 288–290.

Fowler, C., Wiberg, Å., Oreland, L., Marcusson, J., and Winblad, B., *J. Neural Transm.*, **49** (1980), 1–20.

Francis, P. T., Palmer, A. M., Snape, M., and Wilcock, G. K., *J. Neurol. Neurosurg. Psychiatry*, **66** (1999), 137–147.

Frank, E., *J. S. C. Med. Assoc.*, **90** (1994), 417–23.

Giacobini, E., *Pharmacol. Res.*, **50** (2004), 433–440.

Gibson, G. E., and Peterson, C., *J. Neurochem.*, **37** (1981), 978–984.

Grossberg, G. T., Edwards, K. R., and Zhao, Q., *J. Clin. Pharmacol.*, **46** (2006), 17S–26S.

Harel, M., Schalk, I., Ehret-Sabatier, L., Bouet, F., Goeldner, M., Hirth, C., Axelsen, P. H., Silman, I., and Sussman, J. L., *Proc. Natl. Acad. Sci. U. S. A.*, **90** (1993), 9031–9035.

Hebb, C. O., and Whittaker, V. P., *J. Physiol.*, **142** (1958), 187–196.

Hiremathad, A., Chand, K., Esteves, A., Cardoso, S., Ramsay, R., Chaves, S., Keri, R., and Santos, M., *RSC Adv.*, **6** (2016), 53519–53532.

Hosea, N. A., Radić, Z., Tsigelny, I., Berman, H. A., Quinn, D. M., and Taylor, P., *Biochemistry*, **35** (1996), 10995–11004.

Inestrosa, N. C., and Alarcón, R., *J. Physiol.-Paris*, **92** (1998), 341–344.

Inestrosa, N. C., Alvarez, A., Pérez, C. A., Moreno, R. D., Vicente, M., Linker, C., Casanueva, O. I., Soto, C., and Garrido, J., *Neuron*, **16** (1996), 881–891.

Johnson, G., and Moore, S., *Curr. Pharm. Des.*, **12** (2006), 217–225.

Joubert, J., Foka, G., Repsold, B., Oliver, D., Kapp, E., and Malan, S. *Eur. J. Med. Chem.*, **125** (2017), 853–864.

Katzman, R., and Saitoh, T., *Faseb. J.*, **5** (1991), 278–286.

Kawakami, Y., Inoue, A., Kawai, T., Wakita, M., Sugimoto, H., and Hopfinger, A. J., *Bioorganic Med. Chem.*, **4** (1996), 1429–1446.

Kearney, E. B., Salach, J. I., Walker, W. H., Seng, R. L., Kenney, W., Zeszotek, E., and Singer, T. P., *Eur. J. Biochem.*, **24** (1971), 321–327.

Khachaturian, Z. S., *JAMA Neurol.*, **42** (1985), 1097–1105.

Kirksey, T. J., Kwan, S.-W., and Abell, C. W., *Biochemistry*, **37** (1998), 12360–12366.

Knez, D., Sova, M., Košak, U., and Gobec, S., *Future Med. Chem.*, **9** (2017), 811–832.

Košak, U., Knez, D., Coquelle, N., Brus, B., Pišlar, A., Nachon, F., Brazzolotto, X., Kos, J., Colletier, J., and Gobec, S., *Bioorganic Med. Chem.*, **25** (2017), 633–645.

Kreienkamp, H. J., Weise, C., Raba, R., Aaviksaar, A., and Hucho, F *Proc. Natl. Acad. Sci. U. S. A.*, **88** (1991), 6117–6121.

Kryger, G., Silman, I., and Sussman, J. L., *Structure*, **7** (1999), 297–307.

Li, M., Hubálek, F., Newton-Vinson, P., and Edmondson, D. E., *Protein Expr. Purif.*, **24** (2002), 152–162.

Li, F., Wang, Z., Wu, J., Wang, J., Xie, S., Lan, J., Xu, W., Kong, L., and Wang, X., *J. Enzyme Inhib. Med. Chem.*, **31** (2016), 41–53.

Legoabe, L., Petzer, J., and Petzer, A. (2015). *Drug Des. Dev. Ther.*, **9** (2015), 363–365

Lu, C., Guo, Y., Yan, J., Luo, Z., Luo, H., Yan, M., Huang, L., and Li, X., *J. Med. Chem.*, **56** (2013a), 5843–5859.

Lu, C., Zhou, Q., Yan, J., Du, Z., Huang, L., and Li, X., *Eur. J. Med. Chem.*, **62** (2013b), 745–753.

Luo, L., Li, Y., Qiang, X., Cao, Z., Xu, R., Yang, X., Xiao, G., Song, Q., Tan, Z., and Deng, Y. *Bioorganic Med. Chem.*, **25** (2017), 1997–2009.

Mallender, W. D., Szegletes, T., and Rosenberry, T. L., *Biochemistry*, **39** (2000), 7753–7763.

Marco-Contelles, J., Bautista-Aguilera, O., Esteban, G., Chioua, M., Nikolic, K., Agbaba, D., Moraleda, I., Iriepa, I., Soriano, E., Samadi, A., and Unzeta, M. *Drug Des. Devel. Ther.*, (2014), 1893.

Marco-Contelles, J., Unzeta, M., Bolea, I., Esteban, G., Ramsay, R., Romero, A., Martínez-Murillo, R., Carreiras, M., and Ismaili, L., *Front. Neurosci.*, **10** (2016).

Masson, P., Xie, W., Froment, M.-T., Levitsky, V., Fortier, P.-L., Albaret, C., and Lockridge, O., *BBA-Protein Struct. M.*, **1433** (1999), 281–293.

Masson, P., Legrand, P., Bartels, C. F., Froment, M.-T., Schopfer, L. M., and Lockridge, O., *Biochemistry*, **36** (1997), 2266–2277.

Masson, P., Xie, W., Froment, M.-T., and Lockridge, O., *BBA-Protein Struct. Mol. Enzymol.*, **1544** (2001), 166–176.

Massoulié, J., Pezzementi, L., Bon, S., Krejci, E., and Vallette, F.-M., *Prog. Neurobiol.*, **41** (1993), 31–91.

Matos, M., Janeiro, P., González Franco, R., Vilar, S., Tatonetti, N., Santana, L., Uriarte, E., Borges, F., Fontenla, J., and Viña, D., *Future Med. Chem.*, **6** (2014), 371–383.

Mayeux, R., and Sano, M., *N. Engl. J. Med.*, **341** (1999), 1670–1679.

Mehta, M., Adem, A., and Sabbagh, M. *Int. J. Alz. Dis.*, **2012** (2012), pp. 1–8.

Mesulam, M.-M., and Geula, C., *J. Comp. Neurol.*, **306** (1991), 193–220.

Mesulam, M., Guillozet, A., Shaw, P., Levey, A., Duysen, E. G., and Lockridge, O., *Neuroscience*, **110** (2002), 627–639.

Mooser, G., and Sigman, D. S., *Biochemistry*, **13** (1974), 2299–2307.

Munoz-Torrero, D., and Camps, P., *Curr. Med. Chem.*, **13** (2006), 399–422.

Newton-Vinson, P., Hubalek, F., and Edmondson, D. E., *Protein Expr. Purif.*, **20** (2000), 334–345.

Nicotra, A., Pierucci, F., Parvez, H., and Senatori, O., *Neurotoxicology*, **25** (2004), 155–165.

Nielsen, J. A., Mena, E. E., Williams, I. H., Nocerini, M. R., and Liston, D., *Eur. J. Pharmacol.*, **173** (1989), 53–64.

Nilsson, L., Nordberg, A., Hardy, J., Wester, P., and Winblad, B. *J. Neural Transm.*, **67** (1986), 275–285.

O'Carroll, A.-M., Fowler, C. J., Phillips, J. P., Tobbia, I., and Tipton, K. F., *Naunyn-Schmiedeberg's Arch. Pharmacol.*, **322** (1983), 198–202.

Ordentlich, A., Barak, D., Kronman, C., Ariel, N., Segall, Y., Velan, B., and Shafferman, A., *J. Biol. Chem.*, **273** (1998), 19509–19517.

Parsons, C. G., Danysz, W., Dekundy, A., and Pulte, I., *Neurotox. Res.*, **24** (2013), 358–369.

Passos, C., Simões-Pires, C., Nurisso, A., Soldi, T., Kato, L., de Oliveira, C., de Faria, E., Marcourt, L., Gottfried, C., Carrupt, P., and Henriques, A., *Phytochemistry*, **86** (2013), 8–20.

Patil, P., Bari, S., Firke, S., Deshmukh, P., Donda, S., and Patil, D., *Bioorganic Med. Chem.*, **21** (2013), 2434–2450.

Pérez V., Marco J. L., Fernández-Álvarez E., Unzeta M., *Br. J. Pharmacol.* **127**(4) (1999), 869–876.

Perry, E. K., Gibson, P. H., Blessed, G., Perry, R. H., and Tomlinson, B. E., *J. Neurol. Sci.*, **34** (1977), 247–265.

Perry, E., Walker, M., Grace, J., and Perry, R., *Trends Neurosci.*, **22** (1999), 273–280.

Piazzi, L., Cavalli, A., Colizzi, F., Belluti, F., Bartolini, M., Mancini, F., Recanatini, M., Andrisano, V., and Rampa, A., *Bioorganic Med. Chem Lett.*, **18** (2008), 423–426.

Pinto, M., Sousa, M., and Nascimento, M., *Curr. Med. Chem.*, **12** (2005), 2517–2538.

Pisani, L., Catto, M., Leonetti, F., Nicolotti, O., Stefanachi, A., Campagna, F., and Carotti, A. (2011), *Curr. Med. Chem.*, **18** (2011), 4568–4587.

Pisani, L., Farina, R., Soto-Otero, R., Denora, N., Mangiatordi, G., Nicolotti, O., Mendez-Alvarez, E., Altomare, C., Catto, M., and Carotti, A., *Molecules*, **21** (2016), 362.

Praticò, D., and Sung, S., *J. Alzheimer's Dis.*, **6** (2004), 171–175.

Prince, M., Wimo, A., Guerchet, M., Ali, G.-C., Wu, Y.-T., and Prina, M., World Alzheimer Report 2015: The Global Impact of Dementia. Available at: https://www.alz.co.uk/research/worldalzheimerreport-2015summary.pdf (2015) (accessed: 3 November 2016).

Radic, Z., Pickering, N. A., Vellom, D. C., Camp, S., and Taylor, P., *Biochemistry*, **32** (1993), 12074–12084.

Reiner, E., Aldridge, N., Simeon, V., Radic, Z., and Taylor, P., in *Cholinesterases: Structure, Function, Mechanism, Genetics, and Cell Biology: Proceedings of the Third International Meeting on Cholinesterases*, an American Chemical Society Publication, Washington DC (1990), 227–228.

Repsold, B., Malan, S., Joubert, J., and Oliver, D. *SAR QSAR Environ. Res.*, **29** (2018), 231–255.

Riederer, P., *Neurotoxicology*, **25** (2004), 271–277.

Robinson, D., and Keating, G., *Drugs*, **66** (2006), 1515–34.

Rodríguez-Franco, M. I., Fernández-Bachiller, M. I., Pérez, C., Castro, A., and Martínez, A., *Bioorganic Med. Chem.*, **13** (2005), 6795–6802.

Rosenberry, T. L., in *Advances in Enzymology and Related Areas of Molecular Biology*; John Wiley & Sons, New York **43** (2006), pp. 103–218.

Rylett, R. J., Ball, M. J., and Colhoun, E. H., *J. Neuropathol. Exp. Neurol.*, **42** (1983), 345.

Samadi, A., Chioua, M., Bolea, I., de los Ríos, C., Iriepa, I., Moraleda, I., Bastida, A., Esteban, G., Unzeta, M., Gálvez, E., and Marco-Contelles, J., *Eur. J. Med. Chem.*, **46** (2011), 4665–4668.

Samadi, A., de los Ríos, C., Bolea, I., Chioua, M., Iriepa, I., Moraleda, I., Bartolini, M., Andrisano, V., Gálvez, E., Valderas, C., Unzeta, M., and Marco-Contelles. *J. Eur. J. Med. Chem.*, **52** (2012), 251–262.

Sang, Z., Qiang, X., Li, Y., Yuan, W., Liu, Q., Shi, Y., Ang, W., Luo, Y., Tan, Z., and Deng, Y., *Eur. J. Med. Chem.* **94** (2015), 348–366.

Sang, Z., Wang, K., Wang, H., Wang, H., Ma, Q., Han, X., Ye, M., Yu, L., and Liu, W., *Bioorganic Med. Chem. Lett.*, **27** (2017a), 5046–5052.

Sang, Z., Wang, K., Wang, H., Yu, L., Wang, H., Ma, Q., Ye, M., Han, X., and Liu, W., *Bioorganic Med. Chem. Lett.*, **27** (2017b), 5053–5059.

Saxena, A., Redman, A. M. G., Jiang, X., Lockridge, O., and Doctor, B. P. (1999), *Chem. Biol. Interact.*, 119–120, 61–69.

Selkoe, D. J., *J. Biol. Chem.*, **271** (1996), 18295–18298.

Shi, J., Tai, K., McCammon, J. A., Taylor, P., and Johnson, D. A., *J. Biol. Chem.*, **278** (2003), 30905–30911.

Shih, J. C., Chen, K., and Ridd, M. J. *Ann. Rev. Neurosci.*, **22** (1999), 197–217.

Soto-Ortega, D. D., Murphy, B. P., Gonzalez-Velasquez, F. J., Wilson, K. A., Xie, F., Wang, Q., and Moss, M. A., *Bioorganic Med. Chem.*, **19** (2011), 2596–2602.

Stahl, S. M., *J. Clin. Psychiatry*, **61** (2000), 813– 814.

Sterling, J., Herzig, Y., Goren, T., Finkelstein, N., Lerner, D., Goldenberg, W., Miskolczi, I., Molnar, S., Rantal, F., Tamas, T., Toth, G., Zagyva, A., Zekany, A., Lavian, G., Gross, A., Friedman, R., Razin, M., Huang, W., Krais, B., Chorev, M., Youdim, M.B., and Weinstock, M., *J. Med. Chem.*, **45** (2002), 5260–5279.

Sugimoto, H., Iimura, Y., Yamanishi, Y., and Yamatsu, K., *J. Med. Chem.*, **38** (1995), 4821–4829.

Sun, Y., Chen, J., Chen, X., Huang, L., and Li, X., *Bioorganic Med. Chem.*, **21** (2013), 7406–7417.

Sussman, J. L., Harel, M., Frolow, F., Oefner, C., Goldman, A., Toker, L., and Silman, I., *Science*, **253** (1991), 872–879.

Tarihi, Y., Çokuğraş, N. A., and Adresi, Y., *Turk. J. Biochem.*, **28** (2003), 54–61.

Tougu, V., *Cent. Nerv. Syst. Agents Med. Chem.*, **1** (2001), 155–170.

Tripathi, R., Rai, G., and Ayyannan, S., *ChemMedChem*, **11** (2016), 1145–1160.

Tripathi, R. M., Sasi, V., Gupta, S., Krishnamurthy, S., and Ayyannan, S., *J. Enzyme Inhib. Med. Chem.*, **33** (2017), 37–57.

Walker, M. C., and Edmondson, D. E., *Biochemistry*, **33** (1994), 7088–7098.

Wang, L., Esteban, G., Ojima, M., Bautista-Aguilera, O., Inokuchi, T., Moraleda, I., Iriepa, I., Samadi, A., Youdim, M., Romero, A., Soriano, E., Herrero, R., Fernández, A., Ricardo-Martínez-Murillo, Marco-Contelles, J., and Unzeta, M., *Eur J. Med. Chem.*, **80** (2014), 543–561.

Wang, Y., Sun, Y., Guo, Y., Wang, Z., Huang, L., and Li, X., *J. Enzyme Inhib. Med. Chem.* (2015) 1–9.

Watkins, P. B., Zimmerman, H. J., Knapp, M. J., Gracon, S. I., and Lewis, K. W., *JAMA*, **271** (1994), 992–998.

Weinreb, O., Amit, T., Bar-Am, O., and Youdim, M. B. H., *Curr. Drug Targets*, **13** (2012), 483– 492.

Weinstock, M., Bejar, C, Wang, R. H., Poltyrev, T., Gross, A., Finberg, J. P., Youdim, M. B., *J. Neural. Transm* [Suppl] **60** (2000), S157-S70.

Weyler, W., Hsu, Y.-P. P., and Breakafield, X. O., *Pharmacol. Ther.*, **47** (1990), 391–417.

Wu, M., Esteban, G., Brogi, S., Shionoya, M., Wang, L., Campiani, G., Unzeta, M., Inokuchi, T., Butini, S., and Marco-Contelles, J., *Eur J. Med. Chem.*, **121** (2016), 864–879.

Xie, S., Lan, J., Wang, X., Wang, Z., Jiang, N., Li, F., Wu, J., Wang, J., and Kong, L. *Bioorg. Med. Chem.*, **24** (2016), 1528–1539.

Xie, S., Wang, X., Jiang, N., Yu, W., Wang, K., Lan, J., Li, Z., and Kong, L. *Eur. J. Med. Chem*, **95** (2015), 153–165.

Xu, Y., Wang, H., Li, X., Dong, S., Liu, W., Gong, Q., Wang, T., Tang, Y., Zhu, J., Li, J., Zhang, H., and Mao, F. *Eur. J. Med. Chem.*, **143** (2018), 33–47.

Yang, X., Qiang, X., Li, Y., Luo, L., Xu, R., Zheng, Y., Cao, Z., Tan, Z., and Deng, Y., *Bioorg. Chem.*, **71** (2017), 305–314.

Youdim, M. B. H., Edmondson, D., and Tipton, K. F., *Nat. Rev. Neurosci.*, **7** (2006), 295–309.

Youdim, M. B. H., Finberg, J. P. M., and Tipton, K. F. in *Catecholamines I: Handbook of Experimental Pharmacology*; Springer-Verlag, Berlin (1988) pp. 119–165.

Zeller, E. A., *Ann. N. Y. Acad. Sci.*, **107** (1963), 811–821.

Zhang, H.-Y., *FEBS Lett.*, **579** (2005), 5260–5264.

Zindo, F., Joubert, J., and Malan, S., *Future Med. Chem.*, **7** (2015), 609–629.

Zoheir, N., Lappin, D., and Nile, C., *Inflamm. Res.*, **61** (2012), 915–926.

Chapter 12

The Neurodegenerative Characteristics of Alzheimer's Disease and Related Multi-Target Drug Design Studies

Hayrettin Ozan Gülcan[a] and Ilkay Erdogan Orhan[b]

[a]*Eastern Mediterranean University, Faculty of Pharmacy, Department of Pharmaceutical Chemistry, Famagusta, TRNC, via Mersin 10 Turkey*
[b]*Gazi University, Faculty of Pharmacy, Department of Pharmacognosy, Etiler, Ankara, Turkey*

ozan.gulcan@emu.edu.tr

12.1 Introduction

Alzheimer's disease (AD) falls under the category of neurodegenerative diseases. It is defined as a progressive disease accompanied with neuronal loss linked to the appearances of dementia, cognitive decline, mood swings, and personal character loss symptoms (Reisberg et al., 1987, Kumar et al., 2015). Based on this, a total AD cure, therefore, is significant not only for AD patients but also for health care providers, particularly including patient relatives. Imagining a close relative having cognition problems with a continuous and increasing manner would be just

Pharmaceutical Biocatalysis: Fundamentals, Enzyme Inhibitors, and Enzymes in Health and Diseases
Edited by Peter Grunwald

ISBN 978-981-4800-61-7 (Hardcover), 978-0-429-29503-4 (eBook)
www.jennystanford.com

enough to estimate the burden of the disease to patient relatives (Small et al., 1997). Unfortunately, the incidence of the disease is another terrifying point. Statistics indicate that the number of current AD patients is around 50 million worldwide. It is estimated that this will be beyond a hundred million people just in around 30 years. Furthermore, the incidence of the disease is close to 20% above age 65, and 50% for those 85 or older (Hebert et al., 2003).

The disease also provides a staggering economic effect to the developed or developing countries. This is not attributed to the cost of drugs, but to the total expenditures on health care, hospitalization, and other social needs. This is quite normal once remembered that the disease has devastating effects on cognition resulting in the need for continuous patient care particularly at the intermediate and end stages of the disease (Meek et al., 1998; Getsios et al., 2012). Based on this point, developed countries also face with costs that they need to save for funding AD research grants. Indeed, the NIH's estimated funding for AD research studies (e.g., acquired cognitive impairment, AD, and related dementias) in 2019 for US is around $5 billion. Just for detail, it is 10 times higher than the grants saved for each microbial resistance, stroke, human non-embryonic stem-cell research topics (https://report.nih.gov/categorical_spending.aspx).

Although there has been more than a century following the first description of the disease, there is no total cure, simply because of the unknowns waiting for the clarification within the pathophysiology of AD. AD is one of those life-threatening diseases. It is the sixth leading cause of death across all ages, while it is the fifth leading cause of death for those above 65 (https://www.brightfocus.org/alzheimers/article/alzheimers-disease-facts-figures).

For many years, the main research topic on AD was the dementia component of disease. Therefore, it is not surprising to see the first drug tacrine, a cholinesterase inhibitor, reaching to the market in the 1980s as a cognition enhancer agent (Crismon, 1994). Following years have witnessed the approval of couple more cholinesterase inhibitors (i.e., galantamine, donepezil, and rivastigmine). Besides, memantine, an NMDA (*N*-methyl-D-aspartate) receptor antagonist also has obtained license for dementia problems associated with the disease. As it is clearly expressed, these agents only provide symptomatic treatment and have no effect on the progressive neurodegenerative character of the disease (Cummings et al., 2014). AD treatment is looking for better validated pharmacological targets for the radical treatment of the disease.

In this chapter, the validated targets and the current drugs used for the treatment of the disease, the progressive neurodegeneration characteristics of AD concomitant to its pathophysiology, and approaches on the multi-target ligand design studies based on the neurodegenerative characteristics of the disease are provided to summarize the current approaches for alternative treatment options.

12.2 Validated Targets and the Current Drugs Used for the Treatment of AD

Dementia is referred to a declined mental state. It is not always a part of a disease state. The normal stress conditions of life may also result in mild depression states with temporary dementia. In most cases, it is not severe and no drug treatment is necessary. AD-related dementia is quite different (McKhann et al., 2011). In fact, AD is the most common form of dementia. Since the disease has a progressive neurodegenerative characteristic behind, the dementia associated with the disease must be treated. Currently, each drug utilized in the treatment of AD focuses on mainly the dementia symptoms associated with AD development. These drugs are in two categories: cholinesterase inhibitors and NMDA receptor antagonists (Tan et al., 2014).

Although there have been a quite a lot number of targets linked to the development of dementia, the most acknowledged one among them is related to the cholinergic system activation (Giacobini, 1990). Throughout the development of AD, neuronal loss is associated with the decline of a variety of neurotransmitters, including but not limited to acetylcholine, serotonin, and even dopamine at further steps (Reinikainen et al., 1990). Research studies indicated that acetylcholine interaction through with some of the subtypes of both muscarinic and nicotinic receptors is a critical component of routine cognition-related actions (Hasselmo, 2006). Since AD development also provides insufficiency in the amounts of both acetylcholine available in CNS and the related cholinergic receptors, it is practical to consider targeting cholinergic system. One of the key points here is the lack of our current skills on the design of molecules selectively interacting with the receptor subtypes of specific muscarinic and nicotinic receptors (Fisher et al., 2003; Francis et al, 1999; Levin, 1992). Although a variety of molecules have been designed and synthesized to selectively interact with specific muscarinic and nicotinic receptors involved in cognition, none of those agents has been successful in drug development studies (Messer, 2002). From this point of view, only the cholinesterase enzymes (i.e., acetylcholinesterase (AChE) and butyrylcholinesterase (BuChE) have been assigned as validated targets, to increase the amount and the action of acetylcholine within the CNS of AD patients. These enzymes are hydrolytic enzymes both showing selectivity for the hydrolysis of acetylcholine to choline and acetate. BuChE has mainly peripheral distribution; however, it is shown that its activity in CNS is increasing with age and the development of AD (Chatonnet et al., 1989).

Almost all the current drugs used for the treatment of AD (i.e., besides memantine) are inhibitors of cholinesterase enzyme (Fig. 12.1).

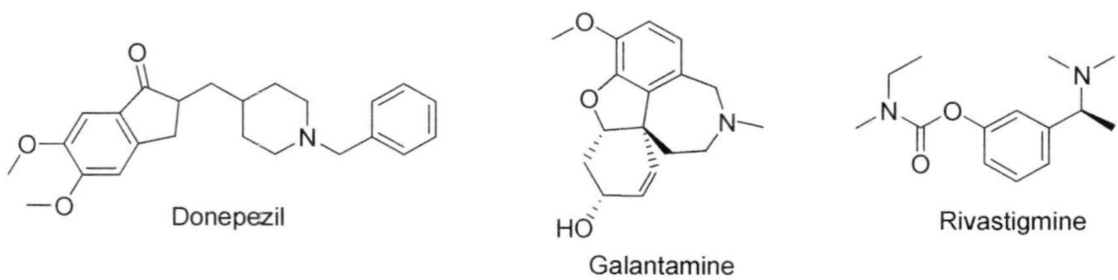

Figure 12.1 Cholinesterase inhibitors used in the treatment of AD.

Tacrine was the first developed analogue, but it is no longer used with respect to its severe hepatotoxicity (Fig. 12.2) (Watkins et al., 1994). Donepezil and galantamine are AChE selective inhibitors, while rivastigmine is a selective BuChE inhibitor (Takeda et al., 2006). Their inhibitory characteristics also show diversity. First, donepezil, a totally synthetic compound, is the most potent AChE inhibitor with a low nM IC_{50} (Sugimoto, 2001). Galantamine is obtained from Amaryllidaceae species (Dall'Acqua

2013). Both of them interact with the active site of AChE to prevent binding of Acetylcholine. Rivastigmine was designed on the basis of a natural alkaloid physostigmine and it is a mechanism-based inhibitor. Through its interaction with BuChE, it is capable of carbamylating the enzyme (Polinsky, 1998). The hydrolysis based regeneration of a carbamate group takes longer duration in comparison to the time required for the hydrolysis of an ester group (Auriacombe et al., 2002).

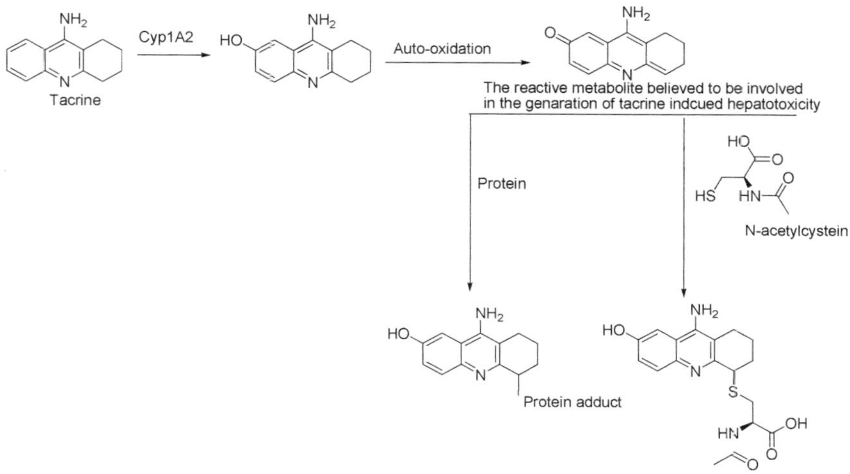

Figure 12.2 A proposed mechanism for the toxicity of tacrine (Spaldin et al., 1994).

The number of research publications on the design of novel cholinesterase inhibitor agents have reached to more than a satisfactory point in the last 2–3 decades. With respect to some deficiencies of current cholinesterase inhibitor molecules (e.g., pharmacokinetic-based, pharmacodynamics-based, or cholinergic side effect-based), thousands of novel designs in diverse structures have been provided. Regarding the effects of these drugs only available in the first and moderate steps of AD concomitant to their lacking properties on the progressive neurodegenerative characteristics of AD, pure cholinesterase inhibition is no longer utilized in drug design studies. However, it is still a validated target for cognitive enhancement and employed in multi-target ligand design studies (Fig. 12.3).

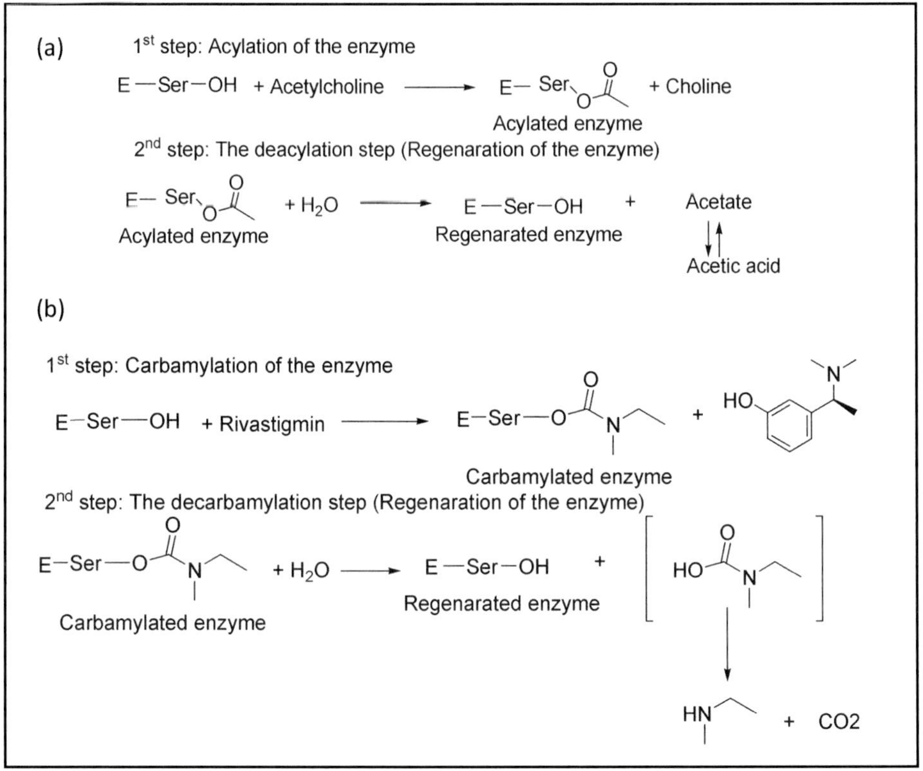

Figure 12.3 The mechanism of cholinesterase activity (a) and carbamylation of a cholinesterase with rivastigmine (b) (Zhou et al., 2010; Jann et al., 2000).

Memantine is an NMDA receptor antagonist and it is the only representative drug of this category employed in AD treatment. Its advantageous part in comparison to cholinesterase inhibitors is its availability to be employed in the moderate and advanced stages of AD treatment (Johnson et al., 2006). Research studies pointed out the excessive activation of NMDA receptors during the development of AD (Lipton, 2006). At one side, this over-activation is linked to glutamate toxicity–related neurodegeneration. Indeed, excessive glutamate activity in CNS results in particularly extensive NMDA activation that is, in turn, stimulating excessive Ca influx into postsynaptic nerves. Excess Ca influx triggers the activity of some enzymes damaging to cellular macromolecules, leading to neuronal loss (Hynd et al., 2004). Therefore, memantine action, in one way, is expressed as a neuroprotective activity. On the other hand, memantine is known to be acting as a cognitive enhancer. This has been proven both in human clinical trials and in some animal experimental models. The exact mechanism of this action, however, is not quite understood. It is postulated that glutamate based excitotoxicity itself generates decline in cognition (Martinez et al., 2010). Therefore, prevention of excessive glutamate activity through NMDA receptors is a way of terminating this excessive neuronal excitation–related cognition symptoms. In fact, memantine is a partial antagonist of NMDA receptors. Compounds showing higher affinity to NMDA receptors were all found to be unsuccessful for the treatment of AD (Parsons et al., 1999). The reason for this is attributed to the regular physiology of NMDA receptor activity (Lipton, 2004)). NMDA receptors action, under normal physiology, is critical for controlling synaptic plasticity. Therefore, it is believed that more potent antagonist agents are not useful for the treatment of AD (Youdim et al., 2005). Memantine action is limited with the blockade of the excessive activation of these receptors (Zheng et al., 2014). Unfortunately, more explanations are required for the mechanism of action of memantine. In one way, no other drug has been at least shown to mimic the activity of memantine in order to display enhancement on cognition. On the other hand, memantine is not an NMDA selective molecule, since it has affinity to some serotoninergic, cholinergic, dopaminergic, and sigmaergic receptor subtypes (Zheng et al., 2014). Therefore, there is debate on the exact mechanism of action of this molecule (Fig. 12.4).

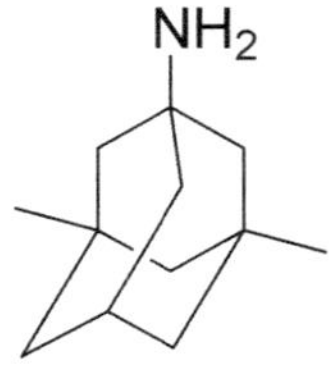

Figure 12.4 The NMDA receptor antagonist drug memantine.

As the name implies neurodegeneration is a process ending up with neuronal death. It involves progressive change within the structure and function of neurons (Halliwell, 2006). Neurodegeneration is not specific to AD, since other disease states such as Parkinson Disease and Huntington Disease also display neurodegeneration characteristics. Although there are varieties in the subcellular organization and pathophysiology mechanisms of neurodegeneration among different CNS diseases states, the progressive character and neuronal loss as well as the alterations in neurotransmitter levels are apparent, independent from the type of the disease (Forman et al., 2004).

In the pathophysiology of AD, senile plaque formation, related to the aggregation of amyloid beta (Aβ) peptides (i.e., derived from amyloid precursor protein), is the most observed change (Haass et al., 2007). The unusual point is the excessive formation of Aβ plaques, since it is also known to be regular part of physiology during aging (Becker et al., 2011). Among those proposed regular physiological functions of Aβ plaques, the role in oxidative stress, the activation of kinase enzymes, and the effect on cholesterol transport are important; however, there is still debate on these topics (Hardy, 2009). On the other hand, the toxicity of excessive Aβ formation has been clearly shown in various types of cell-based studies. The toxicity is also dependent on the type of Aβ; therefore, it is significant to clearly express the cascade leading to the biosynthesis of Aβ peptides (Seilheimer et al., 1997).

Amyloid beta precursor protein (APP) processing generates Aβ peptides with a range of 36 to 43 amino acid length (Mann et al., 1996). It was shown that APP, a transmembrane protein with 695 amino acids, is involved in the regulation of some physiological functions. APP is sequentially cleaved by β-secretase

(BACE-1) and γ-secretase to yield out Aβ peptides (Johnson et al., 1997). This pathway is referred to as amyloidogenic, since there is also a non-amyloidogenic pathway, involving an enzyme (i.e., α-secretase) responsible to cleave APP within the Aβ protein; therefore, its function is critical to prevent Aβ generation (Soriano et al., 2001).

Under normal conditions, $A\beta_{40}$ constitutes the major form of Aβ peptides, while throughout AD development the formation of other $A\beta_{36\text{-}43}$ increases and mainly $A\beta_{42}$ becomes the major biosynthesized form (Kuperstein et al., 2010). The reason that is generating the aggregation of Aβ plaques in AD brain is associated to the unusual peptide misfolding with a nucleation growth mechanism. Accumulation of Aβ leads to several toxicity mechanisms (Chiti et al., 2006).

One of the mechanisms suggested for Aβ plaque toxicity involves the interaction of Aβ with some metals that are inducing the oxidative stress. Indeed, some ions such as cupper and iron in different oxidation states were found co-accumulated with Aβ peptides (House et al., 2004). This co-accumulation is a source of reactive oxygen species (ROS) in such a way that Aβ has the ability to reduce Cu^{+2} to Cu^{+1} as well as Fe^{+3} to Fe^{+2}, and these reduced metals then interact with molecular oxygen to generate superoxide anions (Cuajungco et al., 2000). Superoxide anion then becomes a source to produce H_2O_2 and other hydroxyl radicals involving enzymatic and non-enzymatic pathways. The ROS formed generates lipid and protein oxidations, a well-known mechanism leading to neuronal death.

Another approach suggested for the toxicity of Aβ aggregation is on the loss of synaptic function (Kamenetz et al., 2003). Aβ plaques are able to bind to receptors on various synaptic clefts. This is thought to be involved in blockage of neuronal signaling. In particular, the interaction of Aβ deposits with key receptors (e.g., some glutamatergic and cholinergic receptors), critical to regulate cognitive functions is also considered to be involved within the generation of cognitive decline throughout the development of AD (Shankar et al., 2007).

Aβ deposits are not the only aggregated peptides observed in the development of AD, since tau protein hyper-phosphorylation and related tangle formation are another outcome. Tau is a brain-specific, axon-enriched microtubule-associated protein (De Felice et al., 2008). Hyper-phosphorylation of tau generates insoluble neurofibrillary tangles, also referred to as one of the biomarkers of AD. Neurofibrillary tangle formations are also toxic and lead to neuronal cell loss (Jack et al., 2011). However, the relation of Aβ plaque formation and neurofibrillary tangles is still a topic of debate, since the study results are controversial to depict whether the mechanisms to generate the formations of Aβ plaque formation and neurofibrillary tangles are independent or affecting each other.

As summarized above, the progressive neurodegeneration characteristics of AD is mainly attributed to the two abnormal structures: extracellular amyloid plaques and intraneuronal neurofibrillary tangles, highly insoluble and densely aggregated peptides and proteins. It is currently believed in that the toxic mechanisms are the outcomes of these formations.

Based on these findings, various types of treatment options for AD have been proposed. Regarding the Aβ toxicity, BACE-1 inhibitors and the γ-secretase inhibitors have been designed to prevent Aβ formation. Furthermore, vaccination against Aβ plaques was also tried (Masliah et al., 2005). Regarding the formation of ROS triggered by Aβ toxicity, metal binding ligands and CNS penetrable antioxidant agents as well as MAO inhibitors have also been designed and screened for their activities. There are also agents designed to target the cascades leading to the formation of neurofibrillary tangles. It is noteworthy to state that, none of these agents currently has been found successful in drug development phase stages. Although some of them failed related to the side effects, the main reason of the failures is related to the deficiency of these molecules in the improvement of cognition or prevention of cognitive decline. Therefore, still the cholinesterase enzymes and NMDA receptor are the only validated targets for the treatment of the disease. However, as stated previously, cholinesterase inhibitors do not prevent the progress of toxic mechanisms.

Regarding this situation, the majority of the approaches in the last 2 decades employed multi-target based drug design studies. In some them, at least two of the mechanisms described above have been employed disregarding the failure of agents targeting Aβ and tau cascades. The most preferred strategy has been based on the employment of cholinesterase inhibition concomitant to another target, which is involved in the toxic pathology of AD clarified so far. In this approach, cognitive enhancement is offered via cholinesterase inhibition, while the prevention of progressive neurodegeneration is aimed via acting on Aβ/tau cascades, and ROS formation mechanisms.

12.3 Multi-Target Ligand Design Studies Based on the Neurodegenerative Characteristics of the Disease

The currently employed cholinesterase inhibitor drugs for the treatment of AD have almost no effect on the progress of AD. In other words, they do not have promising effect on one of those pathophysiological mechanisms discovered so far. Therefore agents which target those aforementioned pathological cascades (e.g., mainly related to the pathways leading to neurodegeneration) are referred to as disease modifying drug candidates (Citron et al., 2010). As described, single target approach employing these disease modifying drug candidates has not successfully presented promising results in terms of cognitive enhancement. However, combining disease modifying and cognitive enhancement activities appears to be superior to the currently utilized cholinesterase inhibitor drugs (Leon et al., 2013). Below is the summary of the outcome of these multi-target ligand design and screening studies based on this strategy. It is noteworthy to state that the NMDA receptor partial antagonism is also considered to be an approach as disease modifying. The reason is related to the toxic effects of high innervation (i.e., excitation) of glutamergic receptors within the development of AD. Furthermore, it is also clinically proven that NMDA partial antagonism is another validated strategy to relieve dementia symptoms.

12.3.1 Dual BACE1 and ACHE Inhibitors

BACE-1 enzyme is a member of aspartic protease enzymes. It cleaves APP both at the N-terminus and between E10 and V11

(Hills et al., 2007). There is another BACE enzyme (i.e., BACE-2) that also cleaves APP from different sides. What makes BACE-1, but not BACE-2, as a target is that BACE-1 inhibition absolutely abolishes Aβ production, while BACE-2 inhibition does not display a difference (Fluhrer et al., 2002). Following the initial identification of BACE-1 as a critical enzyme in amyloidogenic pathway, many compounds have been designed and tested for their BACE-1 inhibitory potential. Some of them even have been employed in clinical trials. Although majority of the compounds were found very successful both in human and animal experiments, none of them has been reached to the market yet (Hills et al., 2007). The basic reason is their deficiency in cognitive enhancement (Vassar, 2014). Some of these BACE-1 inhibitors used in clinical trials are shown in Table 12.1.

Table 12.1 Representative BACE-1 inhibitors used in clinical trials (Stamford et al., 2013; Vassar, 2014).

Compound	Status
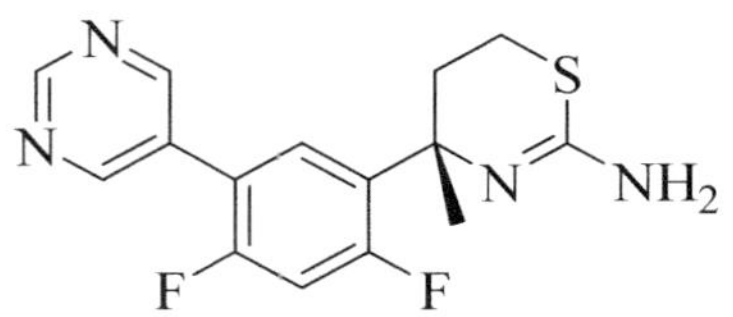 LY2811376 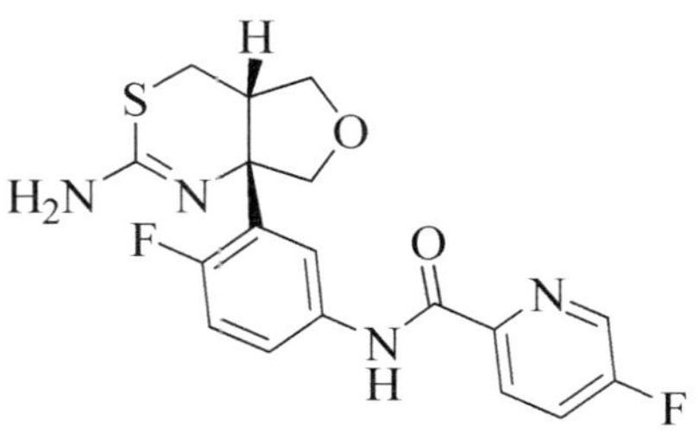 LY2886721	Both developed by Eli Lilly. Terminated clinical trials with respect to unexpected side effects

On the other hand, there are limited numbers of studies conducted so far to discover or design dual BACE-1 and cholinesterase inhibitors. Representative examples are shown in Table 12.2.

This approach is relatively new; therefore, there is no drug candidate, sourced on this perspective, used in clinical trials.

Table 12.2 Representative dual inhibitors of BACE-1 and AChE

Compound	AChE IC_{50} (μM)	BACE-1 IC_{50} (μM)	Reference
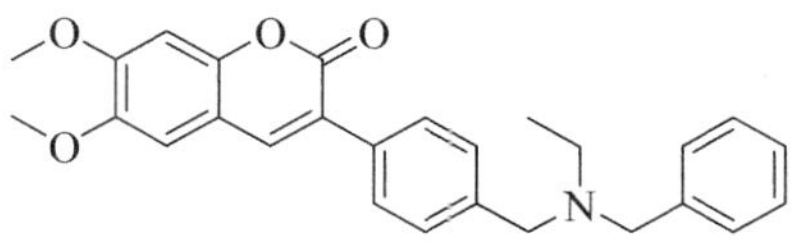 AP2243	0.018	0.23	Piazzi et al., 2008
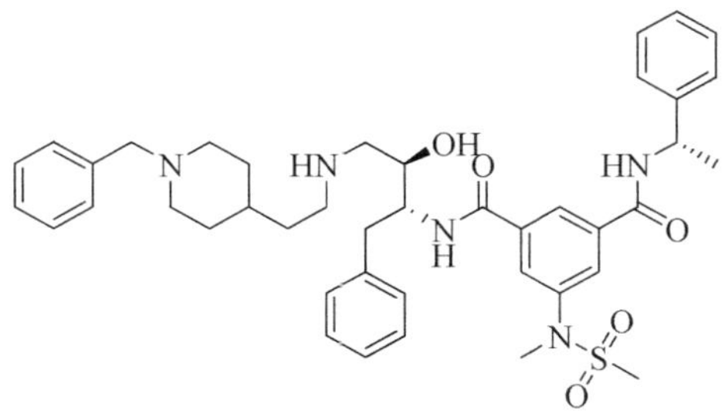	1.83	0.56	Zhu et al., 2009
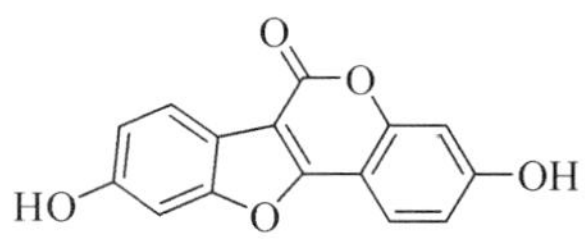 Coumestrol	42.3	51.4	Koirala et al., 2018

12.3.2 Dual Inhibitors of Monoamine Oxidase and Cholinesterase

Monoamine oxidase (MAO) enzyme activity is one of the critical steps in the control of the function of several neurotransmitters (e.g., serotonin, dopamine, and norepinephrine). In other words, through the formation of aldehyde derivatives of neurotransmitters via MAO catalyzed oxidative deamination reactions, the amine neurotransmitters are deactivated (Shih et al., 1999). There are two types of MAO enzymes in human, as referred to as MAO-A and MAO-B. They display different tissue distribution and ratios depending on age. Although they are capable of employing a diverse amine substrates, there is some selectivity difference. MAO-A shows selectivity for serotonin, melatonin, norepinephrine, and epinephrine, while MAO-B possesses substrate selectivity for benzylamine and phenethylamine. On the other hand, both enzymes display similar characteristics to catalyze dopamine, tyramine, and tryptamine oxidative deamination reactions (Kalgutkar et al., 2001). Depending on the substrate specificity difference, MAO-A inhibition is mainly preferred as an option in the treatment of some form depression (e.g., to increase the available levels of serotonin, norepinephrine, and dopamine) (Meyer et al., 2006). Parkinson disease–related symptoms are generally related to deficiency of dopamine. Regarding the side effects of MAO-A inhibitors, MAO-B inhibitors (e.g., rasagiline, selegiline) are preferred to increase the available amount of dopamine in CNS (Youdim, 2006).

The idea of the employment of MAO inhibition within the treatment of AD shows fundamental differences in comparison to its application in Parkinson Disease. The first point is the finding on the increased levels of MAO-B activity within the CNS throughout the development of AD (Gulyás et al., 2011). Another point is the loss of various types of neurotransmitter activity with respect to the progressive neurodegenerative characteristics of the disease (Crow et al., 1984). A final point is the oxidative stress–related (e.g., mainly reactive oxygen species sourced) damage to cellular macromolecules, including but not limited to DNA, protein, and lipid oxidations. Indeed, the studies indicated the too low levels of defense mechanisms (e.g., glutathione) within the CNS, particularly at the final stage of the disease (Lovell et al., 1998). Furthermore, it is important to remember that MAO catalyzed reactions also produce hydrogen peroxide which can be converted to hydroxyl radical via Fenton type reactions. It is very clear that hydroxyl radical is a source of lipid peroxyl radical formations, one of the products of oxidative damage. From this point of view, MAO-B inhibition is also believed to be a critical approach to eliminate ROS based neurodegeneration (Zheng et al., 2005).

There are quite a lot number of studies employed for the design of dual cholinesterase and MAO inhibitor molecules. The strategy is based on the employment of both more cholinergic activity within CNS to relieve cognitive symptoms and more dopaminergic activity to particularly prevent one of the sources creating ROS. The representative molecules deigned and screened for activity under this category in recent years is shown in Table 12.3. Since the cholinesterase and MAO enzymes display quite different structures in terms of substrate selectivity, and protein organization, the design studies are challenging to discover dual MAO and cholinesterase inhibitors. It is important to note that majority of the studies within this category employs chromene analogues (Erdogan et al., 2015).

12.3.3 Dual Inhibitors of Cholinesterase and Aβ Aggregation

One of the alternative applications of preventing Aβ toxicity and related neurodegeneration is the inhibition of Aβ deposit formation. This is particularly significant once considered that insoluble Aβ oligomers are the main constituents of Aβ plaques and the majority of Aβ toxicity is directly relevant to these deposit formations. Therefore, the design of compounds capable of interacting with Aβ peptides has been considered to prevent Aβ aggregation (Hawkes et al., 2009). Although the initial attempts were on the investigation of the development of pure Aβ aggregation inhibitors (e.g., this is also relevant to the idea of development of vaccines against Aβ), later studies have focused on enlarged strategies in which dual inhibitors of cholinesterase and Aβ aggregation are considered. This idea is mainly based on two parameters. At first, the initial attempts with either vaccines or pure Aβ aggregation inhibitor small molecules have not proven a statistically significant improvement in cognitive decline seen in AD (Meilandt et al., 2009). Secondly, there are factors accelerating Aβ deposit formation, including but not limited to metals and cholinesterase enzymes within CNS. Indeed, even in vitro based assays systems display that Aβ aggregation is accelerated in the presence of acetylcholinesterase enzyme (Inestrosa et al., 2005).

Table 12.3 Representative examples for Dual Cholinesterase and MAO Inhibitors

Structure	AChE IC_{50} (μM)	BuChE IC_{50} (μM)	MAO-A IC_{50} (μM)	MAO-B IC_{50} (μM)	12.1 Refs
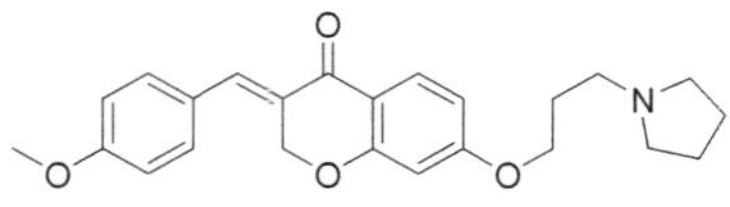	2.83	12.2 9.10 12.3	12.4 NA	12.5 6.93	12.6 Wang et al., 2016
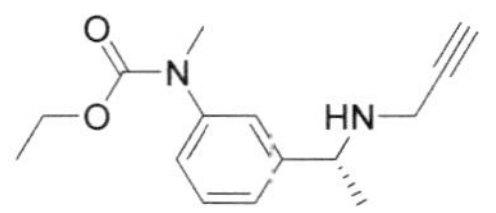	7.7 12.7	12.8 10.3 12.9	12.10 1.04 12.11	12.12 > 1000	12.13 Liu et al., 2016
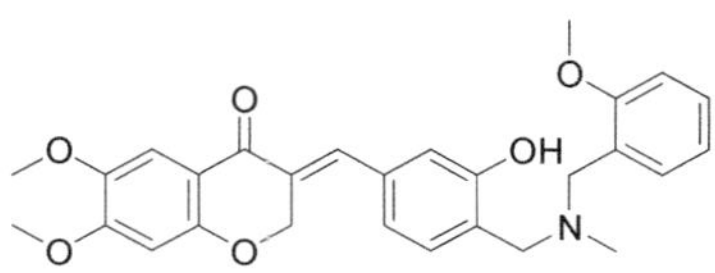	16.4 12.14	12.15 NA	NA	12.16 1.03 12.17	12.18 Li et al., 2017
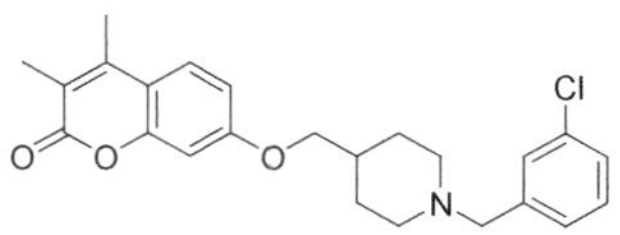	1.03	12.19 6.07	12.20 0.7	12.21 0.089	12.22 Pisani et al., 2016
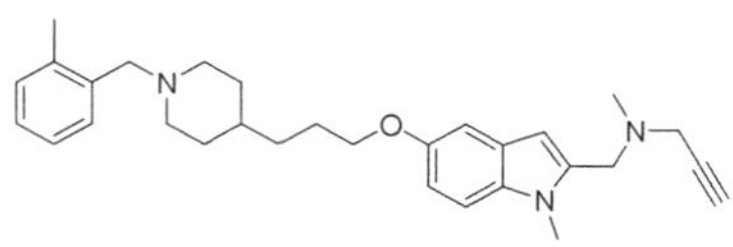	2.8	12.23 4.9	12.24 0.006	12.25 0.183	12.26 Bautista et al., 2014
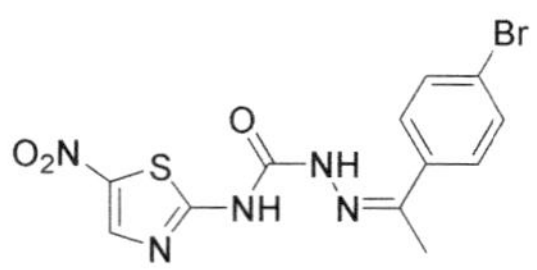	0.34	12.27 0.49	12.28 70.19	12.29 0.212	12.30 Tripathi et al., 2018
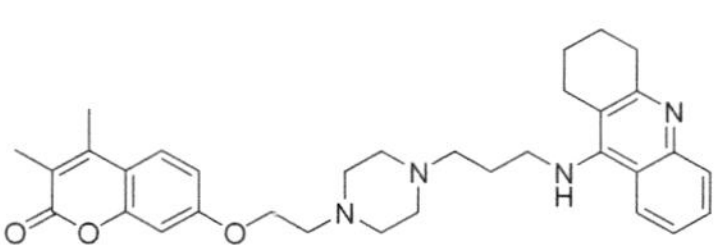	0.033	12.31 0.080	12.32 15.07	12.33 0.24	12.34 Xie et al., 2015

Following the identification of a function of cholinesterase enzymes as an accelerator of Aβ deposit formation, the screening of compounds for their capacity to inhibit both Aβ aggregation and cholinesterase enzymes has become ordinary. The basic thought relied on the idea that a compound with a potential to inhibit cholinesterase activity might also provide inhibitory features to prevent AChE induced- or self- Aβ aggregation (Bolognesi et al., 2005). Following the initial attempts, it has been realized that a cholinesterase inhibitory profile for a drug candidate does not guarantee an extra inhibitory characteristics (Norouzbahari et al., 2018). Indeed, the most potent AChE inhibitor licensed drug, donepezil, has weak inhibitory potential for the prevention of Aβ self- or cholinesterase induced aggregation. Now, it is well recognized that the design of molecules to inhibit both cholinesterase activity and Aβ aggregation is different than pure cholinesterase inhibitor molecule design. More research results are required at the molecular level to understand the factors involved in the self- or inducer-based aggregation of Aβ.

Among those multi-target drug design approaches for the treatment of AD, apparently dual cholinesterase and amyloid beta aggregation inhibitor drug design is the most applied methodology (Castro et al., 2006). Although the previous activity screening studies were based on the employment of Aβ peptides with a Thioflavin-T based fluorometric method, nowadays assays kits are preferred (Solomon et al., 1996). The basic reason relies on the variability of the results in Thioflavin-T based fluorometric method. However, these assay systems are expensive and percent inhibition results are generally presented rather than IC_{50} determinations with respect to cost issues. Some of the representative compounds known to dually inhibit cholinesterase activity and Aβ aggregation is shown in Table 12.4.

12.3.4 Cholinesterase Inhibitors and Antioxidants

In one way or another, neurodegenerative characteristic of AD has been generally linked to the toxicity of ROS, regardless of the mechanistic approach. Although they are not classified as disease modifying agents, antioxidants are always in mind as an auxiliary tool in the design of molecules targeting different biochemical cascades. The idea is mainly on the direct scavenging of excessive ROS. It is noteworthy to mention that single antioxidant application has never been assessed as satisfactory neither in cognitive enhancement nor preventing the related toxic pathways (e.g., amyloidogenic cascade) (Persson et al., 2014).

Table 12.4 Representative dual cholinesterase and amyloid beta aggregation inhibitors

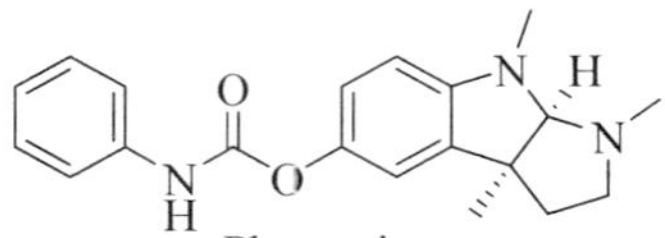 Phenserine Greig et al., 2005	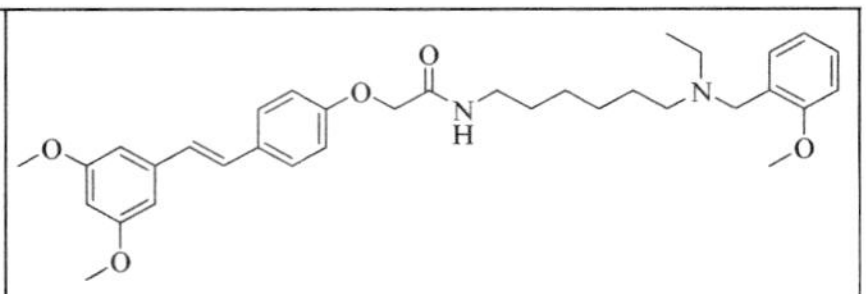 Li et al., 2016
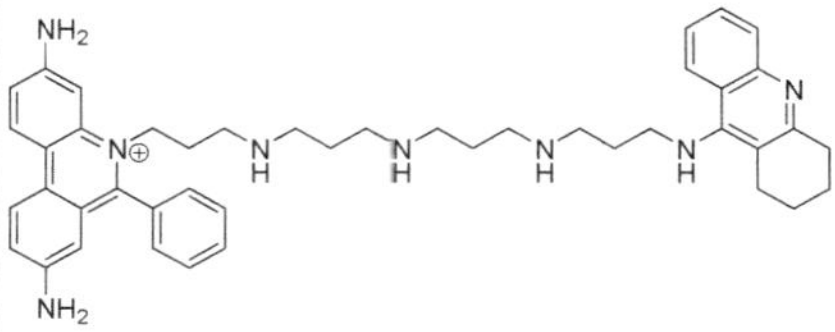 Ma et al., 2004	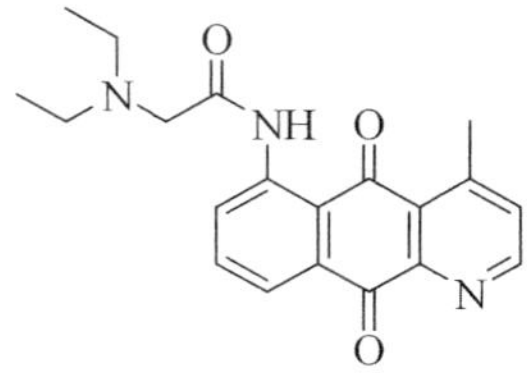 Wu et al., 2018
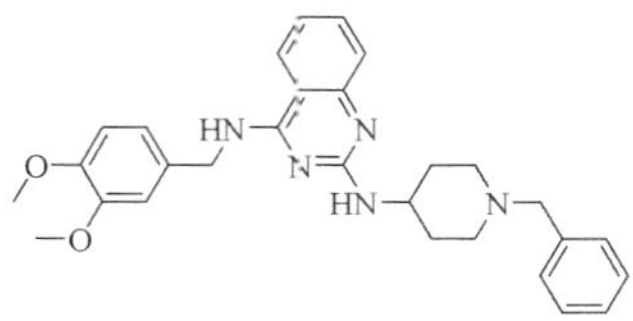 Mohamed et al., 2017	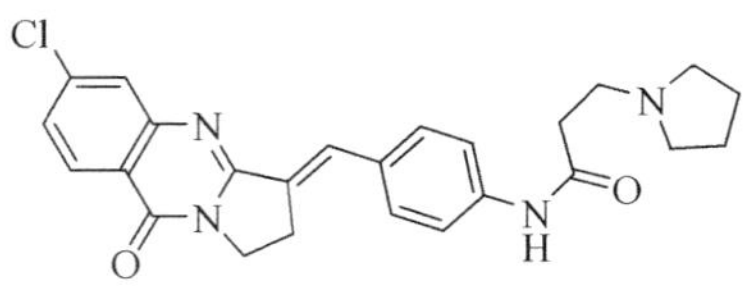 Yan et al., 2012
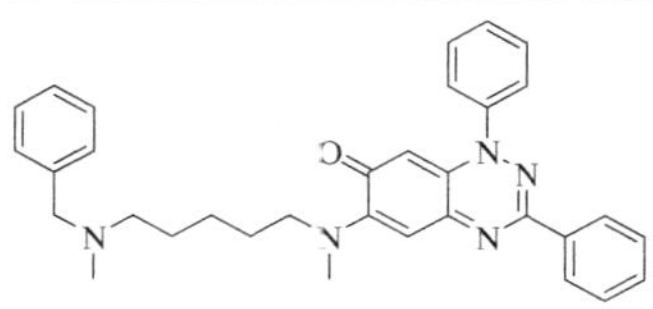 Catto et al., 2012	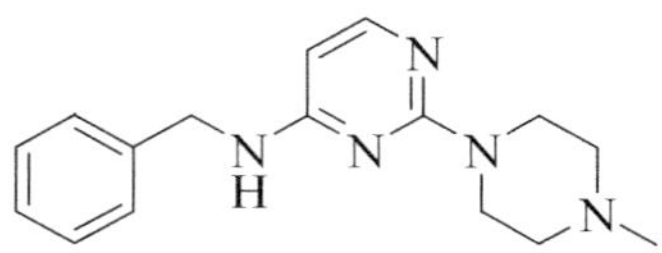 Mohamed et al., 2011
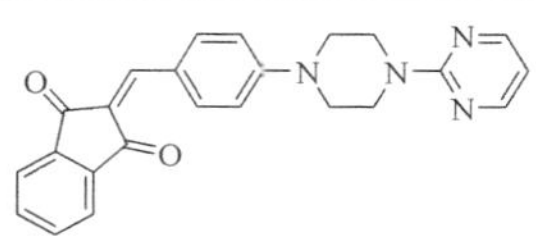 Mishra et al., 2016	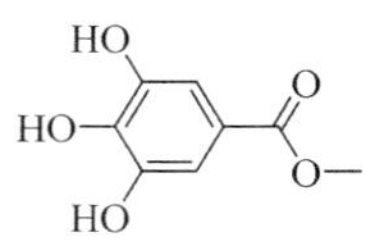 Pugazhendhi et al., 2018

Majority of the investigations on antioxidant drug development have been originated from natural products (Maestri et al., 2006). Since, oxidative stress is not just a part of AD, the attention has been focused on the discovery of natural products providing both cholinesterase inhibition and antioxidant properties (Papandreou et al., 2009). Most of the compounds found active so far have polyphenol structures. Representative natural compounds within this category are provided in Table 12.5. The main drawback of polyphenolic compounds is related to their poor drug-like properties (e.g., poor bioavailability and weak CNS penetrable features) (Yang et al., 2008).

Table 12.5 Representative natural cholinesterase inhibitors and antioxidants

Classification	Representative examples	Reference
Curcuminoids	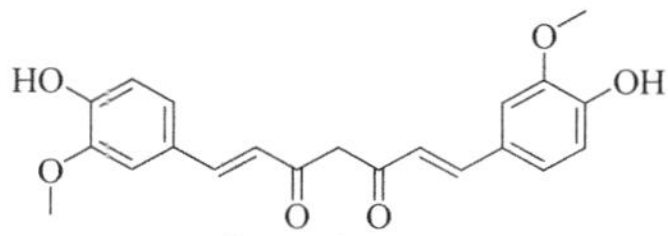 Curcumin 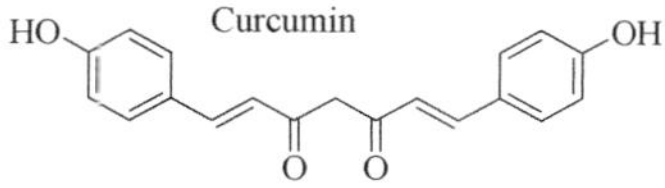 Bisdemethoxycurcumin 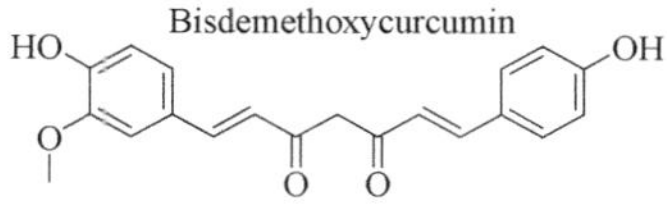 Demethoxycurcumin	Ahmed et al., 2009
Coumarines	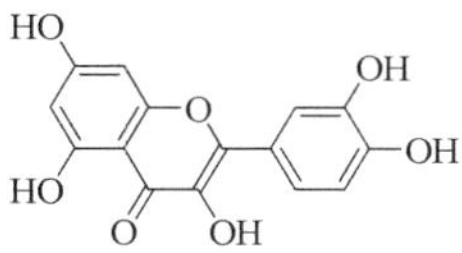 Quercetin	Erdogan et al., 2015
Isothiocyanates	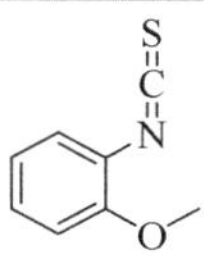	Burčul et al., 2018
Urolithins	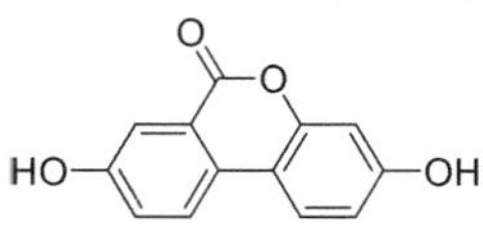 Urolithin A	Gulcan et al., 2014; Ishimoto et al., 2011
Terpenoids	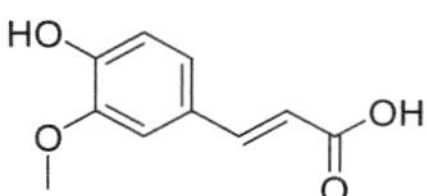 Ferulic acid	de Oliveira et al., 2017; Ahmed et al., 2013

Some medicinal chemistry originated research studies also include the insertion of antioxidant features to design of molecules, besides other main properties such as cholinesterase inhibition and/or disease modifying. Regarding the drawbacks of polyphenols, synthetic compounds are designed on the employment of different functional groups. Some of the representative analogues providing both cholinesterase inhibition and antioxidant properties are displayed in Table 12.6.

Unfortunately, the number of methodologies on the determination of antioxidant properties of compounds generates standardization problems (Frankel et al., 2000). This is reasonable up to some extent once it is considered that there is more than one mechanism to display an antioxidant feature. However, the results can become meaningless when a compound shows varying results in those assays where similar mechanisms are utilized. Therefore, a general comparison of antioxidant activities of different compounds from different sources is almost impossible.

There is also debate on the employment of antioxidant feature in a drug molecule for various disease states including AD. The source of this controversial originates from the regular physiology. Indeed, reactive oxygen or nitrogen species formations are regular pieces of some biochemical cascades (Ježek et al., 2005). Therefore, excessive antioxidant employment (either in CNS or peripheral) might have undesired effects to these systems.

Finally, it is important to state another aspect of antioxidants that is also a topic of debate. Through dietary sources, we are routinely exposed to all different types of xenobiotics and some of them are antioxidants. Even food supplements contain some of these antioxidant compounds. Herbal teas are rich sources of antioxidants as well. Regarding the dosage and frequency to these diets we exposed to, some authorities discuss that they are not worth for the treatment of symptoms of AD. Controversial ideas focus on the beneficial aspects of these antioxidants on the prevention of disease states in which oxidative stress is involved (Luchsinger et al., 2004).

Table 12.6 Representative recent compounds with diverse structures designed to provide cholinesterase inhibition and antioxidant activity

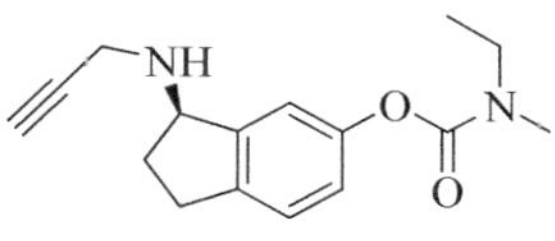

Ladostigil

Bar-Am et al., 2009

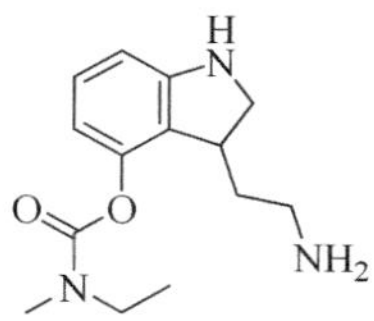

Yanovsky et al., 2012

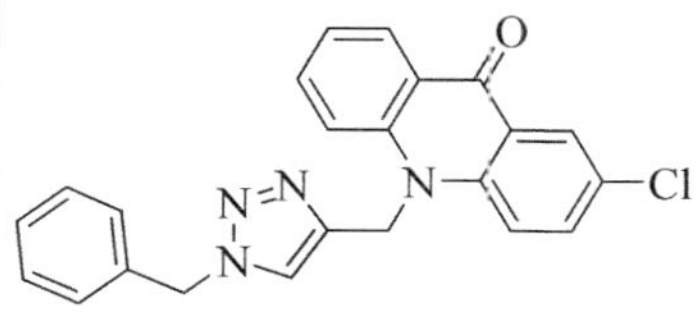

Mohammadi et al., 2015

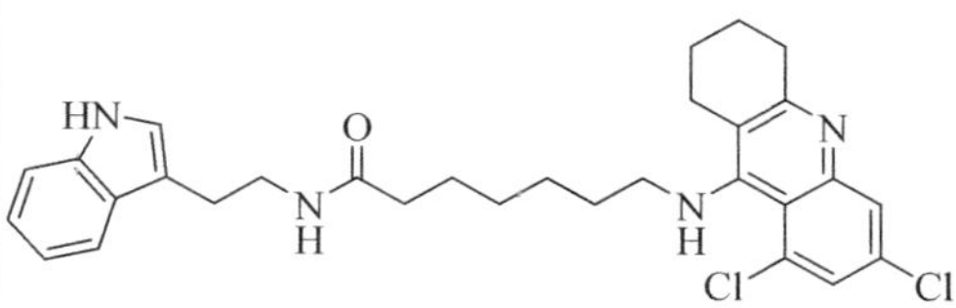

Rodríguez-Franco et al., 2006

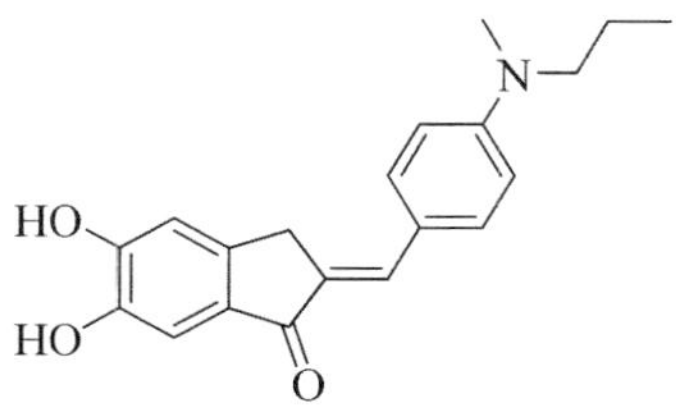

Huang et al., 2012

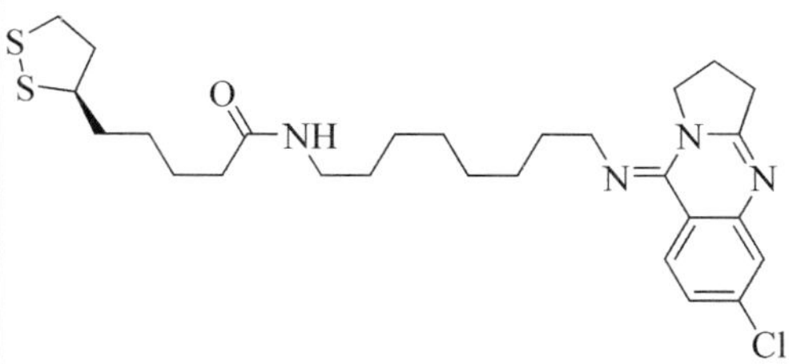

Decker et al., 2008

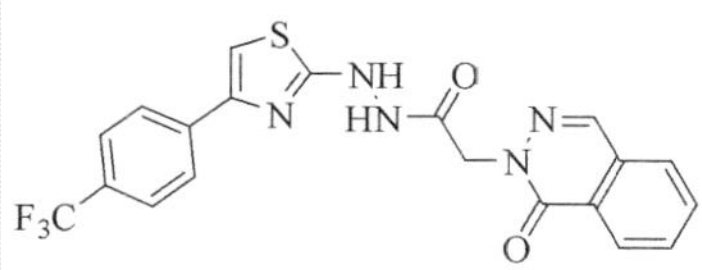

Yamalý et al., 2015

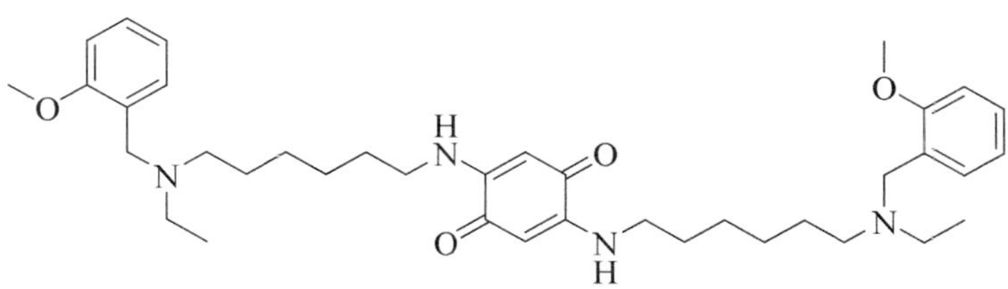

Memoquin

Bolognesi et al., 2009

12.3.5 Gamma Secretase Inhibitors

Gamma-secretase is a membrane bound proteolytic enzyme. APP is one of the substrates of this enzyme. Different than BACE-1, gamma-secretase cleaves APP any of multiple sites producing $A\beta_{39\text{-}42}$ species. The first results on animal models in which gamma secretase inhibitors were utilized displayed promising results in terms of lowered Aβ levels (Dovey et al., 2001). Similar to the results obtained for BACE-1 inhibitors, these initial evaluation on gamma-secretase inhibitors led to the design of phase trials. Unfortunately, serious side effects were observed with gamma-secretase inhibitors. No cognitive improvement was observed (De Strooper, 2014). Very interestingly, long-term exposure to gamma-secretase inhibitors led to an increase in Aβ plasma levels (i.e., a rebound effect) (Siemers et al., 2007). The mechanism for this outcome has not been clarified yet. However, the disappointing results obtained after clinical trials almost totally abolished studies on the design of novel gamma secretase inhibitors (Sogorb et al., 2018).

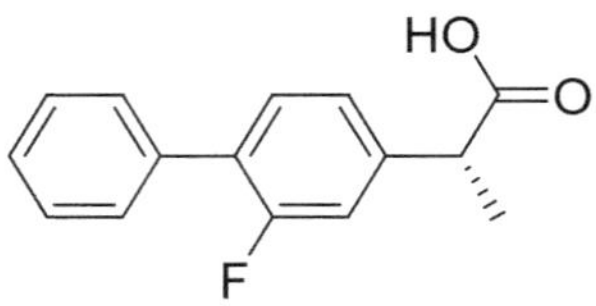

Tarenflurbil

It is the R enantiomer of flurbiprofen. No significant outcome was obtained after phase trials. A speculation on inadequate capacity of the molecule to penetrate CNS was also stated (Sabbagh, 2009; Wilcock et al., 2008).

Figure 12.5 Tarenflurbil, a gamma secretase inhibitor.

12.3.6 Dual Cholinesterase Inhibitors and Metal Chelators

It has been known that Aβ accumulation is also triggered with the presence of metals, particularly with cupper, iron, and zinc (Mantyh et al., 1993). The results indicated that Aβ plaques are able to co-aggregate with these metal ions. Even, these ions are able to accelerate aggregation process. The co-aggregation process also results in the induced cross-linked Aβ plaques which are very stable oligomers and highly aqueous insoluble (Maynard et al.,

2005). As mentioned previously, insoluble oligomer aggregation particularly in synaptic clefts are highly toxic on the continuation of neurotransmission, particularly at the cholinergic and glutamergic (i.e., NMDA) receptors, important for cognitive functions. One other finding within the development of AD is the excessive levels of some metals, including iron ions (e.g., higher than the levels in normal brains) (Bush et al., 2008). Besides the effects of ions on increasing the toxicity of amyloid plaques, they are also involved in oxidation reactions (e.g., Fenton like reactions) to generate ROS. Furthermore, excessive levels of Cu^{+2}, Zn^{+2}, Ca^{+2}, and Fe^{+2} can promote hyper-tau phosphorylation and they are able to increase APP expression which triggers more Aβ synthesis.

On the other hand, there is a metal homeostasis within the CNS. Therefore, a compound with a direct metal binding (i.e., chelator) capacity can also give damage to regular physiological functions wherein these metals are required (e.g., the enzymatic activities of metalloenzymes, direct activities of metals as secondary messengers). From this point of view, a "Metal Hypothesis of Alzheimer's Disease" has been recently proposed (Bush et al., 2008). The term mainly describes metal exchangers or ionophores rather than chelators. Ideally, these compounds are expected to target Aβ bound metals to give less side effect on the regular physiology wherein metals are required (Adlard et al., 2011). This approach can provide agents targeting those metals acting in excessive redox chemistry and upcoming ROS formation. Otherwise, brain metal scavenging or removal via CNS penetrable non-specific agents are expected to be disrupting on brain metal homeostasis. Particularly, studies, followed by discovery of clioquinol (i.e., an antiprotozoal dug) displaying inhibition on Aβ oligomer assembly, have increased the number of studies on this particular area (Schimmer et al., 2012) (Fig. 12.6).

There has been quite acquired number of studies based on the employment of both metal chelation and cholinesterase inhibition approaches. Indeed, pure metal chelators, or metal exchangers as later redefined, are categorized as disease modifying compounds, regarding the effects of metal binding on both Aβ aggregation and upcoming toxic pathways. Therefore, with a dual approach on cholinesterase inhibition and metal binding, a both radical and symptomatic therapy is aimed. Some of the representative examples of this category published in recent years are provided in Table 12.7. It is noteworthy to mention that there are also quite a lot number of controversies regarding metal chelation via CNS penetrable agents. The debates are on the homeostasis of metal levels in brain. Even, metal included food supplements employment in geriatrics is questioned as a disadvantageous or an advantageous approach in elderly.

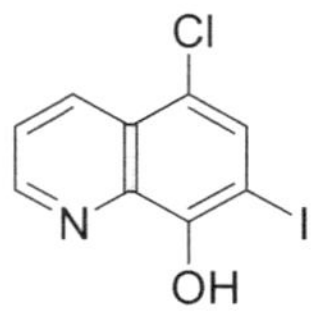

Clioquinol

A copper/zinc ionophore shown to restore cognition in mouse models of Alzheimer's disease. A pilot phase II clinical trial was conducted employing clioquinol. Although promising results were obtained in terms of improved cognition and lowered plasma levels of Aβ42, no phase III trial has been performed (Ritchie et al., 2003).

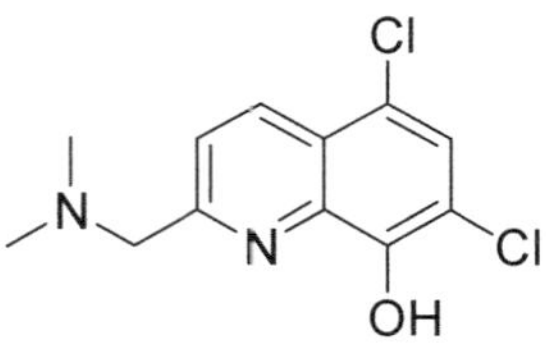

1. PBT2

2. A second-generation 8-hydroxyquinoline analogue. PBT2 is a metal-protein attenuating compound (MPAC) that affects the Cu^{2+}-mediated and Zn^{2+}-mediated toxic oligomerization of Aβ seen in Alzheimer's disease. Although phase II trials pointed out promising results, no further drug development studies were not conducted (Lannfelt et al., 2008).

Figure 12.6 Representative metal chelators in phase trials.

12.4 Conclusion

The pathogenesis of Alzheimer's disease is still very complex. The majority of the attention is focused on the amyloidogenic pathway, since it is the main identified neurodegenerative biochemical cascade induced throughout the development of the disease. Besides, other mechanisms leading to oxidative stress are also considered as alternative targets for the treatment of the disease. Some of the phase trials employing drug candidates purely designed to target one of these neurodegenerative mechanisms (e.g., amyloidogenic pathway) have provided promising results in terms of the prevention of toxic mechanisms. However, none of them has been found validated in terms of the improvement of cognition that is gradually lost throughout the development of AD. Therefore, for the last two decades, a multi-target drug design approach is mainly followed for molecule design in AD research studies. Based on the targets that have been found critical at least two of them are selected. Among those approaches, the employment of cholinesterase inhibition as the main constituent is still valid, since, besides NMDA receptor partial antagonism, cholinesterase inhibition is the only alternative providing cognitive enhancement. The near future will display the outcomes of the clinical results of these multi-target oriented drug candidates.

Table 12.7 Representative molecules designed to act as both cholinesterase inhibitor and metal chelator

Compound	AChE IC_{50} (μM)	BuChE IC_{50} (μM)	Description	Refs
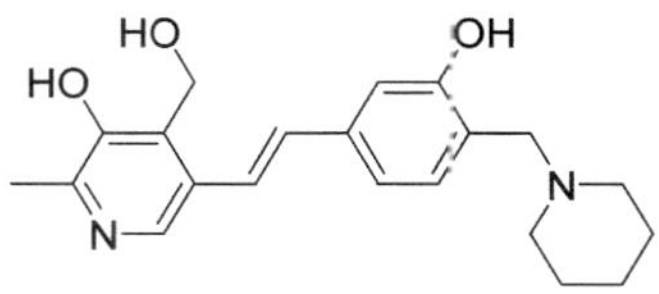	2.1	NA	The compound was shown to be iron and cupper chelator	Yang et al., 2017
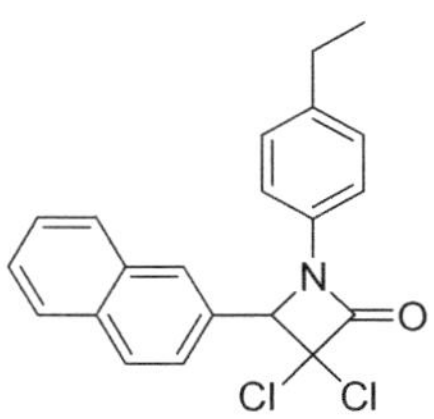	1.18	NA	The compound was shown to be iron chelator	Turan et al., 2016
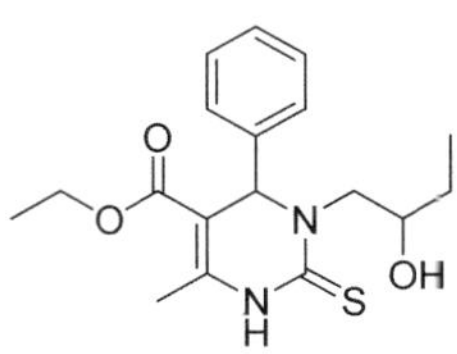	0.12	0.32	The compound was shown to be iron chelator	Sujayev et al., 2016
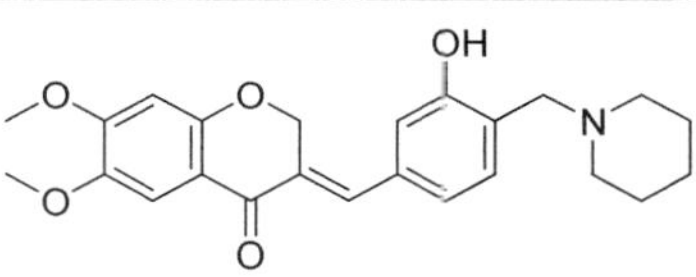	0.002	NA	The compound was shown to be Fe, Al, and Cu chelator	Li et al., 2017

The amyloidogenic pathway, on the other hand, appears to be the outcome of the disease, rather than its original cause. From this point of view, novel approaches are still sought to find out the exact mechanisms underlying the excessive activity of amyloidogenic pathway throughout the development of AD.

References

Adlard P. A., Bica L., White A. R., Nurjono M., Filiz G., Crouch P. J., Bush A. I., *PloS One*, **6** (2011), e17669.

Ahmed F., Ghalib R. M., Sasikala P., Ahmed K. M., *Pharmacog. Rev.*, **7** (2013), 121.

Ahmed, T., Gilani A. H., *Pharmacol. Biochem. Behav.*, **91** (2009), 554–559.

Auriacombe S., Pere J. J., Loria-Kanza Y., Vellas B., *Curr. Med. Res. Opin.*, **18** (2002), 129–138.

Bar-Am O., Weinreb O., Amit T., Youdim M. B., *J. Mol. Neurosci.*, **37** (2009), 135–145.

Bautista-Aguilera O. M., Samadi A., Chioua M., Nikolic K., Filipic S., Agbaba D., Ramsay R. R., *J. Med. Chem.*, **57** (2014), 10455–10463.

Becker J. A., Hedden T., Carmasin J., Maye J., Rentz D. M., Putcha D., et al., *Ann. Neurol.*, **69** (2011), 1032–1042.

Bolognesi M. L., Andrisano V., Bartolini M., Banzi R., Melchiorre C., *J. Med. Chem.*, **48** (2005), 24–27.

Bolognesi M. L., Cavalli A., Bergamini C., Fato R., Lenaz G., Rosini M., et al., *J. Med. Chem.*, **52** (2009), 7883–7886.

Burčul F., Generalić Mekinić I., Radan M., Rollin P., Blažević I., *J. Enz. Inhib. Med. Chem.*, **33** (2018), 577–582.

Bush A. I., Tanzi R. E., *Neurother.*, **5** (2008), 421–432.

Castro A., Martinez A., *Curr. Pharmmaceut Design*, **12** (2006), 4377–4387.

Catto M., Berezin A. A., Re D. L., Loizou G., Demetriades M., De Stradis A., et al., *Eur. J. Med. Chem.*, **58** (2012), 84–97.

Chatonnet A., Lockridge O., *Biochem. J.*, **260** (1989), 625–634.

Chiti F., Dobson C. M., *Annu. Rev. Biochem.*, **75** (2006), 333–366.

Citron M., *Nat. Rev. Drug Discov.*, **9** (2010), 387–398.

Crismon M. L., *Ann. Pharmacother.*, **28** (1994), 744–751.

Crow T. J., Cross A. J., Cooper S. J., Deakin J. F. W., Ferrier I. N., Johnson J. A., et al., *Neuropharmacol.*, **23** (1984), 1561–1569.

Cuajungco M. P., Goldstein L. E., Nunomura A., Smith M. A., Lim J. T., Atwood C. S., et al., *J. Biol. Chem.*, **275** (2000), 19439–19442.

Cummings J. L., Morstorf T., Zhong K., *Alzheimer's Res. Ther.*, **6** (2014), 37.

Dall'Acqua S., *Botanics: Targ. Ther.*, **3** (2013), 19–28.

De Felice F. G., Wu D., Lambert M. P., Fernandez S. J., Velasco P. T., Lacor P. N., et. al., *Neurobiol. Aging*, **29** (2008), 1334–1347.

de Oliveira Silva E., Batista R., *Compreh. Rev. Food Sci. Food Safe.*, **16** (2017), 580–616.

De Strooper B., *Cell*, **159** (2014), 721–726.

Decker M., Kraus B., Heilmann J., *Bioorg. Med. Chem.*, **16** (2008), 4252–4261.

Dovey H. F., John V., Anderson J. P., Chen L. Z., de Saint Andrieu P., Fang L. Y., Hu K. L., *J. Neurochem.*, **76** (2001), 173–181.

Erdogan Orhan, I. Ozan Gulcan H., *Curr. Top. Med. Chem.*, **15** (2015), 1673–1682.

Fisher A., Pittel Z., Haring R., Bar-Ner N., Kliger-Spatz M., Natan N., Brandeis R., *J. Mol. Neurosci.*, **20** (2003), 349–356.

Fluhrer R., Capell A., Westmeyer G., Willem M., Hartung B., Condron M. M., Walter J., *J. Neurochem.*, **81** (2002), 1011–1020.

Forman M. S., Trojanowski J. Q., Lee V. M., *Nat. Med.*, **10** (2004), 1055–1063.

Francis, P. T., Palmer, A. M., Snape, M., Wilcock, G. K., *J. Neurol. Neurosurg. Psychiatry*, **66** (1999), 137–147.

Frankel E. N., Meyer A. S., *J. Sci. Food Agricult.*, **80** (2000), 1925–1941.

Getsios D., Blume S., Ishak K. J., Maclaine G., Hernández L., *Alzheimer's Dem.*, **8** (2012), 22–30.

Giacobini E., The cholinergic system in Alzheimer disease, in *Progress in Brain Research* (Dunnet S. B., Richards S. J.), vol. 84, pp. 321–332, Elsevier (1990).

Greig N. H., Sambamurti K., Yu Q. S., Brossi A., Bruinsma G. B., Lahiri D. K., *Curr. Alzheimer Res.*, **2** (2005), 281–290.

Gulcan H. O., Unlu S., Esiringu Ý., Ercetin T., Sahin Y., Oz D., Sahin M. F., *Biorg. Med. Chem.*, 22 (2014), 5141–5154.

Gulyás B., Pavlova E., Kása P., Gulya K., Bakota L., Várszegi S., Magyar K., *Neurochem. Int.*, **58** (2011), 60–68.

Haass C., Selkoe D. J., *Nat. Rev. Mol. Cell Biol.*, **8** (2007), 101–112.

Halliwell B., *J. Neurochem.*, **97** (2006), 1634–1658.

Hardy J., *J. Neurochem.*, **110** (2009), 1129–1134.

Hasselmo M. E., *Curr. Opin. Neurobiol.*, **16** (2006), 710–715.

Hawkes C. A., Ng V., McLaurin J., *Drug Devel. Res.*, **70** (2009), 111–124.

Hebert L. E., Scherr P. A., Bienias J. L., Bennett D. A., Evans D. A., *Arch. Neurol.*, **60** (2003), 1119–1122.

Hills I. D., Vacca J. P., *Curr. Opin. Drug Discov. Devel.*, **10** (2007), 383–391.

House E., Collingwood J., Khan A., Korchazkina O., Berthon G., Exley C., *J. Alzheimer's Dis.*, **6** (2004), 291–301.

Huang L., Lu C., Sun Y., Mao F., Luo Z., Su T., et al., *J. Med. Chem.*, **55** (2012), 8483–8492.

Hynd M. R., Scott H. L., Dodd P. R., *Neurochem. Int.*, **45** (2004), 583–595.

Inestrosa N. C., Alvarez A., Perez C. A., Moreno R. D., Vicente M., Linker C., Garrido J., *Neuron*, **16** (1996), 881–891.

Ishimoto H., Shibata M., Myojin Y., Ito H., Sugimoto Y., Tai A., Hatano T., *Bioorg. Med. Chem. Lett.*, **21** (2011), 5901–5904.

Jack Jr. C. R., Albert M. S., Knopman D. S., McKhann G. M., Sperling R. A., Carrillo M. C., et al., *Alzheimer's Dem.*, **7** (2011), 257–262.

Jann M. W., *Pharmacother.: J. Hum. Pharmacol. Drug Ther.*, **20** (2000), 1–12.

Ježek P., Hlavatá L., *Int. J. Biochem. Cell Biol.*, **37** (2005), 2478–2503.

Johnson J. W., Kotermanski S. E., *Curr. Opin. Pharmacol.*, **6** (2006), 61–67.

Johnson-Wood K., Lee M., Motter R., Hu K., Gordon G., Barbour R., Lieberburg I., *Proc. Natl. Acad. Sci. U. S. A.*, **94** (1997), 1550–1555.

Kalgutkar A. S., Dalvie D. K., Castagnoli N., Taylor T. J., *Chem. Res. Toxicol.*, **14** (2001), 1139–1162.

Kamenetz F., Tomita T., Hsieh H., Seabrook G., Borchelt D., Iwatsubo T., Malinow R., *Neuron*, **37** (2003), 925–937.

Koirala P., Seong S. H., Jung H. A., Choi J. S., *Molecules*, **23** (2018), pii: E785.

Kumar A., Singh A., *Pharmacol. Rep.*, **67** (2015), 195–203.

Kuperstein I., Broersen K., Benilova I., Rozenski J., Jonckheere W., Debulpaep M., Braeken D., *EMBO J.*, **29** (2010), 3408–3420.

Lannfelt L., Blennow K., Zetterberg H., Batsman S., Ames D., Harrison J., et al., *Lancet Neurol.*, **7** (2008), 779–786.

Leon R., Garcia A. G., Marco-Contelles J., *Med. Res. Rev.*, **33** (2013), 139–189.

Levin E. D., *Psychopharmacol.*, **108** (1992), 417–431.

Li Y., Qiang X., Li Y., Yang X., Luo L., Xiao G., Deng Y., *Bioorg. Med. Chem. Lett.*, **26** (2016), 2035–2039.

Li Y., Qiang X., Luo L., Yang X., Xiao G., Zheng Y., Deng Y., *Bioorg. Med. Chem.*, **25** (2017), 714–726.

Lipton S. A., *NeuroRx*, **1** (2004), 101–110.

Lipton S. A., *Nat. Rev. Drug Discov.*, **5** (2006), 160–170.

Liu W., Lang M., Youdim M. B., Amit T., Sun Y., Zhang Z., Weinreb O., *Neuropharmacol.*, **109** (2016), 376–385.

Lovell M. A., Xie C., Markesbery W. R., *Neurology*, **51** (1998), 1562–1566.

Luchsinger J. A., Mayeux R., *Lancet Neurol.*, **3** (2004), 579–587.

Ma Y., Luo W., Quinn P. J., Liu Z., Hider R. C., *J. Med. Chem.*, **47** (2004), 6349–6362.

Maestri D. M., Nepote V., Lamarque A. L., Zygadlo J. A., *Phytochem.: Adv. Res.*, **37** (2006), 105–135.

Mann D. M., Iwatsubo T., Ihara Y., Cairns N. J., Lantos P. L., Bogdanovic N., Rossor M. N., *Am. J. Pathol.*, **148** (1996), 1257.

Mantyh P. W., Ghilardi J. R., Rogers S., DeMaster E., Allen C. J., Stimson E. R., Maggio J. E. (1993). *J. Neurochem.*, **61** (1993), 1171–1174.

Martinez-Coria H., Green K. N., Billings L. M., Kitazawa M., Albrecht M., Rammes G., et al., *Am. J. Pathol.*, **176** (2010), 870–880.

Masliah E., Hansen L., Adame A., Crews L., Bard F., Lee C., Schenk D., *Neurology*, **64** (2005), 129–131.

Maynard C. J., Bush A. I., Masters C. L., Cappai R., Li Q. X., *Int. J. Experim. Pathol.*, **86** (2005), 147–159.

McKhann G. M., Knopman D. S., Chertkow H., Hyman B. T., Jack Jr. C. R., Kawas C. H., et al., *Alzheimer's Dem.*, **7** (2011), 263–269.

Meek, P. D., McKeithan, E. K., Schumock, G. T., *Pharmacother. J. Human Pharmacol. Drug Ther.*, **18**(2P2) (1998), 68–73.

Meilandt W. J., Cisse M., Ho K., Wu T., Esposito L. A., Scearce-Levie K., Mucke L., *J. Neurosci.*, **29** (2009), 1977–1986.

Messer W. S., *Curr. Top. Med. Chem. Hilversum*, **2** (2002), 353–358.

Meyer J. H., Ginovart N., Boovariwala A., Sagrati S., Hussey D., Garcia A., Houle S., *Arch. Gen. Psychiat.*, **63** (2006), 1209–1216.

Mishra C. B., Manral A., Kumari S., Saini V., Tiwari M., *Bioorg. Med. Chem.*, **24** (2016), 3829–3841.

Mohamed T., Rao P. P., *Eur. J. Med. Chem.*, **126** (2017), 823–843.

Mohamed T., Zhao X., Habib L. K., Yang J., Rao P. P., *Bioorg. Med. Chem.*, **19** (2011), 2269–2281.

Mohammadi-Khanaposhtani M., Saeedi M., Zafarghandi N. S., Mahdavi M., Sabourian R., Razkenari E. K., et al., *Eur. J. Med. Chem.*, **92** (2015), 799–806.

Norouzbahari M., Burgaz E. V., Ercetin T., Fallah A., Foroumadi A., Firoozpour L., Sahin M. F., Gazi M., Gulcan H. O., *Lett. Drug Design Discov.*, **15** (2018), https://doi.org/10.2174/ 1570180815666180115144608.

Papandreou M. A., Dimakopoulou A., Linardaki Z. I., Cordopatis P., Klimis-Zacas D., Margarity M., Lamari F. N., *Behav. Brain Res.*, **198** (2009), 352–358.

Parsons C. G., Danysz W., Quack G., *Neuropharmacol.*, **38** (1999), 735–767.

Persson T., Popescu B. O., Cedazo-Minguez A., *Oxid. Med. Cell. Longevity*, Article ID 427318 (2014).

Piazzi L., Cavalli A., Colizzi F., Belluti F., Bartolini M., Mancini F., Rampa A., *Bioorg. Med. Chem. Lett.*, **18** (2008), 423–426.

Pisani L., Farina R., Catto M., Iacobazzi R. M., Nicolotti O., Cellamare S., Altomare C. D., *J. Med. Chem.*, **59** (2016), 6791–6806.

Polinsky R. J., *Clin. Ther.*, **20** (1998), 634–647.

Pugazhendhi A., Shafreen R. B., Devi K. P., Suganthy N., *J. Mol. Liquids*, **257** (2018), 69–81.

Reinikainen K. J., Soininen H., Riekkinen P. J., *J. Neurosci. Res.*, **27** (1990), 576–586.

Reisberg B., Borenstein J., Salob S. P., Ferris S. H., *J. Clin. Psychiatry*, **48** (1987), Suppl, 9–15.

Ritchie, C. W., Bush A. I., Mackinnon A., Macfarlane S., Mastwyk M., MacGregor L., et al., *Arch. Neurol.*, **60** (2003), 1685–1691.

Rodríguez-Franco M. I., Fernández-Bachiller M. I., Pérez C., Hernández-Ledesma B., Bartolomé B., *J. Med. Chem.*, **49** (2006), 459–462.

Sabbagh M. N., *Am. J. Geriatric Pharmacother.*, **7** (2009), 167–185.

Schimmer A. D., Jitkova Y., Gronda M., Wang Z., Brandwein J., Chen C., Ackloo S., *Clin. Lymphoma Myeloma Leuk.*, **12** (2012), 330–336.

Seilheimer B., Bohrmann B., Bondolfi L., Müller F., Stüber D., Döbeli H., *J. Struct. Biol.*, **119** (1997), 59–71.

Shankar G. M., Bloodgood B. L., Townsend M., Walsh D. M., Selkoe D. J., Sabatini B. L., *J. Neurosci.*, **27** (2007), 2866–2875.

Shih, J. C., Chen, K., Ridd, M. J., *Polish J. Pharm.*, **51**(1) (1999), 25–29.

Siemers E. R., Dean R. A., Friedrich S., Ferguson-Sells L., Gonzales C., Farlow M. R., May P. C., *Clin. Neuropharmacol.*, **30** (2007), 317–325.

Small G. W., Rabins P. V., Barry P. P., Buckholtz N. S., DeKosky S. T., Ferris S. H., et al., *JAMA*, **278** (1997), 1363–1371.

Sogorb-Esteve A., García-Ayllón M. S., Llansola M., Felipo V., Blennow K., Sáez-Valero J., *Mol. Neurobiol.*, **55** (2018), 5047–5058.

Solomon B., Koppel R., Hanan E., Katzav T., *Proc. Natl. Acad. Sci. U. S. A.*, **93** (1996), 452–455.

Soriano S., Lu D. C., Chandra S., Pietrzik C. U., Koo E. H., *J. Biol. Chem.*, **276** (2001), 29045–29050.

Spaldin V., Madden S., Pool W. F., Woolf T. F., Park B. K., *Br. J. Clin. Pharmacol.*, **38** (1994), 15–22.

Stamford A., Strickland C., *Curr. Opin. Chem. Biol.*, **17** (2013), 320–328.

Sugimoto H., *Chem. Rec.*, **1** (2001), 63–73.

Sujayev A., Garibov E., Taslimi P., Gulçin Ý., Gojayeva S., Farzaliyev V., Supuran C. T., *J. Enz. Inhib. Med. Chem.*, **31** (2016), 1531–1539.

Takeda A., Loveman E., Clegg A., Kirby J., Picot J., Payne E., Green C., *Int. J. Geriatr. Psychiatry*, **21** (2006), 17–28.

Tan C. C., Yu J. T., Wang H. F., Tan M. S., Meng X. F., Wang C., et al., *J. Alzheimer's Dis.*, **41** (2014), 615–631.

Tripathi R. K., M. Sasi V., Gupta S. K., Krishnamurthy S., Ayyannan S. R., *J. Enzyme Inhib. Med. Chem.*, **33** (2018), 37–57.

Turan B., Þendil K., Þengül E., Gültekin M. S., Taslimi P., Gulçin Ý., Supuran C. T., *J. Enzyme Inhib. Med. Chem.*, **31** (2016), 79–88.

Vassar R., *Alzheimer's Res. Ther.*, **6** (2014), 89.

Wang, Y., Sun Y., Guo Y., Wang Z., Huang L., Li X., *J. Enzyme Inhib. Med. Chem.*, **31** (2016), 389–397.

Watkins P. B., Zimmerman H. J., Knapp M. J., Gracon S. I., Lewis K. W., *JAMA*, **271** (1994), 992–998.

Wilcock G. K., Black S. E., Hendrix S. B., Zavitz K. H., Swabb E. A., Laughlin M. A., *Lancet Neurol.*, **7** (2008), 483–493.

Wu Z., Liao W., Chen K., Qin J., Tang H., *Bioorg. Chem.*, 76 (2018), 228–236.

Xie S. S., Wang X., Jiang N., Yu W., Wang K. D., Lan J. S., Kong L. Y., *Eur. J. Med. Chem.*, **95** (2015), 153–165.

Yamalý C., Gülcan H. O., Kahya B., Çobanoðlu S., Þüküroðlu M. K., Doðruer D. S., *Med. Chem. Res.*, **24** (2015), 1210–1217.

Yan J. W., Li Y. P., Ye W. J., Chen S. B., Hou J. Q., Tan J. H., Huang Z. S., *Bioorg. Med. Chem.*, 20 (2012), 2527–2534.

Yang X., Qiang X., Li Y., Luo L., Xu R., Zheng Y., et al., *Bioorg. Chem.*, **71** (2017), 305–314.

Yang C. S., Sang S., Lambert J. D., Lee M. J., *Mol. Nutr. Food Res.*, **52** (2008), S139–S151.

Yanovsky I., Finkin-Groner E., Zaikin A., Lerman L., Shalom H., Zeeli S., Weinstock M., *J. Med. Chem.*, **55** (2012), 10700–10715.

Youdim M. B., *Curr. Alzheimer Res.*, **3** (2006), 541–550.

Youdim M. B., Buccafusco J. J., *Trends Pharmacol. Sci.*, **26** (2005), 27–35.

Zheng H., Fridkin M., Youdim M, *Pharmaceuticals*, **7** (2014), 113–135.

Zheng H., Gal S., Weiner L. M., Bar-Am O., Warshawsky A., Fridkin M., Youdim M. B., *J. Neurochem.*, **95** (2005), 68–78.

Zhou Y., Wang S., Zhang Y., *J. Phys. Chem. B*, **114** (2010), 8817–8825.

Zhu Y., Xiao K., Ma L., Xiong B., Fu Y., Yu H., Li J., *Bioorg. Med. Chem.*, **17** (2009), 1600–1613.

Chapter 13

Aldehyde Oxidases as Enzymes in Phase I Drug Metabolism

Cristiano Mota,[a] Teresa Santos-Silva,[a] Mineko Terao,[b] Enrico Garattini,[b] Maria João Romão,[a] and Silke Leimkühler[c]

[a]*UCIBIO, Departamento de Química, Faculdade de Ciências e Tecnologia, Universidade Nova de Lisboa, 2829-516 Caparica, Portugal*
[b]*Laboratory of Molecular Biology, Istituto di Ricerche Farmacologiche Mario Negri IRCCS, via La Masa 19, 20156 Milano, Italy*
[c]*Department of Molecular Enzymology, Institute of Biochemistry and Biology, University of Potsdam, Karl-Liebknecht-Str. 24-25, 14476 Potsdam, Germany*

mjr@fct.unl.pt sleim@uni-potsdam.de

13.1 Introduction

Aldehyde oxidases (AOX, EC 1.2.3.1) are a group of enzymes with important functions in the context of drug and xenobiotic metabolism. These proteins are molybdoflavoenzymes which contain two [2Fe-2S] clusters, the flavin adenine dinucleotide (FAD) cofactor and the molybdenum cofactor (Moco) as the catalytically active units (Alfaro and Jones, 2008; Manevski et al., 2014; Romao, 2009; Schwarz et al., 2009). AOX enzymes are present

Pharmaceutical Biocatalysis: Fundamentals, Enzyme Inhibitors, and Enzymes in Health and Diseases
Edited by Peter Grunwald

ISBN 978-981-4800-61-7 (Hardcover), 978-0-429-29503-4 (eBook)
www.jennystanford.com

in both prokaryotes and eukaryotes. The mammalian AOXs that have been examined to date are exclusively oxidases; NAD^+ is not used as a substrate at the FAD site (Coelho et al., 2012). Thus, AOX enzymes use exclusively molecular oxygen as oxidizing substrate, producing mainly hydrogen peroxide and, to some extent, superoxide as final products. At present, little is known about the physiological relevance of AOX in humans and other mammals.

The number of AOX isoenzymes varies according to the animal species considered. Rodents and humans represent the two extremes. The genomes of mice and rats contain four genetic loci, clustered on chromosome 1, which synthesize AOX1, AOX2, AOX3 and AOX4, respectively (Garattini et al., 2008, 2009; Kurosaki et al., 2013). In contrast, humans and the majority of primates are endowed with a single active AOX gene, i.e., the orthologue of mouse AOX1. The reasons why mammals other than humans express a multiplicity of tissue-specific AOX enzymes is currently unknown.

The specific substrates of AOXs in mammals are also unknown, since the reactions catalyzed by these enzymes are very diverse (Krenitsky et al., 1972). AOXs generally catalyze oxidation reactions that involve nucleophilic attack of electrophilic carbon centers, including aldehyde groups and carbons of N-heterocycles that are *para* to, or adjacent to, the ring's nitrogen (Beedham, 1985, 1997, 1998), e.g., in pyrazine, purine, pyrimidine, quinoline, pyridine, and pteridine. In addition, they reduce certain amino- and sulfo-groups and hydrolyze amide bonds (Garattini et al., 2009; Mota et al., 2018).

AOX enzymes are better known for oxidizing different types of aldehydes to their corresponding acids being also involved in the hydroxylation of *N*-heterocyclic ring systems possessing –CH=N– moieties (e.g., phthalazine) and purine or aromatic and non-aromatic charged heterocycles with a $-CH=N^+$ group (e.g., N^1-methylnicotinamide and *N*-methylphthalazinium). Thus, the term "aldehyde oxidase" might not always be appropriate, since these enzymes do not necessarily act on substrates containing an aldehyde functionality (Coelho et al., 2012; Garattini and Terao, 2011).

Human AOX1 is a cytosolic protein and it is recognized to play an important role in phase I drug metabolism, being involved in the oxidation of several antiviral, hypnotic, and antiepileptic agents (Baral et al., 2014; Bolis et al., 2017; Coelho et al., 2012; Fratelli et al., 2013; Garattini and Terao, 2011; Pryde et al., 2010; Pryde et al., 2012). AOX1 substrates are often intermediates or products of cytochrome-P450 dependent metabolism (Barr and Jones, 2011; Garattini and Terao, 2012; Kitamura and Sugihara, 2014; Obach et al., 2004). Furthermore, hAOX1 is inhibited by various types of drugs, including estrogen receptor modulators (raloxifene) and atypical antipsychotic agents (phenothiazines such as thioridazine and loxapine), which supports the involvement of the enzyme in drug interactions (Barr and Jones, 2011; Obach et al., 2004; Sanoh et al., 2015).

In this chapter, we focus on the significance of aldehyde oxidases for the design and development of new drugs and discuss associated problems.

13.2 Gene Organization and Phylogeny of AOX Enzymes

AOXs and xanthine oxidoreductases (XORs) are structurally and evolutionarily linked molybdoflavoproteins. AOXs and XORs are widely distributed throughout all kingdoms of life and homologous proteins have been identified in many eukaryotic and prokaryotic organisms (Garattini et al., 2003; Hesberg et al., 2004; Mendel et al., 2007 Pritsos, 2000; Yasuhara et al., 2005; Zhang and Gladyshev, 2008). There is a close structural relationship between AOX and XOR proteins of both eukaryotic and prokaryotic origin. The catalytically active forms of eukaryotic AOXs and XORs are dimers consisting of identical subunits. In eukaryotes, the size of AOXs and XOR monomeric subunits is very similar (around 150 kDa) and the corresponding amino acid sequences are highly conserved. For instance, the overall similarity between the various vertebrate AOX isoenzymes and XORs is around 50%. Thus, the amino acid sequences of AOXs and XORs from different animal species can be easily aligned along their entire length (Garattini et al., 2008; Garattini et al., 2003). Vertebrate AOX and XOR proteins are characterized by high levels of similarity not only as far as their amino acid sequence, but also in terms of their 3D structures. This is accompanied by remarkable resemblance in the structure and organization of the corresponding genes.

Up to four different AOX genes, coding for an equivalent number of AOX isoenzymes, have been described in insects (Marelja et al., 2014), plants (Kurosaki et al., 2013) and vertebrates (Garattini et al., 2008; Garattini et al., 2003; Kurosaki et al., 2013), while a single XOR gene is known. All vertebrate AOX and XOR genes consist of 35 and 36 exons, respectively, and the position of the common 35 exon-intron junctions are strictly conserved (Garattini et al., 2008; Kurosaki et al., 2013). The human genes encoding AOX1 and XOR are each approximately 80 kb long. The major difference between the two human genes is suppression of intron 26 in XOR. As a result, exon 26 in the AOX1 gene is identical to exons 26 and 27 in the XOR gene (Terao et al., 1998). A very similar situation is observed in the case of the mouse/rat AOX1, AOX2, AOX3 and AOX4 genes as well as the various orthologous genes present in other vertebrate species (Kurosaki et al., 2013). In humans, the AOX1 and XOR genes are found on different arms of the same chromosome. XOR maps to chromosome 2:p22 (Berger et al., 1995; Ichida et al., 1993; Xu et al., 1994), while AOX1 is located on chromosome 2:q33 (Berger et al., 1995; Terao et al., 1998; Wright et al., 1997). In addition to this functional AOX gene, two human AOX pseudogenes, consisting of 25 and 16 residual exons, respectively, are located on chromosome 2q (Garattini et al., 2003; Kurosaki et al., 2013; Terao et al., 2006). Based on their exon sequences, the two human pseudogenes are likely to represent the orthologs of the functional rodent AOX3 and AOX2 genes (Garattini et al., 2008; Kurosaki et al., 2004; Terao et al., 2006; Terao et al., 2001).

Analysis of the genomic sequencing data available for AOXs and XOR has generated a first draft of the evolutionary history of both types of enzymes (Kurosaki et al., 2013) (Fig. 13.1). A single XOR gene is present in practically all vertebrate species, while the situation of AOX genes is much more complex. In fact, vertebrate genomes are characterized by the presence of one or more AOX genes, each coding for a distinct isoenzyme. Mammalian species present with a maximum of four active AOX genes coding for an equivalent number of proteins (Kurosaki et al., 2013). As already mentioned, the two extremes are represented by humans, which express a single protein (AOX1), and rodents which synthesize four AOX isoenzymes. In mammals, the AOX loci form small gene clusters, as they map to a short region of the same chromosome (Kurosaki et al., 2013) (Fig. 13.2).

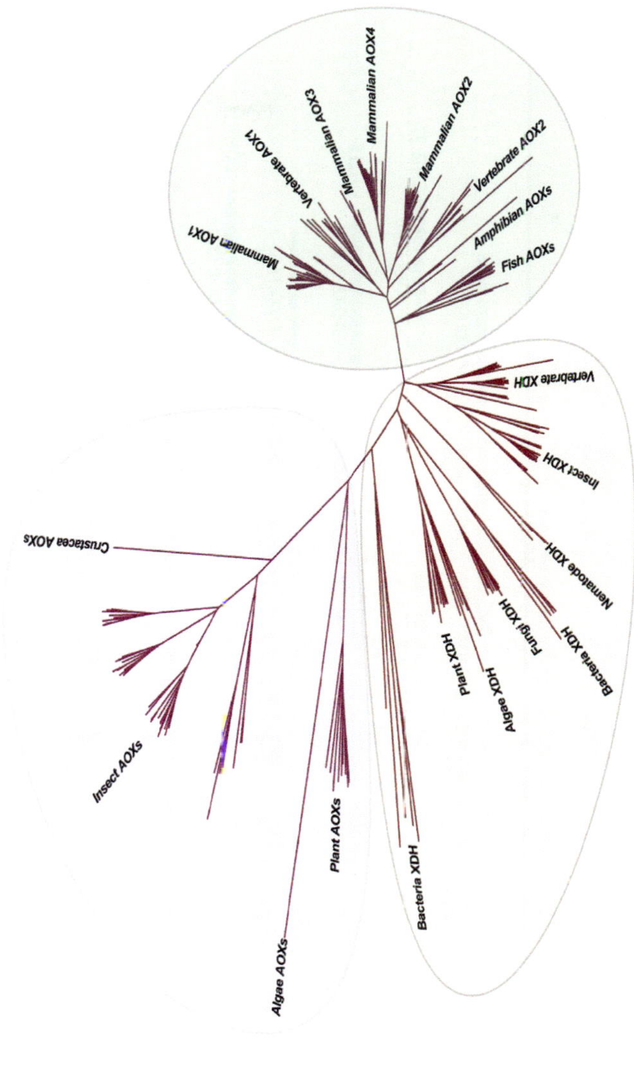

Figure 13.1 Unrooted phylogenetic tree of AOXs and XDHs. Phylogenesis of AOX isoenzymes and XOR (XDH) proteins in eukaryotes and prokaryotes. The phylogenetic tree is constructed aligning a large number of publicly available sequences using the CLUSTAL omega algorithm.

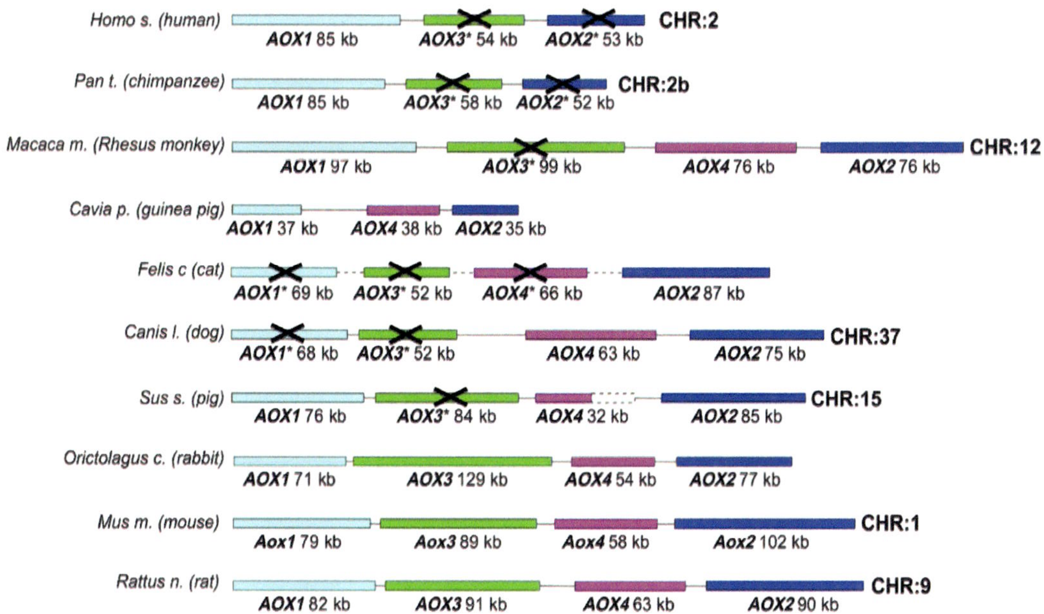

Figure 13.2 AOX genes in humans and selected experimental animals. A schematic representation of *AOX* genes and pseudogenes in humans and selected primates or mammals used for studies on drug and xenobiotic metabolism. Orthologous *AOX* genes and pseudogenes are marked with the same color. Pseudogenes are crossed and marked with an asterisk. When determined, the chromosomal location is indicated on the right: CHR, chromosome.

These observations suggest that vertebrate AOXs and XORs are characterized by a common evolutionary origin. In fact, the present complement of vertebrate AOX and XOR loci originate from gene-duplication and gene-deletion/inactivation events which occurred at different stages during the evolution of the various animal species (Garattini et al., 2008; Kurosaki et al., 2013; Rodriguez-Trelles et al., 2003; Terao et al., 2006). It is likely that prokaryotic, plant, algae, insect and vertebrate AOXs have different origins. In fact, they are likely to be the result of a primordial gene-duplication event involving an ancestral XOR gene (Kurosaki et al., 2013). Subsequently, two distinct lines of AOXs evolved, as indicated by the phylogenetic tree shown in Fig. 13.1. Bacteria, algae, plant and insects are characterized by a similar complement of AOXs, which map to the purple cluster of the tree, and whose center (yellow cluster) is occupied by XORs. In contrast, vertebrate AOXs gather in the green cluster of the dendrogram. This supports the idea that bacteria, algae, plant, and insect AOX genes derive from an original XOR gene-duplication event which is distinct and predates a similar duplication event generating vertebrate AOXs. This analysis suggests that the first phase of vertebrate AOXs evolution is characterized by a species-specific increase in the number of iso-enzymatic forms. In fact, the most ancient vertebrate for which genomic data are available (lamprey) is characterized by a typical XOR gene and the absence of sequences with similarity to AOXs. Ray-finned fishes are the first vertebrates endowed with AOX genes (AOXα and AOXβ (Garattini et al., 2008)). AOXα is likely to be the product of the first duplication event from vertebrate XOR, while AOXβ is the result of a later gene duplication involving AOXα (Garattini et al., 2008; Kurosaki et al., 2013). The subsequent evolution of vertebrate AOXs is characterized by at least two other gene duplications occurring before or concomitant with the appearance of rodents. In mammals, this was followed by a second evolutionary phase resulting in a progressive and species-specific inactivation of one or more AOX genes, which is particularly evident in primates. The last process is the result of gene-deletion or pseudogenization events and led to the maintenance of a single AOX isoform (AOX1) in humans.

The existence of a single AOX in humans and a variable number of AOX isoforms in other mammals may be related to the fact that each isoenzyme exerts a different systemic or tissue-specific function yet to be defined. With respect to this, human AOX1 and possibly other mammalian AOXs are predicted to be involved in specific metabolic pathways, as illustrated by the data available in the KEGG (*Kyoto Encyclopedia of Genes and Genomes*). In addition, mouse AOX1, AOX3 and AOX4 have been demonstrated to catalyze the oxidation of physiological substrates, such as retinaldehyde (Garattini et al., 2008; Terao et al., 2016b). However, the physiological function of human AOX1 and other mammalian AOXs is still largely obscure. At present, the only mouse enzyme for which significant functional data are available is AOX4, given the existence of a specific AOX4 knockout animal (Terao et al., 2016a; Terao et al., 2009). Indeed, phenotypic analysis of these genetically engineered mice is consistent with a systemic action on fat deposition in the adipose tissue, and a tissue-specific role in the control of the epidermal homeostasis. The first observation is in line with the fact that the largest amounts of AOX4 are observed in the Harderian gland, a multifunctional structure located in the orbital cavity, which is not present in humans or higher primates (Terao et al., 2016a). The second finding is compatible with the much lower levels of AOX4, which is found in keratinized epithelia, with particular reference to the skin, esophagus, and stomach. No significant information is available on mouse AOX1 and AOX3, which are predominantly expressed in liver and lung, or mouse AOX2 whose expression is restricted to the epithelial component of the nasal mucosa (Terao et al., 2000). Interestingly, the only functional AOX gene present in humans is the mouse AOX1 ortholog, which is likely to be the most primitive vertebrate AOX isoform. This observation suggests a nonredundant and common physiological role of AOX1 in vertebrates.

13.3 Overall Structure and Domain Architecture of hAOX

hAOX1 is encoded by a single gene in humans and shares 50% amino acid identity with human xanthine oxidoreductase (XOR). In both enzymes, the functional homodimer (2 × 150 kDa) adopts a butterfly shape with approximate dimensions of 150 Å × 90 Å × 65 Å (Fig. 13.3). Each subunit can be further divided into three conserved domains, connected by two linker regions: a small

N-terminal domain I (20 kDa), which harbors two [2Fe-2S] clusters; the FAD domain II (40 kDa); and the large catalytic domain III (90 kDa) containing the molybdenum cofactor (Moco) and the substrate-binding site located at the C-terminus (Coelho et al., 2015; Coelho et al., 2012).

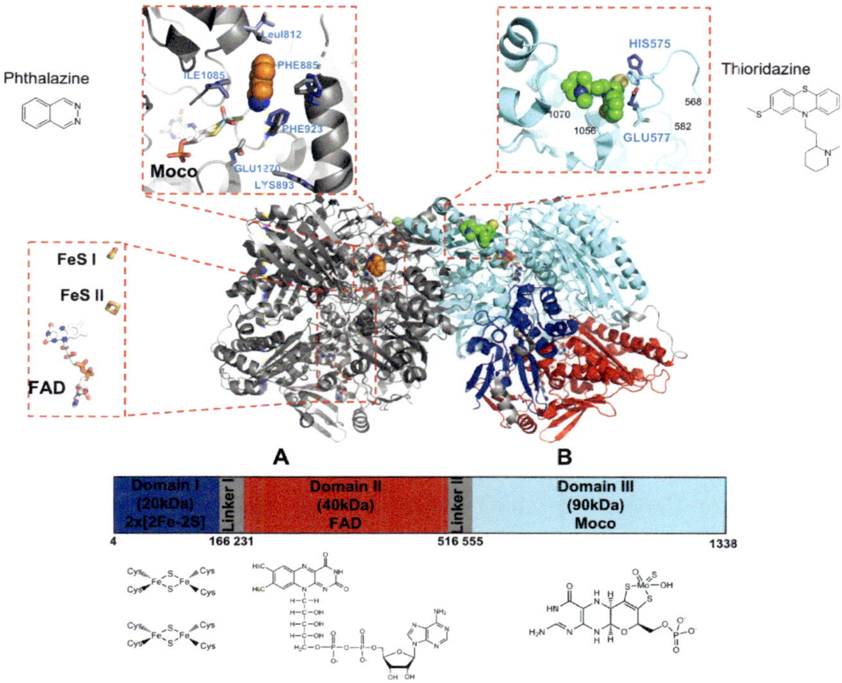

Figure 13.3 Structure of hAOX1. Overall structure of the hAOX1 homodimer in complex with phthalazine (in orange, in space filling) and thioridazine (in green, in space filling) (PDB ID: 4UHX). The three domains represented in monomer B are colored as: domain I in blue, domain II in red and domain III in cyan. Domain I harbors the two [2Fe-2S] centers, domain II contains the FAD cofactor and in domain III Moco is buried. Left: close-up view of phthalazine binding site with relevant amino acid residues from free hAOX1 (in blue, PDB ID: 4UHW) superimposed. Right: close-up view of thioridazine binding pocket with relevant amino acid residues from free hAOX1 (in blue) superimposed. This figure was created using PyMOL v1.7.6.4.

The *N*-terminal domain I contains two distinct [2Fe-2S] clusters linking the Moco and FAD sites in a common arrangement that is also present in other AOX and XOR enzymes (Fig. 13.3). The cofactor disposition is related to the flow of electrons—generated at the Mo active site in the oxidation reactions—passed through the two [2Fe-2S] centers to the FAD, from where they are released to the terminal electron acceptor.

The central domain II of hAOX1 harbors the FAD-binding site that reacts with the terminal electron acceptor: molecular oxygen. The main product of O_2 reduction revealed to be H_2O_2, with a rate of superoxide radical formation of around 10% as compared to the overall reaction (Foti et al., 2017). This value, however, is lower than the reported rates of 16–20% for bovine XO and significantly lower than the rates of 32% for the rat AOXs (Kundu et al., 2012). Relevant structural differences between AOX and other members of the xanthine oxidase family, especially the well-characterized bovine XOR, are at the FAD site of these enzymes (Coelho et al., 2015; Coelho et al., 2012). hAOX1 in general is localized in the cytosol and uses molecular oxygen as sole electron acceptor. Surprisingly, the highly homologous enzyme xanthine dehydrogenase (XDH) uses NAD^+ as physiological oxidizing substrate, although the enzyme can be converted to the oxidase (XO) form that uses molecular oxygen as electron acceptor (D/O conversion) (Nishino et al., 2008). The reversible D/O conversion involves the formation of a new disulfide bond, between two cysteine residues (C535 and C992 in bovine XOR), which are not conserved in AOXs. Additionally, the inter-conversion of XDH into XO requires the structural rearrangement of an eleven-residue loop, so-called the FAD variable loop A (Q423-K433 in bovine XOR), which is in close proximity to the FAD cofactor (Enroth et al., 2000). The loop movement leads to a change in the electrostatic environment around the FAD cofactor (Nishino et al., 2005) and, in the oxidase form, the access of NAD^+ to its binding site is blocked (Enroth et al., 2000). All these structural elements involved in the D/O conversion are absent in the mouse and human AOXs, which might be related to the inability of this inter-conversion mechanism by AOX in general (Terao et al., 2016b). Consistent with this, the amino acid composition of the variable loop A ($Q_{423}ASRREDDIAK_{433}$, in bovine XOR) is very divergent in human AOX1, mouse AOX3 and XORs of different origins. Furthermore, although the loop A orientation of the human AOX1, mouse AOX3 and the XDH form is the same and the loops are superimposable, the structure of human AOX1 reveals a unique conformation of a second variable loop—loop B—close to the FAD site ($T_{1230}RGPDQ_{1235}$). Loop B is flipped by almost 180° relative to mouse AOX3 and bovine XOR. Although most of the loop B sequence is conserved among AOXs, in hAOX1, the poorly conserved Asp1234 establishes a hydrogen bond with the Nε of Trp1334, stabilizing loop B in this new orientation (Coelho et al., 2015). Interestingly, the conformation of this loop, so far unique to human AOX1, occupies nearly the same position as the nicotinamide ring of the electron acceptor NADH, in the structure of the bXDH-NADH complex (PDB ID: 3UNI) (Ishikita et al., 2012). Loop B might be blocking NAD^+ access to the isoaloxazine ring of the FAD, supporting the idea that oxygen is the unique electron acceptor of hAOX1 oxidation reactions. In addition to its role in the electron transfer chain, the FAD cofactor can also be involved in catalysis and there are several reports of reduction reactions proposed to be carried out at this site, instead of the Mo active site as detailed below (for thorough discussion, see review (Mota et al., 2018)).

The C-terminal domain III is binding the Moco in its sulfido-containing form. At the catalytic site, the Moco adopts an approximately square pyramidal geometry. The Mo ion is coordinated by an oxo group at the apical position and two sulfur ligands from the pterin moiety (MPT), one hydroxyl group and one sulfido ligand in the equatorial plan (Fig. 13.3). The sulfido-ligand is essential for the catalytic activity of the enzyme (Okamoto et al., 2004) and is a typical feature of enzymes like AOX and XOR (apart from a few exceptions as AOR from *Desulfovibrio gigas* (Santos-Silva et al., 2009). At the Moco center, the substrate is bound and converted to its product under the consumption of a water molecule. Structural features along the substrate access funnel were identified that might account for differences in substrate specificities between AOX and XOR enzymes. Moreover, differences found among AOX from different species and isoforms could be related with the different active site pockets (Cerqueira et al., 2015; Mota et al., 2018). In general, in AOX enzymes the funnel is wider and more anionic that might imply a better access of larger and bulkier substrates for AOXs enzymes in comparison to XORs. The crystal structure of hAOX1 indicates that several residues of domain III at the entrance of the funnel are flexible and define two putative gates controlling the access of substrates and product release to the Moco active site: Gate 1 comprises residues S_{652}FCFFTEAEK_{661}, several of which are disordered and could not be assigned in the electron density map of the hAOX1 crystal structure; Gate 2 comprises fewer residues (L_{880}DESL_{884}), and also no electron density could be assigned for the two acidic residues in the crystal structure. In the active site of AOXs, some residues (Glu1270, Phe923, Lys893 and Gln776 in hAOX1) are conserved, while many others are species specific and non-conserved among human and mouse isoforms. Glu1270 is the essential catalytic residue present in all enzymes of the xanthine-oxidase family, while Lys893 is specific to AOXs and may be involved in the catalytic cycle, as indicated by preliminary computational data (Coelho et al., 2012) which was confirmed by kinetic studies (Mahro et al., 2013).

13.4 Structural Basis of Substrate and Inhibitor Binding

It is well documented that AOXs can accept a wide range of substrates, catalyzing diverse reactions, from the hydroxylation of aldehydes and various aza-, oxo- and sulfo-heterocycles to the oxidation of iminium functions to cyclic lactams (Mota et al., 2018). In hypoxic conditions, AOXs can also reduce substrates, such as *N*-oxides, sufoxides, nitro-compounds and heterocycles (Dick et al., 2006; Kitamura and Tatsumi, 1984a, 1984b). More recently it was shown that, in addition to oxidation and reduction reactions, AOXs are also able to catalyze amide hydrolysis, a reaction first reported for the drug candidate GDC-0834 (Sodhi et al., 2015) and later demonstrated for a large set of amides (Lepri et al., 2017). Considering the reactions diversity, different catalytic mechanisms at different active sites can be anticipated, with oxidation/hydroxylation reactions occurring at the Moco site and reductions at the FAD site.

The crystal structure of the hAOX1-phthalazine-thioridazine ternary complex (Coelho et al., 2015) provided details on substrate binding at the Mo active site and comparisons with the hAOX free form showed clear changes in the orientations of Leu812 side chain and the phenyl rings of Phe885 and Phe923 (Fig. 13.3).

The structures also revealed that two putative gates condition the access of substrates and inhibitors to the active site cavity. Interestingly, Phe885 is located at one of these gates.

A novel and unexpected inhibitor-binding site was disclosed in the 3D structure of the ternary complex. Thioridazine is an antipsychotic drug belonging to the phenothiazine group which possesses a ring system similar to that of other antidepressants known to inhibit AOX1 activity (Obach and Walsky, 2005; Rani Basu et al., 2005). The thioridazine molecule is bound near the dimer interface in a groove at the enzyme surface and stacked between two loops (residues 568–582 and residues 1056–1070) of each AOX1 monomer (Fig. 13.3). Residues 570–571 are disordered in the substrate-free human AOX1. Upon thioridazine binding, these residues become better structured and define one helical turn close to the inhibitor-binding site. In addition, His575 and Glu577 side chains move, rendering the binding pocket more accessible to thioridazine.

Many studies have been reported on different types of AOX inhibitors that we recently reviewed (Mota et al., 2018). Considering the structural resemblances between substrates and inhibitors and taking into account the results from the crystal structure of the thioridazine-phthalazine-hAOX1 complex as well as the kinetic data, one can postulate that hAOX1 inhibitors can bind at the I) oxidation site (Moco); II) reduction site (FAD) and III) noncompetitive site (thioridazine site). According to this proposal, inhibitors, which structures are similar to aza-heterocycles, will bind at the oxidation site, interacting with the Moco and hindering substrate binding through competitive inhibition. Finally, we suggest that phenothiazine drugs and analogues will bind to the substrate-free site—uncompetitive inhibition—which prevents oxidation reactions probably due to an allosteric effect.

13.5 The Catalytic Mechanism of Human AOX1

Heterologous expression systems for mammalian AOXs, including the human enzyme, in *Escherichia coli* have been developed, allowing site directed mutagenesis approaches to study the role of distinct amino acid residues at the active site (Foti et al., 2016; Hartmann et al., 2012; Mahro et al., 2011; Schumann et al., 2009).

In the active site of AOXs, some residues (Glu1270, Phe923, Lys893 and Gln776, hAOX1 numbering) are highly conserved, while many others are species-specific and non-conserved among human/mouse isoforms (Cerqueira et al., 2015). Glu1270 is the essential catalytic residue present in all enzymes of the xanthine-oxidase family (Coelho et al., 2012), while Phe923 is one of the aromatic amino acids responsible for orienting the substrate in the hAOX1 complex structure.

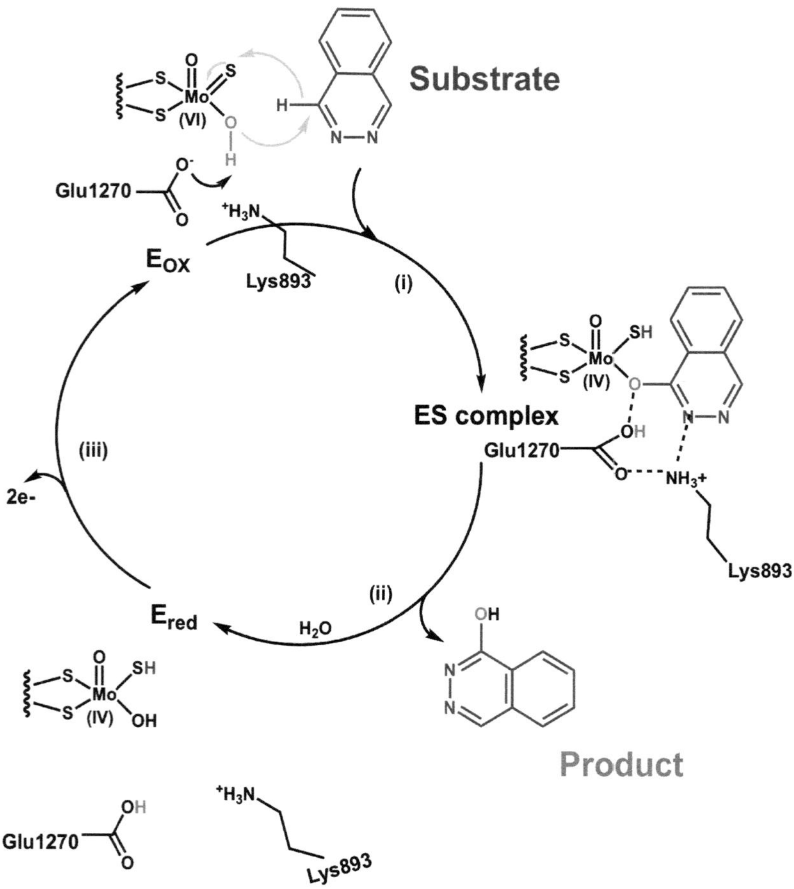

Figure 13.4 Proposed oxidation mechanism for AOXs. Representation of the currently proposed oxidation mechanism of aza-aromatic compounds by AOX, exemplified for the conversion of the substrate phthalazine to hydroxyphthalazine.

The reaction mechanism that has been proposed for AOXs is a base-catalyzed mechanism starting with the initial attack of the deprotonated Glu1270 to the –OH ligand of the Mo-atom, thereby generating a nucleophilic $-O^-$ group (Fig. 13.4). The activated $Mo–O^-$ ligand then attacks the carbon atom of the substrate, adjacent to an aromatic nitrogen atom in the case of *N*-heterocycles oxidation. Concomitantly, hydride transfer occurs to the sulfido ligand and an intermediate species is formed (Fig. 13.4). This intermediate is stabilized by hydrogen bonding interactions with residues of the active site, namely Val811, Met889, and Lys885. This concerted mechanism is favored by experimental evidence of a variety of substituted N-heterocycles (Lepri et al., 2017) as well as by DFT calculations (Alfaro and Jones, 2008). In the following step, the product is released from the reduced Mo site and a water molecule replenishes the vacant coordination position (Fig. 13.4). The reaction cycle is completed once Mo is re-oxidized and the two reducing equivalents are transferred to molecular oxygen via the two [2Fe-2S] and the FAD cofactor.

13.6 Genetic Diversity Provided by Single-Nucleotide Polymorphisms (SNPs)

The identification and characterization of genetic variants of enzymes involved in drug metabolism is highly relevant for the definition of the effectiveness and toxicity of pharmaceutical treatments. In the human population, a major source of inter-individual variability in AOX1 enzymatic activity is represented by the existence of allelic variants characterized by single-nucleotide polymorphisms (SNPs). Numerous SNPs of the human *AOX1* gene are available in the NCBI dbSNP database (http://www.ncbi.nlm.nih.gov/snp), with most of them being uncharacterized to date. Missense SNPs in the coding region of the human AOX1 gene may affect the catalytic function of the enzyme in either a positive or negative fashion (Hartmann et al., 2012). So far, some of the SNP-based base-pair exchanges present in the genome of human individuals were introduced into the hAOX1 gene for heterologous expression and purification in *E. coli* (Foti et al., 2017; Foti et al.,

2016; Hartmann et al., 2012). Some of these SNPs are translated into AOX1 protein variants with increased catalytic activity. Thus, individuals bearing these alleles can be classified as fast AOX1 substrate metabolizers. Other SNPs result in an inactive protein or AOX1 enzymatic variants with reduced catalytic activity(Hartmann et al., 2012). In both cases, individuals characterized by the presence of such alleles are likely to be compromised in their ability to metabolize and potentially inactivate drugs which are known to be AOX1 substrates. Also, SNP-based AOX variants were identified which completely abolished hAOX1 activity (Foti et al., 2016). However, detailed information on the allelic frequency of the SNPs is still missing in the NCBI dbSNP database in addition to information whether the SNPs are homozygous or heterozygous. Only the homozygous SNPs would consequently have an effect on the metabolizing effect of AOX1 on phase I drug metabolism. Therefore, detailed information of AOX1 function is of great clinical relevance in terms of predicting the efficacy and toxicity of new drugs, and further studies are necessary to evaluate the influence on enzyme activity exerted by these and other non-synonymous SNPs in the human population worldwide.

13.7 Relevance of AOX in Drug Metabolism and Cancer

The role of human AOX1 in phase I metabolism of xenobiotics is established and the interest in AOX enzymes is increasing given their emerging significance in drug-development programs (Garattini and Terao, 2011, 2012, 2013). In this last context, a major problem is represented by the animal models used during the preclinical phases of drug development aimed at evaluating the metabolism, pharmacokinetics and pharmacodynamics of new therapeutic agents. As already mentioned, unlike humans, some of the most popular animal models are indeed characterized by expression of other AOXs besides AOX1. As described above, AOXs, including human AOX1, recognize a variety of organic molecules regardless of the presence of an aldehyde function.

All this, along with the high levels of human AOX1 and rodent AOX1/AOX3 in the liver is the basis for the role played by AOXs in xenobiotic and drug metabolism (Coelho et al., 2012; Sanoh et al., 2015; Zientek and Youdim, 2015). In hepatic cells, the main function of AOXs and cytochrome P-450 mono-oxygenases (CYP450s) seems to exert overlapping activities in the biotransformation/inactivation of therapeutic agents and toxicants. While human AOX1 and other AOXs are located in the cytosol, the sub-cellular localization of CYP450s is in the endoplasmic reticulum. Occasionally, AOXs and CYP450s act in concert, as exemplified by the selective serotonin reuptake inhibitors (SSRI), citalopram, whose tertiary amino group is oxidized to an aldehyde by CYP450 (Fig. 13.5) and subsequently bio-transformed into a carboxylic acid by AOXs. AOXs metabolize a variety of anti-tumor, immunosuppressive, anti-malarial and anti-viral drugs as well as molecules acting in the central nervous system. As for the anti-tumor and immunosuppressive agents metabolized by AOXs, methotrexate and 6-mercaptopurine are two prominent examples. Human AOX1, rabbit, rat, hamster and mouse AOXs hydroxylate methotrexate into 7-hydroxy-methotrexate (Fig. 13.5) (Chladek et al., 1997; Kitamura et al., 1999). AOX-dependent hydroxylation affects the pharmacokinetics and it underlays some of methotrexate side effects, as 7-hydroxy-metotrexate is endowed with cytotoxicity. In humans, generation of 7-hydroxy-methotrexate shows inter-individual variability, which correlates with the levels of AOX1 enzymatic activity, namely derived from single-nucleotide polymorphisms (Kitamura et al., 1999). The 6-mercaptopurine metabolite of the azathioprine immunosuppressant is also a long-known AOX substrate (Beedham et al., 1987; Ding and Benet, 1979; Rooseboom et al., 2004). The imidazole ring of 6-mercaptopurine undergoes two successive hydroxylation steps which are catalyzed by XOR and AOXs, respectively (Fig. 13.5). AOX-dependent hydroxylation leads to metabolic inactivation and contributes to the pharmacokinetic profile of 6-mercaptopurine in humans and experimental animals. A last example of immunosuppressant for which some of the toxic metabolites are generated by AOX1 is thiopurine (Chouchana et al., 2012).

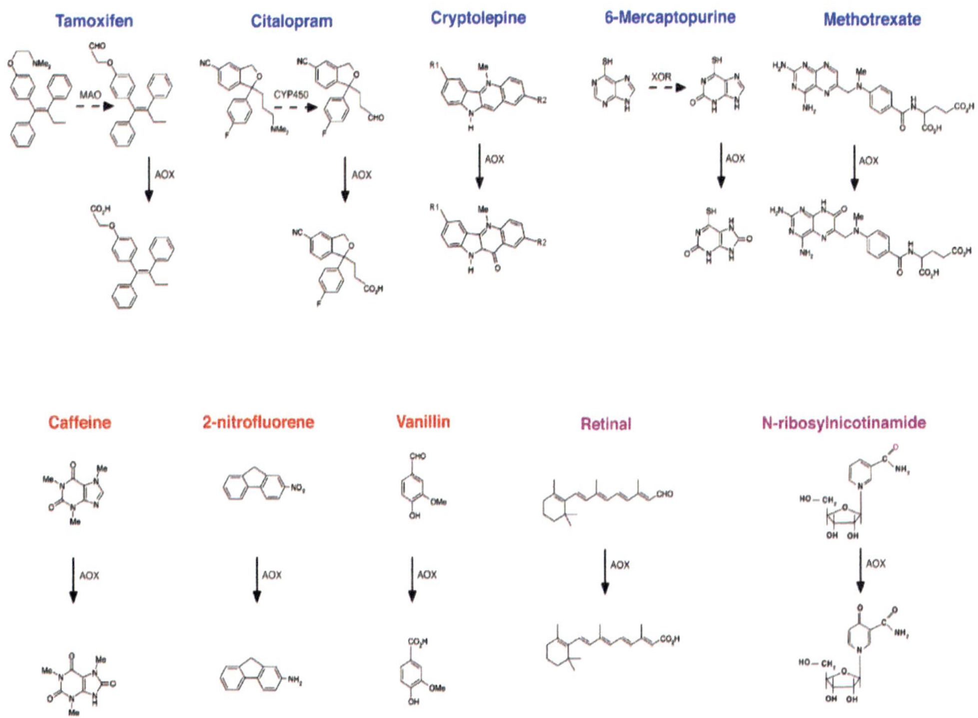

Figure 13.5 Endogenous molecules, drugs and xenobiotics metabolized by AOXs. The panel illustrates the activity of AOXs upon endogenous compounds (purple), drugs (blue), xenobiotics of toxicological interest (red) and AOX-dependent metabolites. When AOXs act on metabolic intermediates of the parental drug, the enzymes involved in the primary metabolic step are also indicated: MAO, monoamine oxidase; CYP450, cytochrome P-450–dependent monooxygenase; XOR, xanthine oxidoreductase.

Other well-known drugs for which there is evidence of AOX-dependent metabolism are some antimalarial and anti-viral agents. The anti-malarial cryptolepine (Fig. 13.5) is oxidized to cryptolepine-11-one (Stell et al., 2012), while quinine is bio-transformed into the 2-quinone derivative by liver AOXs (Beedham et al., 1992). Interestingly, dog and rat livers contain undetectable and low levels of cytosolic quinine oxidase activity, respectively. The first observation is in line with the absence of AOX1 and AOX3 in dog liver (Terao et al., 2006) because of pseudogenization/inactivation of the corresponding genes (Kurosaki et al., 2013). Conversely, the observation in rat liver suggests that quinine is a better substrate of AOX1 than AOX3. This last hypothesis is supported by the results obtained in baboon guinea pig and marmoset livers, which contain quinine oxidizing activity despite pseudogenization/inactivation of the *AOX3* gene (Kurosaki et al., 2013).

There are at least three issues that should be considered in discussing the role played by human AOX1 and other AOXs in drug metabolism. First, AOX-dependent metabolic inactivation of drugs does not simply involve oxidation, but may also involve reduction of appropriate substrates (see above). A relevant example of the point raised is ziprasidone, a drug used for the treatment of different psychiatric disorders. In humans, AOX1 reduces ziprasidone and the resulting metabolite is subsequently methylated by thiol-methyl-transferase (Obach et al., 2012). Second, AOX-dependent drug metabolism does not necessarily lead to metabolic inactivation. Indeed, there are examples of pro-drugs which can be metabolically activated by human AOX1 to avoid pharmacokinetic problems associated with the parental drug. A relevant example is represented by 5-iodo-2-deoxyuridine precursor, 5-iodo-2-pyrimidinone-2-deoxyribose (Kinsella et al., 1994; Kinsella et al., 1998; Kinsella et al., 2000). In the future, human AOX1-dependent activation of pro-drugs may represent a viable approach to augment the therapeutic index of anti-tumor agents via an increase in tumor selectivity. Indeed, in situ activation of pro-drugs by the tumor cell is a possibility, as different cancer types, such as glioblastoma, esophageal and gastric carcinomas (Garattini and Terao, 2011) are deemed to express detectable amounts of human AOX1. A last point to be mentioned relates to the fact that AOX-dependent drug metabolism may be relevant in organs other than liver. For instance, human skin explants metabolize two AOX substrates, like carbarazepan and zoniporide (Manevski et al., 2014). In addition, the presence of AOXs (Terao et al., 2016a) in the skin may represent a problem for the development and the use of drugs or cosmetics to be administered topically.

The relevance of AOXs in drug development is expanding due to the synthesis of an ever-increasing number of organic molecules designed to avoid CYP450-dependent metabolism. In fact, this has resulted in a significant enrichment of chemical structures which are human AOX1 substrates (Barr et al., 2014; Fratelli et al., 2013; Hutzler et al., 2013), paving the way to new medicinal chemistry approaches aimed at avoiding AOX-dependent metabolism (Pryde et al., 2012). The enrichment in potential AOX substrates underscores the necessity to develop in vitro screening assays to predict and validate AOX-dependent metabolism (Garattini and Terao, 2011, 2013). A further issue in drug development is represented by the definition of the most appropriate in vivo model for pharmacokinetics and pharmacodynamics studies of drug candidates metabolized by human AOX1, given the differences in the complement of active *AOX* genes in humans and other mammals (Fig. 13.2). For instance, dogs and cats are not good experimental tools for metabolic studies involving human AOX1 substrates (Garattini and Terao, 2013), because of the already mentioned inactivation of the *AOX1* and *AOX3* genes by pseudogenization (Kurosaki et al., 2013; Terao et al., 2006). Thus, it is not surprising that canine studies have failed to detect metabolites of AOX substrates either in vitro or in vivo (Kaye et al., 1984; Klecker et al., 2006).

AOXs metabolize not only drugs, but also various molecules of toxicological interest, such as the environmental pollutants, phthalazines, which are classical AOX substrates. Different AOXs oxidize phthalazine into 1-hydroxy-phthalazine, which undergoes irreversible isomerization into 1-phthalazinone (Beedham et al., 1990; Coelho et al., 2015; Stubley et al., 1979). Phthalazine administration increases rabbit liver AOX activity, demonstrating that the compound is not only a substrate but also an AOX inducer (Johnson et al., 1984). Another example of toxicologically significant AOX substrate is caffeine (Castro et al., 2001) (Fig. 13.5). Vanillin, a sweetener, is an aromatic aldehyde which is oxidized into the corresponding carboxylic acid by AOXs (Fig. 13.5).

A final AOX substrate is acetaldehyde, the primary ethanol metabolite, which is the reason as to why this class of enzymes has been implicated in the liver toxicity associated with alcohol consumption. The major mechanism deemed to be at the basis of AOX-dependent liver toxicity is the production of reactive oxygen species during acetaldehyde oxidation (Shaw and Jayatilleke, 1990). However, the results obtained in AOX deficient *DBA/2* and AOX proficient *CD1* mice are not consistent with this hypothesis. In fact, *CD1* and *DBA/2* livers do not contain detectable levels of AOX3 and show markedly lower amounts of AOX1 relative to *CD1* animals. Yet *CD1* and *DBA/2* mice were endowed with the same amounts of liver acetaldehyde after ethanol administration (Vila et al., 2004). This supports the concept that neither liver AOX1 nor AOX3 contributes to acetaldehyde metabolism in mice. The finding is also in line with the fact that acetaldehyde is a poor AOX1/AOX3 substrate.

Another class of toxic agents with the potential to be metabolized by mammalian AOXs is represented by nitro compounds. Molecules containing a nitro functionality exemplify the ability of AOXs to act not only as oxidases, but also as reductases, in the presence of electron donors, such as 2-hydroxypyrimidine (Tatsumi et al., 1986; Ueda et al., 2003; Ueda et al., 2005). AOXs have the potential to complete the nitro-reduction of 2-nitrofluorene, 1-nitro-pyrene and 4-nitrobiphenyl, three widespread pollutants, into the corresponding amines. Interestingly, AOX-dependent nitro-reduction is observed not only in the liver (Tatsumi et al., 1986) but also in the skin of experimental animals, such as mice (Ueda et al., 2003; Ueda et al., 2005). As mouse liver synthesizes AOX1 and AOX3, while mouse skin expresses predominantly AOX4, nitro-reduction may not be a characteristic of a specific AOX isoenzyme. The ability of AOX1 to carry out nitro-reduction and the presence of AOXs in the skin has implications for the human situation. In fact, human skin contains AOX1 enzymatic activity (Manevski et al., 2014) and the reduction of environmental nitro-compounds may represent an important defense reaction from pollutants. Another example of AOX-dependent nitro-reductive metabolism is represented by the neonicotinoid insecticide, imidacloprid. The relative contribution of liver AOXs and CYP450s in the metabolism of imidacloprid has been recently assessed (Swenson and Casida, 2013). Administration of the AOX inhibitors, tungsten or hydralazine, to C57BL/6J mice causes a selective reduction in liver AOX activity and no change in CYP450 activity. In tungsten and hydralazine treated mice, CYP450-dependent imidacloprid metabolism is unaffected, whereas the AOX-generated nitrosoguanidine metabolite is significantly decreased. The study provides direct in vivo evidence that liver AOX1 and/or AOX3 play a substantial role in imidacloprid metabolism.

13.8 Conclusions

AOX is a molybdoflavoenzyme, catalyzing oxidation, reduction, and hydrolysis reactions of very different types of substrates. The versatility of this enzyme is quite surprising, especially considering that most of the proteins belonging to the xanthine oxidase family are highly substrate specific. AOX substrates and inhibitors are extremely diverse and, in many cases, discovered by serendipity/misfortune during the last stages of drug development. There are several examples of failures in clinical trials due to the unexpected activity of AOX towards the leading drug or one of its metabolites, giving rise to drug inactivation, toxicity and/or side effects. So far, AOX has not earned the status of CYP450 family of proteins, and, until then, many more drug candidates may lead to a dead-end with increasing costs for the pharma industry.

Over the last years, structural, (bio)chemical, and genetic studies contributed to bring AOX to the drug development agenda, allowing to better understand the mode of action of the enzyme. Efficient in silico, in vitro, and in vivo models need to be further developed, for predicting the putative drug metabolism by AOX, or its inhibition—either by substrate competition (at the Moco site or at the FAD site) or via non-competitive inhibition (at the surface of the protein, where thioridazine was found). Furthermore, when the true function of AOX is revealed, scientists will have better reasoning for activating/inactivating the enzyme, while conceiving new drugs.

It is well known that AOX derives from a gene duplication event and that, in humans, suffered a gene deletion or pseudogenization event. This brings us to a second hurdle regarding AOX metabolism—it is interspecies and inter-individual dependent, making it hard to use animal models to predict activity. The presence of several tissue-dependent isoforms of AOX in mammals usually used in phase I of clinical trials also may misguide the subsequent steps of the drug development cycle. Single-nucleotide polymorphism also affects AOX activity, making individuals fast or slow metabolizers, impacting ongoing therapeutics.

Altogether, it is clear that in spite of our current knowledge, further studies on AOX are required so that new safe and efficient drugs can emerge.

References

Alfaro J. F., Jones J. P., *J. Org. Chem.*, **73** (2008), 9469–9472.

Baral P. K., Swayampakula M., Rout M. K., Kav N. N., Spyracopoulos L., Aguzzi A., James M. N., *Structure*, **22** (2014), 291–303.

Barr J. T., Choughule K., Jones J. P., *Methods Mol. Biol.*, **1113** (2014), 167–186.

Barr J. T., Jones J. P., *Drug Metab. Dispos.*, **39** (2011), 2381–2386.

Beedham C., *Drug Metab. Rev.*, **16** (1985), 119–156.

Beedham C., *Pharm. World Sci.*, **19** (1997), 255–263.

Beedham C., *Biom. Hlth. R.*, **25** (1998), 39–52.

Beedham C., al-Tayib Y., Smith J. A., *Drug Metab. Dispos.*, **20** (1992), 889–895.

Beedham C., Bruce S. E., Critchley D. J., al-Tayib Y., Rance D. J., *Eur. J. Drug. Metab. Pharmacokinet.*, **12** (1987), 307–310.

Beedham C., Bruce S. E., Critchley D. J., Rance D. J., *Biochem. Pharmacol.*, **39** (1990), 1213–1221.

Berger R., Mezey E., Clancy K. P., Harta G., Wright R. M., Repine J. E., Brown R. H., Brownstein M., Patterson D., *Somat. Cell Mol. Gen.*, **21** (1995), 121–131.

Bolis M., Garattini E., Paroni G., Zanetti A., Kurosaki M., Castrignano T., Garattini S. K., Biancardi F., Barzago M. M., Gianni M., et al., *Ann. Oncol.*, **28** (2017) 611–621.

Castro G. D., Delgado de Layno A. M., Costantini M. H., Castro J. A., *Toxicology*, **160** (2001), 11–18.

Cerqueira N. M., Coelho C., Bras N. F., Fernandes P. A., Garattini E., Terao M., Romao M. J., Ramos M. J., *J. Biol. Inorg. Chem.*, **20** (2015), 209–217.

Chladek J., Martinkova J., Sispera L., *Physiol. Res.*, **46** (1997), 371–379.

Chouchana L., Narjoz C., Beaune P., Loriot M. A., Roblin X., *Aliment. Pharmacol. Ther.*, **35** (2012), 15–36.

Coelho C., Foti A., Hartmann T., Santos-Silva T., Leimkühler S., Romao M. J., *Nat. Chem. Biol.*, **11** (2015), 779–783.

Coelho C., Mahro M., Trincao J., Carvalho A. T., Ramos M. J., Terao M., Garattini E., Leimkühler S., Romao M. J., *J. Biol. Chem.*, **287** (2012), 40690–40702.

Dick R. A., Kanne D. B., Casida J. E., *Chem. Res. Toxicol.*, **19** (2006), 38–43. 15.

Ding T. L., Benet L. Z., *Drug Metab. Dispos.*, **7** (1979), 373–377.

Enroth C., Eger B. T., Okamoto K., Nishino T., Nishino T., Pai E. F., *Proc. Natl. Acad. Sci. U. S. A.*, **97** (2000), 10723–10728.

Foti A., Dorendorf F., Leimkühler S., *PLoS One*, **12** (2017), e0182061.

Foti A., Hartmann T., Coelho C., Santos-Silva T., Romao M. J., Leimkühler S., *Drug Metab. Dispos.*, **44** (2016), 1277–1285.

Fratelli M., Fisher J. N., Paroni G., Di Francesco A. M., Pierri F., Pisano C., Godl K., Marx S., Tebbe A., Valli C., et al., *Eur. J. Cancer*, **49** (2013), 1491–1500.

Garattini E., Fratelli M., Terao M., *Cell Mol. Life Sci.*, **65** (2008), 1019–1048.

Garattini E., Fratelli M., Terao M., *Hum. Genomics*, **4** (2009), 119–130.

Garattini E., Mendel R., Romao M. J., Wright R., Terao M., *Biochem. J.*, **372** (2003), 15–32.

Garattini E., Terao M., *Drug Metab. Rev.*, **43** (2011), 374–386.

Garattini E., Terao M., *Expert Opin. Drug Metab. Toxicol.*, **8** (2012), 487–503.

Garattini E., Terao M., Aldehyde oxidase and its importance in novel drug discovery: Present and future challenges. *Expert Opin Drug Discov.*, **8**(6) (2013), 641–654.

Hartmann T., Terao M., Garattini E., Teutloff C., Alfaro J. F., Jones J. P., Leimkühler S., *Drug Metab. Dispos.*, **40** (2012), 856–864.

Hesberg C., Hansch R., Mendel R. R., Bittner F., *J. Biol. Chem.*, **279** (2004), 13547–13554.

Hutzler J. M., Obach R. S., Dalvie D., Zientek M. A., *Expert Opin. Drug Metab. Toxicol.*, **9** (2013), 153–168.

Ichida K., Amaya Y., Noda K., Minoshima S., Hosoya T., Sakai O., Shimizu N., Nishino T., *Gene*, **133** (1993), 279–284.

Ishikita H., Eger B. T., Okamoto K., Nishino T., Pai E. F., *J. Am. Chem. Soc.*, **134** (2012), 999–1009.

Johnson C., Stubley-Beedham C., Stell J. G., *Biochem. Pharmacol.*, **33** (1984), 3699–3705.

Kaye B., Offerman J. L., Reid J. L., Elliott H. L., Hillis W. S., *Xenobiotica*, **14** (1984), 935–945

Kinsella T. J., Kunugi K. A., Vielhuber K. A., McCulloch W., Liu S. H., Cheng Y. C., *Cancer Res.*, **54** (1994), 2695–2700.

Kinsella T. J., Kunugi K. A., Vielhuber K. A., Potter D. M., Fitzsimmons M. E., Collins J. M., *Clin. Cancer Res.*, **4** (1998), 99–109.

Kinsella T. J., Schupp J. E., Davis T. W., Berry S. E., Hwang H. S., Warren K., Balis F., Barnett J., Sands H., *Clin. Cancer Res.*, **6** (2000), 3670–3679.

Kitamura S., Sugihara K., *Xenobiotica*, **44** (2014), 123–134.

Kitamura S., Sugihara K., Nakatani K., Ohta S., Ohhara T., Ninomiya S., Green C. E., Tyson C. A., *IUBMB Life*, **48** (1999), 607–611.

Kitamura S., Tatsumi K., *Biochem. Biophys. Res. Commun.*, **120** (1984a), 602–606.

Kitamura S., Tatsumi K., *Biochem. Biophys. Res. Commun.*, **121** (1984b), 749–754.

Klecker R. W., Cysyk R. L., Collins J. M., *Bioorg. Med. Chem.*, **14** (2006), 62–66.

Krenitsky T. A., Neil S. M., Elion G. B., Hitchings G. H., *Arch. Biochem. Biophys.*, **150** (1972), 585–599.

Kundu T. K., Velayutham M., Zweier J. L., *Biochemistry*, **51** (2012), 2930–2939.

Kurosaki M., Bolis M., Fratelli M., Barzago M. M., Pattini L., Perretta G., Terao M., Garattini E., *Cell Mol. Life Sci.*, **70** (2013), 1807–1830.

Kurosaki M., Terao M., Barzago M. M., Bastone A., Bernardinello D., Salmona M., Garattini E., *J. Biol. Chem.*, **279** (2004), 50482–50498.

Lepri S., Ceccarelli M., Milani N., Tortorella S., Cucco A., Valeri A., Goracci L., Brink A., Cruciani G., *Proc. Natl. Acad. Sci. U. S. A.*, **114** (2017), E3178–E3187.

Mahro M., Bras N. F., Cerqueira N. M., Teutloff C., Coelho C., Romao M. J., Leimkühler S., *PloS One*, **8** (2013), e82285.

Mahro M., Coelho C., Trincao J., Rodrigues D., Terao M., Garattini E., Saggu M., Lendzian F., Hildebrandt P., Romao M. J., et al., *Drug. Metab. Dispos.*, **39** (2011), 1939–1945.

Manevski N., Balavenkatraman K. K., Bertschi B., Swart P., Walles M., Camenisch G., Schiller H., Kretz O., Ling B., Wettstein R., et al., *Drug Metab. Dispos.*, **42** (2014), 2049–2057.

Marelja Z., Dambowsky M., Bolis M., Georgiou M. L., Garattini E., Missirlis F., Leimkühler S., *J. Exp. Biol.*, **217** (2014), 2201–2211.

Mendel R. R., Smith A. G., Marquet A., Warren M. J., *Nat. Prod. Rep.*, **24** (2007), 963–971.

Mota C., Coelho C., Leimkühler S., Garattini E., Terao M., Santos-Silva T., Romao M. J., *Coordin. Chem. Rev.*, **368** (2018), 35–59.

Nishino T., Okamoto K., Eger B. T., Pai E. F., *FEBS J.*, **275** (2008), 3278–3289.

Nishino T., Okamoto K., Kawaguchi Y., Hori H., Matsumura T., Eger B. T., Pai E. F., Nishino T., *J. Biol. Chem.*, **280** (2005), 24888–24894.

Obach R. S., Huynh P., Allen M. C., Beedham C., *J. Clin. Pharmacol.*, **44** (2004), 7–19.

Obach R. S., Prakash C., Kamel A. M., *Xenobiotica*, **42** (2012), 1049–1057.

Obach R. S., Walsky R. L., *J. Clin. Psychopharmacol.*, **25** (2005), 605–608.

Okamoto K., Matsumoto K., Hille R., Eger B. T., Pai E. F., Nishino T., *Proc. Natl. Acad. Sci. U. S. A.*, **101** (2004), 7931–7936.

Pritsos C. A., *Chem. Biol. Interact.*, **129** (2000), 195–208.

Pryde D. C., Dalvie D., Hu Q., Jones P., Obach R. S., Tran T. D., *J. Med. Chem.*, **53** (2010), 8441–8460.

Pryde D. C., Tran T. D., Jones P., Duckworth J., Howard M., Gardner I., Hyland R., Webster R., Wenham T., Bagal S., et al., *Bioorg. Med. Chem. Lett.*, **22** (2012), 2856–2860.

Rani Basu L., Mazumdar K., Dutta N. K., Karak P., Dastidar S. G., *Microbiol. Res.*, **160** (2005), 95–100.

Rodriguez-Trelles F., Tarrio R., Ayala F. J., *Proc. Natl. Acad. Sci. U. S. A.*, **100** (2003), 13413–13417.

Romao M. J., *Dalton Trans.* (2009), 4053–4068.

Rooseboom M., Commandeur J. N., Vermeulen N. P., *Pharmacol. Rev.*, **56** (2004), 53–102.

Sanoh S., Tayama Y., Sugihara K., Kitamura S., Ohta S., *Drug Metab. Pharmacokinet.*, **30** (2015), 52–63.

Santos-Silva T., Ferroni F., Thapper A., Marangon J., Gonzalez P. J., Rizzi A. C., Moura I., Moura J. J., Romao M. J., Brondino C. D., *J. Am. Chem. Soc.*, **131** (2009), 7990–7998.

Schumann S., Terao M., Garattini E., Saggu M., Lendzian F., Hildebrandt P., Leimkühler S., *PloS One*, **4** (2009), e5348.

Schwarz G., Mendel R. R., Ribbe M. W., *Nature*, **460** (2009), 839–847.

Shaw S., Jayatilleke E., *Biochem. J.*, **268** (1990), 579–583.

Sodhi J. K., Wong S., Kirkpatrick D. S., Liu L., Khojasteh S. C., Hop C. E., Barr J. T., Jones J. P., Halladay J. S., *Drug Metab. Dispos.*, **43** (2015), 908–915.

Stell J. G., Wheelhouse R. T., Wright C. W., *J. Pharm. Pharmacol.*, **64** (2012), 237–243.

Stubley C., Stell J. G., Mathieson D. W., *Xenobiotica*, **9** (1979), 475–484.

Swenson T. L., Casida J. E., *Toxicol. Sci.*, **133** (2013), 22–28.

Tatsumi K., Kitamura S., Narai N., *Cancer Res.*, **46** (1986), 1089–1093.

Terao M., Barzago M. M., Kurosaki M., Fratelli M., Bolis M., Borsotti A., Bigini P., Micotti E., Carli M., Invernizzi R. W., et al., *Sci. Rep.*, **6** (2016a), 30343.

Terao M., Kurosaki M., Barzago M. M., Fratelli M., Bagnati R., Bastone A., Giudice C., Scanziani E., Mancuso A., Tiveron C., et al., *Mol. Cell Biol.*, **29** (2009), 357–377.

Terao M., Kurosaki M., Barzago M. M., Varasano E., Boldetti A., Bastone A., Fratelli M., Garattini E., *J. Biol. Chem.*, **281** (2006), 19748–19761.

Terao M., Kurosaki M., Demontis S., Zanotta S., Garattini E., *Biochem. J.*, **332** (Pt 2) (1998), 383–393.

Terao M., Kurosaki M., Marini M., Vanoni M. A., Saltini G., Bonetto V., Bastone A., Federico C., Saccone S., Fanelli R., et al., *J. Biol. Chem.*, **276** (2001), 46347–46363.

Terao M., Kurosaki M., Saltini G., Demontis S., Marini M., Salmona M., Garattini E., *J. Biol. Chem.*, **275** (2000), 30690–30700.

Terao M., Romao M. J., Leimkuhler S., Bolis M., Fratelli M., Coelho C., Santos-Silva T., Garattini E., *Arch. Toxicol.*, **90** (2016b), 753–780.

Ueda O., Kitamura S., Ohashi K., Sugihara K., Ohta S., *Drug Metab. Dispos.*, **31** (2003), 367–372.

Ueda O., Sugihara K., Ohta S., Kitamura S., *Drug Metab. Dispos.*, **33** (2005), 1312–1318.

Vila R., Kurosaki M., Barzago M. M., Kolek M., Bastone A., Colombo L., Salmona M., Terao M., Garattini E., *J. Biol. Chem.*, **279** (2004), 8668–8683.

Wright R. M., Weigel L. K., Varella-Garcia M., Vaitaitis G., Repine J. E., *Redox. Rep.*, **3** (1997), 135–144.

Xu P., Zhu X. L., Huecksteadt T. P., Brothman A. R., Hoidal J. R., *Genomics*, **23** (1994), 289–291.

Yasuhara A., Akiba-Goto M., Aisaka K., *Biosci. Biotechnol. Biochem.*, **69** (2005), 2435–2438.

Zhang Y., Gladyshev V. N., *J. Mol. Biol.*, **379** (2008), 881–899.

Zientek M. A., Youdim K., *Drug Metab. Dispos.*, **43** (2015), 163–181.

Chapter 14

Cytochrome P450 Enzymes for the Synthesis of Novel and Known Drugs and Drug Metabolites

Sanjana Haque,* Yuqing Gong,* Sunitha Kodidela, Mohammad A. Rahman, Sabina Ranjit, and Santosh Kumar

Department of Pharmaceutical Sciences, College of Pharmacy, University of Tennessee Health Science Center, 881 Madison Ave, Rm 456, Memphis, Tennessee 38163, USA

ksantosh@uthsc.edu

14.1 Introduction

As a large group of biocatalysts, cytochrome P450 (CYP) family of enzymes play an indispensable role in the metabolism of both endogenous and exogenous compounds. The ubiquity of CYPs all through the nature with characteristic molecular structure, has allowed these enzymes to become molecules of interest in the development of drugs and drug metabolites. CYPs in their natural form can catalyze a variety of oxidative reactions much

*Both SH and YG are the primary authors.

Pharmaceutical Biocatalysis: Fundamentals, Enzyme Inhibitors, and Enzymes in Health and Diseases
Edited by Peter Grunwald

ISBN 978-981-4800-61-7 (Hardcover), 978-0-429-29503-4 (eBook)
www.jennystanford.com

more efficiently than a chemical reaction. Modification of CYP enzymes by various engineering methods has been resolving their inherent limitations and potentially improving their industrial application for over a decade. In this chapter, we first provide a general background on CYP enzymes followed by (1) wild-type CYP enzymes in the synthesis of novel drugs and drug metabolites, (2) limitations, (3) CYP engineering for enhanced activity, stability, and specificity to overcome the limitations, (4) advantages of using CYP enzymes for drugs and drug metabolites synthesis, and (5) conclusion and future prospects.

14.2 General Background on Cytochrome P450 Enzymes

Cytochrome P450 (CYP) enzymes are a membrane-bound, superfamily of heme containing enzymes with more than 21,000 members throughout nature (Nelson, 2009; Burkina et al., 2017). They are present in abundance in almost all living organisms. Since the discovery of CYPs, they have been drawing attention of researchers, including biochemists, micro- and plant-biologists, pharmacologists, and toxicologists (Urlacher and Girhard, 2012). This heightened interest in CYP is due to their remarkable capacity to catalyze thousands of biological reactions.

Apart from few exceptions, CYP enzymes are mostly monooxygenase proteins with a heme in the center (Correia et al., 2011). In general, CYPs are mostly responsible for metabolizing endogenous or exogenous substances, detoxifying the organism from the harmful substances. These enzymes catalyze a multitude of reactions. They participate in oxidation reactions like activation of C atoms, epoxidation, aromatic hydroxylation, *N*-oxidation, deamination, dehalogenation, and dealkylation (N, O and S) (Vail et al., 2005; Urlacher and Girhard, 2012). In addition, under appropriate reaction conditions, CYP enzymes can catalyze reactions like desaturation, oxidative cleavage of ester and ether, dehydration, and various couplings (C–C and C–O phenol) and rearrangements (e.g., ring formation and oxidative aryl migration) (Vail et al., 2005; Urlacher and Girhard, 2012). As such, CYP enzymes have a diverse substrate selection. Among endogenous compounds, CYP enzymes are responsible for the metabolism of fatty acids, steroids, prostaglandins, terpenes, and poly and heteroatomic compounds. CYP enzymes catalyze reactions for exogenous substrates like drugs, carcinogens, toxins, organic solvents, etc. (Urlacher and Girhard, 2012).

CYP enzymes have been targeted for many years for drugs and drug metabolites development process. Naturally, some of the CYPs also participate in synthesis of medicinally useful products. For example, in bacteria, CYPs are essential for metabolic processes as well as synthesis of antibiotics. In plants, CYPs catalyze biosynthesis of compounds, which can be modified into drug-like compounds that can be used in humans. Mammalian CYPs mainly oxidize drugs and foreign intoxicants. The major drug metabolizing human CYP enzymes are 3A4, 1A2, 2D6, 2C9, and 2C19. Among these enzymes, CYP3A4 is responsible for the metabolism of approximately 60% of marketed drugs (Preissner et al., 2013).

In addition to having the ability to influence a wide variety of reaction and substrate selection, CYP enzymes are highly chemo-, regio- and stereo-selective. These unique characteristics make CYP enzymes a well favorable biocatalyst in different stages of drug discovery and development. For example, CYPs can be used to make minute structural changes like adding functional groups in potential drug candidates (Guengerich, 2002), produce drug metabolites for studying molecular structure and toxicology of the candidates, drug–drug interactions, etc. (Mackenzie et al., 2017). Attaining these goals by only chemical reactions is difficult in terms of time, cost and technicality (Vail et al., 2005).

In the past, scientists used wild-type CYPs as biocatalyst. However, due to their low stability and high cost, they are now engineering the structures of CYP enzymes to overcome their inherent limitations. Attempts have also been made to modify the bacterial/plant host system and integrate CYP enzymes, from other microbial, plant, or human source to increase the yield. In this chapter, we will mostly provide current advances in the industrial drug synthesis using wild-type and engineered CYPs along with the limitations, advantages, and future directions in this field.

14.3 Wild-Type CYP Enzymes in the Synthesis of Novel Drugs and Drug Metabolites

Despite being a challenge to synthesize large quantities of products, use of CYP enzymes in drug development, especially at the initial phase, remains highly useful. Wild-type CYPs, particularly, bacterial CYP enzymes have shown impressive potential in this regard. Compared to mammalian CYP enzymes, bacterial CYPs have better turnover and are much more easily expressed in *Escherichia coli* (*E. coli*), with better NADPH coupling efficiency (Kumar, 2010). Human CYP enzymes are utilized to study the metabolic pathways producing unwanted by-products from endogenous chemicals or carcinogenic drugs. Libraries of metabolites are generated for identifying potential drug candidates by using natural or randomly mutated CYP enzymes. CYP enzymes have also shown prospects in the research for activating prodrugs in the body. Recently, CYP enzymes have been used in transgenic plants in order to develop resistance against pesticides, as well as improve production of beneficial materials for the plant itself, and for industrial applications (Guengerich, 2002).

14.3.1 Bacterial CYP Enzymes

Utilization of bacterial CYP enzymes in drug synthesis dates back to early 1950s when cortisone was converted from progesterone by introducing hydroxyls into sterols, in Rhizopus culture (Peterson et al., 1952). One of the most prevalently used bacterial CYPs, CYPBM3 (also known as CYP102A1), is isolated from a Gram-positive bacteria *Bacillus megaterium.* Using catalytically self-sufficient CYPBM3 is advantageous due to its solubility and overexpression capability in industrial hosts (Munro et al., 2013). Zhang et al. have recently shown that, functional CYPBM3 is a globular protein, with the fusion of heme domain and reductase domain to a single polypeptide chain (Chefson and Auclair, 2006; Urlacher and Eiben, 2006; Zhang et al., 2018). It has relatively more flexible heme domains, and a higher efficiency to transfer electron from NADPH, conforming to the high efficiency in catalytic reactions of a wide range of substrates (Zhang et al., 2018). Reactions catalyzed by CYPBM3 include hydroxylation and/or epoxidation of fatty acids, fatty acid amides, alcohol, etc. (Miura and Fulco, 1975; Narhi and Fulco, 1986). CYP102A has the capability to metabolize the substrates of human CYPs 3A4, 2E1, and 1A2 (Di Nardo et al., 2007). In addition to the substrate diversity, CYP102A shows extremely high reaction rate. For example, CYP102A has long chain fatty acid hydroxylation rate of over 3000 min^{-1} (Narhi and Fulco, 1986), as well as arachidonate oxidation rate of ~17,100 min^{-1} (Noble et al., 1999). Keeping the superior enzymatic activity of CYP102A in mind, this enzyme has been undergoing massive engineering modification to overcome the inherent limitations of wild-type CYPs and being used in industrial synthesis. The importance of this enzyme in industrial drugs and drug metabolites synthesis will be discussed in detail later in this chapter.

CYP106 obtained from *B. megaterium*, is a potential candidate for the preparative scale biotechnological production of hydroxysteroids (Rauschenbach et al., 1993; Schmitz et al., 2014). Chul-Ho Yun and his group in their study with CYP106A1 from *B. megaterium* ATCC 14581 strain showed that the enzyme can hydroxylate steroids (e.g., testosterone, progesterone, 17α-hydroxyprogesterone, 11-deoxycorticosterone, corticosterone, and 11-deoxycortisol). Monohydroxylated compounds are produced as major metabolites by hydroxylation (Lee et al., 2015). Hydroxylation by CYPs can be useful to produce intermediates or to structurally modify steroids for therapeutic applications (Sedlaczek and Smith, 1988). Enzymatic reactions by CYP106A1 could replace several chemical reaction steps in the industrial manufacturing of natural hormones and steroidal drugs (Lee et al., 2015). Recently, studies conducted on *B. megaterium* DSM319 strain have identified the potential of CYP106A1 to convert pentacyclic triterpene 11-keto-β-boswellic acid (KBA), a constituent of resin extracts, used to treat inflammatory disorders and arthritis. Conversion rate of KBA by CYP106A1 whole system was found to be 100 min^{-1} with 80% selectivity towards the main product. The hydroxylated KBA derivatives obtained from this enzymatic reaction are assumed to possess enhanced bioavailability, improved pharmacological activities, or can be modified by chemical methods (Brill et al., 2014). CYP106A2 hydroxylases tricyclic diterpene and pentacyclic triterpene acids (Bleif et al., 2012). A research group led by Rita

Bernhardt, screened a steroid library to identify new substrates for CYP106A2, which included dehydroepiandrosterone (DHEA) and pregnenolone. At a preparative scale, they were able to hydroxylate DHEA and pregnenolone primarily at 7β position. Conversion rate of DHEA to 7β-OH-DHEA using *B. megaterium* was high with a volumetric yield of 103 mg/L/h (Schmitz et al., 2014).

Among other bacterial CYP enzymes, CYP105A1 has been shown to produce active vitamin D3. In fact, CYP105A1 expressed in *E. coli* can hydroxylate both vitamin D2 and D3 (Sawada et al., 2004). CYP107A obtained from a mutant bacteria *Saccharopolyspora erythraea* is used in the synthesis of erythromycin derivative, 6-deoxyerythromycin A (Weber et al., 1991). CYPs from Streptomyces have been used in the synthesis of antibiotics like erythromycin and tetracenomycin by the group led by Hutchinson (Shafiee and Hutchinson, 1988; Andersen and Hutchinson, 1992; Andersen et al., 1993; Shen and Hutchinson, 1993; Shen and Hutchinson, 1994; Guengerich, 2002). CYP genes are also useful in the biosynthesis of epothilone polyketides, which are effective in treatment against taxol-resistant cancers. The CYP epoxidase gene, along with a gene cluster required for this reaction is expressed in *Streptomyces coelicolor* CH999, which successfully produced epothilone (Tang et al., 2000). Another anti-cancer drug perillyl alcohol is reported to be produced in *Pseudomonas Putida* expressing bacterial CYP153 (Notonier, 2016). Harboring CYP153 in *E. coli* resulted in increased bioconversion rate of perillyl aldehyde to perillyl alcohol (Cornelissen et al., 2013).

14.3.2 Yeast and Plant CYP Enzymes

Targeting CYP51, a sterol demethylase isolated from fungi, is a well-known biosynthetic pathway to synthesize sterol drug candidates (Zhu et al., 2006). Antifungal agents like azoles, anti-tuberculosis drugs, drugs against Chagas disease, triterpenes can be synthesized using the CYP51 clan (Zhu et al., 2006; Geisler et al., 2013; Anusha et al., 2015; Lepesheva et al., 2015). The widely used anti-cancer agent taxol, which is the prime bioactive component of the Taxus species, is biosynthesized with CYP enzymes. A 2017 study focused on classifying the CYP enzymes from *Taxus chinensis* revealed the presence of 52 genes from CYP71 clan. Six genes from CYP725s were also identified, which are potential candidates for biosynthetic application. In addition, two gymnosperm specific families were discovered, namely CYP86 and CYP947. Knowledge on the presence of these CYPs can be important in developing new and cost effective industrial biosynthetic pathway for taxol synthesis (Liao et al., 2017). Research has been undertaken to find the biosynthetic pathway for drug synthesis, where CYPs act as a catalytic enzyme in the intermediary step. This will allow better understanding of the process, leading to optimize and accelerate the biotechnological production of complex natural products. For example, production of ingenol mebutate for actinic keratosis could be obtained from *Euphorbia lathyris L.*, utilizing CYP71D445 and CYP726A27 (Luo et al., 2016). Production of ingenol metabutate from plant extraction or chemical synthesis is inefficient and expensive, limitations, which can be overcome by utilizing the newly found biosynthetic pathway (Luo et al., 2016).

14.3.3 Mammalian CYP Enzymes

Similar to bacterial and plant CYPs, human CYP enzymes have also been investigated for potential use in drug synthesis. Libraries of chemical metabolites are generated by using wild-type or randomly mutated CYPs. Human CYPs are an attractive target for this as majority of the drugs and endogenous products are the substrates of human CYPs (Zanger and Schwab, 2013). Through this drug discovery process, we can learn about prodrug activation in the body (Guengerich, 2002). However, incorporating human CYPs in the drug discovery process is still in its very early stages. Few of the examples of human CYPs in industrial drug discovery process include CYPs19, 3A4, and 1B.

CYP19 is a mammalian aromatase, which can be targeted to develop drugs against hormone-responsive cancers. This enzyme is capable of transformation of androgens to estrogens, and reduction of the level of estrogen is an effective way to treat cancers (Guengerich, 2002). Letrozole and anastrazole are drugs that are manufactured based on this mechanism (Brodie, 1994). CYP3A4 can be targeted for developing drugs and prophylactic agents. One example for potent drug candidate development is based on inhibiting CYP3A4 to slow down the metabolism of expensive drugs. The goal is to use lower dose of the drug and to achieve better exposure in target organs (Guengerich, 2002). CYP3A4 is also a promising target for chemotherapeutic agents, e.g., gemcitabine, cisplatin, carboplatin, docetaxel, and paclitaxel used against non-small cell lung cancer (NSCLC). In-silico drug target interaction studies revealed better combination treatment regimen for advanced NSCLC using gemcitabine and carboplatin (Chun et al., 2001; Peter Guengerich et al., 2003; Subhani and Jamil, 2015). Same principle has been applied to CYP1B1. Inhibitors are developed targeting this enzyme, to block the activation of exogenous carcinogens, including estrogens 23, 24, and 25 (Langouet et al., 1995; Chun et al., 2001; Guengerich, 2002; Peter Guengerich et al., 2003). The applications of wild-type CYPs in production of drugs and drug metabolites are summarized in Table 14.1.

Table 14.1 Summary of wild type CYPs used in drug and drug metabolite synthesis

Source	CYP	Applications	References
Bacteria	CYPBM3	Hydroxylation and/or epoxidation of fatty acids, fatty acid amides, alcohol	Miura and Fulco, 1975; Narhi and Fulco, 1986
	CYP106	Hydroxylation of steroids, e.g., testosterone, progesterone, 17α-hydroxyprogesterone, 11-deoxycorticosterone, corticosterone, dehydroepiandrosterone and 11-deoxycortisol; conversion of KBA; hydroxylation of terpenes and terpene acids	Rauschenbach et al., 1993; Bleif et al., 2012; Brill et al., 2014; Schmitz et al., 2014; Lee et al., 2015
	CYP105A1	Hydroxylation of vitamin D2 and D3; Synthesis of erythromycin, tetranomycin and erythromycin derivative	Weber et al., 1991; Shen and Hutchinson, 1994; Sawada et al., 2004
	CYP450 epoxidase	Production of epithilone	Tang et al., 2000
	CYP153	Production of perillyl alcohol	Notonier et al., 2016a

Source	CYP	Applications	References
Yeast and Plant	CYP51	Synthesis of sterols, anti-fungal, anti-tuberculosis, drugs against Chagas disease, triterpenes	Zhu et al., 2006; Geisler et al., 2013; Anusha et al., 2015; Lepesheva et al., 2015
	CYP71D445 CYP726A27	Production of ingenol mebutate	Luo et al., 2016
Mammals	CYP19	Conversion of androgens to estrogens; production of Letrozole and anastrazole	Brodie, 1994; Guengerich, 2002
	CYP3A4	Development of chemotherapeutic agents, e.g., gemcitabine, cisplatin, carboplatin, docetaxel, and paclitaxel	Subhani and Jamil, 2015

14.4 Limitations

CYP is the nature's most versatile biocatalyst, which makes it a promising candidate for biotechnological applications. However, their substrate diversity, low catalytic activity, instability, and need for redox partners for electron transfer hinder their utility as a suitable biocatalyst in industrial applications.

14.4.1 Substrate Diversity

A CYP enzyme can metabolize various class of drugs or produce various metabolites from a parent compound. For example, CYP3A4 metabolizes cyclosporine (immunosuppressant), docetaxel (chemotherapeutic), and amitriptyline (antidepressant), etc. In certain instances, such broad substrate specificity poses drug resistance, e.g., CYP3A4/5 contributes to resistance of osteosarcomas to chemotherapy as it can metabolize several anticancer drugs used in the treatment of osteosarcomas (ifosfamide, vinblastine, etoposide, and doxorubicin) (McFadyen et al., 2004).

14.4.2 Activity

Activity of most of the CYP enzymes varies with co-administered drugs/food. For example, due to the interaction between midazolam (CYP3A4 substrate) and the herbal antidepressant St. John's Wort (CYP3A4 inducer), midazolam AUC and C_{max} decreased by 79% and 65%, respectively (Mueller et al., 2006). CYPs also undergo evolution. During the evolution process, mutations can occur in CYP enzymes leading to decreased/increased activity.

14.4.3 Stability

Enzymes are often poorly stable under different conditions they encounter during production, storage, or use. CYPs are relatively unstable because they comprise multiple subunits and contain thermolabile cofactors.

In addition to all these, complexity of CYP systems due to the need of electron transfer from NADPH via redox partner proteins for the initial hydroxylation step limits a broader technical implementation of CYP enzymes.

Moreover, as discussed earlier, bacterial, mammalian, and plant CYP enzymes differ from each other in their physical and chemical properties. These contrasting features of bacterial and mammalian CYP enzymes clearly suggest that both enzymes have advantages and disadvantages in using them as biocatalyst.

14.5 CYP Engineering for Enhanced Activity, Stability, and Specificity to Overcome the Limitations

To overcome the limitations of wild-type CYP enzymes in industrial synthesis, CYP enzymes have been engineered to enhance the selectivity, catalytic activity, stability, solubility, overcoming need for redox partners, H_2O_2 utilization as well as H_2O_2 production, and NADPH consumption rates of CYP enzymes. In general, engineering of CYP enzymes has typically been achieved by rational site-directed mutagenesis, random mutagenesis, directed evolution approaches, or by a combination of these approaches (Girvan and Munro, 2016). Other than these traditional engineering approaches, gene-fusion approaches and enzyme-mediated DNA shuffling approaches have also been used in engineering plant and mammalian CYP enzymes, respectively (Gillam, 2008; Schuckel et al., 2012).

14.5.1 Bacterial CYP Enzymes

Bacterial CYP enzymes metabolize limited number of drugs (Girvan and Munro, 2016). To expand the use of bacterial CYP enzymes in industrial synthesis, these enzymes have been vastly studied and engineered. CYPs including BM3, 105, 106A2, 153A, and 101A1 have shown most promising effect after engineering, which will be discussed in the following sections.

14.5.1.1 CYPBM3

Arnold's group is the pioneer in engineering CYPBM3. Using engineered CYPBM3, they have generated reactive metabolites from diverse types of substrates, including octane, buspirone, *R*-6-hydroxybuspirone, propranolol, naproxen, verapamil, and astemizole (Farinas et al., 2001; Landwehr et al., 2006; Otey et al., 2006; Sawayama et al., 2009). They also showed that engineered CYPBM3 can be used to enhance enzyme activity, substrate recognition, and thermostability (Whitehouse et al., 2012).

According to recent FDA guidelines, if >10% of a parent drug forms a metabolite, that metabolite will need to undergo separate toxicity testing. For this purpose, large quantities of the drug metabolite is needed (FDA, 2016). Therefore, recent studies highlighted the broad applications of engineered CYPBM3 in the production of human metabolites (Girvan and Munro, 2016). For example, the Munro group engineered CYPBM3 by rational mutagenesis to facilitate the binding of the proton pump inhibitor, omeprazole. The mutations of F87V and A82F resulted in a conformational alternation and a decreased free energy barrier, ultimately changing the substrate recognition. These mutants of CYPBM3 were able to transform omeprazole to the human CYP2C19-type metabolite (Butler et al., 2013). The same group further demonstrated the efficient transformation of several proton pump inhibitors (e.g., lansoprazole, esomeprazole) to human CYP-type metabolites by CYPBM3 F87V/A82F mutants (Butler et al., 2014). In a recent work, a library of CYPBM3 variants carrying multiple mutations that were identified from directed evolutions showed a high oxidation activity for a wide range of drugs. Few of these drugs include, non-steroidal anti-inflammatory drugs/NSAIDs (diclofenac and naproxen), muscle relaxant (chlorzoxazone), antidepressant (amitriptyline), anesthetic, antiarrhythmic drug (lidocaine), and steroid hormones (testosterone). Importantly, CYPBM3 mutants from this library converted diclofenac with 91–100% conversion, and produced human CYP2C9-type and CYP3A4-type metabolites, 4′-hydroxylated and 5′-hydroxylated diclofenac with 34% yield and 47% yield, respectively (Ren et al., 2015). In another recent example, CYPBM3 mutant M11, has been proven to produce human metabolites of fenamic acid NSAIDs including mefenamic acid, meclofenamic acid, and tolfenamic acid (Venkataraman et al., 2014; Capoferri et al., 2016). Mutant M11 was generated by using random mutagenesis and had 90-fold higher initial activities compared with human CYP2D6 (van Vugt-Lussenburg et al., 2007). This mutant was able to synthesize benzylic or aromatic hydroxylation metabolites of the fenamic acid NSAIDs with high substrate conversions and high turnover numbers (2000–6000). Because of the high total turnover numbers, M11 can be used as a biocatalytic tool for large-scale production of fenamic acid metabolites to perform characterization study, activity study, and toxicological evaluation of NSAIDs (Venkataraman et al., 2014). Similarly, mutant M11 was capable of metabolizing anticancer drugs cyclophosphamide and ifosfamide, producing active metabolites. It can also be used for extracellular bioactivation, and for providing a catalytically efficient alternative to liver S9 fraction for toxicity evaluation (Vredenburg et al., 2015). In addition, a set of BM3 mutants, generated using both rational and random mutagenesis, was able to produce human metabolites of 17β-estradiol. This engineered CYPBM3 enzyme can catalyze same reactions as human CYP1A1, CYP1A2, and CYP1B1 with high catalytic efficiency (k_{cat}/K_m) and total turnover numbers between 1040 and 1210, producing human metabolites for industrial applications (Cha et al., 2014).

Engineered CYPBM3 can be used to improve the biocatalytic oxidation, producing valuable industrial chemicals (Wang et al., 2014; Dezvarei et al., 2018a), while wild-type enzymes only catalyze small amount of industrially relevant chemicals (Huisman and Collier, 2013). For example, an engineered CYPBM3 was able to synthesize the chemical core of levomilnacipran, which can be used to treat depression clinically. This engineered enzyme, BM3-Hstar, was able to catalyze the cyclopropanation of *N*,*N*-diethyl-2-phenylacrylamide with 1000 turnovers per minute and can be used under aerobic conditions, while wild-type BM3 has very low cyclopropanation activity. Compared with the traditional transition metal catalytic approaches, the enzymatic method is more attractive because of its stereo-selective and non-toxic profile. Importantly, BM3-Hstar was able to provide the desired product at 92% yield with 92% enantioselectivity and 2:98 diastereoselectivity (Wang et al., 2014). Another report has also shown the enhanced cyclopentane and cyclohexane oxidation by an engineered CYPBM3 with decoy molecules. This engineered CYPBM3 was able to oxidize cycloalkanes up to 8000-fold higher than the wild-type enzyme, at product formation rates up to 1700 nmol·nmol-CYP^{-1}·min^{-1} (Dezvarei et al., 2018a). Moreover, engineered CYPBM3 mutants can be used as biocatalysts to overcome the challenge on selective C-H oxidation of terpenoid scaffolds. Several CYPBM3 variants generated from direct evolution were used to catalyze tobacco cembranoid β-cembrenediol. The F87A/I263L mutant and the L75A/V78A/F87G mutant were able to hydroxylate the neighboring positions C-9 (100% regioselectivity, diastereomeric ratio of 89:11) and C-10 (97% regioselectivity, diastereomeric ratio of 74:26) of β-cembrenediol. Since no chemical synthesis is available to perform this late-stage C-H oxyfunctionlization, this study showed the importance of engineered CYPBM3 as a biocatalyst in chemoenzymatic synthesis of larger complex molecules (Le-Huu et al., 2015).

CYPBM3 can also be engineered to enhance thermal stability (Eiben et al., 2007; Geronimo et al., 2016) and H_2O_2 utilization (Whitehouse et al., 2009; Lim and Sikes, 2015). One approach is to replace the CYPBM3 reductase domain with FAD/FMN reductase domain to improve temperature catalytic tolerance. The maximum activity of this engineered CYPBM3 was obtained at 51°C, and the half-life of the engineered CYPBM3 was ten times longer than the wild-type CYPBM3 (Eiben et al., 2007). Additionally, a recent study showed a targeted region on CYPBM3, which can be used to improve stability. They showed the structural data for wide-type CYPBM3 and R47L/F87V/L188Q/E267V/F81I pentuple mutant (PM) with the native substrate palmitic acid. Then they performed a high-temperature molecular dynamics simulations, and identified labile regions in PM that can be targeted for mutation to improve stability (Geronimo et al., 2016). In addition, engineering of CYPBM3 can also optimize the production of H_2O_2, which is an important element in redox biology. For example, I401P, an engineered CYPBM3, was able to alter NADPH consumption rates and had more reduction potential than the wild-type CYPBM3 (Whitehouse et al., 2009). Recently, the production of H_2O_2 by this engineered CYPBM3 and wild-type CYPBM3 was measured using a genetically encoded H_2O_2 sensor. The result showed that engineered CYPBM3 has an enhanced H_2O_2 production compared to wild-type CYPBM3 (Lim and Sikes, 2015).

14.5.1.2 CYP105 family

The CYP105 family is predominantly found in bacteria, which belongs to the phylum Actinobacteria and the order Actinomycetales (Moody and Loveridge, 2014). CYP105 family is an industrially important enzyme family because CYP105A3 can stereoselectively produce an anti-hyperlipidaemia drug, pravastatin (Watanabe et al., 1995). CYP105 enzyme has large substrate-binding pockets, and can be used in biotransformation and biodegradation of xenobiotics and biosynthesis of bioactive molecules (Yasuda et al., 2018). Recently, many researchers have performed protein engineering of CYP105 family. A mutant library of CYP105A3 (generated from random mutagenesis) performed a 4.1-fold pravastatin yield in whole cells extracts, and a 3.8-fold active CYP expression of the wild-type enzyme (Ba et al., 2013). Further mutagenic study was done in this CYP105A3 mutant library. Five sites located on the substrate-binding pocket, substrate access entrance, and presumed Pdx interaction interface were chosen, followed by a site-directed saturation mutagenesis and three rounds of iterative saturation mutagenesis. The best mutant obtained showed a 7.1-fold of whole cell biotransformation activity of the starting mutant library (Ba et al., 2013a). Recently, pravastatin was produced successfully with a single-step fermentative method with recombinant *Penicillium chrysogenum* expressing a CYP105AS1 mutant. Unlike the wild-type CYP105AS1, which produces mainly the ineffective epimer of pravastatin, the mutant generated from random mutagenesis was able to produce the pharmacologically active pravastatin. Expression of this mutant

CYP105AS1 on chrysogenum yielded more than 6 g/L pravastatin at a pilot production scale, providing the capability of efficient industrial-scale statin drug production (McLean et al., 2015).

The Sakaki group developed double mutants of CYP105A1, R73A/R84A, and R73V/R84A. This enzymatic method can be used industrially to produce 1α, 25-dihydroxy vitamin D3, a therapeutic product for the treatment of osteoporosis and chronic renal failure. The double mutants, R73A/R84A and R73V/ R84A, were generated by using site-directed mutagenesis approaches focusing on CYP105A1 residues, Arg73 and Arg84. These double mutants have a high capability to convert vitamin D3 to its biologically active forms and exhibited approximately 400-fold higher k_{cat}/K_m for 25-hydroxylation, and 100-fold higher k_{cat}/K_m for 1α-hydroxylation than the wild-type CYP105A1 (Hayashi et al., 2008). In their recent study, they also compared the hydroxylation activities of these double mutants towards vitamin D2. Compared with the wild-type enzyme, the double mutants remarkably enhanced hydroxylation of vitamin D2, producing 25- and 26-hydroxylated D2 (Hayashi et al., 2016). Since both double mutants R73A/R84A and R73V/R84A of CYP105A1 were not able to produce the active form of vitamin D2, the same group developed a triple variant R73A/R84A/M239A of CYP105A1 using site-directed mutagenesis approach. This variant showed a 22-fold higher 1α-hydroxylation activity towards 25-(OH)D2 to produce the active form of vitamin D2 (Yasuda et al., 2017).

14.5.1.3 CYP106A2

CYP106A2 is isolated from *B. megaterium* and identified as steroid hydroxylase. However, the wild-type CYP106A2 is not capable to hydroxylate stereoselectively, resulting in the formation of undesired side-products. To overcome the unspecific hydroxylation, protein engineering of CYP106A2 has been successfully performed to optimize the stereoselectivity (Schmitz et al., 2018). The Bernhardt group created a saturation mutagenesis library of CYP106A2 using random mutagenesis approach. Compared with wild-type CYP106A2, 2100 mutants were identified to express higher 11-deoxycortisol hydroxylation activities, and 1900 mutants were identified to have elevated progesterone hydroxylating activities. They further selected the most promising mutant and generated the quadruple mutant A106T/Q109K/T399S/R409L by directed evolution. The quadruple mutant was able to hydroxylate 11-deoxycortisol with a fourfold increase in k_{cat}/K_m, and progesterone with a 1.4-fold higher conversion (Virus and Bernhardt, 2008). Based on previous achievements, the same group generated improved mutants A106T/A395I, A106T/A395I/R409L, and T89N/A395I with an additional directed evolution. Compared with the wild-type CYP106A2, the mutants from this study provided up to 14.3-fold increase in selectivity and up to 108-fold increase in k_{cat}/K_m on 11α-hydroxylation. The best mutant (T89N/A395I) also demonstrated an impressive regioselectivity that increased the production of 11α-hydroxyprogesterone from 27.7 to 80.9%, compared with wild-type enzyme (Nguyen et al., 2012). In their recent work, they generated CYP106A2 mutants with a higher regioselectivity towards hydroxylation at positions 6 and 9 of the steroid compounds. Compared with the wild-type CYP106A2, the mutant A243S, which was generated through site-directed mutagenesis showed more than 10-fold improvement in 6β-selectivity of progesterone. Site-directed mutant F165L/A395E/G397V showed a 10-fold increase in 9α-hydroxylation selectivity of progesterone compared with the wild-type enzyme. Importantly, the metabolites 6β-OH-progesterone and 9α-hydroxyprogesterone are useful pharmaceutical intermediates. 6β-OH-progesterone can be used to produce drug against breast cancer and 9α-hydroxyprogesterone can be used to produce active substances with progestational activity. Moreover, these selected mutants were also able to display 9α- or 6β-hydroxylation of 11-deoxycorticosterone (DOC), testosterone, and cortisol in vivo (Nikolaus et al., 2017).

14.5.1.4 CYP153A

CYP153A has been identified as a fatty acid ω-hydroxylase with a broad substrate range (Honda Malca et al., 2012). CYP153As can be obtained from variety of bacteria including *Marinobacter aquaeolei*, *Alcanivorax borkumensis*, and *Gordonia alkanivorans* (Jung et al., 2016). Recently, the CYP153A enzyme from *M. aquaeolei* was engineered to a double mutant G307A/S233G based on semi-rational protein design to improve the activity of terminal hydroxylation of medium- and long-chain fatty acids. The most active mutant showed a 3.7-fold improvement over the wild-type activity. Additionally, the heme domain of CYP153A mutant was constructed with the reductase domain of CYPBM3 to improve the catalytic activity for its future industrial applications (Notonier et al., 2016b). Moreover, a crystal structure of CYP153A from *M. aquaeolei* revealed a crucial target for engineering CYP153A. CYP153A residues at the extended F-helix and Ω-loop were targeted using site-directed mutagenesis. Importantly, mutant P135A showed an enhanced activity on hydroxylation of hexadecanoic acid with 20% more conversion than the wild-type enzyme (Hoffmann et al., 2016).

14.5.1.5 CYP101A1

CYP101A1 (also known as CYPcam) is part of the metabolic pathway of *P. putida* for hydroxylation of camphor. The engineered CYP101A1 enzymes were often used to enhance catalytic activities of alkanes, such as butane, pentane, hexane, and heptanes (Bell et al., 2002), and produce compounds for the flavor and fragrance industries (Bell et al., 2003). The engineered CYP101A1 was also utilized to produce renewable chemicals from small-chain alkanes, such as alcohols (Xu et al., 2005). A recent report showed that engineered CYP101A1 was able to regio- and stereo-selectively hydroxylate isophorone to (*R*)-4-hydroxyisophorone, which is a synthetic intermediate for pigments and drug molecules. The F87W/Y96F/L244A/V247L mutant of CYP101A1 (generated by site-directed mutagenesis) can selectively yield 98% of (*R*)-4-hydroxyisophorone, while the wild-type enzyme mainly produces 7-hydroxyisophorone (Dezvarei et al., 2018b).

14.5.2 Plant CYP Enzymes

Plants produce more than 200,000 metabolites to interact with and adapt to their environment. Plant specific metabolites, such as terpenoids, phenylpropanoids, and nitrogen-containing compounds are the major source of new drugs and bioactive compounds. Plant CYP enzymes have the potential to be used for industrial hydroxylation reactions. However, plant CYP enzymes typically have low turnover rates (ten to a few hundred per minute range), poor solubility, less suitable expression systems, and requirement of redox proteins (Hamberger and Bak, 2013; Renault et al., 2014). Therefore, gene-fusion approaches have been created for plant CYP enzymes to maximize enzymatic activity. Recently, the plant P450 isoflavone synthase 1 (IFS) from Glycine max (soybean) was used to create a redox-self-sufficient CYP by fusing with the bacterial reductase Rhf-RED. The IFSn-Rhf-RED fusion enzyme can be expressed in *E. coli*, and transform the flavonoid substrate naringenin to genistein, which is an angiogenesis inhibitor and a phytoestrogen. They also created an active CYP fusion enzyme for CYP73A5, a cinnamate-4-hydroxylase (C4H) from *Arabidopsis thaliana*. The enzyme fusion C4H-Rhf-RED was expressed in *E. coli* and can be used for the hydroxylation of cinnamic acid to p-coumaric acid. The fusion system expresses both plant and bacterial CYP enzymes, and has the potential to be used in industrial biocatalysis (Schuckel et al., 2012).

14.5.3 Mammalian CYP Enzymes

In contrast to bacterial CYP enzymes, mammalian CYP enzymes can metabolize numerous substrates and xenobiotics. However, mammalian CYP enzymes have poor activity and stability, and show less suitable expression on heterologous systems, including *E. coli*. Therefore, engineered mammalian CYP enzymes have been developed to enhance enzyme expression, stability, activity, substrate specificity, and regio-selectivity (Kumar, 2010). For example, Kumar and Halpert group have engineered several mammalian CYP enzymes, such as CYP2B and CYP3A4, using rational (site-directed mutagenesis) and directed evolution (random mutagenesis) approaches. These engineered mammalian CYP enzymes showed enhanced catalytic activity, *E. coli* expression, thermostability, solubility, and specificity (Kumar et al., 2003; Kumar et al., 2005; Kumar and Halpert, 2005; Kumar et al., 2007; Talakad et al., 2010). Additionally, the Gillam group has also used an enzyme-mediated DNA shuffling approach to engineered CYP1A and CYP2C enzyme to show novel activities (Gillam, 2007; Huang et al., 2007; Gillam, 2008).

In the past five years, engineering mammalian CYP enzymes has been continually studied to produce pharmaceutical related products. Human CYP3A4 is responsible for about half of the CYP-mediated drug metabolism in the body. Recently, an engineered human CYP3A4 (generated by site-directed mutagenesis) was studied using carbamazepine, a medication used in treatment of epilepsy and neuropathic pain. The mutants I369F, I369L, A370V, and A370L provided enhanced affinity and increased turnover of carbamazepine. The engineered CYP3A4 expressed higher carbamazepine binding and can be used as a CYP3A4-based electrochemical biosensor for determining carbamazepine in clinical samples (Muller et al., 2015). Human CYP3A4 was also engineered to alter its regio- and stereoselectivity. A CYP3A4 mutant A370I generated from random mutagenesis can be used to synthesize (*R*)-lisofylline, which cannot be generated by wild-type enzyme. The active metabolite (*R*)-lisofylline generated by engineered CYP3A4 is an anti-inflammatory substance, which shows efficacy on both type 1 diabetes treatment and hematopoiesis recovery. In addition, the engineered mutants I369F/L482F, I301F/I369F/L482F, and L211F/I301F/I369F/G481W/L482F generated from random mutagenesis approach showed the selectivity of oxidation in a theobromine derivative and sex hormone progesterone (Schiavini et al., 2017). Other human CYP enzymes have also been engineered to produce pharmaceutical products. For example, human CYP4A11, a fatty acid hydroxylase, was engineered to produce a blood pressure regulator 20-hydroxyeicosatetraenoic acid (20-HETE). A random mutagenesis was performed on CYP4A11 and yielded three improved CYP4A11 mutants, CP2600 (A24T/T263A), CP2601 (T263A), and CP2616 (A24T/T263A/V430E). The CYP4A11 mutant showed a threefold increase of expression in whole cells and >10-fold increase in purified proteins on the luminescence assay (Choi et al., 2013).

Other than human CYP enzymes, engineering mammalian CYP enzymes has been successfully developed over a broad range of species. For example, CYP1A9, a piscine CYP enzyme originally from Japanese eel (*Anguilla japonica*), was engineered using site-directed mutagenesis approach. Mutant F263A showed higher α-naphthoflavone (ANF) and 7-ethoxycoumarin hydroxylase activities than wild-type CYP1A9. Mutant F263A, V387, and I391A showed higher catalytic activities of progesterone than wild-type CYP1A9. In this study, engineering of CYP1A9 enhanced the catalytic activity, introduced hydroxyl groups into the steroid ring of progesterone, which would further increase the water solubility of the steroid molecule (Uno et al., 2015). In a recent study, CYP2B from the desert woodrat (*Neotoma lepida*) was engineered for site-directed mutagenesis. Desert woodrats are exposed to a number of toxins with their regular diet. Thus, the CYPs they develop have the capability to detoxify a variety of toxic materials, which makes them highly useful in toxicity studies. CYP enzymes involved in mammalian detoxification generally metabolize a broad range of substrates. Engineering of CYP2B35 changed the enzyme specificity and presented novel catalytic activity, which were not present in the wild-type enzyme. The mutant of CYP2B35 showed maximal activity with 7-butoxycoumarin, which is opposed to the wild-type enzyme. A CYP2B35 mutant I104V/I114V/M209I/A3631/A367V/I477F presented a hybrid activity profile of CYP2B35 and CYP2B37, metabolizing both short-chain and long chain 7-alkoxycoumarins (Huo et al., 2017). In addition, engineered porcine CYP3A46 has also been developed recently. Site-directed mutagenesis was performed on residues Phe-108, Ser-119, Phe-215, Phe-304, and Thr-309 of porcine CYP3A46. They found that porcine CYP3A46 can convert Aflatoxin B1 to a highly reactive epoxide intermediate AFB1-8, 9-epoxide (AFBO), exhibiting similar catalytic procedure as human CYP3A4. The product AFBO can cause hepatotoxicity and hepatocarcinoma in human. The engineered CYP3A46 demonstrated that residues Phe-108, Ser-119, Phe-215, Phe-304, and Thr-309 play important roles in AFB1-8, 9-epoxidation. These results suggested that engineered CYP3A46 can be used for toxicity evaluation studies for industrial use (Jiang et al., 2018). Summary of the engineered CYP enzymes being used in the synthesis of drugs and drug metabolites is presented in Table 14.2.

14.6 Advantages of Using CYP Enzymes for Drugs and Drug Metabolites Synthesis

CYP enzymes have a tremendous advantage for the synthesis of various drugs due to its catalytic function in oxygenation of wide range of chemical as well as biological substrates. Following are some of the advantages of CYPs, which justifies the use of CYPs to be beneficial for drug synthesis.

Table 14.2 Summary of engineered CYPs used in drug and drug metabolite synthesis

Source	CYP	Engineering approaches	Applications	References
Bacteria	CYPBM3	Directed/random/ rational mutagenesis	Generate reactive metabolites from diverse types of substrates; Enhancement of enzyme activity, substrate recognition, and thermo stability	(Farinas et al., 2001; Landwehr et al., 2006; Otey et al., 2006; Sawayama et al., 2009; Whitehouse et al., 2012)
		Rational mutagenesis	Convert omeprazole to the human CYP2C19-type metabolite	(Butler et al., 2013)
		Rational mutagenesis	Convert lansoprazole, esomeprazole to human P450-type metabolites	(Butler et al., 2014)
		Directed mutagenesis	High oxidation activity for diclofenac, naproxen, chlorzoxazone, amitriptyline, lidocaine, and testosterone; production of human CYP2C9-type and CYP3A4-type metabolites of diclofenac	(Ren et al., 2015)
		Random mutagenesis	Produces human metabolites of fenamic acid NSAIDs with high activity; 90-fold higher initial activities compared with human CYP2D6	(van Vugt-Lussenburg et al., 2007; Venkataraman et al., 2014; Capoferri et al., 2016)
		Random mutagenesis	Produces active metabolites of cyclophosphamide and ifosfamide	(Vredenburg et al., 2015)
		Rational and random mutagenesis	Catalyzes same reactions as human CYP1A1, CYP1A2 and CYP1B1 with high catalytic efficiency (k_{cat}/K_m) and total turnover numbers; produce human metabolites of 17 β -estradiol	(Cha et al., 2014)

(*Continued*)

Table 14.2 (*Continued*)

Source	CYP	Engineering approaches	Applications	References
		Rational mutagenesis	Synthesizes the chemical core of levomilnacipran selectively, which can be used to treat depression clinically	(Wang et al., 2014)
		Rational mutagenesis	Oxidizes cycloalkanes up to 8000-fold higher than the wild-type enzyme	(Dezvarei et al., 2018a)
	CYPBM3	Direct mutagenesis	Perform late-stage C-H oxyfunctionlization; catalyzes tobacco cembranoid β-cembrenediol stereoselectively	(Le-Huu et al., 2015)
		Replace the CYPBM3 reductase domain	Enhance thermal stability	(Eiben et al., 2007)
		Rational mutagenesis	Identify the regions for improving stability	(Geronimo et al., 2016)
		Rational mutagenesis	Alter NADPH consumption	(Whitehouse et al., 2009)
		Rational mutagenesis	Enhance H_2O_2 production	(Lim and Sikes, 2015)
	CYP105	Random mutagenesis	High pravastatin yield, and high CYP expression	(Ba et al., 2013)
		Rational mutagenesis	High biotransformation activity of pravastatin	(Ba et al., 2013)
		Random mutagenesis	Produces pharmacologically active pravastatin; yield more than 6 g/L pravastatin	(McLean et al., 2015)
		Rational mutagenesis	Enhances catalytic activity, turnover of vitamin D_3; produces active form of vitamin D_3	(Hayashi et al., 2008)
		Rational mutagenesis	Enhances hydroxylation of vitamin D_2; produces the active form of vitamin D2	(Hayashi et al., 2016; Yasuda et al., 2017)

Source	CYP	Engineering approaches	Applications	References
	CYP106A2	Random mutagenesis	Hydroxylates 11-deoxycortisol with a higher activity; hydroxylates progesterone with a higher conversion	(Virus and Bernhardt, 2008)
		Random mutagenesis + Rational mutagenesis	Higher selectivity, catalytic activity on 11α-hydroxylation	(Nguyen et al., 2012)
		Random mutagenesis + Rational mutagenesis	Higher regioselectivity towards hydroxylation at positions 6 and 9 of the steroid compounds; produces useful pharmaceutical intermediate	(Nikolaus et al., 2017)
	CYP153A	Rational mutagenesis	Improves catalytic activity of terminal hydroxylation of medium- and long-chain fatty acids	(Notonier et al., 2016b)
		Rational mutagenesis	Enhanced activity; Higher conversion	(Hoffmann et al., 2016)
	CYP101A1	Directed/random/ rational mutagenesis	Enhances catalytic activities of alkanes, such as butane, pentane, hexane, and heptanes; produces compounds for the flavor and fragrance industries; produce renewable chemical from small-chain alkanes	(Bell et al., 2002; Bell et al., 2003; Xu et al., 2005)
	CYP101A1	Rational mutagenesis	Selectively yield 98% of (R)-4-hydroxyisophorone	(Dezvarei et al., 2018b)
Plant	P450 isoflavone synthase 1 (IFS)	Create a fusion system	Generates genistein	(Schuckel et al., 2012)

(*Continued*)

Table 14.2 (*Continued*)

Source	CYP	Engineering approaches	Applications	References
	CYP73A5	Create a fusion system	Hydroxylates cinnamic acid to p-coumaric acid	(Schuckel et al., 2012)
Mammalian	CYP2B and CYP3A4	Rational mutagenesis + direct mutagenesis	Enhances catalytic activity, *E. coli* expression, thermostability, solubility, and specificity	(Kumar et al., 2003; Kumar et al., 2007; Talakad et al., 2010)
	CYP1A and CYP2C	Enzyme-mediated DNA shuffling approach	Shows novel activities	(Gillam, 2008)
	CYP3A4	Rational mutagenesis	Enhances affinity and turnover of carbamazepine	(Muller et al., 2015)
	CYP3A4	Random mutagenesis	Synthesizes an anti-inflammatory substance (R)-lisofylline; selective oxidize a theobromine derivative and progesterone	(Schiavini et al., 2017)
	CYP4A11	Random mutagenesis	Increase whole cells expression, produce a blood pressure regulator 20-hydroxyeicosatetraenoic acid (20-HETE)	(Choi et al., 2013)
	CYP1A9	Rational mutagenesis	Enhances catalytic activity; introduce hydroxyl groups into the steroid ring of progesterone; increase solubility	(Uno et al., 2015)
	CYP2B35	Rational mutagenesis	Changes enzyme specificity and present novel catalytic activity	(Huo et al., 2017)
	CYP3A46	Rational mutagenesis	Exhibit similar catalytic procedure as human CYP3A4	(Jiang et al., 2018)

14.6.1 Active Catalysts

CYPs are gaining their value in drug synthesis over chemical methods because of their ability to catalyze oxygenation at specific positions of substrates, which is difficult to perform using synthetic methods. In addition, CYPs have capability to oxidize various compounds in a chemo-, regio, and stereo-selective manner. Further, CYPs can catalyze thousands of reactions such as esterification, rearrangements, epoxidation, aromatic hydroxylation, deamination, dehalogenation, dealkylation (Guengerich, 1993). This wide range of catalysis makes CYPs highly efficient catalysts to functionalize chemicals into drugs.

14.6.2 Easy Availability and Diverse Substrate Selection

CYPs are ubiquitously present in plants, animals, and microorganism (bacteria and fungi), which make them a readily available source for drug synthesis. Further, CYPs have ability to catalyze diverse range of substrates. Several wild-type CYPs from different sources have been used for the synthesis of wide variety of drugs including steroids, terpenes, estrogens, anticancer, antifungal, and anti-tuberculosis drugs (Rauschenbach et al., 1993; Guengerich, 2002; Zhu et al., 2006; Lepesheva et al., 2015; Subhani and Jamil, 2015).

14.6.3 Generation of Higher Yields of Drugs

Isolation of chemotherapeutic drugs (e.g., Taxol) and antibiotics (e.g., Erythromycin D) from the natural source is insufficient. Several chemical, semisynthetic, microbial, and cellular culture synthesis approaches have been applied to produce these drugs in bulk. However, these methods have not achieved the total synthesis of the compounds due to their complex structure, multiple steps involved in their synthesis, low yield, and cost ineffectiveness. Recombinant expression of CYPs in microorganisms has overcome these limitations, producing higher yields of the drugs.

14.6.4 Development of New Antibiotic Drugs

New antibiotics can also be developed through random mutagenesis of the targeted CYPs in antibiotic- producing microorganisms.

The most effective recombinant antibiotic producers can then be screened and cultured in bulk to identify specific mutation in the CYP gene (Guengerich, 2002).

14.6.5 Generation of Drug Metabolites

Studies of drug metabolites are important to evaluate the bioavailability, efficacy, clearance, and toxicity of the drugs. Certain drug metabolites, mainly which require regio- and stereo-selective transformations are very difficult to synthesize using synthetic methods. As CYPs can catalyze wide variety of such transformations, they can be produced recombinantly in large-scale within bacterial cells, where they convert these drugs into metabolites. The recombinant CYPs can further be mutated to improve the binding of drugs and increase the range of substrate recognition. Recently, a more advanced method using immobilized enzyme reactors has been developed for the synthesis of drug metabolites (Lamb et al., 1998; Nicoli et al., 2008). In this method, the recombinant human CYPs immobilized on the gold electrodes catalyze the bioconversion of drugs in presence of NADPH and CPR. This method generates relatively higher yield of drug metabolites with higher purity and stability than the method using microorganisms.

14.7 Conclusion and Future Prospects

Use of CYP enzymes in drug synthesis is gaining popularity over chemical method due to their highly efficient oxygenation capability, in a stereo- and regio-selective manner. Based on the literature evidences, it is clear that manipulating human, fungal, bacterial, and plant CYP catalytic systems to have better yield of drug and drug metabolites has great future potential. Wild-type CYPs can generate drug products, but the molecular structure limits their utility in industrial biosynthesis. Researchers are focusing on engineering these CYPs by rational and random mutagenesis as well as directed evolution processes, which have tremendously improved the catalytic power of CYP enzymes (Girvan and Munro, 2016). CYPs have come a long way from the commercial production of steroids in a CYP dependent *Curvularia lunata* (Sonomoto et al., 1983) to the industrial production of artemisinin by Sanofi using an engineered strain expressing CYP (Paddon et al., 2013).

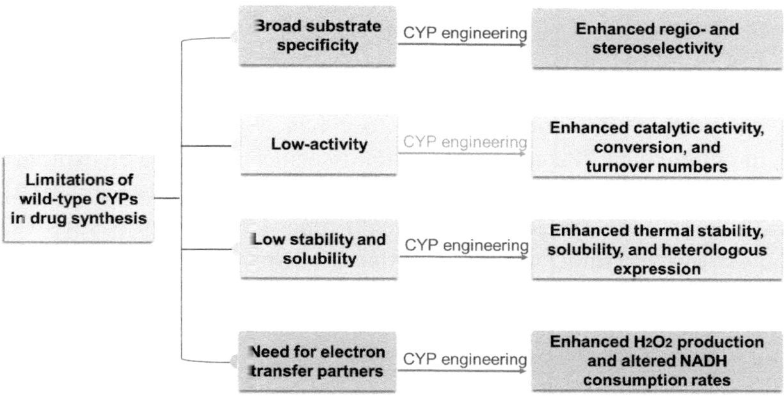

Figure 14.1 Limitations of CYP enzymes and CYP engineering to overcome it.

To apply CYP enzymes in broad scale commercial manufacturing, further investigations need to be conducted on screening and protein engineering of CYPs. Mutagenesis of CYPs along with enzymatic coupling may improvise the already known pharmacophores, moreover, can generate libraries of novel compounds. CYP enzymes are being modified to produce active metabolites, natural lead products, and drug targets, which may reduce the difficulties to manufacture these by means of chemical synthesis (Kandel et al., 2014). Several groups are currently working on generating rich libraries of engineered CYPs from bacteria, plant, and mammals, continuously revealing newer and improved CYP enzyme systems. As most of the CYPs require an electron transfer system for the enzymatic reactions, improving this by introducing fusion proteins and CYP reductase will vastly enhance the enzymatic efficiency (Sakaki et al., 1990; Fisher et al., 1992). Another way to overcome this hurdle is to develop an efficient 'shunt pathway' for the CYP catalytic reactions, which will deliver active oxygen molecule. Using carboxylic acids as decoy molecule in this case may provide necessary power to the enzymatic reaction (Shoji et al., 2010). Future investigations on laboratory formulated strains of bacteria, transgenic yeasts, or plants as host system and optimizing the reaction process might open the door to identify newer CYP isoforms as well as enable higher expression of CYPs (Martinez and Rupashinghe, 2013; Grunwald, 2014). Another obstacle in the way of industrial CYP usage is the structural instability. Use of immobilized CYPs on gold electrodes is showing promising effect (Lamb et al., 1998; Nicoli et al., 2008). A nanodisc system comprising CYP and CYP reductase has demonstrated potential in enhancing CYP activity and stability, which may also be explored for further possibilities (Denisov and Sligar, 2011; Yoshida, 2012). To conclude, by engineering wild-type CYPs, improvising the host system and catalytic reaction pathways, CYP enzymes can be efficiently employed in industrial drug and drug metabolite synthesis.

References

Andersen J. F., Hutchinson C. R., *J. Bacteriol.*, **174** (1992), 725–735.

Andersen J. F., Tatsuta K., Gunji H., Ishiyama T., Hutchinson C. R., *Biochemistry,* **32** (1993), 1905–1913.

Anusha S., B. Cp, Mohan C. D., Mathai J., Rangappa S., Mohan S., Chandra, Paricharak S., Mervin L., Fuchs J. E. M. M., Bender A., Basappa., Rangappa K. S., *PLoS One,* **10** (2015), e0139798.

Ba L., Li P., Zhang H., Duan Y., Lin Z., *Biotechnol. J.*, **8** (2013), 785–793.

Ba L., Li P., Zhang H., Duan Y., Lin Z., *Biotechnol. Bioeng.*, **110** (2013a), 2815–2825.

Bell S. G., Chen X., Sowden R. J., Xu F., Williams J. N., WongL. L., Rao Z., *J. Am. Chem. Soc.*, **125** (2003), 705–714.

Bell S. G., Stevenson J. A., Boyd A. D., Campbell S., Riddle A. D., Orton E. L., Wong L. L., *Chem. Commun. (Camb),* **5** (2002), 490–491.

Bleif S., Hannemann F., Zapp J., Hartmann D., Jauch J., Bernhardt R., *Appl. Microbiol. Biotechnol.,* **93** (2012), 1135–1146.

Brill E., Hannemann F., Zapp J., Bruning G., Jauch J., Bernhardt R., *Appl. Microbiol. Biotechnol.*, **98** (2014), 1701–1717.

Brodie A. M. H., J. *Steroid Biochem. Mol. Biol.*, **49** (1994), 281–287.

Burkina V., Rasmussen M. K., Pilipenko N., Zamaratskaia G., *Toxicology,* **375** (2017), 10–27.

Butler C. F., Peet C., Mason A. E., Voice M. W., Leys D., Munro A. W., *J. Biol. Chem.*, **288** (2013), 25387–25399.

Butler C. F., Peet C., McLean K. J., Baynham M. T., Blankley R. T., Fisher K., Rigby S. E., Leys D., Voice M. W., Munro A. W., *Biochem. J.*, **460** (2014), 247–259.

Capoferri L., Leth R., ter Haar E., Mohanty A. K., Grootenhuis P. D., Vottero E., Commandeur J. N., Vermeulen N. P., Jorgensen F. S., Olsen L., Geerke D. P., *Proteins*, **84** (2016), 383–396.

Cha G. S., Ryu S. H., Ahn T., Yun C. H., *Biotechnol. Lett.*, **36**, (2014), 2501–2506.

Chefson A., Auclair K., *Mol. Biosyst.*, **2** (2006), 462–469.

Choi S., Han S., Lee H., ChunY. J., Kim D., *Biomol. Ther. (Seoul)*, **21** (2013), 487–492.

Chun Y. J., Kim S., Kim D., Lee S. K., Guengerich F. P., *Cancer Res.*, **61** (2001), 8164–8170.

Cornelissen S., Julsing M. K., Volmer J., Riechert O., Schmid A., Buhler B., *Biotechnol. Bioeng.*, **110** (2013), 1282–1292.

Denisov I. G., Sligar S. G., *Biochim. Biophys. Acta*, **1814** (2011), 223–229.

Dezvarei S., Lee J. H. Z., Bell S. G., *Enzyme Microb. Technol.*, **111** (2018b), 29–37.

Dezvarei S., Onoda H., Shoji O., Watanabe Y., Bell S. G., *J. Inorg. Biochem.*, **183** (2018a), 137–145.

Di Nardo G., Fantuzzi A., Sideri A., Panicco P., Sassone C., Giunta C., Gilardi G., *JBIC J. Biol. Inorg. Chem.*, **12** (2007), 313–323.

Eiben S., Bartelmas H., Urlacher V. B., *Appl. Microbiol. Biotechnol.*, **75** (2007), 1055–1061.

Farinas Edgardo T., Schwaneberg U., Glieder A., Arnold Frances H., *Adv. Synth. Catal.*, **343** (2001), 601–606.

FDA, Center for Drug Evaluation and Research (CDER), "Safety Testing of Drug Metabolites Guidance for Industry" (2016).

Fisher C. W., Shet M. S., Caudle D. L., Martin-Wixtrom C. A., Estabrook R. W., *Proc. Natl. Acad. Sci. U. S. A.*, **89** (1992), 10817–10821.

Geisler K., Hughes R. K., Sainsbury F., Lomonossoff G. P., Rejzek M., Fairhurst S., Olsen C. E., Motawia M. S., Melton R. E., Hemmings A. M., Bak S., Osbourn A., *Proc. Natl. Acad. Sci. U. S. A.*, **110** (2013), E3360–E3367.

Geronimo I., Denning C. A., Rogers W. E., Othman T., Huxford T., Heidary D. K., Glazer E. C., Payne C. M., *Biochemistry*, **55** (2016), 3594–3606.

Gillam E. M., *Arch. Biochem. Biophys.*, **464** (2007), 176–186.

Gillam E. M., *Chem. Res. Toxicol.*, **21** (2008), 220–231.

Girvan H. M., Munro A. W., *Curr. Opin. Chem. Biol.*, **31** (2016), 136–145.

Grunwald P., *Industrial Biocatalysis*, Pan Stanford Publishing (2014).

Guengerich F. P., In: Schenkman J. B., Greim H. (eds.), *Cytochrome P450. Handbook of Experimental Pharmacology*, **105** (1993), Springer, Berlin, Heidelberg.

Guengerich F. P., *Nat. Rev. Drug Discov.*, **1** (2002), 359–366.

Hamberger B., Bak S., *Philos. Trans. R. Soc. Lond B Biol. Sci.*, **368** (2013), 20120426.

Hayashi K., Sugimoto H., Shinkyo R., Yamada M., Ikeda S., Ikushiro S., Kamakura M., Shiro Y., Sakaki T., *Biochemistry*, **47** (2008), 11964–11972.

Hayashi K., Yasuda K., Yogo Y., Takita T., Yasukawa K., Ohta M., Kamakura M., Ikushiro S., Sakaki T., *Biochem. Biophys. Res. Commun.*, **473** (2016), 853–858.

Hoffmann S. M., Danesh-Azari H.-R., Spandolf C., Weissenborn M. J., Grogan G., Hauer B., *ChemCatChem*, **8** (2016), 3234–3239.

Honda Malca, S., Scheps D., Kuhnel L., Venegas-Venegas E., Seifert A., Nestl B. M., Hauer B., *Chem. Commun. (Camb)*, **48** (2012), 5115–5117.

Huang W., Johnston W. A., Hayes M. A., De Voss J. J., Gillam E. M., *Arch. Biochem. Biophys.*, **467** (2007), 193–205.

Huisman G. W., Collier S. J., *Curr. Opin. Chem. Biol.*, **17** (2013), 284–292.

Huo L., Liu J., Dearing M. D., Szklarz G. D., Halpert J. R., Wilderman P. R., *Biochemistry*, **56** (2017), 2238–2246.

Jiang H., Wu J., Zhang F., Wen J., Jiang J., Deng Y., *Biochem. Pharmacol.*, **156** (2018), 177–185.

Jung E., Park B. G., Ahsan M. M., Kim J., Yun H., Choi K. Y., Kim B. G., *Appl. Microbiol. Biotechnol.*, **100** (2016), 10375–10384.

Kandel S. E., Wienkers L. C., Lampe J. N., *Annu. Rep. Med. Chem.*, **49** (2014), 347–359.

Kumar S., *Expert Opin. Drug Metab. Toxicol.*, **6** (2010), 115–131.

Kumar S., Chen C. S., Waxman D. J., Halpert J. R., *J. Biol. Chem.*, **280** (2005), 19569–19575.

Kumar S., Halpert J. R., *Biochem. Biophys. Res. Commun.*, **338** (2005), 456–464.

Kumar S., Scott E. E., Liu H., Halpert J. R., *J. Biol. Chem.*, **278** (2003), 17178–17184.

Kumar S., Zhao Y., Sun L., Negi S. S., Halpert J. R., Muralidhara B. K., *Mol. Pharmacol.*, **72** (2007), 1191–1199.

Lamb S. B., Lamb D. C., Kelly S. L., Stuckey D. C., *FEBS Lett.*, **431** (1998), 343–346.

Landwehr M., Hochrein L., Otey C. R., Kasrayan A., Backvall J. E., Arnold F. H., *J. Am. Chem. Soc.*, **128** (2006), 6058–6059.

Langouet S., Coles B., Morel F., Becquemont L., Beaune P., Guengerich F. P., Ketterer B., Guillouzo A., *Cancer Res.*, **55** (1995), 5574–5579.

Le-Huu P., Heidt T., Claasen B., Laschat S., Urlacher V. B., *ACS Catalysis*, **5** (2015), 1772–1780.

Lee G.-Y., Kim D.-H., Kim D., Ahn T., Yun C.-H., *Arch. Pharm. Res.*, **38** (2015), 98–107.

Lepesheva G. I., Hargrove T. Y., Rachakonda G., Wawrzak Z., Pomel P., Cojean S., Nde P. N., Nes W. D., Locuson C. W., Calcutt M. W., Waterman M. R., Daniels J. S., Loiseau P. M., Villalta F., *J. Infect. Dis.* **212**, (2015), 1439–1448.

Liao W., Zhao S., Zhang M., Dong K., Chen Y., Fu C., Yu L., *Front. Plant Sci.*, **8** (2017), 1468.

Lim J. B., Sikes H. D., *Protein Eng. Des. Sel.*, **2** 8(2015), 79–83.

Luo D., Callari R., Hamberger B., Wubshet S. G., Nielsen M. T., Andersen-Ranberg J., Hallstrom B. M., Cozzi F., Heider H., Lindberg Moller B., Staerk D., Hamberger B., *Proc. Natl. Acad. Sci. U. S. A.*, **113** (2016), E5082–E5089.

Mackenzie P. I., Somogyi A. A., Miners J. O., *Pharmacol. Res.*, **116** (2017), 7–19.

Martinez C. A., Rupasinghe S. G., *Curr. Top. Med. Chem.*, **13** (2013), 1470–1490.

McFadyen M. C., Melvin W. T., Murray G. I., *Mol. Cancer Ther.*, **3** (2004), 363–371.

McLean K. J., Hans M., Meijrink B., van Scheppingen W. B., Vollebregt A., Tee K. L., van der Laan J. M., Leys D., Munro A. W., van den Berg M. A., *Proc. Natl. Acad. Sci. U. S. A.*, **112** (2015), 2847–2852.

Miura Y., Fulco A. J., *Biochim. Biophys. Acta*, **388** (1975), 305–317.

Moody S. C., Loveridge E. J., *J. Appl. Microbiol.*, **117** (2014), 1549–1563.

Mueller S. C., Majcher-Peszynska J., Uehleke B., Klammt S., Mundkowski R. G., Miekisch W., Sievers H., Bauer S., Frank B., Kundt G., Drewelow B., *Eur. J. Clin. Pharmacol.*, **62** (2006), 29–36.

Muller C. S., Knehans T., Davydov D. R., Bounds P. L., von Mandach U., Halpert J. R., Caflisch A., Koppenol W. H., *Biochemistry*, **54** (2015), 711–721.

Munro A. W., Girvan H. M., Mason A. E., Dunford A. J., McLean K. J., *Trends Biochem. Sci.*, **38** (2013), 140–150.

Narhi L. O., Fulco A. J., *J. Biol. Chem.*, **261** (1986), 7160–7169.

Nelson D. R., *Hum. Genomics*, **4** (2009), 59–65.

Nguyen K. T., Virus C., Gunnewich N., Hannemann F., Bernhardt R., *ChemBioChem*, **13** (2012), 1161–1166.

Nicoli R., Bartolini M., Rudaz S., Andrisano V., Veuthey J. L., *J. Chromatogr. A*, **1206** (2008), 2–10.

Nikolaus J., Nguyen K. T., Virus C., Riehm J. L., HutterM., Bernhardt R., *Steroids*, **120** (2017), 41–48.

Noble M. A., Miles C. S., Chapman S. K., Lysek D. A., MacKay A. C., Reid G. A., Hanzlik R. P., Munro A. W., *Biochem. J.*, **339** (1999), 371–379.

Notonier S., Alexander M., Jayakody L. N., *Enz.* Eng., **5** (2016), 152.

Notonier S., Alexander M., Jayakody L. N., *Enz. Eng.*, **5** (2016a), 152.

Notonier S., Gricman L., Pleiss J., Hauer B., *ChemBioChem*, **17**(2016b), 1550–1557.

Otey C. R., Bandara G., Lalonde J., Takahashi K., Arnold F. H., *Biotechnol. Bioeng.*, **93** (2006), 494–499.

Paddon C. J., Westfall P., Pitera D. J., Benjamin K., Fisher K., McPhee D., Leavell M. D., Tai A., Main A., Eng D., Polichuk D. R., Teoh K. H., Reed D. W., et al., *Nature*, **496** (2013), 528–532.

Peter Guengerich F., Chun Y. J., Kim D., Gillam E. M., Shimada T., *Mutat. Res.*, **523–524** (2003), 173–182.

Peterson D. H., Murray H. C., Eppstein S. H., Reineke L. M., Weintraub A., Meister P. D., Leigh H. M., *J. Am. Chem. Soc.*, **74** (1952), 5933–5936.

Preissner S. C., Hoffmann M. F., Preissner R., Dunkel M., Gewiess A., Preissner S., *PLoS ONE*, **8** (2013), e82562.

Rauschenbach R., Isernhagen M., Noeske-Jungblut C., Boidol W., Siewert G., *Mol. Gen. Genet.*, **241** (1993), 170–176.

Ren X., Yorke J. A., Taylor E., Zhang T., ZhouW., Wong L. L., *Chemistry*, **21** (2015), 15039–15047.

Renault, H., Bassard J. E., Hamberger B., Werck-Reichhart D., *Curr. Opin. Plant. Biol.*, **19** (2014), 27–34.

Sakaki T., Shibata M., Yabusaki Y., Murakami H., Ohkawa H., *DNA Cell Biol.*, **9** (1990), 603–614.

Sawada N., Sakaki T., Yoneda S., Kusudo T., Shinkyo R., Ohta M., Inouye K., *Biochem. Biophys. Res. Commun.*, **320** (2004), 156–164.

Sawayama A. M., Chen M. M., Kulanthaivel P., Kuo M. S., Hemmerle H., Arnold F. H., *Chemistry*, **15** (2009), 11723–11729.

Schiavini P., Cheong K. J., Moitessier N., Auclair K., *ChemBioChem*, **18** (2017), 248–252.

Schmitz D., Janocha S., Kiss F. M., Bernhardt R., *Biochim. Biophys. Acta Proteins Proteom*, **1866** (2018), 11–22.

Schmitz D., Zapp J., Bernhardt R., *Microb. Cell Fact.*, **13** (2014), 81.

Schuckel J., Rylott E. L., Grogan G., Bruce N. C., *ChemBioChem*,**13** (2012), 2758–2763.

Sedlaczek L., Smith L. L., *Crit. Rev. Biotechnol.*, **7** (1988), 187–236.

Shafiee A., Hutchinson C. R., *J. Bacteriol.*, **170** (1988), 1548–1553.

Shen B., Hutchinson C. R., *Biochemistry*, **32** (1993), 6656–6663.

Shen B., Hutchinson C. R., *J. Biol. Chem.*, 269(48) (1994), 30726–30733.

Shoji O., Fujishiro T., Nagano S., Tanaka S., Hirose T., Shiro Y., Watanabe Y., *J. Biol. Inorg. Chem.*, **15** (2010), 1331–1339.

Sonomoto K., Hoq M. M., Tanaka A., Fukui S., *Appl. Environ. Microbiol.*, **45** (1983), 436–443.

Subhani S., Jamil K., *Biomed. Pharmacother.*, **73** (2015), 65–74.

Talakad J. C., Wilderman P. R., Davydov D. R., Kumar S., Halpert J. R., *Arch. Biochem. Biophys.*, **494** (2010), 151–158.

Tang L., Shah S., Chung L., Carney J., Katz L., Khosla C., Julien B., *Science*, **287** (2000), 640–642.

Uno T., Izumi C., Takenaka S., Yanase T., Imaishi H., Kanamaru K., Yamagata H., Kaminishi Y., Itakura T., *Environ. Toxicol. Pharmacol.*, **40** (2015), 360–368.

Urlacher V. B., Eiben S., *Trends Biotechnol.*, **24** (2006), 324–330.

Urlacher V. B., Girhard M., *Trends Biotechnol.*, **30** (2012), 26–36.

Vail R. B., Homann M. J., Hanna I., Zaks A., *J. Industr. Microbiol. Biotechnol.*, **32** (2005), 67–74.

van Vugt-Lussenburg B. M., Stjernschantz E., Lastdrager J., Oostenbrink C., Vermeulen N. P., Commandeur J. N., *J. Med. Chem.*, **50** (2007), 455–461.

Venkataraman H., Verkade-Vreeker M. C., Capoferri L., Geerke D. P., Vermeulen N. P., Commandeur J. N., *Bioorg. Med. Chem.*, **22** (2014), 5613–5620.

Virus C., Bernhardt R., *Lipids*, **43** (2008), 1133–1141.

Vredenburg G., den Braver-Sewradj S., van Vugt-Lussenburg B. M., Vermeulen N. P., Commandeur J. N., Vos J. C., *Toxicol. Lett.*, **232** (2015), 182–192.

Wang Z. J., Renata H., Peck N. E., Farwell C. C., Coelho P. S., Arnold F. H., *Angew. Chem. Int. Ed.*, **53** (2014), 6810–6813.

Watanabe I., Nara F., Serizawa N., *Gene*, **163** (1995), 81–85.

Weber J. M., Leung J. O., Swanson S. J., Idler K. B., McAlpine J. B., *Science*, **252** (1991), 114–117.

Whitehouse C. J., Bell S. G., Wong L. L., *Chem. Soc. Rev.*, **41** (2012), 1218–1260.

Whitehouse C. J., Bell S. G., Yang W., Yorke J. A., Blanford C. F., Strong A. J., Morse E. J., Bartlam M., Rao Z., Wong L. L., *ChemBioChem*, **10** (2009), 1654–1656.

Xu F., Bell S. G., Lednik J., Insley A., Rao Z., Wong L. L., *Angew. Chem. Int. Ed.*, **44** (2005), 4029–4032.

Yasuda K., Sugimoto H., Hayashi K., Takita T., Yasukawa K., Ohta M., Kamakura M., Ikushiro S., Shiro Y., Sakaki T., *Biochim. Biophys. Acta Proteins Proteom*, **1866** (2018), 23–31.

Yasuda K., Yogo Y., Sugimoto H., Mano H., Takita T., Ohta M., Kamakura M., Ikushiro S., Yasukawa K., Shiro Y., Sakaki T., *Biochem. Biophys. Res. Commun.*, **486** (2017), 336–341.

Yoshida Y., *Biol. Pharm. Bull.*, **35** (2012), 811.

Zanger U. M., Schwab M., *Pharmacol. Ther.*, **138** (2013), 103–141.

Zhang H., Yokom A. L., Cheng S., Su M., Hollenberg P. F., Southworth D. R., Osawa Y., *J. Biol. Chem.*, **293** (2018), 7727–7736.

Zhu J., Lu J., Zhou Y., Li Y., Cheng J., Zheng C., *Bioorg. Med. Chem. Lett.*, **16** (2006), 5285–5289.

Chapter 15

Cytochromes P450, Cardiovascular Homeostasis and Disease

Chin Eng Ong,[a] Amelia Dong,[b] Boon Hooi Tan,[c] and Yan Pan[d]

[a]*School of Pharmacy, International Medical University, 126, Jalan Jalil Perkasa 19, Bukit Jalil, 57000 Kuala Lumpur, Malaysia*
[b]*School of Pharmacy, Monash University Malaysia, Jalan Lagoon Selatan, 47500 Bandar Sunway, Selangor, Malaysia*
[c]*School of Health Sciences, International Medical University, 126, Jalan Jalil Perkasa 19, Bukit Jalil, 57000 Kuala Lumpur, Malaysia*
[d]*Department of Biomedical Science, The University of Nottingham Malaysia Campus, Jalan Broga, 43500 Semenyih, Selangor, Malaysia*

ceong98@hotmail.com

15.1 Introduction

Cytochrome P450 (CYP) superfamily is one group of important enzymes that catalyze phase I oxidations of numerous endogenous compounds and xenobiotics, including commonly used drugs (Di, 2014). Endogenous compounds metabolized by CYPs include steroids, bile acids, fatty acids, fat-soluble vitamins, and many other

Pharmaceutical Biocatalysis: Fundamentals, Enzyme Inhibitors, and Enzymes in Health and Diseases
Edited by Peter Grunwald

ISBN 978-981-4800-61-7 (Hardcover), 978-0-429-29503-4 (eBook)
www.jennystanford.com

lipophilic compounds. While their role in xenobiotic metabolism is well established, the role of CYP enzymes in extra-hepatic tissues including heart, kidneys, and brain in endogenous tissue function has emerged only in recent decades (Nebert et al., 2013). These enzymes are now known to play an important role in the homeostasis of a number of organ systems including blood, cardiovascular, renal, and nervous systems.

This chapter provides an overview of the role of CYP enzymes in organ homeostasis with special focus on vasculature and cardiovascular systems. The CYP eicosanoid metabolism pathways will first be described followed by a description of the role of CYP enzymes (and their genetic polymorphisms) in vascular biology and pathophysiology. Therapeutic strategies aimed at preventing or reducing the mortality resulting from vascular and cardiovascular diseases will be reviewed in the final part of the chapter.

15.2 CYP-Mediated Biosynthetic Pathways of Arachidonic Acid

CYP enzymes play crucial role in eicosanoid metabolism. CYP-derived eicosanoid metabolites are now well known to play important blood and cardiovascular functions. Epoxyeicosatrienoic acids (EETs) and hydroxyeicosatetraenoic acids (HETEs) play a critical role in the maintenance of cardiovascular biological function, while two prostanoids, thromboxane A_2 (TxA_2) and prostacyclin (PGI_2), are crucial for atherogenesis.

All eicosanoid metabolites are derived from arachidonic acid (AA), a 20-carbon, omega-6 polyunsaturated fatty acid (PUFA) released from the cell membrane phospholipids by the cytosolic enzyme phospholipase A_2 (Sudhahar et al., 2010). Cell injury, shear stress, and hormones such as bradykinin, acetylcholine, angiotensin II, and adenosine are among the phospholipase activators that initiate eicosanoid synthesis. Figure 15.1 depicts three major pathways involved in eicosanoid synthesis: the cyclooxygenase (COX) pathway, which generates prostanoids; lipoxygenase (LOX) pathway, which produces leukotrienes and HETEs; and lastly, CYP pathways (CYP epoxygenase and CYP ω-hydroxylase) that generate EETs (Oni-Orisan et al., 2014;

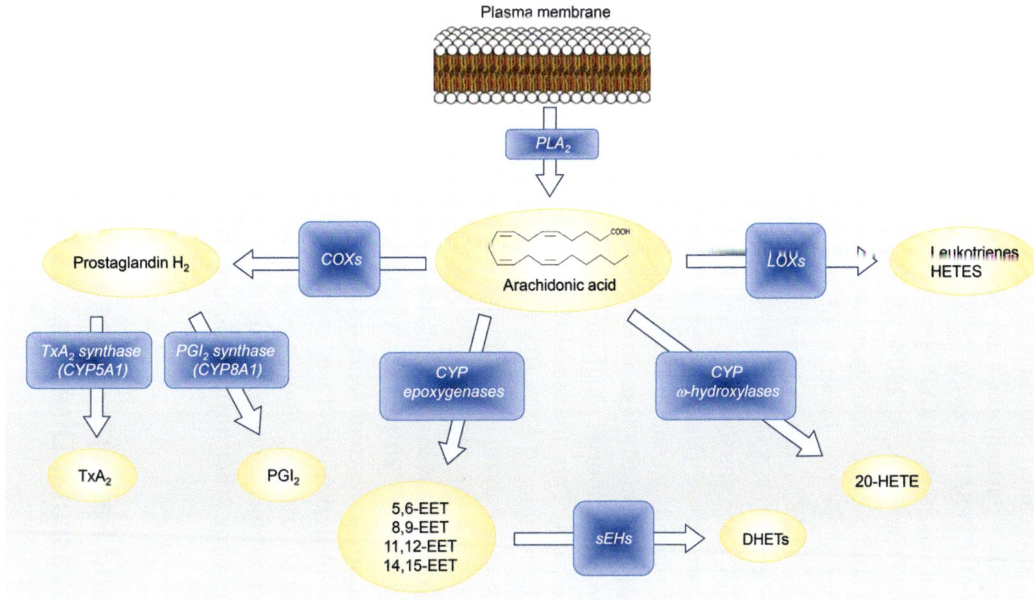

Figure 15.1 The eicosanoid biosynthetic pathways. Arachidonic acid (AA) is a polyunsaturated omega-6 fatty acid which is released from the membrane phospholipids by the activity of phospholipases such as cytosolic phospholipase A_2 (PLA_2). Free AA can be metabolized to eicosanoids through three major pathways: the cyclooxygenase (COX) pathway, the lipoxygenase (LOX) pathway, and the cytochrome P450 (CYP) epoxygenase and ω-hydroxylase pathways. In COX pathway, thromboxane A_2 (TxA_2) and prostacyclin (PGI_2) are further generated by the TxA_2 synthase (CYP5A1) and PGI_2 synthase (CYP8A1) respectively. In the CYP pathways, AA is converted to epoxyeicosatrienoic acids (EETs) and 20-hydroxyeicosatetraenoic acid (20-HETE) by CYP epoxygenases and CYP ω-hydroxylases, respectively. All EETs are then further metabolized by soluble epoxide hydrolase (sEH) forming the less active dihydroxyeicosatrienoic acids (DHETs).

Wang and Dubois, 2012). Two CYP isoforms, CYP5A1 and CYP8A1, are involved in generation of prostanoids in the COX pathway. More popularly known as thromboxane A_2 synthase (Wang and Kulmacz, 2002), CYP5A1 is involved in platelet aggregation, whereas CYP8A1, also called prostacyclin (PGI_2) synthase (Ullrich and Hecker, 1990), participates in platelet disaggregation. Thus, CYP5A1 and CYP8A1 provide a Yin-Yang mechanism for regulation of blood coagulation. CYP epoxygenases are involved in production of EETs (Spector et al., 2004) through olefin epoxidation with both CYP2J and CYP2C epoxygenases generating four EET regioisomers defined by the location of the appended oxygen atom (5,6-, 8,9-, 11,12-, and 14,15-EETs). Each regioisomer consists of a mixture of S/R and R/S enantiomers. CYP2J2, CYP2C8, and CYP2C9 are constitutively and abundantly expressed in the human heart tissue (DeLozier et al., 2007). All EETs are subsequently hydrolyzed by soluble epoxide hydrolase (sEH) (Wang and Dubois, 2012) to form the less active dihydroxyeicosatrienoic acids (DHETs) with a shorter half-life, effectively terminating their biological effects. This makes sEH a target for increasing and prolonging the actions of EETs (Harris and Hammock, 2013). CYP ω-hydroxylases catalyze ω-terminal hydroxylation to form C16–C20 alcohols of AA (16-, 17-, 18-, 19-, and 20-HETEs). Among the different ω-hydroxylases available, CYP4A and CYP4F enzymes are involved in generating 20-HETE, a vasoactive and natriuretic eicosanoid important for vascular function (Fan et al., 2016).

In addition to AA-derived products, members from the omega-6 PUFA family (most notably linoleic acid) and the omega-3 PUFA family, such as docosahexaenoic acid and eicosapentaenoic acid, play a role in many other physiological processes (Baum et al., 2012).

15.3 CYPs and Pathophysiology of Cardiovascular Diseases

15.3.1 Atherothrombosis

Atherosclerosis is a major cause of morbidity and mortality that leads to coronary artery and cerebrovascular diseases (Rosamond et al., 2008). Atherogenesis involves chronic and progressive vascular inflammation accompanied by complicated pathological events including endothelial dysfunction, subendothelial penetration and oxidation of plasma lipids, exposure of adhesion and chemotactic molecules on the endothelial surface, recruitment of leucocytes, platelet activation, leading eventually to formation of arteriosclerotic lesions (Vanhoutte, 2009). The bioactive eicosanoids are essential in modulating atherosclerotic processes, and among these lipids, TxA_2 and PGI_2 and their receptors play a central role (Praticò and Dogné, 2009). A main feature of endothelial dysfunction and inflammation is the increased production of these prostanoids in endothelial cells and activated platelets.

As illustrated in Fig. 15.1, TxA_2 and PGI_2 are produced from AA by the action of thromboxane A_2 and PGI_2 synthase (or CYP5A1 and CYP8A1), respectively in the COX pathway. TxA_2 and PGI_2 act through binding to specific G protein-coupled receptors known as TP and IP receptors, respectively. Both are membrane bound, seven transmembrane receptors found abundantly in the vasculature and cardiovascular tissues (Huang et al., 2004). TxA_2 is mainly produced in blood platelets but can also be synthesized by monocytes and vascular smooth muscle cells (Dogne et al., 2004). It serves as a strong activator of platelets, a vasoconstrictor and smooth muscle cell mitogen. PGI_2, generated predominantly in vascular endothelial cells, acts as a biological opposite to TxA_2 and has potent vasodilating and platelet-inhibitory properties (de Leval et al., 2004). Accumulating evidence demonstrates alteration of their formation, enzymatic machinery, and receptors in atherosclerotic vascular events. For instance, TP receptor expression was found to significantly increase in the aorta and epicardial coronary arteries from ischemic atherosclerotic vessels (Katugampola and Davenport, 2001). TxA_2, acting through TP receptors, triggers various endothelial and inflammatory responses, including enhanced expression of adhesion molecules (e.g., P-selectin, intracellular adhesion molecule [ICAM-1], vascular cell adhesion molecule [VCAM]), vascular smooth muscle contraction and hypertrophy, homing, internalization of monocytes and activation to macrophages, release of inflammatory mediators, matrix metalloproteinases (MMPs), and various tissue factors that collectively form subendothelial necrotic lipid core, finally resulting in advanced atherosclerotic lesions (Praticò and Dogné, 2009). In addition to endothelial dysfunction, platelet activation significantly contributes to plaque formation and atherogenesis (Kleiman, 2004). Platelets increase the expression of adhesion molecules, further providing the reactive surface for chemotaxis of lymphocytes. They also stimulate production of growth factors (i.e., platelet-derived growth factor), and proinflammatory cytokines and chemokines that perpetuate the inflammatory milieu within the plaque.

Due to biological opposite nature of TxA_2 and PGI_2, the balance and regulation of these two mediators is important in both healthy and diseased vasculature. PGI_2 modulates in vivo platelet-vascular interactions in an opposite manner to that of TxA_2 (Praticò and Dogné, 2009). Factors that alter their biological activities would therefore cause changes in vascular function. A clear example is the Ghosal hemato-diaphyseal syndrome, a rare autosomal recessive trait, where the *CYP5A1* gene is mutated (Geneviève et al., 2008). The dysfunction of the gene impacts negatively on platelet aggregation and atherothrombosis progression. In addition to altered platelet aggregation, patients with this syndrome also demonstrate altered bone density, progressive sclerosing dysplasia in the midsection of long bones, defective hematopoiesis and refractory anemia, all of which are associated with hematologic and vascular dysfunction. Mutations in *CYP8A1*gene, on the other hand, are associated with essential hypertension and cerebral infarction (Nakayama et al., 2000; Nakayama et al., 2002). Japanese families with *CYP8A1* polymorphism were found to display a history of hypertension and cerebral infarction, suggesting that abnormal CYP8A1 activity may lead to defective vasculature and platelet function. Another study demonstrated that C1117A polymorphism in *CYP8A1* was linked to development of left main coronary artery disease, a rare but hazardous form of coronary artery disease (Bousoula et al., 2012).

Similar to TxA_2 and PGI_2, other eicosanoid metabolites including EETs and 20-HETE are also critical in atherogenesis and pathophysiology of cardiovascular diseases. Their role is further described below.

15.3.2 Coronary Artery Disease

The most susceptible targets for formation of atheromatous plaques are the aorta and the cerebral and coronary arteries. Coronary artery disease (CAD), is a result of plaque formation in the coronary arteries. More serious consequences of CAD may include myocardial infarction and brain infarction (stroke), depending on the location of the arterial lesions. Different CYP isoforms are known to play a role in the pathophysiology CAD. CYP epoxygenases, in particular, members from the CYP2C and 2J subfamilies, are mainly involved in converting AA into EETs in cardiovascular tissues. EETs exert a number of cardioprotective effects, including vasodilatory, anti-inflammatory, antiapoptotic, thrombolytic and natriuretic actions (Deng et al., 2010).

The impaired endothelial function and dysregulated coronary blood flow in CAD eventually contribute to the onset and progression of atherosclerosis and thrombosis (Kinlay et al., 2001). Substantial evidence has demonstrated that CYP-generated EETs serve as endothelial-derived hyperpolarizing factor (EDHF) in the coronary circulation (Campbell et al., 1996). EDHF activates the large-conductance calcium-activated potassium (BK) channels and the Na^+-K^+-ATPase, and therefore induces hyperpolarization of vascular smooth muscle cells. This results in smooth muscle relaxation, therefore reducing contractile force and oxygen utilization of the muscle, leading to vasodilation (Fleming, 2014; Schinzari et al., 2017). Plasma concentrations of EETs have been demonstrated to be elevated in patients with CAD compared to healthy controls (Theken et al., 2012). It has been suggested that EETs may become up-regulated to compensate for the overall endothelial dysfunction. The key role of CYPs and EDHF is further supported from their expression modulation by the dihydropyridine calcium channel blocker nifedipine. This drug was able to up-regulate expression of CYP2C, but not 2J or 2B, in coronary endothelium. Furthermore, it stimulated EET biosynthesis and potentiated hyperpolarization of coronary artery smooth muscle cells mediated through bradykinin and EDHF (Fisslthaler et al., 2000). The cardioprotective role of EETs has also been demonstrated in transgenic animal model. Transgenic mice with selective overexpression of CYP 2J2 in cardiomyocyte exhibited better post-ischemic recovery of left ventricular function after prolonged ischemia. This protective effect was however abolished in the presence of the CYP epoxygenase inhibitor *N*-methylsulfonyl-6-(2-proparglyloxphenyl) hexamide (Seubert et al., 2004). This was further supported by studies in both adult mongrel dogs and rats (Gross et al., 2011) using the same inhibitor. These findings are in agreement with the reduced formation of atherosclerotic injury in apoplipoprotein E knockout mice observed following sEH inhibition (Ulu et al., 2008). Some other studies demonstrated beneficial effects of EETs on long-term endpoints post-infarction. A study using male rats treated with an sHE inhibitor, GSK2188931B (*N*-(4-bromo-2-[(trifluoromethyl)oxy]phenyl)methyl)-1-[4-methyl-6-(methylamino)-1,3,5-triazin-2-yl]-4 piperidine carboxamide) immediately following permanent ligation of the left anterior descending coronary artery exhibited anti-remodeling effects such as reduced hypertrophy, lower inflammation, decreased ventricular fibrosis and sustained systolic function 5 weeks post-infarction (Kompa et al., 2013). More recently, in vivo studies demonstrated that a selective sHE inhibitor, 4-[[trans-4 [[(tricyclo [3.3.1.13,7]dec-1-ylamino)carbonyl] amino] cyclohexyl]oxy]-benzoic acid (t-AUCB) conferred protection against ischemic injury by preserving cardiomyocyte function and maintaining mitochondrial efficiency (Akhnokh et al., 2016). In another study, improved coronary endothelial function and attenuated cardiac remodeling and diastolic dysfunction were observed in obese insulin-resistant mice following treatment with t-AUCB (Roche et al., 2015).

The role of 20-HETE is not as established in the case of CAD. No difference in 20-HETE levels was observed in patients with CAD compared to healthy controls (Rosamond et al., 2008). This suggests that the CYP-mediated eicosanoid metabolism is dysregulated in some patients with CAD, and demonstrates that the biomarkers of CYP epoxygenase and sEH, but not CYP ω-hydroxylase metabolism are altered in patients with CAD.

EETs also play a key role in ischemia-reperfusion injury (IRI), a condition of exacerbated myocardial damage induced by tissue reperfusion and blood flow restoration after an infarction event (Yellon and Hausenloy, 2007). Studies in recent years demonstrate the protective role of EETs on the heart from IRI (see Fig. 15.2). EETs, when injected directly to the coronary circulation, significantly decreased infarct size and myocardial remodeling in animal IRI models (Gross et al., 2007). Another study showed that EETs or a dual-acting compound possessing EET mimetic and sEH inhibitory properties was able to reduce infarct size and preserve left ventricular developed pressure (LVDP) when administered to isolated mouse hearts (Batchu et al., 2011). The attenuation of deleterious electrocardiographic changes (such as ST elevation and QT prolongation) and mitochondrial damage (increased mitochondrial T-tube fragmentation and swelling) were also noted following ischemia-reperfusion of the isolated perfused heart (Batchu et al., 2009; Katragadda et al., 2009). Acutely, EETs attenuate apoptosis, preserve mitochondrial structure and function, and promote pro-survival signaling. Chronically, EETs are able to prevent adverse remodeling and help maintain systolic function. These protective effects were blocked by the EET antagonist 14,15-epoxyeicosa-5(Z)-enoic acid. Multiple pathways have been associated with cardioprotective effects of EETs including those of sarcolemmal and mitochondrial potassium channels, p42/p44 mitogen-activated protein kinase (MAPK), and phosphoinositide-3-kinase–protein kinase B (PI3K-AKT) signaling (Katragadda et al., 2009; Seubert et al., 2004), all of which are important components of cardioprotective mechanisms.

Recent evidence points to a different role of CYP2 isoforms in IRI of the heart. While over-expression of CYP2J2 and high EET level improve the functional recovery from IRI, CYP2C9 and CYP2C8, have been linked to poor outcome of ischemia-reperfusion (Doggrell, 2004). Both CYP2C isoforms have been demonstrated to generate reactive oxygen species (ROS), which induce deleterious effects, including mitochondrial damages and apoptosis, and accelerated ischemic damage in tissues (see Fig. 15.2). This can be explained by the fact that a CYP epoxygenase is not always just an epoxygenase because superoxide anions (O_2^-), hydrogen peroxide, and hydroxyl radicals can also be generated during the CYP catalysis when electrons for the reduction of the central heme iron are transferred on to the activated bound oxygen molecule (Puntarulo and Cederbaum, 1998). ROS serve as both a cause and an effect of endothelial dysfunction. In IRI, the combination of ROS, exacerbated inflammation, and insufficient response to

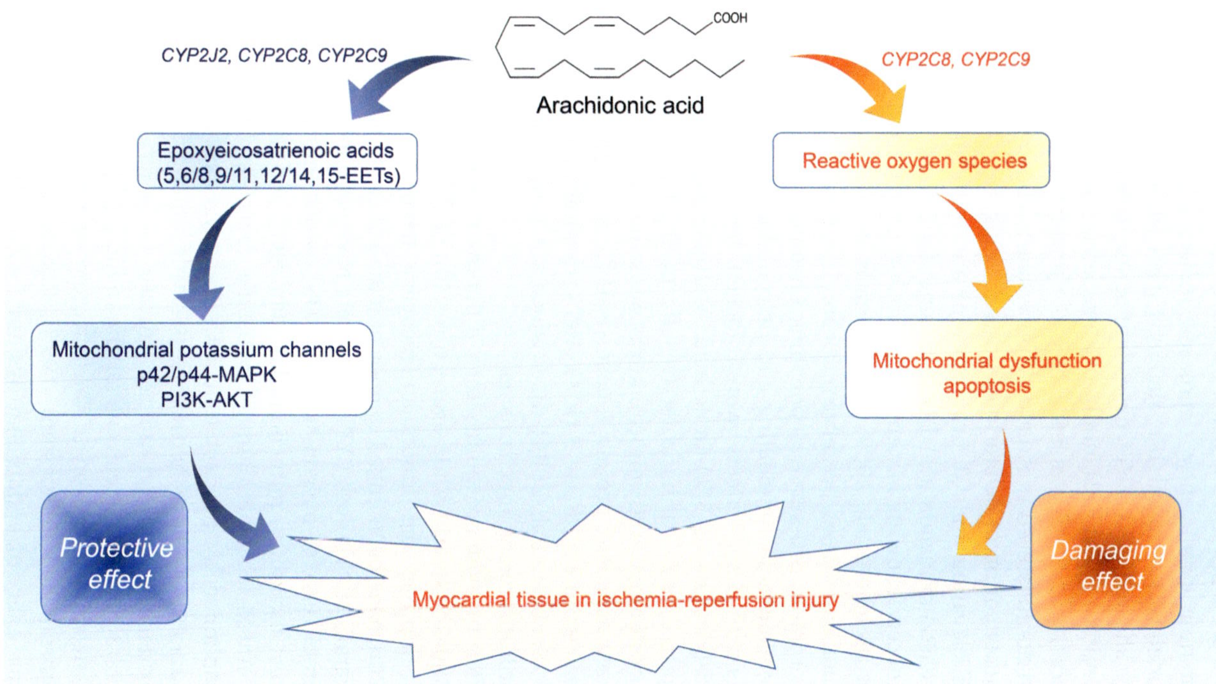

Figure 15.2 Role of CYP epoxygenases in ischemia-perfusion injury of human myocardial tissue.

load results in dilatation, hypertrophy, fibrosis and scar formation following coronary occlusion, all of which result in cardiac failure. Increased ROS formation in endothelial cells was associated with overexpression of human CYP2C9, and this was completely inhibited by sulfaphenazole, the CYP2C9 specific inhibitor (Fichtlscherer et al., 2004; Fleming et al., 2001). Moreover, in an in vivo rat IRI model, intravenous administration of sulfaphenazole was found to reduce the myocardial infarct size and improve the cardiac function during the time of reperfusion (Ishihara et al., 2009). In another study, mice with selective increased endothelial expression of CYP2C8 showed larger infarct size and decreased functional recovery after ischemia-reperfusion (Edin et al., 2011). The above data indicate that type of isoform is critical for ROS generation as well as EET production in myocardial tissue. The nature and severity of IRI would therefore depend on the relative activity and expression profile of the CYP2C and CYP2J isoforms in the affected vascular and cardiac tissues.

CYP genetic polymorphism is expected to have a pathological impact on CAD in view of its important role in eicosanoid metabolism. Patients who are carriers of CYP allelic variants would thus demonstrate altered ability to generate EETs, influencing their vulnerability to CAD (Oni-Orisan et al., 2014). Several association studies have provided proof for this vulnerability. *CYP2C9*2* and *CYP2C9*3*, were associated with the risk of acute myocardial infarction in female patients (Yasar et al., 2003). A higher risk of myocardial infarction was observed in both males and females who were *CYP2C8*3* carriers (Yasar et al., 2003). Furthermore, *CYP2C8*3* was found to increase the CAD incidence in smokers (Lee et al., 2007). This same study also demonstrated that the *CYP2J2* G-50T polymorphism variant-50T allele was associated with significantly lower risk of ischemic heart disease in African Americans. In another study, identification of a mutation in the proximal promoter of human CYP 2J2, *CYP2J2*7*, leading to reduced gene expression, was associated with an increased risk for CAD (Spiecker et al., 2004). In addition, *CYP2J2* polymorphism may be an independent risk factor for the premature onset (below 45 years old) of myocardial infarction in the Chinese Han population (Liu et al., 2006a), However, the positive findings above are discordant with more recent studies, which indicated no association for *CYP2C8* and *CYP2C9* (Haschke-

Becher et al., 2010; Kaur-Knudsen et al., 2009). Further studies that include more ethnic groups or additional independent population cohorts are necessary to clarify these associations. Careful consideration of the potential role of clinical and demographic confounders such as phenotypic definition, population stratification and ethnic admixture is also warranted in the design of such studies.

15.3.3 Hypertension

There is substantial evidence for the role of enzymes in the EET and HETE pathways, including CYPs and sEHs, in fluid and electrolyte homeostasis and blood pressure control. The best evidence was the report that male animals lacking sEH demonstrated high level of EETs, reduced renal production of DHETs (EET metabolites) and a significantly lower blood pressure (Sinal et al., 2000). In humans, blood EET level is reduced in women with pregnancy-induced hypertension (Catella et al., 1990) and in subjects with renovascular hypertension (Minuz et al., 2008). Moreover, significant drop in blood pressure was noted in different hypertensive models in rodents after the administration of a sEH inhibitor (Manhiani et al., 2009; Sinal et al., 2000). Substantial evidence in animals has also demonstrated that transgenic or overexpression of CYP epoxygenase lowered blood pressure in a number of hypertensive models. For example, recombinant adenovirus vector-mediated overexpression of CYP2J2 blocked blood pressure increase in spontaneously hypertensive rats (Xiao et al., 2010). In addition, the in vivo delivery of the homologous gene of *CYP2J2* expressed mainly in rats, *CYP2J3*, enhanced EET level and reduced blood pressure in fructose-treated rats (Xu et al., 2010). EETs can potentially affect blood pressure through various mechanisms. These include modulation of the expression and activity of ion channels directly involved in the vasculature function and regulation of renal sodium excretion such as inwardly rectifying K channels, BK channels, or epithelial sodium channels ($E_{Na}Cs$) (Wang et al., 2008). BK channels, when activated, have been reported to cause hyperpolarization and relaxation of vascular smooth muscle cells (Ye et al., 2002). In addition to modulating the basal blood pressure, EETs can act via the modulation of the renin-angiotensin system. For example, sEH inhibitors showed their efficacy in hypertension associated with activation of the renin-angiotensin system (Imig, 2012). This is related to angiotensin II being a potent vasoconstrictor and a powerful inducer of sEH expression in the rat vasculature (Ai et al., 2007) as well as the rat heart (Ai et al., 2009). Recently, direct administration of EET analogs has been shown to demonstrate cardiovascular therapeutic potential when administered in vivo. One such compound is an 11,12-EET analog that was shown to lower blood pressure in spontaneously hypertensive rats (Sudhahar et al., 2010). Another compound is disodium (*S*)-2-(13-(3-pentyl)ureido)-tridec-8(*Z*)-enamido) succinate (EET-A), which demonstrated a direct vasodilatory effect and attenuated the development of hypertension in angiotensin II-infused rats (Červenka et al., 2018; Sporkova et al., 2017). Furthermore, EET-A has proven its effectiveness in experimental angiotensin II-dependent malignant hypertension, and attenuation of end-organ damage as a result of elevated pressure (Jichova et al., 2016).

Contrary to the vasodilatory effect of EETs, 20-HETE plays a prohypertensive role. This is evidenced from increase in the blood pressure of animal models by infusion of the angiotensin II (Imig, 2013). It has been established that angiotensin II increases 20-HETE production in preglomerular microvessels and reduces EET levels by down regulating epoxygenases and enhancing their degradation by sHE (Alonso-Galicia et al., 2002; Croft et al., 2000). In studies with CYP4F2 transgenic mice, the renal overexpression of CYP4F2 resulted in high 20-HETE in urine and blood, and accounted for the elevation in blood pressure (Lai et al., 2012). More recently, human CYP4A11 transgenic mice were demonstrated to develop a 20-HETE-dependent hypertension associated with high sodium chloride cotransporter expression in the distal tubules (Savas et al., 2016). Release of 20-HETE has been shown to stimulate smooth muscle cell proliferation, migration, inflammation and angiogenesis, and to result in endothelial dysfunction (Cheng et al., 2010; Ishizuka et al., 2008). This is also consistent with finding that inhibiting either 20-HETE generation or angiotensin II receptors significantly ameliorated hypertension mediated by 20-HETE (Sodhi et al., 2010). This agrees with findings of another study that simultaneous inhibition of

20-HETE formation, using the CYP ω-hydrolylase inhibitor, *N*-methylsulfonyl-12,12-dibromododec-11-enamide (DDMS), and metabolism of EETs, using the sEH inhibitor *N*-cyclohexyl-*N*-dodecyl urea (NCND), attenuated cardiac hypertrophy in rats transgenic for the mouse Ren-2 renin gene, the Ren-2 transgenic rats, which is a unique angiotensin II-dependent hypertensive model (Certikova Chabova et al., 2010).

CYP2J2 studies best exemplify clinically relevant roles of CYP polymorphism in the development of hypertension. The presence of *CYP2J2*7*, a high-activity variant, is linked to lower incidence of hypertension in Caucasian males. Nonetheless, this study failed to demonstrate a protective effect of the variant in African-Americans (King et al., 2005). A more recent study demonstrated that *CYP2C8*3*, an allele associated with poor metabolizer phenotype, significantly increased risk for essential hypertension but not CAD and myocardial infarction (Tzveova et al., 2015). All in all, clinical observational studies to date have failed to demonstrate the clinical relevance of many other CYP polymorphisms in hypertension including the *CYP2C8*2*, *CYP2C9*2*, and *CYP2C9*3* alleles (Arun Kumar et al., 2015; Dreisbach et al., 2005; King et al., 2005). Clearly more detailed and exhaustive studies that include a wider range of ethnic groups and possibly other CYP isoforms are warranted.

15.3.4 Renovascular Diseases

The role of 20-HETE and EETs in vascular and tubular abnormalities of renovascular disease (RVD) has been extensively investigated. As a form of secondary hypertension, RVD is one of the major causes of end-stage renal failure, and has been linked to increased cardiovascular mortality (Fatica et al., 2001; Johansson et al., 1999). 20-HETE is a potent vasoconstrictor and an inhibitor of sodium transport in the proximal tubule and thick ascending loop of Henle, whereas EETs induce dilatation of arterioles and inhibit sodium transportation in the proximal tubule and collecting duct, hence promote sodium excretion (Alonso-Galicia et al., 2002). 20-HETE acts by regulating MAPKs, protein kinase C (PKC), tyrosine kinase and Rho kinase pathways. These result in depolarization of vascular smooth muscle cells following blockade of BK channel (Fan et al., 2013), and through inhibition of calcium-sensitive potassium (K_{Ca}) channels (Obara et al., 2002) as well as increase in L-type calcium channel activity (Zeng et al., 2010). The increased 20-HETE level in the renal microcirculation will therefore enhance nephrovascular resistance, which in turn results in decrease in the glomerular filtration rate and capillary pressure. The pressure in the postglomerular circulation is lowered at the same time thus shifting the pressure natriuresis relationship to higher pressures (Williams et al., 2007), resulting in volume retention and hypertension. EETs have been demonstrated to increase the cAMP levels via binding to cell-surface receptors that activate the BK channel, leading to vasodilation. This is also partially due to the stimulation of protein phosphatase 2A and activation of the BK channel in preglomerular arterioles (Imig, 2013). EETs additionally activate small and intermediate calcium-activated potassium (K_{Ca}) channels in the endothelium which alter calcium entry, and possibly increase level and action of nitric oxide. Moreover, EETs enhance sensitivity of the vasodilator response of the renal arterioles to bradykinin, acetylcholine and adenosine (Imig, 2013). Besides changing the vascular tone, 20-HETE inhibits sodium-hydrogen exchanger 3 in the proximal tubule and sodium-potassium-chloride (NKCC2) transporter in the loop of Henle therefore altering sodium and water reabsorption (Alonso-Galicia et al., 2002). On the other hand, EETs' action on tubular transport of sodium is mediated via regulation of $E_{Na}Cs$ activity (Wang et al., 2014).

Changes in the renal 20-HETE and EET levels have been noted in many hypertensive models, and drugs targeting these pathways have been shown to influence pathogenesis of RVD in preclinical studies. In RVD, there is enhanced angiotensin II activity due to activation of the renin-angiotensin system, and this results in elevated 20-HETE and decreased EET levels. These contribute to extracellular matrix deposition, increased endothelin release, vascular remodeling, vasoconstriction, and enhanced glomerulosclerosis and atherogenesis (Kim and Iwao, 2000). For example, in mice with induced renovascular hypertension, sEH inhibition restored the role of EETs in endothelium-dependent relaxation, and eventually attenuated increased blood pressure, cardiac hypertrophy and coronary endothelial dysfunction (Gao et al., 2011). In recent years, allelic polymorphisms in *CYP4A11* and *CYP4F2* and sEHs have been associated with altered 20-HETE levels and variations in blood pressure in rodent models as well as human studies (Capdevila and Wang, 2013; Ward et al., 2014). More specifically, a nucleotide change (T to C) at position 8590 in the human *CYP4A11* that reduces 20-HETE formation has been linked to the development of hypertension due to loss of 20-HETE-mediated natriuresis (Laffer et al., 2008). The G421C polymorphism in the *CYP4F2* in Chinese subjects has additionally been linked to hypertension (Liu et al., 2006b).

EETs and 20-HETE have modulating actions in renal IRI and chronic kidney disease (CKD) (Chen et al., 2012b; Regner and Roman, 2012). The renoprotection of EETs is evidenced from the fact that protection against proteinuria and glomerular injury in angiotensin II and deoxycorticosterone acetate (DOCA)-salt hypertensive rats and mice was demonstrated from chronic administration of an sHE inhibitor (Imig, 2013). Studies with sEH knockout mice showed lowered renal inflammation, glomerular injury, and blood pressure after the induction of hypertension with DOCA-salt or angiotensin II (Imig, 2013). Another evidence is demonstrated by a novel orally active EET analog, *N*-(5-((2-acetamidobenzo[d]thiazol-4-yl)oxy) pentyl)-*N*-isopropylheptanamide) (EET-B). EET-B protected against kidney injury in salt-sensitive hypertensive rats by reducing oxidative stress, endoplasmic reticulum stress and inflammation (Khan et al., 2013). There was markedly reduced urinary albumin along with significant reductions in glomerular injury, kidney fibrosis, and intratubular cast formation (Khan et al., 2013). Another recent study showed that EET-B was able to reduce blood pressure, protect the kidney and markedly lower renal inflammation in angiotensin II-induced hypertension (Khan et al., 2014).

Recent findings indicate impaired vascular responses of tubular arterioles, raised pressure in postglomerular circulation, and increased permeability of the glomerulus to albumin due to 20-HETE deficiency, all of which have resulted in the development of proteinuria and renal injury (Ren et al., 2014; Williams et al., 2012). Animal studies have implicated involvement of 20-HETE in renal IRI. This is supported by a study where a 20-HETE mimetic was shown to protect the injury induced by bilateral renal IRI through prevention of the delayed postischemic fall in medullary blood flow following reperfusion (Regner et al., 2009). Another study showed that pretreatment with a 20-HETE inhibitor, 20-hydroxyeicosa-6(*Z*),15(*Z*)-dienoic acid (6,15,20-HEDE), conferred protection against renal ischemia injury in a uninephrectomized model designed to mimic renal transplant injury (Hoff et al., 2011). More recent studies have demonstrated the role of 20-HETE in the pathogenesis of polycystic kidney disease (PKD) (Klawitter et al., 2014; Park et al., 2009). Renal 20-HETE level was enhanced in a well-characterized rat model of PKD and blocking 20-HETE formation attenuated cystic disease (Park et al., 2009). The same study also demonstrated enhanced expression of *CYP4A1*, *CYP4A2*, *CYP4A3*, and *CYP4A8* mRNA in polycystic kidney, and that daily administration of a 20-HETE inhibitor, dibromododec-11-enoic acid (DBDD), and *N*-hydroxy-*N'*-(4-butyl-2methylphenyl)formamidine (HET0016), significantly reduced kidney size in the PKD rats (Park et al., 2009). Considered together, all the above findings suggest that 20-HETE has great potential to serve as a new biomarker and a therapeutic target for PKD.

15.4 Promising Therapeutic Strategies in Vascular and Cardiovascular Diseases

15.4.1 Cyclooxygenase Inhibitors and Thromboxane Modulators in Atherosclerosis

Numerous lines of evidence support the importance of COX-derived eicosanoids in atherothrombosis as well as the benefit of COX inhibition in decreasing atherosclerosis (Huber et al., 2014). The use of aspirin for secondary prevention of cerebrovascular and cardiovascular disorders has long been established, and action is attributed to its irreversible inhibition of platelet COX-1 activation and subsequent TxA_2 biosynthesis (Ridker et al., 2005), therefore preventing the activation of TP receptor-mediated endothelial and inflammatory responses.

Aspirin is able to inhibit formation of TxA_2 but does not have any effect on the production of other TP receptor ligands such as isoprostanoids and hydroxyeicosanoic acid, which are also pivotal in preserving atherosclerotic vascular injury. Aspirin's action is also limited to platelets, thus blockage of TP receptor can additionally inhibit the effect of macrophage-derived TxA_2 that contributes to the inflammatory status, therefore offering additional therapeutic advantage. A direct inhibition of TP-receptor may also exert better antiplatelet effect than aspirin in high-risk conditions such as diabetes that are associated with increased prostanoid biosynthesis (Giannarelli et al., 2010). These lines of evidence have led to a new research area focusing on TP receptor antagonism as a viable treatment for atherosclerotic diseases. The possibility of combining antiplatelet activity (using aspirin) with an antiatherosclerotic effect via selective inhibition of TP receptor represents an attractive treatment modality because this will allow tackling diseases via multiple mechanisms and thus greatly enhance treatment outcome.

Beneficial effects of TP receptor inhibition have been attested by many animal studies. Daltroban, a TP receptor antagonist, was found to decrease the progression of atherosclerosis in cholesterol-fed rabbits (Metz et al., 1991). Terutroban, another antagonist, exhibited antithrombotic action by inhibiting fibrinogen and platelet deposition in a porcine model (Osende et al., 2004; Vilahur et al., 2007). The regression of atherosclerotic lesions has been observed in a rabbit atherosclerosis model after 6 months of terutroban treatment (Viles-Gonzalez et al., 2005). In an earlier study, terutroban was found to reduce serum levels of the stable metabolite of TxA_2, TxB_2, and significantly decreased serum levels of ICAM-1 and aortic root-lesion formation in mice (Cayatte et al., 2000).

While antiatherogenesis effects of TP receptor blockage have been clearly shown in pre-clinical models, the clinical studies in human subjects have however not been convincing. In the Prevention of cerebrovascular and cardiovascular Events of ischemic origin with teRutroban in patients with a history oF ischemic strOke or tRansient ischeMic attack (PERFORM) study conducted recently, terutroban failed to show superiority over aspirin in prevention of stroke among patients with ischemic cerebrovascular disease (Bots et al., 2014; Bousser et al., 2011). This implies that the pathophysiology of ischemic stroke and CAD is somewhat complicated and more studies comparing other novel antiplatelet drugs with aspirin in recurrent stroke prevention are warranted. Moreover, prevention approaches may require more vigorous and multi-pronged intervention, and the selection of appropriate therapy should also be optimized according to efficacy of the selected agents, patient comorbidities and tolerability to drugs.

15.4.2 EET Modulators in Vascular Dysfunctions

Considering the important role of sEH in EET bioinactivation, various selective and potent sEH inhibitors have been developed in recent years and offer promising value for the treatment of vascular and cardiovascular diseases (Liu et al., 2013; Shen and Hammock, 2011). An additional strategy actively pursued involves increasing CYP-derived EETs using the EET analogs as increased endothelial EET level has proven to confer various cardioprotective effects.

sEH inhibitors have shown proven antihypertensive action in the angiotensin II-dependent animal models (Bellien et al., 2011; Imig, 2012). Besides their antihypertensive effect, sEH inhibitors were able to improve endothelial function and reduce hypertension-induced renal injury and cardiac hypertrophy or dysfunction (Alsaad et al., 2013). Hence, they have been found effective in various cardiovascular disorders, including IRI and heart failure (Bellien et al., 2011; Merabet et al., 2012). The compelling preclinical efficacy supports the prospect of advancing sEH inhibitors to human clinical trials for vascular diseases. An example is AR9281 [1-(1-acetyl-piperidin-4-yl)-3-adamantan-1-yl-urea], a synthetic sEH inhibitor, that has undergone phase I and II clinical trials (Anandan et al., 2011; Chen et al., 2012a). The compound was demonstrated to improve endothelial function in mice models of hypertension, diabetes, and obesity and to significantly lower fasting blood glucose (Zhang et al., 2011). In humans, AR9281 showed a good safety profile and it demonstrated dose-dependent and direct inhibition of blood sEH activity in healthy subjects. Its effectiveness in patients with type 2 diabetes and hypertension has also been investigated. Another example is GSK2256294 ((1R,3S)-N-(4-cyano-2-(trifluoromethyl)benzyl)-3-((4-methyl-6-(methylamino)-1,3,5-triazin-2yl)amino) cyclo hexane-1-carboxamide), a novel potent sEH inhibitor in phase I clinical development (Lazaar et al., 2016). The clinical trials have evaluated GSK2256294 tolerability, safety, pharmacokinetics, and pharmacodynamics and its effect on pulmonary artery systolic pressure under hypoxic conditions in healthy, smokers, obese, elderly, or young subjects. The drug was well tolerated with no reported serious adverse events. There was also no significant impact of food, age or gender on its pharmacokinetic parameters (Lazaar et al., 2016). A recent report indicated GSK2256294 attenuated, in both in vitro and in vivo models, smoking-related endothelial dysfunction mediated by EET in human resistance vessels. This indicates potential benefits of using GSK2256294 in patients with systemic and pulmonary endothelial dysfunctions as seen in many common pulmonary and cardiovascular diseases (Yang et al., 2017).

An important consideration during development of EET modulators is the potential adverse effects resulting from the pharmacological modulation of EET biosynthesis and degradation. Studies have shown that sEH inhibitors can exacerbate hypoxic pulmonary vasoconstriction and hypoxia, induce pulmonary vascular remodeling, and attenuate vascular responsiveness in the respiratory tracts (Keserü et al., 2008). In addition, due to their inhibitory effect on platelet aggregation (Heizer et al., 1991; Krötz et al., 2004), EETs can inactivate platelets via inhibition of adhesion molecule expression, potentially causing bleeding and hemorrhage in patients. Another adverse effect of great concern is that sEH inhibitors and EET analogs can enhance angiogenesis and tumorigenesis (Panigrahy et al., 2011; Ye et al., 2002). EETs have been shown to sustain survival of endothelial cell by pro-angiogenetic and anti-apoptotic mechanisms (Dhanasekaran et al., 2008; Fleming, 2007). They also contribute to VEGF (vascular endothelial growth factor)-mediated stimulation of angiogenesis (Yang et al., 2009). This is supported by study of Panigrahy et al. (2012) that demonstrated the potent stimulatory effect of EETs on primary tumor growth and angiogenesis. Furthermore, elevated EETs enhanced tumorigenesis and multiorgan metastasis in various models of cancer in mice (Tacconelli and Patrignani, 2014). Therefore, future research should focus on prudent medicinal chemistry approach in developing EET modulators with a higher degree of target tissue specificity and decreased angiogenic and mitogenic properties. As an example, a new series of sEH inhibitors have been modified to possess potent and selective sEH inhibition but with significantly reduced proangiogenic properties (Hwang et al., 2013). In fact, data suggest that EET-mediated cancer cell proliferation is regioisomer-specific (Mitra et al., 2011). Thus, sEH inhibitors and EET analogs devoid of carcinogenesis property can potentially be developed using isomer-guided drug design approach.

15.5 Conclusion

Over the past two decades, substantial works have demonstrated the roles played by CYP enzymes in production of a spectrum of bioactive lipid mediators from endogenous substrates. The functional role of CYP5A1 and CYP8A1 in atherogenesis and the cardiovascular actions of members belonging to CYP2C, CYP2J and CYP4 families are know well characterized. Both CYP5A1 and CYP8A1 play the role of Yin and Yang in blood coagulation function via their gene products TxA_2 and PGI_2. The CYP2C and CYP2J epoxygenases produce EETs, which possess cardioprotective properties. The CYP4A and CYP4F ω-hydroxylases generate 20-HETE that is vital to renal microvascular function. Association of genetic polymorphism of some of these *CYP* members with vascular pathology and cardiovascular disorders point to their important roles in human diseases. Potentially feasible treatment modalities are emerging by the way of enhancing EET availability via EET analogs or sEH inhibitors in the clinical management of cardiovascular patients. Similarly, blocking TxA_2 action using TP receptor antagonists has demonstrated promising anti-atherosclerotic effects. These novel classes of agents however are yet to be routinely used due to potential adverse effects including tumorigenesis and angiogenesis. Appropriate drug design studies should be carried out to identify promising agents with acceptable safety profile before they can be considered for clinical use in humans. Future research should also be directed towards defining further the precise signaling mechanisms affected by eicosanoids in different vascular diseases. In addition, more pharmacogenetic studies are needed to determine the correlation between CYP polymorphisms and vasculature and cardiovascular dysfunctions in humans.

References

Ai, D., Fu, Y., Guo, D., Tanaka, H., Wang, N., Tang, C., Hammock, B. D., Shyy, J. Y. J., Zhu, Y., *Proc. Natl. Acad. Sci. U. S. A.*, **104** (2007), 9018–9023.

Ai, D., Pang, W., Li, N., Xu, M., Jones, P. D., Yang, J., Zhang, Y., Chiamvimonvat, N., Shyy, J. Y. J., Hammock, B. D., Zhu, Y., *Proc. Natl. Acad. Sci. U. S. A.*, **106** (2009), 564–569.

Akhnokh, M. K., Yang, F. H., Samokhvalov, V., Jamieson, K. L., Cho, W. J., Wagg, C., Takawale, A., Wang, X., Lopaschuk, G. D., Hammock, B. D., Kassiri, Z., Seubert, J. M., *Front. Pharmacol.*, **7** (2016), 133.

Alonso-Galicia, M., Maier, K. G., Greene, A. S., Cowley, A. W. Jr., Roman, R. J., *Am. J. Physiol. Regul. Integr. Comp. Physiol.*, **283** (2002), R60–R68.

Alsaad, A. M. S., Zordoky, B. N. M., Tse, M. M. Y., El-Kadi, A. O. S., *Drug Metab. Rev.*, **45** (2013), 173–195.

Anandan, S. K., Webb, H. K., Chen, D., Wang, Y. X., Aavula, B. R., Cases, S., Cheng, Y., Do, Z. N., Mehra, U., Tran, V., Vincelette, J., Waszczuk, J., White, K., Wong, K. R., Zhang, L. N., Jones, P. D., Hammock, B. D., Patel, D. V., Whitcomb, R., MacIntyre, D. E., Sabry, J., Gless, R., *Bioorg. Med. Chem. Lett.*, **21** (2011), 983–988.

Arun Kumar, A. S., Kumar, S. S., Umamaheswaran, G., Kesavan, R., Balachandar, J., Adithan, C., *Pharmacol. Rep.*, **67** (2015), 97–101.

Batchu, S. N., Law, E., Brocks, D. R., Falck, J. R., Seubert, J. M., *J. Mol. Cell. Cardiol.*, **46** (2009), 67–74.

Batchu, S. N., Lee, S. B., Qadhi, R. S., Chaudhary, K. R., El-Sikhry, H., Kodela, R., Falck, J. R., Seubert, J. M., *Br. J. Pharmacol.*, **162** (2011), 897–907.

Baum, S. J., Kris-Etherton, P. M., Willett, W. C., Lichtenstein, A. H., Rudel, L. L., Maki, K. C., Whelan, J., Ramsden, C. E., Block, R. C., *J. Clin. Lipidol.*, **6** (2012), 216–234.

Bellien, J., Joannides, R., Richard, V., Thuillez, C., *Pharmacol. Ther.*, **131** (2011), 1–17.

Bots, M. L., Ford, I., Lloyd, S. M., Laurent, S., Touboul, P. J., Hennerici, M. G., *Stroke*, **45** (2014), 2348–2353.

Bousoula, E., Kolovou, V., Vasiliadis, I., Karakosta, A., Xanthos, T., Johnson, E. O., Skandalakis, P., Kolovou, G. D., *Angiology*, **63** (2012), 461–465.

Bousser, M. G., Amarenco, P., Chamorro, A., Fisher, M., Ford, I., Fox, K. M., Hennerici, M. G., Mattle, H. P., Rothwell, P. M., de Cordoüe, A., Fratacci, M. D., *Lancet*, **377** (2011), 2013–2022.

Campbell, W. B., Gebremedhin, D., Pratt, P. F., Harder, D. R., *Circ. Res.*, **78** (1996), 415–423.

Capdevila, J., Wang, W., *Curr. Opin. Nephrol. Hypertens.*, **22** (2013), 163–169.

Catella, F., Lawson, J. A. Fitzgerald, D. J., FitzGerald, G. A., *Proc. Natl. Acad. Sci. U. S. A.*, **87** (1990), 5893–5897.

Cayatte, A. J., Du, Y., Oliver-Krasinski, J., Lavielle, G., Verbeuren, T. J., Cohen, R. A., *Arterioscler. Thromb. Vasc. Biol.*, **20** (2000), 1724–1728.

Certíková Chábová, V., Walkowska, A., Kompanowska-Jezierska, E., Sadowski, J., Kujal, P., Vernerová, Z., Vanourková, Z., Kopkan, L., Kramer, H. J., Falck, J. R., Imig, J. D., Hammock, B. D., Vanecková, I., Cervenka, L., *Clin. Sci. (Lond)*, **118** (2010), 617–632.

Červenka, L., Husková, Z., Kopkan, L., Kikerlová, S., Sedláková, L., Vaňourková, Z., Alánová, P., Kolář, F., Hammock, B. D., Hwang, S. H., Imig, J. D., Falck, J. R., Sadowski, J., Kompanowska-Jezierska, E., Neckář, J., *J. Hypertens.*, **36** (2018), 1326–1341.

Chen, D., Whitcomb, R., MacIntyre, E., Tran, V., Do, Z. N., Sabry, J., Patel, D. V., Anandan, S. K., Gless, R., Webb, H. K., *J. Clin. Pharmacol.*, **52** (2012a), 319–328.

Chen, G., Xu, R., Wang, Y., Zhao, G., Xu, X., Gruzdev, A., Zeldin, D. C., Wang, D. W., *Am. J. Physiol. Endocrinol. Metab.*, **303** (2012b), E563–E575.

Cheng, J., Wu, C. C., Gotlinger, K. H., Zhang, F., Falck, J. R., Narsimhaswamy, D., Schwartzman, M. L., *J. Pharmacol. Exp. Ther.*, **332** (2010), 57–65.

Croft, K. D., McGiff, J. C., Sanchez-Mendoza, A., Carroll, M. A., *Am. J. Physiol. Renal. Physiol.*, **279** (2000), F544–F551.

de Leval, X., Hanson, J., David, J. L., Masereel, B., Pirotte, B., Dogné, J. M., *Curr. Med. Chem.*, **11** (2004), 1243–1252.

DeLozier, T. C., Kissling, G. E., Coulter, S. J., Dai, D., Foley, J. F., Bradbury, J. A., Murphy, E., Steenbergen, C., Zeldin, D. C., Goldstein, J. A., *Drug Metab. Dispos.*, **35** (2007), 682–688.

Deng, Y., Theken, K. N., Lee, C. R., *J. Mol. Cell. Cardiol.*, **48** (2010), 331–341.

Dhanasekaran, A., Gruenloh, S. K., Buonaccorsi, J. N., Zhang, R., Gross, G. J., Falck, J. R., Patel, P. K., Jacobs, E. R., Medhora, M., *Am. J. Physiol. Heart Circ. Physiol.*, **294** (2008), H724–H735.

Di, L., *Expert Opin. Drug Metab. Toxicol.*, **10** (2014), 379–393.

Doggrell, S. A., *Expert Opin. Ther. Targets*, **8** (2004), 491–493.

Dogne, J. M., de Leval, X., Hanson, J., Frederich, M., Lambermont, B., Ghuysen, A., Casini, A., Masereel, B., Ruan, K. H., Pirotte, B., Kolh, P., *Curr. Med. Chem.*, **11** (2004), 1223–1241.

Dreisbach, A. W., Japa, S., Sigel, A., Parenti, M. B., Hess, A. E., Srinouanprachanh, S. L., Rettie, A. E., Kim, H., Farin, F. M., Hamm, L. L., Lertora, J. J., *Am. J. Hypertens.*, **18** (2005), 1276–1281.

Edin, M. L., Wang, Z., Bradbury, J. A., Graves, J. P., Lih, F. B., DeGraff, L. M., Foley, J. F., Torphy, R., Ronnekleiv, O. K., Tomer, K. B., Lee, C. R., Zeldin, D. C., *FASEB J.*, **25** (2011), 3436–3447.

Fan, F., Ge, Y., Lv, W., Elliott, M. R., Muroya, Y., Hirata, T., Booz, G. W., Roman, R. J., *Front. Biosci. (Landmark Ed)*, **21** (2016), 1427–1463.

Fan, F., Sun, C. W., Maier, K. G., Williams, J. M., Pabbidi, M. R., Didion, S. P., Falck, J. R., Zhuo, J., Roman, R. J., *PLoS One*, **8** (2013), e82482.

Fatica, R. A., Port, F. K., Young, E. W., *Am. J. Kidney Dis.*, **37** (2001), 1184–1190.

Fichtlscherer, S., Dimmeler, S., Breuer, S., Busse, R., Zeiher, A., Fleming, I., *Circulation*, **109** (2004), 178–183.

Fisslthaler, B., Hinsch, N., Chataigneau, T., Popp, R., Kiss, L., Busse, R., Fleming, I., *Hypertension*, **36** (2000), 270–275.

Fleming, I., *Prostaglandins Other Lipid Mediat.*, **82** (2007), 60–67.

Fleming, I., *Pharmacol. Rev.*, **66** (2014), 1106–1140.

Fleming, I., Michaelis, U. R., Bredenkotter, D., Fisslthaler, B., Dehghani, F., Brandes, R. P., Busse, R., *Circ. Res.*, **88** (2001), 44–51.

Gao, J., Bellien, J., Gomez, E., Henry, J. P., Dautreaux, B., Bounoure, F., Skiba, M., Thuillez, C., Richard, V., *J. Hypertens.*, **29** (2011), 1128–1135.

Geneviève, D., Proulle, V., Isidor, B., Bellais, S., Serre, V., Djouadi, F., Picard, C., Vignon-Savoye, C, Bader-Meunier, B., Blanche, S., de Vernejoul, M. C., Legeai-Mallet, L., Fischer, A. M., Le Merrer, M., Dreyfus, M., Gaussem, P., Munnich, A., Cormier-Daire, V., *Nat. Genet.*, **40** (2008), 284–286.

Giannarelli, C., Zafar, M. U., Badimon, J. J., *Thromb. Haemost.*, **104** (2010), 949–954.

Gross, G. J., Baker, J. E., Moore, J., Falck, J. R., Nithipatikom, K., *Cardiovasc. Drugs Ther.*, **25** (2011), 517–522.

Gross, G. J., Hsu, A., Falck, J. R., Nithipatikom, K., *J. Mol. Cell. Cardiol.*, **42** (2007), 687–691.

Harris, T. R., Hammock, B. D., *Gene*, **526** (2013), 61–74.

Haschke-Becher, E., Kirchheiner, J., Trummer, O., Grunbacher, G., Kainz, A., Boehm, B. O., Marz, W., Renner, W., *Pharmacogenomics*, **11** (2010), 1359–1365.

Heizer, M. L., McKinney, J. S., Ellis, E. F., *Stroke*, **22** (1991), 1389–1393.

Hoff, U., Lukitsch, I., Chaykovska, L., Ladwig, M., Arnold, C., Manthati, V. L., Fuller, T. F., Schneider, W., Gollasch, M., Muller, D. N., Flemming, B., Seeliger, E., Luft, F. C., Falck, J. R., Dragun, D., Schunck, W. H., *Kidney Int.*, **79** (2011), 57–65.

Huang, J. S., Ramamurthy, S. K., Lin, X., Le Breton, G. C., *Cell Signal.*, **16** (2004), 521–533.

Huber, K., Bates, E. R., Valgimigli, M., Wallentin, L., Kristensen, S. D., Anderson, J. L., Lopez Sendon, J. L., Tubaro, M., Granger, C. B., Bode, C., Ohman, E. M., Steg, P. G., *Am. Heart J.*, **168** (2014), 611–621.

Hwang, S. H., Wecksler, A. T., Zhang, G., Morisseau, C., Nguyen, L. V., Fu, S. H., Hammock, B. D., *Bioorg. Med. Chem. Lett.*, **23** (2013), 3732–3737.

Imig, J. D., *Physiol. Rev.*, **92** (2012), 101–130.

Imig, J. D., *Prostaglandins Other Lipid Mediat.*, **104–105** (2013), 2–7.

Ishihara, Y., Sekine, M., Nakazawa, M., Shimamoto, N., *Eur. J. Pharmacol.*, **611** (2009), 64–71.

Ishizuka, T., Cheng, J., Singh, H., Vitto, M. D., Manthati, V. L., Falck, J. R., Laniado-Schwartzman, M., *J. Pharmacol. Exp. Ther.*, **324** (2008), 103–110.

Jíchová, Š., Kopkan, L., Husková, Z., Doleželová, Š., Neckář, J., Kujal, P., Vernerová, Z., Kramer, H. J., Sadowski, J., Kompanowska-Jezierska, E., Reddy, R. N., Falck, J. R., Imig, J. D., Červenka, L., *J. Hypertens.*, **34** (2016), 2008–2025.

Johansson, M., Herlitz, H., Jensen, G., Rundqvist, B., Friberg, P., *J. Hypertens.*, **17** (1999), 1743–1750.

Katragadda, D., Batchu, S. N., Cho, W. J., Chaudhary, K. R., Falck, J. R., Seubert, J. M., *J. Mol. Cell. Cardiol.*, **46** (2009), 867–875.

Katugampola, S. D., Davenport, A. P., *Br. J. Pharmacol.*, **134** (2001), 1385–1392.

Kaur-Knudsen, D., Bojesen, S. E., Nordestgaard, B. G., *Pharmacogenomics J.*, **9** (2009), 327–332.

Keserü, B., Barbosa-Sicard, E., Popp, R., Fisslthaler, B., Dietrich, A., Gudermann, T., Hammock, B. D., Falck, J. R., Weissmann, N., Busse, R., Fleming, I., *FASEB J.*, **22** (2008), 4306–4315.

Khan, A. H., Falck, J. R., Manthati, V. L., Campbell, W. B., Imig, J. D., *Front. Pharmacol.*, **5** (2014), 216.

Khan, M. A. H., Neckář, J., Manthati, V., Errabelli, R., Pavlov, T. S., Staruschenko, A., Falck, J. R., Imig, J. D., *Hypertension*, **62** (2013), 905–913.

Kim, S., Iwao, H., *Pharmacol. Rev.*, **52** (2000), 11–34.

King, L. M., Gainer, J. V., David, G. L., Dai, D., Goldstein, J. A., Brown, N. J., Zeldin, D. C., *Pharmacogenet. Genomics*, **15** (2005), 7–13.

Kinlay, S., Libby, P., Ganz, P., *Curr. Opin. Lipidol.*, **12** (2001), 383–389.

Klawitter, J., Klawitter, J., McFann, K., Pennington, A. T., Abebe, K. Z., Brosnahan, G., Cadnapaphornchai, M. A., Chonchol, M., Gitomer, B., Christians, U., Schrier, R. W., *J. Lipid Res.*, **55** (2014), 1139–1149.

Kleiman, N. S., *J. Am. Coll. Cardiol.*, **43** (2004), 1989–1991.

Kompa, A. R., Wang, B. H., Xu, G., Zhang, Y., Ho, P. Y., Eisennagel, S., Thalji, R. K., Marino, J. P. Jr., Kelly, D. J., Behm, D. J., Krum, H., *Int. J. Cardiol.*, **167** (2013), 210–219.

Krötz, F., Riexinger, T., Buerkle, M. A., Nithipatikom, K., Gloe, T., Sohn, H. Y., Campbell, W. B., Pohl, U., *Arterioscler. Thromb. Vasc. Biol.*, **24** (2004), 595–600.

Laffer, C. L., Gainer, J. V., Waterman, M. R., Capdevila, J. H., Laniado-Schwartzman, M., Nasjletti, A., Brown, N. J., Elijovich, F., *Hypertension*, **51** (2008), 767–772.

Lai, G., Liu, X., Wu, J., Liu, H., Zhao, Y., *Int. J. Mol. Med.*, **29** (2012), 107–112.

Lazaar, A. L., Yang, L., Boardley, R. L., Goyal, N. S., Robertson, J., Baldwin, S. J., Newby, D. E., Wilkinson, I. B., Tal-Singer, R., Mayer, R. J., Cheriyan, J., *Br. J. Clin. Pharmacol.*, **81** (2016), 971–979.

Lee, C. R., North, K. E., Bray, M. S., Couper, D. J., Heiss, G., Zeldin, D. C., *Pharmacogenet. Genomics*, **17** (2007), 349–358.

Liu, P. Y., Li, Y. H., Chao, T. H., Wu, H. L., Lin, L. J., Tsai, L. M., Chen, J. H., *Atherosclerosis*, **195** (2006a), 199–206.

Liu, J. Y., Lin, Y. P., Qiu, H., Morisseau, C., Rose, T. E., Hwang, S. H., Chiamvimonvat, N., Hammock, B. D., *Eur. J. Pharm. Sci.*, **48** (2013), 619–627.

Liu, H., Zhao, Y. Y., Gong, W., Shi, J. P., Fu, L. Y., Wang, J., Li, G. Y., Lü, J. Y., *Zhongguo Yi Xue Ke Xue Yuan Xue Bao*, **28** (2006b), 143–147.

Manhiani, M., Quigley, J. E., Knight, S. F., Tasoobshirazi, S., Moore, T., Brands, M. W., Hammock, B. D., Imig, J. D., *Am. J. Physiol. Renal Physiol.*, **297** (2009), F740–F748.

Merabet, N., Bellien, J., Glevarec, E., Nicol, L., Lucas, D., Remy-Jouet, I., Bounoure, F., Dreano, Y., Wecker, D., Thuillez, C., Mulder, P., *J. Mol. Cell. Cardiol.*, **52** (2012), 660–666.

Metz, J., Wolf, O., Schmelz, A., Pill, J., Stegmeier, K. H., Hartig, F., *Exp. Pathol.*, **41** (1991), 57–69.

Minuz, P., Jiang, H., Fava, C., Turolo, L., Tacconelli, S., Ricci, M., Patrignani, P., Morganti, A., Lechi, A., McGiff, J. C., *Hypertension*, **51** (2008), 1379–1385.

Mitra, R., Guo, Z., Milani, M., Mesaros, C., Rodriguez, M., Nguyen, J., Luo, X., Clarke, D., Lamba, J., Schuetz, E., Donner, D. B., Puli, N., Falck, J. R.,

Capdevila, J., Gupta, K., Blair, I. A., Potter, D. A., *J. Biol. Chem.*, **286** (2011), 17543–17559.

Nakayama, T., Soma, M., Rehemudula, D., Takahashi, Y., Tobe, H., Sato, M., Uwabo, J., Kunimoto, M., Kanmatsuse, K., *Am. J. Hypertens.*, **13** (2000), 1263–1267.

Nakayama, T., Soma, M., Watanabe, Y., Hasimu, B., Sato, M., Aoi, N., Kosuge, K., Kanmatsuse, K., Kokubun, S., Marrow, J. D., Oates, J. A., *Biochem. Biophys. Res. Commun.*, **297** (2002), 1135–1139.

Nebert, D. W., Wikvall, K., Miller, W. L., *Philos. Trans. R. Soc. Lond. B Biol. Sci.*, **368** (2013), 20120431.

Obara, K., Koide, M., Nakayama, K., *Br. J. Pharmacol.*, **137** (2002), 1362–1370.

Oni-Orisan, A., Alsaleh, N., Lee, C. R., Seubert, J. M., *J. Mol. Cell. Cardiol.*, **74C** (2014), 199–208.

Osende, J. I., Shimbo D., Fuster, V., Dubar, M., Badimon, J. J., *J. Thromb. Haemost.*, **2** (2004), 492–498.

Panigrahy, D., Edin, M. L., Lee, C. R., Huang, S., Bielenberg, D. R., Butterfield, C. E., Barnés, C. M., Mammoto, A., Mammoto, T., Luria, A., Benny, O., Chaponis, D. M., Dudley, A. C., Greene, E. R., Vergilio, J. A., Pietramaggiori, G., Scherer-Pietramaggiori, S. S., Short, S. M., Seth, M., Lih, F. B., Tomer, K. B., Yang, J., Schwendener, R. A., Hammock, B. D., Falck, J. R., Manthati, V. L., Ingber, D. E., Kaipainen, A., D'Amore, P. A., Kieran, M. W., Zeldin, D. C., *J. Clin. Invest.* **122** (2012), 178–191.

Panigraphy, D., Greene, E. R., Pozzi, A., Wang, D. W., Zeldin, D. C., *Cancer Metastasis Rev.*, **30** (2011), 525–540.

Park, F., Sweeney, W. E., Jia, G., Akbulut, T., Mueller, B., Falck, J. R., Birudaraju, S., Roman, R. J., Avner, E. D., *Am. J. Physiol. Renal Physiol.*, **296** (2009), F575–F582.

Praticò, D., Dogné, J. M., *Expert Rev. Cardiovasc. Ther.*, **7** (2009), 1079–1089.

Puntarulo, S., Cederbaum, A. I., *Free Radic. Biol. Med.*, **24** (1998), 1324–1330.

Regner, K. R., Roman, R. J., *Curr. Opin. Nephrol. Hypertens.*, **21** (2012), 33–38.

Regner, K. R., Zuk, A., Van Why, S. K., Shames, B. D., Ryan, R. P., Falck, J. R., Manthati, V. L., McMullen, M. E., Ledbetter, S. R., Roman, R. J., *Kidney Int.*, **75** (2009), 511–517.

Ren, Y., D'Ambrosio, M. A., Garvin, J. L., Peterson, E. L., Carretero, O. A., *Am. J. Physiol. Renal Physiol.*, **307** (2014), F533–F538.

Ridker, P. M., Cook, N. R., Lee, I. M., Gordon, D., Gaziano, J. M., Manson, J. E., Hennekens, C. H., Buring, J. E., *N. Engl. J. Med.*, **352** (2005), 1293–1304.

Roche, C., Besnier, M., Cassel, R., Harouki, N., Coquerel, D., Guerrot, D., Nicol, L., Loizon, E., Remy-Jouet, I., Morisseau, C., Mulder, P., Ouvrard-Pascaud, A., Madec, A. M., Richard, V., Bellien, J., *Am. J. Physiol. Heart Circ. Physiol.*, **308** (2015), H1020–H1029.

Rosamond, W., Flegal, K., Furie, K., Go, A., Greenlund, K., Haase, N., Hailpern, S. M., Ho, M., Howard, V., Kissela, B., Kittner, S., Lloyd-Jones, D., McDermott, M., Meigs, J., Moy, C., Nichol, G., O'Donnell, C., Roger, V., Sorlie, P., Steinberger, J., Thom, T., Wilson, M., Hong, Y., *Circulation*, **117** (2008), e25–e146.

Savas, U., Wei, S., Hsu, M. H., Falck, J. R., Guengerich, F. P., Capdevila, J. H., Johnson, E. F., *J. Biol. Chem.*, **291** (2016), 16904–16919.

Schinzari, F., Tesauro, M., Cardillo, C., *Acta Physiol. (Oxf)*, **219** (2017), 124–137.

Seubert, J., Yang, B., Bradbury, J. A., Graves, J., Degraff, L. M., Gabel, S., Gooch, R., Foley, J., Newman, J., Mao, L., Rockman, H. A., Hammock, B. D., Murphy, E., Zeldin, D. C., *Circ. Res.*, **95** (2004), 506–514.

Shen, H. C., Hammock, B. D., *J. Med. Chem.*, **55** (2011), 1789–1808.

Sinal, C. J., Miyata, M., Tohkin, M., Nagata, K., Bend, J. R., Gonzalez, F. J., *J. Biol. Chem.*, **275** (2000), 40504–40510.

Sodhi, K., Wu, C. C., Cheng, J., Gotlinger, K., Inoue, K., Goli, M., Falck, J. R., Abraham, N. G., Schwartzman, M. L., *Hypertension*, **56** (2010), 871–878.

Spector, A. A., Fang, X., Snyder, G. D., Weintraub, N. L., *Prog. Lipid Res.*, **43** (2004), 55–90.

Spiecker, M., Darius, H., Hankeln, T., Soufi, M., Sattler, A. M., Schaefer, J. R., Node, K., Borgel, J., Mugge, A., Lindpaintner, K., Huesing, A., Maisch, B., Zeldin, D. C., Liao, J. K., *Circulation*, **110** (2004), 2132–2136.

Sporková, A., Husková, Z., Škaroupková, P., Rami Reddy, N., Falck, J. R., Sadowskim, J., Červenka, L., *Physiol. Res.*, **66** (2017), 29–39.

Sudhahar, V., Shaw, S., Imig, J. D., *Curr. Med. Chem.*, **17** (2010), 1181–1190.

Tacconelli, S., Patrignani, P., *Front. Pharmacol.*, **5** (2014), 239.

Theken, K. N., Schuck, R. N., Edin, M. L., Tran, B., Ellis, K., Bass, A., Lih, F. B., Tomer, K. B., Poloyac, S. M., Wu, M. C., Hinderliter, A. L., Zeldin, D. C., Stouffer, G. A., Lee, C. R., *Atherosclerosis*, **222** (2012), 530–536.

Tzveova, R., Naydenova, G., Yaneva, T., Dimitrov, G., Vandeva, S., Matrozova, Y., Pendicheva-Duhlenska, D., Popov, I., Beltheva, O., Naydenov, C.,

Tarnovska-Kadreva, R., Nachev, G., Mitev, V., Kaneva, R., *Biochem. Genet.*, **53** (2015), 319–333.

Ullrich, V., Hecker, M., *Adv. Prostaglandin Thromboxane Leukot. Res.*, **20** (1990), 95–101.

Ulu, A., Davis, B. B., Tsai, H. J., Kim, I. H., Morisseau, C., Inceoglu, B., Fiehn, O., Hammock, B. D., Weiss, R. H., *J. Cardiovasc. Pharmacol.*, **52** (2008), 314–323.

Vanhoutte, P. M., *Circ. J.*, **73** (2009), 595–601.

Vilahur, G., Casani, L., Badimon, L., *Thromb. Haemost.*, **98** (2007), 662–669.

Viles-Gonzalez, J. F., Fuster, V., Corti, R., Valdiviezo, C., Hutter, R., Corda, S., Anand, S. X., Badimon, J. J., *Eur. Heart J.*, **26** (2005), 1557–1561.

Wang, D., Dubois, R. N., *J. Clin. Invest.*, **122** (2012), 19–22.

Wang, L. H., Kulmacz, R. J., *Prostaglandins Other Lipid Mediat.*, **68–69** (2002), 409–422.

Wang, Z., Wei, Y., Falck, J. R., Atcha, K. R., Wang, W. H., *Am. J. Physiol. Renal Physiol.*, **294** (2008), F1441–F1447.

Wang, W. H., Zhang, C., Lin, D. H., Wang, L., Graves, J. P., Zeldin, D. C., Capdevila, J. H., *Am. J. Physiol. Renal. Physiol.*, **307** (2014), F453–F460.

Ward, N. C., Croft, K. D., Puddey, I. B., Phillips, M., van Bockxmeer, F., Beilin, L. J., Barden, A. E., *J. Hypertens.*, **32** (2014), 1495–1502.

Williams, J. M., Fan, F., Murphy, S., Schreck, C., Lazar, J., Jacob, H. J., Roman, R. J., *Am. J. Physiol. Regul. Integr. Comp. Physiol.*, **302** (2012), R1209–R1218.

Williams, J. M., Sarkis, A., Lopez, B., Ryan, R. P., Flasch, A. K., Roman, R. J., *Hypertension*, **49** (2007), 687–694.

Xiao, B., Li, X., Yan, J., Yu, X., Yang, G., Xiao, X., Voltz, J. W., Zeldin, D. C., Wang, D. W., *J. Pharmacol. Exp. Ther.*, **334** (2010), 784–794.

Xu, X., Zhao, C. X., Wang, L., Tu, L., Fang, X., Zheng, C., Edin, M. L., Zeldin, D. C., Wang, D. W., *Diabetes*, **59** (2010), 997–1005.

Yang, L., Cheriyan, J., Gutterman, D. D., Mayer, R. J., Ament, Z., Griffin, J. L., Lazaar, A. L., Newby, D. E., Tal-Singer, R., Wilkinson, I. B., *Chest*, **151** (2017), 555–563.

Yang, S., Wei, S., Pozzi, A., Capdevila, J. H., *Arch. Biochem. Biophys.*, **489** (2009), 82–91.

Yasar, U., Bennet, A. M., Eliasson, E., Lundgren, S., Wiman, B., De Faire, U., Rane, A., *Pharmacogenetics*, **13** (2003), 715–720.

Ye, D., Zhang, D., Oltman, C., Dellsperger, K., Lee, H. C., VanRollins, M., *J. Pharmacol. Exp. Ther.*, **303** (2002), 768–776.

Yellon, D. M., Hausenloy, D. J., *N. Engl. J. Med.*, **357** (2007), 1121–1135.

Zeng, Q., Han, Y., Bao, Y., Li, W., Li, X., Shen, X., Wang, X., Yao, F., O'Rourke, S. T., Sun, C., *Am. J. Physiol. Heart Circ. Physiol.*, **299** (2010), H1109–H1117.

Zhang, L. N., Vincellette, J., Chen, D., Gless, R. D., Anandan, S. K., Rubanyi, G. M., Webb, H. K., MacIntyre, D. E., Wang, Y. X., *Eur. J. Pharmacol.*, **654** (2011), 68–74.

Chapter 16

Protein Degradation Inducers SNIPERs and Protacs against Oncogenic Proteins

Norihito Shibata, Nobumichi Ohoka, Takayuki Hattori, and Mikihiko Naito

Divisions of Molecular Target and Gene Therapy Products, National Institute of Health Sciences, 3-25-26 Tonomachi, Kawasaki-ku, Kawasaki-shi, Kanagawa 210-9501, Japan

miki-naito@nihs.go.jp

16.1 Introduction

Identification of disease-related proteins is an important step in drug development. Following deep sequencing of the human genome and genome-wide association studies, many oncogenic proteins that drive tumor growth have been identified. Although these oncogenic proteins are attractive targets for development of novel anti-tumor drugs, such oncogenic proteins are not always "druggable." Many molecular targeted drugs recently developed against tumors are small molecule inhibitors and antibodies that inhibit the activities of target proteins. However, it is generally

Pharmaceutical Biocatalysis: Fundamentals, Enzyme Inhibitors, and Enzymes in Health and Diseases
Edited by Peter Grunwald

ISBN 978-981-4800-61-7 (Hardcover), 978-0-429-29503-4 (eBook)
www.jennystanford.com

difficult to develop potent inhibitors of intracellular proteins without enzymatic activity. Therefore, a new drug modality is required to target "undruggable" proteins.

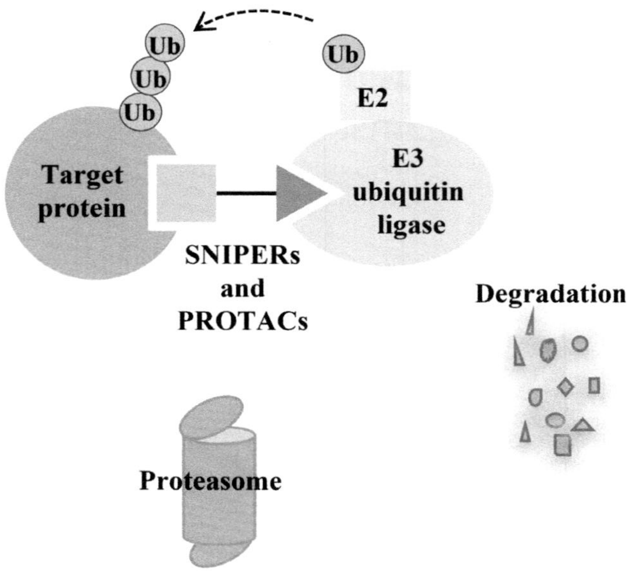

Figure 16.1 Brief illustration of protein degradation by hybrid molecules SNIPERs and PROTACs.

An alternative approach is downregulation of the target protein, which should have a potential therapeutic effect. Recently, we and others have developed a protein knockdown technology that induces degradation of target proteins using hybrid small molecules named SNIPERs [specific and non-genetic inhibitor of apoptosis protein (IAP)-dependent protein erasers] or PROTACs (proteolysis targeting chimeras). They are chimeric molecules consisting of two different ligands connected by a linker. One ligand is for the target protein, while the other is for E3 ubiquitin ligases. Accordingly, these molecules are expected to crosslink the target protein and E3 ubiquitin ligases in cells, resulting in ubiquitylation and subsequent degradation of the target protein via the ubiquitin-proteasome system (UPS) (Fig. 16.1). In this chapter, we present examples of drugs that act by degrading oncogenic proteins, and discuss the features and prospect of chemical degraders.

16.2 Chemical Protein Knockdown Technology

16.2.1 SNIPERs (Specific and Non-Genetic Inhibitor of Apoptosis Protein [IAP]-Dependent Protein Erasers)

In 2008, we reported that (-)-*N*-[(*2S,3R*)-3-amino-2-hydroxy-4-phenyl-butyryl]-L-leucine methyl ester (MeBS) destabilizes cellular IAP1 (cIAP1), an E3 ubiquitin ligase belonging to the IAP family (Sekine et al. 2008). MeBS directly interacts with cIAP1 at the third baculoviral IAP repeat domain and induces auto-ubiquitylation of cIAP1 depending on its RING domain, resulting in proteasomal degradation of cIAP1. A structure–activity relationship study indicated that analogs with a carboxyl-ester reduce the amount of cIAP1, even though the methyl group is substituted to other residues, and other modifications of the bestatin backbone seriously affect activity (Sekine et al., 2008). Based on these observations, we hypothesized that the methyl group can be substituted for a ligand of a target protein without ablating the ubiquitin ligase activity of cIAP1, and we designed hybrid molecules consisting of MeBS and ligands of target proteins connected by a linker. The molecules are expected to crosslink target proteins and cIAP1 in cells, resulting in ubiquitylation and subsequent degradation of the target proteins via the UPS. As a proof-of-concept study, we designed and synthesized a series of hybrid molecules consisting of MeBS and all-trans retinoic acid, and found that they induced proteasomal degradation of cellular retinoic acid binding protein-II (Itoh et al., 2010; Okuhira et al., 2011; Itoh et al., 2012; Okuhira et al., 2017). Next, we developed various hybrid molecules by connecting MeBS and ligands of oncogenic proteins, which induce degradation of estrogen receptor α (ERα) (Demizu et al., 2012; Okuhira et al., 2013), androgen receptor (AR) (Shibata et al., 2018a), transforming acidic coiled-coil containing protein 3 (Ohoka et al., 2014; Ohoka et al., 2017a), and BCR-ABL (Demizu et al., 2016) at 10–100 μM. Thus, selective degradation of target proteins can be attained by hybrid molecules that crosslink the target protein and cIAP1. We termed the hybrid molecules as SNIPERs.

IAPs are a family of anti-apoptotic proteins, and some family members are attractive targets for tumor therapy because IAPs are overexpressed in multiple human malignancies and implicated in promoting tumor progression, treatment failures, and poor prognoses. Thus, many potent and cell-permeable IAP antagonists have been developed, some of which are under evaluation in clinical phase studies as anti-tumor drugs (Cohen et al., 2010; Fulda et al., 2012). Because these IAP antagonists have higher affinity for IAPs than MeBS, we reasoned that novel SNIPERs with a potent protein degradation activity could be developed by incorporating an IAP antagonist into SNIPERs. First, to develop a potent protein degrader of ERα [SNIPER(ER)], we tested various combinations of ERα ligands and IAP antagonists for protein degradation activity of SNIPER(ER) (Ohoka et al., 2017b). The resulting SNIPER(ER)-87, in which ERα ligand 4-hydroxytamoxifen is conjugated to an IAP antagonist LCL161 derivative by a polyethylene glycol × 3 linker, shows potent activity to degrade the ERα protein. Effective protein degradation of ERα by SNIPER(ER)-87 was observed at concentrations of ≥3 nM. Furthermore, SNIPER(ER)-87 reduced ERα levels in tumor xenografts and suppressed the growth of ERα-positive breast tumors in mice (Ohoka et al., 2017b; Ohoka et al., 2018). Using this IAP antagonist, potent SNIPERs were also developed against other oncogenic proteins, BCR-ABL (Ohoka et al., 2017b; Shibata et al., 2017; Shimokawa et al., 2017) and bromodomain-containing protein 4 (BRD4) (Ohoka et al., 2017b). In addition, by derivatization of the IAP ligand molecule, we have developed novel SNIPER(ER)s whose protein degradation and anti-tumor activities are more potent than those of SNIPER (ER)-87 (Ohoka et al., 2018). Because IAP ligands with higher binding affinities for IAP proteins were incorporated into SNIPER(ER)s, incorporating these IAP ligands may be a promising strategy to develop SNIPERs against other oncogenic proteins with better protein degradation and anti-tumor activities.

16.2.2 PROTACS (Proteolysis Targeting Chimeras)

Cullin RING E3 ubiquitin ligases (CRLs) are multi-subunit complexes that catalyze the transfer of ubiquitin to specific substrate proteins. The complex of cullin1 (CUL1), S-phase kinase associated protein 1 (SKP1), and mammalian F-box protein β-transducin repeat-containing protein (β-TRCP) is a CRL known as $CRL1^{\beta\text{-TRCP}}$. The β-TRCP of $CRL1^{\beta\text{-TRCP}}$ binds to IκBα and promotes its ubiquitination and degradation (Yaron et al., 1998). A 10 amino acid phosphopeptide segment of IκBα is both necessary and sufficient to mediate its binding to $CRL1^{\beta\text{-TRCP}}$ and subsequent ubiquitination and degradation (Yaron et al., 1997; Yaron et al., 1998). Dr. Deshaies' and Crews' groups designed and synthesized a series of hybrid molecules named PROTACs consisting of the IκBα phosphopeptide and a ligand of target proteins such as methionine aminopeptidase 2 (Sakamoto et al., 2001), ERα, and AR (Sakamoto et al., 2003). Although the β-TRCP-recruiting PROTACs induce ubiquitination and degradation of target proteins in vitro, they have low activity in cells, presumably because of the poor cell permeability of the employed peptide. To overcome this problem, the IκBα phosphopeptide was replaced by a hydroxyproline-containing pentapeptide from hypoxia-inducible factor α (HIF-1α), which is recognized by von Hippel-Lindau (VHL). VHL is a substrate recognition component of an E3 ubiquitin ligase complex containing cullin 2 (CUL2), RING-box protein 1 (RBX1), elongin B (ELOB), and elongin C (ELOC). The CUL2-RBX1-ELOB-ELOC-VHL complex (known as $CRL2^{VHL}$) directs the ubiquitylation and subsequent proteasomal degradation of HIF-1α under normoxic conditions. VHL-recruiting PROTACs, consisting of the HIF-1α pentapeptide and a ligand of ERα or AR, are cell permeable and induce significant degradation of ERα and AR at 12.5 and 50 μM, respectively (Rodriguez-Gonzalez et al., 2008).

To improve the protein knockdown activity of PRORACs, Dr. Crews' group developed hydroxyproline-based small molecules targeting VHL (Buckley et al., 2012a; Buckley et al., 2012b; Galdeano et al., 2014; Frost et al., 2016). They synthesized all small molecule PROTACs consisting of VHL032 (a VHL ligand) and a ligand of estrogen-related receptor α (ERRα) or receptor-interacting serine-threonine kinase 2 (RIPK2) (Bondeson et al., 2015). These novel VHL-recruiting PROTACs induce significant degradation of ERRα and RIPK2 at 100 and 3 nM, respectively. In addition, the VHL-recruiting PROTAC against ERRα provides broad tissue distribution and knockdown of ERRα protein in a tumor xenograft model. Using the VHL ligand or its derivative, potent PROTACs were also developed against oncogenic proteins such as BRD4 (Raina et al., 2016), TRIM24 (Gechijian et al., 2018), and receptor tyrosine kinases (epidermal growth factor receptor, HER2, and c-MET) (Burslem et al., 2018).

In addition to a VHL ligand, the PROTAC platform uses thalidomides to recruit Cereblon (CRBN) that is part of the cullin 4 (CUL4) containing E3 ubiquitin ligase complex CUL4-RBX1-DDB1-CRBN (known as $CRL4^{CRBN}$). Thalidomide-based drugs were developed initially as a sedative, but were withdrawn owing to teratogenicity (Bartlett et al., 2004). Nevertheless, thalidomides have been repurposed for the treatment of leprosum, multiple myeloma, and 5-q-deletion-associated dysplasia. By ligand-affinity chromatography, CRBN was identified as the target of thalidomides (Ito et al., 2010). Moreover, several groups, including Dr. Bradner's group, have reported that thalidomides modulate the activity of E3 ubiquitin ligase in $CRL4^{CRBN}$ and promote ubiquitylation of IKAROS family transcription factors (IKZF1 and IKZF3) (Fischer et al., 2014; Kronke et al., 2014; Lu et al., 2014), casein kinase 1α (Kronke et al., 2015; Petzold et al., 2016), translation termination factor GSPT1 (Matyskiela et al., 2016), and zinc finger protein ZFP91 (An et al., 2017). Subsequently, Dr. Bradner's group hypothesized that a rational design of bifunctional thalidomides-conjugated ligands could induce CRBN-dependent degradation of target proteins. To test the hypothesis, they developed a hybrid molecule named dBET consisting of a thalidomide derivative and JQ1, a ligand of bromodomain and extra-terminal (BET) proteins (Winter et al., 2015; Winter et al., 2017). dBET shows potent and CRBN-dependent degradation activity of BET protein in vitro at 100 nM. In addition, dBET induces BET protein degradation and delays leukemia progression in a tumor xenograft model. Dr. Crews' group also developed a similar hybrid molecule named ARV-825 that shows potent activity to degrade BRD4 protein (Lu et al., 2015). After this achievement, many groups have developed CRBN-recruiting PROTACs against various oncogenic proteins such as BCR-ABL (Lai et al., 2016), BRD9 (Remillard et al., 2017), EML4-ALK (Powell et al., 2018; Zhang et al., 2018), fms related tyrosine kinase 3 (FLT3) (Huang et al., 2018), Bruton's tyrosine kinase (BTK) (Buhimschi et al., 2018), and CDK9 (Olson et al., 2018).

16.3 Features of Protein Knockdown Technology

16.3.1 Comparison with Enzyme Inhibitors

Identification and understanding of oncogenic proteins have led to the development of several successful cancer therapies. From this viewpoint, small molecule inhibitors against protein kinases (e.g., imatinib, crizotinib, gefitinib, and sorafenib) are clinically successful. These small molecule inhibitors are based on an "occupancy strategy" enzyme inhibitors must bind to a target enzyme to inhibit their activity (Cromm et al., 2017; Salami et al., 2017) (Fig. 16.2A). Therefore, high concentrations of these inhibitors should be maintained to ensure active site occupancy and sustain their inhibitory effect. The achievement of sustained high drug levels in vivo is one of the major hurdles for drug development. However, such high concentrations might cause undesirable off-target effects. In contrast, SNIPERs and PROTACs are based on an "event-driven strategy." These degraders suppress protein function by reducing the target protein level (Cromm et al., 2017; Salami et al., 2017) (Fig. 16.2B). Therefore, in the case of oncogenic proteins, tumor cells could not undergo proliferation following degrader treatment until sufficient amounts of oncogenic protein accumulate in cells by de novo synthesis of the protein, which occasionally results in apoptosis of the cells. Consistently, short-term treatment with SNIPERs and PROTACs results in long-lasting suppression of tumor cell proliferation, whereas inhibitors slightly retard the growth of cells after drug removal (Lu et al., 2015; Shibata et al., 2018b). In addition, short-term treatment with SNIPERs and PROTACs markedly reduces the target protein level, which is maintained thereafter.

In contrast to enzyme inhibition, the protein knockdown activity is rather attenuated at high concentrations of SNIPERs and PROTACs. This phenomenon is known as the high dose hook effect, where a certain pharmaceutical activity is interfered with higher concentration of drugs. This effect is explained by the formation of a ternary complex consisting of target protein-SNIPERs or PROTACs-E3 ubiquitin ligase required for the protein knockdown activity that would be suppressed by higher concentrations of the hybrid molecules. Therefore, the protein knockdown activity is attenuated (Fig. 16.3).

As mentioned above, there are many small molecule inhibitors that are clinically successful, and some of them are used as the first-line therapy for patients. However, a significant number of patients develop resistance that is commonly attributable to point mutations in the enzyme domain of the target oncogenic protein. Therefore, novel enzyme inhibitors should be constantly developed in an effort to overcome drug resistance. SNIPERs and PROTACs require a ligand of the target protein, which is sufficient to bind to the target protein. If the target protein has an additional domain other than the enzyme domain, novel ligands might be developed, which bind to such a domain. Incorporation of these ligands into SNIPERs or PROTACs would allow induced degradation of target proteins resistant to enzyme inhibitors.

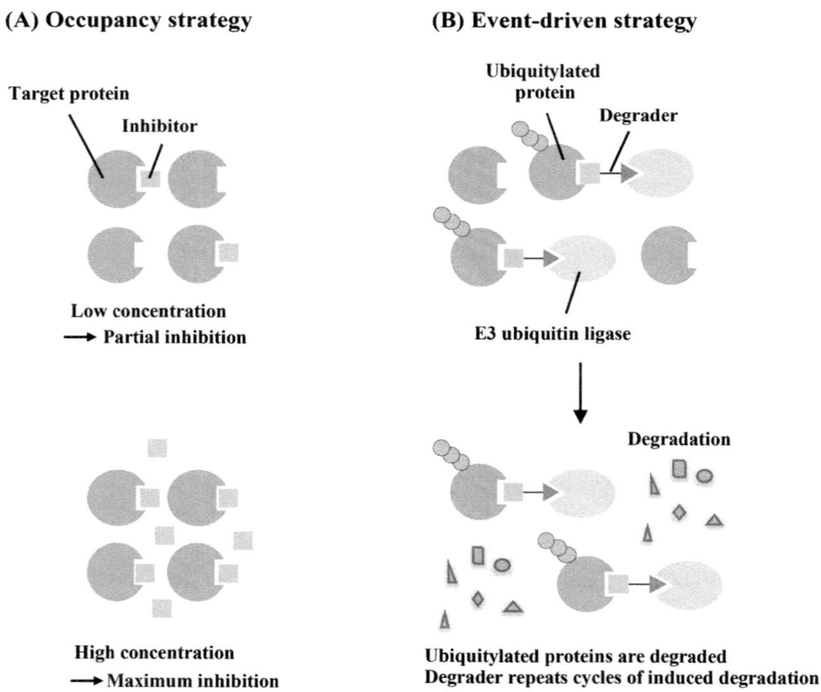

Figure 16.2 Comparison between the occupancy strategy (A) and event-driven strategy (B).

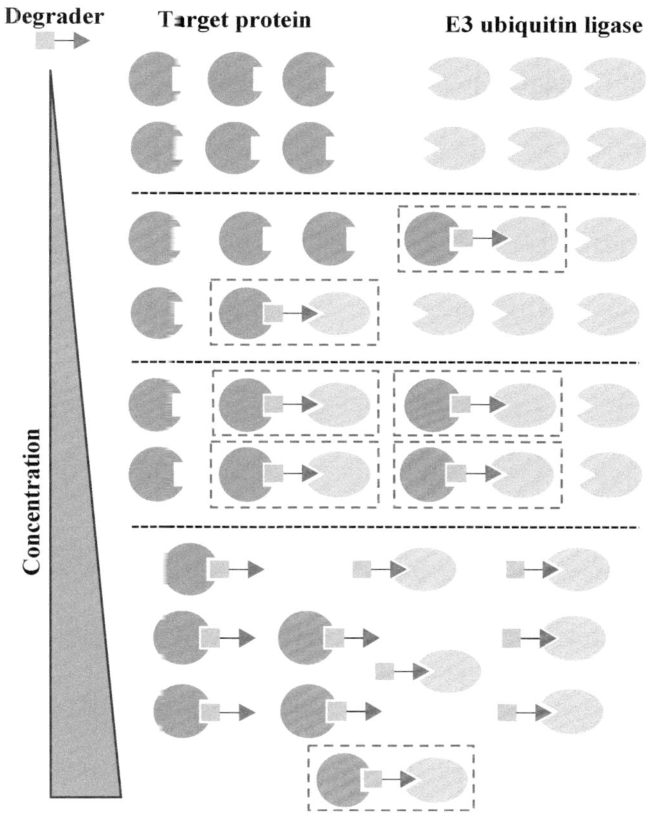

Figure 16.3 The hook effect on formation of a ternary complex.

16.3.2 Comparison with Genetic Knockdown Technologies

In addition to SNIPERs and PROTACs, downregulation of oncogenic proteins can be achieved by genetic knockdown technologies such as antisense oligonucleotides, RNA interference, and genome editing (CRISPR-CAS). Currently, oligonucleotide therapeutics have been developed extensively and are emerging as a new platform for drug development (Yoshida et al., 2018). The U.S. Food and Drug Administration (FDA) have approved five oligonucleotide therapeutics (fomivirsen, pegaptanib, mipomersen, eteplirsen, and nusinersen) for clinical treatment (Stein et al., 2017). Despite its great therapeutic potential, genetic knockdown remains clinically challenging because oligonucleotide therapeutics are not efficiently taken up by cells, and delivery to the desired target tissues is technically difficult. SNIPERs and PROTACs are small molecules, so their uptake by cells and delivery to desired target tissues are expected to be easier compared with genetic knockdown technologies. As mentioned above, several SNIPERs and PROTACs against oncogenic proteins have been successfully developed, and some of them suppress the growth of tumor cells in vivo (Winter et al., 2015; Raina et al., 2016; Ohoka et al., 2017b; Winter et al., 2017).

16.3.3 Comparison with "Mono-Ligand" Type Protein Degraders

Certain small molecule ligands are reported to destabilize proteins. Such "mono-ligand type" small molecules have been developed to induce degradation of target proteins. Fulvestrant is an example of this type of degrader, which binds to its target protein, ERα. Fulvestrant was approved for treatment of ERα-positive breast cancers. Similarly, canertinib (Citri et al., 2002) and arsenic trioxide (Zhang et al., 2001) are clinically approved degraders. Canertinib binds to HER2, and arsenic trioxide binds to PML-RARα.

Thalidomides are another type of protein degrader with a mono-ligand type structure. They interact with CRBN and modulate recognition of substrates by $CRL4^{CRBN}$ E3 ubiquitin ligase. Recently, another modulator of an E3 ubiquitin ligase was reported (Han et al., 2017; Uehara et al., 2017). Anti-tumor sulfonamide indisulam binds to DCAF15 (DDB1- and CUL4-associated factor 15) and modulates recognition of substrates by $CRL4^{DCAF15}$ (CUL4-RBX1-DDB1-DCAF15 complex) E3 ubiquitin ligase, leading to proteasomal degradation of RBM39 (RNA binding motif protein 39).

Although the success of these degraders has attracted much attention for the development of degraders against new target proteins, it is not possible to rationally design a molecule to target any specific protein with mono-ligand type E3-ligase modulators. In contrast, SNIPERs and PROTACs are chimeric molecules containing target ligand modules. Therefore, it is possible to rationally design a degrader against any protein of interest.

16.3.4 Comparison with Therapeutic Antibodies

Currently, therapeutic antibodies are well established as one of the major drug modalities, and some therapeutic antibodies have become very well known (Elvin et al., 2013). The FDA has approved more than 30 therapeutic antibodies for cancer treatment. Among them, antibodies against the PD-1/PD-L1 pathway, such as nivolumab, pembrolizumab, and atezolizumab, are the most successful drugs. The efficacy of therapeutic antibodies stems from neutralization, antibody-dependent cellular cytotoxic activity, complement-dependent cytotoxic activity, or application as a drug delivery carrier (Suzuki et al., 2015). The crucial difference that can be targeted by therapeutic antibodies and chemical protein knockdown technology is the localization of target proteins, i.e., antibodies interact with extracellular proteins, whereas SNIPERs and PROTACs degrade intracellular proteins. Thus, the two drug modalities should be differentially applied depending on the localization of target proteins.

16.4 Prospect of Chemical Protein Knockdown Technology

16.4.1 Rational Design and Optimization of SNIPERs and PROTACs

SNIPERs and PROTACs require a ligand of target proteins. However, there are many proteins for which such ligands have not been found, and it is generally difficult to identify these ligands. In addition, there is no guarantee that all proteins can be degraded by protein knockdown technology, because it is dependent on various factors such as protein stability and the position of lysine residues in the proteins. To address these issues, we and others have developed a protein degrader that recognizes a tag fused to a target protein, such as the HaloTag (Neklesa et al., 2013; Tomoshige et al., 2015), hexa-His (Hattori et al., 2017; Okitsu et al., 2018), and mutated 12 kDa FK506-binding protein (FKBP12^{F36V}) (Nabet et al., 2018). This strategy theoretically allows all target proteins to be recognized by a particular degrader without a specific ligand, and provides an easy method to validate the susceptibility of target proteins to UPS-mediated degradation by protein knockdown technology. Recently, the tag system elucidated the consequence of degrading one of the prevalent oncogenic proteins, KRAS (Nabet et al., 2018). Although the RAS family, including KRAS, is an attractive target for development of anti-tumor drugs, inhibitors targeting RAS are far from clinical translation (Cox et al., 2014). The consequence provides validation that degraders based on protein knockdown technology are a potentially new approach for RAS-targeted therapeutics, as well as the motivation to screen ligands that potently and "silently" bind to KRAS.

After the validation and identification of a specific ligand of target proteins, the optimal combination of an E3 ligase ligand and target ligand needs to be determined for degradation of target proteins. Recently, we and other groups have developed degraders against the oncogenic BCR-ABL protein (Lai et al., 2016; Shibata et al., 2017). When dasatinib was incorporated as a ligand of BCR-ABL protein, IAP- and CRBN-recruiting degraders showed effective protein knockdown activity against the BCR-ABL protein, whereas a VHL-recruiting degrader showed negligible activity (Lai et al., 2016; Shibata et al., 2017; Shibata et al., 2018b). However, when HG-7-85-01 (another inhibitor of BCR-ABL kinase) was incorporated, the VHL-recruiting degrader showed effective protein knockdown activity, whereas IAP- and CRBN-recruiting degraders showed weak or no activity, respectively (Shibata et al., 2018b). Most likely, effective degraders recruit E3 ligases to an appropriate position, so that the lysine residues on the surface of BCR-ABL can be ubiquitylated. Accordingly, development of VHL- and CRBN-recruiting PROTACs based on the promiscuous kinase inhibitor ferotinib revealed that the degradation profiles of the two PROTACs were not identical, and that there was no correlation between the binding affinity of either PROTAC to the kinases and the extent of PROTAC-induced degradation of the kinases (Bondeson et al., 2018). Thus, a suitable combination of an E3 ligase ligand and target ligand is critically important to develop a potent degrader.

Optimization of the degrader is needed for efficient and selective degradation of target proteins. Useful information is the crystal structure of the ternary complex, the target protein-SNIPER, or PROTAC-E3 ubiquitin ligase, although there is room for discussion on how these data should be used for development of SNIPERs and PROTACs. Recently, Dr. Ciulli's group solved the crystal structure of MZ1 (a VHL-recruited PROTAC against BRD4)-bound ELOB-ELOC-VHL-BRD4^{BD2} (the second bromodomain of BRD4) complexes, and found that cooperative binding of VHL to BRD4^{BD2} contributes to effective and specific degradation of BRD4^{BD2} by the PROTAC (Gadd et al., 2017). Based on the crystal structure, they developed a highly selective and novel degrader of BRD4 (Gadd et al., 2017). Similarly, Dr. Fischer's group resolved the crystal structures of dBET (CRBN-recruited PROTAC against BRD4)-bound DD1ΔB [harboring an internal deletion of the flexible BPB propeller (Petzold et al., 2016)]-CRBN-BRD4^{BD1} (the first bromodomain of BRD4) complexes. However, X-ray crystal structures together with biochemical, cellular, and computational characterizations showed that the effective activity and selectivity of the degraders did not correlate with the strong binding or positive cooperativity between CRBN and BRD4 (Nowak et al., 2018). They demonstrated the plasticity between CRBN and BRD4, which results in several distinct low energy binding conformations, leading to the effective activity and selectivity of the degrader-induced protein degradation. In addition, they demonstrated that in silico protein docking can facilitate the design of a BRD4-selective degrader (Nowak et al., 2018). Thus, the effect of cooperativity between the target protein and E3 ubiquitin ligase on the activity and selectivity of degradation remains unclear. It might be dependent on the combination of the E3 ubiquitin ligase and target protein, and the issue should be resolved in the future. Because crystal structure analysis could be helpful to design and develop highly selective degraders, such analysis together with in silico protein docking should be performed to optimize SNIPERs and PROTACs.

16.4.2 Drug-Like Properties

Lipinski's rule of five is a guideline for oral bioavailability and has probably been the most influential concept in preclinical drug development (Lipinski et al., 2001). It shows that poor absorption or permeability of a compound are more likely when there are more than five hydrogen bond donors, the molecular mass is more than 500 Da, the lipophilicity is high, and the sum of nitrogen and oxygen atoms is more than 10. Although the rules do not cover some drugs, such as natural products, SNIPERs and PROTACs violate at least the condition of molecular mass. Therefore, it is necessary to improve the bioavailability of SNIPERs and PROTACs for clinical use. It is noteworthy that an orally bioavailable AR PROTAC (ARV-110) that demonstrated efficacy against prostate cancer was presented at the 2018 Genitourinary Cancers Symposium (Neklesa et al., 2018). More optimization of degraders at the linker or the combination of ligands would be more effective to improve their bioavailability. A cell-penetrating peptide might also be helpful.

16.4.3 Need for Development of a Novel E3 Ubiquitin Ligase Ligand

Here, we described that SNIPERs, and VHL- and CRBN-recruiting PROTACs achieve potent activity to induce degradation of target proteins in vitro and in vivo. So far, only three E3 ubiquitin ligases (IAPs, CRL2^{VHL}, and CRL4^{CRBN}) can be used in chemical protein knockdown technology. Currently, an obvious issue is whether other E3 ubiquitin ligases can be used for the protein knockdown platform. The answer is probably yes. However, we need more potent small molecules. Previously, researchers attempted to use ligands of other E3 ligases to recruit them, such as MDM2 (Schneekloth et al., 2008) and CRL3^{KEAP1} (Lu et al., 2018). The attempts were conceptually successful, but the MDM2-recruiting degrader showed low degradation activity of the target protein, and the KEAP1-recruiting degrader required peptide sequences for KEAP1 recognition. Therefore, it is necessary to identify a more potent small molecule binder for E3 ubiquitin ligases.

Although incorporating ligands of IAPs, VHL, and CRBN leads to successful development of various potent degraders, it should not be limited to these ligands. To expand chemical protein knockdown technology as a more promising novel drug modality, it is necessary to identify a novel small molecule that binds to other E3 ubiquitin ligases. Hopefully, ligands of disease-specific, tissue-specific, or cellular localization-specific E3 ubiquitin ligases can be developed, which would be helpful to minimize off-target effects and maximize on-target effects of degraders.

16.5 Conclusion

Owing to the discovery of novel and potent small molecules that bind to E3 ubiquitin ligases (IAPs, CRL2^{VHL}, and CRL4^{CRBN}), chemical protein knockdown technology has recently progressed, and it is now possible to develop degraders with a potent activity in vivo. These recent advances shed light on various features and issues of the technology, which should be addressed. Protein knockdown technology may become a promising drug modality for development of novel drugs and could be applied to "undruggable" oncogenic proteins.

Acknowledgment

This study was supported, in part, by Japan Society for the Promotion of Science (KAKENHI Grant Numbers 26860050 and 18K07311 to N.S., 26860049 and 18K06567 to N.O., 16H05090 to T.H. and M.N., and 16K15121 to N.O. and M.N.), by the Project for Cancer Research And Therapeutic Evolution (P-CREATE) (JP17cm0106522j0002 to N.S. and JP17cm0106124j0002 to N.O.) and the Research on Development of New Drugs (JP16ak0101029j1403 to M.N.) from the Japan Agency for Medical Research and Development (AMED), by Takeda Science Foundation (to N.O.) and by Takeda Pharmaceutical Co. Ltd. (to M.N.). The authors thank M. Seki for measurement of the protein knockdown activities, and the Mitchell Arico from Edanz Group (www.edanzediting.com/ac) for editing a draft of this manuscript.

References

An, J., Ponthier, C. M., Sack, R., Seebacher, J., Stadler, M. B., Donovan, K. A., Fischer, E. S., *Nat. Commun.*, **8** (2017), 15398.

Bartlett, J. B., Dredge, K., Dalgleish, A. G., *Nat. Rev. Cancer*, **4** (2004), 314–322.

Bondeson, D. P., Mares, A., Smith, I. E., Ko, E., Campos, S., Miah, A. H., et al., *Nat. Chem. Biol.*, **11** (2015), 611–617.

Bondeson, D. P., Smith, B. E., Burslem, G. M., Buhimschi, A. D., Hines, J., Jaime-Figueroa, S., Wang, J., Hamman, B. D., Ishchenko, A. Crews, C. M., *Cell. Chem. Biol.*, **25** (2018), 78–87.

Buckley, D. L., Gustafson, J. L., Van Molle, I., Roth, A. G., Tae, H. S., Gareiss, P. C., Jorgensen, W. L., Ciulli, A. Crews, C. M., *Angew. Chem. Int. Ed. Engl.*, **51** (2012a), 11463–11467.

Buckley, D. L., Van Molle, I., Gareiss, P. C., Tae, H. S., Michel, J., Noblin, D. J., Jorgensen, W. L., Ciulli, A. Crews, C. M., *J. Am. Chem. Soc.*, **134** (2012b), 4465–4468.

Buhimschi, A. D., Armstrong, H. A., Toure, M., Jaime-Figueroa, S., Chen, T. L., Lehman, A. M., Woyach, J. A., Johnson, A. J., Byrd, J. C., Crews, C. M., *Biochemistry*, **57** (2018), 3564–3575.

Burslem, G. M., Smith, B. E., Lai, A. C., Jaime-Figueroa, S., McQuaid, D. C., Bondeson, D. P., Toure, M., Dong, H., Qian, Y., Wang, J., Crew, A. P., Hines, J. Crews, C. M., *Cell. Chem. Biol.*, **25** (2018), 67–77.

Citri, A., Alroy, I., Lavi, S., Rubin, C., Xu, W., Grammatikakis, N., Patterson, C., Neckers, L., Fry, D. W., Yarden, Y., *EMBO J.*, **21** (2002), 2407–2417.

Cohen, P. Tcherpakov, M., *Cell*, **143** (2010), 686–693.

Cox, A. D., Fesik, S. W., Kimmelman, A. C., Luo, J., Der, C. J., *Nat. Rev. Drug Discov.*, **13** (2014), 828–851.

Cromm, P. M., Crews, C. M., *Cell Chem. Biol.*, **24** (2017), 1181–1190.

Demizu, Y., Okuhira, K., Motoi, H., Ohno, A., Shoda, T., Fukuhara, K., Okuda, H., Naito, M. Kurihara, M., *Bioorg. Med. Chem. Lett.*, **22** (2012), 1793–1796.

Demizu, Y., Shibata, N., Hattori, T., Ohoka, N., Motoi, H., Misawa, T., Shoda, T., Naito, M. Kurihara, M., *Bioorg. Med. Chem. Lett.*, **26** (2016), 4865–4869.

Elvin, J. G., Couston, R. G., van der Walle, C. F., *Int. J. Pharm.*, **440** (2013), 83–98.

Fischer, E. S., Bohm, K., Lydeard, J. R., Yang, H., Stadler, M. B., Cavadini, S., et al., *Nature*, **512** (2014), 49–53.

Frost, J., Galdeano, C., Soares, P., Gadd, M. S., Grzes, K. M., Ellis, L., Epemolu, O., Shimamura, S., Bantscheff, M., Grandi, P., Read, K. D., Cantrell, D. A., Rocha, S. Ciulli, A., *Nat. Commun.*, **7** (2016), 13312.

Fulda, S., Vucic, D., *Nat. Rev. Drug Discov.*, **11** (2012), 109–124.

Gadd, M. S., Testa, A., Lucas, X., Chan, K. H., Chen, W., Lamont, D. J., Zengerle, M., Ciulli, A., *Nat. Chem. Biol.*, **13** (2017), 514–521.

Galdeano, C., Gadd, M. S., Soares, P., Scaffidi, S., Van Molle, I., Birced, I., Hewitt, S., Dias, D. M., Ciulli, A., *J. Med. Chem.*, **57** (2014), 8657–8663.

Gechijian, L. N., Buckley, D. L., Lawlor, M. A., Reyes, J. M., Paulk, J., Ott, C. J., Winter, G. E., Erb, M. A., Scott, T. G., Xu, M., Seo, H. S., Dhe-Paganon, S.,

Kwiatkowski, N. P., Perry, J. A., Qi, J., Gray, N. S., Bradner, J. E., *Nat. Chem. Biol.*, **14** (2018), 405–412.

Han, T., Goralski, M., Gaskill, N., Capota, E., Kim, J., Ting, T. C., Xie, Y., Williams, N. S., Nijhawan, D., *Science*, **356** (2017) eaal3755.

Hattori, T., Okitsu, K., Yamazaki, N., Ohoka, N., Shibata, N., Misawa, T., Kurihara, M., Demizu, Y., Naito, M., *Bioorg. Med. Chem. Lett.*, **27** (2017), 4478–4481.

Huang, H. T., Dobrovolsky, D., Paulk, J., Yang, G., Weisberg, E. L., Doctor, Z. M., Buckley, D. L., Cho, J. H., Ko, E., Jang, J., Shi, K., Choi, H. G., Griffin, J. D., Li, Y., Treon, S. P., Fischer, E. S., Bradner, J. E., Tan, L. Gray, N. S., *Cell Chem. Biol.*, **25** (2018), 38–99 e86.

Ito, T., Ando, H., Suzuki, T., Ogura, T., Hotta, K., Imamura, Y., Yamaguchi, Y., Handa, H., *Science*, **327** (2010), 1345–1350.

Itoh, Y., Ishikawa, M., Kitaguchi, R., Okuhira, K., Naito, M. Hashimoto, Y., *Bioorg. Med. Chem. Lett.*, **22** (2012), 4453–4457.

Itoh, Y., Ishikawa, M., Naito, M. Hashimoto, Y., *J. Am. Chem. Soc.*, **132** (2010), 5820–5826.

Kronke, J., Fink, E. C., Hollenbach, P. W., MacBeth, K. J., Hurst, S. N., Udeshi, N. D., et al., *Nature*, **523** (2015), 183–188.

Kronke, J., Udeshi, N. D., Narla, A., Grauman, P., Hurst, S. N., McConkey, M., et al., *Science*, **343** (2014), 301–305.

Lai, A. C., Toure, M., Hellerschmied, D., Salami, J., Jaime-Figueroa, S., Ko, E., Hines, J. Crews, C. M., *Angew. Chem. Int. Ed. Engl.*, **55** (2016), 807–810.

Lipinski, C. A., Lombardo, F., Dominy, B. W., Feeney, P. J., *Adv. Drug Deliv. Rev.*, **46** (2001) 3–26.

Lu, M., Liu, T., Jiao, Q. Ji, J., Tao, M., Liu, Y., You, Q., Jiang, Z., *Eur. J. Med. Chem.*, **146** (2018), 251–259.

Lu, G., Middleton, R. E., Sun, H., Naniong, M., Ott, C. J., Mitsiades, C. S., Wong, K. K., Bradner, J. E., Kaelin, W. G., Jr., *Science*, **343** (2014), 305–309.

Lu, J., Qian, Y., Altieri, M., Dong, H., Wang, J., Raina, K., Hines, J., Winkler, J. D., Crew, A. P., Coleman, K., Crews, C. M., *Chem. Biol.*, **22** (2015), 755–763.

Matyskiela, M. E., Lu, G., Ito, T., Pagarigan, B., Lu, C. C., Miller, K., Fang, W., Wang, N. Y., Nguyen, D., Houston, J., Carmel, G., Tran, T., Riley, M., Nosaka, L. Lander, G. C., Gaidarova, S., Xu, S., Ruchelman, A. L., Handa, H., Carmichael, J., Daniel, T. O., Cathers, B. E., Lopez-Girona, A., Chamberlain, P. P., *Nature*, **535** (2016), 252–257.

Nabet, B., Roberts, J. M., Buckley, D. L., Paulk, J., Dastjerdi, S., Yang, A., Leggett, A. L., Erb, M. A., Lawlor, M. A., Souza, A., Scott, T. G., Vittori, S., Perry,

J. A., Qi, J., Winter, G. E., Wong, K. K., Gray, N. S., Bradner, J. E., *Nat. Chem. Biol.*, **14** (2018), 431–441.

Neklesa, T. K., Noblin, D. J., Kuzin, A. P., Lew, S., Seetharaman, J., Acton, T. B., Kornhaber, G. J., Xiao, R., Montelione, G. T., Tong, L., Crews, C. M., *ACS Chem. Biol.*, **8** (2013), 2293–2300.

Neklesa, T., Snyder, L. B., Willard, R. R., Vitale, N., Raina, K., Pizzano, J., Gordon, D. A., Bookbinder, M., Macaluso, J., Dong, H., Liu, Z., Ferraro, C., Wang, G., Wang, J., Crews, C. M., Houston, J., Crew, A. P., Taylor, I., *J. Clin. Oncol.*, **36**(6)**_suppl** (2018), 381–381.

Nowak, R. P., DeAngelo, S. L., Buckley, D., He, Z., Donovan, K. A., An, J., Safaee, N., Jedrychowski, M. P., Ponthier, C. M., Ishoey, M., Zhang, T., Mancias, J. D., Gray, N. S., Bradner, J. E., Fischer, E. S., *Nat. Chem. Biol.*, **14** (2018), 706–714.

Ohoka, N., Morita, Y., Nagai, K., Shimokawa, K., Ujikawa, O., Fujimori, I., Ito, M., Hayase, Y., Okuhira, K., Shibata, N., Hattori, T., Sameshima, T., Sano, O., Koyama, R., Imaeda, Y., Nara, H., Cho, N. Naito, M., *J. Biol. Chem.*, **293** (2018), 6776–6790.

Ohoka, N., Nagai, K., Hattori, T., Okuhira, K., Shibata, N., Cho, N., Naito, M., *Cell Death. Dis.*, **5** (2014), e1513.

Ohoka, N., Nagai, K., Shibata, N., Hattori, T., Nara, H., Cho, N., Naito, M., *Cancer Sci.*, **108** (2017a), 1032–1041.

Ohoka, N., Okuhira, K., Ito, M., Nagai, K., Shibata, N., Hattori, T., Ujikawa, O., Shimokawa, K., Sano, O., Koyama, R., Fujita, H., Teratani, M., Matsumoto, H., Imaeda, Y., Nara, H., Cho, N. Naito, M., *J. Biol. Chem.*, **292** (2017b), 4556–4570.

Okitsu, K., Hattori, T., Misawa, T., Shoda, T., Kurihara, M., Naito, M., Demizu, Y., *J. Med. Chem.*, **61** (2018), 576–582.

Okuhira, K., Demizu, Y., Hattori, T., Ohoka, N., Shibata, N., Nishimaki-Mogami, T., Okuda, H., Kurihara, M., Naito, M., *Cancer Sci.*, **104** (2013), 1492–1498.

Okuhira, K., Ohoka, N., Sai, K., Nishimaki-Mogami, T., Itoh, Y., Ishikawa, M., Hashimoto, Y., Naito, M., *FEBS Lett.*, **585** (2011), 1147–1152.

Okuhira, K., Shoda, T., Omura, R., Ohoka, N., Hattori, T., Shibata, N., Demizu, Y., Sugihara, R., Ichino, A., Kawahara, H., Itoh, Y., Ishikawa, M., Hashimoto, Y., Kurihara, M., Itoh, S., Saito, H., Naito, M., *Mol. Pharmacol.*, **91** (2017), 159–166.

Olson, C. M., Jiang, B., Erb, M. A., Liang, Y., Doctor, Z. M., Zhang, Z., Zhang, T., Kwiatkowski, N., Boukhali, M., Green, J. L., Haas, W., Nomanbhoy,

T., Fischer, E. S., Young, R. A., Bradner, J. E., Winter, G. E., Gray, N. S., *Nat. Chem. Biol.*, **14** (2018), 163–170.

Petzold, G., Fischer, E. S., Thoma, N. H., *Nature*, **532** (2016), 127–130.

Powell, C. E., Gao, Y., Tan, L., Donovan, K. A., Nowak, R. P., Loehr, A., Bahcall, M., Fischer, E. S., Janne, P. A., George, R. E., Gray, N. S., *J. Med. Chem.*, **61** (2018), 4249–4255.

Raina, K., Lu, J., Qian, Y., Altieri, M., Gordon, D., Rossi, A. M., Wang, J., Chen, X., Dong, H., Siu, K., Winkler, J. D., Crew, A. P., Crews, C. M., Coleman, K. G., *Proc. Natl. Acad. Sci. U. S. A.*, **113** (2016), 7124–7129.

Remillard, D., Buckley, D. L., Paulk, J., Brien, G. L., Sonnett, M., Seo, H. S., Dastjerdi, S., Wuhr, M., Dhe-Paganon, S., Armstrong, S. A., Bradner, J. E., *Angew. Chem. Int. Ed. Engl.*, **56** (2017), 5738–5743.

Rodriguez-Gonzalez, A., Cyrus, K., Salcius, M., Kim, K., Crews, C. M., Deshaies, R. J., Sakamoto, K. M., *Oncogene*, **27** (2008), 7201–7211.

Sakamoto, K. M., Kim, K. B., Kumagai, A., Mercurio, F., Crews, C. M., Deshaies, R. J., *Proc. Natl. Acad. Sci. U. S. A.*, **98** (2001), 8554–8559.

Sakamoto, K. M., Kim, K. B., Verma, R., Ransick, A., Stein, B., Crews, C. M., Deshaies, R. J., *Mol. Cell Proteomics*, **2** (2003), 1350–1358.

Salami, J., Crews, C. M., *Science*, **355** (2017), 1163–1167.

Schneekloth, A. R., Pucheault, M., Tae, H. S., Crews, C. M., *Bioorg. Med. Chem. Lett.*, **18** (2008), 5904–5908.

Sekine, K., Takubo, K., Kikuchi, R., Nishimoto, M., Kitagawa, M., Abe, F., Nishikawa, K., Tsuruo, T., Naito, M., *J. Biol. Chem.*, **283** (2008), 8961–8968.

Shibata, N., Miyamoto, N., Nagai, K., Shimokawa, K., Sameshima, T., Ohoka, N., Hattori, T., Imaeda, Y., Nara, H., Cho, N., Naito, M., *Cancer Sci.*, **108** (2017), 1657–1666.

Shibata, N., Nagai, K., Morita, Y., Ujikawa, O., Ohoka, N., Hattori, T., Koyama, R., Sano, O., Imaeda, Y., Nara, H., Cho, N., Naito, M., *J. Med. Chem.*, **61** (2018a), 543–575.

Shibata, N., Shimokawa, K., Nagai, K., Ohoka, N., Hattori, T., Miyamoto, N., Ujikawa, O., Sameshima, T., Nara, H., Cho, N., Naito, M., *Sci. Rep.*, **8** (2018b), 13549.

Shimokawa, K., Shibata, N., Sameshima, T., Miyamoto, N., Ujikawa, O., Nara, H., Ohoka, N., Hattori, T., Cho, N. Naito, M., *ACS Med. Chem. Lett.*, **8** (2017), 1042–1047.

Stein, C. A., Castanotto, D., *Mol. Ther.*, **25** (2017), 1069–1075.

Suzuki, M., Kato, C. Kato, A., *J. Toxicol. Pathol.*, **28** (2015), 133–139.

Tomoshige, S., Naito, M., Hashimoto, Y., Ishikawa, M., *Org. Biomol. Chem.*, **13** (2015), 9746–9750.

Uehara, T., Minoshima, Y., Sagane, K., Sugi, N. H., Mitsuhashi, K. O., Yamamoto, N., Kamiyama, H., Takahashi, K., Kotake, Y., Uesugi, M., Yokoi, A., Inoue, A., Yoshida, T., Mabuchi, M., Tanaka, A. Owa, T., *Nat. Chem. Biol.*, **13** (2017), 675–680.

Winter, G. E., Buckley, D. L., Paulk, J., Roberts, J. M., Souza, A., Dhe-Paganon, S., Bradner, J. E., *Science*, **348** (2015), 1376–1381.

Winter, G. E., Mayer, A., Buckley, D. L., Erb, M. A., Roderick, J. E., Vittori, S., Reyes, J. M., di Iulio, J., Souza, A., Ott, C. J., Roberts, J. M., Zeid, R., Scott, T. G., Paulk, J., Lachance, K., Olson, C. M., Dastjerdi, S., Bauer, S., Lin, C. Y., Gray, N. S., Kelliher, M. A., Churchman, L. S., Bradner, J. E., *Mol. Cell*, **67** (2017), 5–18 e19.

Yaron, A., Gonen, H., Alkalay, I., Hatzubai, A., Jung, S., Beyth, S., Mercurio, F., Manning, A. M., Ciechanover, A., Ben-Neriah, Y., *EMBO J.*, **16** (1997), 6486–6494.

Yaron, A., Hatzubai, A., Davis, M., Lavon, I., Amit, S., Manning, A. M., Andersen, J. S., Mann, M., Mercurio, F., Ben-Neriah, Y., *Nature*, **396** (1998), 590–594.

Yoshida, T., Naito, Y., Sasaki, K., Uchida, E., Sato, Y., Naito, M., Kawanishi, T., Obika, S., Inoue, T., *Genes Cells*, **23** (2018), 448–455.

Zhang, T. D., Chen, G. Q., Wang, Z. G., Wang, Z. Y., Chen, S. J., Chen, Z., *Oncogene*, **20** (2001), 7146–7153.

Zhang, C., Han, X. R., Yang, X., Jiang, B., Liu, J., Xiong, Y., Jin, J., *Eur. J. Med. Chem.*, **151** (2018), 304–314.

Chapter 17

Resistance Mechanisms of Tumor Cells

Rolf Marschalek

Johann Wolfgang Goethe-University,
Department of Biochemistry, Chemistry and Pharmacy,
Institute of Pharmaceutical Biology/Diagnostic Center of Acute Leukemia,
Max-von-Laue-Str. 9, D-60438 Frankfurt/Main, Germany

rolf.marschalek@em.uni-frankfurt.de

17.1 Introduction

The average risk of women and men developing a malignant disease during their lifetime is about 37.7% and 39.7%, respectively, and the risk of dying from that disease is 18.8% and 22.3%, respectively (SEER's database). Therefore, the topic of tumor cells and their potential resistance mechanisms is probably quite important for many of us. Tumors have been historically classified according to their tissue origin into carcinomas (epithelial tumors: breast, prostate, lung, pancreas, and colon), sarcomas (mesenchymal origin: bone, cartilage, fat and nerve), lymphomas and leukemias (hematopoietic cells), germ cell tumors (pluripotent cells: testicle

Pharmaceutical Biocatalysis: Fundamentals, Enzyme Inhibitors, and Enzymes in Health and Diseases
Edited by Peter Grunwald

ISBN 978-981-4800-61-7 (Hardcover), 978-0-429-29503-4 (eBook)
www.jennystanford.com

and ovary) or blastomas (embryonic tissue and precursor cells). These classifications are usually further subspecified by organ names (e.g., colon carcinoma).

Cancer cells have the capability to grow infinitely. In addition, results obtained during the last years suggest that cancer cells strongly modulate their cellular environment for their own benefit. Finally, cancer cells need to cope with the systemic response of the host that tries to kill the cancer cells with immune cells. If a tumor is successful in coping with all these adverse situations, it will start to disseminate into the body and finally kill the host— if no therapeutic intervention occurs.

17.2 Development of Solid Tumors

In general, somatic cells in a multicellular organism are usually in the G0 phase of their cell cycle because of outside signals, usually exerted by the extracellular matrix or by direct cell–cell contacts. Therefore, a cancer cell derives from a somatic cell that has acquired a series of gene mutations which allowed this cell to overcome its normal, growth inhibitory environment. In solid tumors, this usually occurs in a nearly fixed sequel of consecutive mutations that have been identified by studying the different stages and gradings of various tumors, as, e.g., described by Vogelstein and Kinzler (1993). In recent years, these data have been confirmed independently by whole genome sequencing experiments of different human tumor types (see Fig. 17.1, upper right panel A). The picture that is emerging from these big data indicates that solid tumors display a narrow spectrum of gene mutations that are highly cancer-specific, and partly overlapping with similar spectra of mutations in other tumor types. Beside the numerous subclonal or patient-specific mutations, most of the identified mutations can be summarized in a "mutational landscape," resembling a relatively small number of genes. Within these genes that can be annotated to different cellular pathways, a subgroup belongs to the class of either "tumor suppressor genes" or "oncogenes." Genes of both groups are either deleted/mutated on both alleles (tumor suppressors), or represent monoallelic mutated versions of so-called "proto-oncogenes" that are necessary for normal cell growth.

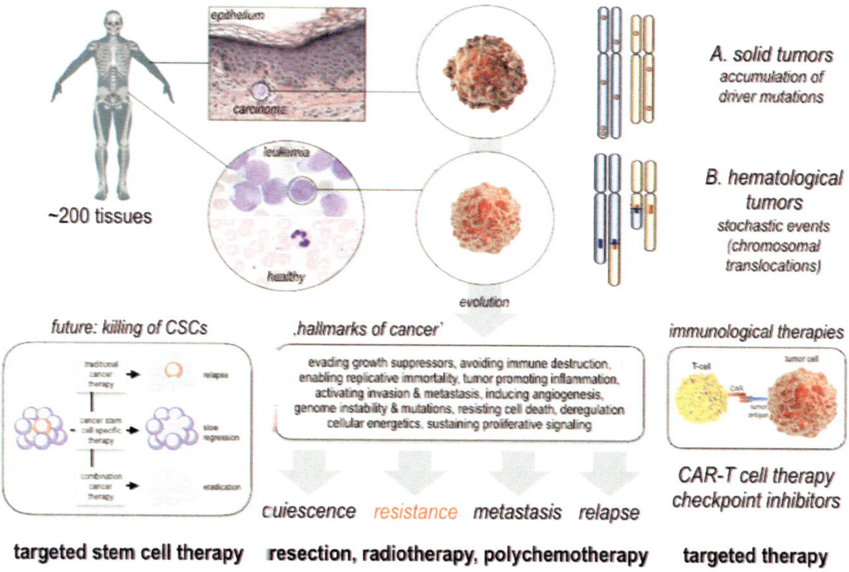

Figure 17.1 Mechanism of cancer development and treatment options. Human beings exhibit about 200 different tissues. Basically, we can divide tumors into solid and hematological tumors. These two tumor types are genetically different, as solid tumor accumulate over time many “driver mutations,” while hematological tumors mostly occur after a single genetic event (chromosomal translocations). All tumor cells display a genetic evolution to manifest certain properties important for their survival (hallmarks of cancer). These acquired features help the tumor cells to become resistant against therapy, to metastasize, to relapse, or to become quiescent. Targeted therapies against cancer stem cells (CSCs) will be important for the future, while CAR-T cell therapy against hematological tumors and checkpoint inhibitors against solid tumors have been quite successfully tested in clinical trials.

Sequencing of cancer genomes has also led to the identification of “driver mutations,” necessary to convert a normal cell into a premalignant or malignant cell, and “passenger mutations” which are useful for an individual tumor but not necessarily important in general. The “mutational landscapes” of different tumor types (Stransky et al., 2011; Hodis et al., 2012; Stephens et al., 2012; Kandoth et al., 2013; Vogelstein et al., 2013; Wang et al., 2014; Choi et al., 2015; Suzuki et al., 2015; Nik-Zainal et al., 2016; Northcott et al., 2017) has been published and allows to define common cancer pathways, which fit perfectly to the already described "hallmarks in cancer" (as described by Hanahan and Weinberg, 2000, 2011). These "hallmarks in cancer" are describing the all necessary changes to convert a somatic cell into a tumor cell (evading growth suppressors, avoiding immune destruction, enabling replicative immortality, tumor promoting inflammation, activating invasion and metastasis, inducing angiogenesis, genome instability and mutations, resisting cell death, deregulation cellular energetics, sustaining proliferative signaling (see Fig. 17.1, lower middle panel).

17.3 Development of Hematological Tumors

Almost all hematological tumors do not exhibit such a mutational landscape rather than display recurrent chromosomal translocations that result in "chimeric fusion genes" (see Fig. 17.1, upper right panel B). These fusions genes are highly specific and prototypic for leukemia disease subtypes (ALL, AML, CML, and CLL). The term "chromosomal translocation" describes an illegitimate exchange of chromosome material between two or more non-homologous chromosomes. In most cases, these illegitimate events are caused by a DNA damage situation where a subsequent DNA repair process results in wrongly fused "derivative chromosomes" (Reichel et al., 1998; Richardson et al., 1998). Chromosomal translocations can be subcategorized by the results of their illegitimate recombination event: (1) overexpression of proto-oncogenes by fusing them to strong enhancers, or (2) creation of chimeric fusion genes. Since these chromosomal changes occur recurrently at nearly precise points of our genomes, one might argue that these genes represent recombination hot spots. DNA damage situation may occur more frequently due to endogenous events (like, e.g., early apoptosis, torsional stress, DNase I hypersensitive sites, etc.) or exogenous xenobiotic stress conditions (Strick et al., 2000).

Hematological tumors can be categorized into leukemias and lymphomas, due to their site of origin (onset in primary or secondary lymphatic organs) which perfectly correlate with their aggressiveness. In general, leukemias derive from hematological stem or precursor cells, while lymphomas derive from mature cells. Therefore, lymphomas represent per se benign tumors—growing over years in many patients. These tumors are mostly diagnosed in patients with grade II or III and in stage III or IV due to organ infiltration (splenomegaly or swollen lymph nodes). By contrast, acute leukemias always start in the bone marrow (B-ALL, AML) or thymus (T-ALL), are highly aggressive, and quickly infiltrate many organs when starting to disseminate out of the bone marrow. Acute leukemias are quite dangerous and can kill a patient within a few weeks if no adequate treatment is applied.

Since chromosomal translocations are the hallmark of nearly all hematological tumors (a few leukemias are associated with either hyperdiploid of hypodiploid genomes), very few secondary mutations were usually identified in whole genome sequencing projects. Secondary aberrations could be diagnosed only in subpopulations, indicating that they are not necessarily driving tumor development rather than prepare the ground for selection processes (Dobbins et al., 2013; Martelli et al., 2013; Anderson et al., 2015; Bodini et al., 2015; Lindqvist et al., 2015; Stieglitz et al., 2015; Ding et al., 2017; Bolouri et al., 2018). Based on our current knowledge, the chimeric fusion genes are sufficient to cause the onset and maintenance of the different disease types (AML, ALL). Thus, these chimeric fusion genes are highly malignant when expressed in the appropriate hematopoietic (stem- or precursor cell) compartment. These fusion genes change either the epigenetic set-up and/or transcriptional profiles (e.g., MLL-r leukemias), or convert existing "physiological" into "oncogenic signaling pathways" (e.g., BCR-ABL). The latter cases can be quite successfully treated by the available TKIs (tyrosine kinase inhibitors such as Imatinib) and KIs (kinase inhibitors). The success of these novel therapies for certain patient groups has led to the term "personalized medicine," which is based on the rational treatment of patient based on their diagnosed gene mutations (or present 'fusion gene').

Note: Specific chromosomal translocations are also diagnosed in solid tumors. Today, 360 out of about 600 cancer genes listed in the Sanger Cancer Census Database are recurrently involved in chromosomal translocations (60%), indicating the general importance of this genetic mechanism for all human cancers (see http://cancer.sanger.ac.uk/census).

17.4 General Treatment of Tumors and Patient Outcome

The different tumors have been treated over decades by three major strategies: surgical resection, radiotherapy, and polychemotherapy. Treatment regimens (timing, doses, and the right sequel of combinations of chemotherapeutics) have been evolved in many clinical trials over nearly 7 decades in order to maximize the survival rates of cancer patients. Survival of patients is usually visualized by Kaplan-Meier curves, displaying "event free survival" (EFS) and "overall survival" (OS) over a time frame of 5–10 years. Despite this clinical success, roughly 50% of all cancer patients are still dying from their relapses (18.8% out of 37.7% female patients, and 22.3% out of 39.7% male patients). This is mainly due to therapy resistance and the presence of cancer (stem) cells which allow survival and regrowth of tumor cells even after very long latency.

17.5 Concept of Cancer Stem Cells

It is worth mentioning that the concept of "cancer stem cells" in solid tumors was not known for a very long time; rather it became adapted from conceptional studies performed with hematological tumor models (as described by Bonnet and Dick, 1997), which were subsequently validated also for solid tumors (Jordan, 2004). The novel concept of cancer stem cells has changed the mind of many researchers and physicians, because all the treatment schedules that have been established in clinical studies over decades have been optimized for killing of the large tumor burden, rather than targeting specifically the very small compartment of cancer stem cells. Importantly, this stem cell population is not only responsible for the creation of the tumor bulk but also most likely the source for relapses in tumor patients.

The ability to selectively target and kill cancer stem cells will be the next big challenge in tumor medicine. Proof-of-concept data already exist for several tumor types. One of them is the t(15;17) translocation which is creating the PML-RARA fusion protein typically found in acute promyelocytic leukemia (APML).

This type of leukemia is a rare subform of AML (FAB type M3) which was successfully treated for the first time by high doses of all-trans retinoic acid (ATRA) to induce differentiation of cancer cells into normal cells (Hoe et al., 1992). The implementation of the ATRA treatment regimen into the normal chemotherapeutic backbone resulted immediately in survival rates up to 70–80% in later studies. However, treatment was even better when the ATRA treatment schedule was accompanied by the addition of arsenic trioxide (ATO, AS_2O_3; Chen et al., 1996). It turned out that arsenite was selectively killing the cancer stem cells, because ATO causes the selective degradation of the oncogenic PML-RARA fusion protein (Zheng et al., 2007). Ongoing research over the next decade(s) will help to uncover more examples of such strategies which are coping with the stem cell problem in other tumor entities (see Fig. 17.1, lower left panel).

17.6 Immunotherapies

Another possible solution is to use of immunological therapies that could be easily combined with conventional therapies. Targeted therapies using CAR-T (or CAR-NK) cells are quite successful, because those cells may have the power to systematically eliminate cancer stem cells, bulk tumor cells and to generate "memory" against a given tumor type in patients (Hurton et al., 2016; Song et al., 2017). This type of cellular treatment has lead to the novel term "therapy with living drugs" (see Fig. 17.1, lower right panel). The observed memory effect in some patients could be of great importance, because it may prevent potential relapses. Another immunotherapy treatment strategy, namely the use of checkpoint inhibitors, will be discussed below.

17.7 Therapy Resistance Mechanisms

Besides all these developments in the past decades, resistance against a diverse panel of therapies is still the major challenge for clinicians when treating tumor patients. Tumor cells usually develop evading strategies to survive a hostile environment. This is particularly true in multicellular organisms with an active immune system. In principle, five different mechanisms contribute to the resistance of tumor cells against modern therapies: (1) They simply express certain ABC transporters that pump chemotherapeutic drugs to the outside of the tumor cell; (2) they change their DNA repair capacity—mostly by inflicting with the p53 pathway; (3) tumor cells acquire features of stem cells and go into quiescence, most likely in a niche where oxygen and other supplies are rare; (4) tumor cells change their (epi)genetic program and manipulate their environment (by exosomes and cytokines) in order to maintain a growth advantage; (5) finally the tumor cells directly express surface ligands (e.g., PD-L1) that calm down any hostile activity of immune cells. All these mechanisms are important features for tumor cells. They need to be addressed when trying to combat a given tumor. All of these mechanisms will be highlighted in Sections 17.7.1 to 17.7.5 to understand the large variety of tumor cell resistance mechanisms.

17.7.1 Multidrug Resistance

Multidrug resistance is a major problem for many therapeutic interventions, because the expression of distinct multidrug transporter proteins (ATP-driven ABC transporters) is a natural defense mechanism of our cells and, obviously, also part of the detoxifying capacity of cancer cells. Several ABC transporter families are known to confer this ability. However, clinically most relevant were only very few, which are ABCB1 (MDR1, PGP), ABCC1 (MRP1), and ABCG2 (MXR, ABCP). They represent enzymes that lower the adverse effects of poisons by simply pumping them in an ATP-dependent manner to the outside (efflux transporters). They are ubiquitously expressed, but certain organs or tissues have additional ABC transporters for protection. MDR1 was the first efflux transporter ever described because of its ability to confer drug resistance to cancer cells (Roninson et al., 1986). It is preferentially pumping uncharged or positively charged hydrophobic compounds, while MRP1 pumps organic anions including drug-conjugates (glutahione, glucuronate, or sulfate). MXR pumps several drugs like, e.g., topoisomerase inhibitors, anthracyclines, camptothecin, TKI, antimetabolites mitoxanthrones, anthracyclines, topothecan, flavopiridol, and methotrexate. Noteworthy, most ABC transporters can be inactivated by a distinct set of drugs.

Drugs known to inhibit MDR1 (Amiodaron, Azithromycin, Captopril, Chinidin, Chinin, Clarithromycin, Ciclosporin, Elacridar, Piperin, Quercetin, Reserpin, Ritonavir, Tariquidar, and Verapamil) can be used to manipulate tumor cells to make them more sensitive against chemotherapy. Similarly, nucleoside reverse transcriptase inhibitors (NRTIs: Zidovudin, Stavudin, Lamivudin, Didanosin, Abacavir) or non-nucleoside reverse transcriptase inhibitors (NNRTIs: Nevirapin, Efavirenz, and Delavirdin)—usually used for HIV treatment—have the capability to inhibit MRP1 in a dose-dependent fashion (Weiss et al., 2007). MRP1 is also very important for the maintenance of the blood–brain barrier and thus counteracts brain tumor treatment. In those cases, both Reversan (inhibitor of MDR1 and MRP1; Burkhardt et al., 2009) and MK571 (inhibitor of MRP1 and MRP4; Gekeler et al., 1995) can be used to enhance treatment efficacy (Tivnan et al., 2015). The fungal toxin Fumitremorgin C (Rabindran et al., 2000) and Curcumin (Zhou et al., 2015) act both as inhibitors of ABCG2 and breast cancer resistance protein (BCRP: mediates resistance to mitoxantrone and doxorubicin); however, clinical studies are yet missing to use this compound alone or in combination with the above-mentioned chemotherapeutic drugs for the treatment of patients (Hasanabady and Kalalina, 2016).

17.7.2 TP53 Mutations

TP53 has been discovered in 1981 as a tumor-specific phosphoprotein (Crawford et al., 1981). The literature of p53—the guardian of our genome—is largely overwhelming with nearly 67,000 publications. However, the key points about how p53 becomes activated and functionally coordinated in healthy cells can be briefly summarized: The p53 protein is usually bound to MDM2 which itself represents a p53 specific E3-ligase that cause a permanent degradation of p53 (Momand et al., 1992; Wu et al., 1993). Upon DNA damage, MDM2 becomes inactivated by phosphorylation through the ATM-CHK2 and ATR-CHK1 signaling pathways (reviewed in Rotman and Shiloh, 1997; Tibbetts et al., 1999), or directly by ARF (CDKN2A product; Zhang et al., 1998). In addition, both signaling pathway, ATM-CHK2 and ATR-CHK1, activate p53 protein by phosphorylation at several serine residues (e.g., Ser-15, Ser-20, Ser-37, Ser-313, Ser-314, Ser-377, Ser-378).

These multiple phosphorylations stabilize p53 by disrupting the p53::MDM2 interaction and allow p53 to tetramerize. The resulting p53 tetramers are able to bind to target gene promoters of several cell cycle control factors (e.g., CDKN1A), or to promoters the expression of pro-apoptotic proteins (like PUMA, NOXA, and BAX). Under physiological conditions, this leads to a cell cycle arrest, and—if not repaired in an appropriate time window—to the induction of apoptosis.

Although the human genome is designed in a way that important gene functions have a certain redundancy, e.g., by gene families, this is not the case for TP53: the human genome comprises only a single copy of this important gene. Genetic changes in TP53 are therefore of great importance (Bullock and Fersht, 2001) and found in various degrees in all diagnosed tumor types (lung 80–95%; ovary: 90%; head and neck: 70–80%; breast: 15–80%; esophagus: 40–60%; bladder: 50%; colon and brain: 20–60%). Missense mutations in TP53 are mainly found in the DNA binding domain (~95%), and therefore the presence of a single mutated TP53 allele in a cell causes a 75% drop of p53 activity. This is due to the fact that only a p53 tetramer, consisting of four intact p53 proteins, remains functional. In addition, mutant p53 protein is more stable than wild-type p53 protein, and interacts with a large variety of other transcription factors (NF-Y, E2Fs, NFkBp65, NfkBp50, SREBP, YAP, VDR, or NRF2) to steer proliferation, invasion, and metastasis. Binding of mutant p53 to p63 inhibits p63-mediated metastasis suppressor functions (Muller et al., 2014), and mutant p53 can regulate the expression of SHARP1 and Dicer and thus influences a complete network of miRNAs (Wei et al., 2014; Luo et al., 2018).

Haploinsufficiency of TP53—either by a point-mutation or allele deletion—has the negative side effect that tumor cells become prone to accumulate mutations because of the missing cell cycle control and pro-apoptotic function. This "mutator phenotype" helps the tumor cell to progress into an evolutionary phase where the "most effective mutations" become positively selected. The resulting pattern of "secondary mutations" can be recurrently found in cancer patients but differs slightly when different tumor types are analyzed.

Fortunately, there are several drugs available (e.g., CP-31398, PRIMA-1, ARP-246, MIRA-1, STIMA-1, a.o.) that selectively bind to distinct TP53 mutants and partially restore their activity, or, inhibit at least the TP53 interactions with other proteins (e.g., TP73; Blandino and Di Agostino, 2018; Bykov et al., 2018). For most of these drugs, it has been demonstrated by in vitro—and in some cases also by in vivo experiments—that they are highly potent to re-sensitize tumor cells that bear different types of p53 mutations. Future studies will unravel the power of restoring TP53 activity in tumor cells, but all experimental evidence looks quite promising, and hopefully these drugs will quickly enter larger clinical trials to circumvent the deleterious effects of TP53 mutations.

17.7.3 Acquiring Stem Cell Features and Quiescence

Tumor cells are quite dangerous because of their infinite cell growth in conjunction with their ability to accumulate secondary mutations and to metastasize in other organs. Moreover, tumor cells often acquire "stem cell like features." This concept was first described in 1959 (Makino, 1959), with several decades of FACS-phenotyping experiments that allowed to isolate and characterize cancer stem cells (Grunt, 2001; Kopper and Hajdú, 2004). From 2004 onward, the concept of stochastic and hierarchic cancer stem cell compartments have been investigated in different tumors, mostly driven by technical advances that helped to classify and genetically analyze those cells. In 2007, the first publications about "tumor-initiating cells"—equivalent to cancer stem cells—became evident (Lee and Herlyn, 2007). In these experiments, isolated cancer stem cells were transplanted into animal model systems in very low numbers (~10 cells) to generate new tumors.

Nowadays, it is pretty clear that cancer stem cells (termed CSCs in the literature) are the important drivers for the relapses observed in patients. Therefore, the question arises how these stem cell-like features can be acquired. In times of the iPS technology (Takahashi and Yamanaka, 2006), everyone understands that this is presumably not a too complicated task, because stem cell-like features can be acquired by the expression of a specific set of proteins. These can be either the naturally expressed triumvirate of embryonic stem cell genes (NANOG, OCT4, and SOX2) or other factors that provide features of immortality. One important factor is the TERT gene which is usually inactive in somatic cells, while active in stem cells. It is interesting to note that a single nucleotide exchange in the promoter of one TERT allele (C>T at −124 upstream of TSS) allows to reactivate this gene in order to gain immortality (Stern et al., 2015). Others have investigated this in a more comprehensive way to find similar mutations (C228T and C250T) in different tumor entities (Bell et al., 2016), which suggests that it is quite easy to reactivate this important gene which is absolutely necessary for the maintenance of tumor cells.

The above-mentioned factors are able to reprogram a cell back to the state of their origin. Expression of SOX2 and OCT4 factor together with KLF4 and c-MYC seems to be sufficient, as it is the standard technology to create iPS cells (Takahashi and Yamanaka, 2006). These factors steer a small genetic program that allow to reset chromatin back to "start," like in a fertilized egg cell. Tumor cells are using the same or similar mechanism to gain "immortality." Every tumor type has a certain set of such proteins which can be found in the isolated cancer stem cells fraction (e.g., CD133 in different tumor entities). CD133 is upregulated by IGF2 and promote cancer stemness (Xu et al., 2018). However, CD133 is also an important tumor migration marker that can be secreted by tumor cells via exosomes to other tumor cells. This vesicular transfer of CD133 to non-stem cells looks like a "transplantation strategy" of the tumor to generate more stem cells (Yang et al., 2018). CD133 plays also an important role for stemness in many brain tumors.

Similarly, breast cancer stem cells with an $ALDH^+/CD44^+/CD24^-$ signature display a high activity of phosphorylated STAT3 which is representing an Achilles heel of breast cancer stem cells. Inhibiting P-STAT3 by different methods led to an inhibition of stem cell features and impaired the successful engraftment in mammary fat pad mouse models (Lin et al., 2013).

In the hematopoietic system, other proteins are also able to exhibit these features. The EGR1 transcription factor (as well as EGR2 or EGR3) allows stem cells to go into homeostasis or dormancy to maintain them (Min et al., 2008; reviewed in Kühn et al., 2016). EGR1 has initially identified as activator of p21 and as gatekeeper of the TP53 (Krones-Herzig et al., 2003). Cells (over)expressing EGR1 protein can potentially escape treatment, and are presumably one of the reasons for relapses. A dormant cell needs only to switch back from this dormant state to the normal growth program. Recently, it has been shown that CDK4 but mainly overexpressed CDK6 (under certain stress conditions) re-activates dormant stem cells and causes tumor cell formation (Scheicher et al., 2015). Similarly, other factors have been described, such as HOXB4 (an OCT4 and GATA2 downstream target gene; Huang et al., 2016), which is capable of inducing sufficient amounts of the RUNX1 transcription factor to maintain hematopoietic stem cells (Teichweyde et al., 2017).

In summary, the ability of different tumors to create their own stem cell compartment makes it harder to treat them. A "dormant state" allows these cells to become insensitive for certain treatments by chemotherapeutic agents and thus represents a natural reservoir for relapsing cells. Future studies on these cancer stem cells and how they can be therapeutically addressed (e.g., by mobilization out of their niche) will hopefully improve the outcome of patients.

17.7.4 Manipulating the Niche

It is becoming more and more evident that "stem cells" are crosstalking with their "niche" where they are residing in. This form of communication is life sustaining for the cancer stem cells. Exemplarily, the molecular "crosstalk" of leukemic stem cells with their "niche" allows the production of cytokines necessary for the leukemic stem cells to grow and survive, while concomitantly inhibiting normal HSCs (as outlined in Chen et al., 2016). This microenvironment includes stromal cells (cancer-associated fibroblasts, mesenchymal cells), infiltrating immune cells (B, T, DC and Macrophages) and vascular structures, including extracellular matrix (ECM) proteins. Tumor cells use cell-to-cell contacts, secrete soluble factors, and release exosomes to manipulate the gene expression of surrounding cells (Plaks et al., 2015; Kumar et al., 2016). These events are necessary to initiate and propagate the tumor cells and influence the outcome of neoplastic transformation processes.

In particular, exosome production by cancer stem cells has been recognized as a crucial element for tumor initiation, development and protection (Wu et al., 2017; Sharma, 2018). Exosomes are created by a very special pathway in our cells, namely by the ESCRT-system (Juan and Fürthauer, 2018), the same system which is being used by Hepatitis C virus for its replication and packaging (Ariumi et al., 2011). In healthy cells, the ESCRT system allows to build a big vacuole by folding the ER membrane. This resulting structure contains many smaller vesicles (exosomes) with distinct membrane proteins (Alix, Tetraspanins) and of distinct size (30–100 nm) which is then termed "multi-vesicular body" (MVB). All the vesicles contain "cargo molecules" which consists of proteins, mRNAs, and/or miRNA or lncRNA, etc. This MVB can subsequently be fused with the outer membrane in order to release all exosomes. Exosomes are then distributed via body fluids, but mainly act in cis by fusing these cargo-loaded exosomes directly with cells in their neighborhood. In case of cancer stem cells, this could be the above-mentioned stromal cells, cancer-associated fibroblast, dendritic cells, or immune cells. This way, a cancer stem cell directly influences the surrounding tissue and provokes the production of molecules for their benefit. These niche interactions also allow the change in the metabolism of surrounding cells (e.g., increase fatty acid production) in order to feed tumor cells, or the secretion of ligands that favor tumor cell growth (Cruz et al., 2018). Therefore, we can assume that acquiring novel insights into the process of exosome formation in different cancer types will also allow in the future the use of self-made, artificial exosomes as novel drugs to disrupt these cancer stem cell/niche interactions. In addition, several drugs are known that influence the formation of exosomes or other extracellular vesicles (EVs). They might be used in the future to disrupt this "molecular reprogramming" of niche cells by cancer (stem) cells and thus may interfere with the persistence of cancer stem cells.

17.7.5 Triggering Checkpoints

Another interesting mechanism of cancer (stem) cells is the manipulation of the body's immune cells to guarantee their own survival. For this purpose, tumor cells express molecules on their cell surface (B7, PD-L1, or PD-L2), which allows them to functionally control T-cells and other immune cells (Iwai et al., 2002). In most cases, PD-L1 is the key molecule expressed on solid tumors. This ligand binds to PD-1 on T-cells, a functional homolog of CTLA-4, and represses any T-cell activity directed towards the tumor cell. Basically, T cells are initially very strongly activated by several signaling pathways, including TCR/CD3, CD28, and IL2R. However, T-cell effector functions are usually active only for a couple of days, until the surface receptor CD28 is being exchanged into CTLA-4. CTLA-4 (like PD-1) signaling is blocking T cell effector functions and initiates the important homeostatic phase of any immune reactions. When tumor cells express PD-L1 on their surface, they simply trigger down any T-cell activity which could be harmful for them (Norde et al., 2011). Thus, a tumor cell can be surrounded by hundreds of T-cells, but none of them will attack the corresponding tumor cell when expressing PD-L1. This knowledge has led to the development of "checkpoint inhibitors," specific antibodies blocking the PD-1::PD-L1 interaction, as novel drugs to re-activate T-cell functions directed against tumor cells (as described by Suzuki et al., 2014; Callahan and Wolchok, 2013). Interestingly, this new type of therapy was quite successful in those cancer types that express many neoantigens, like melanoma, different lung cancers, and different gastro-intestinal tumors (Schumacher and Schreiber, 2015). This is an interesting point to acknowledge, because if a tumor is too similar to normal host cells, it may prevent the attack of its own immune system (see Fig. 17.1, lower right panel).

17.8 Summary

Taking together all this information, it becomes clear that tumor cells have the capacity to evade different treatment regimens and also the endogenous defense strategies provided by the host immune system. Tumor cells manipulate their surrounding tissue, while transporters help to stay away from therapeutic drugs. Finally, if all fail-safe mechanisms are not working efficiently, there is still the dormancy state as a final escape strategy to remain alive. Of interest, analyses of the localization of leukemic cells after chemotherapy revealed another interesting defense mechanism: AML cells survived chemotherapy when residing next to the adipose tissue in the bone marrow (Shafat et al., 2017). It could also be shown that the tumor cells secrete molecules that induced lipolysis in the fat cells in order to get fed. This again indicates how pivotal these tumor-niche interactions really are. In that study, feeding of the tumor cells by fat cells of the bone marrow provided the pro-tumoral microenvironment necessary to survive the selective pressure of the applied chemotherapy.

This makes clear how complicated it is to cure an individual cancer in a given patient. Cancer cells have developed all kinds of mechanisms to survive diverse treatment situations. A successful treatment will only be possible if we address different defense mechanisms concomitantly by different treatment strategies. The recent success in the treatment of different tumors by checkpoint inhibitors (solid tumors) or CAR-T cells (hematological disorders) is only one possibility. Drugs that inhibit exosomes production/secretion could be important to interfere with the niche/cancer (stem) cell crosstalk. Alternatively, the use of self-made exosomes which are targeting the niche/cancer (stem) cell interaction could be of great benefit. Similarly, drugs that manipulate ABC transporters or reactivate mutant p53 will sensitize tumor cells to make standard therapies more effective. Taken together, the intelligent combination of targeted (molecular) therapies with conventional treatments will improve the outcome of cancer patients in the future.

References

Andersson, A. K., Ma, J., Wang, J., Chen, X., Gedman, A. L., Dang, J., et al., *Nat. Genet.*, **47** (2015), 330–337.

Ariumi, Y., Kuroki, M., Maki, M., Ikeda, M., Dansako, H., Wakita, T., et al., *PLoS One*, **6** (2011), e14517.

Bell, R. J., Rube, H. T., Xavier-Magalhães, A., Costa, B. M., Mancini, A., Song, J. S., et al., *Mol. Cancer Res.*, **14** (2016), 15–23.

Blandino, G., Di Agostino, S., *J. Exp. Clin. Cancer Res.*, **37** (2018), 30.

Bodini, M., Ronchini, C., Giacò, L., Russo, A., Melloni, G. E., Luzi, L., et al., *Blood*, **125** (2015), 600–605.

Bolouri, H., Farrar, J. E., Triche, T. Jr., Ries, R. E., Lim, E. L., Alonzo, T. A., et al., *Nat. Med.*, **24** (2018), 103–112.

Bonnet, D., Dick, J. E., *Nat. Med.*, **3** (1997), 730–737.

Bullock, A. N., Fersht, A. R., *Nat. Rev. Cancer*, **1** (2001), 68–76.

Burkhart, C. A., Watt, F., Murray, J., Pajic, M., Prokvolit, A., Xue, C., et al., *Cancer Res.*, **69** (2009), 6573–6580.

Bykov, V. J. N., Eriksson, S. E., Bianchi, J., Wiman, K. G., *Nat. Rev. Cancer*, **18** (2018), 89–102.

Callahan, M. K., Wolchok, J. D., *J. Leukoc. Biol.*, **94** (2013), 41–53.

Chen, G. Q., Zhu, J., Shi, X. G., Ni, J. H., Zhong, H. J., Si, G. Y., et al., *Blood*, **88** (1996),1052–1061.

Choi, J., Goh, G., Walradt, T., Hong, B. S., Bunick, C. G., Chen, K., et al., *Nat. Genet.*, **47** (2015), 1011–1019.

Crawford, L. V., Pim, D. C., Gurney, E. G., Goodfellow, P., Taylor-Papadimitriou, J., *Proc. Natl. Acad. Sci. U. S. A.*, **78** (1981), 41–45.

Cruz, L., Arevalo Romero, J. A., Brandão Prado, M., Santos, T. G., Hohmuth Lopes, M., *Stem Cell Rev.*, **14** (2018), 262–276.

Ding, L. W., Sun, Q. Y., Tan, K. T., Chien, W., Mayakonda, A., Yeoh, A. E. J., et al., *Cancer Res.*, **77** (2017), 390–400.

Dobbins, S. E., Sherborne, A. L., Ma, Y. P., Bardini, M., Biondi, A., Cazzaniga, G., et al., *Genes Chrom. Cancer*, **52** (2013), 954–960.

Gekeler, V., Ise, W., Sanders, K. H., Ulrich, W. R., Beck, J., *Biochem. Biophys. Res. Commun.*, **208** (1995), 345–352.

Grunt, T. W., *Methods Mol. Med.*, **39** (2001), 161–168.

Hanahan, D., Weinberg, R. A., *Cell*, **100** (2000), 57–70.

Hanahan, D., Weinberg, R. A., *Cell*, **144** (2011), 646–674.

Hasanabady, M. H., Kalalinia, F., *J. Biosci.*, **41** (2016), 313–324.

Hodis, E., Watson, I. R., Kryukov, G. V., Arold, S. T., Imielinski, M., et al., *Cell*, **150** (2012), 251–263.

Hoe, T. S., Cheong, S. K., Hussin, N. H., Chum, K. W., *Med. J. Malaysia*, **47** (1992), 225–227.

Huang, X., Lee, M. R., Cooper, S., Hangoc, G., Hong, K. S., Chung, H. M., et al., *Leukemia*, **30** (2016), 144–153.

Hurton, L. V., Singh, H., Najjar, A. M., Switzer, K. C., Mi, T., Maiti, S., et al., *Proc. Natl. Acad. Sci. U. S. A.*, **113** (2016), E7788–E7797.

Iwai, Y., Ishida, M., Tanaka, Y., Okazaki, T., Honjo, T., Minato, N., *Proc. Natl. Acad. Sci. U. S. A.*, **99** (2002), 12293–12297.

Jordan, C. T., *Curr. Opin. Cell Biol.*, **16** (2004), 708–712.

Juan, T., Fürthauer, M., *Semin. Cell. Dev. Biol.*, **74** (2018), 66–77.

Kandoth, C., McLellan, M. D., Vandin, F., Ye, K., Niu, B., Lu C., et al., *Nature*, **502** (2013), 333–339.

Kopper, L., Hajdú, M., *Pathol. Oncol. Res.*, **10** (2004), 69–73.

Krones-Herzig, A., Adamson, E., Mercola, D., *Proc. Natl. Acad. Sci. U. S. A.*, **100** (2003), 3233–3238.

Kühn, A., Löscher, D., Marschalek, R., *Oncotarget*, **7** (2016), 35341–35352.

Kumar, B., Garcia, M., Murakami, J. L., Chen, C. C., *Biochim. Biophys. Acta*, **1863** (2016), 464–470.

Lee, J. T., Herlyn, M., *J. Cell. Physiol.*, **213** (2007), 603–609.

Lin, L., Hutzen, B., Lee, H. F., Peng, Z., Wang, W., Zhao, C., et al., *PLoS One*, **8** (2013), e82821.

Lindqvist, C. M., Nordlund, J., Ekman, D., Johansson, A., Moghadam, B. T., Raine, A., et al., *Hum. Mutat.*, **36** (2015), 118–128.

Luo, Z., Cui, R., Tili, E., Croce, C., *Cancer Lett.*, **419** (2018), 96–102.

Makino, S., *Acta Unio Int. Contra Cancrum*, **15** (1959), 196–198.

Martelli, M. P., Sportoletti, P., Tiacci, E., Martelli, M. F., Falini, B., *Blood Rev.*, **27** (2013), 13–22.

Min, I. M., Pietramaggiori, G., Kim, F. S., Passegué, E., Stevenson, K. E., Wagers, A. J., *Cell Stem Cell*, **2** (2008), 380–391.

Momand, J., Zambetti, G. P., Olson, D. C., George, D., Levine, A. J., *Cell*, **69** (1992), 1237–1245.

Muller, P. A., Trinidad, A. G., Caswell, P. T., Norman, J. C., Vousden, K. H., *J. Biol. Chem.*, **289** (2014), 122–132.

Nik-Zainal, S., Davies, H., Staaf, J., Ramakrishna, M., Glodzik, D., Zou, X., et al., *Nature*, **534** (2016), 47–54.

Norde, W. J., Maas, F., Hobo, W., Korman, A., Quigley, M., Kester, M. G., et al., *Cancer Res.*, **71** (2011), 5111–5122.

Northcott, P. A., Buchhalter, I., Morrissy, A. S., Hovestadt, V., Weischenfeldt, J., Ehrenberger, T., et al., *Nature*, **547** (2017), 311–317.

Plaks, V., Kong, N., Werb, Z., *Cell Stem Cell*, **16** (2015), 225–238.

Rabindran, S. K., Ross, D. D., Doyle, L. A., Yang, W., Greenberger, L. M., *Cancer Res.*, **60** (2000), 47–50.

Reichel, M., Gillert, E., Nilson, I., Siegler, G., Greil, J., Fey, G. H., et al., *Oncogene*, **17** (1998), 3035–3044.

Richardson, C., Moynahan, M. E., Jasin, M., *Genes Dev.*, **12** (1998), 3831–3842.

Roninson, I. B., Chin, J. E. Choi, K. G., Gros, P., Housman, D. E., Fojo, A., et al., *Proc. Natl. Acad. Sci U. S. A.*, **83** (1986), 4538–4542.

Rotman, G., Shiloh, Y., *Cancer Surv.*, **29** (1997), 285–304.

Scheicher, R., Hoelbl-Kovacic, A., Bellutti, F., Tigan, A. S., Prchal-Murphy, M., Heller, G., et al., *Blood*, **125** (2015), 90–101.

Schumacher, T. N. Schreiber, R. D., *Science*, **348** (2015), 69–74.

Shafat, M. S., Oellerich, T., Mohr, S., Robinson, S. D., Edwards, D. R., Marlein, C. R., et al., *Blood*, **129** (2017), 1320–1332.

Sharma A., *Int. J. Cancer*, **142** (2018), 1086–1092.

Song, Y., Tong, C., Wang, Y., Gao, Y., Dai, H., Guo, Y., et al., *Protein Cell*, 2017 Mar 10. doi: 10.1007/s13238-017-0384-8. [Epub ahead of print].

Stephens, P. J., Tarpey, P. S., Davies, H., Van Loo, P., Greenman, C., Wedge, D. C., et al., *Nature*, **486** (2012), 400–404.

Stern, J. L., Theodorescu, D., Vogelstein, B., Papadopoulos, N., Cech, T. R., *Genes Dev.*, **29** (2015), 2219–2224.

Stieglitz, E., Taylor-Weiner, A. N., Chang, T. Y., Gelston, L. C., Wang, Y. D., Mazor, T., et al., *Nat. Genet.*, **47** (2015), 1326–1333.

Stransky, N., Egloff, A. M., Tward, A. D., Kostic, A. D., Cibulskis, K., et al., *Science*, **333** (2011), 1157–1160.

Strick, R., Strissel, P. L. Borgers, S., Smith, S. L., Rowley, J. D., *Proc. Natl. Acad. Sci. U. S. A.*, **97** (2000), 4790–4795.

Suzuki, H., Aoki, K., Chiba, K., Sato, Y., Shiozawa, Y., Shiraishi, Y., et al., *Nat. Genet.*, **47** (2015), 458–468.

Suzuki, H., Owada, Y. Watanabe, Y., Inoue, T., Fukuharav, M., Yamaura, T., et al., *Hum. Vaccin Immunother.*, **10** (2014), 352–357.

Takahashi, K., Yamanaka, S., *Cell*, **126** (2006), 663–676.

Teichweyde, N., Horn, P. A., Klump, H., *Transfus. Med. Hemother.*, **44** (2017), 128–134.

Tibbetts, R. S., Brumbaugh, K. M., Williams, J. M., Sarkaria, J. N., Cliby, W. A., Shieh. S. Y., et al., *Genes Dev.*, **13** (1999), 152–157.

Tivnan, A., Zakaria, Z., O'Leary, C., Kögel, D., Pokorny, J. L., Sarkaria, J. N., et al., *Front. Neurosci.*, **9** (2015), 218.

Vogelstein, B., Kinzler, K. W., *Trends Genet.*, **9** (1993), 138–141.

Vogelstein, B., Papadopoulos, N., Velculescu, V. E., Zhou, S., Diaz, L. A. Jr., Kinzler, K. W., *Science*, **339** (2013), 1546–1558.

Wang, K., Yuen, S. T., Xu, J., Lee, S. P., Yan, H. H., Shi, S. T., et al., *Nat. Genet.*, **46** (2014), 573–582.

Wei, S., Wang, H., Lu, C., Malmut, S., Zhang, J., Ren, S., et al., *J. Biol. Chem.*, **289** (2014), 8947–8959.

Weiss, J., Theile, D., Ketabi-Kiyanvash, N., Lindenmaier, H., Haefeli, W. E., *Drug Metab. Dispos.*, **35** (2007), 340–344.

Wu, X., Bayle, J. H., Olson, D., Levine, A. J., *Genes Dev.*, **7** (1993), 1126–1132.

Wu, J., Qu, Z., Fei, Z. W., Wu, J. H., Jiang, C. P., *Oncol. Lett.*, **13** (2017), 2855–2866.

Xu, W. W., Li, B., Zhao, J. F., Yang, J. G., Li, J. Q., Tsao, S. W., et al, *Cancer Lett.*, **425** (2018), 88–100.

Yang, F., Xing, Y., Li, Y., Chen, X., Ai, Z., Wei, Y., et al., *Mol. Cell. Biol.*, 2018 May 14. pii: MCB.00024-18. doi: 10.1128/MCB.00024-18. [Epub ahead of print].

Zhang, Y., Xiong, Y., Yarbrough, W. G., *Cell*, **92** (1998), 725–734.

Zheng, X., Seshire, A., Rüster, B., Bug, G., Beissert, T., Puccetti, E., et al., *Haematologica*, **92** (2007), 323–331.

Zhou, Q., Ye, M., Lu, Y., Zhang, H., Chen, Q., Huang, S., et al., *PLoS One*, **10** (2015), e0136694.

Chapter 18

Biocatalytic Nanoreactors for Medical Purposes

Oscar González-Davis, Kanchan Chauhan, and Rafael Vazquez-Duhalt

Centro de Nanociencias y Nanotecnología, Universidad Nacional Autónoma de México, Km 107 carretera Tijuana-Ensenada, Ensenada, Baja California 22860, México

rvd@cnyn.unam.mx

18.1 Introduction

Nanotechnology is a rapidly expanding field, in which materials at the nanoscale are synthesized to take advantage of enhanced properties such as higher strength, lighter weight, increased electrical conductivity, and chemical reactivity compared to their larger-scale equivalents. Nanomaterials are already revolutionizing several industrial fields by the introduction of new processes, and unique materials for aeronautics, energy generation, biomedical and environmental applications, and coating, among others. Nanotechnology has significantly impacted the development of medicine. Nanomedicine is a branch of medicine that applies

Pharmaceutical Biocatalysis: Fundamentals, Enzyme Inhibitors, and Enzymes in Health and Diseases
Edited by Peter Grunwald

ISBN 978-981-4800-61-7 (Hardcover), 978-0-429-29503-4 (eBook)
www.jennystanford.com

the nanotechnological tools for the prevention and treatment of disease. Nanomedicine involves the use of nanomaterials, such as biocompatible nanoparticles and nanodevices, for diagnosis, delivery, sensing, or actuation purposes in a living organism.

A significant research effort is currently focused on the use of nanoparticles (NPs) for drug delivery. The design of nanocarriers (NCs) containing a specific drug that could be targeted to specific tissues, reducing the doses of drug needed by increasing the treatment efficiency is the primary goal of these smart drug delivery systems. Nevertheless, the NPs could also be used as NCs for enzymatic activity. The treatment of disorders originated from enzymatic activity deficiencies using therapeutic enzymes was first described 50 years ago (Duve, 1966). There are many diseases originated from the lack of one or more enzymatic activities, and because of this, in some cases, the administration of exogenous enzymes has been successfully used as enzyme replacement therapy (ERT).

Enzyme replacement therapy has been recently reviewed (Yari et al., 2017; Grabowski and Hopkin, 2003; Solomon and Muro, 2017). The success of ERT is mainly based on enzyme biochemistry, local enzyme concentrations, the ability of treated cells to turn over and be replaced by normal cells, and the substrate movement, redistribution, and storage. All these factors could be modulated and improved by the use of enzyme functionalized NCs.

In this chapter, the potential use of NPs as carriers for enzymes with medical relevance is critically reviewed and discussed.

18.2 Enzymes of Medical Importance

18.2.1 Enzyme Deficiency Disorders

Adenosine Deaminase (ADA) deficiency prevents the degradation of adenosine and deoxyadenosine into inosine and deoxyinosine, which are ultimately degraded into uric acid, leading to a severe combined immunodeficiency (Blackburn and Kellems, 2005). The most notable effect is on lymphocytes. The affected patients show persistent infections and pulmonary insufficiency since early infancy (Treem et al., 1999). Bone marrow transplantation and enzyme therapy have proved successful in treating ADA deficiency and other genetic disorders (Buckley et al., 1997). Polyethylene glycol (PEG) covered ADA was the first commercial PEGylated protein for ERT. The results of ERT have shown lower mortality and morbidity than those from haploidentical bone marrow transplantation (Hershfield, 1995).

Congenital sucrose-isomaltase deficiency (CSID) is a rare inherited metabolic disorder characterized by the deficiency or absence of the enzymes sucrase and isomaltase. These enzymes are involved in the digestion of certain sugars such as sucrose and dextrin, and their deficiency causes ingested sucrose and starch not to be able to be adequately absorbed from the gut (Gericke et al., 2017; Puntis and Zamvar, 2015) Symptoms are variable and include watery diarrhea, abdominal distension, and discomfort, bloating sometimes noticeably worse after ingestion of fruit. It is well known that adding microorganisms, especially *Saccharomyces cerevisiae*, which shows high sucrase activity, allows patients to maintain a regular diet. Administration of sacrosidase (a liquid preparation produced from *S. cerevisiae* containing sucrase and isomaltase activities) has produced a marked reduction in symptoms with no adverse events (Puntis and Zamvar, 2015).

Fabry disease is a disorder linked to glycosphingolipid metabolism caused by α-galactosidase A (α-Gal) deficiency. The successful use of ERT for Fabry disease has been reported (Alfadhel and Sirrs, 2011). Recombinant agalsidase alfa and agalsidase beta have been compared for Fabry disease treatment (Arends et al., 2018). The authors found no difference between patients treated with either agalsidase, and a more pronounced reduction of storage materials was found with agalsidase beta.

Gaucher's disease is an inherited disorder caused by deficient activity of the enzyme β-glucocerebrosidase, found mainly in lysosomes. The lack of glucocerebrosidase activity results in an accumulation of glucocerebroside in the lysosomes of macrophages, especially in the reticuloendothelial system. The accumulation of glucocerebroside leads to hepatomegaly, splenomegaly, anemia, and thrombocytopenia causing fatigue, discomfort, infections, bleeding and bruising. In addition, it induces bone related problems such as pain, bone crises, and avascular necrosis. Other problems such as lung disease, impaired growth, and delayed puberty are also associated with Gaucher's disease (Grabowski et al., 1998; Stirnemann et al., 2017). The clinical effectiveness of the ERT has been reviewed (Grabowski et al., 1998; Connock et al., 2006). A recombinant and glycoengineered glucocerebrosidase containing mannosyl-ended oligosaccharides has been designed in order to be recognized by the specific receptors for α-mannosyl on macrophages (Oh, 2015). Also, a plant-derived variant of glucocerebrosidase has been mannosyl-targeted to disease-affected cells (Tekoah et al., 2013). Thus, the targeted delivery of enzyme for the specific treatment of Gaucher's disease is feasible.

Glycogen storage disease is an inherited metabolic disorder caused by the absence or deficiency of one of the enzymes responsible for glycogen synthesis, breakdown or glycolysis, mainly within muscles and liver cells (Özen, 2007). The incidence is estimated at 1 case per 20000–43000 live births. There are several types of glycogen storage disease according to the enzyme deficiency. The enzymes involved are glycogen synthase, glucose-6-phosphatase, acid α-glucosidase, glycogen debranching enzyme, glycogen branching enzyme, glycogen phosphorylase, and others.

Hypophosphatasia is osteomalacia caused by mutations in the gene that encodes the alkaline phosphatase (ALP) that affects the development of bones and teeth (Whyte, 2010; Mornet, 2018). The ALP deficiency disrupts a process called mineralization, in which minerals such as calcium and phosphorus are deposited in developing bones and teeth. The ERT is now available for hypophosphatasia. Asfotase alfa is a recombinant glycoprotein that contains the catalytic domain (the active site) of tissue-nonspecific alkaline phosphatase. Asfotase alfa therapy improved skeletal, respiratory, and physical symptoms with only a few severe side effects in patients with hypophosphatasia, supporting the safety and efficacy of ERT (Kitaoka et al., 2017; Whyte, 2017).

Insufficiency of the pancreatic exocrine gland is the primary consequence of some diseases including chronic pancreatitis, cystic fibrosis and pancreatic tumors (Domínguez-Muñoz, 2011, 2018). In order to maintain a good nutritional status, avoiding digestion problems, it is crucial to administer a mixture of enzymes containing lipases, proteases, and amylases. Several commercial products are available from porcine, bovine, and microbial sources (Berry, 2014). The enzyme preparations are formulated to be taken orally, and thus they should be protected from a low pH until they enter the intestinal lumen. Several preparations are enzymes embedded into microspheres that could be coated with polyacrylic acid, which dissolves at a pH higher than 5.5. It seems essential that the enzymatic activity reach the small intestine and thus enzyme encapsulation is desirable to protect the protein crossing the digestive pathway when orally administered.

Mitochondrial neurogastrointestinal encephalomyopathy is a progressive metabolic disorder caused by thymidine phosphorylase enzyme deficiency. Clinical features include mental regression, ophthalmoplegia, and fatal gastrointestinal complications (Yadak et al., 2017). Thymidine phosphorylase was encapsulated into polymeric NPs (De Vocht et al., 2009) and erythrocytes (Moran et al., 2008) in order to prolong half-life and reduce the immunogenic reactions.

Mucopolysaccharidosis (MPS) is a related lysosomal storage disease originated by the absence of enough enzyme to metabolize mucopolysaccharides. The different subtypes of MPS share some clinical features and show different degrees of severity. The symptoms are not evident at birth, but progress with the storage of glycosaminoglycans (also called mucopolysaccharides) induces a compression of nerves in the spinal cord or the peripheral nervous system. The accumulation of mucopolysaccharides leads to affections in bone, skeletal structure, connective tissues, and organs. Neurological complications may include damage to neurons as well as pain and impaired motor function. Additionally, cognitive impairments, developmental delay, or severe behavioral problems may be encountered (Coutinho et al., 2012).

Mucopolysaccharidosis type I (MPS I) is a rare disease in which the body is missing or does not have enough of α-L-iduronidase needed to break down long chains of glycosaminoglycans. There are several other types of MPS, including MPS II or Hunter syndrome originated by the absence of iduronate-2-sulfatase; MPS III or Sanfilippo syndrome derived by low activity of *N*-acetylglucosaminidase; MPS IV or Morquio syndrome originated by the absence of galactose-6-sulfate sulfatase activity, and MPS VI derived from the absence of *N*-acetylgalactosamine-4-sulfatase.

The ERT has been studied for most of MPS subtypes (Dornelles et al., 2017; Whiteman and Kimura, 2017; Gilkes and Heldermon, 2014; Tomatsu et al., 2015; Vairo et al., 2015).

Phenylketonuria is an inborn error of amino acid metabolism caused by phenylalanine hydroxylase deficiency (Kim et al., 2004). Phenylalanine hydroxylase is a non-heme, iron-containing protein that catalyzes the conversion of phenylalanine to tyrosine and is responsible for the catabolism of most of the dietary phenylalanine. In addition to phenylalanine hydroxylase, phenylalanine ammonia-lyase from plants is being developed for the treatment of Phenylketonuria. This plant enzyme has some advantages when compared with phenylalanine hydroxylase: It does not require cofactors for degrading phenylalanine, the product trans-cinnamic acid is converted to benzoic acid in the liver, which is then excreted via the urine, and it is more stable (Hosking and Gray, 1982).

Pompe disease also called glycogen storage disease type II (GSD II), is an autosomal recessively inherited disease caused by a deficiency of acid α-glucosidase (GAA). Pompe disease patients present hypotonia, feeding problems, hepatosplenomegaly, and cardiomyopathy in early infancy. These patients show progressive weakness due to deposition of glycogen from skeletal muscle (Kishnani and Beckemeyer, 2014) and ERT is the only established drug therapy targeting this enzyme deficiency (Reuser et al., 2002). Recombinant human acid α-glucosidase is administered as an intravenous infusion twice a week. An important factor that can limit the bioavailability of recombinant enzyme is the development of anti-GAA neutralizing antibodies that occur after the start of the enzyme replacement treatment. In the case of infantile Pompe disease, the probability of developing high titers of anti-GAA antibodies is dependent of the cross-reactive immunologic material (Tarnopolsky et al., 2016).

18.2.2 Enzymes for Cancer Therapy

Acute lymphoblastic leukemia, also known as *acute lymphocytic leukemia,* is a type of blood and bone marrow cancer and is the most common type of cancer in children. This disease progresses rapidly and creates immature blood cells that are auxotrophic for asparagine. Thus, L-asparaginase is an essential therapeutic enzyme widely used for the treatment of this specific cancer (Galluzzi et al., 2013; Ghoshoon et al., 2015). The treatment goal is to reduce the concentration of asparagine in blood by the transformation of this amino acid to aspartic acid and therefore induce selective growth inhibition of sensitive tumor cells. The FDA first approved an enzymatic preparation of asparaginase from *Escherichia coli* in 1978, and 15 years later PEGylated asparaginase was approved to avoid undesirable immune responses. The main problem with asparaginase from *E. coli* is that it exhibits a high glutaminase activity, which could render the treatment toxic. Site-directed mutagenesis has been performed in order to reduce the glutaminase activity while maintaining high asparaginase activity (Ramya et al., 2011). Also, in several cases, asparaginase from *E. coli* has had a reduced lifetime due to inactivation. Thus, immunologically distinct asparaginase from *Erwinia chrysanthemi* is being tested (Pieters et al., 2011).

Arginine deprivation (see also Chapter 8) has been proposed for the treatment of different cancers such as acute myeloid leukemia (Miraki-Moud et al., 2015), pancreatic cancer (Liu et al., 2014), and prostate cancer (Hsueh et al., 2012). Arginine deiminase, isolated from Mycoplasma, degrades arginine, which is involved in several essential biosynthetic pathways. PEGylated arginine deiminase has been investigated in clinical trials (Ott et al., 2013) and directed evolution has been performed to improve its catalytic performance for medical applications (Zhu et al., 2010; 2014).

Cytochrome P450 (CYP). So far, chemotherapy is the most used therapy for cancer (Al-Lazikani et al., 2012). There are almost 200 different cytotoxic compounds that are currently available for this use (Huttunen et al., 2008), most of them administered as prodrugs, which have to be transformed into the active anticarcinogenic drug. This transformation is mainly performed by the cytochrome P450 superfamily (Huttunen et al., 2008). When administered, most of these prodrugs are activated in the liver, where CYPs are overexpressed (Bièche et al., 2007). On the other hand, CYP expression varies significantly in the different tissues (Choudharya et al., 2003), and even more, in some cases healthy cells near the tumor cells express higher CYP activity (Zhao et al., 2012), in which the cytotoxic agent is preferentially activated affecting these healthy cells.

In order to increase the generation of the active drug form in situ within the tumor, different strategies have been proposed. The increase of CYP activity in the tumor tissue could induce a more efficient chemotherapeutic treatment and consequent reduction of the amount of drug administered, thereby reducing the toxic systemic side effects. Genetic therapy (see also Vol. 6 of this Series on Biocatalysis) such as gene-directed enzyme prodrug therapy (GDEPT) (Chen et al., 2012) and virus-directed enzyme prodrug therapy (VDEPT) (Palmer et al., 2004; Tychopoulos et al., 2005) have been proposed to increase the CYP activity and sensitize tumor cells to anticancer prodrugs.

18.2.3 Other Medically Relevant Enzymes

Cystic fibrosis is a genetic disorder that mainly affects the lungs. It provokes difficult breathing and coughing up mucus as a result of frequent lung infections. Patients show high concentrations of DNA in the secretions causing an obstruction of the airways. Human DNase and some engineered variants are used to cleave DNA in order to facilitate mucus clearance in the lungs (Konstan and Ratjen, 2012). A DNase variant produced a 35-fold increase in the degradation of DNA when compared to the wild type (Pan and Lazarus, 1997).

Peyronie's disease is caused by the formation of scar tissue inside the penis producing difficulties in obtaining a straight erection. Collagenase is an enzyme that selectively dissolves collagen and is used for the clinical treatment of Peyronie's disease (Hay et al., 2013). The use of NCs has improved enzyme delivery through the dense extracellular matrices (Villegas et al., 2015).

Thrombosis is a major cause of morbidity and mortality in developed countries. Fibrinolytic enzymes transform plasminogen to plasmin which cleaves fibrin causing clot dissolution and blood flow restoration. Streptokinase from *Streptococcus sp.* has been used as a thrombotic agent (Perler, 2005). On the other hand, recombinant human tissue plasminogen activator, which is a serine protease, also transforming plasminogen to plasmin, is currently used in thrombosis treatment. Several variants of tissue plasminogen activator obtained by protein engineering are available in the market (Nordt and Bode, 2003).

18.3 Nanobioreactors

Nanobioreactors are NPs carrying enzymatic activity; they are the combination of the sophisticated catalytic mechanism of enzymes with the exceptional properties of nanomaterials which can be readily functionalized with ligands for specific cell receptors.

As stated by Vriezema et al. (2005), the coupling of reactions in time and space begins with the design and construction of a confined environment within which the conversion reaction will take place. The design of nanobioreactors is based on the attachment, encapsulation, or insertion of enzymes on polymeric supramolecular assemblies in which they can act in situ.

There are many challenges associated with the delivery and retention of enzymatic activity and because of this, biocompatible and inert environments are sought to prevent interference with the native structure of enzymes, which can be highly sensitive to the environment (Holcenberg, 1982; Mitchell et al., 2002). These environments can be modulated by external parameters such as temperature, pH or ionic strength and allow the control of both NPs stability and catalytic activity (Koyani et al., 2017).

The premise of using biomaterial-based nanoparticles for enzyme immobilization and delivery is that they intrinsically possess high reactivity due to their small size and large specific surface area, which tend to facilitate the exposure of active sites for triggering catalytic reactions. Additionally, they can be easily modified with functional groups to covalently bond enzymes, they reduce diffusion limitations while increasing stability and targeting capabilities, they protect them from proteolysis, and most importantly they are compatible with cellular components (Koyani et al., 2017). Moreover, the NC modification with "stealth" moieties, such as PEG, can reduce the immune response and further prevent enzyme degradation increasing their circulation in the bloodstream (Lin et al., 2018; Xie et al., 2009; Zhao et al., 2011). Few examples of nanobioreactors for enzyme delivery are highlighted in Fig. 18.1.

The use of NPs for the transport of therapeutic enzymes has been reviewed (Bosio et al., 2016; Rostro-Alanis et al., 2016; Dean et al., 2017; Mukerabigwi et al., 2018). Nevertheless, most of these works are focused on cross-linked enzymatic aggregates (CLEA), metal-, silica-, and carbon-based nanoparticles, including synthetic polymers. Recently, Koyani et al. (2017) reviewed, discussed, and proposed the use of biomaterials (materials from a biological origin) for enzyme delivery as a better alternative for the design and production of NCs for medical purposes. Among these biomaterials, liposomes, proteins, polysaccharides, and DNA are most mentioned in the literature (Fig. 18.1).

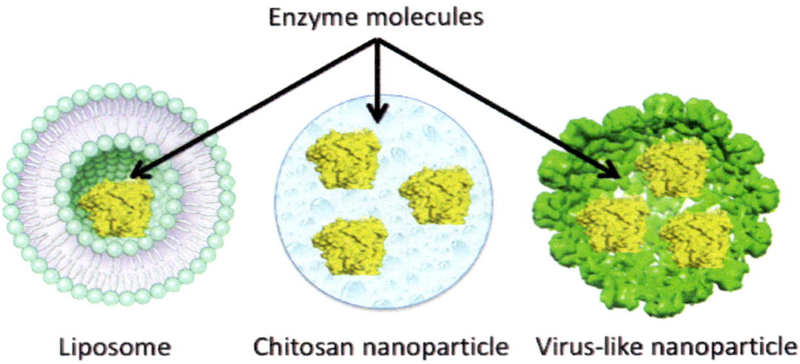

Figure 18.1 Different biomaterials for enzyme nanocarriers. Lipid bilayer in liposomes, gel nanoparticles in chitosan and enzyme-containing virus capsids.

Despite the promising medical applications of nanobioreactors, their use as carriers for enzymatic activity has had less attention than those used for drug delivery. Nonetheless, they are potential platforms to mimic the spatial organization of metabolic enzymes. Furthermore, the current knowledge in DNA technology and recombinant proteins has allowed researchers to begin investigating their use in novel therapeutic approaches and even though they are only in the early stages of research, they represent a viable candidate to improve the efficiency of various therapeutic domains.

18.3.1 Nanobioreactor Design Strategies

18.3.1.1 Liposomes and micelles

Lipid-based NPs can be further divided into liposome and micellar NPs; the primary difference being that liposomes are comprised of a lipid bilayer that encapsulates an aqueous core whereas micelles do not possess a bilayer and have a hydrophobic core made up of the lipid tails (Alves et al., 2015). Liposomes have been one of the most traditional NCs with a significant role in the biomedical field. These structures are formed by the spontaneous well-ordered arrangement of natural or synthetic lipid molecules placed in aqueous solution (Fu et al., 2014).

Liposomes are possibly the most-studied therapeutic delivery platforms, in large part because of their biocompatibility, biodegradability, low toxicity, and aptitude to trap both hydrophilic and lipophilic molecules. Reports on the use of liposomes for enzyme delivery go back a few decades; Belchetz et al. (1977) injected liposomes containing human β-glucocerebrosidase into an adult with Gaucher's disease. After 13 months of treatment, there was a reduction in the size of the liver, which was associated with pressure symptom relief in the lower abdomen.

Liposomes can be tuned through the use of different lipid monomers, combining lipid molecules and functionalizing the exterior surface. However, inconsistent size, difficulties in cargo loading, physical stability, the risk of cargo damage during production, and degradation prior to targeted therapeutical application has restricted their biomedical use (Dean et al., 2017; Xu et al., 2012).

In 1996, Martins et al. characterized an L-asparaginase bioconjugate obtained by linkage of palmitic acid chains to the native enzyme. Comparisons between the isoelectric points, hydrophobicity, pH, and temperature profiles for the bioconjugate and the native enzyme were performed. A 10-fold increase in hydrophobicity was obtained; however, both bioconjugated- and native-enzymes produced similar behaviors instability in human serum and on storage. Catalytic activity retention was dependent on the presence of micelles, showing 65% of initial activity for preparations without surfactant, and 98–100% activity retention in the presence of surfactant micelles. The catalytic properties of the bioconjugated- and native-enzyme preparations in micelles of different hydrophobicities were compared. The K_m of the native enzyme was found to be 0.030 mM, independently of the presence of a surfactant, while the K_m of the bioconjugate varied with the surfactant concentration from 0.036 to 0.046 mM.

Zelphati et al. (2001) developed a lipid-mediated delivery system composed of trifluoroacetylated lipopolyamine (TFA-DODAPL) and dioleoyl phosphatidylethanolamine (DOPE) able of successfully delivering β-galactosidase (β-Gal) and caspase 3 (CP3), caspase 8 (CP8), and granzyme B into cultured cell lines as demonstrated by intracellular fluorescence intensity measurements. In order to prove the functionality of their delivery system, the authors tested the ability of TFA-DODAPL:DOPE to deliver functional granzyme B into primary human AML cells by using a fluorogenic cell permeable substrate, CaspaTag, and measured the activation of endogenous caspases by flow cytometry. Similarly, CP3, CP8 and granzyme B activity was evaluated in Jurkat cells. Granzyme B and CP3 induced apoptosis in ~60% of Ki-Ras 267β1 cells.

Kim et al. (2012) produced DOTAP/DOPE and apolipoprotein (APOA-1) liposome NPs to mediate the targeted delivery of cytochrome C (cyt C) as an inducer of apoptosis, into non-small cell lung tumors. Cytochrome C was conjugated with a membrane permeable sequence (MPS) peptide and mixed with the NPs formulations, then, protein-loaded NPs were modified with DSPE-PEG-anisamide to enable specific NP targeting. The H460 cells treated with the PEGylated-NPs exhibited massive apoptosis, and when these NPs were intravenously administered in H460 tumor-bearing mice, a specific tumor targeting effect with low NP accumulation in the liver was observed. Furthermore, this treatment provoked tumor growth retardation in H460 xenograft mice.

By utilizing a combination of oligonucleotides and lipid-like molecules, Eltoukhy et al. (2014) demonstrated that intracellular horseradish peroxidase (HRP) NPs delivery is dramatically enhanced. Protein was azide-functionalized using an NHS-ester cross-linker with a short PEG spacer. This azide-modified HRP was then conjugated to an oligonucleotide and finally made to react with different lipid formulations. The lipid-covered protein NPs were incubated with HeLa cells and the results showed that oligonucleotide conjugation was required to achieve effective intracellular delivery. Likewise, immunohistochemistry data revealed substantial intracellular HRP staining within cells treated with lipid-DNA conjugates as opposed to controls with naked HRP-DNA.

Zuris et al. (2015) complexed proteins with a polyanionic molecule and mixed the complexes with cationic liposomes to produce an efficient intracellular protein delivery system. By fusing proteins to a supernegatively charged GFP they delivered functional nM protein concentrations into cultured mammalian cells. Furthermore, by fusing CRISPR-associated protein 9 (Cas9) to a polyanionic single guide RNA (sgRNA), up to 80% of cultured mammalian cells were modified from a single treatment.

18.3.1.2 DNA nanostructures

Recently DNA nanostructures have emerged as programmable, sequence-driven, self-assembled scaffolds. One of the most appealing applications for these nanostructures is the encapsulation of external moieties for biologically targeted delivery. The emergence of DNA origami propelled this field and allowed researchers to manipulate DNA into specific structures from a long, single strand scaffold by the addition of short complementary staple strands. Nonetheless, this technique represents only a fraction among the strategies being used to create such nanocages (Chandrasekaran and Levchenko, 2016).

In 2006, Erben et al. encapsulated cyt C within a tetrahedral cage made out of DNA. The protein was conjugated to one of the component strands via a surface amine before assembly of the tetrahedron, which resulted in the protein being encapsulated within the cage; however, the authors reported that by modifying the sequence in the strand conjugating the protein, the later could be positioned at specific sites on the tetrahedron.

Brodin et al. (2015) reported a strategy for protein transfection composed of a chemically modified functional protein core with a dense shell of oligonucleotides termed ProSNAs for *protein spherical nucleic acids*. The large homotetrameric enzyme core of β-galactosidase (β-Gal) was reacted with a fluorophore to allow the monitoring of cellular uptake and was then covered with PEG-modified DNA strands. Remarkably, the oligonucleotide shell enhanced cellular uptake by up to ~280-fold while maintaining β-Gal catalytic functionality with working concentrations as low as 100 pM enzyme.

DNA origami has been used to produce modular nanotubes encapsulating glucose oxidase (GOx) and horseradish peroxidase. Linko et al. (2015) separately produced and purified DNA origami units containing these enzymes and then glued them together via a programmable base pairing. By controlling the combination of both enzymes inside the nanotubes, an enzymatic cascade reaction was produced. Although this particular enzymatic cascade reaction was not meant as a therapeutic process, their modular nanoplatform concept could be applied to many other enzymes.

Finally, Zhao et al. (2016) reported the encapsulation of metabolic enzymes including glucose-6-phosphate dehydrogenase (G6PDH), lactic dehydrogenase (LDH) and β-Gal into DNA nanocages. Their assemblies showed well-controlled stoichiometry and architecture that enabled a systematic study of the impact of both encapsulation and proximal polyanionic surfaces. Activity assays demonstrated an increased substrate turnover number for DNA nanocage-encapsulated enzymes.

18.3.1.3 Protein-based nanoparticles

Zhao et al. (2011) reported the preparation of redox-responsive protein-based nanocapsules for intracellular delivery of active caspase (CP3). The targeted enzyme was non-covalently encapsulated into a polymeric shell that was interconnected by disulfide-containing cross-linkers via in situ interfacial polymerization. The polymeric shell of these reactors could be dissociated under reducing conditions and release its cargo in the presence of glutathione (GSH) and therefore, delivering the enzymatic cargo to the cytosol. With nanocapsule treatment, apoptosis was induced in a variety of human cancer cell lines. This core-shell concept was further utilized to reversibly encapsulate a maltose-binding protein fused with apoptin (MBP-APO) into water-soluble polymer shells. These sub-100 nm nanocapsules were efficiently internalized by mammalian cell lines with the selective accumulation of rhodamine-labeled apoptin in the nuclei of cancer cells. Intracellular release of apoptin induced tumor-specific apoptosis in several cancer cell lines and inhibited tumor growth in vivo (Zhao et al., 2013).

Herrera Estrada et al. (2014) used a desolvation and cross-linking process to produce 283 nm β-Gal-enhanced green fluorescent protein (eGFP) nanoparticles. Over 85% of β-Gal used was incorporated into the eGFP nanoparticles and enzymatic activity retained over 70% of native protein. Additionally, flow cytometry analysis showed that these NPs exhibited higher internalization in multiple cell lines in vitro compared to the soluble enzyme while maintaining enzymatic activity inside target cells. Furthermore, upon lyophilization in PBS buffer, these nanobioreactors were stable for six months retaining 84% of β-Gal enzymatic activity after resuspension and their size reduced by 5% to 35% depending on the cross-linker and cross-linker time utilized.

Virus-like particles (VLPs) are multimeric proteins derived from virus capsids, which form symmetric protein shells with a hollow interior space. A diverse array of recombinant VLPs have been produced and used as enzyme NCs providing them with a layer of protection against proteases, pH shifts, and raised temperatures (Wilkerson et al., 2018). Furthermore, recombinant VLPs can be easily purified and loaded with cargo using disassembly-assembly in vitro protocols or through the genetic fusion of protein cargo to components that direct localization into de particle interior, making this type of particles very versatile for the delivery of therapeutic enzymes.

Comellas-Aragonès et al. (2007), developed a virus-based single-enzyme nanoreactor. Horseradish peroxidase was encapsulated in the chlorotic mottle virus (CCMV) and enzymatic behavior of the developed nanoreactor was studied. It was the first report that showed the permeability of virus capsid for substrate/product and the alteration of permeability by the change of pH. A breakthrough in VLP based nanoreactors was achieved by Patterson et al. (2014) who constructed a densely packed multienzyme system able to perform a coupled cascade of reactions. In this work, a 160 kDa protein was produced by the fusion of β-glucosidase, galactokinase, and glucokinase and encapsulated into bacteriophage P22 derived capsids. In addition, their results showed that intermediate channeling between sequential enzymes is dependent in both inter-enzyme distance and a balance between the kinetic parameters of the enzymes involved.

Sánchez-Sánchez et al. (2014) effectively encapsulated a cytochrome P450 (CYP) variant from *Bacillus megaterium* into VLPs constituted of coat protein from cowpea chlorotic mottle virus (CCMV) (Fig. 18.2A). Up to ten CYP molecules were encapsulated inside each 28 nm diameter capsid. These catalytic NPs were able to transform the chemotherapeutic prodrug tamoxifen into active products similar to those obtained with human CYP. In a subsequent study, Sánchez-Sánchez et al. (2015) encapsulated CYP into bacteriophage P22 derived capsids. Nanobioreactors contained 109 molecules of CYP per capsid and enzyme stability towards protease degradation, and acidic pH was increased.

Furthermore, these nanobioreactors were internalized into human cervix carcinoma cells, and as expected, P22-CYP transfected cells showed a 10-fold higher CYP activity when compared to non-treated cells. PEGylation of the VLP capsids strongly reduced or eliminated the immunogenic response and by functionalizing specific ligands to the free end of the PEG molecules, the virus-derived nanoreactors could be targeted to receptors on tumor cells (Tapia-Moreno et al., 2017; Chauhan et al., 2018a). Ligand–receptor–mediated cell internalization increased CYP activity inside cells and thus increased tamoxifen cell sensitivity of both human cervix and breast tumor cells, reducing the dose needed to kill these cells by 50% (Fig. 18.2B) (Tapia-Moreno et al., 2017).

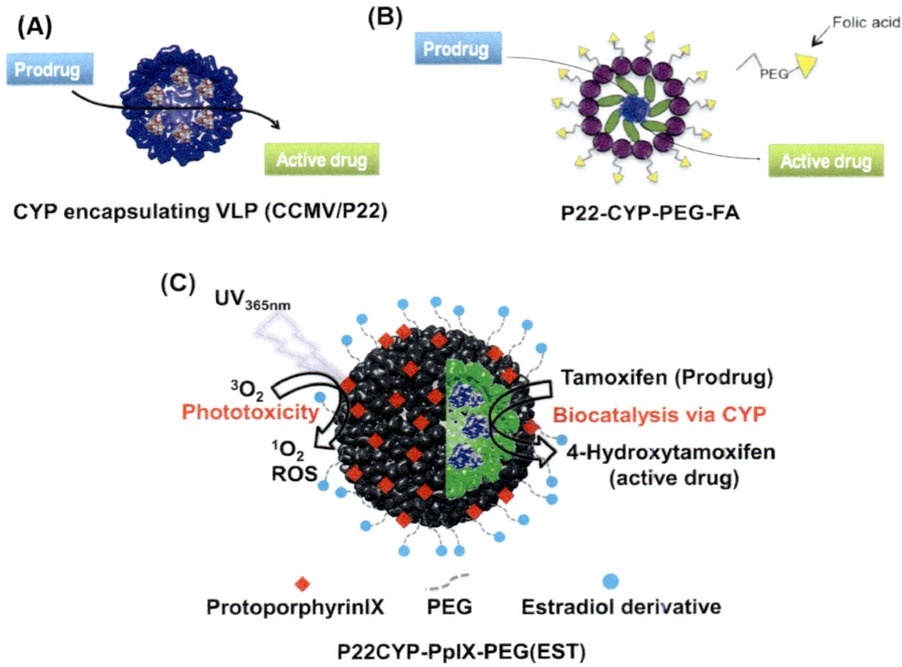

Figure 18.2 CYP encapsulating VLPs for cancer treatment: (A) CYP encapsulation in CCMV and P22 VLPs by Sanchez-Sanchez et al. (2014) and (2015); (B) Folic acid functionalized biocatalytic P22 for targeted cancer treatment by Tapia-Moreno et al. (2017); (C) Multifunctionalized biocatalytic P22 nanoreactor integrating enzyme prodrug and photodynamic therapy by Chauhan et al. (2018).

The multivalency and versatility of virus-derived nanovehicles were also used for targeted enzyme delivery in conjunction with the synergistic effect of photodynamic therapy (PDT) by Chauhan et al. (2018a). The P22 bacteriophage encapsulating CYP activity was multifunctionalized with a porphyrin-based photosensitizer for PDT effect and estradiol-based ligand for targeted delivery to ER+ breast tumor cells (Fig. 18.2C). The system was able to generate reactive oxygen species upon illumination with light that in synergy with CYP activity transformed tamoxifen (prodrug) to 4-hydroxytamoxifen (active drug) and showed ~3-fold higher toxicity than non-targeted VLPs (Fig. 18.2C). Additionally, the PEG-coat of the multifunctional nanoreactors rendered them invisible to macrophages and treatment of tumor cells with these nanoreactors significantly reduced prodrug dosage. Results obtained indicate that a drastic reduction of chemotherapy side effects and an increase in treatment effectiveness could be expected. The study is the first example to show the applicability of the VLP platform as a combinatory treatment modality and inhibiting tumor cells by multiple ways at once, which will be advantageous to overcome tumor heterogeneity and reoccurrence issues.

Schoonen et al. (2017) encapsulated T4 lysozyme (T4L) inside elastin-like polypeptide (EPL) stabilized CCMV capsids. The activity of T4L is highly dependent on the salt concentration and pH of the environment under which these capsids are not able to form; however, by adding metal ions, the system was stabilized. Four T4L molecules per CCMV capsid were encapsulated and remained catalytically active. Their work opened the possibility of utilizing shielded T4L in antibiotic applications.

18.3.1.4 Carbohydrate-based nanoparticles

Chitosan is widely used as an enzyme immobilization support. Due to its biodegradability, biocompatibility, and renewability, chitosan has been used in a wide range of applications. Giannotti et al. (2011) developed chitosan-based NCs containing α-galactosidase (α-Gal). To have a suitable molecule able to form these complexes at near neutral pH conditions, they prepared partially quaternized chitosan with methyl groups generating trimethyl-chitosan (TMC). The NPs displayed 62% of specific activity as compared to free enzyme preparations and were able to dissociate α-Gal from TMC at acidic pH values as those encountered inside lysosomes. Furthermore, these nanobioreactors were internalized by endothelial cells and mostly destined to lysosomal compartments. In a subsequent study, they evaluated the intracellular activity of these complexes and introduced RGD motifs to enhance their delivery into endothelial cell lysosomes. RGD-modified nanobioreactors presented significantly higher in vitro internalization, and intracellular activity was enhanced making them an ideal system for delivery of lysosomal proteins in living cells (Giannotti et al., 2016).

Klein et al. (2012) investigated the use of macro- and nanoparticles of chitosan for β-Gal immobilization. Their research evaluated the effect of the support size on the properties of the immobilized enzyme. They observed that different particle sizes and porosities modified the enzyme load, as well as the activity and thermal stability of the immobilized biocatalyst. Nonetheless, immobilized β-Gal exhibited excellent operational stability at 37°C. Although their focus was on the food industry, modifications could be made to accommodate this system for a therapeutic approach.

Bahreini et al. (2014) assessed the in vitro preparation of L-asparaginase II-loaded chitosan NPs cross-linked with tripolyphosphate (TPP) to evaluate their efficacy for the entrapment and controlled release of protein. The immobilized enzyme had an increased half-life in comparison with the free enzyme and release from the NC could be influenced by pH and ionic strength of the medium.

Fernandes and Gregoriadis (1997) covalently coupled a low molecular weight polysialic acid and colominic acid to the surface of the anti-leukemic enzyme L-asparaginase from the *Erwinia carotovora*. Polysialylated constructs retained between 82 and 86% of activity depending on the molar ratio of colominic acid-asparaginase used, and K_m values remained the same when compared with the native enzyme. Upon a 6 h exposure to mouse blood plasma at 37°C, constructs exhibited up to 83% of the initial enzyme activity in contrast to native enzyme inactivation under the same conditions. Similarly, Vỹna et al. (2001) utilized the biologically active fructose polysaccharide levan from *Zymomonas mobilis* to covalently bond L-asparaginase from *E. carotovora*. They found that glycosylation of L-asparaginase widened the range of pH stability, temperature tolerance and storage in aqueous solutions as compared to the native enzyme, representing a feasible approach to biocompatible stabilization for improved use in therapeutics.

Using the "polymer masked-unmasked protein therapy" concept, Ferguson and Duncan (2009) designed and synthesized a dextrin-phospholipase A2 (PLA2) conjugate for the targeting of deregulated lipid metabolism in cancer. This concept took advantage of the presence of up to 85-fold increase of α-amylase in some tumors, which could be used to degrade the dextrin NC and expose PLA2 to the cellular environment. The conjugated nonreactors decreased to 36% the enzyme activity compared to native PLA2; however, upon incubation with α-amylase, the activity was almost completely restored. Whereas dextrin conjugation caused a marked reduction in PLA2's hemolytic activity, the conjugate was cytotoxic toward MCF-7, HT29, and B16F10 cells at a level that was comparable to that observed for free PLA2, showing the potential of conjugation to diminish nonspecific toxicity.

Perez et al. (2005) coupled end-group activated dextran to the surface of superoxide dismutase (SOD). These NPs retained 81% of initial enzymatic activity after surface glycosidation and showed improved resistance to oxidation by H_2O_2 after a 4 h incubation period. This stability was further reflected by a 10.8-fold increase in SOD-dextran half-life when compared to the native counterpart. SOD-dextran conjugates after intravenous administration showed a significant lower body clearance in rats (6.4 U h^{-1}) when compared with native SOD (62.5 U h^{-1}).

18.3.2 Tissue Targeting

One of the important aspects for the utilization of nanobioreactors in medical application is the tissue targeting efficiency. There has been a significant improvement in the delivery of therapeutic cargo using NCs, however; the clinical translation of the majority of them has been hindered by poor targeting efficiency and lower uptake in the intended tissue. Targeting involves the structural manipulations of nanoparticles for specific localization at the diseased site and is broadly categorized into two strategies.

18.3.2.1 Passive targeting

The process of NC uptake by the influence of physicochemical properties of NCs (size and surface charge) and the pathophysiological conditions of diseased tissue (leaky vasculature) is referred to as passive targeting (Himri and Guaadaoui, 2018). It details the targeting of long-circulating NCs to the pathological sites with compromised vasculature. For instance, cancer cells have a great extent of vascularization that results in the formation of leaky vessels which promote the internalization of NPs in the tumor by "enhanced permeation and retention effect" (EPR) (Matsumura and Maeda, 1986). However, the percentage accumulation of NCs even in high EPR regions is minimal owing to the several biological barriers. Recent reviews highlight the measures and challenges to overcome these barriers and promote NC extravasation (Blanco et al., 2015; Rosenblum et al., 2018).

Precise size, monodispersity, shape, and surface charge are important features in NC design significantly influencing their circulation and biodistribution (Fig. 18.3). Various biological parameters have discrete size cut-offs. Ideally, NCs in a 20–100 nm size manage the renal and endothelium filtration and extravasate through a leaky vasculature; NCs larger than 150 nm show non-specific uptake. Particles in the micrometer size range (>2 μm) are usually entrapped in lung capillaries and can be beneficial for targeting the predominant metastatic site (Soo Choi et al., 2007; Longmire et al., 2008). Variable NP geometries show a distinct flow that governs their circulation time, interaction with biological entities and macrophage uptake and hence, affects their overall biodistribution (Champion and Mitragotri, 2006). Neutral or slightly negatively charged NCs prevent/minimize interactions with serum proteins resulting in prolonged circulation time and higher uptake via the EPR effect (Xiao et al., 2011; Yuan et al., 2012). Particle elasticity, biodegradability, and high stability are other essential features that govern therapeutic delivery (Anselmo et al., 2015). Nanocarriers should be stable enough to hold the therapeutic cargo during circulation to avoid the uptake in healthy tissue; however, NPs are susceptible to phagocytosis leading to premature clearance. The coating of NCs with hydrophilic polymers provide steric brush creating the "stealth" effect and thus, enhancing longevity and stability (Gamucci et al., 2014; Qie et al., 2016). Emerging bioinspired stealth strategies are the alternate approach for safer delivery (White et al., 2012; Lee et al., 2014; Jutz et al., 2015; Jiang et al., 2016; Pitek et al., 2016a; Hill et al., 2017).

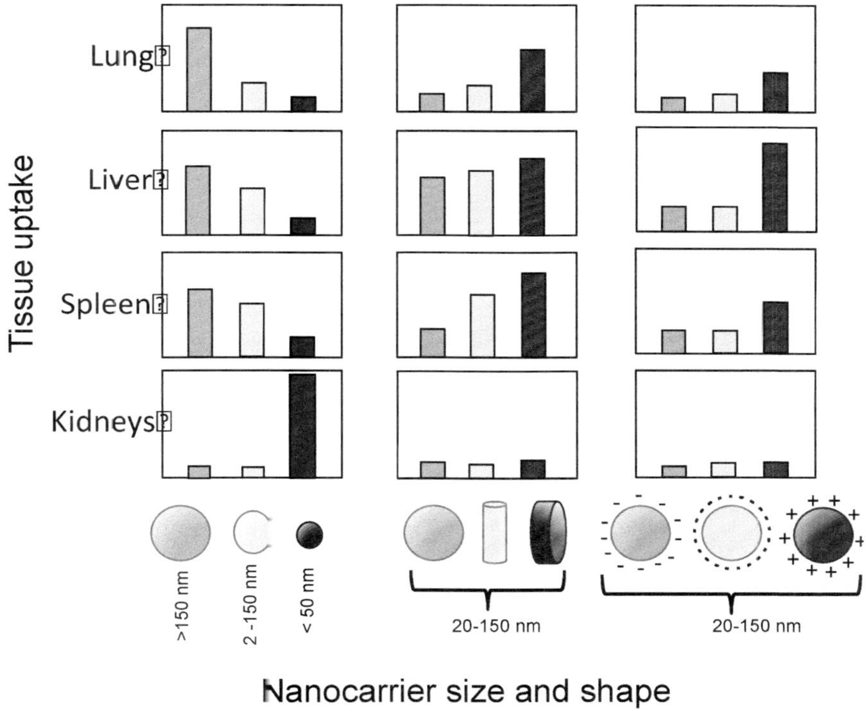

Figure 18.3 The influence of size, shape and surface charge of NCs on macrophage uptake present in lungs, liver, and spleen that affect overall biodistribution. Modified from Blanco et al. (2015).

18.3.2.2 Active targeting

The main obstacle in the passive targeting of NCs is the non-specific distribution resulting in the lower uptake due to which, higher doses should be administered that cause off-target toxicity. Thus, active targeting via ligand–receptor interaction was introduced. Complimentary combination of passive targeting with the high-affinity ligands specific towards receptors (CD19, folate receptor, HER2, CD44, among others) overexpressed on the diseased tissue, and cell surfaces are used to enhance the NC localization via receptor-mediated endocytosis. Several diagnostic and therapeutic agents are benefitted from active targeting to minimize the non-specific retention in normal cells (Chauhan et al., 2016; Quester et al., 2016; Chauhan et al., 2018b; Sengar et al., 2018). The targeting ligands can be based on small molecules, antibodies, proteins or peptides. Different groups have used active targeting to enhance the enzymatic payload at the target site for maximum prodrug transformation, thus enhancing the therapeutic efficacy in comparison to the EPR effect (Papademetriou et al., 2013; Yata et al., 2014; Acosta et al., 2015; Chen et al., 2015; Condori et al., 2016; Tapia-Moreno et al., 2017; Chauhan et al., 2018a). A list of targeting ligands and their cognate receptors are summarized in Table 18.1.

Table 18.1 List of targeting ligands and their cognate receptors (Dean et al., 2017)

Ligand	Targeted receptor	Example of disease
Folate	Folate	Prostate cancer
Transferrin	Transferrin	Breast cancer
RGD peptide	Integrin	Lymphoma
LL-37 antibody	Bacterial membrane	Antibacterial
Galactose	Asialogycoprotein receptor (ASGPR)	Hepatocellular carcinoma
CD-19 Antibody	CD-19 antigen	Lymphoma
VEGF peptide	VEGF	Breast cancer
Epidermal growth factor (EGF)	EGF	Head and neck cancer

As the biocatalytic NCs carry therapeutic enzymes, the covalent conjugation of targeting ligands should involve simple chemistry under ambient conditions that do not impact the enzymatic activity. Moreover, the controlled delivery of therapeutic payload may enhance the overall outcome and avoid the premature release of cargo, one of the limitations of NCs (Che and van Hest, 2016). For instance, in cancer, NCs can take advantage of tumor microenvironment (TME) conditions such as acidic pH, redox species, enzymes, among others, for the triggered release of therapeutic cargo. The commonly used conjugation approach for linking targeting vectors on the NCs and introducing TME responsive vectors is listed in Table 18.2. Recent reviews briefly highlight the bioconjugation chemistry involved in NPs functionalization (Sapsford et al., 2013; Schoonen and van Hest, 2014; Werengowska-Ciećwierz et al., 2015).

Table 18.2 Common conjugation strategies used for the attachment of affinity ligands and the introduction of TME responsive function to the NCs

Conjugation strategy	Bonding	Reaction conditions	Stability
Amine/ Carboxylic	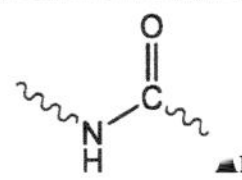 Amide	EDC, NHS, pH 7–8.5, 2–24 h, RT or 4°C	Stable
Thiol/ Maleimide	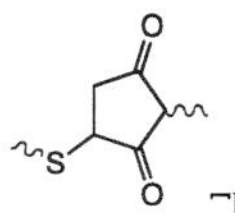 Thio-ether	pH > 7, 4–24 h, RT	Stable
Thiol/Thiol	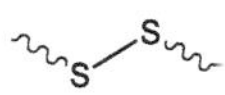 Disulfide	pH 8, 2-24 h, RT or 4°C	Labile in reducing environment
Azide/Alkyne	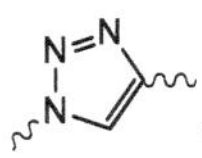 Triazole ring	Cu(I), 2–3 h, RT	Stable
Hydrazine/ Aldehyde	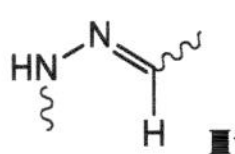 Hydrazone	24 h, 5–6°C	Acid labile

Abbreviations: EDC, 1-Ethyl-3-(3-dimethylaminopropyl)carbodiimide; NHS, N-hydroxysuccinimide; RT, room temperature.

There are impressive advances in specific targeting, even if recent studies stating that the targeting function enhances cellular internalization but has little effect on the overall tumor localization raise concerns (Kirpotin et al., 2006; Schmidt and Wittrup, 2009). Similar to passive targeting, the systemically administered actively targeted NCs could be lost before reaching the targeted tissue thus, the conditions for NC design in passive targeting should also be considered in active targeting. Additional benefits may arise thanks to the multifunctionality of NCs as multitargeting to various receptors overexpressed on the diseased tissue can avoid low targeting efficiency problems due to diseased tissue heterogeneity.

In a recent review by Rosenblum et al. (2018), the existing challenges and the critical considerations in NC design for the targeted delivery of cancer therapeutics are briefly summarized. These considerations could also be applied to the targeting of various other diseases, nevertheless, the variable distribution of sequentially administered bionanocatalysts and the prodrug also hamper the therapeutic outcomes thus, co-administration of enzyme and prodrug using a single targeted nanoplatform could be a beneficial approach. For instance, a block copolymer prodrug (camptothecin) based polymersome nanoreactor encapsulating glucose oxidase (GOx) was constructed for orchestrated oxidation/chemotherapy of cancer (Li et al., 2017). These nanoreactors remained inactive in healthy tissues upon systemic administration; however, tumor acidity triggered the enhanced permeability of polymersomes and promoted catalysis of diffused glucose by GOx. The catalytic reaction generated H_2O_2, which further activated the polymer to release the active drug. Thus, the high oxidative stress and chemotherapy synergistically suppress the tumor growth.

18.4 Immunogenic Response and Masking

The NPs are known to induce, inhibit and/or modify the innate immune system (Boraschi et al., 2012). Macrophages or neutrophils are the proficient phagocytic cells that are part and parcel in innate/acquired immunity. These cells are the key factors in encapsulation and invasion of foreign substances including NPs that fall within the size range of viruses and are thus, proficiently recognized and processed. Both natural and synthetic nanomaterials are known to interact with the immune cells and produce long-term effects (Pitek et al., 2016b; Liu et al., 2017). For example, NCs can be active immunosuppressants either by the killing or by impairment of the immune cells, by cascading the down-regulatory mechanisms or by stimulating the inflammatory response leading to the acquired immunogenic response.

In some cases, the NPs can act as immunomodulators by inducing an unexpected immune response (Barbero et al., 2017).

The immune response of NPs has also been exploited in cancer therapy by localized induction of immunogenic reactions (He et al., 2016; Lizotte et al., 2016). The immune response cascade by nanomaterials begins with protein opsonization involving the formation of a corona which is mostly dependent on the size, shape and surface charge of NPs. The plasma protein is comprised of serum albumin, complement components, apolipoproteins, and immunoglobulins that endow the NPs with a specific biological identity that is recognized by the immune system. The main features of the immune system are recently reviewed in brief (Gamucci et al., 2014; Barbero et al., 2017; Liu et al., 2017).

Opsonization causes non-specific uptake in the mononuclear phagocyte system like the liver, lung, and spleen, and is also unfavorable for active-targeting as the adhered protein envelop masks the affinity ligands reducing their specificity significantly (Salvati et al., 2013; Docter et al., 2014). A similar approach involves the use of hydrophilic polymers to coat the surface of NCs creating a "stealth" effect that prevents protein accumulation. PEGylation is the most widely utilized strategy (Harris and Chess, 2003); however, the effect is greatly dependent on the length and conformation of the polymer chain (Steinmetz and Manchester, 2009; Lee et al., 2015).

In spite that PEG is widely used and has been approved by the regulatory agencies worldwide, the recent recognition of PEG-specific antibodies in ~25% of the human population and the rapid clearance of PEG-conjugates from blood on repetitive doses has raised concerns and produced the need for a paradigm shift (Armstrong et al., 2007; Garay et al., 2012). Nevertheless, it is still challenging to design a synthetic NC with prolonged blood circulation and native protein compatibility matching. One alternative is the self-assembly of peptides and proteins to camouflage NCs (White et al., 2012). These emerging bioinspired stealth strategies are a step towards the next generation NCs and could open new avenues for safer delivery with prolonged retention (Gulati et al., 2018).

Rodriguez et al. (2013) investigated the active stealth approach by camouflaging NPs with "don't eat me" marker CD47 peptides which provided the "self" property and thus, avoided phagocytosis. Pitek et al. (2016a) used a serum albumin camouflage to enhance the pharmacokinetics of tobacco mosaic virus (TMV) by preventing antibody recognition. Huang et al. (2015) utilized T-cell-conjugated nanocapsules for therapeutic delivery to disseminated tumors by using their natural trafficking potential all over the lymphatic system. The cell membranes extracted from red blood cells (Hu et al., 2011), leukocytes (Parodi et al., 2013), stem cells (Gao et al., 2016) and platelets (Hu et al., 2015) also provide a biomimetic camouflage and are found to significantly enhance the NC circulation in vivo.

18.5 Conclusions

Nanobiotechnology as an emerging technological field has opened a vast number of potential applications, in which biomedical applications are included. The combination of enzymatic activity and nanosized biomaterials has a promissory future for the design of new therapeutic strategies in Enzyme Replacement Therapies. The diseases originated by the lack or low enzymatic activity reviewed here could be treated with smarter ERT approaches in which the specific enzyme could be specifically targeted to the diseased tissue, avoiding systemic administrations. Biomaterial-based NPs carrying enzymatic activity could be effectively functionalized with different ligands to be recognized and cell-internalized for tissues containing the specific receptor. Moreover, the co-administration of the prodrug and the enzymatic activity via a single nanoplatform could produce even better treatment outcomes by overcoming the different distribution and pharmacokinetic profile of both cofactors. In addition, the recent advancements in the bioinspired "active" stealth covers combining the property to reduce or eliminate the immunogenic response and active targeting could be beneficial for treating diseased tissue via a bio-mimicry approach with lower side-effects". We can conclude so far, that the use of biocatalytic nanoreactors for biomedical and therapeutic purposes is still in its infancy but shows enormous potential and thus research efforts are still needed.

References

Acosta W., Ayala J., Dolan M. C., Cramer C. L., *Sci. Rep.*, **5** (2015), 14144.

Alfadhel M., Sirrs S., *Ther. Clin. Risk Manag.*, **7** (2011), 69–82.

Al-Lazikani B., Banerji U., Workman P., *Nat. Biotechnol.*, **30** (2012), 679–692.

Alves N. J., Turner K. B., Medintz I. L., Walper S. A., *Ther. Deliv.*, **6** (2015), 873–887.

Anselmo A. C., Zhang M., Kumar S., Vogus D. R., Menegatti S., Helgeson M. E., Mitragotri S., *ACS Nano.*, **9** (2015), 3169–3177.

Arends M., Biegstraaten M., Wanner C., Sirrs S., Mehta A., Elliott P. M., Oder D., Watkinson O. T., Bichet D. G., Khan A., Iwanochko M., Vaz F. M., van Kuilenburg A. B. P., West M. L., Hughes D. A., Hollak C. E. M., *J. Med. Genet.*, **55** (2018), 351–358.

Armstrong J. K., Hempel G., Koling S., Chan L. S., Fisher T., Meiselman H. J., Garratty G., *Cancer*, **110** (2007), 103–111.

Bahreini E., Aghaiypour K., Abbasalipourkabir R., Mokarram A. R., Goodarzi M. T., Saidijam M., *Nanoscale Res. Lett.*, **9** (2014), 1–13.

Barbero F., Russo L., Vitali M., Piella J., Salvo I., Borrajo M. L., Busquets-Fité M., Grandori R., Bastús N. G., Casals E., Puntes V., *Semin. Immunol.*, **34** (2017), 52–60.

Belchetz P. E., Crawley J. C. W., Braidman I. P., Gregoriadis G., *Lancet*, (1977), 116–117.

Berry A., *J. Nutr. Clin. Pract.*, **29** (2014), 312–321.

Bièche I., Narjoz C., Asselah T., Vacher S., Marcellin P., Lidereau R., Beaune P., de Waziers I., *Pharmacogenet. Genomics*, **17** (2007), 731–742.

Blackburn M. R., Kellems R. E., *Adv. Immunol.*, **86** (2005), 1–41.

Blanco E., Shen H., Ferrari M., *Nat. Biotechnol.*, **33** (2015), 941–951.

Boraschi D., Costantino L., Italiani P., *Nanomedicine*, **7** (2012), 121–131.

Bosio V. E., Islan G. A., Martínez Y. N., Durán N., Castro G. R., *Crit. Rev. Biotechnol.*, **36** (2016), 447–464.

Brodin J. D., Sprangers A. J., McMillan J. R., Mirkin C. A., *J. Am. Chem. Soc.*, **137** (2015), 14838–14841.

Buckley R. H., Schiff R. I., Schiff S. E., Markert M. L., Williams L. W., Harville T. O., Roberts J. L., Puck J. M., *J. Pediatr.*, **130** (1997), 378–387.

Champion J. A., Mitragotri S., *Proc. Natl. Acad. Sci.*, **103** (2006), 4930–4934.

Chandrasekaran A. R., Levchenko O., *Chem. Mater.*, **28** (2016), 5569–5581.

Chauhan, K., Arun A., Singh S., Manohar M., Chuttani K., Konwar R., Dwivedi A., Soni R., Singh A. K., Mishra A. K., Dutta A., *Bioconjug. Chem.*, **27** (2016), 961–972.

Chauhan K., Hernandez-Meza J. M., Rodríguez-Hernández A. G., Juarez-Moreno K., Sengar P., Vazquez-Duhalt R, *J. Nanobiotechnology*, **16** (2018a), 1.

Chauhan K., Tiwari A. K., Chadha N., Kaul A., Singh A. K., Datta A., *Mol. Pharm.*, **15** (2018b), 1515–1525.

Che H., van Hest J. C. M., *J. Mater. Chem. B*, **4** (2016), 4632–4647.

Chen H., Tian J., He W., Guo Z., *J. Am. Chem. Soc.*, **137** (2015), 1539–1547.

Chen L., Waxman D., *J. Curr. Pharm. Des.*, **8** (2012), 1405–1416.

Choudharya D., Janssona I., Schenkmana J. B., Sarfarazib M., Stoilovb I., *Arch. Biochem. Biophys.*, **414** (2003), 91–100.

Comellas-Aragonès M., Engelkamp H., Claessen V. I., Sommerdijk N. A. J. M., Rowan A. E., Christianen P. C. M., Maan J. C., Verduin B. J. M., Cornelissen J. J. L. M., Nolte A. R. J. M., *Nat. Nanotechnol.*, **2** (2007), 635–639.

Condori J., Acosta W., Ayala J., Katta V., Flory A., Martin R., Radin J., Cramer C. L., Radin D. N., *Mol. Genet. Metab.*, **117** (2016), 199–209.

Connock M., Burls A., Frew E., Fry-Smith A., Juarez-Garcia A., McCabe C., Wailoo A., Abrams K., Cooper N., Sutton A., O'Hagan A., Moore D., *Health Technol. Assessment.*, **10** (2006), ix–136.

Coutinho M. F., Lacerda L., Alves S., *Biochem. Res. Int.*, **2012** (2012), 1.

De Vocht C., Ranquin A., Willaert R., Van Ginderachter J. A., Vanhaecke T., Rogiers V., Versées W., Van Gelder P., Steyaert J., *J. Control. Release*, **137** (2009), 246–254.

Dean S. N., Turner K. B., Medintz I. L., Walper S. A., *Ther. Deliv.*, **8** (2017), 577–595.

Docter D., Distler U., Storck W., Kuharev J., Wünsch D., Hahlbrock A., Knauer S. K., Tenzer S., Stauber R. H., *Nat. Protoc.*, **9** (2014), 2030–2044.

Domínguez-Muñoz J. E., *J. Gastroenterol. Hepatol.*, **26** (2011), 12–16.

Domínguez-Muñoz J. E., *Curr. Opin. Gastroenterol.*, **34** (2018), 349–354.

Dornelles A. D., Artigalás O., da Silva A. A., Ardila D. L. V., Alegra T., Pereira T. V., Vairo F. P. E., Schwartz I. V. D., *PLoS One*, **12** (2017), e0184065.

Duve C. D., *Proc. Inst. Med. Chicago*, **26** (1966), 73–76.

Eltoukhy A. A., Chen D., Veiseh O., Pelet J. M., Yin H., Dong Y., Anderson D. G., *Biomaterials*, **35** (2014), 6454–6461.

Erben C. M., Goodman R. P., Turberfield A., *Angew. Chem. Int. Ed.*, **45** (2006), 7414–7417.

Ferguson E. L., Duncan R., *Biomacromolecules*, **10** (2009), 1358–1364.

Fernandes A. I., Gregoriadis G., *Biochim. Biophys. Acta Protein Struct. Mol. Enzymol.*, **1341** (1997), 26–34.

Fu A., Tang R., Hardie J., Farkas M. E., Rotello V. M., *Bioconjug. Chem.*, **25** (2014), 1602–1608.

Galluzzi L., Kepp O., Vander Heiden M. G., Kroemer G., *Nat. Rev. Drug Discov.*, **12** (2013), 829–846.

Gamucci O., Bertero A., Gagliardi M., Bardi G., *Coatings*, **4** (2014), 139–159.

Gao C., Lin Z., Jurado-Sánchez B., Lin X., Wu Z., He Q., *Small*, **12** (2016), 4056–4062.

Garay R. P., El-Gewely R., Armstrong J. K., Garratty G., Richette, P., *Expert Opin. Drug Deliv.*, **9** (2012), 1319–1323.

Gericke B., Amiri M., Scott C. R., Naim H. Y., *Biochim. Biophys. Acta*, **1863** (2017), 817–826.

Ghoshoon M. B., Berenjian A., Hemmati S., Dabbagh F., Karimi Z., Negahdaripour M., Ghasemi Y., *Int. J. Pept. Res. Ther.*, **21** (2015), 487–495.

Giannotti M. I., Abasolo I., Oliva M., Andrade F., García-Aranda N., Melgarejo M., Pulido D., Corchero J. L., Fernández Y., Villaverde A., Royo M., García-Parajo M. F., Sanz F., Schwartz S. Jr., *ACS Appl. Mater. Interfaces.*, **8** (2016), 25741–25752.

Giannotti M. I., Esteban O., Oliva M., García-Parajo M. F., Sanz F., *Biomacromolecules*, **12** (2011), 2524–2533.

Gilkes J. A., Heldermon C. D., *Pediatr. Endocrinol. Rev.*, **12** (2014), 133–140.

Grabowski G. A., Hopkin R. J., *Annu. Rev. Genom. Hum. Gen.*, **4** (2003), 403–436.

Grabowski G. A., Leslie N., Wenstrup R., *Blood Rev.*, **12** (1998), 115–133.

Gulati N. M., Stewart P. L., Steinmetz N. F., *Mol. Pharm.*, **15** (2018), 2900–2909.

Harris J. M., Chess R. B., *Nat. Rev. Drug Discov.*, **2** (2003), 214–221.

Hay D., Louie D., Earp B., Kaplan F., Akelman E., Blazar P., *J. Hand Surg. Br.*, **39** (2013), 463–465.

He C., Duan X., Guo N., Chan C., Poon C., Weichselbaum R. R., Lin W., *Nat. Commun.*, **7** (2016), 12499.

Herrera Estrada L., Chu S., Champion J. A., *J. Pharm. Sci.*, **103** (2014), 1863–1871.

Hershfield M. S., *Clin. Immunol. Immunopathol.*, **76** (1995), S228–S232.

Hill B. D., Zak A., Khera E., Wen F., *Curr. Protein Pept. Sci.*, **19** (2017), 112–127.

Himri I., Guaadaoui A., Cell and organ drug targeting: Types of drug delivery systems and advanced targeting strategies, in *Nanostructures for the Engineering of Cells, Tissues and Organs From Design to Applications*, Holban A. M., eds.; Elsevier Inc. (2018).

Holcenberg J. S., *Annu. Rev. Biochem.*, **51** (1982), 795–812.

Hoskins J. A., Gray J., *Res. Commun. Chem. Pathol. Pharmacol.*, **35** (1982), 275–282.

Hsueh E. C., Knebel S. M., Lo W.-H., Leung Y.-C., Cheng P. N.-M., Hsueh C.-T., *J. Hematol. Oncol.*, **5** (2012), 1.

Hu C. M. J., Fang R. H., Wang K. C., Luk B. T., Thamphiwatana S., Dehaini D., Nguyen P., Angsantikul P., Wen C. H., Kroll A. V., Carpenter C., Ramesh M., Qu V., Patel S. H., Zhu J., Shi W., Hofman F. M., Chen T. C., Gao W., Zhang K., Chien S., Zhang L., *Nature*, **526** (2015), 118–121.

Hu C. M. J., Zhang L., Aryal S., Cheung C., Fang R. H., Zhang L., *PNAS*, **108** (2011), 10980–10985.

Huang B., Abraham W. D., Zheng Y., Bustamante López S. C., Luo S. S., Irvine D. J., *Sci. Transl. Med.*, **7** (2015), 291.

Huttunen K. M., Mahonen N., Raunio H., Rautio J., *Curr. Med. Chem.*, **15** (2008), 2346–2365.

Jiang Y., Lu H., Chen F., Callari M., Pourgholami M., Morris D. L., Stenzel M. H., *Biomacromolecules*, **17** (2016), 808–817.

Jutz G., van Rijn P., Miranda B. S., Bo A., *Chem. Rev.*, **115** (2015), 1653–1701.

Kim W., Erlandsen H., Surendran S., Stevens R. C., Gamez A., Michols-Matalon K., Tyring S. K., Matalon R., *Mol. Ther.*, **10** (2004), 220–224.

Kim S. K., Foote M. B., Huang L., *Biomaterials*, **33** (2012), 3959–3966.

Kirpotin D. B., Drummond D. C., Shao Y., Shalaby M. R., Hong K., Nielsen U. B., Marks J. D., Benz C. C., Park J. W., *Cancer Res.*, **66** (2006), 6732–6740.

Kishnani P. S., Beckemeyer A. A., *Pediatr. Endocrinol. Rev.*, **12** (2014), 114–124.

Kitaoka T., Tajima T., Nagasaki K., Kikuchi T., Yamamoto K., Michigami T., Okada S., Fujiwara I., Kokaji M., Mochizuki H., Ogata T., Tatebayashi K., Watanabe A., Yatsuga S., Kubota T., Ozono K., *Clin. Endocrinol. (Oxf).*, **87** (2017), 10–19.

Klein M. P., Nunes M. R., Rodrigues R. C., Benvenutti E. V., Costa T. M. H., Hertz P. F., Ninow J. L., *Biomacromolecules*, **13** (2012), 2456–2464.

Konstan M. W., Ratjen F., *J. Cyst. Fibros.*, **11** (2012), 78–83.

Koyani R., Pérez-Robles J., Cadena-Nava R. D., Vazquez-Duhalt R., *Nanotechnol. Rev.*, **6** (2017), 405–419.

Lee H. J., Park H. H., Kim J. A., Park J. H., Ryu J., Choi J., Lee J., Rhee W. J., Park T. H., *Biomaterials*, **35** (2014), 1696–1704.

Lee K. L., Shukla S., Wu M., Ayat N. R., El Sanadi C. E., Wen A. M., Edelbrock J. F., Pokorski J. K., Commandeur U., Dubyak G. R., Steinmetz N. F., *Acta Biomater.*, **19** (2015), 166–179.

Li J., Li Y., Wang Y., Ke W., Chen W., Wang W., Ge Z., *Nano Lett.*, **17** (2017), 6983–6990.

Lin H., Chen Y., Shi J., *Chem. Soc. Rev.*, **47** (2018), 1938–1958.

Linko V., Eerikäinen M., Kostiainen M. A., *Chem. Commun.*, **51** (2015), 5351–5354.

Liu Y., Hardie J., Zhang X., Rotello V. M., *Semin. Immunol.*, **34** (2017), 25–32.

Liu J., Ma J., Wu Z., Li W., Zhang D., Han L., Wang F., Reindl K. M., Wu E., Ma Q., *BMC Cancer*, **14** (2014), 686.

Lizotte P. H., Wen A. M., Sheen M. R., Fields J., Rojanasopondist P., Steinmetz N. F., Fiering S., *Nat. Nanotechnol.*, **11** (2016), 295–303.

Longmire M., Choyke P. L., Kobayashi H., *Nanomedicine*, **3** (2008), 703–717.

Martins M. B. A. F., Gonçalves A. P. V., Cruz M. E. M., *Bioconjug. Chem.*, **7** (1996), 430–435

Matsumura Y., Maeda H., *Cancer Res.*, **46** (1986), 6387–6392.

Miraki-Moud F., Ghazaly E., Ariza-McNaughton L., Hodby K. A., Clear A., Anjos-Afonso F., Liapis K., Grantham M., Sohrabi F., Cavenagh J., *Blood*, **125** (2015), 4060–4068.

Mitchell D. T., Lee S. B., Trofin L., Li N., Nevanen T. K., Söderlund H., Martin C. R., *J. Am. Chem. Soc.*, **124** (2002), 11864–11865.

Moran N. F., Bain M. D., Muqit M. M., Bax B. E., *Neurology*, **71** (2008), 686–688.

Mornet E., *Metabolism*, **82** (2018), 142–155.

Mukerabigwi J. F., Ge Z., Kataoka K., *Chem. Eur. J.*, **24** (2018), 1–20.

Nordt T. K., Bode C., *Heart*, **89** (2003), 1358–1362.

Oh D.-B., *BMB Rep.*, **48** (2015), 438–444.

Ott P. A., Carvajal R. D., Pandit-Taskar N., Jungbluth A. A., Hoffman E. W., Wu B. W., Bomalaski J. S., Venhaus R., Pan L., Old L. J., Pavlick A. C., Wolchok, J. D., *Invest. New Drugs*, **31** (2013), 425–34.

Özen H., *World J. Gastroenterol.*, **13** (2007), 2541–2553.

Palmer D. H., Mautner V., Mirza D., Oliff S., Gerritsen W., van der Sijp J. R., Hubscher S., Reynolds G., Bonney S., Rajaratnam R., Hull D., Horne M.,

Ellis J., Mountain A., Hill S., Harris P. A., Searle P. F., Young L. S., James N. D., Kerr D. J., *J. Clin. Oncol.*, **22** (2004), 1546–1552.

Pan C. Q., Lazarus R. A., *Biochemistry*, **36** (1997), 6624–6632.

Papademetriou J., Garnacho C., Serrano D., Bhowmick T., Schuchman E. H., Muro S., *J. Inherit. Metab. Dis.*, **36** (2013), 467–477.

Parodi A., Quattrocchi N., van de Ven A. L., Chiappini C., Evangelopoulos M., Martinez J. O., Brown B. S., Khaled S. Z., Yazdi I. K., Enzo M. V., Isenhart L., Ferrari M., Tasciotti E., *Nat. Nanotechnol.*, **8** (2013), 61–68.

Patterson D. P., Schwarz B., Waters R. S., Gedeon T., Douglas T., *ACS Chem. Biol.*, **9** (2014), 359–365.

Perez Y., Valdivia A., Gomez L., Simpson B. K., Villalonga R., *Macromol. Biosci.*, **5** (2005), 1220–1225.

Perler B., *J. Endovasc. Ther.*, **12** (2005), 224–232.

Pieters R., Hunger S. P., Boos J., Rizzari C., Silverman L., Baruchel A., Goekbuget N., Schrappe M., Pui C. H., *Cancer*, **117** (2011), 238–249.

Pitek A. S., Jameson S. A., Veliz F. A., Shukla S., Steinmetz N. F., *Biomaterials*, **89** (2016a), 89–97.

Pitek A. S., Wen A. M., Shukla S., Steinmetz N. F., *Small*, **12** (2016b), 1758–1769.

Puntis J. W., Zamvar V., *Arch. Dis. Child.*, **100** (2015), 869–871.

Qie Y., Yuan H., Von Roemeling C. A., Chen Y., Liu X., Shih K. D., Knight J. A., Tun H. W., Wharen R. E., Jiang W, Kim B. Y., *Sci. Rep.*, **6** (2016), 1–11.

Quester K., Juarez-Moreno K., Secundino I., Roseinstein Y., Alejo K. P., Huerta-Saquero A., Vazquez-Duhalt R., *Macromol. Biosci.*, **17** (2016), 1–9.

Ramya L., Doble M., Rekha V., Pulicherla K., *J. Pediatr. Hematol. Oncol.*, **33** (2011), 617–621.

Reuser A. J. J., Van den Hout H., Bijvoet A. G. A., Kroos M. A., Verbeet M. P., Van der Ploeg A. T., *Eur. J. Pediat.*, **161** (2002), S106–S111.

Rodriguez P. L., Harada T., Christian D. A., Pantano D. A., Tsai R. K., Discher D. E., *Science*, **339** (2013), 971–975.

Rosenblum D., Joshi N., Tao W., Karp J. M., Peer D., *Nat. Commun.*, **9** (2018), 1410.

Rostro-Alanis M. de J., Mancera-Andrade E. I., Patiño M. B. G., Arrieta-Baez D., Cardenas B., Martinez-Chapa S. O., Saldívar R. P., *Biocatalysis*, **2** (2016), 1–24.

Salvati A., Pitek A. S., Monopoli M. P., Prapainop K., Bombelli F. B., Hristov D. R., Kelly P. M., Åberg C., Mahon E., Dawson K. A., *Nat. Nanotechnol.*, **8** (2013), 137–143.

Sánchez-Sánchez L., Cadena-Nava R., Palomares L., Ruiz-Garcia J., Koay M., Cornelissen J., Vazquez-Duhalt R., *Enzyme Microb. Technol.*, **60** (2014), 24–31.

Sánchez-Sánchez L., Tapia-Moreno A., Juarez-Moreno K., Patterson D. P., Cadena-Nava R. D., Douglas T., Vazquez-Duhalt R., *J. Nanobiotechnol.*, **13** (2015), 1.

Sapsford K. E., Algar W. R., Berti L., Gemmill K. B., Casey B. J., Oh E., Stewart M. H., Medintz I. L., *Chem. Rev.*, **113** (2013), 1904–2074.

Schmidt M. M., Wittrup, K. D. A., *Mol. Cancer Ther.*, **8** (2009), 2861–2871.

Schoonen L., Maassen S., Nolte R. J. M., van Hest J. C. M., *Biomacromolecules*, **18** (2017), 3492–3497.

Schoonen L., van Hest J. C. M., *Nanoscale*, **6** (2014), 7124–7141.

Sengar P., Juárez P., Verdugo-Meza A., Arellano D. L., Jain A., Chauhan K., Hirata G. A., Fournier P. G. J., *J. Nanobiotechnology*, **16** (2018), 1–19.

Solomon M., Muro S., *Adv. Drug Deliv. Rev.*, **118** (2017), 109–134.

Soo Choi H., Liu W., Misra P., Tanaka E., Zimmer J. P., Itty Ipe B., Bawendi M. G., Frangioni J. V., *Nat. Biotechnol.*, **25** (2007), 1165–1170.

Steinmetz N. F., Manchester M., *Biomacromolecules*, **10** (2009), 784–792.

Stirnemann J., Belmatoug N., Camou F., Serratrice C., Froissart R., Caillaud C., Levade T., Astudillo L., Serratrice J., Brassier A., Rose C., Billette de Villemeur T., Berger M. G., *Int. J. Mol. Sci.*, **18** (2017), 441.

Tapia-Moreno A., Juarez-Moreno K., Gonzalez-Davis O., Cadena-Nava R. D., Vazquez-Duhalt R., *Biotechnol. J.*, **12** (2017), 1600706.

Tarnopolsky M., Katzberg H., Petrof B. J., Sirrs S., Sarnat H. B., Myers K., Dupré N., Dodig D., George A., Venance S. L., Korngut L., Raiman J., Khan A., *Can. J. Neurol. Sci.*, **43** (2016), 472–485.

Tekoah Y., Tzaban S., Kizhner T., Hainrichson M., Gantman A., Golembo M., Aviezer D., Shaaltiel Y., *Biosci. Rep.*, **33** (2013), e00071.

Tomatsu S., Sawamoto K., Alméciga-Díaz C. J., Shimada T., Bober M. B., Chinen Y., Yabe H., Montaño A. M., Giugliani R., Kubaski F., Yasuda E., Rodríguez-López A., Espejo-Mojica A. J., Sánchez O. F., Mason R. W., Barrera L. A., Mackenzie W. G., Orii T., *Drug Des. Devel. Ther.*, **9** (2015), 1937–1953.

Treem W. R., McAdams L., Stanford L., Kastoff G., Justinich C., Hyams J., *J. Pediat. Gastroenterol. Nutrit.*, **28** (1999), 137–142.

Tychopoulos M., Corcos L., Genne P., Beaune P., de Waziers I., *Cancer Gene Ther.*, **12** (2005), 497–508.

Vairo F., Federhen A., Baldo G., Riegel M., Burin M., Leistner-Segal S., Giugliani R., *Appl. Clin. Genet.*, **8** (2015), 245–255.

Villegas M. R., Baeza A., Vallet-Regí M., *ACS Appl. Mater. Interfaces.*, **7** (2015), 24075–24081.

Vŷna I., Karsakevich A., Bekers M., *J. Mol. Catal. B Enzym.*, **11** (2001), 551–558.

Vriezema D. M., Comellas Aragonès M., Elemans J. A. A. W., Cornelissen J. J. L. M., Rowan A. E., Nolte R. J. M., *Chem. Rev.*, **105** (2005), 1445–1490.

Werengowska-Ciećwierz K., Wiśniewski M., Terzyk A. P., Furmaniak S., *Adv. Condens. Matter Phys.*, (2015), 198175.

White A. D., Nowinski A. K., Huang W., Keefe A. J., Sun F., Jiang S., *Chem. Sci.*, **3** (2012), 3488–3494.

Whiteman D. A., Kimura A., *Drug Des. Devel. Ther.*, **11** (2017), 2467–2480.

Whyte M. P., *Ann. N. Y. Acad. Sci.*, **1192** (2010), 190–200.

Whyte M. P., *J. Bone Miner. Res.*, **32** (2017), 667–675.

Wilkerson J. W., Yang S. O., Funk P. J., Stanley S. K., Bundy B. C., *N. Biotechnology*, **44** (2018), 59–63.

Xiao K., Li Y., Luo J., Lee J. S., Xiao W., Gonik A. M., Agarwal R. G., Lam K. S., *Biomaterials*, **32** (2011), 3435–3446.

Xie T., Wang A., Huang L., Li H., Chen Z., Wang Q., *African J. Biotechnol.*, **8** (2009), 4724–4733.

Xu X., Costa A., Burgess D. J., *Pharm. Res.*, **29** (2012), 1919–1931.

Yadak R., Sillevis Smitt P., van Gisbergen M. W., van Til N. P., de Coo I. F. M., *Front. Cell. Neurosci.*, **11** (2017), 1.

Yari M., Ghoshoon M. B., Vakili B., Ghasemi Y., *Curr. Pharm. Biotechnol.*, **18** (2017), 531–540.

Yata V. K., Banerjee S., Ghosh S. S., *Adv. Sci. Eng. Med.*, **6** (2014), 388–392.

Yuan Y. Y., Mao C. Q., Du X. J., Du J. Z., Wang F., Wang J., *Adv. Mater.*, **24** (2012), 5476–5480.

Zelphati O., Wang Y., Kitada S., Reed J. C., Felgner P. L., Corbeil J., *J. Biol. Chem.*, **276** (2001), 35103–35110.

Zhao M., Biswas A., Hu B., Joo K. Il., Wang P., Gu Z., Tang Y., *Biomaterials*, **32** (2011), 5223–5230.

Zhao Z., Fu J., Dhakal S., Johnson-Buck A., Liu M., Zhang T., Woodbury N. W., Liu Y., Walter N. G., Yan H., *Nat. Commun.*, **7** (2016), 1–9.

Zhao M., Hu B., Gu Z., Joo K., Wang P., Tang Y., *Nano Today*, **8** (2013), 11–20.

Zhao Y.-N., Zhang W., Chen Y.-C., Fang F., Liu X.-Q., *Maturitas*, **72** (2012), 139–145.

Zhu L., Cheng F., Piatkowski V., Schwaneberg U., *ChemBioChem.*, **15** (2014), 276–83.

Zhu L., Tee K. L., Roccatano D., Sonmez B., Ni Y., Sun Z.- H., Schwaneberg U., *ChemBioChem.*, **11** (2010), 691–697.

Zuris J. A., Thompson D. B., Shu Y., Guilinger J. P., Bessen J. L., Hu J. H., Maeder M. L., Joung J. K., Chen Z. Y., Liu D. R., *Nat. Biotechnol.*, **33** (2015), 73–80.

Chapter 19

Computer-Aided Drug Design for the Identification of Multi-Target Directed Ligands (MTDLs) in Complex Diseases: An Overview

Pravin Ambure and Kunal Roy

Drug Theoretics and Cheminformatics Laboratory, Department of Pharmaceutical Technology, Jadavpur University, Kolkata 700 032, West Bengal, India

kunalroy_in@yahoo.com, kunal.roy@jadavpuruniversity.in

19.1 Introduction to Multi-Target Drug Designing

19.1.1 Multi-Target Drug Designing and Its Role in Complex Diseases

Complex diseases such as cancer, multiple sclerosis, osteoporosis, and neurodegenerative diseases, whose origin is multi-factorial in nature, are caused by a combination of several factors associated

Pharmaceutical Biocatalysis: Fundamentals, Enzyme Inhibitors, and Enzymes in Health and Diseases
Edited by Peter Grunwald

ISBN 978-981-4800-61-7 (Hardcover), 978-0-429-29503-4 (eBook)
www.jennystanford.com

with genetic, environment, and lifestyle (Craig, 2008). Thus, in the present scenario, drugs acting on a single target enzyme or receptor are inadequate for the treatment of complex diseases. An emerging approach, i.e., multi-target drug designing (MTDD), focuses on the development of a single ligand that can act on multiple targets simultaneously. Such ligands are termed as multi-target directed ligands (MTDLs). Polypharmacology is a branch of pharmaceutical science that deals with the development of MTDLs that can hit multiple targets for same or multiple disease pathways (Jalencas and Mestres, 2013). It has emerged as the next paradigm in rational drug design (RDD) under drug discovery, particularly for the complex diseases (Ambure and Roy, 2017; Roy, 2019).

19.1.2 Multi-Target Directed Ligands over Single-Target Directed Ligands, Multiple-Medication Therapy, and Multiple-Compound Medication

Multi-target drug designing approach is not easy to accomplish; however, it is still better than conventional approaches such as multiple-medication therapy (MMT), i.e., consuming more than one medication at a time or multiple-compound medication (MCM), i.e., different drugs into the same formulation. The benefits of MTDLs over the conventional approaches are that MTDLs avoid any likely drug–drug interactions and resolve the problems associated with the usage of multiple drugs simultaneously such as managing the different rate of solubility, absorption, bioavailability, distribution, and metabolism. More importantly, the clinical trial of any MTDL should be similar to any single lead and thus have easy regulatory acceptance compared to MMT or MCM (Cavalli et al., 2008).

19.1.3 MTDD Trend in Research and Development

The potential of MTDD or polypharmacology in the pharmaceutical field has been already demonstrated in a published study (Lu et al., 2012). In this study, the authors compared the drug targets and the market sales of the new molecular entities approved by the U.S. Food and Drug Administration from January 2000 to December 2009. The study concludes that there is a significant progress in the development of multi-targeted drugs, which have a substantial potential against complex diseases.

In another report (Jalencas and Mestres, 2013), based on the collected drug–target interaction data, the authors have suggested to change the biased perception of drug selectivity. Further, it was stated that most of the therapeutically effective drugs interact with multiple proteins to show their actual pharmacological effect, while selective drugs that targeting only a single target are just few exceptions. Based on their observation, currently only 15% of drugs are known to interact with a single target, whereas over 50% of them interact with more than five targets.

19.1.4 Role of Computer-Aided Drug Discovery (CADD) in MTDD

Computational or in silico studies play an important role in the drug discovery and drug design paradigm by providing guidance towards the lead identification, lead optimization, prioritization of chemicals before initiating the synthetic or experimental analysis and designing superior analogues. Notably, in silico techniques complement the 3Rs approach, i.e., replacement, refinement, and reduction of animals in research as well as these are cost-effective and time saving. There are several available in silico techniques that can be divided into two major application areas, i.e., structure-based drug design and ligand-based drug design (Huang et al., 2010). Structure-based drug design relies on three-dimensional (3D) structural knowledge of the target protein (enzyme or receptor) and its binding sites to investigate vital inter-molecular interactions as well as their corresponding binding energy. On the other hand, ligand-based drug design relies on knowledge of ligands with known binding information with the target of interest. Notably, both type of techniques together become a powerful in silico tool to design potential ligands against one or multiple targets.

19.1.5 In silico Testing Prior to Experimental Validation

The early stage experimental validation of selected MTDLs is important for post-optimization. However, prior to experimental validation, a few in silico testing or filters can be applied like drug-likeness (Lipinski et al., 1997) or central nervous system (CNS) drug-likeness (Lipinski, 2004), or finding colloidal aggregators

(Irwin et al., 2015), etc. Once the compounds are screened and prioritized, further experimental testing can be done to confirm the in silico results on selected compounds.

19.2 Computational Techniques Involved in MTDD

19.2.1 Preliminary Studies for Initial Selection of Potential Ligands and Target Proteins

A preliminary study is certainly helpful for initial selection of possible MTDLs and/or target proteins. In such study, the potential ligand as MTDLs can be selected by finding a set of structurally similar ligands with known binding affinity to different protein targets (from same or different protein families). Here, the similar ligands can be identified based on a similarity/distance-based metric like Tanimoto similarity coefficient, Euclidean distance, etc. The potential target proteins can be selected via a binding site similarity study, where, the binding site can be estimated using several available computational tools, such as LIGSITE, Q-SiteFinder, SURFNET, or PASS (Henrich et al., 2010). Proteins that belong to the same gene family, or proteins showing complete sequence homology are more likely to have similar binding sites. Ligand binding-site similarities among unrelated proteins are interesting viewpoint in MTDD approach (Haupt et al., 2013). Another approach is drug-repurposing or drug re-profiling studies, where, the goal is to find already known and approved drugs as potential candidate for MTDD via studying the side effects or beneficial effects caused by binding to multiple off-target proteins (Zhang et al., 2017). Moreover, network pharmacology (Hopkins, 2008) can also be employed to create a chemical network that relates proteins that bind to similar ligand.

19.2.2 In silico Approaches in MTDD

The in silico approaches utilized in MTDD can be classified into two broad categories (see Fig. 19.1):

(a) Combinatorial
(b) Fragment-based or de novo design

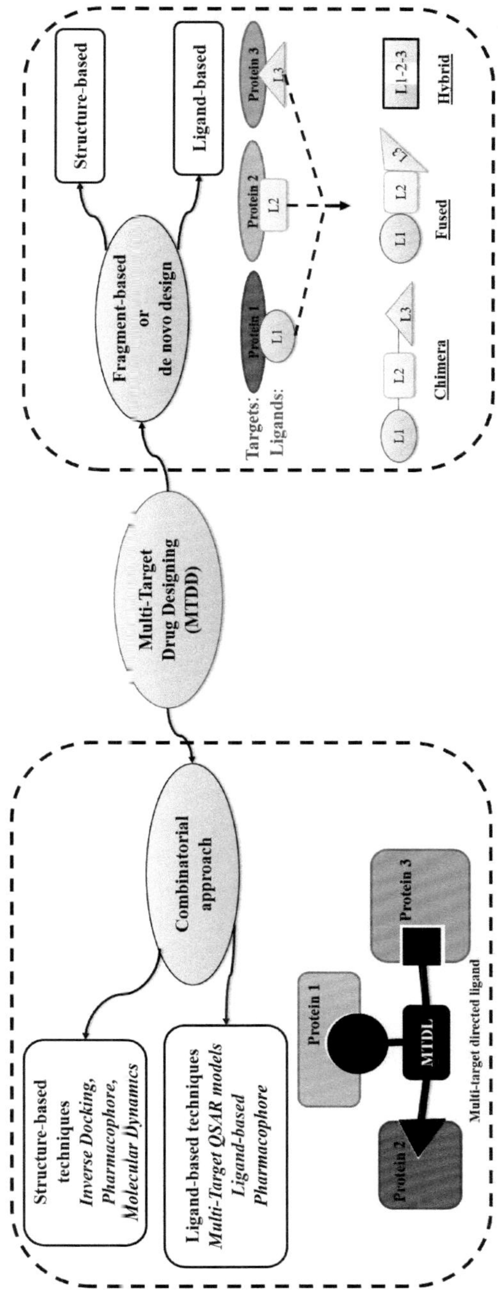

Figure 19.1 Computational approaches in Multi-target drug designing (MTDD).

19.2.2.1 Combinatorial approaches

In this approach, as the name suggests, the results obtained against individual targets are combined to find MTDLs that can simultaneously interact with multiple targets (illustrated in Fig. 19.1).

There are two types of techniques under combinatorial approaches:

(i) Ligand-based techniques

These techniques rely on the knowledge of diverse ligands that interact with the biological targets of interest. For instance, similarity-based screening (2D and 3D) like ligand-based pharmacophore mapping, machine learning techniques, are ligand-based approaches. Further, we will discuss these techniques, which are as follows:

Ligand-based pharmacophore model

A pharmacophore can be defined as the basic structural features (such as H-bond acceptors and donors, hydrophobic aliphatic, hydrophobic aromatic, and ring aromatic) along with the proper geometrical constraints (distance, and dihedral angle), which are essential in a ligand for showing biological activity or interactions against a biological target. In silico pharmacophore development technique can be classified into ligand-based (*by overlapping a set of known active and inactive compounds for identifying the common structural features that are required for their biological activity*) and structure-based (*by finding the potential interaction points between the ligands and target protein*) pharmacophore modeling. Pharmacophore modeling is a highly useful as well as time-saving technique for identifying MTDLs, where the simplest approach is to develop a set of pharmacophore models against the protein targets of interest and then screen the available chemical (synthetic and/or natural compounds) databases from each of the developed and validated pharmacophore. The compounds that successfully pass (possessing minimum required features) through more than one pharmacophore models can be explored further for confirming activity against the respective targets. Here, the screened compounds can be ranked based on their geometric fit values. Another way is to derive a common pharmacophore from a set of pharmacophore models against different targets and then the single pharmacophore model is used to screen and identify MTDLs (Ambure and Roy, 2017). Ligand-based pharmacophore modeling can be done using several available software—such as HipHop and HypoGen (Li et al., 2000) (by Accelrys Inc., http://www.accelrys.com), PHASE (Dixon et al., 2006) (by Schrödinger Inc., http://www.schrodinger.com), DISCO (Martin, 2000), GASP (Jones et al., 2000), and GALAHAD (Poptodorov et al., 2006) (by Tripos Inc., http://www.tripos.com).

Multi-Target QSAR Models

Quantitative structure-activity relationship (QSAR) is a widely used computational technique to develop a quantitative relationship between the descriptors representing chemical features and the biological activity/property/response variable for a series of compounds (Roy, 2017). There are two types of QSAR models, namely, (a) classification-based that defines the relationship between the descriptors and categorical values of the response variable, and (b) regression-based that defines the relationship between the descriptors and quantitative values of the response variable. QSAR models are usually developed for a set of compounds against a single target. However, in MTDD, multiple QSAR models are developed employing several different data sets aiming at different biological targets of interest, as also reported in a recent study (Ambure et al., 2018) for identifying multi-target directed leads against Alzheimer's disease. The developed multiple QSAR models are then appropriately validated, and then utilized to screen chemical databases. Compounds that are found active against more than one target are of special interest and are then studied further for confirming their biological activities experimentally. Another approach is to develop a single multi-target QSAR model that can simultaneously predict activity or classify compounds as actives or in-actives against respective targets. To develop a single multi-target model, specific descriptors are introduced that are characteristics of a set of active compounds against the same target including auxiliary variables for indicating which response one wants to predict. Once the model is built with the aid of such descriptors, the developed model can be used to predict whether a compound is active or not against multiple targets using the same QSAR equation. For instance, several studies (Kleandrova et al., 2016; Luan et al., 2014; Speck-Planche and Cordeiro, 2017; Speck-Planche et al., 2012; Speck-Planche et al., 2013) have reported single multi-target QSAR models using QSAR-perturbation approach (Gonzalez-Diaz et al., 2013) or Box–Jenkins approach (Hill et al., 2006), which can perform simultaneous classification and prediction of chemicals against multiple different target proteins or multiple different experimental conditions.

(ii) Structure-based methods

As mentioned before, structure-based techniques rely on 3D structural knowledge of the target protein and its active/binding site which helps in investigating the key ligand–protein interactions as well as to predict the binding energy. Now we will discuss some key structure-based techniques that are employed in MTDD.

Structure-based pharmacophore modeling

Structure-based pharmacophore (SBP) model development is only possible when the 3D structural information of target macromolecule is available. Here, the methodology involves an analysis of the complementary chemical features of the active/binding site along with their spatial arrangement. Subsequently from this analysis, a pharmacophore model with selected features is built. Further, the (SBP) modeling technique can be further divided into two sub-categories, namely, target–ligand complex-based and target (only)-based. The target–ligand complex-based approach is more suitable as it helps in finding the ligand-binding site of the macromolecular target and for understanding the key interaction points between ligands and target. LigandScout (Wolber and Langer, 2005), Pocket v.2 (Chen and Lai, 2006) and GRID-based pharmacophore model (GBPM) (Ortuso et al., 2006) are some tools to perform target–ligand complex-based pharmacophore modeling. However, in cases when no compounds are known that are targeting the binding site of interest, then one can perform pharmacophore modeling using the target (only)-based approach. The SBP method implemented in Discovery Studio software (Discovery Studio, 2009) can be used to perform target (only)-based approach (Yang, 2010). Further, to identify the MTDLs, the parallel screening is performed on a set of chemicals obtained from several available databases through the multiple pharmacophore models representing different targets. Another faster way to perform pharmacophore-based virtual screening is to use the available online platform such as PharmMapper (http://lilab.ecust.edu.cn/pharmmapper/) (Liu et al., 2010), which is an integrated pharmacophore matching platform with statistical method for potential target identification and it includes a database (i.e., PharmTargetDB) of over 7000 target-based pharmacophore models. Another database "PharmaDB" (Meslamani and Rognan, 2015), which is included in Discovery Studio software (Discovery Studio, 2009), comprises about 140,000 receptor-ligand pharmacophore models, which are built from and validated using the scPDB (an annotated database of druggable binding sites from the Protein Data Bank) (Kellenberger et al., 2006). Finally, pharmacophore-based techniques are not computationally expensive when compared to molecular docking (or inverse docking) or molecular dynamics studies. However, identifying a good-quality pharmacophore hypothesis is a big task (Lavecchia and Cerchia, 2016).

Inverse docking

Molecular docking is a computational technique to find and measure the inter-molecular interactions between the ligand and the binding site of the target protein. Additionally, it helps in predicting the bioactive conformation as well as the probable binding energy (or docking score) that is computed using several available scoring functions (Ambure and Roy, 2016). Molecular docking technique is extensively used to screen big chemical databases to identify a set of ligands that can bind aptly to the target of interest and thus such study involves several ligands, but one target protein. However, a MTDD study usually involves a few or more ligands against multiple target proteins and since this approach is opposite to the usual molecular docking, thus it is termed as "inverse docking." Initially inverse docking was employed to understand or find the side effects of known drugs by finding the "off-targets," but the same approach can be successfully employed to find off-targets that will improve the desired pharmacological response (Rognan, 2010). Although inverse docking is a powerful technique to identify MTDLs, it should be noted that inverse docking is considerably time-consuming technique as it requires exhaustive protein structure preparation and usually the target flexibility issue is not addressed. Further, results can be affected based on the poor performance of scoring functions, so selection of appropriate scoring function (Perola et al., 2004) is important.

Molecular dynamics (MD) simulation study

The scoring functions of existing docking methodologies help in ranking or prioritization of the compounds but are not accurate enough to compute reliable binding free energy values. Molecular dynamics simulation study helps in investigating the ligand–protein interactions in detail, which includes understanding the most stable or key interactions and to find significantly accurate binding free energy. Usually Molecular Mechanics/Poisson–Boltzmann Surface Area (MM/PBSA) or Molecular Mechanics/Generalized Born Surface Area (MM/GBSA) simulation approaches are employed to compute the binding free energy between ligand and target protein (Hou et al., 2011). However, the molecular dynamics studies should be performed on the finally screened compounds with potential multi-target characteristics, as this technique is highly time consuming (Ambure et al., 2018). Moreover, the issue of protein flexibility and false-positive hits sometimes observed in molecular docking and (SBP) technique can be resolved by performing molecular dynamics analysis.

19.2.2.2 Fragment-based or de novo design approaches

In this approach, multiple fragments with known binding capability against individual targets of interest are combined to design a MTDL. Here, according to the "rule-of-three," fragments are defined as molecules with a molecular weight < 300, H-bond acceptors and donors ≤ 3, and a log $P \leq 3$ (Congreve et al., 2003). Additionally, number of rotatable bonds ≤ 3 and a polar surface area ≤ 60 can be used as extra parameters. Fragment-based approach for designing MTDLs can be performed using any one of the following strategies (Bottegoni et al., 2012):

- Chimera: linking of the key pharmacophoric functions by means of suitable spacers
- Fusion: blending of the key functions without adding or missing any substructure

- Hybrid: amalgamation of the requisite pharmacophoric features into one molecule

These strategies are well illustrated in Fig. 19.1. Moreover, the fragment-based or de novo design study can also be categorized as ligand-based or structure based. In ligand-based de novo MTDD, the known active ligands for the targets of interest are collected to build a virtual ligand library. Using such library, one may perform QSAR studies against each target of interest to analyze relationship between fragment/functional group descriptors and biological response against the target of interest. Finally, essential fragments are identified for each target and are then combined using the above-mentioned strategies to develop optimized MTDLs.

Using the structure-based de novo MTDD, an easy way to find essential fragments against individual targets is to use the molecular docking technique. This approach seems more rational, since it directly considers the interactions between the ligand and multiple targets. Here, the initial seeds are first placed into the known binding/active sites and a pool of molecules is incubated following a genetic algorithm-based growth scheme. Here, the degree of inter-molecular interactions with targets of interest can be used as fitness function. All the generated candidate molecules are tested against the fitness function and other optimal objectives. The top-ranked molecules are then selected for the next generation of optimization and this cycle is continued until one or more termination criteria is fulfilled (Zhang et al., 2017; Fechner and Schneider, 2007).

Many de novo drug design tools have been developed and are available, such as LUDI (Böhm, 1992), SMoG (DeWitte and Shakhnovich, 1996), LigBuilder 2 (Yuan et al., 2011), FLUX (Fechner and Schneider, 2007), Pharmacophore-based De Novo Design (PhDD) (Huang et al., 2010), etc.

19.3 Conclusion

In this chapter, we have provided an overview of the computational techniques for carrying out multi-target drug designing (MTDD). We have briefly discussed both combinatorial and fragment-based or de novo designing, approaches which are employed for performing MTDD. The computational techniques are further divided into structure-based and ligand-based studies. The most promising way is to combine both types of studies, which significantly improves the quality of results. However, the selection of techniques is completely based on the availability as well as the diversity and the quality of available input data. Furthermore, MTDD strategies play an important role in rational designing of new MTDLs in complex diseases. However, it is still challenging task to design and optimize MTDLs while retaining the drug-like properties and regulating the unwanted off-target side effects.

References

Ambure P., Bhat J., Puzyn T., Roy K., *J. Biomol. Struct. Dynamics* (2018), 1–25. https://doi.org/10.1080/07391102.2018.1456975.

Ambure P., Roy K., *Curr. Drug Targets,* **18** (2017), 522–533.

Ambure P., Roy K., *Methods and Algorithms for Molecular Docking-Based Drug Design and Discovery* (Dastmalchi S., Hamzeh-Mivehroud, M., eds.) IGI Global, PA, USA (2016), pp. 54–98.

Böhm H.-J., *J. Computer-Aided Mol. Des.,* **6** (1992), 61–78.

Bottegoni G., Favia A. D., Recanatini M., Cavalli A., *Drug Disc. Today,* **17** (2012), 23–34.

Cavalli A., Bolognesi M. L., Minarini A., Rosini M., Tumiatti V., Recanatini M., Melchiorre C., *J. Med. Chem.,* **51** (2008), 347–372.

Chen J., Lai L., *J. Chem. Inform. Model.,* **46** (2006), 2684–2691.

Congreve M., Carr R., Murray C., Jhoti H., *Drug Disc. Today,* **19** (2003), 876–877.

Craig J., *Nat. Educ.,* **1** (2008), 184.

DeWitte R. S., Shakhnovich E. I., *J. Am. Chem. Soc.,* **118** (1996), 11733–11744.

Discovery Studio, Accelrys Inc.: San Diego, CA, USA (2009).

Dixon S. L., Smondyrev A. M., Knoll E. H., Rao S. N., Shaw D. E., Friesner R. A., *J. Computer Aided Mol. Des.,* **20** (2006), 647–671.

Fechner U., Schneider G., *J. Chem. Inform. Model.,* **47** (2007), 656–667.

Gonzalez-Diaz H., Arrasate S., Gomez-SanJuan A., Sotomayor N., Lete E., Besada-Porto L., Ruso J. M., *Curr. Topics Med. Chem.,* **13** (2013), 1713–1741.

Haupt V. J., Daminelli S., Schroeder M., *PloS One,* **8** (2013), e65894.

Henrich S., Salo-Ahen O. M., Huang B., Rippmann F. F., Cruciani G., Wade R. C., *J. Mol. Recogn.: An Interdisc. J.*, **23** (2010), 209–219.

Hill T., Lewicki P., Lewicki P., *Statistics: Methods and Applications: A Comprehensive Reference for Science, Industry, and Data Mining*, StatSoft, Inc. (2006).

Hopkins A. L., *Nat. Chem. Biol.*, **4** (2008), 682.

Hou T., Wang J., Li Y., Wang W., *J. Comput. Chem.*, **32** (2011), 866–877.

Huang H.-J., Yu H. W., Chen C.-Y., Hsu C.-H., Chen H.-Y., Lee K.-J., Tsai F.-J., Chen C. Y.-C., *J. Taiwan Institute Chem. Eng.*, **41** (2010), 623–635.

Huang Q., Li L.-L., Yang S.-Y., *J. Mol. Graphics Model*, **28** (2010), 775–787.

Irwin J. J., Duan D., Torosyan H., Doak A. K., Ziebart K. T., Sterling T., Tumanian G., Shoichet B. K., *J. Med. Chem.*, **58** (2015), 7076–7087.

Jalencas X., Mestres J., *MedChemComm*, **4** (2013), 80–87.

Jones G., Willet P., Glen R., *Pharmacophore Perception, Development, and Use in Drug Design* (2000), pp. 85–106, International University Line: La Jolla, CA.

Kellenberger E., Muller P., Schalon C., Bret G., Foata N., Rognan D., *J. Chem. Inform. Model.*, **46** (2006), 717–727.

Kleandrova V. V., Ruso J. M., Speck-Planche A., Dias Soeiro Cordeiro M. N. l., *ACS Combinat. Sci.*, **18** (2016), 490–498.

Lavecchia A., Cerchia C., *Drug Discov. Today*, **21** (2016), 288–298.

Li H., Sutter J., Hoffmann R., *Pharmacophore perception, Development, and Use in Drug Design*, **2** (2000), p. 171, International University Line: La Jolla, CA.

Lipinski C. A., *Drug Discov. Today: Technol.*, **1** (2004), 337–341.

Lipinski C. A., Lombardo F., Dominy B. W., Feeney P. J., *Adv. Drug Deliv. Rev.*, **23** (1997), 3–25.

Liu X., Ouyang S., Yu B., Liu Y., Huang K., Gong J., Zheng S., Li Z., Li H., Jiang H., *Nucleic Acids Res.*, **38** (2010), W609–W614.

Lu J.-J., Pan W., Hu Y.-J., Wang Y.-T., *PloS One*, **7** (2012), e40262.

Luan F., Kleandrova V. V., González-Díaz H., Ruso J. M., Melo A., Speck-Planche A., Cordeiro M. N. D., *Nanoscale*, **6** (2014), 10623–10630.

Martin Y., *Pharmacophore Perception, Development, and Use in Drug Design*, **2** (2000), 49–68, International University Line: La Jolla, CA.

Meslamani J., Rognan D., *CICSJ Bull.*, **33** (2015), 27.

Ortuso F., Langer T., Alcaro S., Bioinformatics, **22** (2006), 1449–1455.

Perola E., Walters W. P., Charifson P. S., *Proteins: Struct., Funct. Bioinform.*, **56** (2004), 235–249.

Poptodorov K., Luu T., Hoffmann R. D., *Meth. Principles Med. Chem.*, **32** (2006), 17.

Rognan D., *Mol. Inform.*, **29** (2010), 176–187.

Roy K. (ed), *Advances in QSAR Modeling*. Applications in Pharmaceutical, Chemical, Food, Agricultural and Environmental Sciences., Springer (2017).

Roy K. (ed.), *Multi-Target Drug Design Using Chem-Bioinformatic Approaches*, Springer (2019).

Speck-Planche A., Cordeiro M. N. D., *Mol. Divers.*, **21** (2017), 511–523.

Speck-Planche A., Kleandrova V. V., Luan F., Cordeiro M. N. D., *Bioorg. Med. Chem.*, **20** (2012), 4848–4855.

Speck-Planche A., Kleandrova V., Luan F. Cordeiro N. D., *Curr. Alzheimer Res.*, **10** (2013), 117–124.

Wolber G., Langer T., *J. Chem. Inform. Model.*, **45** (2005), 160–169.

Yang S.-Y., *Drug Disc. Today*, **15** (2010), 444–450.

Yuan Y., Pei J., Lai L., *J. Chem. Inform. Model.*, **51** (2011), 1083–1091.

Zhang W., Pei J., Lai L., *J. Chem. Inform. Model.*, **57** (2017), 403–412.

Chapter 20

The Development of Improved Therapeutics through a Glycan-"Designer" Approach

Natalia G. Bednarska

London School of Hygiene and Tropical Medicine, Keppel Street, London WC1E 7HT, UK

natalia.bednarska@lshtm.ac.uk

20.1 Introduction

Glycosylation is a natural post-translational modification, being currently explored in the field of biotechnology to improve activity and pharmacological propensities of many therapeutics. Most of the functional proteins in the human body are glycosylated (~80%); moreover, any mis-glycosylation or changes in the glycan structure are linked to numerous diseases (Apweiler et al., 1999). The analysis of human serum showed that 96% of the total glycome consists of complex *N*-glycans, while hybrid and high-mannose type of glycans account to the remaining 4% (An et al., 2009).

Pharmaceutical Biocatalysis: Fundamentals, Enzyme Inhibitors, and Enzymes in Health and Diseases
Edited by Peter Grunwald

ISBN 978-981-4800-61-7 (Hardcover), 978-0-429-29503-4 (eBook)
www.jennystanford.com

Depending on the linkage between protein and saccharide, the glycosylation can be classified as follows:

- *N*-linked glycosylation, where glycan is attached via Asparagine amino acid
- *S*-linked, where glycans are attached through Serine or Tyrosine residue of the protein
- *C*-linked

Throughout the years of research on glycans and their lectin receptors, it has become clear that the glycans not only help to maintain protein intrinsic properties (folding, stability, structure) but also affect the biological pathways and cell-mediated immune processes.

In mammalian cells, glycosylation occurs in the lumen of endoplasmic reticulum and in the Golgi apparatus. There are around 200 glycosyltrasferases encoded in mammalian genome, which along with glycosidases contribute to the unique glycocode of each cell specific for cell lineages and cellular activation states (Marth and Grewal, 2008). Many of these are now being explored to serve as rapid diagnostics and prognostic markers for cancers, autoimmune and genetic diseases (Maverakis et al., 2015; An et al., 2009). Glycosylation plays important role in the immune system, where any changes in glycosylation type or level have an impact on leukocyte migration, apoptosis, B-cell and T-cell receptors, IgG function and antigen presentation among many other unknown regulatory functions. Changes in physiological levels of glycosylation can occur and may lead to abnormal glycan production and pathologies. This is because the glycosylation influences protein interactions, structure, folding, and various pathways and signaling (e.g., tyrosine phosphorylation). There are sufficient findings in autoimmunity field demonstrating that each autoimmune disease can be linked to a unique glycan structure responsible for an abnormal function and recognition of immune cells, so-called "altered glycan theory" (Maverakis et al., 2015). This theory confirms that a new method for the development of therapeutics should include vigilant glycan-modifications approach, because any small differences in glycan composition may affect the immune system recognition, leading to a disease. Since glycans have either pro-inflammatory and/or anti-inflammatory effect, it could be worth to consider using adequate glycosylation to up/downregulate the immune system accordingly to a given therapeutic demand. Glycobiology research is rapidly expanding with many biophysical and bioinformatics tools under development to study glycan structures, diversity, and glycan influence on protein interactions. In this chapter, current application of protein–glycan coupling technology will be introduced showing how to modify protein/peptide immunogenicity through glycosylation in the prospective of enhancing the antimicrobial and immunogenic properties.

20.2 Types of Glycosylation

There are four known types of glycosylation based on the type of carbohydrate-peptide linkage: *N*-linked (Asparagine), *O*-linked (Threonine, Serine) *C*-linked (via Carbon atom) and *S*-glycosylation (Cysteine). Glycosylation can be further diversified on the base of glycosidic linkage, glycan composition, structure, and length. The *N*-type glycosylation in which glycan is attached to the amine group of asparagine residue resulting in formation of amide bond, is the most prevalent naturally occurring glycosylation type. The most common glycan linkage in human body among *N*-glycans is *N*-acetylglucosamine (GlcNAcβ1-Asn) found abundant in human serum. Overall different *N*-glycans are prevalent structures in prokaryotic surface proteins with a unique species-specific oligosaccharide structure; for that reason the glycoproteins derived from pathogens are excellent immunogens. The synthesis of *N*-glycans is well described for both prokaryotes and eukaryotes. While glycosylation in prokaryotes can be easily predicted and controlled the synthesis of all eukaryotic *N*-glycans is more complexed.

O-linked glycosylation on the contrary is a dynamically explored field due to its potent role in mammalian pathophysiological processes. *O*-linked glycosylation is characterized by a covalent attachment of glycan through an oxygen atom. The *O*-linked consensus is initiated by an attachment of N-Acetylgalactosamine (GalNAc) to Ser/Thre but may also be composed of *O*-linked β-*N*-acetylglucosamine; hence, classification of *O*-glycans is according to their initiating monosaccharide. There are around five classes of *O*-glycosyl modifications known in mammalian cells, these are: *O*-*N*-acetylgalactosamine, *O*-fucose, *O*-glucose, *O*-

N-acetylglucosamine and *O*-mannose. Oligosaccharide attachment is catalyzed by *O*-glycosyltransferases (OGTs) upon recognition of the Asn-X-Ser/Thr (where X is any amino acid). There are many bacterial glycosyltransferases already employed for in vitro controlled glycosylation, opening very successful field of protein glycoengineering. The glycan polymer may vary in heterogeneity, which makes the prediction of monosaccharide building blocks variable in vivo. Additionally, the extra branching of *O*-glycans involves multiple glycosyltransferases.

Protein *C*-linked glycosylation differs fundamentally from *N*- and *O*-glycosylation and defines a type of glycosylation in which carbohydrate is linked to a protein via carbon atom. In eukaryotes the *C*-glycosylation was detected in multiple cell lines: mucins, erythropoietin receptor, human RNase 2, IL-12, rat liver microsomes, and proteins of complement system (Hartmann and Hofsteenge, 2000). The biological role of *C*-glycosylation is still unclear; however, the high interest toward *C*-glycosylation of protein in pharmaceutical industry is majorly related to its unique resistance to metabolic hydrolysis. For the *C*-mannosylation the acceptor sequence consists of W-x-x-W where the first Trp becomes *C*-mannosylated with α-mannopyranose. The alteration of hyaluronidase and lipoprotein lipase enzymes with *C*-mannosylation showed to be crucial for enzyme secretion-efficiency and overall enzymatic activity (Okamoto et al., 2017; Wang, 2012). The synthesis of *C*-mannosylated thrombospondin type 1 repeat-derived peptides showed to have a stimulatory effect on lipopolysaccharide-induced signaling in macrophage (Ihara et al., 2010).

The *S*-linked glycosylation is most recently discovered type of glycosylation in which a hexose residue is linked to the thiol group of Cysteine. The *S*-linked glycosylation occurs post-translationally at the Cysteine residue allowing sulfur to α-carbon bridges being formed and giving conformational strength similarly to disulphide bonds. In vitro study of the glycotransferase SunS showed that the enzyme can transfer glucose as well as any hexose to multiple cysteine residues. Therefore *S*-glycotransferase relaxed substrate specificity suggests its possible application in protein/peptide engineering. The *S*-linked glycosylation could potentially improve peptide half-life in serum due to its increased resistance to proteolysis (Stepper et al., 2011). The peptide glycocin-F, from the family of bacteriocins, is an excellent example for an improved stability by *S*-linked saccharide modification. Glycocin-F has a great stability in a wide pH range (2–10) and is resistant to boiling for up to 2 h (Stepper et al., 2011).

20.3 The Effect of Glycosylation on Peptides, Proteins and Other Therapeutics

The expanding field of rational design of antimicrobial peptides utilizes various glycan databases and software that allows visualization of the saccharide structure. The availability of glycan structures allows one to design a library targeting desired ligands in form of lectins. Glycosylation of proteins or incorporation into nanocarriers showed to be very effective way to target tumor cells/tissue selectively. For example, gold glyconanoparticles incorporating antigens were part of the polyvalent anticancer vaccine candidate progressing rapidly thorough clinical trials with excellent results (Cai et al., 2018).

The knowledge about glycan receptors allows a design of a tailored conjugate. An example here would be peptide with mannose glycan capable of binding to mannose-binding lectin (MBL) and leading to activation of complement system, macrophages, and dendritic cells. Conjugation of peptides or proteins with mannose is a very simple process and requires C-2 epimer of glucose as a substrate or a D-mannosamine. *C*-mannosylation consist of a unique C-C bond linking C1 of Man to C2 of the indole ring of tryptophan and its post-translational modification found in mammalian cells.

Other newly discovered lectins that could be used to enhance the innate immune system are intelectins—soluble lectins with capacity of binding to galactofuranose saccharides. The galactofuranose is naturally occurring in various bacteria, but not in humans, therefore making the saccharide a potent immunomodulator (Tsuji et al., 2001). Intelectins were showed to be expressed in the heart, small intestine, colon and thymus with much lesser amount expressed in human spleen and uterus (Tsuji et al., 2001). It was suggested that the intelectin serves as a defense receptor against pathogens causing endocarditis (e.g., Streptococci group).

Peptides are also excellent to use as drugs for CNS due to a low toxicity of their metabolites. Glycosylation of neuropeptides allows blood–brain barrier passage, enhanced biodistribution and improved peptide/receptor interactions. An example of effective modification of enkephalin opioid with β-D-glucose at different position shows how important is not only the choice of the suitable glycan but also the glycosylation site on the peptide (Egleton et al., 2001). The Serine monoglycosylated enkephalin showed excellent receptor binding kinetics profile along with improved metabolic stability.

20.3.1 Glycosylation of Antimicrobial Peptides

Antimicrobial drug development no longer aims to only fight infections, but also to improve systemic health and reduce general drug toxicity/adverse effects. The interdisciplinary approach to new drug development aims at bringing all aspects of the immune system and linking pharmacological knowledge with biology of antigen processing and immunoregulation. In the past decade it became clear that glycosylation of both natural and synthetic antimicrobial peptides can influence antimicrobial activity, ability to enhance host immunity, improve target specificity, and peptide's biological stability (Bednarska et al., 2017). Glycosylation had been described in five naturally occurring insect-derived AMPs (antimicrobial peptides/proteins), all *O*-linked, and proline-rich: diptericin, drosocin, formaecin, lebocin, and phyrrorricin. These observations and the availability of synthetic glycosylation allowed rapid synthetic antimicrobial glycopeptide development. Currently the glycobiology field is rapidly expanding with many synthetic approaches being implemented to generate large libraries of improved synthetic AMPs (Fig. 20.1).

The database of AMP describes small fraction of peptides carrying glycosylation motifs, most of which are insect- or bacteria-derived. The proline rich formaecin I is a natural *O*-linked glycopeptide derived from *Myrmecia gulosa*, showing strong antibacterial properties. The study comparing glycosylated form of synthetic formaecin I and its de-glycosylated counterpart (removed GalNAc) showed significant loss of antimicrobial propensities against Gram-negative bacteria (Lele et al., 2015; Talat et al., 2011).

Glycocin-F (GccF) is another glyco-active 43 amino acids long peptide, belonging to the bacteriocin family. GccF is glycosylated by β-linked-GlcNAc moieties attached on the Serine residue via oxygen atom and second glycan is attached via sulfur atom (Stepper et al., 2011). Similar glycosylation has been recorded for the peptide thurandacin and sublancin. Sublancin is antimicrobial peptide produced by *Bacillus subtilis* that contains *S*-linked HexNAc at the C-terminus. The glycosylation showed to be essential for the peptide's antimicrobial activity but also the *S*-glycosylation improves the stability of peptides in general (Oman et al., 2011). Any variation in the conformation of the glycan chain and length affects the stereo isometry of AMP that can influence stability. For example, depending on the position of the glycosylated amino acid, the linkage of glycan α-D-GalNAc-Thr in drosocin, alters not only the structure of that peptide but also its antibacterial potency against Gram-negative bacteria (Gobbo et al., 2002) An excellent study by Lele et al. (2015) shows how different glycan modifications affect internalization of AMPs into bacterial cells. The group showed that not only the length of a saccharide chain but also the interglycosidic linkage influence the antimicrobial properties of proline-rich AMPs.

Lanthipeptides are ribosomally synthetized peptides belonging to lantibiotics family, few were shown to be post-translationally modified with *N*-linked deoxyhexose moiety linked to tryptophan residue (Iorio et al., 2014). Lanthipeptides showed to have a broad-spectrum activity ranging from antibactierial, antifungal to immunomodulatory effect. As discussed in the next section, glycans have broad effect on immune modulation; hence, some gAMPs are more effective than others in activation of immune responses in addition to their antibacterial properties.

Apart of their antimicrobial activities, AMP and host defense peptides (HDPs) display range of functions such as: immunomodulatory functions, promotion of cytokine production, stimulatory effect on dendritic cells, and macrophage differentiation, angiogenesis and inhibition of inflammatory responses. These are also termed innate defense regulator peptides (IDRs) as they have shown the ability to stimulate both B- and T-cell responses (Werdelin et al., 2002). Host defense peptides interact with both MHCI and MHCII, depending on their length, glycosylation and other post-translational modifications. Appropriate glycan choice

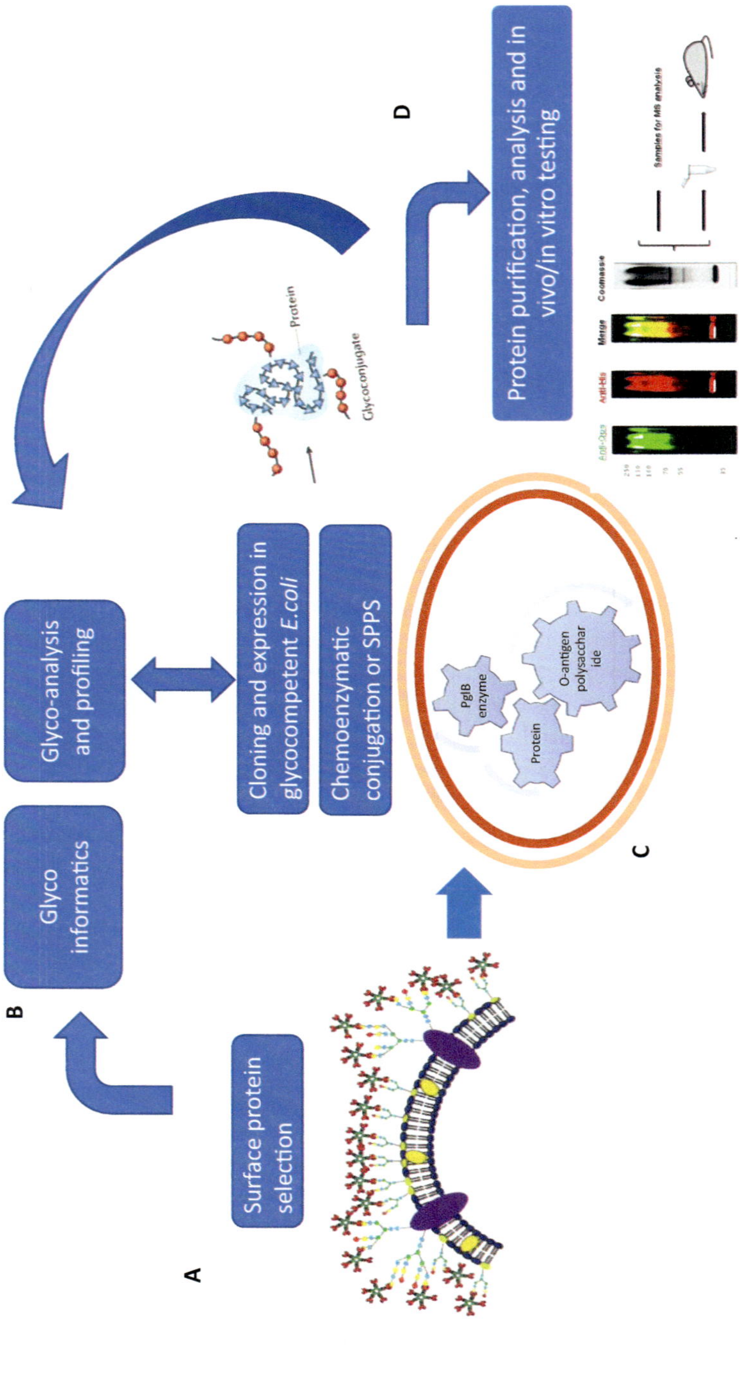
A
Surface protein selection
B
Glyco informatics
Glyco-analysis and profiling
Cloning and expression in glycocompetent *E.coli*
Chemoenzymatic conjugation or SPPS
C
PglB enzyme
Protein
O-antigen polysacchar ide
Protein
Glycoconjugate
D
Protein purification, analysis and in vivo/in vitro testing
Anti-His
Merge
Coomassie
Samples for MS analysis

Figure 20.1 Smart glycodesign approach for development of improved therapeutics. An example of a glycotherapeutic design strategy using PGC technology. (A) An example shows selection of immunogen from a cell surface, followed by analysis of the glycan structure and visualization of the protein:glycan structure (B). One may use the KEGG Cancer database to begin with cancer-associated carbohydrates exploration. Other repositories may serve for saccharide structure and glycosyltransferase selection. (C) For different applications one may consider different glycosylation methods or changing the enzymes. For PGCT methodology the PglB enzyme is used to drive the periplasmic glycosylation onto an effector sequence of the protein. Protein needs to either be engineered or contain glycosylation tag sequence. To glycosylate peptides, it is recommended to use plasmid encoding sequences interlinked with protease cleavage sites and purification tags. GlycoSNAP (Ollis et al., 2015) method can be implemented to optimize glycosylation enzymes. (D) Purification of protein/peptides and analysis/verification of the glycan structure. Varieties of in vitro assays can be applied to evaluate immunogenicity and receptor binding. Multiple assays (immunoassays, SPR, ELISA based assays) should be applied prior to evaluation in vivo models. The overall strategy should be tested for reproducibility for clinical application.

(from humanized sialic acid to zwitterionic bacterial glycans) can render the peptides either non-recognizable or strongly immunogenic. Another beneficial role of glycan additions on therapeutic peptides is the protection from both neutralization antibodies and protease cleavage. Ability to modify peptides with saccharides opens further application of AMPs as immunomodulatory agents in addition to their antimicrobial activity (e.g., adjuvant system, wound healing, receptor binding, etc.). An excellent example of tailored modification is alteration of a glycopeptide with Thomsen–Friedenreich (TF) antigen (β-Gal-[1→3]-α-GalNAc-*O*-serine) to generate specific cytotoxic T-Cell response against various tumors (Xu et al., 2004).

The relatively fast and cheap process of phenolic *O*-glycosylation showed to improve the stability and membrane permeability of many large cyclic peptides. The most recent method development for synthetic *O*-glycosylation of peptides allows the attachment of glycan in water solution with no need for a group-protection. This is a breakthrough in the field of synthetic glycosylation, since until this discovery most of the methods were relaying on protection of glycosyl donors and organic solvents had to be used otherwise glycosyltransferases were needed to catalyze the reaction (Wadzinski et al., 2018). The process of phenolic *O*-glycosylation is very fast (approximately 10 min at the room temperature) and gives very high yields of peptides with high anomeric selectivity. This method tolerates macrocyclic rings and side-chain disulphide bridges, adding up to the possibility of glycosylation of macrocyclic peptides. The authors, however, aim to further optimize the method to increase specificity towards an addition of glycan to -OH position on Tyrosine instead of competitive addition to the -SH group on the Tyrosine residue.

Knowing that the benefit of glycosylation serves tailored immunogenicity and activity of antimicrobial peptides, there is need for a fast, cheap, and reliable technique to attach glycan at the precise location. For this purpose, many techniques have been developed, including recombinant Protein–glycan coupling technology (PGCT) and recently published similar technique for *O*-linked glycan attachment, so-called site-directed glycosylation. The site-directed glycosylation similarly to PGCT uses *E. coli* as a host. However, it consists of two steps: Introduction of *O*-linked *N*-Acetylglucosamine (O-Glc-NAc) priming monosaccharide on the target protein/peptide followed by a ligation of eukaryotic N-glycan onto the O-GlcNAc (Wu et al., 2016). The first step requires catalyzing enzyme oligosaccharyltransferase (OGT) that transfers *O*-linked GlcNAc onto the accessible Serine/Threonine residue of a protein through β-linkage. In the method of site directed glycosylation both protein and the OGTase are expressed on separate plasmids in vivo *E. coli*. In the second step of transferring the *N*-glycan onto the primed protein researchers used an optimized enzyme endo-β-N-acetylglucosaminidase (Endo MN175Q) (Wu et al., 2016; Umekawa et al., 2010), which showed to have promiscuous substrate tolerance for GlcNAc-*O*-linked peptides/proteins in addition to its GlcNAc-*N*-linked natural preference. The efficiency of the *O*-linked substrate transfer was much lower (21%) then the *N*-linked one (73%) nonetheless with future improvements and optimizations this technique could be viable for tailoring glycans of interest to a monosaccharide primed protein.

The most well-established method for recombinant glycoprotein production for vaccine development is Protein Glycan Coupling Technology (Fig. 20.1). The method uses safe *E. coli* strains with various mutations to optimize the level of glycosylation and reduce production of endotoxin. To reduce *E. coli* natural glycoprotein production and natural synthesis of toxic lipopolysaccharide the PGCT uses WaaL (O-antigen ligase) and WecA (transferase) knockout *E. coli* mutants. WecA enzyme catalyzes the transfer of the GlcNAc-1-phosphate from undecaprenyl-GlcNAc onto the carrier lipid moiety yielding a first lipid-linked intermediate involved in LPS synthesis. The engineered strain is able to produce a glycoconjugate with high efficiency due to a lack of competition for either the undecaprenyl-phosphate lipid carrier by WecA or the OPS II-Und-PP substrate from WaaL (Garcia-Quintanilla et al., 2014).

The technology is based on expression of *C. jejuni* oligosaccharyltransferase (PglB) locus on either a plasmid or chromosome of *E. coli* strain. The PglB locus consists of 12 clustered genes. The PglB enzymes synthesize a heptasaccharide, GalNAc-α1,4-GalNAc-α1,4-(Glc-β1,3)-GalNAc-α1,4-GalNAc-α1,4-GalNAc-α1,3-Bac on a lipid carrier undecaprenylpyrophosphate (Kowarik et al., 2006). The *E. coli* host also has to carry two more plasmids: plasmid with a protein tagged with the glycosylation sequence (D/E-X-N-X-S/T) and one encoding the polysaccharide synthesis locus. The PglB enzyme has broad specificity and many mutations already described in the literature showed to improve this propensity to the extent that the PglB system is being used on an industrial scale (Cuccui et al., 2013). The enzyme recognizes the lipid-linked reducing end sugar (so-called lipid-linked oligosaccharide LLOs) and transfers the entire polysaccharide onto the acceptor protein sequence. The major limitation of this method is the fact that a wild type PglB enzyme only recognizes glycans with acylated sugar at the reducing end. In addition, the first two sugars cannot be linked by β1–4 linkage for the PglB to transfer. Another limiting factor of this technique is the fact that PglB operates on the inner membrane, flipping the glycosylated protein into a periplasmic space, limiting that way the amount of protein that can be produced per cell. Additional concerns for using this method are the batch variation and inability to control the extent of glycosylation and mixture heterogeneity. Using this method for future therapeutic production would need to be combined with several purification steps and glycan analysis of the final product to ensure homogeneity of the preparation.

With the availability of many oligosaccharyltransferases and multiple glycosyltransferases a chemoenzymatic glycosylation is becoming the method of choice. This is because the glycosylation can be controlled efficiently and is suitable for complex carbohydrates. Glycosyltransferases extend the sugar chain by the attachment of one monosaccharide at a time. For example, conjugation of GlcNAc to the lactose moiety of enkephalin was done using acetylglucosaminyltransferase derived from *Neisseria* species. With the current dynamic research in exploring new glycosyltransferases (e.g., database GlyTouCan) from different species, the chemoenzymatic glycosylation would be one of the most precise methods for glycosylation of therapeutics.

The synthesis of glycopeptides with *N*-linked and *O*-linked glycans can be achieved using solid phase peptide synthesis (SPPS), where the glycosylated amino acid is coupled to the elongating peptide using either fluorenylmethyloxycarbonyl (Fmoc) or tert-butyloxycarbonyl (Boc) chemistry (Cudic and Burstein, 2008; Comegna et al., 2015). For longer peptides (>50 amino acids) the convergent fragment-condensation method is a method of preference for *N*-glycosylation, where Lansbury asparylation is applied to eliminate formation of possible side products (Marcaurelle and Bertozzi, 2002). The SPPS method is widely used for conjugation of peptides with mucins based on a well-established protocol in which GalNAc-α-O-Ser/Thr building blocks are used for the assembly of the full-length glycopeptides.

Other well described method for chemoenzymatic glycosylation employ ENGases (Endo-β-*N*-acetylglucosaminidases), glycosyltransferases (GTases) and oligosaccharyltransferases (OGTases). Different ENGases range allows transfer of high-mannose, hybrid, and complex *N*-glycans to Glc-NAc (Tomabechi et al., 2014). For example, a 37-residue sublancin was successfully synthesized using SPPS method couples with a convergent ligation methodology whilst thurandacin and enterocin 96 peptides were prepared using chemoenzymatic methodology (Stepper et al., 2011)

20.4 Optimizing Immunogenicity

Glycans have multiple effects on immune cells, varying from anti-inflammatory properties (e.g., sialic acids) to pro-inflammatory effect (various bacterial *N*-glycans, viral capsules, etc.). To select for the most beneficial glycan-modification one should carefully study the effect of their modification on both cellular level and systemic level. For example, until very recent no research has been done on a subset of glycan recognizing T-cells, which shows to play a major role in immunorecognition and generation of strong pro-inflammatory feedback to antigen presenting cells, allowing for generation of long-lasting immunoglobulins (Avci et al., 2013). An excellent example of tailored modification is alteration of a glycopeptide with Thomsen–Friedenreich (TF) antigen (β-Gal-[1→3]-α-GalNAc-*O*-serine) to generate specific cytotoxic T-cell response against various tumors (Xu et al., 2004).

Abnormal *O*-glycosylation is often associated with various malignancies. For example, Tn antigen is adenocarcinoma-associated carbohydrate antigen built of D-GalNac unit α-linked to a Serine or Threonine amino acids (α-GalNAc-*O*-Ser/Thr glycan). One of the cancer vaccine constructs aimed to use the Tn glycan as a conjugate to Mucin-6 protein. Mucin-6 is a secretory *O*-glycosylated protein normally found at high levels in stomach but also aberrantly expressed on different carcinoma types and metastatic tumor cells. The immunology data showed that glycosylation of MU6 protein with the Tn glycan modulates immunogenicity by inducing IL-17 responses and by an efficient presentation of the antigen to CD4+T cells (Freire et al., 2006). The research conducted by Freire et.al (Freire et al., 2006) is an excellent example of strategic glycan designer approach, where different constructs were synthesized using in vitro enzymatic method (Fig. 20.1). To obtain MUC6 proteins researchers used *E. coli* expression system which allowed them to introduce various modifications in glycosylation sites of the protein. The protein was purified using NTA-agarose and C18 column separation methods followed by verification of the recombinant glycoprotein using high performance liquid chromatography (HPLC). To conjugate MUC6 with Tn antigen in vitro glycosylation was the method of choice using ppGAlNac transferase enzyme, with subsequent analysis of immunological properties in various in vitro and in vivo assays.

Other glycans that either mimic or are derived from pathogen-associated molecular patterns (PAMPs) are recognized by mannan-binding lectins. These can activate innate immune system via mannan-binding lectin (MBL) complement pathway and drive pro-inflammatory responses.

20.4.1 T Cell Signaling

The small subset of T cells, called T_{carb} are able to specifically recognize glycopeptides complexed with MHCI and MHCII receptors (Cobb et al., 2004). Their specificity towards glycans showed to be dependent upon amino acid composition of T cell receptors, where a specific glycan-recognizing region within αβTCR CDR3 regions was described (Speir et al., 1999). However, the $\alpha\beta T_{carb}$ cells cannot recognize large and complex glycan structures; therefore naturally any branched sugars attached to an antigen would be hidden from recognition of $\alpha\beta T_{carb}$ cell (Werdelin et al., 2002). Yet the bulkier glycans are being recognized by a B-cell subset, which showed to uptake the glycoprotein into endosomes where the glycoproteins are hydrolyzed into shorter glycopeptides followed by presentation on the MHCII complex for TCR recognition (Fig. 20.2) (Dintzis et al., 1976). For the MHCII complex presentation, a pH-dependent endolysosomal cysteine protease, asparaginyl endopeptidase (AEP) cleaves antigens into peptides. The N-glycosylation plays an important role in AEP cleavage, as it was shown that N-glycans can hide the cleavage motifs for that enzyme so the glycosylation in that case would affect the epitope-exposure on MHCII complex (Manoury et al., 2003). For that reason, protein engineering where specific glycotags are added close to epitopes that needs to be masked is currently researched.

The larger glycans can be recognized by gamma-delta T cell type (CD3+). These T-cell types are important in activation of innate immune system and their cytotoxic effect showed to be promising anti-tumor therapies, wound healing, or antiparasitic therapy (Wu et al., 2014). The γδT cells do not require MHC complex and are able to display antigen on their surface thus stimulate B-cell responses, DC and macrophages independently. Knowing T_{carb} receptor's affinity and with availability of glycan structure databases one may utilize this information to design tailored immunogenic or immunosenescent peptides and screen large peptide libraries. If, however, one aims to generate more T-cell specific immune responses the interest should be directed towards bacterial zwitterionic carbohydrates as these showed to be capable of inducing CD4+ T cell responses and regulate inflammatory responses (Cobb et al., 2004). Several studies have shown that glycopeptides are the key modulators of CD4+ T cells and some of the *N*-glycans showed to be essential for MHC class II-restricted CD4+T cell recognition (Malaker et al., 2017). For example, *N*-linked glycans on the tyrosinase activate the CD4+ T cells and any point mutations introduced in the glycosites yield the protein immunosenescent (Housseau et al., 2001). Figure 20.2 represents intricate effect of glycoconjugates on cell-mediated immune responses.

It is now clear that bacterial glycopeptides are recognized by both CD4+ and CD8+ T cells. Furthermore, the study on glycoconjugated proteins shows that the protein is being endocytosed by B-cell and displayed on MHCII in form of glycopeptide followed by recognition via T-cell receptor (Avci et al., 2011, 2013). One of the recent studies showed an excellent workflow to design and

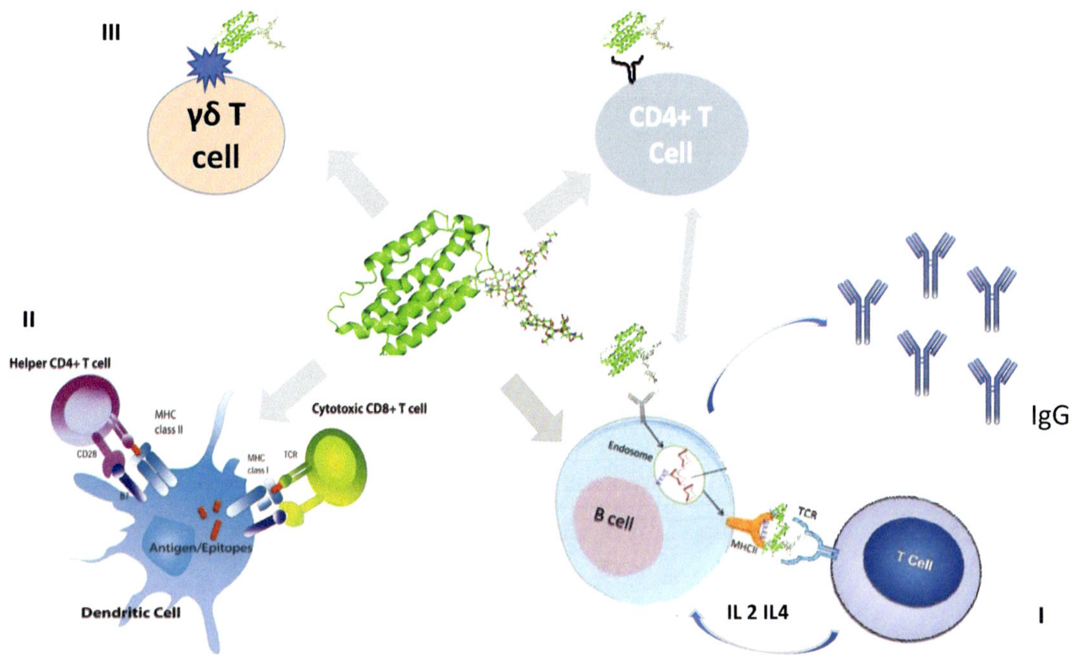

Figure 20.2 The effect of glycoconjugate on cells of the immune system. (I) Glycoprotein binds to B-cell receptor and it is processed through an endosome followed by being presented in form of glycopeptides on MHCII complex. The specific T_{carbs} detect the B-cell presented glycopeptides and release IL2 and IL4 cytokines. (II) Glycoproteins may also be detected by dendritic cell (MHCI or MHCII) switching tyrosine kinase pathway and antigen presentation. T_{carb} cells recognize glycans complexed with MHCI or MHCII. (III) Gamma-delta T cells (CD3+) detect larger glycans and do not require MHC complex. γδ T cells can display antigen on their surface and stimulate other immune cells (B-cells, dendritic cells).

model glycopeptides for specific HLA type. The workflow for generating leukemic glycopeptides involved purification of surface glycan neoantigens from cancer patients, characterization of the neoantigens, synthesis of glycopeptides mimetics and evaluation of the peptide immune responses in T-cells derived from healthy donors. This study also shows that not only the presence of GlcNAc residues but also their position affects TCR recognition. Furthermore, the same group used the Immune Epitope Database and Analysis Resource to predict the MHC class II binding site, followed by a molecular modeling (using the 2014 Molecular Operating Environment MOE software suit) to visualize the HLA/TCR binding (Malaker et al., 2017). These findings again prompt to the need of designer approach of glycotherapeutics which would undoubtedly lead to a very specific immune stimulation and tailored cancer-therapy.

The T-cell receptor glycan expression is also being regulated to tune the adaptive immune responses. The gene encoding for mannoside *N*-acetylglucosaminyltransferase 5 (*Mgat*5) showed to be upregulated in T-cells, generating *N*-linked glycan branches with repeated Gal-GlcNAc disaccharide which bound galectin-3 receptors with a high-affinity. The Gal-GlcNA—rich TCRs drive the formation of immune complexes that lower the overall T-cell activation threshold. The immune regulatory effect of Gal-GlcNAc was demonstrated in mouse models of arthritis, where oral administration of this glycan showed to reduce symptoms of the experimental autoimmunity (Grigorian et al., 2007).

20.4.2 Antigen Presenting Cell Signaling and Activation

The protein–glycan conjugates are very efficient in stimulatory effect on B-cells due to their ability to cross-link the B-cell receptor (BCR). The larger carbohydrates on their own showed to cluster B-cell receptors in the "Immunon"-like manner (Dintzis et al., 1976) inducing IgM type immunoglobulins with low avidity and short half-life. A widely accepted theory for immunogenicity enhancement by proteins and larger polymers is the additional T cell-independent activation of B cells. This occurs through cross-linking mechanisms by the repetitive and ordered structures that are typical for protein aggregates or capsular saccharides (Baker et al., 2010; Sauerborn et al., 2010). Larger molecules with repeating structure may resemble "immunons," continuous clusters of haptens on a linear polymer (>100 kD in size with >10–20 haptens spaced in 100 Å). Immunons can directly activate B cells through receptor clustering (Dintzis et al., 1976). Interestingly immunon-like polymers could also inhibit the B-cell activation if binding to BCRs is non-stimulatory. However, for most glycopeptides or protein–glycan conjugates B-cell epitopes were shown to co-localize with T-helper epitopes showing therefore ability to induce specific IgGs. The larger glycans showed to lose the ability to bind MHCII class receptors and instead activate directly B cells or were recognized by previously described γδT -cells.

Another very important glycan detecting receptor family that need consideration when designing a glycotherapeutic are Siglec receptors (Sialic acid-binding Ig-like lectins). Siglecs are lectins receptors expressed mostly on the cells of the immune system. Siglecs bind and recognize different glycans made of sialic acids. Until now there are 16 human Siglec receptors described (Schnaar, 2016). Not all of Siglec's functions are known; however, ongoing research shows that Siglec-1 is able to recognize pathogenic sialic acids and drive the inflammatory immune response. Other CD33-related Siglecs are associated with tyrosine-activation pathway of the immune cells. Many human pathogens mimic sialic acids and utilize Siglec receptor for their advantage.

The key antigen presenting cells of the immune system able to bind and process glycosylated antigens are Dendritic cells (DCs). DCs can present the antigens on both MHCI and MHCII complex, DCs are able to either prime or suppress T-cell mediated immune responses. DCs express C-type lectin receptors (CTLs) allowing Ca^{2+} dependent glycan binding (Taylor and Drickamer, 2003). From the perspective of immune modulation these receptors are very good target for activation of myeloid cell via spleen tyrosine kinase pathway.

Dendritic cell CTL, called DC-SIGN, is an excellent example of C-type lectin receptor which recognizes high-mannose glycans and Lewis-type antigens. These receptors are used for targeted delivery to dendritic cells. Overall lectins serve as detectors for both self and non-self-molecular patterns, playing an important role in both pathogen defense system and autoimmune protection. There are many proteins of complement system that act as potential glycan receptors allowing prompt immune detection and signaling. Most of the natural complement proteins are synthesized in liver and the saccharide composition and expression levels vary between complement components, also may vary between individuals and life span of the cells. Nonetheless, most of the complement components synthesized in the liver contain sialylated, biantennary glycans, while component C3 contains only oligo-mannose type glycans.

Other important glycoreceptors involved in important immunoregulation and signaling are selectins, which are responsible to leukocyte rolling and extravasation into a tissue. These can be explored for a design of therapeutics targeting blood–brain barrier and body compartments.

20.5 Rational Design of Glycotherapeutics

The bioinformatics for glycan prediction, analysis, and visualization is a growing area of extensive research. Using both prediction software and available databases with already described glycans, it is possible to start with in silico design of glycomodifications without the need for labor-intensive methods. However, the high throughput prediction workflow is a major challenge since the glycosylation is a complexed, condition-dependent PTM controlled by gene expression responsible for glycan synthesis. Therefore, to decipher glycobiology of a cell one needs to investigate gene expression, glycan structure, glycan binding receptors and their role in a cell signaling and regulation. For this purpose, the field of glycobiology developed a range of open databases and repositories.

Designing human therapeutics appears to be simple since there are nine known building blocks of human glycome: glucose, galctose, mannose, *N*-acetylglucosamine, *N*-acetylgalactosamine, sialic acid, fucose, xylose, and glucoronic acid, which are substrates of around 200 different glycosyltransferases, and glycosidases. However, due to a high specificity of glycan-binding lectins any differences in linkages between the monosaccharides or even linkage position or configuration could have a large impact on recognition and signaling (Fig. 20.3). Each cell might change its glycopatterns at different time points and build glycan trees that are impossible to predict simply from gene expression analysis.

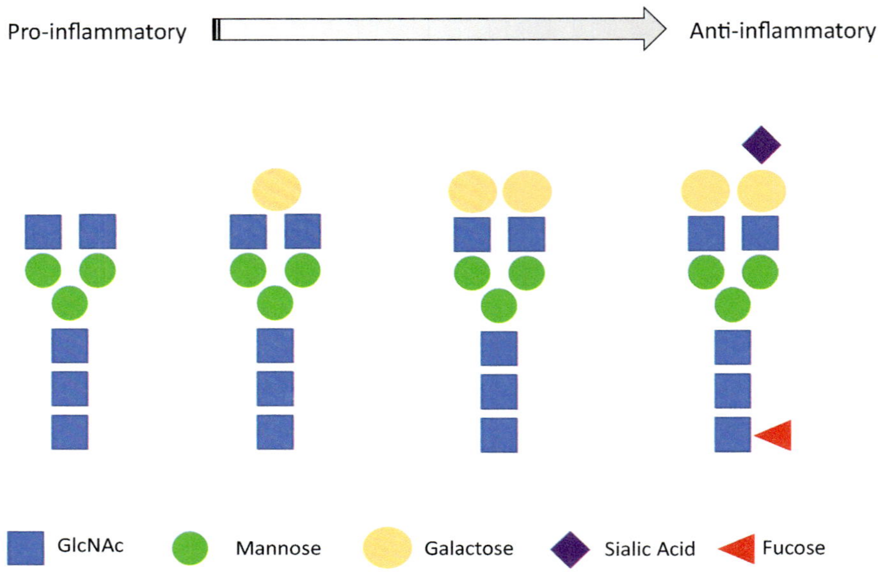

Figure 20.3 The immunogenic effect of different types of N-glycans.

For in silico glycan-design one may start with a search for a precise glycan using free, standardized glycan repository database. **GlyTouCan** is an excellent free glycan repository with over 100,000 glycans stored in the database, including a graphical representation of most of the structures, where monosaccharides are assigned in multiple nomenclatures (with the CFG set as standard).

For a more specific investigation of cancer-associated carbohydrates and their pathways in cancer research, the free database **KEGG Cancer** can be explored (https://www.genome.jp/kegg/glycan/). KEGG mapping allows one to view the glycan biosynthesis pathway, obtaining additional information of the genetics underlying certain diseases including references. The KEGG glycan structure database relies on the previously mentioned CarBank database. KEGG database is one of the first databases that implement cross-disciplinary data (genes, disease pathways, structures, glycan structures).

To start exploring natural glycoproteins and their possible glycosylation sites different algorithms can be used or free online glycosite prediction calculators. The **GlycoMine** online predictor of *N*-glycosylation and *O*-glycosylation sites provides the probability scores for each site of glycosylation. It is a good start-point for mapping the protein possible glycosylation sites taking into consideration both protein-sequence and its structural features.

NGlycPred is a pioneering server run by NIH that can score *N*-linked glycosylation sites effectively, taking into consideration structural properties and sequence of the protein. Both predictors need PDB format of the protein and a selected chain of interest. NGlycPred also gives a decision score which is calculated by the Random Forrest Classifier. For both servers the probability score >0.5 is a positive threshold.

One of the well-explored for therapeutic purposes human glycan, which structure and functions are well studied is Hyaluronan (HA). HA is a natural human glycosaminoglycan used for immunomodulation and immune signaling, composed of an altering sequence of β1,3 glucuronic acid (GlcA) and β1,4 *N*-acetylglucosamine (GlcNAc). Hyaluronan has a dynamic turnover in human cells with different species circulating in tissues ranging from a high molecular weight HA, a lower molecular weight and oligosaccharides. The most recent data shows that HA plays a major role in cell signaling, inflammation, and tumorigenesis (Misra et al., 2015). The low-molecular-weight hyaluronan shows to interact with CD44 receptor and a prolonged interaction was linked with pathological conditions, including inflammation, cancer and angiogenesis. The CD44 receptor showed to be widely distributed on hematopoietic cells, tumor cells, immune cells, and placenta cells; hence, binding of CD44 to HLA depends on the activation state of the receptor (Misra et al., 2015). The high molecular weight HA showed to promote malignant cell proliferation and development of cancer, while agonist to HA-CD44 interactions proved to be excellent inhibitors of tumor cell growth (Lokeshwar et al., 2014). Hyaluronans are very successful conjugates of many anti-inflammatory, anti-cancer and surgery therapies due to their selective cell/tissue targeting and improved solubilization in water, e.g., ONCOFID-S, ONCOFID-P) (Montagner et al., 2015). Overall, the approaches of inhibiting the HA-CD44 interaction have been made in many different cancer types, in which CD44 has regulating activities (e.g., colon cancer, melanoma, breast, and prostate cancer). For this purpose, many peptides were designed to block different hyaluronans, as an example here Pep-1 conjugate against melanoma invasion and migration. The 12 amino acid long Pep-1 inhibitor of HA showed to block expression of contact hypersensitivity responses in mouse models and later used to treat inflammatory disorders (Mummert et al., 2002).

Other very interesting from the perspective of medicinal chemistry glycan groups (that should be rather avoided or carefully evaluated prior to usage) are sialic acids (Sias). Sialic acids are negatively charged, 9-carbon chain backbone monosaccharides, abundant in human cells and tissues. Their role ranges from cellular processes, regeneration to regulation of immune system. The key enzyme in the synthesis of sialic acids in human body is GNE (UDP-*N*-acetylglucosamine 2-epimerase/*N*-acetylmannosamine-kinase. Sialic acids are one of the most diverse glycans. The variety arises from diverse α-linkages between second carbon C-2 of the sialic acids and underlying sugars. The variety of natural modifications adds up to a secondary level of their diversity. The combinations of different glycosidic linkages with multitude of possible modifications (e.g., Neu5Ac, Neu5Gc, addition of hydroxyl groups on C-4, C7, C-8, or C9) generate huge variety among sialic acids and account to distinctive glycocode of different cell types. Sialic acids modifications were linked to particular types of malignancies; for example, there is an association of *N*-glycolyl-neuraminic acid (Neu5Gc) prevalence on cancer tissues, yielding an inflammatory response (Pearce and Laubli, 2016). Some of the sialic acids in conjugation with highly immunogenic proteins/peptides are explored for their cancer protective vaccine application. Research in the field of sialic acids and their interaction with Siglec receptors could open many new ideas on immune regulation and cell signaling.

20.6 Conclusions

Without any doubt, glycosylation is a powerful post-translational modification. Exploring the effect of glycans to improve therapeutic potency will help to optimize receptor binding, lower toxic effects of immunogenic epitopes, and reduce clearance rate, among many other possible benefits. Carbohydrate modifications of peptides using simple hexoses like glucose or maltose resulted in a higher renal clearance; hence, glycosylation plays a role at both cellular and systemic level. The large impact of small glycans prompts one to consider employing a range of glycotools combined with in vitro/in vivo testing for variations of glycotherapeutics before proceeding into the clinical evaluations. The host for production of a recombinant protein/peptide should also be carefully evaluated in terms of its glycosylation modifications. For example, for humanized glycan one may chose glycoengineered *Pichia pastoris*, while for immunogenic glycan production/vaccine the bacterial glycosylation system would be optimal. Finally, this chapter gives an important insight into immune regulatory effect of different glycans, which could be selectively tailored to maximize therapeutic effect of drugs.

References

An H. J., Kronewitter S. R., de Leoz M. L. A., Lebrilla C. B., *Curr. Opin. Chem. Biol.*, **13** (2009), 601–607.

Apweiler R., Hermjakob H., Sharon N., *Biochim. Biophys. Acta,* **1473** (1999), 4–8.

Avci, F. Y., Li X., Tsuji M., Kasper D. L., *Nat. Med.,* **17** (2011), 1602–1609.

Avci, F. Y., Li X., Tsuji M., Kasper D. L., *Semin. Immunol.,* **25** (2013), 146–151.

Baker M. P, Helen M. R., Brooke L., and Christine J. B., *Self/nonself,* **1**(4) (2010), 314–322.

Bednarska N. G., Wren B. W., Willcocks S. J., *Drug Discov. Today,* **22** (2017), 919–926.

Cai L., Gu Z., Zhong J., Wen D., Chen G., He L., Wu J., Gu Z., *Drug Discov. Today,* **23** (2018), 1126–1138.

Cobb B. A, Wang Q., Tzianabos A. O., and Kasper D. L., *Cell,* **117**(5) (2004), 677–687.

Comegna D., de Paola I., Saviano M., Del Gatto A., Zaccaro L., *Org. Lett.,* **17** (2015), 640–643.

Cuccui J., Thomas R. M., Moule M. G., D'Elia R. V., Laws T. R., Mills D. C., Williamson D., Atkins T. P., Prior J. L., Wren B. W., *Open Biol.*, **3** (2013), 130002.

Cudic M., Burstein G. D., *Methods Mol. Biol.,* **494** (2008), 187–208.

Dintzis H. M., Dintzis R. Z., Vogelstein B., *Proc. Natl. Acad. Sci. U. S. A.*, **73** (1976), 3671–3675.

Egleton R. D., Mitchell S. A., Huber J. D., Palian M. M., Polt R., Davis T. P., *J. Pharmacol. Exp. Ther.*, **299** (2001), 967–972.

Freire T., Lo-Man R., Piller F., Piller V., Leclerc C., Bay S., *Glycobiology*, **16** (2006), 390–401.

Garcia-Quintanilla F., Iwashkiw J. A., Price N. L., Stratilo C., Feldman M. F., *Front. Microbiol.*, **5** (2014), 381.

Gobbo M., Biondi L., Filira F., Gennaro R., Benincasa M., Scolaro B., Rocchi R., *J. Med. Chem.*, **45** (2002), 4494–4504.

Grigorian A., Lee S. U., Tian W., I. J. Gao G., Mendelsohn R., Dennis J. W., Demetriou M., *J. Biol. Chem*, **282** (2007), 20027–20035.

Hartmann S., Hofsteenge J., *J. Biol. Chem.*, **275** (2000), 28569–28574.

Housseau F., Moorthy A., Langer D. A., Robbins P. F., Gonzales M. I., Topalian S. L., *Eur. J. Immunol.*, **31** (2001), 2690–2701.

Ihara Y., Manabe S., Ikezaki M., Inai Y., Matsui I. S., Ohta Y., Muroi E., Ito Y., *Glycobiology*, **20** (2010), 1298–1310.

Iorio M., Sasso O., Maffioli S. I., Bertorelli R., Monciardini P., Sosio M., Bonezzi F., Summa M., Brunati C., Bordoni R., Corti G., Tarozzo G., Piomelli D., Reggiani A., Donadio S., *ACS Chem. Biol.*, **9** (2014), 398–404.

Kowarik M., Young N. M., Numao S., Schulz B. L., Hug I., Callewaert N., Mills D. C., et al., *EMBO J.*, **25** (2006), 1957–1966.

Lele D. S., Talat S., Kumari S., Srivastava N., Kaur, K. J., *Eur. J. Med. Chem.*, **92** (2015), 637–647.

Lokeshwar V. B, Mirza S., Jordan A., *Adv. Cancer Res.*, **123** (2014), 35–65.

Malaker S. A,. Ferracane M. J., Depontieu F. R., Zarling A. L., Shabanowitz J., Bai D. L., Topalian S. L., Engelhard V. H., Hunt D. F., *J. Proteome Res.*, **16** (2017), 228–237.

Manoury B., Mazzeo D., Li D. L., Billson J., Loak K., Benaroch P., Watts C., *Immunity*, **18** (2003), 489–498.

Marcaurelle L. A., Bertozzi C. R., *Glycobiology*, **12** (2002), 69R–77R.

Marth J. D., Grewal P. K., *Nat. Rev. Immunol.*, **8** (2008), 874–887.

Maverakis E., Kim K., Shimoda M., Gershwin M. E., Patel F., Wilken R., Raychaudhuri S., Ruhaak L. R., Lebrilla C. B., *J. Autoimmun.*, **57** (2015), 1–13.

Misra S., Hascall V. C., Markwald R. R., Ghatak S., *Front. Immunol.*, **6** (2015), 201.

Montagner I. M., Merlo A., Carpanese D., Zuccolotto G., Renier D., Campisi M., Pasut G., Zanovello P., Rosato A., *Oncoscience,* **2** (2015), 373–381.

Mummert M. E., Mummert D., Edelbaum D., Hui F., Matsue H., Takashima A., *J. Immunol.,* **169** (2002), 4322–31.

Okamoto S., Murano T., Suzuki T., Uematsu S., Niwa Y., Sasazawa Y., Dohmae N., Bujo H., Simizu S., *Biochem. Biophys. Res. Commun.,* **486** (2017), 558–563.

Ollis A. A., Yi Chai Y., DeLisa M. P., *Methods Mol. Biol.,* **1321** (2015), 37–47.

Oman T. J., Boettcher J. M., Wang H., Okalibe X. N., van der Donk W. A., *Nat. Chem. Biol.,* **7** (2011), 78–80.

Pearce O. M. T., Laubli H., *Glycobiology,* **26** (2016), 111–128.

Sauerborn M., Brinks V., Jiskoot W., and Schellekens H., *Trends Pharmacol. Sci.,* **31**(2) (2010), 53–59.

Schnaar R. L., *J. Leukocyte Biol.,* **99** (2016). 825–838.

Speir J. A., Abdel-Motal U. M., Jondal M., and Wilson I. A., *Immunity,* **10**(1) (1999), 51–61.

Stepper J., Shastri S., Loo T. S., Preston J. C., Novak P., Man P., Moore C. H., Havlicek V., Patchett M. L., Norris G. E., *FEBS Lett.,* **585** (2011), 645–650.

Talat S., Thiruvikraman M., Kumari S., Kaur K. J., *Glycoconj. J.,* **28** (2011), 537–555.

Taylor M. E., Drickamer K., *Methods Enzymol.,* **363** (2003), 3–16.

Tomabechi Y., Squire M. A., Fairbanks A. J., *Org. Biomol. Chem.,* **12** (2014), 942.

Tsuji S., Uehori J., Matsumoto M., Suzuki Y., Matsuhisa A., Toyoshima K., Seya T., *J. Biol. Chem.,* **276** (2001), 23456–23463.

Umekawa M., Li C., Higashiyama T., Huang W., Ashida H., Yamamoto K., Wang L.-X., *J. Biol. Chem.,* **285** (2010), 511–521.

Wadzinski T. J., Steinauer A., Hie L., Pelletier G., Schepartz A., Miller S. J., *Nat. Chem.,* **10** (2018), 644–652.

Wang G., *Curr. Biotechnol.,* **1** (2012), 72–79.

Werdelin O., Meldal M., Jensen T., *Proc. Natl. Acad. Sci. U.S.A.,* **99** (2002), 9611–9613.

Wu Z., Jiang K., Zhu H., Ma C., Yu Z., Li L., Guan W., et al., *Bioconj. Chem.,* **27**(9) (2016), 1972–1975.

Xu Y., Gendler S. J., and Franco A., *J. Exp. Med.,* **199**(5), (2004), 707–716.

Chapter 21

On Biocatalysis as Resourceful Methodology for Complex Syntheses: Selective Catalysis, Cascades and Biosynthesis

Andreas Sebastian Klein,[a] Thomas Classen,[b] and Jörg Pietruszka[a,b]

[a]*Institute for Bioorganic Chemistry, Heinrich Heine University Düsseldorf, D-52426 Jülich, Germany*
[b]*Institute for Bio- and Geosciences I: Bioorganic Chemistry, Forschungszentrum Jülich GmbH, D-52426 Jülich, Germany*

j.pietruszka@fz-juelich.de

21.1 Introduction

Natural compounds have been used as pharmaceuticals since ancient times. While fragrances or dyes were substituted by chemical compounds in virtually all cases, natural compounds still are important as active agents. Feher and Schmidt (2003) analyzed a plethora of compounds with respect to their physico-chemical properties such as molecular weight, number of stereogenic centers etc. for thousands of combinatorial chemicals, natural

Pharmaceutical Biocatalysis: Fundamentals, Enzyme Inhibitors, and Enzymes in Health and Diseases
Edited by Peter Grunwald

ISBN 978-981-4800-61-7 (Hardcover), 978-0-429-29503-4 (eBook)
www.jennystanford.com

compounds, and pharmaceuticals. The principal component analysis revealed that the chemical space of pharmaceuticals and natural compounds are more congruent than the chemical space of combinatorial compounds. However, natural compounds are rather complex in terms of numbers of functional moieties as well as stereogenic elements (see Fig. 21.1), which renders their synthesis often as particular challenge. Thus, it is important to have selective synthetic tools for these jobs. Besides human-made synthetic tools, biocatalytic methods acquired importance during the last decades (Bornscheuer, 2018; Classen and Pietruszka, 2017; Gröger, 2018; Patel, 2018; Sun et al., 2018).

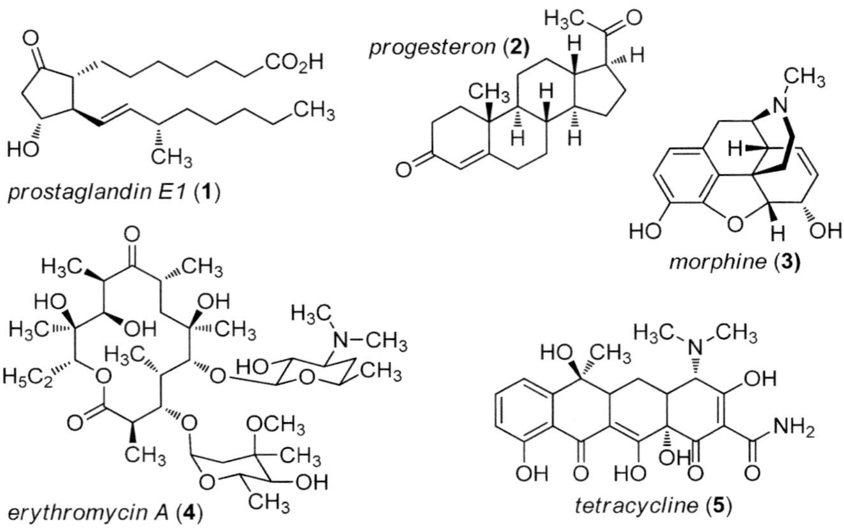

Figure 21.1 Selection of natural compounds with pharmaceutical use illustrating the complex structures.

In this chapter, we want to illustrate various modes of utilizing biocatalysis including biosynthesis, which can be applied for the production of pharmaceuticals or other high value products. We will start with simple single-step conversions, which might be carried-out by isolated enzymes but also by living, starving, or dead—permeabilized—cells (see Fig. 21.2). More complex processes are multi-step conversions combining various steps, which do not necessarily need to be biocatalysts but also chemical catalysts or even uncatalyzed chemical steps in combination. Although the processes might be more complex in terms of optimizing the individual rates of each step as well as finding compatible reaction conditions for all reactants, the main advantage of multi-step conversions is the reduction of downstream processing operations, which tends to be the most time-consuming and expensive part of a process. The most complex type of biocatalysis is biosynthesis. The biocatalyst is here a living being—an organism, a cell. Thus, it could be considered as the perfect situation for the employed enzymes: perfect environment, cofactor-supply, (re-)folding assistance, and a continuous regeneration of the biosynthesis machinery. But also, biosynthesis has to cope with some drawbacks like transport limitations, toxic intermediates, and product metabolization. Nevertheless, biocatalysis offers a wide range of methods for the selective formation of complex molecules, which reduces the overall number of steps and makes such syntheses more effective. This chapter will lay a focus on the biosynthetic methods and will only highlight the principal ideas from single-step conversions to cascade reactions.

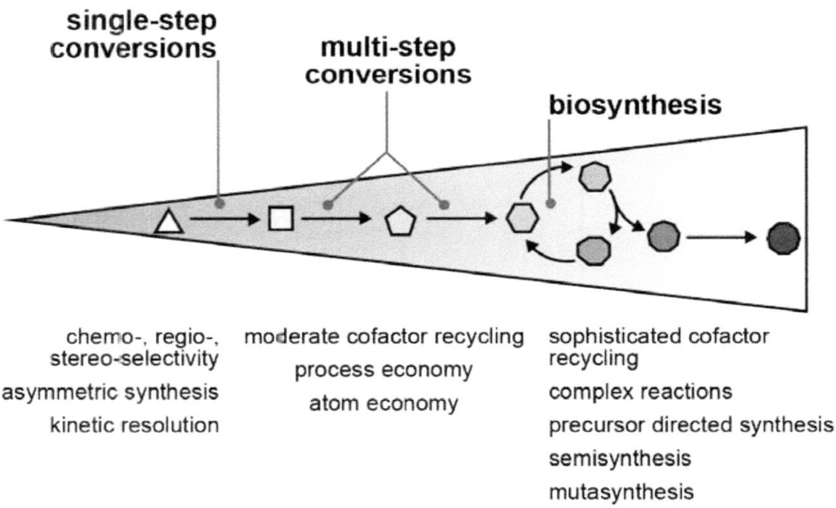

Figure 21.2 Spectrum of biocatalysis. The complexity of the process increases from left to right, but so does the complexity of the addressed chemical products.

21.2 Single-Step Conversions

Enzymes are often associated with the term "green chemistry" because they are natural, biological degradable and so on. However, it must be considered that enzymes are usually used in aqueous solutions causing downstream challenges: energy consuming evaporation or salt/solvent contaminated wastewater. These drawbacks may be taken into account if the synthesis profits from an enzyme's selectivity, which is in many cases much higher than the selectivity of competing chemical catalysts. Enzymes may install chemical moieties by their chemoselectivity and may bypass step-intensive protective strategies. Stereogenic centers might be installed with high selectivity by means of asymmetric synthesis or the enantiomeric excess can be increased to pharmaceutical relevance by means of kinetic resolution.

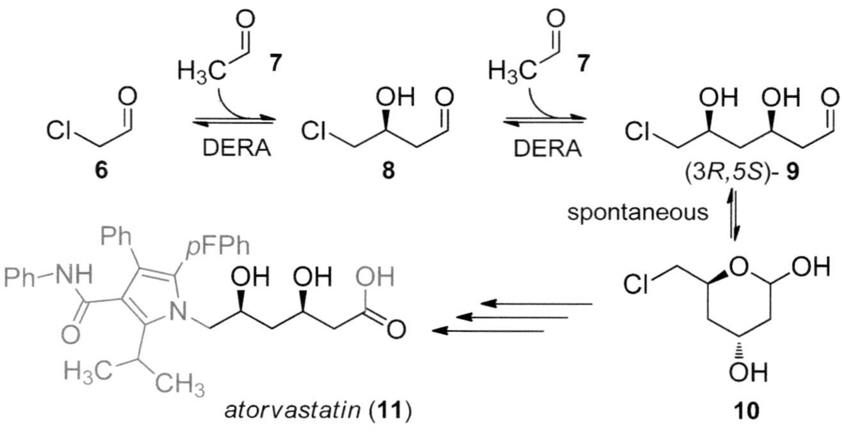

Figure 21.3 Biocatalytic key-step for producing atorvastatin (**11**) (Patel, 2009; Jennewein et al., 2006).

Statins decrease the cholesterol level and were blockbuster par excellence in affluent societies. Especially the chiral side-chain of atorvastatin (**11**) has been synthesized by many groups using biocatalysis with great success (Patel, 2009; Müller, 2004). The chemical company DSM was able to implement a multi-ton scale production based on the 2-deoxy-ribose-5-phosphate aldolase (DERA), which naturally cleaves the name giving substrate into acetaldehyde and glyceraldehyde-3-phosphate (Haridas et al., 2018). Since enzymes accelerate a given reaction but do not influence the thermodynamic equilibrium, the DERA can also be used in the aldol rather than the retro-aldol reaction (see Fig. 21.3). The precursor of the side-chain **9** can be produced in one reaction sequence, where DERA performs two consecutive aldol reactions: the first between acetaldehyde (**7**) and chloroacetaldehyde (**8**), and the second between previously formed aldol (**8**) and second molecule of acetaldehyde (**7**). The formed lactol **10** can be transformed into the respective pharmacophore of atorvastatin (**11**). This process takes advantage of DERA's chemo- and stereoselectivity.

The sequence yields very good diastereomeric and enantiomeric excesses suitable for the final drug and the process is favorable in terms of atom efficiency, because all atoms are included within the product and ostensibly there are no process inherent side-products. But what if the process would not need the (3*R*,5*S*)-configuration? There is no stereocomplementary DERA in nature, which may give the opposite configuration (Bisterfeld et al., 2016). However, even in cases, where an enantioselective enzyme is available but with the undesired enantioselectivity, processes can be developed leading to high enantiomeric excesses. Using kinetic resolution a racemic or enantioenriched sample is converted using a selective enzyme. One enantiomer will be converted while the other enantiomer remains untouched creating a pair of enantiocomplementary and enantiopure compounds, which are often conveniently separable. Although when starting from a racemic mixture, one can maximally obtain 50% yield of each enantiomer, this method offers access to both enantiomers. This renders the method important for research endeavors e.g. compound screenings or toxicity tests in the preclinical phase. Pietruszka and Simon (2009) developed a process for the synthesis of the natural compound irciniastatin A (**14**, Fig. 21.4), where one building block (**12**) was synthesized diastereoselectively by means of classical synthesis. The racemic compound was then subjected to a kinetic resolution employing an esterase. This yielded the nearly enantiopure acid (4*S*,5*S*)-**13** and the ester (4*R*,5*R*)-**12** with almost the achievable yield of 50% each. Though 50% of the starting material would be discarded in a preparative process, the process might be still economically more efficient if the enantioselective, asymmetric synthesis would include more steps or expensive reagents. Even the limitations of the 50% yield have been overcome by advances in dynamic kinetic resolutions, where the remaining enantiomer is racemised in situ (Martín-Matute and Bäckvall, 2007; Steinreiber et al., 2008; Turner, 2010; Verho and Bäckvall, 2015).

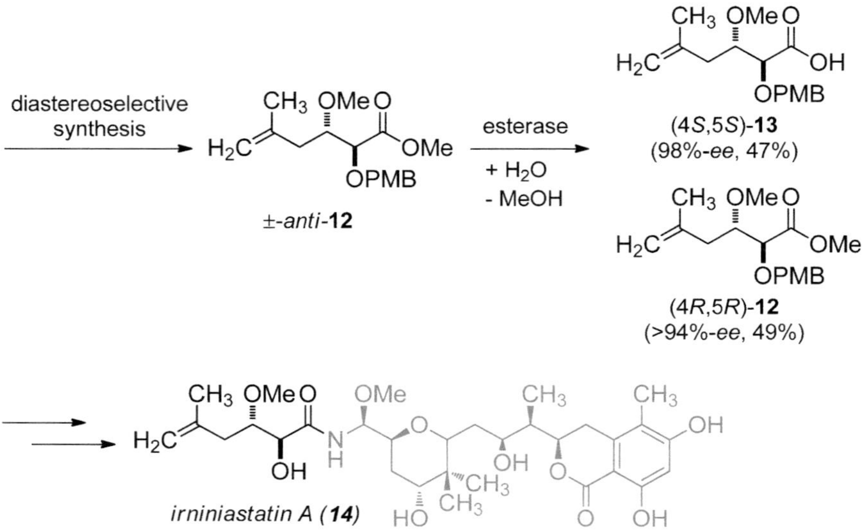

Figure 21.4 Kinetic resolution of racemic compound **12** to yield a building block for the synthesis of irciniastatin A (**14**) (Pietruszka and Simon, 2009).

21.3 Multi-Step Conversions and Cascades

Often enzymes are considered as laborious and expensive catalysts, which is not true. Chemical catalysts must be synthesized as well and their production costs usually increase with their degree of decoration, which are usually present in selective chemical catalysts. Once the cloning and fermentation is established, the reproduction costs of new enzyme charges are relatively low. The most abundant process costs as well as loss of product are located within the downstream processing. Thus, reducing the number of downstream processing steps will increase the effectiveness of a process. This is the main idea of cascade reactions. A detailed review on multi-enzyme cascades and their different architectures can be found in Ricca et al. (2011).

The principle is illustrated on a cascade including three enzymes to produce tetrahydroisoquinolines (**20**) from inexpensive compounds (see Fig. 21.5). Besides their excellent stereoselectivity, the chemoselectivity of the enzymes is exploited. The enzymes can act within one medium because they are acting on their specific substrate. For instance, the transaminase catalyzes the reductive amination of ketone **16** but does not convert aldehyde **15** despite having the higher carbonyl activity, or the Pictet-Spenglerase uses benzyl aldehyde (**19**) but not *meta*-hydroxy benzaldehyde (**15**). Furthermore, the authors could show that the very last step—the Pictet-Spengler-reaction—could be carried-out using the phosphate buffer rather than the enzyme, which leads to the epimer of the newly formed stereogenic centre. This is a nice example illustrating that the combination of chemical steps and biocatalytical steps is a fruitful expansion of the synthetic methodology.

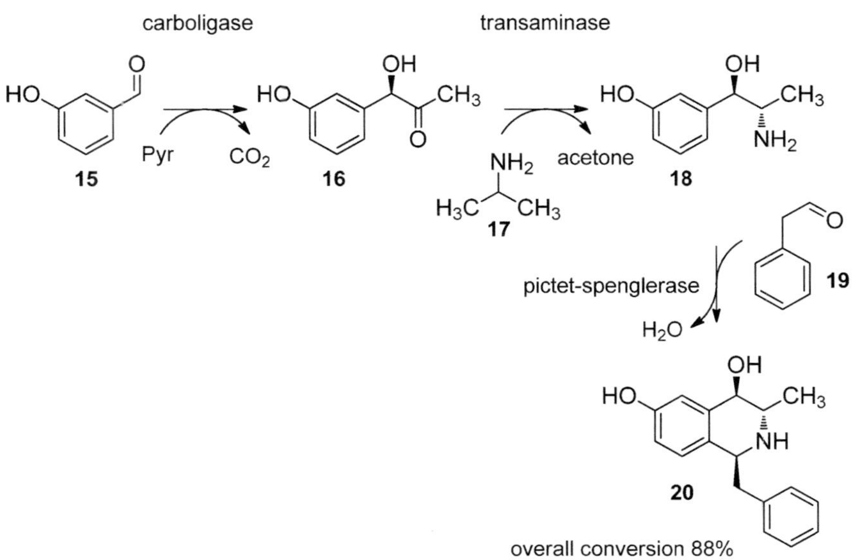

Figure 21.5 Step-efficient cascade employing three enzymes to form tetrahydroisoquinolines (Erdmann et al., 2017).

As we have illustrated, in vitro biocatalysis offers a wide range of options to exploit the extraordinary selectivity of many enzymes. Synthesis routes can be streamlined and thus the biocatalytic routes might be more efficient. In this chapter the emphasis was not on the in vitro catalysis methodology, but provided selected examples, only. For further reading, a recent review (Classen and Pietruszka, 2017) is recommended. In the following sections, we focus on in vivo biocatalysis and highlight these advanced techniques, which may grant access to complex natural target structures as well as derivatives thereof.

21.4 Biosynthesis

In the previous sections, specific reaction pathways and individual reactions for the production of (pharmaceutically active) molecules were presented. However, most of these reactions are only singular steps of the entire production machinery of all chemical compounds in living organisms. This so-called *biosynthesis* of biomolecules is the motor of life as we know it and, through the creativity of millions of years of evolution, represents an almost inexhaustible source of natural substances.

Therefore, it is not surprising that humans have been using biosynthesis since earliest times—often without knowing it. As examples for this *native biosynthesis*, the use of plants as pharmaceutical products (Fig. 21.6), but also the process of fermentation for prolonging the shelf life of food, can be mentioned. The understanding of the molecular reactions of biosynthetic pathways, the isolation and structural elucidation of pure substances as well as the elucidation of their biological activity only goes hand in hand with the strong scientific development since the early 19th century. Today, this knowledge and technical progress form the basis for manipulations in native biosynthetic pathways with the aim to modify biosyntheses and produce natural product derivatives.

Due to this historical background, the *native biosynthesis* is first presented in this chapter, which in the second part is then merged into the production of derivatives using methods developed more recently.

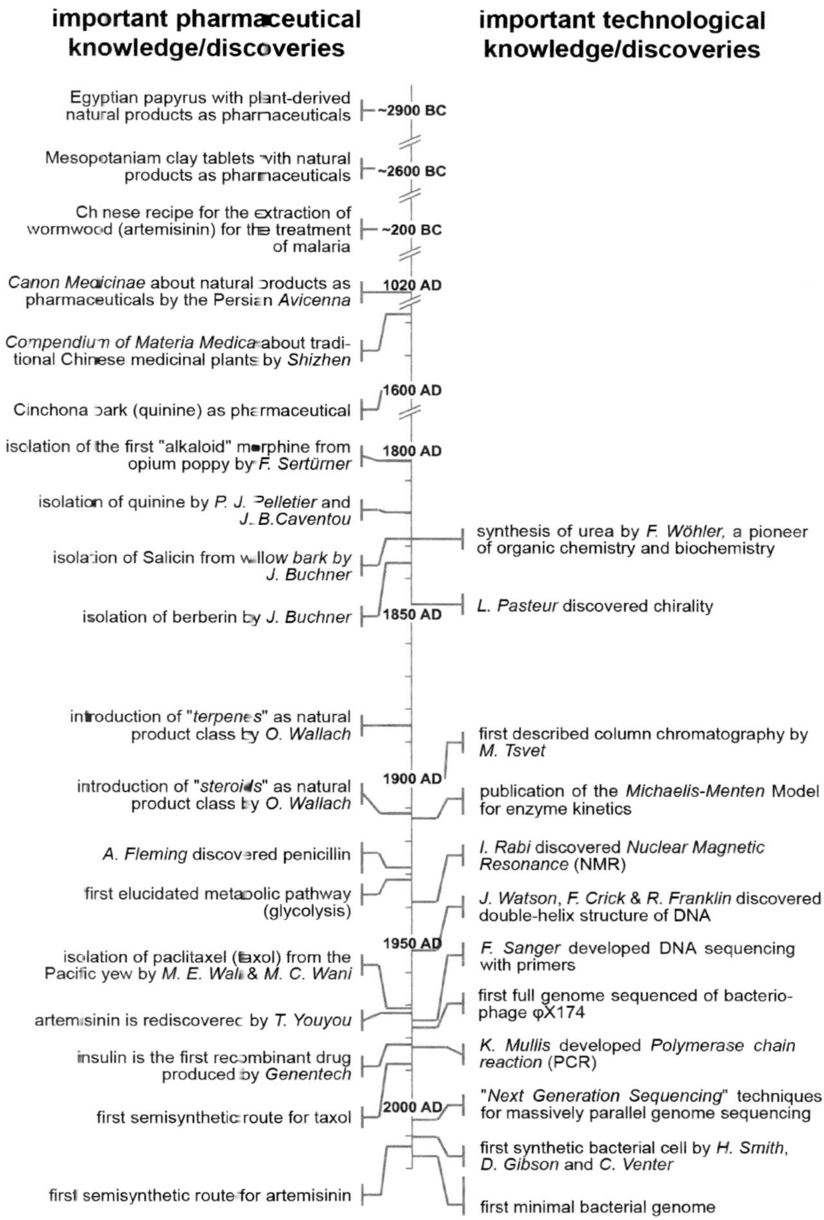

Figure 21.6 Timeline about important pharmaceutical and technological discoveries.

21.4.1 Native Biosynthesis

The use of native biosynthesis is the oldest and easiest way to produce natural products and has been used for thousands of years in the production and use of pharmaceutical products. Fundamentally, no knowledge of biosynthesis and related metabolic pathways or genetic constructs is required; mostly, precursors are provided by primary metabolism and enzymes and their cofactors/coenzymes are made available by the organism itself (Fig. 21.7). The advantage of biosynthesis is therefore the supply of all components involved by the host and the consequently higher economic value for an industrial process (bioeconomy). Examples of industrial biosynthesis products are riboflavin (**21**, vitamin B2), amino acids such as L-glutamate (**22**), L-serine and L-lysine as feed additives, flavorings, alcohols as biofuels and polylactides for the production of polymers (Fig. 21.8) (Ledesma-Amaro et al., 2015; Eggeling and Bott, 2005; Madhavan Nampoothiri et al., 2010; Vink et al., 2004; Izumi et al., 1978; Kubota et al., 1971).

Nonetheless, native biosynthesis is subject to some fundamental disadvantages:

- Especially for natural substances from plant organisms, the climate results in local and seasonal fluctuations in availability. E.g., the cultivation of *Vanilla planifolia* to produce the natural vanilla flavoring **23** only takes place in tropical regions of the earth (Gallage and Møller, 2015; Ramachandra

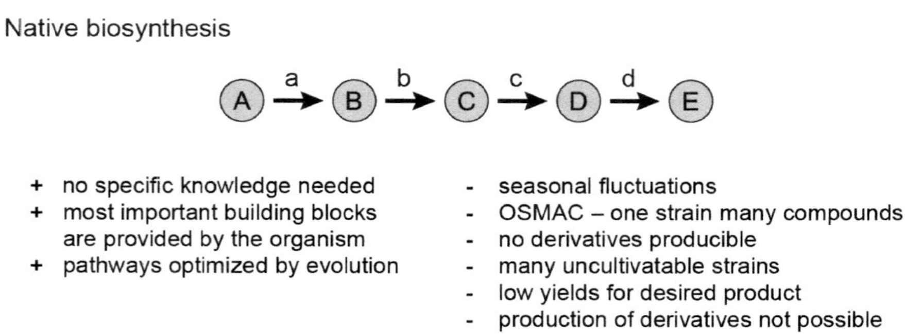

Figure 21.7 Scheme of native biosynthesis. The scheme is based on Kirschning et al. (2007). A, substrate; B–D – biosynthetic intermediate; E, natural product; a–d, enzymatic reactions/enzymes.

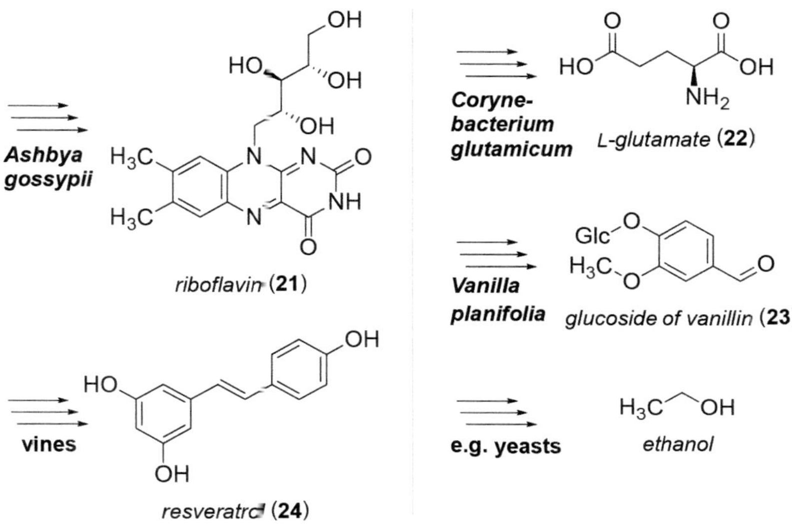

Figure 21.8 Examples for natural products with industrial relevance gained by native biosynthesis.

Rao and Ravishankar, 2000). The grape harvest to produce wine and as the source of the polyphenol resveratrol (**24**) can only be carried out once a year (Fig. 21.9A).

- Many organisms, such as *Streptomyces* bacteria, myxobacteria (slime bacteria), fungi and plants, produce a large variety of natural compounds in a single cell (Fig. 21.9C). Often microorganisms harbor silent pathways, which are not always expressed under standard laboratory conditions. These microorganisms are therefore poorly suited for the production of individual metabolites. This principle is called OSMAC (one strain many compounds) (Höfs et al., 2000; Wang et al., 2014; Scherlach and Hertweck, 2009).
- A large number of the naturally occurring bacteria can either not be cultivated at all or only under special conditions in a microbiological laboratory, which makes the production of natural products more difficult (Lewis et al., 2010). In addition, some organisms can represent a safety risk.
- The production quantities are often too low from an ecological and economic point of view. For example, about 12 trees and 9 weeks of production time are needed to provide enough of

the anti-cancer agent paclitaxel (**25**, taxol) from the Pacific yew (*Taxus brevifolia*) for the treatment of one patient (Pletsch et al., 1995). The content of the antimalarial active ingredient artemisinin (**26**) from *Artemisia annua* is approx. only 1% of the dry weight of the plant (Fig. 21.9B).

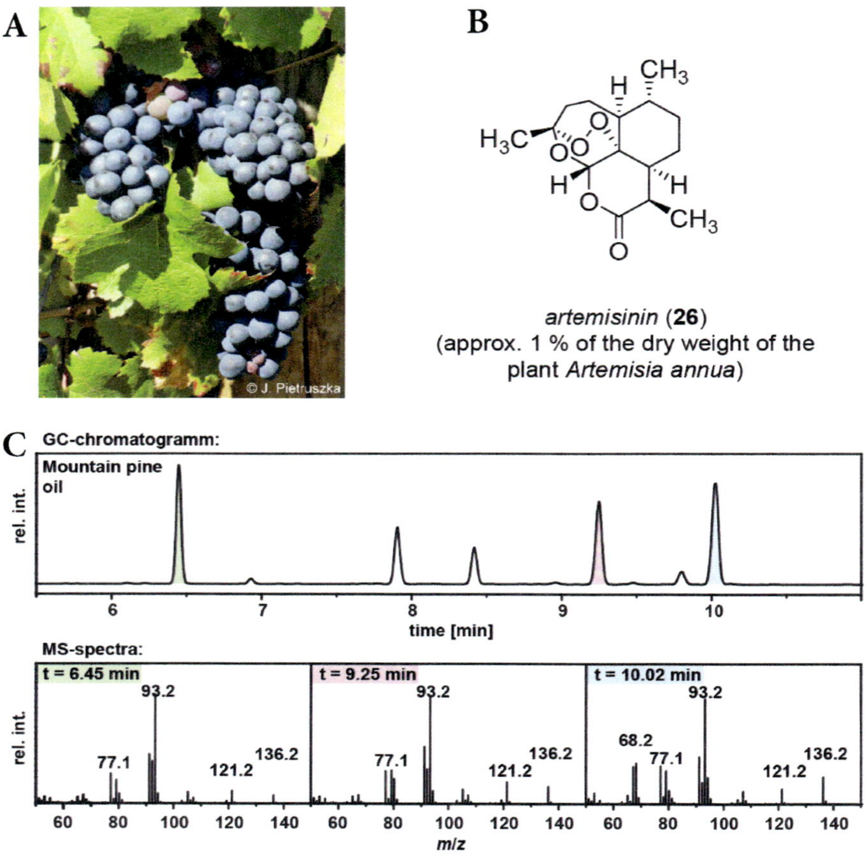

Figure 21.9 Examples of fundamental disadvantages in native biosynthesis. (A) Grapes, as a source for wine and the polyphenol resveratrol (**24**), can only be harvested once a year; (B) Artemisinin (**26**) is only available in low quantities in the native plant *Artemisia annua*; (C) GC-MS trace of Mountain pine (*Pinus mugo*) oil as an extract demonstrates the principle of OSMAC (one strain many compounds). The plant produces various terpenes with the same mass but different retention times, which are related to the different structures of the molecules.

Nevertheless, native biosynthesis can be optimized by additional knowledge about the metabolic pathway, e.g. the supplementation of building blocks and additives into the culture broth or the genetic optimization of metabolism by *metabolic flux analysis* (Bilyk and Luzhetskyy, 2016; King et al., 2016; Pickens et al., 2011). Biosynthesis can also be optimized regarding the host, since the genetic tools allow the transfer of entire gene clusters into heterologous hosts (see Section 21.4.2.2). However, biosynthesis makes hardly any contribution to the production of derivatives, since in most cases only one defined product is produced.

21.4.2 Utilizing Biosynthesis Providing Derivatives

As already mentioned at the beginning of this chapter, natural substances, in particular secondary metabolites, offer an almost inexhaustible source for pharmaceutical applications. Even though most of these components have been optimized for their specific task by evolution, an improvement with regard to their therapeutic use can still be obligatory. Among other things, the specification and improvement of efficacy towards the target molecules, the solubility in water and the metabolic stability should be mentioned. The provision of natural substances and a compound library of their derivatives is therefore unavoidable for medical studies and is therefore considered as a challenge for natural product research.

Especially due to the development of organic chemistry from the second half of the 20th century onwards, the total synthesis of natural products forms the basis for the provision of bioactive compounds (vice versa, the demand for natural products formed the basis for the evolution of organic chemistry to this day). Through the continuous development of new reactions, catalysts and reagents, multi-component and domino reactions and stereospecific reactions—not least through the aforementioned use of biocatalysts—the most complex structures can currently be produced. A major advantage of total synthesis is certainly the control of each step in the synthesis, which simplifies the provision of derivatives. However, with increasing complexity of the molecules, the number of synthetic steps usually also increases, which inevitably reduces the total yield. It is precisely this

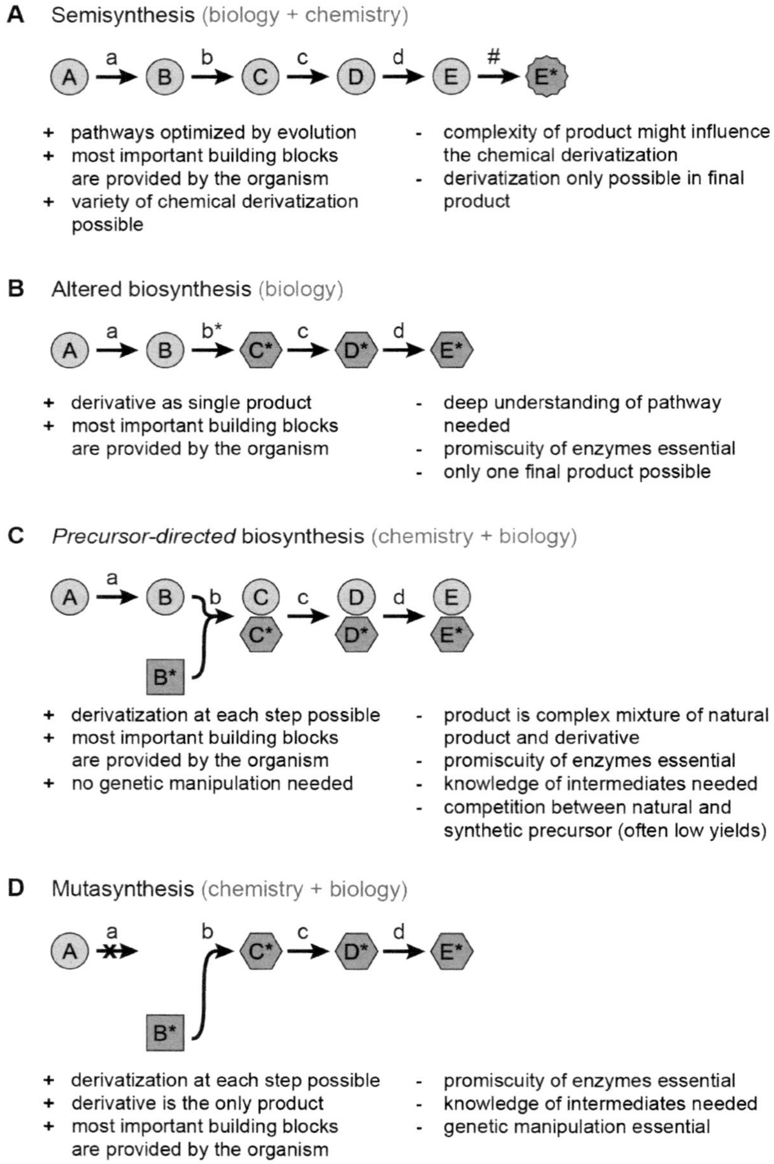

Figure 21.10 Overview on the different methods for the production of natural product derivatives. The scheme is based on Kirschning et al. (2007). A, substrate; B–D, biosynthetic intermediate; E, natural product; a–d, enzymatic reactions/enzymes; *, derivative; #, chemical conversion.

argument that makes total synthesis for an industrial approach in a number of cases unattractive, so that a critical economic consideration is necessary in comparison with the potential of a natural product.

Alternatives for total synthesis can be found in biosynthesis and the synergy potential by combining chemistry and biology. In the following, methodological approaches like *semisynthesis*, *altered biosynthesis*, *precursor-directed biosynthesis*, and *mutasynthesis* (Fig. 21.10) are describing the aim at optimizing biosynthesis and the production of natural product derivatives.

21.4.2.1 Semisynthesis

Semisynthesis is the oldest way of diversifying natural products, as it does not require knowledge of biosynthetic pathways or genetic constructs. Semisynthesis combines classical biosynthesis and organic chemistry and generates derivatives by consecutive chemical modification of the extracted natural substance (Fig. 21.10A) (Von Nussbaum et al., 2006; Eichner et al., 2012; Kennedy, 2008; Mahoney et al., 2014). For this approach, mostly complex natural products or intermediates are used as a starting material, and either these are completed to bioactive natural metabolite or the specific bioactivity is improved by (non-biological) functional groups. However, this approach is hampered by a limited range of subsequent chemical transformations. With structurally highly complex natural products, the synthetic modification must be highly efficient as well as chemo- and regioselective in order to avoid a subsequent separation problem and the associated loss of yield. Well-known examples of semisynthetic products are the penicillin (**27**) derivatives amoxicillin and methicillin, the antitumor therapeutic paclitaxel (**25**, taxol), the antimalarial therapeutic artemisinin (**26**) and the cholesterol-lowering simvastatin (a lovastatin derivative) (Kennedy, 2008; Patel, 1998; Kung et al., 2018; Paddon and Keasling, 2014).

Paclitaxel (25, taxol)

The small biosynthetic amounts of anti-cancer agent paclitaxel (**25**, taxol) obtained from the Pacific yew (*Taxus brevifolia*) led to the development of alternative production strategies (Patel,

1998). Although total syntheses were published, these were not economically justifiable due to the complexity of the molecule. Since an intermediate of the taxol biosynthesis, 10-deacetylbaccatin III (**28**), occurs in larger amounts in the leaves of the European yew (*Taxus baccata*), Holton (1993) developed a semisynthesis approach for the production of taxol (**25**). Starting from 10-deacetylbaccatin III (**28**), the 7-*O*-triethylsilyl baccatin III (**29**) is first obtained by two consecutive protection reactions, which resulted after conversion with the *Ojima* lactone (**30**) and a subsequent deprotection in taxol (Fig. 21.11).

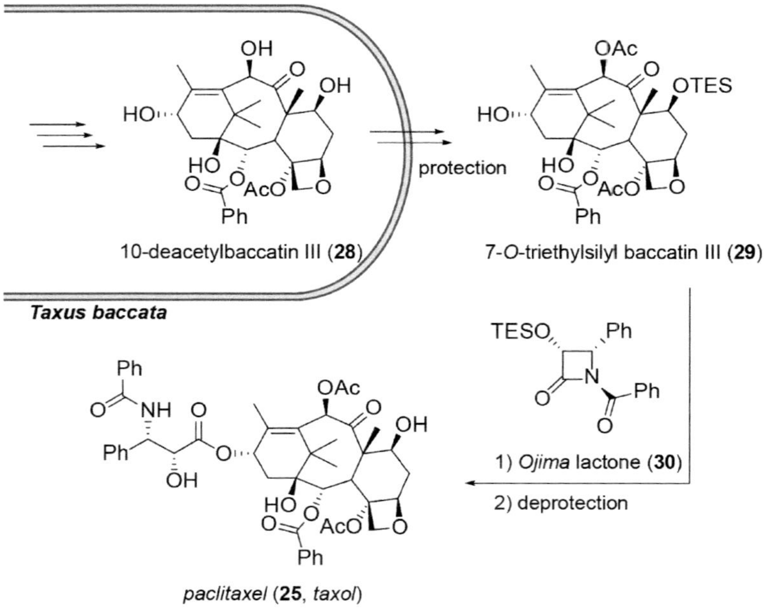

Figure 21.11 Semisynthesis of paclitaxel (**25**, taxol) starting with the extraction of 10-deacetylbaccatin III (**28**) from the leaves of the European yew (*Taxus baccata*) and coupling with the *Ojima* lactone (**30**) published by Holton (1993).

Artemisinin (26)

In analogy to paclitaxel (**25**), an alternative process was found for the industrial production of artemisinin (**26**) using a semisynthesis approach, as the biosynthesis by *Artemisia annua* produces too small quantities of this antimalarial therapeutic. In contrast to paclitaxel (**25**), no native source was used for the biosynthesis intermediate artemisinic acid (**31**), a late-stage precursor of artemisinin (**26**). Artemisinic acid (**31**) was produced by heterologous expression of plant genes in the yeast *Saccharomyces cerevisiae* by fermentation, combining methods of synthetic biology and metabolic engineering (see Fig. 21.12 on page 730; Paddon and Keasling, 2014; Kung et al., 2018). For the subsequent semisynthesis approach the artemisinic acid (**31**) was chemically modified and converted into artemisinin (**26**) in a final photochemical process with a yield of 55%—starting from the extracted artemisinic acid (**31**). This process was designed by *Amyris*, Inc, and implemented and optimized by the pharmaceutical company *Sanofi* (Turconi et al., 2014; Burgard et al., 2016). It is currently the only available industrial-scale process for the semisynthetic production of artemisinin (**26**) and is still being further developed (Kung et al., 2018).

21.4.2.2 Altered Biosynthesis

Today, native biosynthesis in its natural strains is rarely used for the production of natural products. Often biosyntheses are optimized through genetic modifications of the production strains by adapting the metabolism for the production of the desired natural product supported by *metabolic flux analysis* (Park et al., 2018). In early times, random mutation experiments by radiation (*mutation breeding*) or the use of mutagenic agents were carried out and new production strains were selected by large-scale screenings (Van Harten, 1998; Oladosu et al., 2016).

Nowadays, by gaining knowledge of genetic information and elucidating metabolic pathways (see Fig. 21.6), production strains can be specifically mutated at defined regions on the DNA level. In addition to influencing the metabolism, genetic modifications can also be carried out in the biosynthesis cluster and even entire genes for enzymes can be exchanged, which have been investigated in the previously mentioned biocatalytic approaches (see Sections 21.2 and 21.3). Therefore, the combination of defined mutations or gene exchanges with biosynthesis strategies resulted in the term *combinatorial biosynthesis*. Examples of combinatorial biosynthesis approaches are often polyketides and nonribosomal peptides (Reeves, 2003).

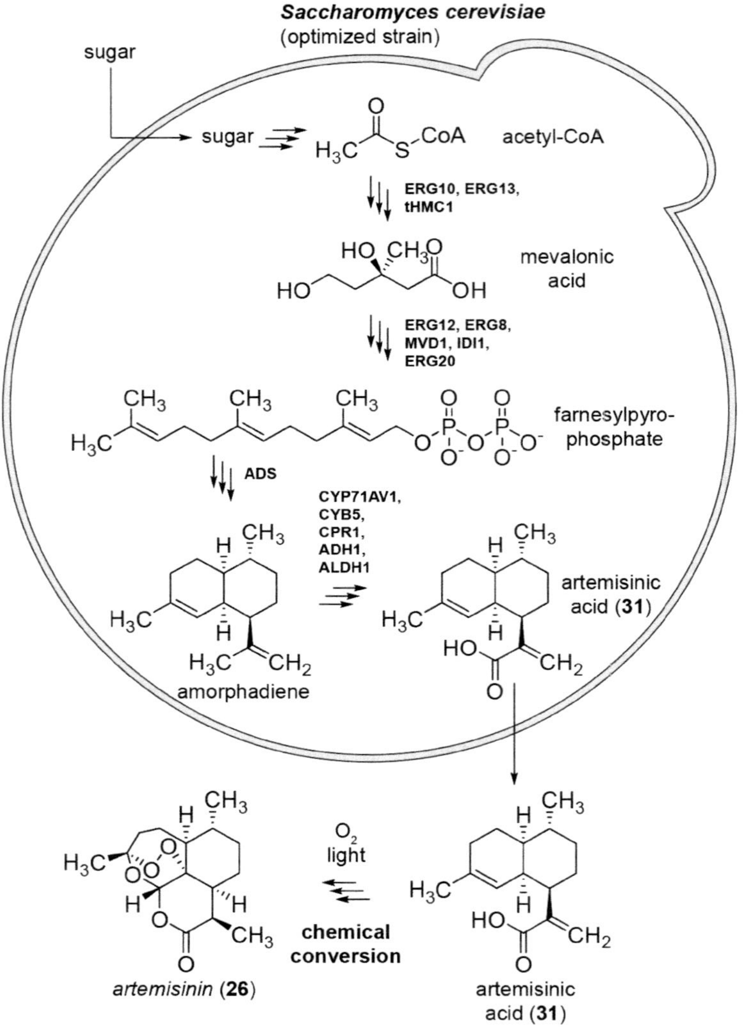

Figure 21.12 Semisynthesis of artemisinin (**26**) starting with the fermentation of sugar in a genetically modified *Saccharomyces cerevisiae* strain and the extraction of artemisinic acid (**31**) followed by a photochemical conversion. This process and the strain were designed by *Amyris,* Inc, and implemented by pharmaceutical company *Sanofi* (Paddon and Keasling, 2014).

A further optimization of native biosynthesis results from the already described problem of the cultivability of microorganisms in microbiological laboratories and production facilities (see Section 21.4.1). In addition, the minimization of the risk by using non-pathogenic production strains is one of the main objectives of current research. The knowledge about the genetic sequences of the gene clusters and the development of genetic tools enables the transfer of entire gene clusters into heterologous safe strains (*horizontal gene transfer*) and thus reduces the infection of employees and the contamination of products (Pfeifer and Khosla, 2001, Wenzel and Müller, 2005). Examples of safe strains are the so-called GRAS-certified (*generally recognized as safe*) organisms, which are approved by the Food and Drug Administration (FDA) for use in the food industry and as food additives and are declared as safe.

Combinatorial biosynthesis: Avermectins 32/Ivermectins 33

The important groups of neurotoxins avermectins **32** and ivermectins **33** are pesticides and are used to treat ectoparasites (lice, mites and ticks) and nematodes (Pitterna et al., 2009). For the discovery and development of these compounds, the Nobel Prize for Physiology or Medicine was awarded in 2015. These natural products are polyketides, which are produced from acyl building blocks by a modularly clustered enzyme machinery. Some avermectins **32** are obtained by fermentation of *Streptomyces avermitilis*, other derivatives are only chemically addressable.

Polyketide synthases (PKS) are particularly suitable for the approach of combinatorial biosynthesis, since the exchange of homologous genes is simplified by the modular character in order to generate new derivatives. For some avermectins **32** it could be shown that an exchange of reducing enzymes from module 2 (DH: dehydratase; KR: ketoreductase) with enzymes from a rapamycin-producing PKS of *Streptomyces hygroscopicus* (DH: dehydratase; KR: ketoreductase; ER: enoyl reductase) led to products with reduced double bond (Fig. 21.13) (Gaisser et al., 2003).

Alternative Host

The production of natural products in heterologous organisms requires the transfer of genetic information for the corresponding enzymes. Tools such as the TREX-system (transfer and expression) developed by Loeschcke et al. (2013) are available that can be used for the horizontal gene transfer of entire gene clusters into the genome of the new heterologous host (Fig. 21.14A). The TREX-system was used to transfer the carotenoid gene cluster from *Pantoea ananatis* (*crt*-cluster, 6 genes, 6.9 kb) and the prodigiosin pathway from *Serratia marcescens* (*pig*-cluster, 15 genes, 21.8 kb) into *Escherichia coli* BL21, *Pseudomonas putida* KT2440 and *Rhodobacter capsulatus* B10S (Figs. 21.14B,C). This enabled the production of the carotenoids *β*-carotene (**34**) and zeaxanthin (**35**) as well as the tripyrrole antibiotic prodigiosin (**36**) (Loeschcke et al., 2013; Loeschcke and Thies, 2015; Domröse et al., 2015).

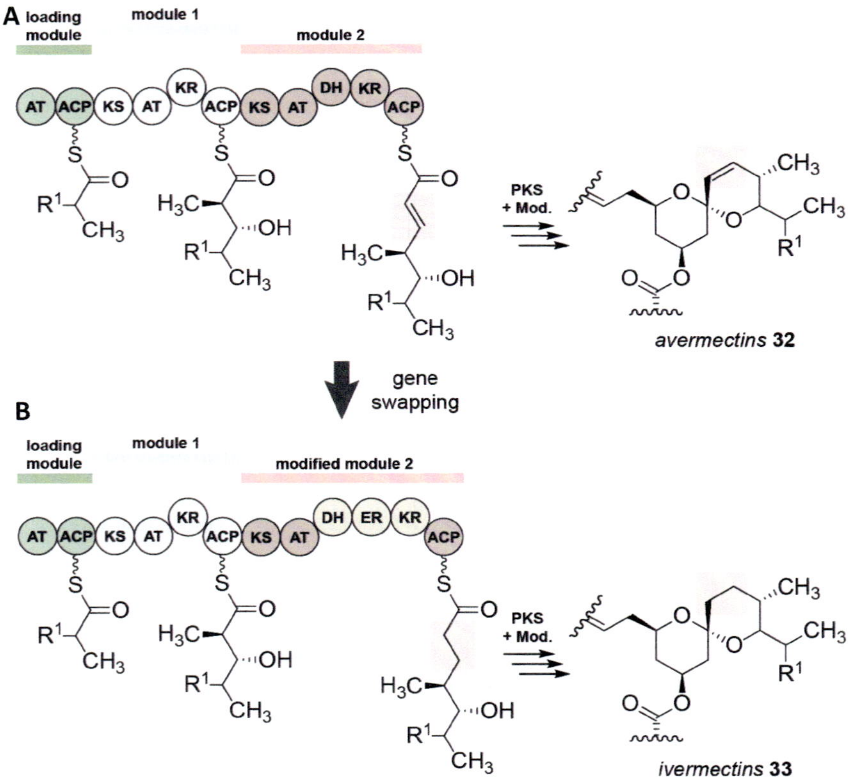

Figure 21.13 Production of ivermectins **33** using combinatorial biosynthesis altering the biosynthesis avermectin polyketide synthase cluster. The reducing enzymes from module 2 (**A:** DH and KR) were exchanged with enzymes from a rapamycin-producing PKS of *Streptomyces hygroscopicus* (**B:** DH, KR and ER) resulting in a reduced double bond. The scheme is based on Weissman and Leadlay (2005). DH, dehydratase; KR, ketoreductase; ER, enoyl reductase.

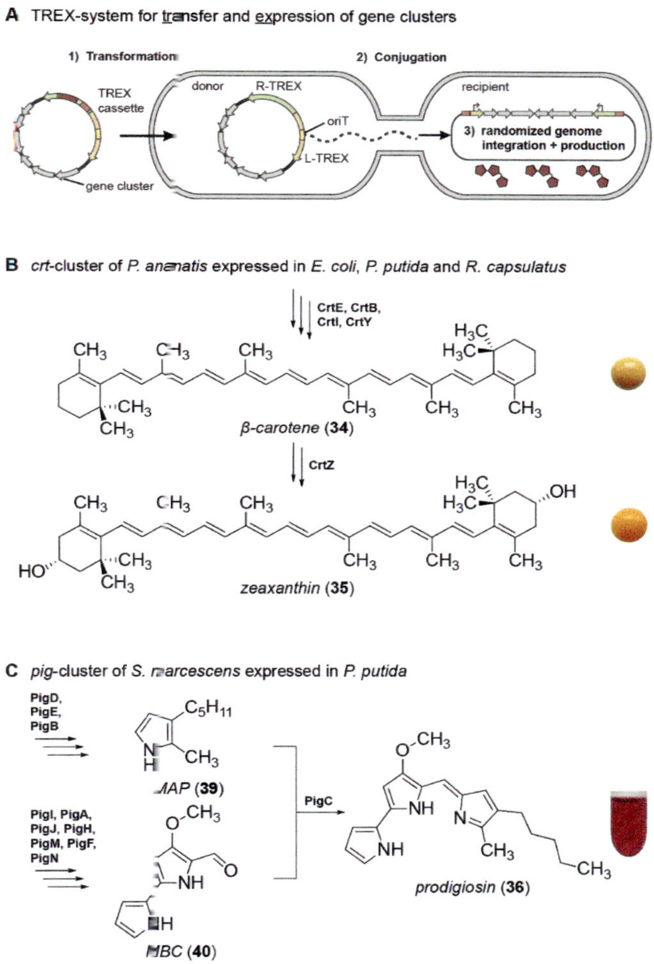

Figure 21.14 Principles of using a heterologous host for the production of natural compounds. (A) The TREX-system published by Loeschcke et al. (2013) as example for the integration of heterologous gene clusters in the genome of a new production host. The target gene cluster is flanked by the TREX cassette and is transferred in a donor strain. After conjugation the gene cluster is randomly integrated into the genome of the recipient strain by a Tn5 transposase. (B) The *crt*-cluster of *Pantoea ananatis* is transferred into the genome of *Escherichia coli* BL21, *Pseudomonas putida* KT2440 and *Rhodobacter capsulatus* B10S using the TREX-system producing *β*-carotene (**34**) and zeaxanthin (**35**). (C) The *pig*-cluster of *Serratia marcescens* is transferred into the genome of *P. putida* KT2440 using the TREX-system producing prodigiosin (**36**).

21.4.2.3 Precursor-directed biosynthesis and mutasynthesis

Precursor-directed biosynthesis

With the increasing knowledge of biosynthetic pathways and the function of individual enzymes, the method of precursor-directed biosynthesis (PDB) was developed, which combines classical biosynthesis with the organic synthesis of precursor molecules (Kirschning et al., 2007; Kennedy, 2008; Mahoney et al., 2014). By supplementing simple unnatural precursor analogues to the production strain, these are incorporated into the biosynthesis and thus incorporated into a natural product derivative. Therefore, derivatives can be produced relatively quickly using wild type biosynthesis without time-consuming genetic manipulations. However, the yields of the analogues are often low because the unnatural precursor has to compete with the preferred natural precursor. By using quite high precursor concentrations, this competition can be shifted in favor of the unnatural building block, but in this case concentration-dependent toxic properties of the compounds have to be considered. A further disadvantage is the generation of a product mixture of structurally and chemically often similar compounds, whereby complex separation processes have to be established. Well-known examples of a PDB are penicillin V (**27a**) (Fig. 21.15) and rapamycin analogues (Halliday and Arnstein, 1956; Boecker et al., 2016; Lowden et al., 2004).

Penicillin V (27a)

Phenoxymethylpenicillin (**27a**, penicillin V) is produced—like penicillin G (**27b**)—fermentative in the fungus *Penicillium chrysogenum*. By the supplementation of phenoxyacetate first the activated CoA-thioester **37** is formed, which competes with the natural precursor benzoyl-CoA (**38**) (Fig. 21.15). At high substrate concentrations, penicillin V (**27a**) is predominantly produced. Due to a higher acid stability, penicillin V (**27a**) can be given orally compared to penicillin G (**27b**). Penicillin V (**27a**) is specifically used for the treatment of bacterial infections, like streptococcal pharyngitis (strep throat), caused only by Gram-positive species. It is further less active against Gram-negative bacteria than penicillin G (**27b**).

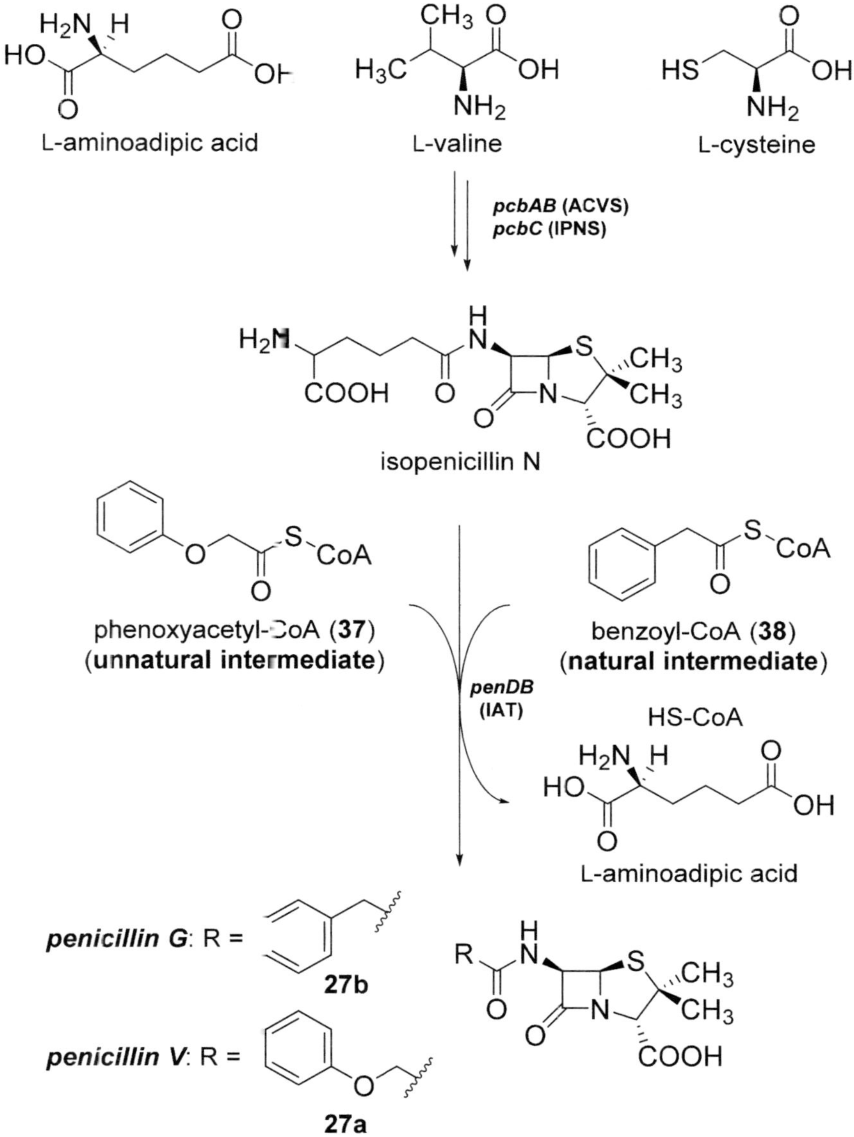

Figure 21.15 Production of penicillin V (**27a**) via a precursor-directed biosynthesis (PDB) in *Penicillium chrysogenum*. The unnatural intermediate phenoxyacetate is supplemented into the fermentation broth and competes as activated thioester **37** with the natural precursor benzoyl-CoA (**38**).

21.4.2.4 Mutasynthesis

Mutasynthesis (*mutational biosynthesis*) is based conceptually on the PDB and reduces the disadvantages to the chemical synthesis and acceptance of (possibly toxic) precursors. In mutasynthesis, unnatural precursor analogues (*mutasynthons*) are also supplemented to a production strain, but the biosynthesis of these organisms is interrupted at a defined step, which prevents the competition with natural precursors. As a result, these strains are no longer able to produce the natural compound and therefore do not form product mixtures (Kirschning et al., 2007; Kirschning and Hahn, 2012; Eichner et al., 2012; Kennedy, 2008).

The basic requirement for a mutasynthesis approach is knowledge of the individual steps of biosynthesis and their gene clusters. Consequently, development only began after the discovery of genetic tools such as sequencing of small DNA fragments in the 1970s and 1980s, the polymerase chain reaction in the 1990s and the sequencing of entire genomes, and is thus the most recent of the concepts presented here (Fig. 21.16).

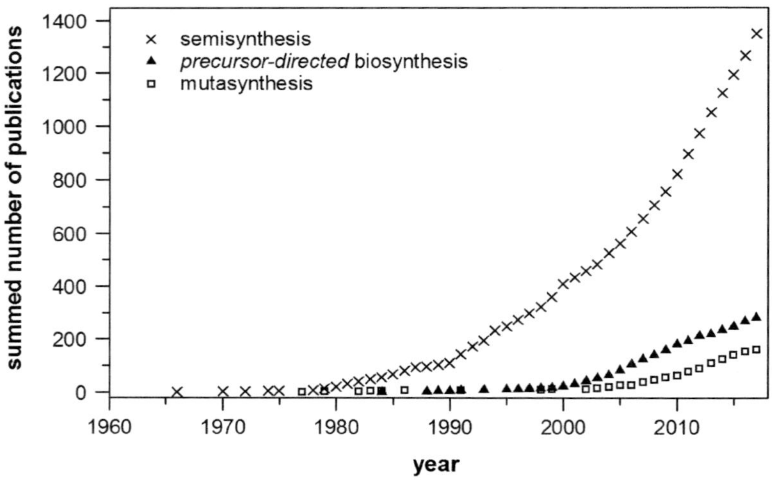

Figure 21.16 Number of total publications on semisynthesis, precursor-directed biosynthesis and mutasynthesis until the end of 2017. The database "Web of Science™" by Thomson Reuters™ was used as data set (status September 2018). The English terms "semisynthesis", "precursor(-)directed biosynthesis" and "mutasynthesis" were chosen for the title and/or topic of the publication.

If the biosynthesis of a natural substance is known, the entry point for mutasynthesis can be specifically chosen in order to increase the efficiency of this methodology. A mutation in one of the early key reactions in biosynthesis enables the production of complex natural product derivatives by supplementing very simple mutasynthons. It should be noted that these early mutasynthons are neither obligatory for the growth of the host nor is the acceptance of the following enzymes too low for the altered intermediates (*enzyme promiscuity*). Therefore, parallel biosynthesis pathways are particularly suitable for the mutasynthesis approach. Examples of mutasynthesis are the production of ansamitocin, geldanamycin, rapamycin, vancomycin, and prodigiosin (**36**) derivatives (Bilyk and Luzhetskyy, 2016; Boecker et al., 2016; Taft et al., 2008; Weissman, 2007; Eichner et al., 2009; Gregory et al., 2005; Weist et al., 2004; Klein et al., 2017; Klein et al., 2018).

Prodigiosin (36) derivatives

The tripyrrole pigment prodigiosin (**36**), which in addition to its characteristic color is also known for its antibiotic effect against pathogenic microorganisms and antitumor activity. The native producer of prodigiosin is the opportunistically pathogenic bacterium *Serratia marcescens*. The *pig*-gene cluster, which comprises the prodigiosin biosynthesis genes *pig*A to *pig*N carries the genetic information for the enzymatic production of the monopyrrole MAP (**39**, 2-methyl-3-amyl-pyrrole) and the dipyrrole MBC (**40**, 4-methoxy-2,2′-bipyrrole-5-carbaldehyde), which are condensed to the tripyrrole prodigiosin (**36**) (Hu et al., 2016).

To generate prodigiosin (**36**) derivatives, the gene cluster was first transferred by Domröse et al. (2015) into the genome of the GRAS-certified heterologous strain *P. putida* KT2440 and afterwards the gene for the first enzyme of the biosynthesis of the monopyrrole MAP (**39**; *pig*D) was deleted (Fig. 21.17). The supplementation of 25 pyrrole mutasynthons **41**, whose spectrum ranged from mono- and dialkylated pyrroles to cyclic alkyl pyrroles and alkene functionalities, led to the production of 19 derivatives **42** (Klein et al., 2018). These derivatives **42** were used in investigations concerning their antibiotic effect and the modulation of autophagy (Klein et al., 2017; Klein et al., 2018).

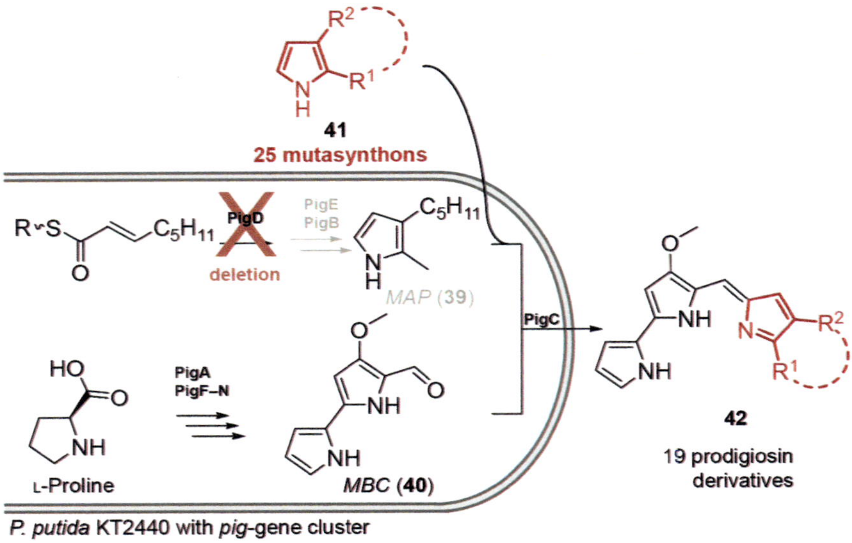

Figure 21.17 Mutasynthesis of prodigiosin derivatives in *Pseudomonas putida* KT2440. The heterologous production *P. putida* strain is blocked in the very first step of the MAP (**39**) biosynthesis by knockout of *pig*D. Different monopyrrole mutasynthons were supplemented to this knockout strain producing a large variety of prodigiosin derivatives **42**.

21.5 Conclusion

This chapter provides an overview on how much influence biocatalysis and biosynthesis have already gained in the production of bioactive natural products. Thus, single-step conversions form the basis for the mechanistic and application-oriented understanding of enzymes in biocatalysis, which in turn merge into multi-step conversions enabling the development of production cascades. In particular, the biosynthesis examples of paclitaxel (**25**) and artemisinin (**26**) illustrate how this knowledge can be applied on an industrial scale for the production of pharmaceuticals by combining the diversity of methods. Even if these examples provide the basis for a future-oriented industrial development, science and industry naturally compete with existing technical processes for every new product. Only by continuously improving biocatalysis and methods of biosynthesis for the production of pharmaceutical compounds in sufficient quantities, this technology will become the major player in the future.

References

Bilyk O., Luzhetskyy A., *Curr. Opin. Biotechnol.,* **42** (2016), 98–107.

Bisterfeld C., Classen T., Küberl I., Henßen B., Metz A., Gohlke H., Pietruszka J., *PLOS ONE,* **11**(6) (2016), e0156525.

Boecker S., Zobel S., Meyer V., Süssmuth R. D., *Fungal Genet. Biol.,* **89** (2016), 89–101.

Bornscheuer U. T., *Philos. Trans. R. Soc. A Math. Phys. Eng. Sci.,* **376** (2018), 2110.

Burgard A., Gieshoff T., Peschl A., Hörstermann D., Keleschovsky C., Villa R., Michelis S., Feth M. P., *Chem. Eng. J.,* **294** (2016), 83-96.

Classen T., Pietruszka J., *Biorg. Med. Chem.,* **26** (7) (2017), 1285–1303.

Domröse A., Klein A. S., Hage-Hülsmann J., Thies S., Svensson V., Classen T., Pietruszka J., Jaeger K.-E., Drepper T., Loeschcke A., *Front. Microbiol.,* **6**(972) (2015), 972.

Eggeling L., Bott M. *Handbook of Corynebacterium Glutamicum* (2005), CRC press.

Eichner S., Floss H. G., Sasse F., Kirschning A., *ChemBioChem,* **10**(11) (2009), 1801–1805.

Eichner S., Knobloch T., Floss H. G., Fohrer J., Harmrolfs K., Hermane J., Schulz A., Sasse F., Spiteller P., Taft F., Kirschning A., *Angew. Chem. Int. Ed.,* **51**(3) (2012), 752–757.

Erdmann V., Lichman B. R., Zhao J., Simon R. C., Kroutil W., Ward J. M., Hailes H. C., Rother D., *Angew. Chem.,* **129**(41) (2017), 12677–12681.

Feher M., Schmidt J. M., *J. Chem. Inf. Model.,* **43**(1) (2003), 218–227.

Gaisser S., Kellenberger L., Kaja A. L., Weston A. J., Lill R. E., Wirtz G., Kendrew S. G., Low L., Sheridan R. M., Wilkinson B., *Org. Biomol. Chem.,* **1**(16) (2003), 2840–2847.

Gallage N. J., Møller B. L., *Mol. Plant,* **8**(1) (2015), 40–57.

Gregory M. A., Petkovic H., Lill R. E., Moss S. J., Wilkinson B., Gaisser S., Leadlay P. F., Sheridan R. M., *Angew. Chem.,* **117**(30) (2005), 4835–4838.

Gröger H., *Biorg. Med. Chem.,* **26**(7) (2018), 1239–1240.

Halliday W. J., Arnstein H. R. V., *Biochem. J.,* **64**(2) (1956), 380.

Haridas M., Abdelraheem E. M. M., Hanefeld U., *Appl. Microbiol. Biotechnol.,* 10.1007/s00253-018-9392-8 (2018).

Höfs R., Walker M., Zeeck A., *Angew. Chem. Int. Ed.,* **39**(18) (2000), 3258–3261.

Holton R. A., Florida State University (1993), U.S. Patent 5,274,124.

Hu D. X., Withall D. M., Challis G. L., Thomson R. J., *Chem. Rev.,* **116**(14) (2016), 7818–7853.

Izumi Y., Chibata I., Itoh T., *Angew. Chem.,* **90**(3) (1978), 187–194.

Jennewein S., Schürmann M., Wolberg M., Hilker I., Luiten R., Wubbolts M., Mink D., *Biotechnol. J.,* **1**(5) (2006), 537–548.

Kennedy J., *Nat. Prod. Rep.,* **25**(1) (2008), 25–34.

King J. R., Edgar S., Qiao K., Stephanopoulos G., *F1000Research,* **5** (2016).

Kirschning A., Hahn F., *Angew. Chem. Int. Ed.,* **51**(17) (2012), 4012–4022.

Kirschning A., Taft F., Knobloch T., *Org. Biomol. Chem.,* **5**(20) (2007), 3245–3259.

Klein A. S., Brass H. U. C., Klebl D. P., Classen T., Loeschcke A., Drepper T., Sievers S., Jaeger K.-E., Pietruszka J., *ChemBioChem,* **19**(14) (2018), 1545–1552.

Klein A. S., Domröse A., Bongen P., Brass H. U. C., Classen T., Loeschcke A., Drepper T., Laraia L., Sievers S., Jaeger K.-E., Pietruszka J., *ACS Synth. Biol.,* **6**(9) (2017), 1757–1765.

Kubota K., Kageyama K., Shiro T., Okumura S., *J. Gen. Appl. Microbiol.,* **17**(2) (1971), 167–168.

Kung S. H., Lund S., Murarka A., McPhee D., Paddon C. J., *Front. Plant Sci.,* **9**(2018), 87.

Ledesma-Amaro R., Serrano-Amatriain C., Jiménez A., Revuelta J. L., *Microb. Cell Fact.,* **14** (2015), 163.

Lewis K., Epstein S., D'Onofrio A., Ling L. L., *J. Antibiot.,* **63**(8) (2010), 468–476.

Loeschcke A., Markert A., Wilhelm S., Wirtz A., Rosenau F., Jaeger K. E., Drepper T., *ACS Synth. Biol.,* **2**(1) (2013), 22–33.

Loeschcke A., Thies S., *Appl. Microbiol. Biotechnol.,* **99**(15) (2015), 6197–6214.

Lowden P. A. S., Böhm G. A., Metcalfe S., Staunton J., Leadlay P. F., *ChemBioChem,* **5**(4) (2004), 535–538.

Madhavan Nampoothiri K., Nair N. R., John R. P., *Bioresour. Technol.,* **101**(22) (2010), 8493–8501.

Mahoney K. P. P., Smith D. R. M., Bogosyan E. J. A., Goss R. J. M., *Synthesis,* **46**(16) (2014), 2122–2132.

Martín-Matute B., Bäckvall J.-E., *Curr. Opin. Chem. Biol.,* **11**(2) (2007), 226–232.

Müller M., *Angew. Chem. Int. Ed.,* **44**(3) (2004), 362–365.

Oladosu Y., Rafii M. Y., Abdullah N., Hussin G., Ramli A., Rahim H. A., Miah G., Usman M., *Biotechnol. Biotechnol. Equip.,* **30**(1) (2016), 1–16.

Paddon C. J., Keasling J. D., *Nat. Rev. Microbiol.,* **12**(5) (2014), 355.

Park S. Y., Yang D., Ha S. H., Lee S. Y., *Adv. Biosyst.,* **2**(1) (2018), 1700190.

Patel R. N., *Annu. Rev. Microbiol.,* **52**(1) (1998), 361–395.

Patel J. M., *J. Mol. Catal. B: Enzym.,* **61**(3) (2009), 123–128.

Patel R. N., *Biorg. Med. Chem.,* **26**(7) (2018), 1252–1274.

Pfeifer B. A., Khosla C., *Microbiol. Mol. Biol. Rev.,* **65**(1) (2001), 106–118.

Pickens L. B., Tang Y., Chooi Y.-H., *Annu. Rev. Chem. Biomol. Eng.,* **2**(2011), 211–236.

Pietruszka J., Simon R. C., *Eur. J. Org. Chem.,* **21** (2009), 3628–3634.

Pitterna T., Cassayre J., Hüter O. F., Jung P. M. J., Maienfisch P., Kessabi F. M., Quaranta L., Tobler H., *Bioorg. Med. Chem.,* **17**(12) (2009), 4085–4095.

Pletsch M., Sant'Ana A. E. G., Charlwood B. V., Secondary compound accumulation in plants—the application of plant biotechnology to plant improvement, in *Chemistry of the Amazon* (Seidel P. R., Gottlieb O. R., Kaplan M. A. C., eds.); American Chemical Society (1995).

Ramachandra Rao S., Ravishankar G. A., *J. Sci. Food Agric.,* **80**(3) (2000), 289–304.

Reeves C. D., *Crit. Rev. Biotechnol.,* **23**(2) (2003), 95–147.

Ricca E., Brucher B., Schrittwieser J. H., *Adv. Synth. Catal.,* **353**(13) (2011), 2239–2262.

Scherlach K., Hertweck C., *Org. Biomol. Chem.,* **7**(9) (2009), 1753–1760.

Steinreiber J., Faber K., Griengl H., *Chem. A Eur. J.,* **14**(27) (2008), 8060–8072.

Sun H., Zhang H., Ang E. L., Zhao H., *Biorg. Med. Chem.,* **26**(7) (2018), 1275–1284.

Taft F., Brunjes M., Floss H. G., Czempinski N., Grond S., Sasse F., Kirschning A., *ChemBioChem,* **9**(7) (2008), 1057–1060.

Turconi J., Griolet F., Guevel R., Oddon G., Villa R., Geatti A., Hvala M., Rossen K., Göller R., Burgard A., *Org. Process Res. Dev.,* **18**(3) (2014), 417–422.

Turner N. J., *Curr. Opin. Chem. Biol.,* **14**(2) (2010), 115–121.

Van Harten A. M. *Mutation Breeding: Theory and Practical Applications,* Cambridge University Press (1998).

Verho O., Bäckvall J.-E., *J. Am. Chem. Soc.,* **137**(12) (2015), 3996–4009.

Vink E. T. H., Rábago K. R., Glassner D. A., Springs B., O'Connor R. P., Kolstad J., Gruber P. R., *Macromol. Biosci.,* **4**(6) (2004), 551–564.

Von Nussbaum F., Brands M., Hinzen B., Weigand S., Häbich D., *Angew. Chem. Int. Ed.,* **45**(31) (2006), 5072–5129.

Wang W.-J., Li D.-Y., Li Y.-C., Hua H.-M., Ma E.-L., Li Z.-L., *J. Nat. Prod.,* **77**(6) (2014), 1367–1371.

Weissman K. J., *Trends Biotechnol.,* **25**(4) (2007), 139–142.

Weissman K. J., Leadlay P. F., *Nat. Rev. Microbiol.,* **3**(12) (2005), 925.

Weist S., Kittel C., Bischoff D., Bister B., Pfeifer V., Nicholson G. J., Wohlleben W., Süssmuth R. D., *J. Am. Chem. Soc.,* **126**(19) (2004), 5942–5943.

Wenzel S. C., Müller R., *Curr. Opin. Biotechnol.,* **16**(6) (2005), 594–606.

Index

T - #0168 - 290819 - C33 - 229/152/37 [39] - CB - 9789814800617